Tutorial: VLSI Testing & Validation Techniques

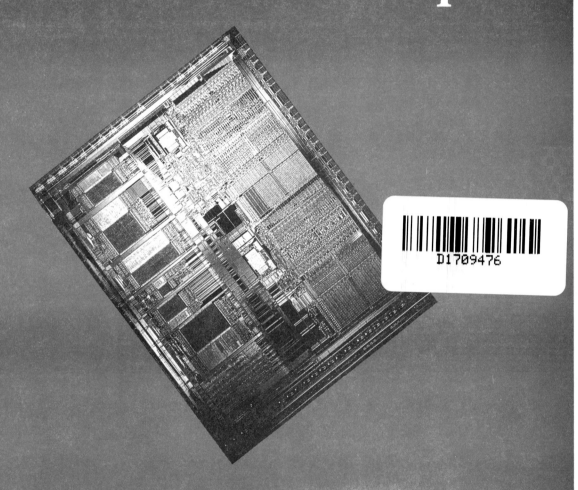

IEEE Catalog Number EH0237-8
Library of Congress Number 85-80876
IEEE Computer Society Order Number 668
ISBN 0-8186-0668-1

Hassan K. Reghbati
Simon Fraser University
Burnaby, British Columbia, Canada

IEEE COMPUTER SOCIETY

THE INSTITUTE OF ELECTRICAL
AND ELECTRONICS ENGINEERS, INC.

Published by IEEE Computer Society Press
1730 Massachusetts Avenue, N.W.
Washington, D.C. 20036-1903

Cover is by courtesy of AT&T Bell Laboratories

Copyright and Reprint Permissions: Abstracting is permitted with credit to the source. Libraries are permitted to photocopy beyond the limits of U.S. copyright law for private use of patrons those articles in this volume that carry a code at the bottom of the first page, provided the per-copy fee indicated in the code is paid through the Copyright Clearance Center, 29 Congress Street, Salem, MA 01970. Instructors are permitted to photocopy isolated articles for noncommercial classroom use without fee. For other copying, reprint or republication permission, write to Director, Publishing Services, IEEE, 345 E. 47 St., New York, NY 10017. All rights reserved. Copyright © 1985 by The Institute of Electrical and Electronics Engineers, Inc.

IEEE Catalog Number EH0237-8
Library of Congress Number 85-80876
IEEE Computer Society Order Number 668
ISBN 0-8186-0668-1 (Paper)
ISBN 0-8186-4668-3 (Microfiche)

Order from: IEEE Computer Society   IEEE Service Center
             Post Office Box 80452   445 Hoes Lane
             Worldway Postal Center   Piscataway, NJ 08854
             Los Angeles, CA 90080

 THE INSTITUTE OF ELECTRICAL AND ELECTRONICS ENGINEERS, INC.

About the Cover

The image on the cover shows the WE 32000 (formerly the Bellmac-32A) 32-bit microprocessor chip under test. The 84 pads around the 1 × 1-cm chip are being contacted by 2-mil diameter tungsten probes connected to the pin-electronics circuits in the VLSI test system. These circuits, enclosed in a box-shaped test head suspended above the automatic, silicon-wafer stepper, are designed to stimulate the chip inputs with high-speed digital signals at frequencies ranging up to 80 MHz. The pin-electronics circuits are also capable of detecting digital output signals from the chip and comparing the results with stored, expected data previously calculated by a remote CAT system.

The test head is connected to the mainframe of the test system, where the digital stimulus and expected one-zero patterns are stored in high-speed memories together with facilities for modulating the signals with pulse delay, width, and amplitude data. The system is also equipped to perform dc voltage and current measurements on the circuits around the periphery of the chip.

Unfortunately for today's silicon chip manufacturers, the cost of a typical VLSI test system is roughly equal to $8000 per pin, or $2 million in the case of a typical, fully equipped 256-pin system. As a result, systems that can simultaneously control two or three test heads are definitely more cost effective. It is also important for the system to provide multiple, high-speed, dc voltage and current measuring circuits, since these tend to restrict the testing throughput in a manufacturing environment. A typical test time for a silicon VLSI chip might be a few seconds, of which only 10 ms might be consumed by the actual digital test patterns.

The main challenge in testing the WE 32000 is to stimulate each of its tens of thousands of circuits and, by observing the outputs, determining whether or not there are any faults. In practice, the number of digital test patterns containing stimuli and expected data is roughly equal to the number of transistors in the microprocessor. This leads to the requirement that the test system must contain an expensive 10M-byte test pattern memory that can be accessed at rates up to 80 MHz. As MOS transistor sizes shrink toward the 1-micron level, providing high fidelity digital signals at rates up to 80 MHz becomes another challenge in testing modern VLSI circuits. However, a major problem arises here due to inductance, capacitance, and transmission line reflections that exist between the silicon chip under test and the pin driver and detection circuits housed in the test head. This problem becomes serious as the number of pin-electronics circuits grows to 128 and beyond. In the case of a 256 pin test head, it may be necessary to place these circuits as far as 50 cm from the chip. The basis of the problem lies in the fact that silicon chip manufacturers have been able to shrink their circuits at a faster rate than those being used to construct test system drivers and detectors.

A major, future challenge awaiting the test industry will be presented by wafer-scale silicon ICs. For them, design for testability will become mandatory, as will tungsten probe arrays that can test at many points across the wafer surface. It may also be necessary to incorporate lasers in the test system in order to connect redundant circuit elements, since it will be exceedingly difficult to produce defect-free circuits containing millions of MOS transistors on wafers exceeding 10 cm in diameter.

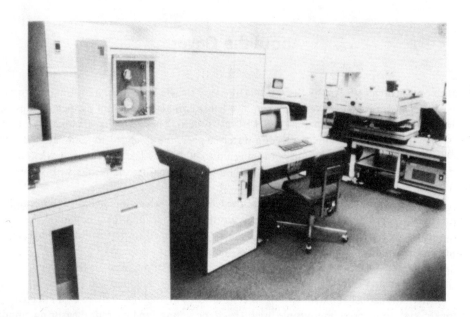

The system used by AT&T Bell Laboratories to test the WE 32000 microprocessor includes (from left to right) a line printer, a disk and computer drive, an on-line terminal (connected to a mainframe (behind)), and a test head and probe. Three off-line units, such as the one at the rear of the photo, can also be used by test personnel. These units are served by a Unix link.

Preface

A very large-scale integrated (VLSI) circuit is commonly defined as a single chip that contains more than 100,000 devices. Generally the MOS technology is being used with a minimum feature size in the 1–2 micron range. The advent of VLSI systems, while making significant contributions to the cost effectiveness of many products, is presenting challenges not previously faced by design and test engineers.

The development of semiconductor devices has depended upon a synergism with computers. This is particularly true for integrated circuits (ICs) whose development was motivated by the computer applications. With each advance in components, computers resulting from IC use reached a wider market, motivating further advances in the semiconductor technology. Furthermore, the development and testing of components depends extensively on computers.

While "computer-aided design and test" (CADT) is an explosively growing area in many engineering disciplines, VLSI-CADT is certainly the most widely used and best developed CADT branch. New CADT tools are continually being developed to cope with the ever increasing problems of VLSI complexity. The bristling richness of this rapidly growing field unfolds in this text with 46 authoritative papers. The material of the text is organized into chapters based on major topic headings. Each major topic is introduced with additional material supplying background, extensions, and annotations to the reprinted articles. The extensive bibliography is indispensable for those persons who want to learn more about the field.

The opening chapter provides a comprehensive overview of the entire field. It provides the necessary background for the reader who might not be proficient in VLSI-CADT. The remaining five chapters review the state of the art of CADT as it relates to VLSI. The second chapter is on VLSI design aids. Chapter Three examines the failures and testing of microelectronic circuits. Functional testing of specific circuits and testing of iterative arrays are discussed in Chapter Four. As the complexity of VLSI increases, testability of designs becomes more crucial. The current state of the art in built-in self-test and designing for testability is the theme of Chapter Five. Finally in Chapter Six, VLSI test systems and the future evolution of VLSI-CADT environments are discussed.

As can be seen from the papers in this book, VLSI-CADT is not a luxury; it is a rapidly evolving field and is an essential ingredient in the design, manufacturing, and use of integrated circuits. This book will serve its purpose if the reader achieves a greater understanding of the problems associated with VLSI-CADT as well as a better insight into the various tools (available and emerging) being developed for VLSI's evolution. It is a suitable text for graduate-level VLSI-CADT courses in computer science and electrical engineering departments. It is also a highly recommended reference book for VLSI designers and test engineers.

This project has benefited immensely from the works of others. Their contribution is highly appreciated. The author wishes to express his thanks to Chip G. Stockton, Dr. William D. Carroll, Dr. Roy L. Russo, and Margaret Brown for their invaluable assistance during the preparation of the manuscript. I am especially grateful to my friends—Professors Jacob A. Abraham, Vinod K. Agrawal, Sheldon B. Akers, Melvin A. Breuer, Nick J. Cercone, Donald A. George, John P. Hayes, Edward J. McCluskey, and Dhiraj K. Pradhan—for their support and encouragement.

This book is dedicated to my brother, Dr. Asghar Karimzadeh Reghbati, and to my mother, Ms. Sakineh Eslami Nasab, who are responsible for all the worthwhile things I do.

Table of Contents

About the Cover... iii

Preface v

Chapter 1: An Introduction to Testing and VLSI Design 1

1.1 LSI Testing Techniques 6
    M.S. Abadir and H.K. Reghbati (IEEE Micro, February 1983, pages 34-51)
1.2 Computer-Aided Design of VLSI Circuits 24
    A.R. Newton (Proceedings of the IEEE, October 1981, pages 1189-1199)
1.3 A CAD System for Design for Testability.. 35
    V.D. Agrawal, S.K. Jain, and D.M. Singer (VLSI Design, October 1984, pages 46-54)

Chapter 2: VLSI Design Validation 41

2.1 The Magic VLSI Layout System.. 47
    J.K. Ousterhout, G.T. Hamachi, R.N. Mayo, W.S. Scott, and G.S. Taylor (IEEE Design & Test, February 1985, pages 19-30)
2.2 Tools for Verifying Integrated Circuit Designs 59
    C.M. Baker and C. Terman (Lambda Magazine, 4th Quarter, 1980, pages 22-30)
2.3 EXCL: A Circuit Extractor for IC Designs 67
    S.P. McCormick (Proceedings of the 21st Design Automation Conference, June 1984, pages 616-623)
2.4 Circuit Analysis, Logic Simulation, and Design Verification for VLSI.......... 75
    A.E. Ruehli and G.S. Ditlow (Proceedings of the IEEE, January 1983, pages 34-48)
2.5 A Survey of Hardware Accelerators Used in Computer-Aided Design 90
    T. Blank (IEEE Design & Test, August 1984, pages 21-39)
2.6 A Switch-Level Timing Verifier for Digital MOS VLSI 109
    J.K. Ousterhout (IEEE Transactions on Computer-Aided Design, July 1985)
2.7 VHSIC Hardware Description Language 150
    M. Shahdad, R. Lipsett, E. Marschner, K. Sheehan, H. Cohen, R. Waxman, and D. Ackley (Computer, February 1985, pages 94-103)

Chapter 3: Failures, Fault Models, and Testing... 161

3.1 Closing the Loop: An Expanding Role for ATE in Semiconductor Manufacturing 165
    M. Mahoney (Proceedings of the International Test Conference, 1982, pages 12-21)
3.2 Fundamentals of Electron Beam Testing of Integrated Circuits..................... 175
    E. Menzel and E. Kubalek (Scanning, 1983, pages 103-122)
3.3 An Automated Laser Prober to Determine VLSI Internal Node Logic States 195
    F.J. Henley (Proceedings of the International Test Conference, 1984, pages 536-542)
3.4 Microelectronic Test Chips for VLSI Electronics 202
    M.G. Buehler (VLSI Electronics: Microstructure Science, Volume 6, 1983, pages 529-576)
3.5 Diagnostic Techniques 249
    R.B. Marcus (VLSI Technology, 1983, pages 507-549)
3.6 Physical Versus Logical Fault Model in MOS LSI Circuits: Impact on Their Testability....... 283
    J. Galiay, Y. Crouzet, and M. Vergniault (IEEE Transactions on Computers, June 1980, pages 527-531)

3.7 Fault Modeling for Digital MOS Integrated Circuits 289
J.P. Hayes (*IEEE Transactions on Computer-Aided Design*, July 1984, pages 200-208)

3.8 Test Generation for MOS Circuits Using D-Algorithm 298
S.K. Jain and V.D. Agrawal (*Proceedings of the 20th Design Automation Conference*, 1983, pages 64-70)

3.9 Robust Tests for Stuck-Open Faults in CMOS Combinational Logic Circuits 305
S.M. Reddy, M.K. Reddy, and V.D. Agrawal (*Proceedings of the 14th Fault-Tolerant Computing Conference*, 1984, pages 44-49)

3.10 Stafan: An Alternative to Fault Simulation 311
S.K. Jain and V.D. Agrawal (*Proceedings of the 21st Design Automation Conference*, 1984, pages 18-23)

Chapter 4: Testing Regular Structures and Specific Circuits 317

4.1 Easy Testable Iterative Systems 320
A.D. Friedman (*IEEE Transactions on Computers*, December 1973, pages 1061-1064)

4.2 Design of Easily Testable Bit-Sliced Systems 324
T. Sridhar and J.P. Hayes (*IEEE Transactions on Computers*, November 1981, pages 842-854)

4.3 Testing and Fault Tolerance of Multistage Inteconnection Networks 337
D.P. Agrawal (*Computer*, April 1982, pages 41-53)

4.4 A Design of Programmable Logic Arrays with Universal Tests 350
H. Fujiwara and K. Kinoshita (*IEEE Transactions on Computers*, November 1981, pages 823-828)

4.5 Built-In Tests for VLSI Finite-State Machines 356
K.A. Hua, J.-Y. Jou, and J.A. Abraham (*Proceedings of the 14th Fault-Tolerant Computing Conference*, 1984, pages 292-297)

4.6 New Techniques for High Speed Analog Testing 362
M.V. Mahoney (*Proceedings of the International Test Conference*, 1983, pages 589-597)

4.7 Codec Testing Using Synchronized Analog and Digital Signals 371
D.K. Shirachi (*Proceedings of the International Test Conference*, 1984, pages 447-454)

Chapter 5: Testable Design and Built-In Self-Test 379

5.1 Design for Testability—A Survey 383
T.W. Williams and K.P. Parker (*Proceedings of the IEEE*, January 1983, pages 98-112)

5.2 Good Controllability and Observability Do Not Guarantee Good Testability 398
J. Savir (*IEEE Transactions on Computers*, December 1983, pages 1198-1200)

5.3 Testability Measures—What Do They Tell Us? 401
V.D. Agrawal and M.R. Mercer (*Proceedings of the International Test Conference*, 1982, pages 391-396)

5.4 Syndrome-Testable Design of Combinational Circuits 407
J. Savir (*IEEE Transactions on Computers*, June 1980, pages 442-451)

5.5 Syndrome-Testability Can Be Achieved by Circuit Modification 419
G. Markowsky (*IEEE Transactions on Computers*, August 1981, pages 604-606)

5.6 Syndrome-Testing of "Syndrome-Untestable" Combinational Circuits 421
J. Savir (*IEEE Transactions on Computers*, August 1981, pages 606-608)

5.7 The Weighted Syndrome Sums Approach to VLSI Testing 425
Z. Barzilai, J. Savir, G. Markowsky, and M.G. Smith (*IEEE Transactions on Computers*, December 1981, pages 996-1000)

5.8 Built-In Testing: State-of-the-Art 430
B. Koenemann, G. Zwiehoff, and R. Bosch (*Proceedings of the Curriculum for Test Technology*, 1983, pages 83-89)

5.9	Built-In Self-Test Techniques	437
	E.J. McCluskey (*IEEE Design & Test*, April 1985, pges 21-28)	
5.10	Built-In Self-Test Structures	445
	E.J. McCluskey (*IEEE Design & Test*, April 1985, pages 29-36)	
5.11	Built-In Testing of One-Dimensional Unilateral Iterative Arrays	453
	E.M. Aboulhamid and E. Cerny (*IEEE Transactions on Computers*, June 1984, pages 560-564)	
5.12	LOCST: A Built-In Self-Test Technique	457
	J.J. LeBlanc (*IEEE Design & Test*, November 1984, pages 45-52)	

Chapter 6: VLSI Test Systems and the Future 465

6.1	VLSI Test Gear Keeps Pace with Chip Advances	467
	H. Bierman (*Electronics*, April 1984, pages 125-128)	
6.2	A Merger of CAD and CAT Is Breaking the VLSI Test Bottleneck	471
	E. Hnatek (*Electronics*, April 1984, pages 129-134)	
6.3	VLSI Testers Ramp up Capabilities for Mixed-Signal Chips and Hybrids	477
	D. Johnson (*Electronics*, April 1984, pages 135-139)	
6.4	Fundamental Timing Problems in Testing MOS VLSI on Modern ATE	482
	M.R. Barbar (*IEEE Design & Test*, August 1984, pages 90-97)	
6.5	Managing VLSI Complexity: An Outlook	490
	C.H. Séquin (*Proceedings of the IEEE*, January 1983, pages 149-166)	
6.6	A Knowledge-Based System Using Design for Testability Rules	508
	P.W. Horstmann (*Proceedings of the 14th Fault-Tolerant Computing Conference*, 1984, pages 278-284)	
6.7	Diagnostic Reasoning Based on Structure and Behavior	515
	R. Davis (*Artificial Intelligence*, 1984, pages 347-410)	

Bibliography 579

Glossary 597

Author Index 599

Subject Index 601

Author's Biography 603

Chapter One: An Introduction to Testing and VLSI Design

This book is about the techniques and tools that may help us solve the VLSI design validation and testing crisis of the 1980s. There are many reasons why a particular design may not work. These range from very low-level problems, such as two signals shorted together because they were too close to each other, to high-level bugs in algorithms. In another dimension, VLSI circuits may fail due to production problems or bonding errors.

The purpose of this chapter is to summarize the VLSI design activities, to review the status of VLSI testing, and to establish the need for diagnosis and reliable design of VLSI systems. The remaining chapters will focus on specific ways for satisfying this need.

1.1 VLSI Circuits

The advances of integrated circuit (IC) technology in the last decade have been staggering. For the past twenty years chip complexity and thus the functional capability of a single IC have roughly doubled every year. And this trend is expected to continue for another decade! New approaches and new tools will be necessary for the design and testing of the next generation of VLSI chips.

The increase in functional density of ICs is mainly due to two factors. First, it is the technological advances that make possible ever smaller feature dimensions. MOS device scaling theory as presented in 1974 [1] offered a methodology of designing devices and processes for high-density circuits. The way in which processes are scaled to make the dimensions smaller involves the careful coordination of several different variables including the supply voltage and the impurity distributions. Because classic scaling [1] can lead to many undesirable properties, particularly in terms of delays along the interconnect, measures have been taken to avoid these difficulties [2]. However, there appears to be a fundamental limit [3,4] of approximately 0.25-micron channel length, where certain physical effects such as the tunneling through the gate oxide and fluctuations in the positions of impurities in the depletion layers begin to make devices having smaller dimensions unworkable. Currently, direct-write electron beam lithography systems can reliably produce 0.5-micron chip features.

The improvement in performance and density of ICs is only partly due to technological advances. An equally important factor is the ingenuity of the circuit and chip designers. For example, the number of individual features (e.g., MOS transistors) necessary to implement a particular function (e.g., storing one bit of information) has been reduced dramatically in the past decade [5]. However, as manufacturers continue their quest for denser memory cells, a number of fundamental limits may come into play. Four of the most critical are thermal stability, thermal noise, quantum mechanical tunneling, and thermal dissipation [4,5].

1.2 Process Monitoring and Yield

The process engineer is generally responsible for producing integrated circuits once he has been provided with proven designs. In many cases, where technology is rapidly changing, he is also responsible for changing the process to meet new design goals. These two demands mean that he needs to have intimate and immediate knowledge of how his processes are working. The devices he uses to test his process and the effort expended on process characterization and assurance will vary considerably depending on whether he is developing a new process or working with a mature one.

Numerical simulation has emerged recently as an important aid to process and device developments. In fact, process and device simulations are now as common as circuit simulation for two major reasons: Computer simulations are less expensive and much faster than experimental approaches.

Many instrumental methods are useful for solving problems that arise in VLSI technology development efforts. Some of the methods are most applicable to the analysis of circuit structures; others are more applicable to problems generated during experiments on the preparation of new materials for VLSI processing programs. The major areas of application of instrumental methods to VLSI problems are the determination of morphology, chemical analysis, the determination of crystallographic structures and mechanical properties, and electrical mapping of sites of device leakage and breakdown.

1.3 Integrated Circuit Fabrication Yield

Many factors contribute to the IC fabrication yield. Usually, 75% to 95% of all wafers have some operating chips. The percentage of functioning chips per wafer varies drastically depending on the complexity of the circuit: It varies from 5% for complex chips to 90% for simple logic circuits. It is important to note that in the early stages of developing a complex circuit, such as a microprocessor, the yield at wafer probe time may be very low. In particular, yields of 5% or even less may be common until the causes of such a low yield are well understood. On the other hand, for a highly optimized part that is in stable production, yields may be at a level of 50% or higher. Obtaining such a high yield, particularly for complex chips, is no easy task and is the result of a great deal of interaction between processing engineers and chip designers.

Chip assembly (i.e., bonding and packaging) is also a source of yield reduction. Generally, 85% to 95% of the chips are successfully assembled. After assembly, the final package undergoes a final test (with or without burn-in). On the average, 60% to 95% of the packaged chips pass this testing stage.

To summarize, the main components of IC fabrication yield (Y) are wafer processing yield (Y_W), wafer probe yield (Y_P), assembly yield (Y_A), and final test yield (Y_T). Mathematically, this can be represented as $Y = (Y_W)(Y_P)(Y_A)(Y_T)$.

1.4 Computer-Aided Design

It has become generally recognized that the computer is an essential tool for designing VLSI circuits. One of the principal goals behind the design aids is to significantly reduce the time between the initial concept of a complex system and the generation of IC masks. A second equally important goal is to allow the designer to efficiently explore design alternatives. Since design aids can consider a large number of design trade-offs per unit time, it is even conceivable that for sufficiently complex designs the computer aids could help produce designs superior to those produced manually. A third purpose of computer-aided design (CAD) systems is to assist designers in verifying the correctness of their designs. We are mainly interested in this last application of CAD tools.

1.5 Testing and Reliable Operation

The three major techniques for achieving reliable operation are testing, concurrent error detection, and fault-tolerance. Testing (off-line) is mainly used for revealing manufacturing defects and permanent failures. To catch transient errors (due to intermittent failures, timing problems, noise, radiation, etc.), concurrent (on-line) error detection is most suitable. Continuous system operation under physical failures (both transient and permanent) can be achieved by fault-tolerance techniques.

Testing and fault-tolerance are not independent. More specifically, testing is a fundamental part of fault-tolerance. Every system must be tested thoroughly during manufacture to ensure that it is defect-free. Moreover, testing should also be performed during the life of the system because the overhead required for fault-tolerance grows very quickly as more faults must be tolerated at one time. Usually, it is assumed that only a single fault exists in the system. This assumption is justifiable only if periodic testing of the system is done in order to scrub the latent faults and ensure that, during system operation, the probability of two failures is extremely small.

1.6 Testing

A VLSI circuit will undergo testing at different phases of development and production, not to mention in-the-field testing.

1.6.1 Prototype Testing

When the first prototype wafers of a new design have been fabricated, it is common practice to visually inspect the circuits under a microscope and package say 20–50 samples for evaluation, since it is difficult to carry out detailed diagnostic tests at the wafer stage. In such prototype circuit testing, it is imperative to be able to identify and localize faults, and this requirement, coupled with device characterization can assist with fault diagnosis and correction. Prototype faults may be process-induced such as those caused by pin-holes, open circuits, and short circuits which are statistically distributed over a wafer, or they may be design faults such as missing contacts, wrong interconnections, wrong transistor ratios, or excessive signal delays on long interconnection lines. Prototype testing is invariably executed by the designer using sophisticated test equipment. Testing of the chip using electron-beam techniques to detect both physical and electrical defects can also be applied. One such useful technique involves the use of voltage contrast in a scanning electron microscope. This voltage contrast technique presents the designer with a "map" of the logic states present in his design and facilitates fault finding. In prototype testing, parametrical tests to evaluate circuit speed, power consumption, and temperature performance are also required. The number and nature of these tests will depend on the type of circuit being tested, and from these parametric test results a data sheet can be produced for the circuit.

1.6.2 Production Testing

Production test requirements are quite different since, here, testing time per circuit becomes a crucial factor. Production tests are required to run on production-oriented test equipment, and short test times commensurate with reasonable fault coverage should be used to provide a go/no-go decision. It is not generally possible to redesign or repair an integrated circuit at this stage.

Since the packaging of devices is a relatively expensive process, as much of the testing as possible is performed at the wafer stage to identify those circuits that have obvious defects. At the wafer test stage, a wafer prober is used to make contact with the individual pads on the circuit and the prober can step over the wafer in x and y directions to test every circuit location. This stepping operation is done under computer control, and at every circuit location, the individual probes can be used to excite or monitor voltages on the pads, again controlled by the computer.

The first tests performed by the wafer probes monitor the quality of the fabrication process. Two alternative approaches can be used with special-purpose test structures: incorporating the structures in every circuit (development and teaching situations) or including them on the wafers as "drop-ins" in approximately five or six die locations distributed over the wafer. Typical test structures monitor process parameters such as gate-to-source threshold voltages and leakage currents for the various types of transistors, drain-to-source punch-through breakdown voltage, gate oxide breakdown voltage, field breakdown voltage, transistor gains, and circuit speeds as measured by a ring oscillator consisting of a string (prime number) of minimum geometry inverter circuits.

Individual wafers which satisfy the production spreads on processing tolerances are then tested for functional integrity by the prober. Here most of the significant states of the circuit are analyzed at DC and low frequencies, but tests at high frequency to the full specifications are normally performed on the packaged devices. The wafer prober can also be used to perform low-frequency parametric testing. During wafer probing, circuits that fail to satisfy any of the test requirements are marked with an ink dot to be rejected at the packaging stage.

Automatic test equipment (ATE) for the production situation normally operates on packaged chips, and computer control is used to define the function of individual pin connections as well as their signal waveforms. Most of these systems operate by applying a sequence of predetermined test stimuli to the input pins and then comparing the output responses at the output pins with the anticipated responses, thereby detecting errors.

1.7 Concurrent Error Detection

Concurrent (on-line) error detection is a technique used to check the results of a computation when it is required that the system produce correct results. Such a technique is imperative when a system is operated in environments that can induce transient errors. Note that this does not test the hardware, since only the symptoms of failure can be observed on-line. Classical work in this area has been concerned with the design of self-checking circuits. Most of these results have been for gate-level designs and gate-level faults. Recently, transistor-level failures have been studied, and techniques for designing self-checking circuits under these have been reported. More work is needed in this area. There have also been some promising results in the use of time redundancy, rather than hardware redundancy, for concurrent error detection.

Except for memories, and some large computer systems, there has been very little application of concurrent error-detection techniques in chips or small systems. In the future, these techniques will become more important as feature sizes shrink with advances in technology, and this will result in a greater likelihood of transient errors.

1.8 Fault-Tolerance

When very high reliability is required of a system, not only do failures have to be detected, but the system should have additional resources to replace the failed parts. Two aspects of fault-tolerance can be identified: 1) techniques to tolerate manufacturing defects owing to the requirements of high yield, and 2) techniques to tolerate failures in the field.

Fault-tolerance techniques for yield improvement have been primarily applied to memory circuits where additional cells are provided to take up the function of failed cells. There has been interest in extending this to logic blocks in wafer-scale circuits. The major problem here is that of testing the blocks and locating the exact points of failure. Built-in tests might provide a solution to this problem, but further work is required in this area.

When there is a requirement for continuous operation of a system in the field, fault-tolerance techniques must be incorporated into the design to ensure that a failure does not cause a crash of the system. Two methods are used to achieve this: fault masking (where the failures are masked by other correctly operating units), and error detection followed by correction and reconfiguration. Fault masking, for example, using triplication and voting, has been used in space and military systems. The problem is the high hardware overhead that is required. The use of self-checking circuits with backup hardware has been proposed, but there are problems in the area of the switching mechanism, as well as in the operating system. Some recent work has concentrated on techniques that use information about the nature of the computation to define appropriate system-level coding techniques to detect faulty units and that, with the use of a few additional computation units, can detect and correct errors. This technique, which seems extremely useful for signal processing systems, shows a great deal of promise and should be explored in the future.

1.9 Reprints

The theory, design, fabrication, and operation of integrated circuits are introduced in the paper by Clark [6]. The classical texts by Mead and Conway [7], Muroga [8], Hodges and Jackson [9], Glasser and Dobberpuhl [10], Seitz [11], Sze [12], Elmasry [13], Gray et al. [14], Newkirk and Mathews [15], Mavor, et al. [16], and Weste and Eshraghian [17] are highly recommended for further reference.

Testing of logic circuits is introduced in the first reprinted tutorial paper by Abadir and Reghbati. Various aspects of testing are discussed in the books by Breuer and Friedman [18], McCluskey [19], Bennetts [20,21], Stover [22], Lenk [23,24], Roth [25], Fee [26], Lala [27], Pradhan [28], Arsenault and Roberts [29], and Siewiorek and Swarz [30]. A good presentation of self-checking circuits is given in Wakerley [31].

Computer-aided design and simulation of VLSI circuits is surveyed in the second reprinted tutorial paper by Newton. The books by Antognetti et al. [32], Rabbat [33], Breuer [34,35], Ullman [36], and Ayres [37], are recommended for further reading.

The third reprinted paper, by Agrawal et al., describes the TITUS design automation system, developed at Bell Labs. This system includes programs for automatic design audits, scan implementation, and test vector generation. Special layout features that minimize the hardware overhead in custom polycell designs have also been developed.

For a discussion of VLSI test systems and the future of testing and VLSI, the reader is referred to Chapter Six. A comprehensive and up-to-date discussion of the current status and future of VLSI and testing appears in Volume 26 of *Advances in Computers* [38]. Application of artificial intelligence techniques to VLSI CADT is discussed in [39].

For further references on a specific topic, the reader is referred to the Bibliography that appears at the end of the book.

References

[1] R.H. Dennard et al., "Design of Ion-Implanted MOSFET's with Very Small Physical Dimensions," *IEEE J. Solid-State Circuits,* Vol. SC-9, No. 5, Oct. 1974, pp. 256–268.

[2] Y.A. El-Mansy, "On Scaling MOS Devices for VLSI," *Proc. of IEEE Int. Conf. on Circuits and Computers,* 1980, pp. 457–460.

[3] B. Hoeneisen and C.A. Mead, "Fundamental Limitations in Micro-Electronics—I . MOS Technology," *Solid-State Electronics,* Vol. 15, 1972, pp. 819–829.

[4] R.W. Keyes, "Physical Limits in Digital Electronics," *Proc. IEEE,* Vol. 63, May 1975, pp. 740–767.

[5] A.V. Pohm, "High-Speed Memory Systems," *IEEE Computer,* Vol. 17, No. 10, Oct. 1984, pp. 162–171.

[6] W.A. Clark, "From Electron Mobility to Logical Structure: A View of Integrated Circuits," *ACM Computing Surveys,* Vol. 12, No. 3, Sept. 1980, pp. 325–356.

[7] C.A. Mead and L. Conway, *Introduction to VLSI Systems,* Addison-Wesley, Reading, MA, 1980.

[8] S. Muroga, *VLSI System Design,* Wiley, New York, NY, 1982.

[9] D.A. Hodges and H.G. Jackson, *Analysis and Design of Digital Integrated Circuits,* McGraw-Hill, New York, NY, 1983.

[10] L.A. Glasser and D.W. Dobberpuhl, *The Design and Analysis of VLSI Circuits,* Addison-Wesley, 1985.

[11] C. Seitz, *Structured VLSI Design,* Addison-Wesley, 1986 (to appear).

[12] S.M. Sze (Ed.), *VLSI Technology,* McGraw-Hill, New York, NY, 1983.

[13] M.I. Elmasry (Ed.), *Digital MOS Integrated Circuits,* IEEE Press, New York, NY, 1981.

[14] Gray et al. (Eds.), *Analog MOS Integrated Circuits,* IEEE Press, New York, NY, 1980.

[15] J. Newkirk and R. Mathews, *The VLSI Designer's Library,* Addison-Wesley, Reading, MA, 1983.

[16] J. Mavor, M.A. Jack, and P.B. Denyer, *Introduction to MOS LSI Design,* Addison-Wesley, Reading, MA, 1983.

[17] N. Weste and K. Eshraghian, *Principles of CMOS VLSI Design: A Systems Perspective,* Addison-Wesley, Reading, MA, 1985.

[18] M.A. Breuer and A.D. Friedman, *Diagnosis and Reliable Design of Digital Systems,* Computer Science Press, Rockville, MD, 1976.

[19] E.J. McCluskey, *Logic Design Principles,* Prentice-Hall, Englewood Cliffs, NJ, 1986 (to appear).

[20] R.G. Bennetts, *Introduction to Digital Board Testing,* Crane-Russak, New York, NY, 1982.

[21] R.G. Bennetts, *Design of Testable Logic Circuits,* Addison-Wesley, Reading, MA, 1984.

[22] A.C. Stover, *ATE: Automatic Test Equipment,* McGraw-Hill, New York, NY, 1984.

[23] J.D. Lenk, *Handbook of Electronic Test Procedures,* Prentice-Hall, Englewood Cliffs, NJ, 1982.

[24] J.D. Lenk, *Handbook of Advanced Troubleshooting,* Prentice-Hall, Englewood Cliffs, NJ, 1983.

[25] J.P. Roth, *Computer Logic, Testing, and Verification,* Computer Science Press, Rockville, MD, 1980.

[26] W.G. Fee (Ed.), *Tutorial: LSI Testing,* IEEE Computer Society, Washington, D.C., 1978.

[27] P. Lala, *Fault-Tolerant & Fault-Testable Hardware Design,* Prentice-Hall, Englewood Cliffs, NJ, 1985.

[28] D.J. Pradhan (Ed.), *Fault-Tolerant Computing,* Prentice-Hall, Englewood Cliffs, NJ, 1986 (to appear).

[29] J.E. Arsenault and J.A. Roberts (Eds.), *Reliability and Maintainability of Electronic Systems,* Computer Science Press, 1980.

[30] D.P. Siewiorek and R.S. Swarz, *The Theory and*

Practice of Reliable System Design, Digital Press, Maynard, MA, 1982.

[31] J.F. Wakerley, *Error Detecting Codes, Self-Checking Circuits and Applications,* North Holland, 1978.

[32] P. Antognetti, et al. (Eds.), *Computer Design Aids for VLSI Circuits,* Sifthoff & Nordhoff, Rockville, MD, 1981.

[33] G. Rabbat (Ed.), *Hardware and Software Concepts in VLSI,* Van Nostrand Reinhold, New York, NY, 1983.

[34] M.A. Breuer (Ed.), *Digital System Design Automation: Languages, Simulation & Data Base,* Computer Science Press, Rockville, MD, 1975.

[35] M.A. Breuer (Ed.), *Design Automation of Digital Systems,* Prentice-Hall, Englewood Cliffs, NJ, 1972.

[36] J.D. Ullman, *Computational Aspects of VLSI,* Computer Science Press, Rockville, MD, 1984.

[37] R.F. Ayres, *Silicon Compilation and the Art of Automatic Microchip Design,* Prentice-Hall, Englewood Cliffs, NJ, 1983.

[38] H.K. Reghbati, "VLSI Testing and Validation Techniques," in *Advances in Computers,* Vol. 26, Academic Press, 1987 (to appear).

[39] H.K. Reghbati and N. Cercone, "Knowledge Representation for Computer-Aided Design and Test of VLSI Systems," in *Knowledge Representation,* Springer Verlag, 1986 (to appear).

Tests good for SSI and MSI circuits can't cope with the complexity of LSI. New techniques for test generation and response evaluation are required.

LSI Testing Tech

M. S. Abadir*

H. K. Reghbati**

University of Saskatchewan

The growth in the complexity and performance of digital circuits can only be described as explosive. Large-scale integrated circuits are being used today in a variety of applications, many of which require highly reliable operation. This is causing concern among designers of tests for LSI circuits. The testing of these circuits is difficult for several reasons:

• The number of faults that has to be considered is large, since an LSI circuit contains thousands of gates, memory elements, and interconnecting lines, all individually subject to different kinds of faults.

• The observability and controllability of the internal elements of any LSI circuit are limited by the available number of I/O pins. As more and more elements are packed into one chip, the task of creating an adequate test becomes more difficult. A typical LSI chip may contain 5000 gates but only 40 I/O pins.

• The implementation details of the circuits usually are not disclosed by the manufacturer. For example, the only source of information about commercially available microprocessors is the user's manual, which details the instruction set and describes the architecture of the microprocessor at the register-transfer level, with some information on the system timing. The lack of implementation information eliminates the use of many powerful test generation techniques that depend on the actual implementation of the unit under test.

• As more and more gates and flip-flops are packed into one chip, new failure modes—such as pattern-sensitivity faults—arise.[1] These new types of faults are difficult to detect and require lengthy test patterns.

• The dynamic nature of LSI devices requires high-speed test systems that can test the circuits when they are operating at their maximum speeds.

• The bus structure of most LSI systems makes fault isolation more difficult because many devices—any of which can cause a fault—share the same bus.

• Solving the problems above increases the number of test patterns required for a successful test. This in turn increases both the time required for applying that test and the memory needed to store the test patterns and their results.

LSI testing is a challenging task. Techniques that worked well for SSI and MSI circuits, such as the D-algorithm, do not cope with today's complicated LSI and VLSI circuits. New testing techniques must be developed. In what follows, we describe some basic techniques developed to solve the problems associated with LSI testing.

---

*M. S. Abadir is now in the Department of Electrical Engineering of the University of Southern California.
**H. K. Reghbati is now in the Department of Computing Science of Simon Fraser University, Burnaby, British Columbia.

Reprinted from *IEEE Micro*, February 1983, pages 34-51. Copyright © 1983 by The Institute of Electrical and Electronics Engineers, Inc.

Testing methods

There are many test methods for LSI circuits, each with its own way of generating and processing test data. These approaches can be divided into two broad categories—*concurrent* and *explicit*.[2]

In concurrent approaches, normal user-application input patterns serve as diagnostic patterns. Thus testing and normal computation proceed concurrently. In explicit approaches, on the other hand, special input patterns are applied as tests. Hence, normal computation and testing occur at different times.

Concurrent testing. Systems that are tested concurrently are designed such that all the information transferred among various parts of the system is coded with different types of error detecting codes. In addition, special circuits monitor these coded data continuously and signal the detection of any fault.

Different coding techniques are required to suit the different types of information used inside LSI systems. For example, m-out-of-n codes (n-bit patterns with exactly m 1's and $n-m$ 0's) are suitable for coding control signals, while arithmetic codes are best suited for coding ALU operands.[3]

The monitoring circuits—*checkers*—are placed in various locations inside the system so that they can detect most of the faults. A checker is sometimes designed in a way that enables it to detect a fault in its own circuitry as well as in the monitored data. Such a checker is called a *self-checking checker*.[3]

Hayes and McClusky surveyed various concurrent testing methods that can be used with microprocessor-based LSI systems.[2] Concurrent testing approaches provide the following advantages:

• Explicit testing expenses (e.g., for test equipment, down time, and test pattern generation) are eliminated during the life of the system, since the data patterns used in normal operation serve as test patterns.
• The faults are detected instantaneously during the use of the LSI chip, hence the first faulty data pattern caused by a certain fault is detected. Thus, the user can rely on the correctness of his output results within the degree of fault coverage provided by the error detection code used. In explicit approaches, on the other hand, nothing can be said about the correctness of the results until the chip is explicitly tested.
• Transient faults, which may occur during normal operation, are detected if they cause any faulty data pattern. These faults cannot be detected by any explicit testing method.

Unfortunately, the concurrent testing approach suffers from several problems that limit its usage in LSI testing:

• The application patterns may not exercise all the storage elements or all the internal connection lines. Defects may exist in places that are not exercised, and hence the faults these defects would produce will not be detected. Thus, the assumption that faults are detected as they occur, or at least before any other fault occurs, is no longer valid. Undetected faults will cause fault accumulation. As a result, the fault detection mechanism may fail because most error detection codes have a limited capability for detecting multiple faults.

- Using error detecting codes to code the information signals used in an LSI chip requires additional I/O pins. At least two extra pins are needed as error signal indicators. (A single pin cannot be used, since such a pin stuck at the good value could go undetected.) Because of constraints on pin count, however, such requirements cannot be fulfilled.

- Additional hardware circuitry is required to implement the checkers and to increase the width of the data carriers used for storing and transferring the coded information.

- Designing an LSI circuit for concurrent testing is a much more complicated task than designing a similar LSI circuit that will be tested explicitly.

- Concurrent approaches provide no control over critical voltage or timing parameters. Hence, devices cannot be tested under marginal timing and electrical conditions.

- The degree of fault coverage usually provided by concurrent methods is less than that provided by explicit methods.

The above-mentioned problems have limited the use of concurrent testing for most commercially available LSI circuits. However, as digital systems grow more complex and difficult to test, it becomes increasingly attractive to build test procedures into the UUT—unit under test—itself. We will not consider the concurrent approach further in this article. For a survey of work in concurrent testing, see Hayes and McCluskey.[2]

Explicit testing. All explicit testing methods separate the testing process from normal operation. In general, an explicit testing process involves three steps:

- *Generating the test patterns.* The goal of this step is to produce those input patterns which will exercise the UUT under different modes of operation while trying to detect any existing fault.

- *Applying the test patterns to the UUT.* There are two ways to accomplish this step. The first is external testing—the use of special test equipment to apply the test patterns externally. The second is internal testing—the application of test patterns internally by forcing the UUT to execute a self-testing procedure.[2] Obviously, the second method can only be used with systems that can execute programs (for example, with microprocessor-based systems). External testing gives better control over the test process and enables testing under different timing and electrical conditons. On the other hand, internal testing is easier to use because it does not need special test equipment or engineering skills.

- *Evaluating the responses obtained from the UUT.* This step is designed with one of two goals in mind. The first is the detection of an erroneous response, which indicates the existence of one or more faults (*go/no-go testing*). The other is the isolation of the fault, if one exists, in an easily replaceable module (*fault location testing*). Our interest in this article will be go/no-go testing, since fault location testing of LSI circuits sees only limited use.

Many explicit test methods have evolved in the last decade. They can be distinguished by the techniques used to generate the test patterns and to detect and evaluate the faulty responses (Figure 1). In what follows, we concentrate on explicit testing and present in-depth discussions of the methods of test generation and response evaluation employed with explicit testing.

Test generation techniques

The test generation process represents the most important part of any explicit testing method. Its main goal is to generate those test patterns that, when applied to the UUT, sensitize existing faults and propagate a faulty response to an observable output of the UUT. A test sequence is considered good if it can detect a high percentage of the possible UUT faults; it is considered good, in other words, if its degree of *fault coverage* is high.

Rigorous test generation should consist of three main activities:

- Selecting a good descriptive model, at a suitable level, for the system under consideration. Such a model

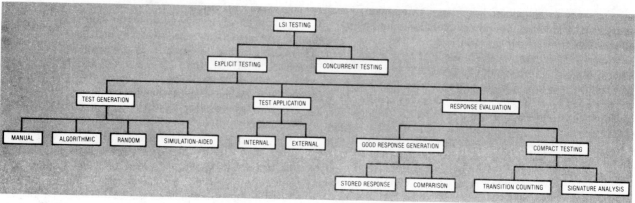

Figure 1. LSI test technology.

> **NP-complete problems**
>
> The theory of NP-completeness is perhaps the most important theoretical development in algorithm research in the past decade.[1] Its results have meaning for all researchers who are developing computer algorithms.
>
> It is an unexplained phenomenon that for many of the problems we know and study, the best algorithms for their solution have computing times which cluster into two groups. The first group consists of problems whose solution is bounded by a polynomial of small degree. Examples include ordered searching, which is $O(\log n)$, polynomial evaluation, which is $O(n)$, and sorting, which is $O(n \log n)$.[2]
>
> The second group contains problems whose best-known algorithms are nonpolynomial. For example, the best algorithms described in Horowitz and Sahni's book[2] for the traveling salesman and the knapsack problems have a complexity of $O(n^2 2^n)$ and $O(2^{n/2})$, respectively. In the quest to develop efficient algorithms, no one has been able to develop a polynomial-time algorithm for any problem in the second group.
>
> The theory of NP-completeness does not provide a method for obtaining polynomial-time algorithms for these problems. But neither does it say that algorithms of this complexity do not exist. What it does show is that many of the problems for which there is no known polynomial-time algorithm are computationally related. In fact, a problem that is NP-complete has the property that it can be solved in polynomial time if all other NP-complete problems can also be solved in polynomial time.
>
> **References**
>
> 1. M. Garey and D. Johnson, *Computers and Intractability: A Guide to the Theory of NP-Completeness*, W. H. Freeman, San Francisco, 1978.
> 2. E. Horowitz and S. Sahni, *Fundamentals of Computer Algorithms*, Computer Science Press, Washington, DC, 1978.

should reflect the exact behavior of the system in all its possible modes of operation.

• Developing a fault model to define the types of faults that will be considered during test generation. In selecting a fault model, the percentage of possible faults covered by the model should be maximized, and the test costs associated with the use of the model should be minimized. The latter can be accomplished by keeping the complexity of the test generation low and the length of the tests short. Clearly these objectives contradict one another—a good fault model is usually found as a result of a trade-off between them. The nature of the fault model is usually influenced by the model used to describe the system.

• Generating tests to detect all the faults in the fault model. This part of test generation is the soul of the whole test process. Designing a test sequence to detect a certain fault in a digital circuit usually involves two problems. First, the fault must be *excited;* i.e., a certain test sequence must be applied that will force a faulty value to appear at the fault site if the fault exists. Second, the test must be *made sensitive to* the fault; i.e., the effect of the fault must propagate through the network to an observable output.

Rigorous test generation rests heavily on both accurate descriptive (system) models and accurate fault models.

Test generation for digital circuits is usually approached either at the gate level or at the functional level. The classical approach of modeling digital circuits as a group of connected gates and flip-flops has been used extensively. Using this level of description, test designers introduced many types of fault models, such as the classical stuck-at model. They also assumed that such models could describe physical circuit failures in terms of logic. This assumption has sometimes restricted the number of physical failures that can be modeled, but it has also reduced the complexity of test generation since failures at the elementary level do not have to be considered.

Many algorithms have been developed for generating tests for a given fault in combinational networks.[1,4,5,6,7] However, the complexity of these algorithms depends on the topology of the network; it can become very high for some circuits. Ibarra and Sahni have shown that the problem of generating tests to detect single stuck-at faults in a combinational circuit modeled at the gate level is an NP-complete problem.[8] Moreover, if the circuit is sequential, the problem can become even more difficult depending on the deepness of the circuit's sequential logic.

Thus, for LSI circuits having many thousands of gates, the gate-level approach to the test generation problem is not very feasible. A new approach—the functional level—is needed.

Another important reason for considering faults at the functional level is the constraint imposed on LSI testing by a user environment—the test patterns have to be generated without a knowledge of the implementation details of the chip at the gate level. The only source of information usually available is the typical IC catalog, which details the different modes of operation and describes the general architecture of the circuit. With such information, the test designer finds it easier to define the functional behavior of the circuit and to associate faults with the functions. He can partition the UUT into various modules such as registers, multiplexers, ALUs, ROMs, and RAMs. Each module can be treated as a "black box" performing a specified input/output mapping. These modules can then be tested for *functional failures;* explicit consideration of faults affecting the internal lines is not necessary. The example given below clarifies the idea.

Consider a simple one-out-of-four multiplexer such as the one shown in Figure 2. This multiplexer can be modeled at the gate level as shown in Figure 2a, or at the functional level as shown in Figure 2b.

A possible fault model for the gate-level description is the single stuck-at fault model. With this model, the fault list may contain faults such as the line labeled with "f" is stuck at 0, or the control line "C_0" is stuck at 1.

At the functional level, the multiplexer is considered a black box with a well-defined function. Thus, a fault model for it may specify the following as possible faults: selection of wrong source, selection of no source, or presence of stuck-at faults in the input lines or in the multiplexer output. With this model, the fault list may contain faults such as source "X" is selected instead of source "Y," or line "Z" is stuck at 1.

Ad hoc methods—which determine what faults are the most probable—are sometimes used to generate fault lists. But if no fault model is assumed, then the tests derived must be either exhaustive or a rather ad hoc check of the functionality of the system. Exhaustive tests are impossible for even small systems because of the enormous number of possible states, and superficial tests provide neither good coverage nor even an indication of what faults are covered.

Once the fault list has been defined, the next step is to find the test patterns required to detect the faults in the list. As previously mentioned, each fault first has to be excited so that an error signal will be generated somewhere in the UUT. Then this signal has to be sensitized at one of the observable outputs of the UUT. The three examples below describe how to excite and sensitize different types of faults in the types of modules usually encountered in LSI circuits.

Consider the gate-level description of the three-bit incrementer shown in Figure 3. The incrementer output $Y_2Y_1Y_0$ is the binary sum of C_i and the three-bit binary number $X_2X_1X_0$, while C_o is the carry-out bit of the sum. Note that $X_0(Y_0)$ is the least significant bit of the incrementer input (output).

Assume we want to detect the fault "line f is stuck at 0." To excite that fault we will force a 1 to appear on line f so that, if it is stuck at 0, a faulty value will be generated at the fault site. To accomplish this both X_0 and C_i must be set to 1. To sensitize the faulty 0 at f, we have to set X_1 to 1; this will propagate the fault to Y_2 independent of the value of X_2. Note that if we set X_1 to 0, the fault will be masked since the AND gate output will be 0, independent of the value at f. Note also that X_2 was not specified in the above test. However, by setting X_2 to 1, the fault will propagate to both Y_2 and C_o, which makes the response evaluation task easier.

Consider a microprocessor RAM and assume we want to generate a test sequence to detect the fault "accessing word i in the RAM results in accessing word j instead." To excite such a fault, we will use the following sequence

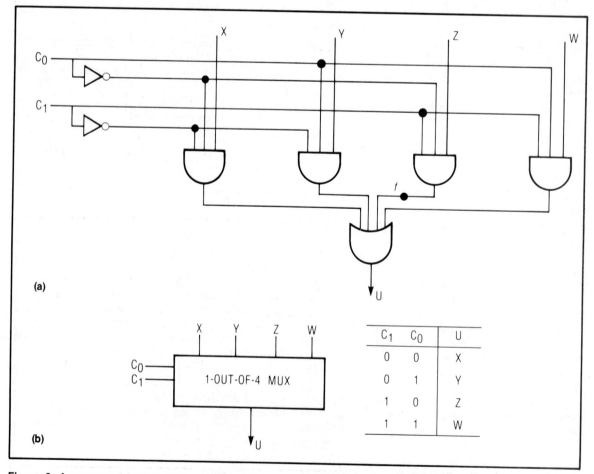

Figure 2. A one-out-of-four multiplexer—gate-level description (a); functional-level description (b).

of instructions (assume a microprocessor with single-operand instructions):

Load the word 00 . . . 0 into the accumulator.
Store the accumulator contents into memory address j.
Load the word 11 . . . 1 into the accumulator.
Store the accumulator contents into memory address i.

If the fault exists, these instructions will force a 11 . . . 1 word to be stored in memory address j instead of 00 . . . 0. To sensitize the fault, we need only read what is in memory address j, using the appropriate instructions. Note that the RAM and its fault have been considered at the functional level, since we did not specify how the RAM is implemented.

Consider the program counter (PC) of a microprocessor and assume we want to generate a test sequence that will detect any fault in the incrementing mode of this PC, i.e., any fault that makes the PC unable to be incremented from x to $x+1$ for any address x. One way to excite this fault is to force the PC to step through all the possible addresses. This can be easily done by initializing the PC to zero and then executing the no-operation instruction $x+1$ times. As a result, the PC will contain an address different than $x+1$. By executing another no-operation instruction, the wrong address can be observed at the address bus and the fault detected. In practice, such an exhaustive test sequence is very expensive, and more economical tests have to be used. Note that, as in the example immediately above, the problem and its solution have been considered at the functional level.

Four methods are currently used to generate test patterns for LSI circuits: manual test generation, algorithmic test generation, simulation-aided test generation, and random test generation.

Manual test generation. In manual test generation, the test designer carefully analyzes the UUT. This analysis can be done at the gate level, at the functional level, or at a combination of the two. The analysis of the different parts of the UUT is intended to determine the specific patterns that will excite and sensitize each fault in the fault list. At one time, the manual approach was widely used for medium- and small-scale digital circuits. Then, the formulation of the D-algorithm and similar algorithms eliminated the need for analyzing each circuit manually and provided an efficient means to generate the required test patterns.[1,5] However, the arrival of LSI circuits and microprocessors required a shift back toward manual test generation techniques, because most of the algorithmic techniques used with SSI and MSI circuits were not suitable for LSI circuits.

Manual test generation tends to optimize the length of the test patterns and provides a relatively high degree of fault coverage. However, generating tests manually takes a considerable amount of effort and requires persons with special skills. Realizing that test generation has to be done economically, test designers are now moving in the direction of automatic test generation.

One good example of manual test generation is the work done by Sridhar and Hayes,[9] who generated test patterns for a simple bit-sliced microprocessor at the functional level.

A bit-sliced microprocessor is an array of n identical ICs called slices, each of which is a simple processor for operands of k-bit length, where k is typically 2 or 4. The interconnections among the n slices are such that the entire array forms a processor for nk-bit operands. The simplicity of the individual slices and the regularity of the interconnections make it feasible to use systematic methods for fault analysis and test generation.

Sridhar and Hayes considered a one-bit processor slice as a simplified model for commercially available bit-sliced processors such as the Am2901.[10] A slice can be modeled as a collection of modules interconnected in a known way. These modules are regarded as black boxes with well-defined input-output relationships. Examples of these functional modules are ALUs, multiplexers, and registers. Combinational modules are described by their truth tables, while sequential modules are defined by their state tables (or state diagrams).

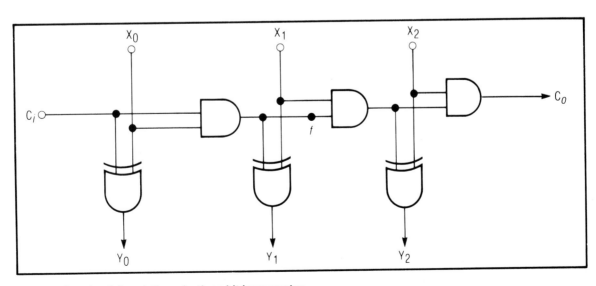

Figure 3. Gate-level description of a three-bit incrementer.

The following fault categories were considered:

• For combinational modules, all possible faults that induce arbitrary changes in the truth table of the module, but that cannot convert it into a sequential circuit.

• For sequential modules, all possible faults that can cause arbitrary changes in the state table of the module without increasing the number of states.

Only one module was assumed to be faulty at any time.

To test for the faults allowed by the above-mentioned fault model, all possible input patterns must be applied to each combinational module (exhaustive testing), and a checking sequence[11] to each sequential module. In addition, the responses of each module must be propagated to observable output lines. The tests required by the individual modules were easily generated manually—a direct consequence of the small operand size ($k=1$). And because the slices were identical, the tests for one slice were easily extended to the whole array of slices. In fact, Sridhar and Hayes showed that an arbitrary number of simple interconnected slices could be tested with the same number of tests as that required for a single slice, as long as only one slice was faulty at one time. This property is called *C-testability*. Note that the use of carry-lookahead when connecting slices eliminates C-testability. Also note that slices with operand sizes equal to 2 or more usually are not C-testable.

The idea of modeling a digital system as a collection of interconnected functional modules can be used in modeling any LSI circuit. However, using exhaustive tests and checking sequences to test individual modules is feasible only for toy systems. Hence, the fault model proposed by Sridhar and Hayes, though very powerful, is not directly applicable to LSI testing.

Algorithmic test generation. In algorithmic test generation, the test designer devises a set of algorithms to generate the 1's and 0's needed to test the UUT. Algorithmic test techniques are much more economical than manual techniques. They also provide the test designer with a high level of flexibility. Thus, he can improve the fault coverage of the tests by replacing or modifying parts of the algorithms. Of course, this task is much

---

Path sensitization and the D-algorithm

One of the classical fault detection methods at the gate and flip-flop level is the D-algorithm[1,2] employing the path sensitization testing technique.[3] The basic principle involved in path sensitization is relatively simple. For an input X_i to detect a fault "line a is stuck-at j, $j = 0, 1$," the input X_i must cause the signal a in the normal (fault-free) circuit to take the value $\bar{j}$. This condition is necessary but not sufficient to detect the fault. The error signal must be propagated along some path from its site to an observable output.

To generate a test to detect a stuck-at fault in a combinational circuit, the following path sensitization procedure must be followed:

• Excitation—The inputs must be specified so as to generate the appropriate value (0 for stuck-at 1 and 1 for stuck-at 0) at the site of the fault.

• Error propagation—A path from the fault site to an observable output must be selected, and additional signal values to propagate the fault signal along this path must be specified.

• Line justification—Input values must be specified so as to produce the signals values specified in the step above.

There may be several possible choices for error propagation and line justification. Also, in some cases there may be a choice of ways in which to excite the fault. Some of these choices may lead to an inconsistency, and so the procedure must backtrack and consider the next alternative. If all the alternatives lead to an inconsistency, this implies that the fault cannot be detected.

To facilitate the path sensitization process, we introduce the symbol D to represent a signal which has the value 1 in a normal circuit and 0 in a faulty circuit, and $\bar{D}$ to represent a signal which has the value 0 in a normal circuit and 1 in a faulty circuit. The path sensitization procedure can be formulated in terms of a cubical algebra[1,2] to enable automatic generation of tests. This also facilitates test generation for more complex fault models and for fault propagation through complex logic elements.

We shall define three types of cubes (i.e., line values specified in positional notation):

• For a circuit element E which realizes the combinational function f, the "primitive cubes" offer a typical presentation of the prime implicants of f and $\bar{f}$. These cubes concisely represent the logical behavior of E.

• A "primitive D-cube of a fault" in a logic element E specifies the minimal input conditions that must be applied to E in order to produce an error signal (D or $\bar{D}$) at the output of E.

• The "propagation D-cubes" of a logic element E specify the minimal input conditions to the logic element that are required to propagate an error signal on an input (or inputs) to the output of that element.

To generate a test for a stuck-at fault in a combinational circuit, the D-algorithm must perform the following:

(1) Fault excitation—A primitive D-cube of the fault under consideration must be selected. This generates the error signal D or $\bar{D}$ at the site of the fault. (Usually a choice exists in this step. The initial choice is arbitrary, and it may be necessary to backtrack and consider another choice).

simpler than modifying the 1's and 0's in a manually generated test sequence.

Techniques that use the gate-level description of the UUT, such as path sensitization[4] and the D-algorithm,[5] can no longer be used in testing complicated LSI circuits. Thus, the problem of generating meaningful sets of tests directly from the functional description of the UUT has become increasingly important. Relatively little work has been done on functional-level testing of LSI chips that are not memory elements.[9,12,13,14,15,16,17] Functional testing of memory chips is relatively simple because of the regularity of their design and also because their components can be easily controlled and observed from the outside. Various test generation algorithms have been developed to detect different types of faults in memories.[1,18] In the rest of this section we will concentrate on the general problem of generating tests for irregular LSI chips, i.e., for LSI chips which are not strictly memory chips.

It is highly desirable to find an algorithm that can generate tests for any LSI circuit, or at least most LSI circuits. One good example of work in this area is the technique proposed by Thatte and Abraham for generating tests for microprocessors.[12,13] Another approach, pursued by the authors of this article, is a test generation procedure capable of handling general LSI circuits.[15,16,17]

The Thatte-Abraham technique. Microprocessors constitute a high percentage of today's LSI circuits. Thatte and Abraham[12,13] approached the microprocessor test generation problem at the functional level.

The test generation procedure they developed was based on

- A functional description of the microprocessor at the register-transfer level. The model is defined in terms of data flow among storage units during the execution of an instruction. The functional behavior of a microprocessor is thus described by information about its instruction set and the functions performed by each instruction.
- A fault model describing faults in the various functional parts of the UUT (e.g., the data transfer function, the data storage function, the instruction decoding and control function). This fault model describes the faulty

---

(2) Implication—In Step 1 some of the gate inputs or outputs may be specified so as to uniquely imply values on other signals in the circuit. The implication procedure is performed both forwards and backwards through the circuit. Implication is performed as follows: Whenever a previously unspecified signal value becomes specified, all the elements associated with this signal are placed on a list B and processed one at a time (and removed). For each element processed, it is determined if new values of 0, 1, D, and $\overline{D}$ are implied, based on the previously specified inputs and outputs. These implied line values are determined by intersecting the test cube (which specifies all the previously determined signal values of the circuit) with the primitive cubes of the element. If any line values are implied, they are specified in the test cube, and the associated gates are placed on the list B. An inconsistency occurs when a value is implied on a line which has been specified previously to a different value. If an inconsistency occurs, the procedure must backtrack to the last point a choice existed, reset all lines to their values at that point, and begin again with the next choice.

(3) D-propagation—All the elements in the circuit whose output values are unspecified and whose input has some signal D or $\overline{D}$ are placed on a list called the D-frontier. In this step, an element from the D-frontier is selected and values are assigned to its unspecified inputs so as to propagate the D or $\overline{D}$ on its inputs to one of its outputs. This is accomplished by intersecting the current test cube describing the circuit signal values with a propagation D-cube of the selected element of the D-frontier, resulting in a new test cube. If such intersection is impossible, a new element in the D-frontier is selected. If intersection fails for all the elements in the D-frontier, the procedure backtracks to the last point at which a choice existed.

(4) Implication of D-propagation—Implication is performed for the new test cube derived in Step 3.

(5) Steps 3 and 4 are repeated until the faulty signal has been propagated to an output of the circuit.

(6) Line justification—Execution of Steps 1 to 5 may result in specifying the output value of an element E but leaving some of the inputs to the element unspecified. The unspecified inputs of such an element are assigned values so as to produce the desired output value. This is done by intersecting the test cube with any primitive cube of the element which has no specified signal values that differ from those of the test cube.

(7) Implication of line justification—Implication is performed on the new test cube derived in Step 6.

(8) Steps 6 and 7 are repeated until all specified element outputs have been justified. Backtracking may again be required.

References

1. J. P. Roth, W. G. Bouricius, and P. R. Schneider, "Programmed Algorithms to Compute Tests to Detect and Distinguish Between Failures in Logic Circuits," *IEEE Trans. Electronic Computers*, Vol. CE-16, No. 5, Oct. 1967, pp. 567-580.

2. M. A. Breuer and A. D. Friedman, *Diagnosis and Reliable Design of Digital Systems*, Computer Science Press, Washington, DC, 1976.

3. D. B. Armstrong, "On Finding a Nearly Minimal Set of Fault Detection Tests for Combinatorial Nets," *IEEE Trans. Electronic Computers*, Vol. EC-15, No. 2, Feb. 1966, pp. 63-73.

behavior of the UUT without knowing its implementation details.

The microprocessor is modeled by a graph. Each register in the microprocessor (including general-purpose registers and accumulator, stack, program counter, address buffer, and processor status word registers) is represented by a node of the graph. Instructions of the microprocessor are classified as being of transfer, data manipulation, or branch type. There exists a directed edge (labeled with an instruction) from one node to another if during the execution of the instruction data flow occurs from the register represented by the first node to that represented by the second. Examples of instruction representation are given in Figure 4.

Having described the function or the structure of the UUT, one needs an appropriate fault model in order to derive useful tests. The approach used by Thatte and Abraham is to partition the various functions of a microprocessor into five classes: the register decoding function, the instruction decoding and control function, the data storage function, the data transfer function, and the data manipulation function. Fault models are derived for each of these functions at a higher level and independently of the details of implementation for the microprocessor. The fault model is quite general. Tests are derived allowing any number of faults, but only in one function at a time; this restriction exists solely to cut down the complexity of test generation.

The fault model for the register decoding function allows any possible set of registers to be accessed instead of a particular register. (If the set is null then no register is accessed.) This fault model is thus very general and independent of the actual realization of the decoding mechanism.

For the instruction decoding and control function, the faulty behavior of the microprocessor is specified as follows—when instruction I_j is executed any one of the following can happen:

• Instead of instruction I_j some other instruction I_k is executed. This fault is denoted by $F(I_j/I_k)$.
• In addition to instruction I_j, some other instruction I_k is activated. This fault is denoted by $F(I_j/I_j+I_k)$.
• No instruction is executed. This fault is denoted by $F(I_j/\Phi)$.

Under this specification, any number of instructions can be faulty.

In the fault model for the data storage function, any cell in any data storage module is allowed to be stuck at 0 or 1. This can occur in any number of cells.

The fault model for the data transfer function includes the following types of faults:

• A line in a path used in the execution of an instruction is stuck at 0 or 1.
• Two lines of a path used in the instruction are coupled; i.e., they fail to carry different logic values.

Note that the second fault type cannot be modeled by single stuck-at faults. The transfer paths in this fault model are logical paths and thus will account for any failure in the actual physical paths.

Since there is a variety of designs for the ALU and other functional units such as increment or shift logic, no specific fault model is used for the data manipulation function. It is assumed that complete test sets can be derived for the functional units for a given fault model.

By carefully analyzing the logical behavior of the microprocessor according to the fault models presented above, Thatte and Abraham formulated a set of algorithms to generate the necessary test patterns. These algorithms step the microprocessor through a precisely defined set of instructions and addresses. Each algorithm was designed for detecting a particular class of faults, and theorems were proved which showed exactly the kind of faults detected by each algorithm. These algorithms employ the excitation and sensitization concepts previously described.

To gain insight into the problems involved in using the algorithms, Thatte investigated the testing of an eight-bit microprocessor from Hewlett-Packard.[12] He generated the test patterns for the microprocessor by hand, using the algorithms. He found that 96 percent of the single stuck-at faults that could affect the microprocessor were detected by the test sequence he generated. This figure indicates the validity of the technique.

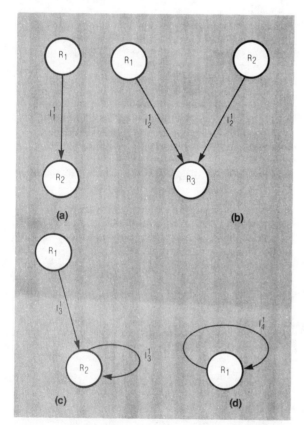

Figure 4. Representations of microprocessor instructions—I_1, transfer instruction, $R_2 \leftarrow R_1$ (a); I_2, add instruction, $R_3 \leftarrow R_1 + R_2$ (b); I_3, or instruction, $R_2 \leftarrow R_1$ OR R_2 (c); I_4, rotate left instruction (d).

The Abadir-Reghbati technique. Here we will briefly describe a test generation technique we developed for LSI circuits.[15,16] We assumed that the tests would be generated in a user environment in which the gate- and flip-flop-level details of the chip were not known.

We developed a module-level model for LSI circuits. This model bypasses the gate and flip-flop levels and directly describes blocks of logic (modules) according to their functions. Any LSI circuit can be modeled as a network of interconnected modules such as counters, registers, ALUs, ROMs, RAMs, multiplexers, and decoders.

Each module in an LSI circuit was modeled as a black box having a number of functions defined by a set of *binary decision diagrams* (see box, next page).[19] This type of diagram, a functional description tool introduced by Akers in 1978, is a concise means for completely defining the logical operation of one or more digital functions in an implementation-free form. The information usually found in an IC catalog is sufficient to derive the set of binary decision diagrams describing the functions performed by the different modules in a device. These diagrams—like truth tables and state tables—are amenable to extensive logical analysis. However, unlike truth tables and state tables, they do not have the unpleasant property of growing exponentially with the number of variables involved. Moreover, the diagrams can be stored and processed easily in a digital computer. An important feature of these diagrams is that they state exactly how the module will behave in every one of its operation modes. Such information can be extracted from the module's diagrams in the form of a set of *experiments*.[15,20] Each of these experiments describes the behavior of the module in one of its modes of operation. The structure of these experiments makes them suitable for use in automatic test generation.

We also developed a functional-level fault model describing faulty behavior in the different modules of an LSI chip. This model is quite independent of the details of implementation and covers functional faults that alter the behavior of a module during one of its modes of operation. It also covers stuck-at faults affecting any input or output pin or any interconnection line in the chip.

Using the above-mentioned models, we proposed a functional test generation procedure based on path sensitization and the D-algorithm.[15] The procedure takes the module-level model of the LSI chip and the functional description of its modules as parameters and generates tests to detect faults in the fault model. The *fault collapsing* technique[1] was used to reduce the length of the test sequence. As in the D-algorithm, the procedure employs three basic operations, namely implication, D-propagation, and line justification. However, these operations are performed on functional modules.

We also presented algorithmic solutions to the problems of performing these operations on functional modules.[16] For each of the three operations, we gave an algorithm which takes the module's set of experiments and current state (i.e., the values assigned to the module inputs, outputs, and internal memory elements) as parameters and generates all the possible states of the module after performing the required operation.

We have also reported our efforts to develop test sequences based on our test generation procedure for typical LSI circuits.[17] More specifically, we considered a one-bit microprocessor slice C that has all the basic features of the four-bit Am2901 microprocessor slice.[10] The circuit C was modeled as a network of eight functional modules: an ALU, a latch register, an addressable register, and five multiplexers. The functions of the individual modules were described in terms of binary decision diagrams or equivalent sets of experiments. Tests capable of detecting various faults covered by the fault model were then generated for the circuit C. We showed that if the fault collapsing technique is used, a significant reduction in the length of the final test sequence results.

The test generation effort was quite straightforward, indicating that the technique can be automated without much difficulty. Our study also shows that for a simplified version of the circuit C the length of the test sequence generated by our technique is very close to the length of the test sequence manually generated by Sridhar and Hayes[9] for the same circuit. We also described techniques for modeling some of the features of the Am2909 four-bit microprogram sequencer[10] that are not covered by the circuit C.

The results of our case study were quite promising and showed that our technique is a viable and effective one for generating tests for LSI circuits.

Simulation-aided test generation. Logic simulation techniques have been used widely in the evaluation and verification of new digital circuits. However, an important application of logic simulation is to interpret the behavior of a circuit under a certain fault or faults. This is known as *fault simulation*. To clarify how this technique can be used to generate tests for LSI systems, we will first describe its use with SSI/MSI-type circuits.

To generate a fault simulator for an SSI/MSI circuit, the following information is needed:[1]

- the gate-level description of the circuit, written in a special language;
- the initial conditions of the memory elements; and
- a list of the faults to be simulated, including classical types of faults such as stuck-at faults and adjacent pin shorts.

The above is fed to a simulation package which generates the fault simulator of the circuit under test. The resulting simulator can simulate the behavior of the circuit under normal conditions as well as when any faults exist.

Now, by applying various input patterns (either generated by hand, by an algorithm, or at random), the simulator checks to see if the output response of the correct circuit differs from one of the responses of the faulty circuits. If it does, then this input pattern detects the fault which created the wrong output response; otherwise the input pattern is useless. If an input pattern is found to detect a certain fault, this fault is deleted from the fault list and the process continues until either the input patterns or the faults are finished. At the end, the faults remaining in the fault list are those which cannot be detected by the input patterns. This directly measures the degree of fault coverage of the input patterns used.

Two examples of this type of logic simulator are LAMP—the Logic Analyzer for Maintenance Planning developed at Bell Laboratories,[21] and the Testaid III fault simulator developed at the Hewlett-Packard Company.[12] Both work primarily at the gate level and simulate stuck-at faults only. One of the main applications of such fault simulators is to determine the degree of fault coverage provided by a test sequence generated by any other test generation technique.

There are two key requirements that affect the success of any fault simulator:

- the existence of a software model for each primitive element of the circuit, and
- the existence of a good fault model for the UUT which can be used to generate a fault list covering most of the actual physical faults.

These two requirements have been met for SSI/MSI circuits, but they pose serious problems for LSI circuits. If it can be done at all, modeling LSI circuits at the gate level requires great effort. One part of the problem is the lack of detailed information about the internal structure of most LSI chips. The other is the time and memory required to simulate an LSI circuit containing thousands of gates. Another severe problem facing almost all LSI test generation techniques is the lack of good fault models at a level higher than the gate level.

The Abadir-Reghbati description model proposed in the previous section permits the test designer to bypass the gate-level description and, using binary decision diagrams, to define blocks of logic according to their functions. Thus, the simulation of complex LSI circuits can take place at a higher level, and this eliminates the large time and memory requirements. Furthermore, the Abadir-Reghbati fault model is quite efficient and is suitable for simulation purposes. In fact, the implication operation[16] employed by the test generation procedure represents the main building block of any fault simulator. It must be noted that fault simulation techniques are very useful in optimizing the length of the test sequence generated by any test generation technique.

Random test generation. This method can be considered the simplest method for testing a device. A random number generator is used to simultaneously apply random input patterns both to the UUT and to a copy of it known to be fault-free. (This copy is called the *golden unit*.) The results obtained from the two units are compared, and if they do not match, a fault in the UUT is detected. This response evaluation technique is known as comparison testing; we will discuss it later. It is important to note that every time the UUT is tested, a new random test sequence is used.

The important question is how effective the random test is, or, in other words, what fault coverage a random

Binary decision diagrams

Binary decision diagrams are a means of defining the logical operation of digital functions.[1] They tell the user how to determine the output value of a digital function by examining the values of the inputs. Each node in these diagrams is associated with a binary variable, and there are two branches coming out from each node. The right branch is the "1" branch, while the left branch is the "0" branch. Depending on the value of the node variable, one of the two branches will be selected when the diagram is processed.

To see how binary decision diagrams can be used, consider the half-adder shown in Figure 1a. Assume we are interested in defining a procedure to determine the value of C, given the binary values of X and Y. We can do this by looking at the value of X. If X = 0, then

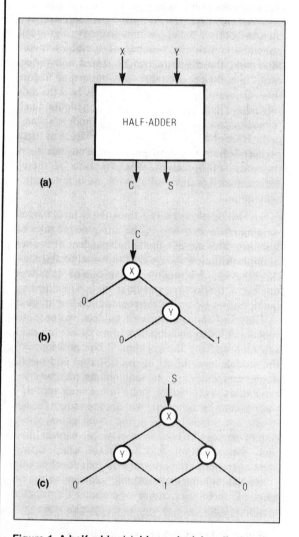

Figure 1. A half-adder (a); binary decision diagram for C = X·Y (b); binary decision diagram for S = X⊕Y (c).

C = 0, and we are finished. If X = 1, we look at Y. If Y = 0, then C = 0, else C = 1, and in either case we are finished. Figure 1b shows a simple diagram of this procedure. By entering the diagram at the node indicated by the arrow labeled with C and then proceeding through the diagram following the appropriate branches until a 0 or 1 value is reached, we can determine the value of C. Figure 1c shows the diagram representing the function S of the half-adder.

To simplify the diagrams, any diagram node which has two branches as exit branches can be replaced by the variable itself or its complement. These variables are called exit variables. Figure 2 shows how this convention is used to simplify the diagrams describing the half-adder.

In the previous discussion, we have considered only simple diagrams in which the variables within the nodes are primary input variables. However, we can expand the scope of these diagrams by using auxiliary variables as the node variables. These auxiliary variables are defined by their diagrams. Thus, when a user encounters such a node variable, say g, while tracing a path, he must first process the diagram defining g to determine the value of g, and then return to the original node and take the appropriate branch. This process is similar to the use of subroutines in high-level programming languages.

For example, consider the full-adder defined by

$$C_{i+1} = E_i C_i + \overline{E_i} A_i$$

$$S_i = E_i \oplus C_i,$$

where $E_i = A_i \oplus B_i$. Figure 3 shows the diagrams for these three equations. If the user wants to know the value of C_{i+1} when the values of the three primary inputs $A_i, B_i,$ and C are all 1's, he enters the C_{i+1} diagram, where he encounters the node variable E_i. By traversing the E_i diagram, he obtains a value of 0. Returning to the original C_{i+1} diagram with $E_i = 0$ will result in taking the 0 branch and exiting with $C_{i+1} = A_i = 1$.

Since node variables can refer to other auxiliary functions, we can simply describe complex modules by breaking their functions into small subfunctions. Thus, the system diagram will consist of small diagrams connected in a hierarchical structure. Each of these diagrams describes either a module output or an auxiliary variable.

Akers[1] described two procedures to generate the binary decision diagram of a combinational function f. The first one uses the truth table description of f, while the other uses the boolean expression of f. A similar procedure can be derived to generate the binary decision diagram for any sequential function defined by a state table.

Binary decision diagrams can be easily stored and processed by a computer through the use of binary tree structures. Each node can be completely defined by an ordered triple: the node variable and two pointers to the two nodes to which its 0 and 1 branches are directed. Binary decision diagrams can be used in functional testing.[2]

References

1. S. B. Akers, "Binary Decision Diagram," *IEEE Trans. Computers*, Vol. C-27, No. 6, June 1978, pp. 509-516.
2. S. B. Akers, "Functional Testing with Binary Decision Diagrams," *Proc. 8th Int'l Symp. Fault-Tolerant Computing*, June 1978, pp. 82-92.

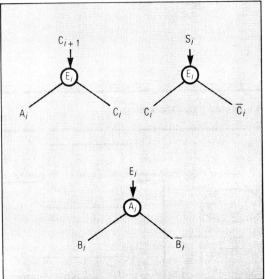

Figure 3. Binary decision diagrams for a full-adder.

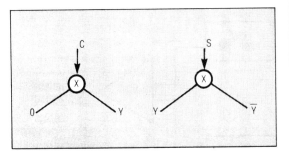

Figure 2. Simplified binary decision diagrams for the half-adder.

test of given length provides. This question can be answered by employing a fault simulator to simulate the effect of random test patterns of various lengths. The results of such experiments on SSI and MSI circuits show that random test generation is most suitable for circuits without deep sequential logic.[1,22,23] However, by combining random patterns with manually generated ones, test designers can obtain very good results.

The increased sequentiality of LSI circuits reduces the applicability of random testing. Again, combining manually generated test patterns with random ones improves the degree of fault coverage. However, two factors restrict the use of the random test generation technique:

- The dependency on the golden unit, which is assumed to be fault-free, weakens the level of confidence in the results.
- There is no accurate measure of how effective the test is, since all the data gathered about random tests are statistical data. Thus, the amount of fault coverage provided by a particular random test process is unpredictable.

Response evaluation techniques

Different methods have been used to evaluate UUT responses to test patterns. We restrict our discussion to the case where the final goal is only to detect faults or, equivalently, to detect any wrong output response. There are two ways of achieving this goal—using a good response generator or using a compact testing technique.

Good response generation. This technique implements an ideal strategy—comparing UUT responses with good response patterns to detect any faulty response. Clearly, the key problems are how to obtain a good response and at what stage in the testing process that response will be generated. In current test systems, two approaches to solving these problems are taken—*stored response testing* and *comparison testing*.

Stored response testing. In stored response testing, a one-shot operation generates the good response patterns at the end of the test generation stage. These patterns are stored in an auxiliary memory (usually a ROM). A flow diagram of the stored response testing technique is shown in Figure 5.

Different methods can be used to obtain good responses of a circuit to a particular test sequence. One way is to do it manually by analyzing the UUT and the test patterns. This method is the most suitable if the test patterns were generated manually in the first place.

The method most widely used to obtain good responses from the UUT is to apply the test patterns either to a known good copy of the UUT—the golden unit—or

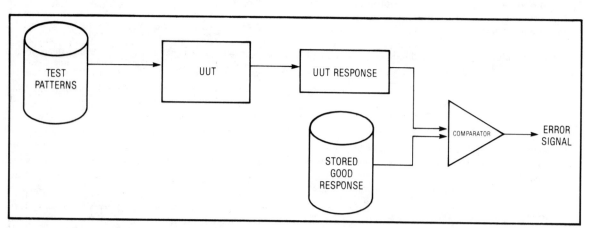

Figure 5. Stored response testing.

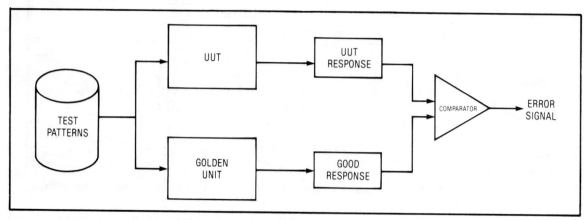

Figure 6. Comparison testing.

to a software-simulated version of the UUT. Of course, if fault simulation techniques were used to generate the test patterns, the UUT's good responses can be obtained very easily as a partial product from the simulator.

The use of a known good device depends on the availability of such a device. Hence, different techniques must be used for the user who wants to test his LSI system and for the designer who wants to test his prototype design. However, golden units are usually available once the device goes into production. Moreover, confidence in the correctness of the responses can be increased by using three or five good devices together to generate the good responses.

The major advantage of the stored response technique is that the good responses are generated only once for each test sequence, thus reducing the cost of the response evaluation step. However, the stored response technique suffers from various disadvantages:

- Any change in the test sequence requires the whole process to be repeated.
- A very large memory is usually needed to store all the good responses to a reasonable test sequence, because both the length and the width of the responses are relatively large. As a result, the cost of the testing equipment increases.
- The speed with which the test patterns can be applied to the UUT is limited by the access time of the memory used to store the good responses.

Comparison testing. Another way to evaluate the responses of the UUT during the testing process is to apply the test patterns simultaneously to both the UUT and a golden unit and to compare their responses to detect any faulty response. The flow diagram of the comparison testing technique is shown in Figure 6. The use of comparison testing makes possible the testing of the UUT at different speeds under different electrical parameters, given that these parameters are within the operating limits of the golden unit, which is assumed to be ideal.

Note that in comparison testing the golden unit is used to generate the good responses every time the UUT is tested. In stored response testing, on the other hand, the golden unit is used to generate the good responses only once.

The disadvantages of depending on a golden unit are more serious here, however, since every explicit testing process requires one golden unit. This means that every tester must contain a golden copy of each LSI circuit tested by that tester.

One of the major advantages of comparison testing is that nothing has to be changed in the response evaluation stage if the test sequence is altered. This makes comparison testing highly desirable if test patterns are generated randomly.

Compact testing. The major drawback of good response generation techniques in general, and stored response testing in particular, is the huge amount of response data that must be analyzed and stored. Compact testing methods attempt to solve this by compressing the response data R into a more compact form $f(R)$ from which most of the fault information in R can be derived. Thus, because only the compact form of the good responses has to be stored, the need for large memory or expensive golden units is eliminated. An important property of the compression function f is that it can be implemented with simple circuitry. Thus, compact testing does not require much test equipment and is especially suited for field maintenance work. A general diagram of the compact testing technique is shown in Figure 7.

Several choices for the function f exist, such as "the number of 1's in the sequence," "the number of 0 to 1 and 1 to 0 transitions in the sequence" (*transition counting*),[24] or "the signature of the sequence" (*signature analysis*).[25] For each compression function f, there is a slight probability that a response R1 different from the fault-free response R0 will be compressed to a form equal to $f(R0)$, i.e., $f(R1) = f(R0)$. Thus, the fault causing the UUT to produce R1 instead of R0 will not be detected, even though it is covered by the test patterns.

The two compression functions that are the most widely accepted commercially are transition counting and signature analysis.

Transition counting. In transition counting, the number of logical transitions (0 to 1 and vice versa) is computed at each output pin by simply running each output of the UUT into a special counter. Thus, the number of counters needed is equal to the number of

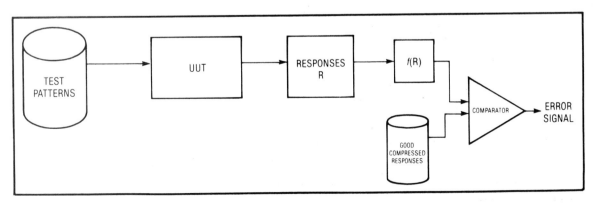

Figure 7. Compact testing.

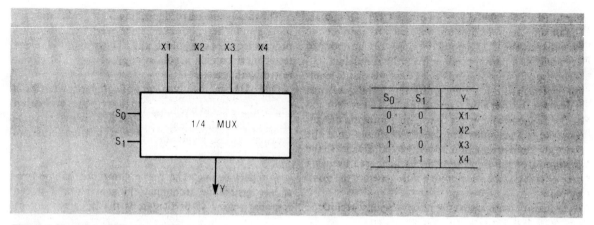

Figure 8. A one-out-of-four multiplexer.

output pins observed. For every m-bit output data stream (at one pin), an n-bit counter is required, where $n = \lceil \log_2 m \rceil$. As in stored response testing, the transition counts of the good responses are obtained by applying the test sequence to a golden copy of the UUT and counting the number of transitions at each output pin. This latter information is used as a reference in any explicit testing process.

In the testing of an LSI circuit by means of transition counting, the input patterns can be applied to the UUT at a very high rate, since the response evaluation circuitry is very fast. Also, the size of the memory needed to store the transition counts of the good responses can be very small. For example, a transition counting test using 16 million patterns at a rate of one MHz will take 16 seconds, and the compressed stored response will occupy only K 24-bit words, where K is the number of output pins. This can be contrasted with the 16 million K-bit words of storage space needed if regular stored response testing is used.

The test patterns used in a transition counting test system must be designed such that their output responses maximize the fault coverage of the test.[24] The example below shows how this can be done.

Consider the one-out-of-four multiplexer shown in Figure 8. To check for multiple stuck-at faults in the multiplexer input lines, eight test patterns are required, as shown in Table 1. The sequence of applying these eight patterns to the multiplexer is not important if we want to evaluate the output responses one by one. However, this sequence will greatly affect the degree of fault coverage if transition counting is used. To illustrate this fact, consider the eight single stuck-at faults in the four input lines X1, X2, X3, and X4 (i.e., X1 stuck-at 0, X1 stuck-at 1, X2 stuck-at 0, and so on). Each of these faults will be detected by only one pattern among the eight test patterns. For example, the fault "X1 stuck-at 0" will be detected by applying the first test pattern in Table 1, but the other seven test patterns will not detect this fault. Now, suppose we want to use transition counting to evaluate the output responses of the multiplexer. Applying the eight test patterns in the sequence shown in Table 1 (from top to bottom) will produce the output response 10101010 (from left to right), with a transition count of seven. Any possible combination of the eight faults described above will change the transition count to a number different from seven, and the fault will be detected. (Note that no more than four of the eight faults can occur at any one time.) Thus, the test sequence shown in Table 1 will detect all single and multiple stuck-at faults in the four input lines of the multiplexer.

Now, if we change the sequence of the test patterns to the one shown in Table 2, the fault coverage of the test will decrease considerably. The output responses of the sequence of Table 2 will be 11001100, with a transition count of three. As a result, six of the eight single stuck-at

Table 1.
The eight test patterns used for testing the multiplexer of Figure 8.

S_0	S_1	X1	X2	X3	X4	Y
0	0	1	0	0	0	1
0	0	0	1	1	1	0
0	1	0	1	0	0	1
0	1	1	0	1	1	0
1	0	0	0	1	0	1
1	0	1	1	0	1	0
1	1	0	0	0	1	1
1	1	1	1	1	0	0

Table 2.
A different sequence of the eight multiplexer test patterns.

S_0	S_1	X1	X2	X3	X4	Y
0	0	1	0	0	0	1
0	1	0	1	0	0	1
0	0	0	1	1	1	0
0	1	1	0	1	1	0
1	0	0	0	1	0	1
1	1	0	0	0	1	1
1	0	1	1	0	1	0
1	1	1	1	1	0	0

faults will not be detected, because the transition count of the six faulty responses will remain three. For example, the fault "X1 stuck-at 1" will change the output response to 11101100, which has a transition count of three. Hence, this fault will not be detected. Moreover, most of the multiple combinations of the eight faults will not change the transition count of the output, and hence they will not be detected either.

It is clear from the above example that the order of applying the test patterns to the UUT greatly affects the fault coverage of the test. When testing combinational circuits, the test designer is completely free to choose the order of the test patterns. However, he cannot do the same with test patterns for sequential circuits. More seriously, because he is dealing with LSI circuits that probably have multiple output lines, he will find that a particular test sequence may give good results at some outputs and bad results at others. One way to solve these contradictions is to use simulation techniques to find the optimal test sequence. However, because of the limitations discussed here, transition counting cannot be recognized as a powerful compact LSI testing method.

Signature analysis. In 1977 Hewlett-Packard Corporation introduced a new compact testing technique called signature analysis, intended for testing LSI systems.[25-28] In this method, each output response is passed through a 16-bit linear feedback shift register whose contents $f(R)$, after all the test patterns have been applied, are called the test *signature*. Figure 9 shows an example of a linear feedback shift register used in signature analysis.

The signature provided by linear feedback shift registers can be regarded as a unique fingerprint—hence, test designers have extremely high confidence in these shift registers as tools for catching errors. To better understand this confidence, let us examine the 16-bit linear feedback shift register shown in Figure 9. Let us assume a data stream of length n is fed to the serial data input line (representing the output response to be evaluated). There are 2^n possible combinations of data streams, and each one will be compressed to one of the 2^{16} possible signatures. Linear feedback shift registers have the property of equally distributing the different combinations of data streams over the different signatures.[27] This property is illustrated by the following numerical examples:

- Assume $n=16$. Then each data stream will be mapped to a distinctive signature (one-to-one mapping).
- Assume $n=17$. Then exactly two data streams will be mapped to the same signature. Thus, for a particular data stream (the UUT good output response), there is only one other data stream (a faulty output response) that will have the same signature; i.e., only one faulty response out of $2^{17}-1$ possible faulty responses will not be detected.
- Assume $n=18$. Then four different data streams will be mapped to the same signature. Hence, only three faults out of $2^{18}-1$ possible faults will not be detected.

We can generalize the results obtained above. For any response data stream of length $n>16$, the probability of missing a faulty response when using a 16-bit signature analyzer is[27]

$$\frac{2^{n-16}-1}{2^n-1} \cong 2^{-16}, \text{ for } n >> 16.$$

Hence, the possibility of missing an error in the bit stream is very small (on the order of 0.002 percent). Note also that a great percentage of the faults will affect more than one output pin—hence the probability of not detecting these kind of faults is even lower.

Signature analysis provides a much higher level of confidence for detecting faulty output responses than that provided by transition counting. But, like transition counting, it requires only very simple hardware circuitry and a small amount of memory for storing the good signatures. As a result, the signatures of the output responses can be calculated even when the UUT is tested at its maximum speed. Unlike transition counting, the degree of fault coverage provided by signature analysis is not sensitive to the order of the test patterns. Thus, it is clear that signature analysis is the most attractive solution to the response evaluation problem.

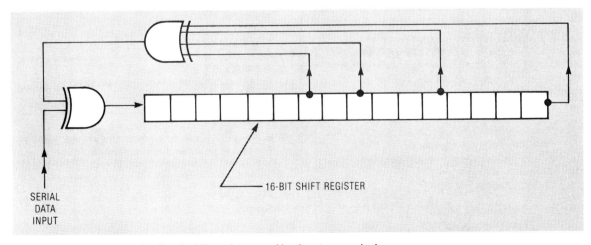

Figure 9. The 16-bit linear feedback shift register used in signature analysis.

The rapid growth of the complexity and performance of digital circuits presents a testing problem of increasing severity. Although many testing methods have worked well for SSI and MSI circuits, most of them are rapidly becoming obsolete. New techniques are required to cope with the vastly more complicated LSI circuits.

In general, testing techniques fall into the concurrent and explicit categories. In this article, we gave special attention to explicit testing techniques, especially those approaching the problem at the functional level. The explicit testing process can be partitioned into three steps: generating the test, applying the test to the UUT, and evaluating the UUT's responses. The various testing techniques are distinguished by the methods they use to perform these three steps. Each of these techniques has certain strengths and weaknesses.

We have tried to emphasize the range of testing techniques available, and to highlight some of the milestones in the evolution of LSI testing. The details of an individual test method can be found in the sources we have cited. ∎

References

1. M. A. Breuer and A. D. Friedman, *Diagnosis and Reliable Design of Digital Systems,* Computer Science Press, Washington, DC, 1976.
2. J. P. Hayes and E. J. McCluskey, "Testing Considerations in Microprocessor-Based Design," *Computer,* Vol. 13, No. 3, Mar. 1980, pp. 17-26.
3. J. Wakerly, *Error Detecting Codes, Self-Checking Circuits and Applications,* American Elsevier, New York, 1978.
4. D. B. Armstrong, "On Finding a Nearly Minimal Set of Fault Detection Tests for Combinatorial Nets," *IEEE Trans. Electronic Computers,* Vol. EC-15, No. 2, Feb. 1966, pp. 63-73.
5. J. P. Roth, W. G. Bouricius, and P. R. Schneider, "Programmed Algorithms to Compute Tests to Detect and Distinguish Between Failures in Logic Circuits," *IEEE Trans. Electronic Computers,* Vol. EC-16, No. 5, Oct. 1967, pp. 567-580.
6. S. B. Akers, "Test Generation Techniques," *Computer,* Vol. 13, No. 3, Mar. 1980, pp. 9-15.
7. E. I. Muehldorf and A. D. Savkar, "LSI Logic Testing—An Overview," *IEEE Trans. Computers,* Vol. C-30, No. 1, Jan. 1981, pp. 1-17.
8. O. H. Ibarra and S. K. Sahni, "Polynomially Complete Fault Detection Problems," *IEEE Trans. Computers,* Vol. C-24, No. 3, Mar. 1975, pp. 242-249.
9. T. Sridhar and J. P. Hayes, "Testing Bit-Sliced Microprocessors," *Proc. 9th Int'l Symp. Fault-Tolerant Computing,* 1979, pp. 211-218.
10. *The Am2900 Family Data Book,* Advanced Micro Devices, Inc., 1979.
11. Z. Kohavi, *Switching and Finite Automata Theory,* McGraw-Hill, New York, 1970.
12. S. M. Thatte, "Test Generation for Microprocessors," PhD thesis, University of Illinois, Urbana, 1979.
13. S. M. Thatte and J. A. Abraham, "Test Generation for Microprocessors," *IEEE Trans. Computers,* Vol. C-29, No. 6, June 1980, pp. 429-441.
14. M. A. Breuer and A. D. Friedman, "Functional Level Primitives in Test Generation," *IEEE Trans. Computers,* Vol. C-29, No. 3, Mar. 1980, pp. 223-235.
15. M. S. Abadir and H. K. Reghbati, "Test Generation for LSI: A New Approach," Tech. Report 81-7, Dept. of Computational Science, University of Saskatchewan, Saskatoon, 1981.
16. M. S. Abadir and H. K. Reghbati, "Test Generation for LSI: Basic Operations," Tech. Report 81-8, Dept. of Computational Science, University of Saskatchewan, Saskatoon, 1981.
17. M. S. Abadir and H. K. Reghbati, "Test Generation for LSI: A Case Study," Tech. Report 81-9, Dept. of Computational Science, University of Saskatchewan, Saskatoon, 1981.
18. M. S. Abadir and H. K. Reghbati, "Functional Testing of Semiconductor Random Access Memories," Tech. Report 81-6, Dept. of Computational Science, University of Saskatchewan, Saskatoon, 1981.
19. S. B. Akers, "Binary Decision Diagram," *IEEE Trans. Computers,* Vol. C-27, No. 6, June 1978, pp. 509-516.
20. S. B. Akers, "Functional Testing with Binary Decision Diagram," *Proc. 8th Int'l Symp. Fault-Tolerant Computing,* June 1978, pp. 82-92.
21. B. A. Zimmer, "Test Techniques for Circuit Boards Containing Large Memories and Microprocessors," *Proc. 1976 Semiconductor Test Symp.,* pp. 16-21.
22. P. Agrawal and V. D. Agrawal, "On Improving the Efficiency of Monte Carlo Test Generation," *Proc. 5th Int'l Symp. Fault-Tolerant Computing,* June 1975, pp. 205-209.
23. D. Bastin, E. Girard, J. C. Rault, and R. Tulloue, "Probabilistic Test Generation Methods," *Proc. 3rd Int'l Symp. Fault-Tolerant Computing,* June 1973, p. 171.
24. J. P. Hayes, "Transition Count Testing of Combinational Logic Circuits," *IEEE Trans. Computers,* Vol. C-25, No. 6, June 1976, pp. 613-620.
25. "Signature Analysis," *Hewlett-Packard J.,* Vol. 28, No. 9, May 1977.
26. R. David, "Feedback Shift Register Testing," *Proc. 8th Int'l Symp. Fault-Tolerant Computing,* June 1978.
27. H. J. Nadig, "Testing a Microprocessor Product Using Signature Analysis," *Proc. 1978 Semiconductor Test Symp.,* pp. 159-169.
28. J. B. Peatman, *Digital Hardware Design,* McGraw-Hill, New York, 1980.

Magdy S. Abadir is a research assistant and graduate student working towards the PhD degree in electrical engineering at the University of Southern California. His research interests include functional testing, design for testability, test pattern generation, and design automation. He received the BSc degree in computer science from Alexandria University, Egypt, in 1978 and the MSc in computer science from the University of Saskatchewan in 1981.

Abadir's address is the Department of Electrical Engineering, University of Southern California, Los Angeles, CA 90007.

**Hassan K. Reghbati** is an assistant professor in the Department of Computing Science at Simon Fraser University, Burnaby, British Columbia. He was an assistant professor at the University of Saskatchewan from 1978 to 1982, where he was granted tenure. From 1970 to 1973 he was a lecturer at Arya-Mehr University of Technology, Tehran, Iran. His research interests include fault-tolerant computing, VLSI systems, design automation, and computer communication. The author or coauthor of over 15 papers, he has published in *IEEE Micro, Computer, Infor,* and *Software—Practice & Experience*. One of his papers has been reprinted in the *Auerbach Annual 1980: Best Computer Papers*. He has served as a referee for many journals and was a member of the program committee of the CIPS '82 National Conference.

Reghbati holds a BSc from Arya-Mehr University of Technology and an MSc in electrical engineering from the University of Toronto, where he is completing requirements for his PhD. He is a member of the IEEE.

His address is the Department of Computing Science, Simon Fraser University, Burnaby, BC V5A 1S6, Canada.

Computer-Aided Design of VLSI Circuits

ARTHUR RICHARD NEWTON, MEMBER, IEEE

Invited Paper

Abstract—With the rapid evolution of integrated circuit (IC) technology to larger and more complex circuits, new approaches are needed for the design and verification of these very-large-scale integrated (VLSI) circuits. A large number of design methods are currently in use. However, the evolution of these computer aids has occurred in an ad hoc manner. In most cases, computer programs have been written to solve specific problems as they have arisen and no truly integrated computer-aided design (CAD) systems exist for the design of IC's. A structured approach both to circuit design and to circuit verification, as well as the development of integrated design systems, is necessary to produce cost-effective error-free VLSI circuits. This paper presents a review of the CAD techniques which have been used in the design of IC's, as well as a number of design methods to which the application of computer aids has proven most successful. The successful application of design-aids to VLSI circuits requires an evolution from these techniques and design methods.

I. INTRODUCTION

THE NUMBER of components which can be implemented on a single-chip integrated circuit (IC) has increased rapidly in recent years [1]. As a result, new approaches to design and verification are necessary for the effective use of very large scale integrated (VLSI) circuits. The evolution of computer aids for IC design has occurred in an ad hoc manner. In most cases, computer programs have written to solve specific problems as they have arisen and very few truly integrated computer-aided design (CAD) systems exist for the design of IC's. Most CAD systems currently in use consist of a loose collection of programs, requiring a large collection of data formats and often requiring manual intervention to move from one program to another.

The use of a particular class of circuit structures is referred to as a *design method*, or design style, and while the development of new algorithms and techniques for CAD continues, the most significant contribution to the design of VLSI circuits will come from the development of new circuit design methods. However, while the implementation of a design method does not *require* the use of computer aids, the most successful design methods will be those designed to take maximum advantage of the computer in both the circuit design and verification phases. The design method must provide the *structure* necessary to use both human and computer resources effectively. For VLSI, this structure also provides the reduction in design complexity necessary to reduce design time and to ensure that the circuit function can be verified and the resulting circuit can be tested. In describing the variety of computer-aids used for IC design, a distinction is made between those techniques used for *design*, or synthesis, of the IC and those techniques used for its *verification*. In both of these categories, a further distinction is made between techniques relating to the *physical*, or topological, aspects of the design process, such as mask layout or the placement of components in a circuit, and *functional* considerations, such as logic description, synthesis, simulation, and test-pattern generation.

Computer aids for design, or synthesis, at both the functional and physical levels, are primarily concerned with the use of *optimization* to improve performance and cost. These design tasks may be formulated as combinatorial optimization problems for operations such as cell placement, routing, logic minimization, and logic state assignment, or as parametric optimization problems for operations such as design at the electrical level. These optimization problems are often too complex to solve directly. Therefore, *partitioning* is often used to reduce the problem to a set of simpler subproblems. The solutions of these subproblems are later combined in a separate step. Both the partitioning task and the solution of each subproblem generally involves the use of heuristics to reduce the complexity further.

This paper presents a review of the CAD techniques which have been used in the design of IC's, as well as a number of design methods to which the application of computer aids has proven most successful. The successful application of design-aids to VLSI requires an evolution from these techniques and design methods. The organization of this paper follows the historical evolution of CAD. Following a brief historical review in Section II, Section III describes the variety of data representations used in the design of VLSI circuits. Sections IV–VII review CAD techniques in use and under development for the design and verification of basic circuits, or cells, circuit building blocks which may consist of many cells, and integrated systems, which may consist of a number of building blocks. After a summary of the major ideas described in this paper, the need for a unified approach to IC design and verification is described in Section VIII.

II. HISTORICAL REMARKS

The first digital IC's were available commercially in the early 1960's. However, it was a number of years before computer-aids were applied to the design and verification of these circuits. In retrospect, it is surprising how little the computer has been used in the design of IC's. Early circuits were sufficiently small that mask patterns could be drawn by hand on rubylith, and then photographically reduced to generate the IC masks directly. However, for the verification of the *function* of the circuit, simulators proved quite useful. Hence initial work in the mid-1960's focused on the development of device analysis [2], [3] and circuit analysis [4]–[9] techniques. These circuit simulators were originally developed for the analysis of nonlinear and radiation effects in discrete circuits and it was not until the early 1970's that circuit simulators suitable for IC analysis became generally available [10]-[17].

As the complexity of the circuits increased, industry turned to the computer to store IC layout data and to produce the masks required for manufacture. Systems for layout digitization and interactive correction found extensive use by the early 1970's. However, it was not until the mid-1970's that programs for the physical layout rule checking (LRC) of the circuit began to find widespread use [18]-[20].

By 1975 it had become clear that computer-aids were a necessity in the design of complex IC's, both for physical and for functional design and verification. Until then, the layout of an IC and its transistor-level schematic diagram had been quite separate. In the late 1970's, computer programs became available for such tasks as connectivity verification [21], extraction of transistor-level schematics from IC artwork data [22]-[25] and even extraction of gate-level schematics from the transistor list [26]. These programs are loosely coupled in general and are often incompatible with one another. In fact, the *only* integrated CAD systems that exist today for the design of complex IC's are those for some highly specialized design approaches, such as standard cell and gate arrays.

In parallel with the development of computer-aids for IC design, a great deal of work has been done to aid the digital system designer, particularly as applied to printed circuit (PC) board design using standard components. In particular, algorithms and programs for the optimal placement and routing of cells [72], [73], logic simulation techniques [27]-[29], and test grading [30], [31] have resulted in sophisticated design packages.

III. Representations of the Design

Throughout the IC design process, a variety of different representations or *views* of the design are used. These representations may reflect a particular level of abstraction, such as the functional specification of the circuit or its mask layout, or they may reflect the view required for a certain application, such as the information required for simulation. The choice of appropriate representations for each level of the design process is a key factor in determining the effectiveness of computer aids since it is via these representations that both the structure of the design as well as specific information relating to a particular design level are expressed. The design process then involves transformations between these representations, both for design and verification. In this section, a classification of representations is presented. This classification is used in the following sections to relate different design aids. One particular representation which promises to provide a significant improvement in the design process over the next few years is symbolic layout. For this reason, a number of symbolic layout representations are described in more detail at the end of the section.

While the particular set of representations used in a design depends on the particular design approach being used, the major categories may be defined as shown in Fig. 1. These representations fall into three major categories: behavioral, schematic, and physical. At the behavioral or algorithmic level, functional intent of the design is described independent of a particular implementation. In some cases, programming languages such as concurrent Pascal [99] have been used to represent the design at this level, as well as providing a simulation capability. Languages specifically designed for this task have also been developed [114], [116], [117].

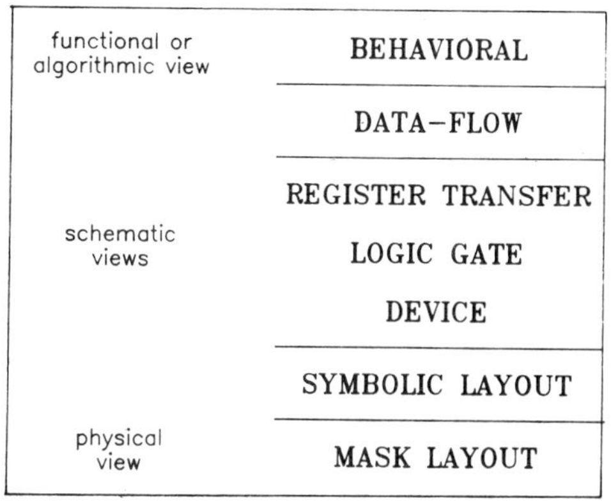

functional or algorithmic view	BEHAVIORAL
schematic views	DATA-FLOW
	REGISTER TRANSFER
	LOGIC GATE
	DEVICE
physical view	SYMBOLIC LAYOUT
	MASK LAYOUT

Fig. 1. Representations of the IC design process and their classification.

Once a functional implementation strategy has been determined, a schematic view may be generated. At its most abstract level, this schematic view consists of a *chip plan*, illustrating the loose physical placement of the major components and busses, or a register transfer level (RTL) description, defining the functional relationships between the major components of the design. As the implementation is refined further, logic gate level and finally transistor level schematics may be generated. While the nature of the information contained at each level is different, each more detailed view may be considered a different level of "zooming in" on the implementation. With each new level of refinement more information concerned with the detailed physical and functional implementation of the circuit is included in the description. The final transformation consists of the generation of detailed, mask-level geometries from a device-level schematic view.

The transition between functional schematic descriptions and between schematic and mask layout may involve the use of additional views. At the higher level, a data-flow description of the circuit function may serve this purpose, as shown in Fig. 1. At the behavioral level, this description may be viewed as the parse tree generated by a compiler operating on the algorithmic description of the intended function. At the RTL level nodes in the data-flow graph represent an initial configuration of circuit building blocks used to implement the function and branches indicate data paths to- and from- these functions.

At each level, these descriptions must express the structure of the design in such a manner that it can be exploited by the design aids. In particular, *regularity* and *hierarchy* must be exploited. For example, regularity in the form of one- to two-dimensional arrays of similar, e.g., RAM, or iterated, e.g., ROM, components can reduce the design time since only a small number of basic circuit types need be designed by hand. The verification time is reduced also since only one example of each possible spatial combination of this small number of cells need be verified to certify the entire array. Hierarchy can aid the verification process in a similar manner. The components of a circuit block, such as the logic gates used to implement an arithmetic-logic unit, need only be checked in detail once. When the composition of these cells is checked, only the relationships *between* the cells need be verified. A detailed check of the internals of the cells is not necessary. If these cells are used a number of times, this process can provide substantial

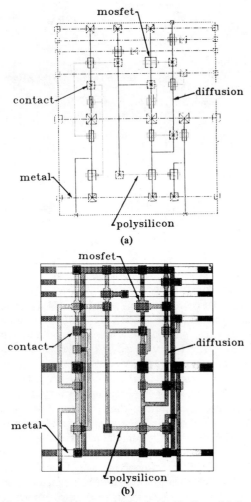

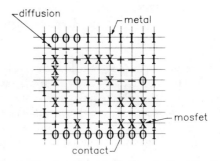

Fig. 3. Simple-grid symbolic layout of an IC cell.

Fig. 2. (a) Loose symbolic layout of a simple flip-flop. (b) Corresponding uncompacted mask-level layout.

savings in computer time. Circuit structure can also be exploited in other areas, such as simulation, circuit synthesis, and testing, as described in the remaining sections of the paper.

Symbolic layout forms a bridge between a schematic view of the circuit and its mask-level layout. A symbolic layout contains explicit connectivity information as well as the relative placement of circuit components, such as transistors, to form a basic circuit cell, cells to form a building-block, and building-blocks to describe the entire circuit. At the transistor level, the symbolic layout is often called a *stick diagram* since interconnections are represented by their center lines and hence resemble sticks. Fig. 2(a) shows the uncompacted, symbolic layout of a cell and Fig. 2(b) shows its corresponding mask level representation [32]. One of the key advantages of a symbolic layout is its ability to maintain explicit electrical connectivity information through to the mask level descriptions. Not only can symbolic layout be used to aid the verification of the circuit, but by separating layout-sensitive cells and interconnections, computer programs can be used to optimize the area utilization of the circuit by modifying only the noncritical interconnections. This is described in Section IV.

There are two common approaches to the use of symbolic layout. In the *fixed grid* or coarse grid method [33]-[35], the layout is drawn on spaced grids. The grid spacing is determined by a set of dominant layout rules in such a way that it is impossible for user input to violate spacing rules. A conversion program may then be used to replace each symbol in the user input with its equivalent mask layout description. The compactness of the resulting layout depends on the quality of the user input, since spaces in the layout left by the user are not removed, and the coarseness of the grid. A simple ASCII format can be used to represent fixed-grid symbolic layout, as shown in Fig. 3 [34].

The second type of symbolic layout representation is called *relative grid* since the loose layout plan indicates only the relative placement and interconnection of element symbols [36]-[39]. Fig. 2(a) is an example of a graphical view of such a relative-grid approach.

Once an appropriate set of representations for a particular design method has been determined, it is essential that a self-consistent approach be used to maintain the data for each design.

IV. DEVICE LEVEL

As IC technology scales down to ever-smaller feature sizes, new processing techniques and the circuit devices constructed in the new processes must be characterized accurately for circuit design. If a process has been developed, it may be characterized by extensive measurement. However, if a process is to be designed for specific device characteristics, computer aids can play a major role. For a clear understanding of the process, not only is accurate modeling and analysis of subsilicon surface effects essential, but the characteristics of the lithographic process itself becomes critical as the fundamental limits of mask making, exposure, and pattern-etching technologies are approached. In this section, a number of techniques used for the analysis and design of MOS processes and devices are described. Fig. 4 lists the type of programs used most often for these tasks.

Process simulators, such as the SUPREM program [40], provide one-dimensional impurity profiles and oxide thicknesses for any sequence of process steps, such as oxidation, ion implantation, diffusion, and epitaxial growth. These programs are finding widespread use while work continues to develop more accurate models for each step in the IC manufacturing process.

With this decrease in device size there is also an increasing need to simulate two-dimensional processes, such as line-edge profiles resulting from the various lithography, growth, and etching processes used during IC manufacturing. Programs such as SAMPLE [41], [42] can be used to simulate and model profiles for lithography and etching. SAMPLE allows the user to mimic a large variety of lithographic and etching processes,

	FUNCTIONAL	PHYSICAL
DESIGN	circuit simulation schematic capture	layout digitization symbolic layout compaction
VERIFICATION	schematic extraction electrical-rule check circuit simulation schematic comparison	layout-rule check connectivity verification

Fig. 4. Computer aids used in the design and verification of IC devices and technologies.

	FUNCTIONAL	PHYSICAL
DESIGN	device analysis parasitic analysis interconnect analysis	process simulation lithography simulation process optimization
VERIFICATION	device analysis parasitic analysis interconnect analysis	process simulation lithography simulation

Fig. 5. Computer aids used in the design and verification of basic circuit cells.

including oxide etching, sputter etching, evaporation, and epitaxial growth, and provides cross sections of the line-edge profiles at various stages in the processing.

Once a set of physical processing steps has been established for the design of a circuit device, such as an MOS transistor, the electrical characteristics such as transconductance, breakdown voltage, and parasitic capacitance values must be determined. One, two, or three-dimensional semiconductor device analysis programs can be used here. Most of these programs use finite-element or finite-difference techniques to solve for the static [43] or dynamic [44] characteristics of the device. It is clear that there is a strong interaction between the processing steps and the device characteristics. Recently, work has been carried out on the coupling of process and device analysis to allow the process to be optimized for required device characteristics [45].

A further result of the reduction in size of IC geometries is that parasitic and distributed effects become increasingly significant. Numerical analysis techniques such as finite-difference and finite-element methods may also be used here to determine the electrical characteristics of complex geometrical patterns [46]. These analyses can then be used to produce equivalent electrical models for this geometry [47]. It may soon be necessary to consider detailed coupled-line effects and microstrip propagation in the design of these circuits.

V. Basic Cell

Once the IC process has been characterized, either by computer analysis or by measurement, basic circuit cells can be designed and verified. These cells may be simple logic gates, such as NAND, NOR, or flip-flops. Alternatively, they may be cells used in regular arrays, such as the one-transistor cells used in a programmable logic array (PLA), which do not perform a complete logic function alone. Computer aids for basic cell design are listed in Fig. 5 and these programs are described in the following.

A. Layout Entry

The most commonly used aid for the physical design of a cell is mask-level digitization and interactive correction program. The user draws the mask layout of each mask layer and enters it into the computer via a digitization process. Once the layout has been entered, it may be plotted and then edited on an interactive graphics display. In some cases the user may enter the initial mask information via the graphics editor directly.

The output from such programs is a file suitable for driving a computer-controlled mask pattern generator. Some of these programs provide rudimentary error detection on input, such as checking for undersized geometry on a particular mask layer or checking the local physical layout rules are not violated as each new piece of geometry is entered [48].

Some programs allow direct symbolic layout entry, using either fixed-grid [33]-[35] or relative-grid [36]-[39] schemes. With the fixed-grid symbolic approach, the grid is designed to ensure all basic layout rules are satisfied upon data entry. For relative-grid schemes, it is necessary to modify the layout such that all layout rules are satisfied. Programs which carry out this operation are often referred to as *compaction* programs since they also attempt to reduce the area occupied by the circuit.

B. Layout Compaction

Once a symbolic layout has been entered into the computer, it may be compacted by adjusting the size of noncritical components, such as interconnections, under the constraints imposed by the physical and electrical layout rules of a given technology. The FLOSS program [36], developed at RCA for the compaction of circuit cells, paved the way for the development of transistor-level compaction programs for IC's. The algorithms used for compaction generally perform x and y axes iterative compaction steps until all layout rules are satisfied and no further area reduction can be achieved. Local modifications to the layout can then be performed to allow further compaction. These modifications generally consist of distortions to interconnect, such as the introduction of "jogs" or "doglegs," or the rotation of transistors and cells [32], [37]. Critical path algorithms and force-directed heuristics are used to determine the best location for the introduction of these layout modifications. To ensure the layout is least sensitive to processing tolerances, noncritical components must then be placed midway between constraints to maximize yield. At each compaction step, a layout-rule analysis must be performed to determine the values of geometrical constraints on the layout. This process is relatively expensive and determines the time complexity of the compaction process. The compaction time for very large layouts can become prohibitive.

The use of an hierarchical description of the circuit can be exploited to reduce the analysis time by compacting the cells independently. The resulting compacted cells may then be combined and compacted to form the circuit. While this may not result in an optimal area utilization, the primary objective

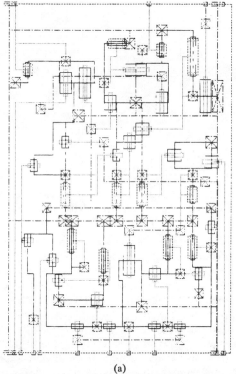

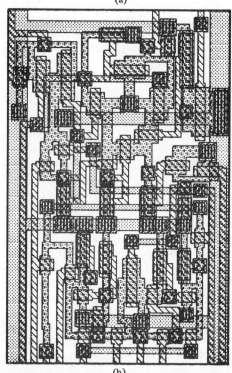

Fig. 6. (a) Loose symbolic layout of a latch-buffer cell in NMOS depletion load technology. (b) Mask layout of the cell after compaction.

of error-free layout is achieved. Fig. 6(a) shows the loose, symbolic layout of an IC building-block and Fig. 6(b) shows its physical layout after compaction with the CABBAGE compaction program [32].

C. Circuit Simulation and Modeling

When accurate circuit models are available, circuit simulators provide precise electrical information, such as frequency response, time-domain waveforms, and sensitivity information, about the circuit under analysis. The majority of circuit simulators currently in use contain models for a wide range of active devices and hence are largely independent of technology. For this reason, these programs must employ general algorithms for the solution of the set of coupled, nonlinear, ordinary differential equations which describe the integrated circuit and hence cannot exploit the special characteristics of a particular technology.

Without models whose accuracy is well matched to the expected accuracy of a simulation, the results of the simulation may not reflect the performance of the circuit under analysis. Recent work on modeling for MOS circuit simulation [49]–[52] has focussed on the development of empirical models for MOS transistors which predict the characteristics of the devices accurately without requiring large amounts of computer time. With small geometries, signal delays and signal degradation caused by interconnect can dominate circuit operation. For this reason, explicit models for interconnect are necessary for accurate simulation. The parameters of such models may be provided by the designer interactively or by design programs directly.

Most circuit simulators are batch-oriented programs and the input to the program consists of a textual description of the transistors and their interconnections. In some cases, an interactive graphics editor is used to capture the schematic diagram and provide simulator input. This description can also be used for comparison with a schematic diagram extracted directly from the layout, as described later.

D. Layout-Rule Check

If hand-generated mask-level layout is used for cell design, compaction programs cannot be used and LRC programs are necessary to verify that all of the physical layout rules, such as minimum spacing, minimum size, and minimum enclosure constraints, are satisfied. Since the introduction of such programs in the mid-1970's [20], considerable improvements have been achieved in time and memory requirement complexity [53], [54]. However, if unconstrained polygonal shapes are permitted in the layout, the computer run times of LRC programs on circuits containing over 10 000 transistors become prohibitive. LRC programs are under development which exploit the structure of a circuit, in particular the circuit hierarchy, to reduce the cost of LRC for complex circuits [42].

E. Layout Analysis

If a transistor-level schematic description of a circuit is available, either from simulator input data or a graphical entry system, a comparison between the circuit elements and their connections in the schematic and the corresponding geometry in the layout can be performed. A layout extraction program is required to extract a schematic diagram from the circuit layout and a connectivity verification, or electrical rules check, program can be used compare it with the intended schematic captured earlier [21]–[24]. The extraction program can also be used to determine parameter values for simulation, such as device and interconnect capacitance, the effective width and length of MOSFET channels, and the area of source and drain diffusions. The extracted schematic can also be checked for simple electrical rule violations, such as a short circuit between power supply and ground or an isolated transistor, without the need for a full schematic diagram entered separately. A large number of errors can be detected by connectivity verification and extraction before any simulations are performed.

VI. BUILDING BLOCKS

Computer aids can be used to combine the cells described earlier and form more complex building blocks for VLSI. While many design methods are used for the implementation of these building blocks, the methods can be classified in four categories: programmable arrays, standard-cell, macrocell, and procedural design.

A VLSI circuit may consist of one large building block or it may consist of a number of building blocks combined either manually, using the techniques described in the previous section, or by a computer program. After a brief description of each of these design approaches, a number of computer aids used for the design and verification of IC building blocks are described in this Section. These programs are listed in Fig. 7.

A. Design Methods

A programmable array is a one- or two-dimensional array of repeated cells which can be customized by adding or deleting geometry from specific mask layers. Since a number of processing steps are completed prior to customization, the locations of components on those layers are independent of a particular circuit implementation. Examples of programmable arrays include the gate array, storage-logic array, PLA, and ROM.

The *gate array* (also referred to as master-slice, or uncommitted logic array) is by far the most common programmable array designed by computer. In this approach, a two-dimensional array of replicated transistors is fabricated to a point just prior to the interconnection levels. A particular circuit function is then implemented by customizing the connections within each local group of transistors, to define its characteristics as a basic cell, and by customizing the interconnections between cells in the array to define the overall circuit. Generally a two-level interconnection scheme is used for signals and, in some approaches, a third, more coarsely defined layer of interconnections is provided for power and ground connections. The interconnections are implemented on a rectilinear grid in the *channels* between the cells. In many cases, channels are also provided which run over the cells themselves and in some arrays, wider channels are provided in the center of the array to alleviate the congestion often found in that area.

Gate arrays are used in many technologies, in particular bipolar and CMOS, and arrays containing many thousands of gates have been used [56], [57]. In the storage logic array (SLA) approach [58], each "gate" consists of a storage element (flip-flop) and a small, uncommitted PLA. This design method has considerable potential for VLSI but effective design-aids for the synthesis of logic functions in SLA form are not yet available.

PLA's may also be used to implement building blocks directly, with storage elements in the feedback path to implement sequential logic in the classical Moore or Mealy style [59]. The PLA consists of a number of transistor arrays which implement logic AND and OR operations. In MOS technology, NOR arrays are used [60]. A conventional PLA consists of two arrays of cells: an input, or lookup, plane followed by an output plane. A *folded* PLA may use an additional plane, since rows and/or columns in the structure may be shared by more than one circuit variable, as described later.

	FUNCTIONAL	PHYSICAL
DESIGN	circuit simulation timing simulation logic simulation mixed—mode simulation logic synthesis testability analysis	cell placement routing direct layout synthesis
VERIFICATION	circuit simulation timing simulation logic simulation mixed—mode simulation gate extraction schematic comparison critical path analysis fault coverage analysis test generation	layout—rule check connectivity verification

Fig. 7. Computer-aids used in the design and verification of circuit building blocks.

The *standard cell* (or polycell) approach refers to a design method where a library of custom-designed cells is used to implement a logic function. These cells are generally of the complexity of simple logic gates or flip-flops and may be restricted to constant height and/or width to aid packing and ease of power distribution. Unlike the programmable array approach, standard cell layout involves the customization of all mask layers. This additional freedom permits variable width channels to be used. While most standard cell systems only permit intercell wiring in the channels between rows of cells or through cells via predetermined "feedthrough" cells, some systems permit over-cell routing. Standard cell systems are also used extensively in a variety of technologies including bipolar and CMOS [61], [62].

It is often relatively inefficient to implement all classes of logic functions in a single design approach. For example, a standard cell approach is inefficient for memory circuits such as RAM and stack. In the *macrocell* method, large circuit blocks, customized to a certain type of logic function, are available in a circuit library. These blocks are of irregular size and shape and may allow functional customization via interconnect, such as a PLA or ROM macro [63], or they can be parameterized with respect to topology as well [64]-[66]. With the parameterized cell, the number of inputs and outputs may be parameters of the cell. In some systems macrocells may also be embedded in gate array or standard-cell designs.

All of the design methods described above may be classified as *data driven*. That is, a description of the required logic function, in the form of equations or an interconnection list, is used as input to a software system which interprets the data and generates the final design. Techniques have been developed recently which can be classified as *procedure*, or program, driven [64]-[66]. In this case, each design is described by a set of procedures and the execution of those procedures is required to generate the design. While this technique provides a flexible and powerful design paradigm, sophisticated verification and debugging techniques must be developed to support it. The use of conditional circuit assembly and loop constructs within the design specification make this verification difficult at other than the final geometrical level.

B. Placement and Routing

All of these techniques involve the choice of an optimal cell *placement* and the interconnection, or *routing*, of the cells to form the final circuit.

The placement problem involves the assignment of specific

locations to building blocks of the layout. This includes the assignment of logic gates within a gate array [67], the placement of cells in a standard cell layout [68], [69] or the placement of macrocells in a nonuniform building-block approach [71], [72]. While a considerable amount of theoretical work has been developed in this area [72], [73], the most successful approaches involve the use of simple, force-directed heuristics and pairwise interchange. Partitioning may be used here to improve the final result by first clustering cells, placing the clusters, and then assigning the cell placements within clusters. In these approaches, either total interconnect length or estimated or exact routing requirements are used to drive the placement algorithms.

Once a cell placement has been determined, either by hand or by computer, an optimal channel routing must be determined. A channel router solves the problem of routing a specified list of nets, or multiport interconnections, between rows and/or columns of terminals across a multilayer channel. In most cases, two layers are available for signal interconnections. Nets are generally routed with horizontal segments on one layer and vertical segments on the other. One objective of channel routing is the minimization of the number of tracks needed to route the nets. Unfortunately, solving the resulting optimization problem exactly, perhaps with a branch-and-bound method [74], may require enormous computing resources for practical problems of even moderate size. This problem can be overcome using heuristic algorithms which generate optimum or near-optimum solutions in a reasonable amount of computer time. As an example, [74] describes new algorithms which allow fast routing for single-channel problems with and without track-sharing by multiple nets.

It is clear that in a methodology which permits only preallocated, fixed-width channels, such as the gate array approach, it is advantageous to determine *a priori* whether a circuit function can be implemented on a given master template and if so what circuit density can be achieved. While techniques are not available to solve this problem exactly, without actually routing the circuit, *estimation* techniques can be used to predict the feasibility of the placement and routing tasks. Estimation algorithms are based on the assumption of equal sized blocks and uniform channel widths [75], or more recent work has extended these techniques to unequal size logic blocks and channels [76].

C. Logic Synthesis

While the above placement and routing techniques involve physical optimization at the topological level, it is also possible to perform optimization at the functional level. Logic synthesis techniques, which carry out this optimization, may be applied to both combinational and sequential circuits.

The generation of a minimal, multiple-output combinational logic implementation of a set of Boolean equations requires the use of heurisitc techniques. For two-level logic, a number of successful programs exist and have been applied extensively to the design of such structures as the PLA [77], [78]. For multilevel logic minimization, a number of heuristic algorithms have been developed for, and applied to, specific classes of problems, such as the implementation of multilevel multiplexor-based logic [79].

The problem of sequential logic design can be divided into the tasks of determining an optimal partitioning at the state-machine level, state assignment, and optimal combinational

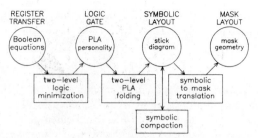

Fig. 8. Data representations (circles) and design aids (boxes) used in the translation of a set of Boolean logic equations to an optimized two-level folded PLA implementation.

logic design techniques. While considerable research has been done in terms of general algorithms for use in this area, these techniques have not yet been successfully applied to the design of integrated digital systems.

Clearly there is often a strong interaction between the topological design of a circuit and its functional counterpart. For example, it may be better to leave additional product terms in a PLA if the resulting increase in sparsity can be used more effectively to increase the overall density of the PLA implementation. A folded PLA is a PLA in which a number of circuit variables share the same row or column in the PLA [80]. In a two-level folding scheme, this is achieved by "splitting" the row or column electrode [81]. This is illustrated in Fig. 9(d) for column folding where the circuit variables A and B share the same input column. Fig. 8 indicates the steps involved in translating from Boolean equations to the IC implementation of a two-level folded PLA and Figs. 9(a)–(d) show the representations of a simple PLA at each stage of the synthesis process. If the objective function of the logic minimizer is simply to provide a minimal set of product terms, the resulting PLA may not be well suited to the folding process. However, if the logic minimization heuristics attempt to increase the sparsity of the PLA with folding in mind, as well as reduce the number of the product terms, the result after folding may be more area-efficient.

D. Simulation

Circuit simulation techniques can provide accurate waveform analysis for circuits of building-block complexity. However, as circuit size increases the time and memory requirements of a circuit simulation become prohibitive. On an IBM 370/168 computer, the average cost of a SPICE [16] analysis is 6 ms/device/clock/timepoint. For a 10 000 device circuit, with 3 clocks and for an analysis of 10 μs at 1 ns steps, the computation time would be in excess of 20 computer days! Nevertheless, the success of circuit simulation in design evaluation has been such that designers wish to continue to simulate large circuits at the level of accuracy provided by this type of program.

By applying node tearing techniques [82], [83] to the interface between cells in the circuit, inactive cells can be bypassed during the equation solution phase. However, it is anticipated that these techniques alone will provide less than an order of magnitude speed improvement. This is not sufficient improvement in performance to permit cost effective device-level analysis of VLSI circuits.

If simulation algorithms are tailored to specific technologies or applications substantial speed improvements can be achieved. Many components of digital MOS or I^2L circuits can be con-

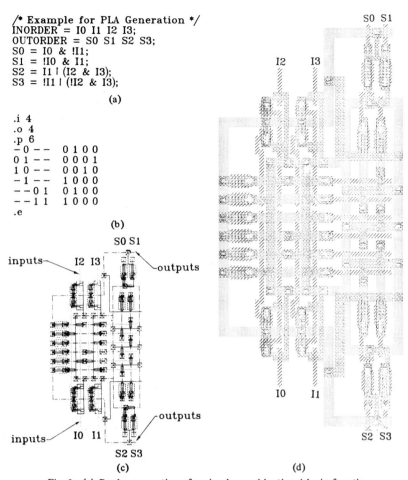

Fig. 9. (a) Boolean equations for simple combinational logic function. (b) PLA personality after logic minimization. (c) Symbolic layout after topological minimization (folding). (d) Final layout of an NMOS implementation of the PLA.

sidered unilateral in nature. This characteristic, as well as the facts that these families are saturating and hence accumulated voltage errors are lost at the extremes of signal swing, and that large digital circuits are relatively inactive at the gate level, are exploited in *timing simulation*. Timing simulators [84]-[86] can improve simulation speed by up to two orders of magnitude while maintaining acceptable waveform accuracy. These savings are achieved by using node decoupling techniques in conjunction with simplified lookup table models for nonlinear devices.

Where a library of cells is used during the design, or when a group of transistors is used to implement a common function, such as a cell or logic gate, it is often possible to exploit the known structure of the circuit and use a simplified representation which maintains the essential characteristics of the cell at reduced computational expense. Such a reduced representation is called a *macromodel* [87]-[89] and macromodels are used in both circuit and timing level analysis.

The speed of timing simulation is insufficient for the analysis of large circuit blocks. Gate-level logic simulators have been used for over twenty years in digital system design [90] however it was not until the late 1970's that logic simulators applied to IC simulation were capable of dealing with many of the problems associated with MOS logic gates [91]. Logic simulation is considerably faster than circuit simulation since the evaluation of a new logic state involves only Boolean operations, the fastest operations available on most computers, and the logic simulator deals with logic gates directly rather than the details of their transistor-level implementation. In a logic simulator, the signals propagated between gates are no longer voltages or currents but rather they are discrete logic states [92]-[94]. The most successful MOS logic simulators use as many as nine static logic states to describe the terminal characteristics of a gate or bus [52], [91].

In some simulations it may be necessary to obtain the waveform accuracy of circuit simulation for parts of the circuit while in other parts of the circuit, a more efficient gate-level logic analysis is sufficient. In this situation, *mixed-mode* analysis can be used [95]-[98]. In a mixed-mode simulator, different levels of abstraction are combined in the single analysis. Circuit, timing, and logic analyses [95]-[98] or even behavioral and RTL analyses [99] can be combined in a single program.

E. Testing

With the increasing complexity of IC's, techniques for the cost-effective evaluation of whether a circuit has been manufactured correctly becomes critical to the success or failure of a design [100]. One approach to this problem is to determine an adequate set of tests, after the circuit has been designed, which can qualify it as a functioning part. Even for LSI circuits, the ability to generate test patterns automatically and perform fault simulation is almost impossible unless the testability issue has been considered during the design phase. While a number of ad hoc techniques have been used to ensure cost-effective testability, two very successful and somewhat similar structured approaches have evolved. They are the *level sensitive scan design* (LSSD) [101] and the *scan path* [102] approaches. These techniques reduce the problem of test generation to one of generating tests for combinational logic, for which powerful algorithms are available [105]. Both approaches involve the addition of circuitry to the design, which may increase manufacturing cost and can reduce performance, but for a major reduction in testing cost. A recent extension of these approaches involves the use of *built-in logic block observation* (BILBO) [103]. The BILBO approach uses the separation of combinational and sequential functions, as in the LSSD and scan path approaches, combined with on-chip linear feedback registers to generate input patterns for a signature analysis [104] form of circuit evaluation.

If an ad hoc approach is used, then programs are available which provide a *testability measure* [106, 107] of the design. These programs are quite fast and provide the designer with a heuristic measure of how "difficult" it may be to observe or control nodes in the circuit. While this approach is useful, further research is required to calibrate these techniques and to extend the heuristics.

Once the circuit has been designed, test patterns must be generated which provide an adequate cover of all possible faults. If a structured design approach was used, or if only combinational logic is involved, this task is relatively inexpensive and the required test sequences are generally quite small. If testability was not considered during design, a great deal of fault simulation [108] will be required to find an adequate set of test patterns, if they can be found at all.

VII. System Design

Very few design aids are available to assist the IC designer at the system level. Design at this level involves the translation of a required behavioral-level specification into a register-transfer-level implementation. Once the functional partitioning of the design is completed, estimates of the layout size, power-supply requirements, and speed of the high-level circuit blocks used to implement the various sub-functions are required. A chip-plan must also be constructed to determine the relative placement of these building blocks. This chip plan is then further refined as the design proceeds. These tasks are often performed manually, perhaps with the computer used to perform bookkeeping tasks such as the storage of interconnection data. Research is in progress to develop aids for the designer in this area.

The Carnegie-Mellon University's Design Automation Project [109] aims at the evaluation of a set of high-level tradeoffs involving such parameters as speed, cost, and power supply requirements as a function of *design style*. For example, increasing the priority for speed of the circuit would result in a new design, probably using more parallelism. Once a satisfactory design style has been selected [110], the system will continue through an *allocation* [111] and *module binding* phase [112], where library components are allocated to implement the building blocks of the design, and finally to produce mask layout. Initial results for a restricted class of implementations look promising.

Recent work in the use of constraint-based design [113] has also shown some significant results. In this approach, the relationships between the components of a circuit are defined by a set of *incremental constraints*. These constraints may be physical, such as the required minimum spacing between mask geometries, as well as electrical, such as the size of a bus as a function of its electrical context. A computer program can then act as the *designer's assistant* and remember the reasons for design choices as well as perform simple deductions with them. One of the tasks controlled by the assistant is the maintenance of the design constraints.

Digital system-level simulation can be performed with the use of register transfer level simulators [114], [115]. These simulators generally operate at the synchronous machine level and hence they cannot be used to detect asynchronous timing problems. As mentioned in Section II, behavioral-level simulation can also be performed using programming languages, such as concurrent Pascal [99], as well as special-purpose languages and associated simulators [116], [117].

VIII. Concluding Remarks

In the design and verification of VLSI circuits extensive use of computer aids is required. The effective use of these computer programs necessitates structured design approaches so that the complexity of the design and verification tasks is reduced to a manageable level. Nevertheless, large amounts of data and a variety of design representations will be used for each circuit. For this reason, it is important that an integrated set of computer aids, coupled with a unified approach to data management, be provided to the IC designer [118].

With the decreasing cost of hardware, it also seems apparent that a dedicated special-purpose *design station* can be provided for VLSI designers on a per-designer basis. It is anticipated that such a design station would have the computation power of a high-performance minicomputer of today, coupled to a high-resolution color graphics display, and interfaced to other design stations and common resources via a high-speed computer network.

In the past, design software has often been tailored to a specific design style. To employ a different approach to design would then require a complete rewrite of the software in the design system. As is clear from the previous sections of this paper, an optimal design methodology for VLSI is not known at this time. In fact, it is clear that different design approaches are necessary for different applications. A design style suitable for low-volume custom circuits would not suit the high-volume general-purpose component designer. Hence it is necessary that the underlying operating system software on such a design station be configured to support a variety of different design methods without the need to rewrite common components used in a number of differing approaches. This software framework would involve the use of flexible data management aids which understand such structural metaphors as *hierarchy* and *regularity*. Other key requirements of such a system are the ability to apply integrity constraints to the various data representations, concurrency control to allow multiple access to the data, and incremental update so that the system can be "backed up" to an earlier version of the design. Early experiments with the use of general-purpose data-base systems for VLSI have met with limited success [119, 120] and further research is required in this area.

VLSI concerns the implementation of systems in silicon and the skills required in the design of a single chip span the entire range, from technology to system architecture. The most successful solutions to the VLSI design problem will come from a synergism of skills in all areas.

Acknowledgment

The author wishes to thank D. O. Pederson and A. L. Sangiovanni-Vincentelli for their continuous encouragement and criticism during the development of this paper. He also gratefully acknowledges the assistance of R. S. Tucker for his careful review of the manuscript.

REFERENCES

[1] W. Lattin, "VLSI design methodology: the problem of the 80's for microprocessor design," in *Proc. 16th Design Automation Conf.* (San Diego, CA) pp. 548-549, June 1979.
[2] H. K. Gummel, "A self-consistent iterative scheme for one-dimensional steady-state transistor calculations," *IEEE Trans. Electron Devices*, vol. ED-11, pp. 455-465, Oct. 1964.
[3] C. N. Gwyn, D. L. Scharfetter, and J. L. Wirth, "The analysis of radiation effects in semiconductor junction devices," *IEEE Trans. Nuclear Sci.*, vol. NS-14, pp. 153-169, Dec. 1967.
[4] R. W. Jensen and M. D. Lieberman, *The IBM Electronic Circuit Analysis Program.* Englewood Cliffs, NJ: Prentice-Hall, 1968.
[5] A. F. Malmberg, F. L. Cornwell, and F. N. Hofer, "NET-1 network analysis program 7090/94 Version," Los Alamos Scientific Lab., Los Alamos, NM, Rep. LA-3119, 1964.
[6] L. D. Milliman, W. A. Massera, and R. H. Dickhaut, "CIRCUS-A digital computer program for transient analysis of electronic circuits," Harry Diamond Lab., Washington, DC, Rep. AD-346-1, 1967.
[7] J. C. Bowers and S. R. Sedore, *SCEPTRE: A Computer Program for Circuit and System Analysis.* Englewood Cliffs, NJ: Prentice-Hall, 1971.
[8] E. D. Johnson et al., "Transient radiation analysis by computer analysis (TRAC)," Harry Diamond Lab., Washington, DC, Contract DAA 639-68-C-0041, 1968.
[9] F. F. Kuo, "Network analysis by digital computer," *Proc. IEEE*, vol. 54, pp. 821-835, June 1966.
[10] W. J. McCalla and W. G. Howard, "BIAS-3—A program for the nonlinear dc analysis of bipolar transistor circuits," in *Dig. Tech. Papers, IEEE Int. Solid State Circuits Conf.* (Philadelphia, PA), pp. 82-83, Feb. 1970.
[11] L. Nagel and R. A. Rohrer, "Computer analysis of nonlinear circuits, excluding radiation (CANCER)," *IEEE J. Solid State Circuits*, vol. SC-6, pp. 166-182, Aug. 1971.
[12] T. E. Idleman, F. S. Jenkins, W. J. McCalla, and D. O. Pederson, "SLIC—A simulator for linear integrated circuits," *IEEE J. Solid-State Circuits*, vol. SC-6, pp. 188-204, Aug. 1971.
[13] F. S. Jenkins and S. P. Fan, "TIME—A nonlinear dc and time-domain circuit simulation program," *IEEE J. Solid State Circuits*, vol. SC-6, pp. 188-192, Aug. 1971.
[14] L. W. Nagel and D. O. Pederson, "Simulation program with integrated circuit emphasis," in *Proc. 16th Midwest Symp. Circuit Theory* (Waterloo, Ont. Canada), Apr. 1973.
[15] W. T. Weeks et al., "Algorithms for ASTAP—A network analysis program," *IEEE Trans. Circuit Theory*, vol. CT-20, pp. 628-634, Nov. 1973.
[16] L. W. Nagel, "SPICE2: A computer program to simulate semiconductor circuits," Univ. California, Berkeley, ERL Memo ERL-M520, May 1975.
[17] E. Cohen, "Program reference for SPICE2, " Electronics Res. Lab., Univ. California, Berkeley, ERL Memo ERL-M592, June 1976.
[18] H. S. Baird, "A survey of computer aids for IC mask artwork verification," in *Proc. IEEE Int. Symp. on Circuit and Systems* (Phoenix, AZ), Apr. 1977.
[19] L. Rosenberg and C. Benbassat, "Critic: An integrated circuit design rule checking program," in *Proc. 11th Design Automation Workshop* (Denver, CO), pp. 14-18, June 1974.
[20] Design Rule Checking System, NCA Corporation, Sunnyvale, CA.
[21] R. M. Allgair and D. S. Evans, "A comprehensive approach to a connectivity audit, or a fruitful comparison of apples and oranges," in *Proc. 14th Design Automation Conf.* (New Orleans, LA), pp. 312-321, June 1977.
[22] I. Dobes and R. Byrd, "The automatic recognition of silicon gate transistor geometries: An LSI design aid program," in *Proc. 13th Design Automation Conf.* (San Francisco, CA), pp. 414-420, June 1976.
[23] J. Le Charpentier, "Computer aided synthesis of an IC electrical diagram from mask data," in *Dig. Tech. Papers, IEEE Int. Solid State Circuits Conf.* (Philadelphia, PA), pp. 84-85, Feb. 1975.
[24] B. T. Preas, B. W. Lindsay, and C. W. Gwyn, "Automatic circuit analysis based on mask information," in *Proc. 13th Design Automation Conf.* (San Francisco, CA), pp. 309-317, June 1976.
[25] L. Szanto, "Network recognition of an MOS integrated circuit from the topography of its masks," *Computer-Aided Design*, vol. 10, no. 2, pp. 135-140, Mar. 1978.
[26] L. Scheffer and R. Apti, "LSI design verification using topology extraction," in *Proc. 12th Asilomar Conf. Circuits, Systems, and Computers* (Asilomar, CA), pp. 149-153, Nov. 1978.
[27] E. B. Eichelberger, "Hazard detection in combinational and sequential switching circuits," *IBM J. Res. Develop.*, Mar. 1965.
[28] L. Bening, "Developments in computer simulation of gate level physical logic," in *Proc. 16th Design Automation Conf.* (San Diego, CA), pp. 561-567, June 1979.
[29] S. A. Szygenda and E. W. Thompson, "Modeling and digital simulation for design verification and diagnosis," *IEEE Trans. Computers*, vol. C-25, pp. 1242-1253, Dec. 1976.
[30] T. W. Williams, "Design for testability," presented at NATO Advanced Study Institute on Computer Design Aids for VLSI Circuits, Sogesta-Urbino, Italy, July 1980.
[31] P. S. Bottorff, "Computer aids to testing—An overview," presented at NATO Advanced Study Institute on Computer Design Aids for VLSI Circuits, Sogesta-Urbino, Italy, July 1980.
[32] M. Y. Hsueh, "Symbolic layout and compaction of integrated circuits," Univ. California, Berkeley, ERL Memo UCB/ERL M79/80, Dec. 1979.
[33] R. P. Larson, "Versatile mask generation technique for custom microelectronic devices," in *Proc. 15th Design Automation Conf.*, pp. 193-198, June 1978.
[34] D. Gibson and S. Nance, "SLIC—Symbolic layout of integrated circuits," in *Proc. 13th Design Automation Conf.*, pp. 434-440, June 1976.
[35] K. Hardage, "ASAP: Advanced symbolic artwork preparation," *LAMBDA*, vol. 1, no. 3, pp. 32-39, Fall 1980.
[36] Y. E. Cho, A. J. Korenjak, and D. E. Stockton, "FLOSS: An approach to automated layout for high-volume designs," in *Proc. 14th Design Automation Conf.*, pp. 138-141, June 1977.
[37] A. E. Dunlop, "SLIP: Symbolic layout of integrated circuits with compaction," *Computer-Aided Design*, vol. 10, no. 6, pp. 387-391, Nov. 1978.
[38] J. D. Williams, "STICKS—A graphical compiler for high-level LSI design," in *Proc. AFIPS Conf.*, vol. 47, pp. 289-295, June 1978.
[39] A. D. Ivannikov and P. P. Sipchuk, "Computer-aided design of MOS integrated circuit layout," in *Proc. IEE-CADMECCS* (Brighton, England), pp. 47-51, July 1979.
[40] D. Antoniadis, S. Hansen, and R. Dutton, "Suprem II—A program for IC process modeling and simulation," Stanford Electron. Lab., Stanford, CA, Tech. Rep. 5019-2, 1978.
[41] W. G. Oldham, S. N. Nandgaodnkar, A. R. Neureuther, and M. O'Toole, "A general simulator for VLSI lithography and etching processes: Part I—Application to projection lithography," *IEEE Trans. Electron Devices*, vol. ED-26, pp. 717-723, Apr. 1979.
[42] W. G. Oldham, A. R. Neureuther, C. Sung, J. L. Reynolds, and S. N. Nandgaonkar, "A general simulator for VLSI lithography and etching processes: Part II—Application to deposition and etching," *IEEE Trans. Electron Devices*, vol. ED-27, no. 8, pp. 1455-1459, Aug. 1980.
[43] S. Liu, B. Hoefflinger, and D. O. Pederson, "Interactive two-dimensional design of barrier-controlled MOS transistors," *IEEE Trans. Electron Devices*, vol. ED-27, pp. 1550-1558, Aug. 1980.
[44] S. Selberherr, A. Schultz, and H. W. Potzl, "MINIMOS—A two-dimensional MOS transistor analyzer," *IEEE J. Solid State Circuits*, vol. SC-15, pp. 605-615, Aug. 1980.
[45] A. Doganis and R. W. Dutton, "Optimization of IC processes using SUPREM," Stanford Electron. Lab., Stanford, CA, Tech. Rep., 1981.
[46] A. E. Rhueli, "Inductance calculations in a complex integrated circuit environment," *IBM J. Res. Develop.*, vol. 16, no. 5, pp. 470-481, Sept. 1972.
[47] A. E. Rhuehli, P. A. Brennan, "A micromodel for the inductance of perpendicular crossing lines," in *Proc. IEEE Int. Symp. Circuits, and Systems* (New York, NY), pp. 781-783, May 1978.
[48] W. J. McCalla, and D. Hoffman, "Symbolic representation and and incremental DRC for interactive layout," in *Proc. IEEE Int. Symp. on Circuits, and Systems* (Chicago, IL), p. 710-715, Apr. 1981.
[49] "MOS models for micro-computer CAD," MOSAID, Inc.
[50] L. M. Dang, "A simple current model for short-channel IGFET and its application to circuit simulation," *IEEE J. Solid State Circuits*, vol. SC-14, pp. 358-367, Apr. 1979.
[51] S. Liu, "A unified CAD model for MOSFETS," Ph.D. dissertation, Univ. California, Berkeley, Dec. 1980.
[52] A. R. Newton, "Timing, logic, and mixed mode simulation for large MOS integrated circuits," presented at NATO Advanced Study Institute on Computer Design Aids for VLSI Circuits, Sogesta-Urbino, Italy, July 1980.
[53] H. S. Baird, "Fast algorithms for LSI artwork analysis," in *Proc. 14th Design Automation Conf.* (New Orleans, LA), pp. 303-311, June 1977.
[54] P. Wilcox, H. Rombeek, and D. M. Caughey, "Design rule verification based on one-dimensional scans," in *Proc. 15th Design Automation Conf.* (Las Vegas, NV), pp. 285-289, June 1978.
[55] T. Whitney, "Hierarchical design-rule checking," M. S. rep., Dep. Computer Science, California Institute of Technology, Pasadena, June 1981.
[56] Y. Horiba et al., "A bipolar 2500-gate subnanosecond masterslice

LSI," in *Dig. Tech. Papers, IEEE Int. Solid State Circuits Conf.* (New York, NY), pp. 228-229, Feb. 1981.
[57] C. M. Davis et al., "IBM System 370 bipolar gate-array microprocessor chip," in *Proc. IEEE Int. Conf. Circuits and Computers*, pp. 669-673, Oct. 1980.
[58] S. H. Patil and T. A. Welch, "A programmable logic approach for VLSI," *IEEE Trans. Computers*, vol. C-28, pp. 594-601, Sept. 1979.
[59] C. H. Roth, *Fundamentals of Logic Design*. St. Paul, MN: West Publ. Co., 1979.
[60] C. Mead and L. Conway, *Introduction to VLSI Systems*. Reading, MA: Addison-Wesley, 1979.
[61] B. T. Murphy et al., "A CMOS 32b single-chip microprocessor," in *Dig. Papers, IEEE Int. Solid State Circuits Conf.* (New York, NY), pp. 230-231, Feb. 1981.
[62] M. Watanabe, "CAD tools for designing VLSI in Japan," in *Dig. Tech. Papers, IEEE Int. Solid State Circuits Conf.* (Philadelphia, PA), pp. 242-243, Feb. 1979.
[63] J. W. Jones, "Array logic macros," *IBM J. Res. Develop.*, pp. 98-109, March 1975.
[64] D. Johansen, "Bristle blocks: A silicon compiler," in *Proc. 16th Design Automation Conf.*, pp. 310-313, June 1979.
[65] J. B. Brinton, "CHAS seeks title to global CAD system," *Electronics*, pp. 100-102, Feb. 10, 1981.
[66] G. B. Goates, "ABLE: A LISP-based layout modeling language with user-definable procedural models for storage logic array design," M.S. Rep., Univ. Utah, Salt Lake City, Dec. 1980.
[67] M. Hanan, P. K. Wolff, and B. J. Agule, "Some experimental results on placement techniques," in *Proc. 13th Design Automation Conf.* (San Francisco, CA), pp. 214-220, June 1976.
[68] R. L. Mattison, "Design automation of MOS artwork," *Computer*, vol. 7, pp. 21-28, Jan 1974.
[69] G. Persky, D. N. Deutsch, and D. G. Schweikert, "LTX—A system for the directed automation design of LSI circuits," in *Proc. 13th Design Automation Conf.*, 1976.
[70] U. Lauther, "A min-cut placement algorithm for general cell assemblies based on a graph representation," in *Proc. 16th Design Automation Conf.* (San Diego, CA), pp. 1-10, June 1979.
[71] B. T. Preas and C. W. Gwyn, "General hierarchical automatic layout of custom VLSI circuit masks," *J. Des. Automat. Fault-Tolerant Comput.*, vol. 3, no. 1, pp. 41-58, Jan. 1979.
[72] M. Hanan and J. Kurtzberg, "Placement techniques," in *Design Automation of Digital Systems: Vol. 1, Theory and Techniques*, M. A. Breuer, Ed. Englewood, Cliffs, NJ: Prentice-Hall, 1972, Ch. 5.
[73] M. Hanan, P. Wolff, and B. Agule, "A study of placement techniques," *J. Des. Automat. Fault-Tolerant Comput.*, vol. 1, no. 1, pp. 28-61, Oct. 1976.
[74] T. Yoshimura and E. S. Kuh, "Efficient algorithms for channel routing," to be published in *IEEE Trans. Circuits and Syst.*
[75] W. R. Heller, W. F. Mikhail, and W. E. Donath, "Prediction of wiring space requirements for LSI," *J. Des. Autom. Fault Tolerant Comput.*, vol. 2, pp. 117-144, 1978.
[76] A. A. El Gamal, "Two-dimensional estimation model for interconnections in master-slice integrated circuits," *IEEE Trans. Circuits, and Syst.*, vol. CAS-28, pp. 127-137, Feb. 1981.
[77] S. J. Hong, R. G. Cain, and D. L. Ostapko, "MINI: A heuristic approach for logic minimization," *IBM J. Res. Develop.*, vol. 18, no. 5, pp. 443-458, Sept. 1974.
[78] The PRESTO program was developed by A. Svoboda and D. Brown, Tektronix, Inc.
[79] E. Porter, STC Microtechnology Corp., Private Commun.
[80] R. A. Wood, "A high density programmable logic array chip," *IEEE Trans. Comput.*, vol. C-28, pp. 602-608, Sept. 1979.
[81] G. D. Hachtel, A. L. Sangiovanni-Vincentelli, and A. R. Newton, "Some results in optimal PLA folding," in *Proc. IEEE Int. Conf. Circuits, and Computers*, pp. 1023-1028, Oct. 1980.
[82] A. L. Sangiovanni-Vincentelli, L-K. Chen, and L. O. Chua, "A new tearing approach—Node-tearing nodal analysis," in *Proc. IEEE Int. Symp. Circuits and Systems*, pp. 143-147, Apr. 1977.
[83] P. Yang, "An investigation of ordering, tearing, and latency algorithms for the time-domain simulation of large circuits," *Tech. Rep.* R-891 Univ. Illinois, Urbana, Aug. 1980.
[84] B. R. Chawla, H. K. Gummel, and P. Kozak, "MOTIS—An MOS timing simulator," *IEEE Trans. Circuits and Syst.*, vol. CAS-22, pp. 901-909, Dec. 1975.
[85] S. P. Fan, M. Y. Hsueh, A. R. Newton, and D. O. Pederson, "MOTIS-C: A new circuit simulator for MOS LSI circuits," in *Proc. IEEE Int. Symp. Circuits Syst.*, pp. 700-703, Apr. 1977.
[86] G. R. Boyle, "Simulation of integrated injection logic," Univ. California, Berkeley, ERL Memo ERL-M78/13, Mar. 1978.
[87] E. B. Kosemchak, "Computer aided analysis of digital integrated circuits by macromodeling," Ph.D. dissertation, Columbia Univ., New York, NY, 1971.
[88] N. B. Rabbat, "Macromodeling and transient simulation of large integrated digital systems," Ph.D. dissertation, The Queen's Univ., Belfast, U.K., 1971.
[89] M. Y. Hsueh, A. R. Newton and D. O. Pederson, "The development of macromodels for MOS timing simulators," in *Proc. IEEE Symp on Circuits Systems* (New York) pp. 345-349, May 1978.
[90] R. C. Baldwin, "An approach to the simulation of computer logic," 1959 AIEE Conf. paper.
[91] The LOGIS logic simulator, Information Systems Design, Inc., Santa Clara, CA.
[92] S. A. Szygenda and E. W. Thompson, "Digital logic simulation in a time-based, table-driven environment, Part 1. Design verification," *IEEE Comput.*, pp. 24-36, Mar. 1975.
[93] M. A. Breuer, "General survey of design automation of digital computers," *Proc. IEEE*, vol. 54, no. 12, Dec. 1966.
[94] P. Wilcox and A. Rombeck, "F/LOGIC—An interactive fault and logic simulation for digital circuits," in *Proc. 13th Design Automation Conf.*, pp. 68-73, 1976.
[95] A. R. Newton, "Techniques for the simulation of large-scale integrated circuits," *IEEE Trans. Circuits Syst.*, vol. CAS-26, pp. 741-749, Sept. 1979.
[96] G. Arnout and H. De Man, "The use of threshold functions and Boolean-controlled network elements for macromodeling of LSI Circuits," *IEEE J. Solid-State Circuits*, vol. SC-13, pp. 326-332, June 1978.
[97] V. D. Agrawal et al., "The mixed mode simulator," in *Proc. 17th Design Automation Conf.*, pp. 618-625, June 1980.
[98] K. Sakallah and S. W. Director, "An activity-directed circuit simulation algorithm," in *Proc. IEEE Int. Conf. on Circuits and Computers*, pp. 1032-1035, Oct. 1980.
[99] D. D. Hill and W. M. Van Cleemput, "SABLE: Multi-level simulation for hierarchical design," in *Proc. IEEE Int. Symp. Circuits and Systems*, pp. 431-434, (Houston, TX), Apr. 1980.
[100] T. W. Williams and K. P. Parker, "Testing logic networks and design for testability," *Computer*, pp. 9-21, Oct. 1979.
[101] E. B. Eichelberger and T. W. Williams, "A logic design structure for LSI testability," *J. Des. Automat. Fault Tolerant Comput.*, vol. 2, no. 2, pp. 165-178, May 1978.
[102] S. Funatsu, N. Wakatsuki, and T. Arima, "Test generation systems in Japan," in *Proc. 12th Design Automation Conf.*, pp. 114-122, June 1975.
[103] B. Koenemann, J. Mucha, and G. Zwiehoff, "Built-in logic block observation techniques," in *Dig. Papers, 1979 IEEE Test Conf.*, pp. 37-41, Oct. 1979.
[104] H. J. Nadig, "Signature analysis: Concepts, examples, and guidelines," *Hewlett-Packard J.*, pp. 15-21, May 1977.
[105] J. P. Roth, W. G. Bouricius, and P. R. Schneider, "Programmed algorithms to compute tests to detect and distinguish between failures in logic circuits," *IEEE Trans. Electron. Comput.*, vol. EC-16, pp. 567-580, Oct. 1967.
[106] J. Grason, "TMEAS, a testability measurement program," in *Proc. 16th Des. Automat. Conf.* (San Diego, CA), pp. 156-161, June 1979.
[107] L. H. Goldstein and E. L. Thigpen, "SCOAP: Sandia controllability/observability analysis program," in *Proc. 17th Design Automation Conf.* (Minneapolis, MN), pp. 190-196, June 1980.
[108] E. G. Ulrich, T. Baker, and L. R. Williams, "Fault test analysis techniques based on simulation," in *Proc. 9th Design Automation Workshop*, pp. 111-115, June 1972.
[109] A. Parker et al., "The CMU design automation system: an example of automated data path design," in *Proc. 16th Design Automation Conf.* (San Diego, CA), pp. 73-80, June 1979.
[110] D. E. Thomas, "The design and analysis of an automated design style selector," Ph.D. dissertation, Dep. Electrical Engineering, Carnegie-Mellon Univ., Pittsburgh, PA, 1977.
[111] L. Hafer, "Data-memory allocation in the distributed logic design style," M.S. Rep., Dep. Electrical Engineering, Carnegie-Mellon Univ., Pittsburgh, PA, 1977.
[112] G. W. Leive, "The binding of modules to abstract digital hardware descriptions," Ph.D. dissertation, Dep. Electrical Engineering, Carnegie-Mellon Univ., Pittsburgh, PA, May 1980.
[113] G. J. Sussman, J. Holloway, and T. F. Knight, Jr., "Design aids for digital integrated systems, an artificial intelligence approach," in *Proc. IEEE Int. Symp. on Circuits and Computers*, pp. 612-615, Oct. 1980.
[114] M. Barbacci, "The ISPL language," Dep. of Computer Science, Carnegie-Mellon Univ., Pittsburgh, PA, 1977.
[115] A number of references may be found in *IEEE Computer*, vol. 7, no. 12, Dec. 1974.
[116] C. Y. Chu, "An ALGOL-like computer design language," *Commun. ACM*, vol. 8, no. 10, pp. 607-615, Oct. 1965.
[117] CASSANDRE and LASCAR systems—User's manual. Grenoble, France: ENSIMAG.
[118] P. M. Carmody et al., "An interactive graphics system for custom physical design," in *Dig. Tech. Papers, IEEE Int. Solid State Circuits Conf.* (Philadelphia, PA), pp. 246-247, Feb. 1979.
[119] L. Rosenberg, "The evolution of design automation to meet the challenges of VLSI," in *Proc. 17th Design Automation Conf.* (Minneapolis, MN), pp. 3-11, June, 1980.
[120] P. B. Weil and L. P. McNamee, "Report of data base workshop," presented at IEEE Computer Science Design Automation Technical Committee, Santa Barbara, CA, Feb. 1980.

A CAD System for Design for Testability

Vishwani D. Agrawal, Sunil K. Jain, and David M. Singer, AT&T Bell Laboratories, Murray Hill, NJ

Because the scan approach permits chip testability to be automated, a design automation system, called TITUS—for Testability Implementation and Test-generation Using Scan—can automatically check the circuit for design rules, implement the testability hardware, and generate tests. Special features in this CAD system include handling of on-chip memory, chip layout techniques to optimize hardware overhead and performance penalties, and test generation in the presence of tristate devices, buses, and bidirectional devices. A 100-percent fault coverage is guaranteed in almost all cases. Because TITUS automates testability implementation and test generation, it frees the designer to concentrate on circuit optimization and functional verification. This approach produces not only a high quality design but also a high-quality product.

Design for testability means designing in such a way that testing can be guaranteed. All of the methods for implementing testability—serial scan, random scan, and built-in test (Williams and Parker 1982)—require hardware beyond that needed to accomplish the given function of the circuit. TITUS uses serial scan because that method adds a minimum amount of hardware hardware and permits the use of the conventional test technology.

To take best advantage of automation, the design process is divided in two stages. In the first stage, the designer implements only the basic function of the circuit, and operates within a set of design rules as guided by a computer-aided design audit tool. Simulation techniques verify the required circuit functions.

On the other hand, the second stage of the design is completely automatic. In this stage, the design automation system implements the testability hardware, re-verifies the design, and generates tests.

This article describes this design automation system, TITUS. The TITUS system includes programs for automatic design audits, scan implementation, and test vector generation. Special layout features that minimize the hardware overhead in custom polycell designs have also been developed. Specially designed on-chip memory blocks support scan testability. Because TITUS performs its test generation at the combinational logic level, special logic models were required for MOS circuits that contain tristate devices, buses, and bidirectional devices with memory states. The tests generated for these special logic models remain valid for the MOS devices.

TITUS fulfills the testability needs of an integrated CAD

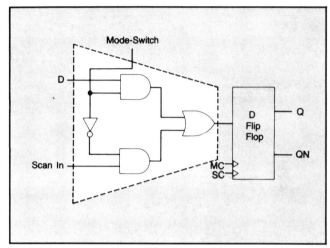

FIGURE 1. Scan modification of D flip-flop.

environment that also includes a MOS simulator, MOTIS (Agrawal *et al.* 1981), a fault simulator (Bose *et al.* 1982) and a layout system, LTX2 (Dunlop 1983).

Scan Design

The scan concept (Williams and Angell 1973) attempts to solve the general problem of test generation for sequential circuits. It takes advantage of the fact that if all the flip-flops (memory elements) can be controlled to any specific value, and if they can be observed with a straightforward operation, then the test generation task can be reduced to that of test generation for a combinational logic network. Scan design permits designers to control and observe the state, or value, of all the internal flip-flops by connecting the flip-flops into one or more shift registers when the circuit is in a test (scan) mode.

The memory elements are implemented using D-type master-slave flip-flops, whose master clock (MC) and slave clock (SC) are two non-overlapping clock signals. Two modifications of this flip-flop can accommodate scan design. In the first modification, as suggested by Williams and Angel, a double-throw switch is inserted in the data path and a mode-switch input selects the normal or scan shift mode (Figure 1). The same master and slave clocks are used in both scan and normal modes. Furthermore, because the slave clock can be generated locally from the master clock, only one clock signal need be routed over the whole chip. This fact is important in chip layouts because routing several clock signals can intro-

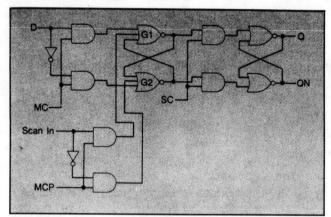

FIGURE 2. Scan register flip-flop (Eichelberger et al. 1977), which requires two clocks, MC and SC, plus a scan clock, MCP.

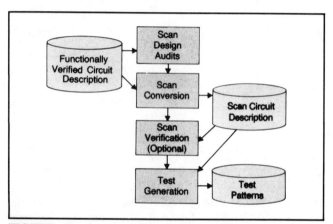

FIGURE 3. Automated scan design and test generation.

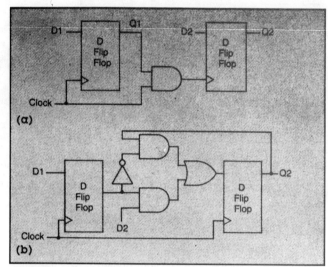

FIGURE 4. An example of a design rule audit: violation of Rule 3a (a), and a circuit modification to avoid violation (b).

duce time skews between them. Even in a single-clock design, designers usually take special care in routing the clock signal. Routing of more than once clock can introduce major layout and timing problems. Hence, from a layout point of view, this scheme has its advantages.

On the other hand, the double-throw switch introduces an additional two-gate delay (actually, it is only one gate delay in a MOS implementation). This gate delay can be kept to a minimum if the switch is implemented within the flip-flop cell in the cell-based design approach. In the flip-flop for the cell-based approach, the actual increase in cell area is only 21%, even though the increase in transistor count is 38.9%.

An alternative modification of the flip-flop (Eichelberger and Williams 1977) is shown in Figure 2. In the normal operation, inputs MC and SC are used as master and slave clocks, respectively, while the scan-clock input MCP is held low. For the scan operation, the scan-clock input and SC are used as active master and slave clocks, while the clock line MC is held low to detach the normal data line D. The area overhead due to the added gates is about the same as in the previous case. In this scheme, there is a smaller increase in delay in the data path (as compared to the previous scheme) due to increased fan-in of gates G1 and G2. However, the slave clock input SC must be a primary input. Hence, in this scheme, three clock signals have to be routed over the whole chip. (In the previous scheme, only two signals, clock and mode switch, need to be routed.) This requirement will result in higher routing overhead. In addition, routing of three clock signals opens the possibility for skews and timing problems. Routing the mode switch signal is easier than routing the clock signal: the problem of skews does not exist, so that more delay can be tolerated on the mode switch signal.

TITUS, which uses the former approach to scan modification, can automatically convert a design without scan into a scan design. Because this conversion is automatic, a designer using TITUS can complete the design verification without worrying about testability. Nevertheless, the designer must still work with a set of design rules.

In the first phase of the scan design implementation and test generation, an audit program checks the circuit for adherence to design rules (Figure 3). If the circuit does not violate the design rules, then the system automatically augments the circuit with scan design, using one of two options discussed below. An automated scan register audit can be performed at this stage. In the final step, automatic test generation covers all stuck-type faults in the scan circuit.

Design Rules for Scan Circuits

In this section, we describe a specific set of design rules that result in a design suitable for scan implementation. The rules are simple to follow and can be checked automatically by a CAD tool. These rules result in a hazard- and race-free sequential design and still provide considerable flexibility to the designer.

Rule 1: All internal memory elements must be implemented in D-type master-slave flip-flops.

Rule 2: One primary input pin must be allocated for specifying the mode (scan or normal mode). Normal input and output pins can be used as scan-in and scan-out signals of the shift register. Hence, even for multiple scan chains, only one extra primary input is required for implementation of scan design.

Rule 3: It must be possible to identify a set of clock primary inputs that control the clock inputs to the flip-flops. Furthermore, clock inputs should not be controlled through logic that is gated by flip-flop outputs or other primary inputs (Rule 3a).

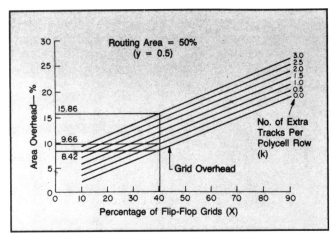

FIGURE 5. Estimated overhead of scan testability.

Nor should clock primary inputs feed the data input to a flip-flop, either directly or through combinational logic (Rule 3b). A common situation that violates Rule 3a is shown in Figure 4, along with circuit modifications that avoid the violation.

Rule 4: The edge triggered flip-flop can have only one asynchronous input (clear/preset). Furthermore, this asynchronous input must be controllable by non-clock primary inputs.

The above design rules can be effectively audited for compliance by a CAD tool, such as that described in the literature (Godoy et al. 1977). That tool checks logic structures for compliance to design rules through logic simulation of a behavioral model, which consists of primitive gates (AND, OR, NAND, NOR) and a shift register latch. However, this type of tool could not be used efficiently for circuit structures described at the MOS gate level.

An audit algorithm was developed that does not require a logic simulator. This algorithm identifies the violation of design rules by analyzing the structure of the circuit, rather than examining its function. As a result, this approach simplifies and speeds up the audit algorithms. A topological analysis of the circuit detects the situations that may violate the scan design rules. The audit is implemented as a depth-first path search algorithm. The audit algorithm initiates the path tracing from the primary clock inputs or other primary inputs of the circuit. During these path tracings, the logical functions of gates in the path are ignored. Often this results in a conservative check. This tool can also quickly check very large logic designs. Hence, it acts as an effective early warning tool for the logic designer and can be used early in the logic design phase.

A sequential circuit designed in accordance with Rules 1 through 4 will be scan-testable. After the scan audits, the flip-flops in the circuit must be connected into one or more shift registers. This can be done automatically by TITUS, or manually by the designer, who should follow the following rules:

Rule 5: All flip-flops should be connected in a shift register. There may be logical inversions in the path of the scan chain.

Rule 6: Each shift register should have a primary input and a primary output in the scan mode. However, normal primary inputs and outputs may be used or shared for this purpose.

Rule 7: When the mode specification line is in scan mode, then the output of a flip-flop or scan-out primary output should be a function of only the preceding flip-flop output or scan-in primary input of the shift register.

A scan verification audit checks compliance with Rules 5 through 7. (When TITUS is used to implement the scan design, the user does not worry about Rules 5-7.) This audit also detects the inversions in the scan chain. The technique used to check for scan chain verification is similar to logic simulation (at the MOS gate level) performed locally on the scan portion of the circuit.

Area and Performance Overheads

The polycell layout style consists of standard cells placed on grids in the rows of the layout. The polycells contain simple Boolean or memory functions. One dimension (height) of the cells is fixed to allow for an arrangement in rows, although the width of the polycells varies. Adjacent rows of polycells are separated by a space for the routing channels. CAD programs facilitate this routing process (Dunlop 1983).

The implementation of scan design increases the area of the chip in two ways. First, the width of a scan flip-flop is greater than that of an ordinary flip-flop—although the height of both types of flip-flops remains the same. The larger flip-flop is reflected in the increase in the width of the polycell row (X-direction). Secondly, the scan design requires at least two additional routing tracks per pair of polycell rows. One of these tracks is for the scan data signal and the other track is for the mode specification line (scan or normal mode). The added tracks will increase the area in the Y-direction.

The increase in area due to larger scan flip-flops depends on the fraction of chip area that is occupied by the flip-flops. The total increase in area can be calculated theoretically. The increase in the X direction (Δ_X) is $(W_r - 1)x$, where x is the combined width of the flip-flop polycells in the pre-scan circuit as a fraction of total polycell width; and W_r is the width of the scan flip-flops, relative to the width of the non-scan flip-flops. The increase in the Y direction (Δ_Y) is $(1 - y)k/T$, where y is the routing area fraction in the pre-scan circuit; k is the number of extra tracks per polycell row required for scan design; and T is the polycell height in terms of number of routing tracks. The area overhead is then $(1 + \Delta_X)(1 + \Delta_Y) - 1$. Figure 5 shows the computed area overhead versus percentage of flip-flops in the circuit. A routing area fraction of 50% and polycell height of 21.5 tracks are assumed. Note that area overhead is a function of both the flip-flop percentage and the extra tracks required for scan routing.

Scan Implementation

In TITUS, the scan design can be implemented automatically in one of two different ways. The circuit designs are described heirarchically through a connectivity language. Primitives in this description are the polycells, pre-defined layouts of commonly used functions having complexities up to 30-40 MOS devices. Flip-flop polycells are automatically replaced by their scan versions.

In the first approach to scan implementation, the flip-flops are connected in the shift register such that the hierarchy in the design is preserved. The shift register connection obtained in this manner is similar to that as implemented by a designer.

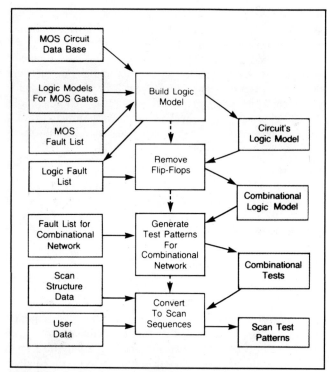

FIGURE 6. Scan test generation system.

With this approach, the impact of scan conversion on a circuit's interconnection description hierarchy is minimal. However, the overhead due to scan wiring cannot be estimated easily. Furthermore, the layout program does not differentiate between normal and scan wiring.

The scan wiring may cause the normal wiring to be stretched out—which translates directly into larger routing capacitances for the normal paths. However, the new interconnection description of the circuit is the same as the original description of the circuit. Therefore, it is more readable and the modified schematics can be obtained easily.

The second approach takes advantage of the fact that the order of flip-flops in the shift register is not important. In this approach, the circuit is first laid out by the layout program without the shift register wiring. The order of the flip-flops in the shift register is then selected to minimize the number of extra routing tracks required for scan wiring. The layout program can differentiate between normal and scan wiring at this point. Hence, normal wiring is optimized before the scan wiring is added.

In this approach, the new circuit interconnection description is not similar to the original hierarchical interconnection description. Actually, the new description does not have any heirarchy and is often difficult to understand. However, the area and performance considerations make this approach attractive.

The overhead due to the increased size of the scan flip-flops is the same in both approaches. The overhead due to extra routing tracks for scan wiring will be different in the two approaches. The second approach gives more optimized scan wiring.

To check this overhead, a 2000-gate CMOS chip (polycell design) was laid out in three different ways. First, the design was laid out without scan. Then, two more layouts, with the approaches discussed, were completed. The actual area overheads were determined and compared with the theoretical values. The results of this experiment are presented in Table 1.

The chip without scan had 47.8% of the total polycell area occupied by flip-flop polycells. From the first layout, the total routing track area was found to be 47.1%. This gives $x = 0.478$ and $y = 0.471$. Taking the polycell height $T = 21.5$ tracks, and $k = 1.5$ tracks per polycell row for scan wiring, the overhead is calculated as 14.05%. The measured overhead in the scan chip with preserved hierarchy was 16.93%. The chip with optimized scan wiring had an overhead of 11.9%, which is even better than the computed value. Evidently, the layout program's flip-flop ordering is better than the ordering used in the hierarchical design.

The operating frequency for the three layouts was determined by a path analysis technique (Agrawal 1982). As expected, the scan design also increased the delay of normal data paths, reducing the speed of operation (clock rate). Two components contribute to the increased path delay in MOS polycell circuits. First, the multiplexer on the data input of the flip-flop introduces an extra gate delay. Second, the output of a flip-flop is connected to the scan input of the next flip-flop in the scan register chain. This results in extra capacitive loading due to scan wiring at the flip-flop outputs.

Although the delay overhead due to the multiplexer will be the same in both of the scan implementation approaches, the delay due to extra capacitances can be minimized if the scan wiring is optimized through proper ordering of flip-flops in the scan register. Table 1 indicates that scan implementation during layout gives better performance than scan implementation during the logic design phase.

Test Generation

In the test generation system (Figure 6), efficient algorithms (Roth et al. 1967) are available for test generation (for stuck-at faults) for combinational networks described with primitive logic gates (AND, OR, NAND, NOR, and NOT). In order to use these algorithms, the MOS circuits should be modeled using only primitive logic gates. Thus, the first step in test generation consists of obtaining the circuit's logic gate model. This is achieved by using a logic gate library for MOS gates.

MOS circuits often contain tristate and bidirectional buffers, and internal buses using tristate drivers. Special logic gate models, discussed below, are substituted for these gates for combinational test generation. In preparing the logic model during this step, the system also prepares a list of stuck-at faults for test generation. It translates a fault of a CMOS gate to an equivalent fault on the logic gate.

The next step in test generation consists of removing the flip-flops to obtain the remaining combinational network. The scan flip-flop consists of a multiplexer and a D flip-flop. While the multiplexer is retained as part of combinational network, the flip-flop is removed. This step produces a circuit logic model for combinational test generation. In this manner, any existing combinational test generator can be used for test generation. The faults related to the flip-flops are removed from the fault list for combinational test generation.

Based on an algorithm similar to the D-algorithm (Roth et al. 1967), an automatic test generator produces test patterns

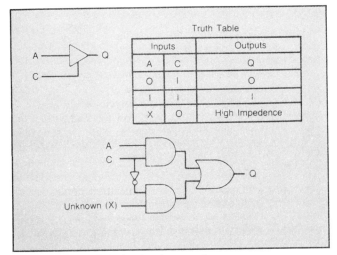

FIGURE 7. A tristate device and its test generation model

for stuck-at faults in the combinational network. The system uses a few random vectors to find "easy" faults. The logic test generation model for a tristate buffer is shown in Figure 7; the model for a tristate bus is shown in Figure 8. Such models will appropriately propagate the faults through the data inputs and set the value of control inputs.

The faults through the control inputs will not be properly propagated in these models. The faults on the control input often result in sequential behavior at the output and therefore cannot be properly represented by using primitive logic gates. Usually, the control input is a primary input and does not have combinational logic in its path. In the model the input UNKNOWN is tied to a common primary input. This is a specially added primary input signal that remains fixed to an unknown (X) value that cannot be changed by the test generator.

In general, a test pattern for a fault may contain don't-care states. For a combinational circuit, two patterns can be merged into one as long as the corresponding bits are identical or align with a don't-care. The test patterns obtained are compacted, so as to obtain a smaller set of patterns. The last step consists of translating the test patterns for a combinational network into scan tests patterns for the sequential circuit. The combinational network inputs corresponding to flip-flop outputs are converted into scan sequences. This step also includes the automatic generation of test patterns for the shift registers to verify the correct shifting operation. The test consists of a shift test in which a 010101 pattern is shifted through every flip-flop of the scan register. Analysis has shown that this test pattern is sufficient to detect faults in the shift path and in the master-slave flip-flops.

Special Features

There are several special situations in circuit designs that need careful handling. Some chips contain RAM or ROM or both. New techniques were developed for implementing scan design around the memory blocks. In order to incorporate scan design around the RAM, the data, address, and read-write (R/W) inputs need to be observed and data outputs need to be controlled. If all these signals are latched into master-slave flip-flops, then there is no problem in incorporating scan design, as these flip-flops can be put in a shift register.

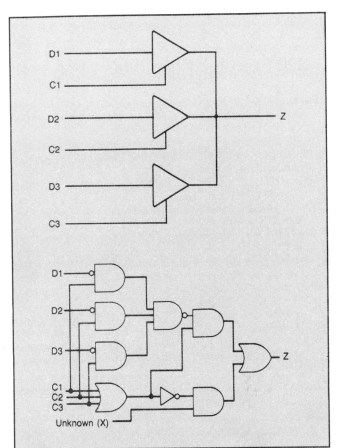

FIGURE 8. A tristate bus and its test generation model.

However, usually the data outputs are not latched into master-slave flip-flops. In a modification that makes data-out controllable, the data inputs are shorted to data outputs in the test mode. This test mode signal for memory is a separate signal from the scan mode signal of the chip. The memory itself is not tested in this mode and only the logic around the memory is tested.

Often a different strategy is adopted for testing RAM or ROM. The designer makes sure that memory inputs are controllable and outputs are observable so that memory tests can be applied directly. This is necessary because memory tests are usually long and their application through a scan register might be inefficient.

Other special conditions apply to those circuits that contain more than one clock. If two flip-flops are driven by different clocks, then they should be put in separate scan chains. The scan implementation program automatically creates separate shift registers for separately clocked flip-flops.

Finally, some circuits contain functional shift registers as part of the chip architecture. Scan design takes advantage of these preconnected shift registers and adds scan only to the first flip-flop of the shift register. This results in very little overhead. □

References

Agrawal, V.D., A.K. Bose, P. Kozak, N.H. Nham, and E. Pacas-Skewes. Winter 1981. "Mixed-mode Simulation in the MOTIS System," *Journal of Digital Systems*.

Agrawal, V.D. 1982. "Synchronous Path Analysis in MOS Circuit Simulator," *19th Design Automation Conference*, Las Vegas, NV.

Bose, A.K., P. Kozak, C-Y. Lo, H.N. Nham, E. Pacas-Skewes, and K. Wu. 1982. "A Fault Simulator for MOS LSI Circuits," *19th Design Automation Conference*, Las Vegas, NV.

Dunlop, A.E. 1983. "Automatic Layout of Gate Arrays," *1983 IEEE Symposium on Circuits and Systems*, Newport Beach, CA.

Eichelberger, E.B., and T.W. Williams. 1977. "A Logic Design Structure for LSI Testability," *14th Design Automation Conference*, New Orleans, LA.

Godoy, H.C., C.B. Franklin, and P. Bottorff. 1977. "Automatic Checking of Logic Design Structures for Compliance with Testability Ground Rules," *14th Design Automation Conference*, New Orleans, LA.

Roth, J.P., W.G. Bouricius, and P.R. Schneider. October 1967. "Programmed Algorithms to Compute Tests to Detect and Distinguish Between Failures in Logic Circuits," *IEEE Transactions on Electronic Computers*.

Williams, M.J.Y. and J.B. Angell. January 1973. "Enhancing Testability of Large Scale Integrated Circuits via Test Points and Additional Logic," *IEEE Transactions on Computers*.

Williams, T.W. and K.P. Parker. January 1982. "Design for Testability—A Survey," *IEEE Transactions on Computers*.

About the Authors

Vishwani D. Agrawal is the supervisor of the Test Aids Group at the AT&T Bell Laboratories, Murray Hill, NJ. He is also currently serving on a special assignment to assess the company's CAE needs. He obtained his Ph.D degree from the University of Illinois and has previously worked at TRW, Indian Institute of Technology, and E.G.&G., Inc. He has published widely and three of his papers have won awards from the IEEE.

Sunil K. Jain received the Bachelor of Technology degree from the Indian Institute of Technology, Kanpur, India in 1980, and the M.S. degree from Lehigh University, Bethlehem, PA. in 1982. Sunil has worked on computer-aided design and testing of VLSI circuits at the AT&T Bell Laboratories, Murray Hill, NJ. His research interests include design for testability, built-in test, high level design verification and logic synthesis. He also received an award for the Best Presentation at the ACM IEEE 21st Design Automation Conference in Albuquerque, NM.

David M. Singer received the A.A.S. degree in 1980, and the B.S. degree in 1982 in electronic technology from the University of Hartford, West Hartford, Conn. Recently, David has been developing computer-aided design tools for VLSI circuit testing at the AT&T Bell Laboratories, Murray Hill, NJ. In addition, he is pursuing a M.S. degree in computer science at the Stevens Institute of Technology, Hoboken, NJ.

Corrections

In "A CAD System for Design for Testability" (October 1984), Table 1 should have appeared as shown:

"First, the design was laid out without scan. Then, two more layouts, with the approaches discussed, were completed. The actual area overheads were determined and compared with the theoretical values. The results of this experiment are presented in Table 1."

Scan Implementation	Predicted Overhead $k = 1.5$	Actual Area Overhead	Normalized Operating Frequency
None	0	0	1.0
Hierarchical	14.05%	16.93%	0.87
Optimized	14.05%	11.9%	0.91

TABLE 1. Overhead of scan design

Number of CMOS gates = 2000
Percentage width of flip-flop polycells = 47.8%
Percentage of routing area = 47.1%.

Chapter Two: VLSI Design Validation

There are many chances for errors to occur in the design of large chips. In addition, there are many different types of errors that can occur, any one of which may cause the whole chip to fail, possibly without the designer having any idea why. Some tools are needed that will help designers debug their chips before they are manufactured, so that the chips have a better chance of working when they are actually implemented [1].

The reasons why a particular design may not work are numerous. They range from very low-level problems, such as two signals shorted together because they were too close to each other, to high-level bugs in algorithms. In another dimension, VLSI circuits may fail due to production problems or bonding errors. In this chapter, we will focus our attention on the errors that can be discovered from the mask descriptions that will be sent for fabrication. Other types of flaws are dealt with in the following chapters.

2.1 VLSI Design Rules

The standard approach to keep the complexity of designing VLSI circuits at a manageable level is to discipline that process by adopting a number of rules to be respected throughout the design. The analysis process can also benefit from a decision to follow a simple, well-defined methodology during the initial stages of synthesis. This is particularly significant if conformity to the chosen rules can be verified by machine.

Different methodologies are better suited to achieving different goals, but they all share a basic characteristic: They help isolate the designer from the details, allowing the design effort to be concentrated on higher-level abstractions. In that sense, the rules that constitute a methodology can be regarded as a syntax for the creation of circuits. Once that syntax is respected, the designer's attention is free to dwell on the semantics, i.e., the circuit functionality.

The rules of a methodology result from the interrelations among a variety of design considerations. It is not surprising, then, that the rules normally refer to distinct types of circuit information. It is convenient to classify them accordingly, as this will help understanding and taking full advantage of any methodology checking tool.

- Topological rules restrict the ways in which circuit elements can be interconnected. They prescribe the configuration of pull-ups, pull-downs, pass structures, logic gates, and more complex modules. The information needed to verify these rules is the transistor-level description of the circuit.

- Timing rules restrict the ways in which the designer can impose precedence of certain events over others to control the dynamics of the circuit. Statements about clock wave-forms and precharging of elements, for example, fall in this group. Verifying these rules requires the information needed for topological rules plus the time behavior of a few important inputs to the circuit.

- Electrical rules establish limits on capacitances, resistances, and other electrical properties of the various components of the circuit. A rule concerning maximum current densities in wires is an example of rules that belong to this category. Worst-case process parameters are needed to check compliance to electrical rules.

- Layout rules refer to geometrical shapes and dimensions on the mask. This class includes, but is not limited to, the important subcategory of geometrical design rules. Performing design rule checking (DRC) requires the complete mask-level description. Some rules in this group, however, can be checked using only vestigial layout information in addition to the circuit topology. Such is the case of restrictions on pull-up-to-pull-down ratios (also known as beta ratios), which depend but on the transistors' nominal channel sizes.

- Special rules account for useful special cases not predicted by the other categories. A special rule might, for instance, legalize a certain configuration in a situation that would otherwise exclude it. To illustrate, in NMOS technology, part of the input protection circuitry includes an enhancement-mode transistor with its gate and source shorted—This topology is only reasonable on chip inputs. By their very nature, special rules invite a pattern-matching verification method.

It is interesting to point out that rules belonging to the same class do not necessarily stem from similar considerations. Layout rules, for example, often include both design rules and beta ratio restrictions. The former are process derived whereas the latter are related to the size of noise margins.

The type of circuit information dealt with by any analysis program is implicit in its circuit description language. Thus, it should be clear from the classification above that, in a methodology verification system, the circuit description language is largely responsible for determining what kinds of syntactic rules can be checked. Furthermore, it is conceivable that a given description language will provide the means to convey some functional information about the circuits too, in which case a bit of semantic analysis can also be performed.*

2.2 Design Rule Checking

Some of the constraints on VLSI designs have a simple syntactic flavor; these are often called design rules [2]. Until recently DRC was often done by eye. Computer programs to check geometric rules are now in widespread use [3]. Since these programs often require substantial computer time to check large designs, special-purpose hardware for DRC has been proposed [4]. Most DRC programs are batch-oriented. They run from two input files; a primary file contains mask data for the design to be checked and a secondary file contains layout rules for the fabrication process under consideration. Evidently, the algebra used to represent the layout rules must be capable of expressing overlap as well as separation and of dealing with rules that apply between multiple mask levels. Typically, a complete set of rules for a silicon-gate NMOS process may be represented in some 40 to 50 logical statements at this level.

For VLSI chips, a batch DRC run may require many hours of computation time. To combat this problem, some checkers operate "incrementally." When the layout is modified, they record which areas have changed and recheck only those areas. While the user continues editing, the checker runs in background and highlights errors as it finds them. Since most changes made with the layout editor are small, the DRC program can usually display errors instantly.

2.3 Circuit Extraction

Whether or not a particular collection of mask shapes will actually realize the function intended is an important design issue. In this context, a range of circuit extraction programs is of interest. The function of these circuit extractors is to automatically interpret the circuit that will actually be realized, when given only the mask geometry data. The important point here is the guarantee that the network to be simulated is a correct interpretation of the network to be fabricated. As with other processes, when this interpretation is done manually the scope for error is considerable.

---

*The dividing line between syntactic and semantic errors in circuits may not be very well defined: In the sense used here, the former are violations of the rules explicitly selected by the user to undergo verification and the latter include connection or potential connection (via pass transistors) of gate outputs to power supplies, rings of inverters, "write-only memories" and so forth.

Generally, two levels of abstraction are of interest: logical (or perhaps switch-level) abstraction, whose primary purpose is to verify the functionality of the circuit, and circuit abstraction (including geometric data such as transistor aspect ratios), whose usual purpose is to verify the electrical performance of the circuit.

Circuit extractors are usually capable of accurate computations of interconnection resistance, internodal capacitance, ground capacitance, and transistor sizes. However, for switch-level abstraction, simpler techniques can be used to improve the execution speed of the circuit extractor.

2.4 Simulation

Simulation is the principal debugging tool of the designer. It may also be used as an aid in engineering the circuit to achieve certain performance parameters, particularly speed and power dissipation. Also, ideally, a simulator is required at each interface in the hierarchy to verify that the detail created by designing the lower level from the higher level is a faithful representation and has not introduced any errors. The most common simulations are performed with software whose input is some form of machine-readable description of the design. These descriptions usually correspond to specifications of the design at different levels in the behavioral hierarchy.

2.4.1 Process Simulation

As VLSI technology moves toward finer line-widths, the topographical features and doping profiles of devices are becoming an increasing concern [5]. Not only is greater line-width control required for high resolution but it must be accomplished in a context where features in the third dimension are as large as the line-width itself. When process models can be established, simulation algorithms can often be used to explore process effects by simulating time evolution of the line-edge profile.

Physical models and algorithms to simulate lithography, etching, deposition, and thermal processes have been developed. For example the SUPREM program [5], developed at Stanford, simulates thermal processes such as diffusion and oxidation, and the SAMPLE program [5], developed at Berkeley, permits the simulation of individual lithography, etching, and deposition processes.

2.4.2 Device Simulation

The reason for modeling devices is twofold. The device designer wants to understand how a device operates and the circuit designer seeks a quantitative description of the terminal behavior of the device. Many programs have been developed for device simulation [6]. For example, for the two-dimensional static analysis of MOS transistors, the TWIST program [7] can be used. The detailed study of problems such as the weak-inversion and weak-injection punch-through phenomena in short-channel MOS devices is facilitated by such a program.

2.4.3 Circuit Simulation

Circuit simulators offer the most detailed level of simulation normally used for design. They model electrical circuits of transistors and other elements to determine the static and transient behavior of node voltages and branch currents within the network. The results are usually presented in the form of plots of voltage (or current) versus time at selected points in the circuit, rather like an oscilloscope trace. The primary input to these circuit simulators is a list of labeled circuit elements and their connections, including definable wave-form generators at input nodes. There is also a secondary input of process parameters (threshold voltage, oxide thickness, etc.) for each type of element to be modeled. For a CMOS process, for example, two sets of parameters would be required, one for the p-channel devices and one for the n-channel devices. For a depletion-load NMOS process, one set of parameters is required for the enhancement devices, and one for the depletion devices.

The transient behavior of the network is then modeled in a series of small time steps (say 1 ns or less). In an elementary way, the voltages present on all of the circuit nodes at the start of the step determine the individual branch currents in the network. When multiplied by the time step, these currents give the net gain or loss of charge at each node which, when divided by the nodal capacitance, gives the new node voltages for the next time step. The transistor currents are computed from internal models using the node voltages, transistor dimensions, and process parameters given for each element. Evidently the accuracy of the simulation depends not only upon these parameters, but also upon the model itself. Generally, the more complex models offer greater simulation accuracy at the cost of longer run times.

The choice of a correct set of process parameters is very important to the accuracy of the simulation. This is a potentially confusing area because of the variety of ways in which a parameter, such as threshold voltage, may be defined. Misunderstandings can easily lead to large simulation errors, perhaps as large as orders of magnitude. It is therefore essential that the parameters be extracted from the process in a manner consistent with the simulation model. Because no perfect model exists, there may still be some unfortunate ambiguity, and in this circumstance only experience can dictate how to set a parameter for the best simulation results.

SPICE2 [8] and ASTAP [9] are two popular circuit simulators. They are essentially based on the exact numerical solution of the network's differential equations. However, even with the increase in speed afforded by the wave-form relaxation method [10], exact numerical solutions are too slow for VLSI circuits.

2.4.4 Timing Simulation

Circuit simulators provide precise electrical wave-form information about the circuit being analyzed. Most circuit simulators contain models for a wide range of active devices and hence are largely independent of technology. For this reason, these programs must employ general algorithms for the solution of the set of coupled, nonlinear, ordinary differential equations that describes the integrated circuit. Consequently, the characteristics of a particular technology cannot be exploited.

Many components of digital MOS or I^2L circuits may be considered to be unilateral. This characteristic, as well as the facts that these families are saturating (accumulated voltage errors are lost at the extremes of signal swing), and that large digital circuits are relatively inactive at the gate level, is exploited in timing simulation. Using table look-up models for nonlinear devices, in conjunction with node decoupling techniques, timing simulators can increase simulation speed by up to two orders of magnitude. Where a library of cells is used during the design, simplified macromodels of the cells may be used to reduce computation further.

Specialized MOS timing simulators like MOTIS-C [11] and SPLICE [12] rely on the table look-up of device characteristic for speed, and they save additional time by terminating a Newton-Raphson or similar iteration before attaining convergence. In both of these programs, the termination of an iterative step prior to convergence saves time at the cost of accuracy, and in some instances at the cost of numerical stability [13].

Some recent timing simulators such as CRYSTAL [14], TV [15], and RSIM [16] are based on a highly simplified electrical description of the network, and fall at the opposite end of the speed-accuracy trade-off curve from SPICE2. In these simulators a MOSFET is typically represented by a resistor in series with a switch; a polysilicon or diffusion line is represented by a lumped capacitance in RSIM or by a delay in TV (which is obtained by simply averaging the upper and lower delay bounds.) Although these programs are very fast, there are no absolute known limits to the error in their total delay estimates. Recent attempts [17,18] try to combine the computational speed that results from avoiding the numerical solution of differential equations with the user confidence in the results that comes from rigorous uncertainty bounds. The reader is referred to the references for details.

2.4.5 Switch-Level Simulation

To increase the simulation speed, or potential circuit complexity, it is necessary to abandon analog wave-form analysis and instead use some type of gate-level logic simulation. However, elementary (1,0) logic simulators are unsuitable for several reasons. Firstly, MOS logic elements, pass-transistors in particular, are not always unidirectional, and secondly, a high impedance state is possible, with memory of the previous state. Finally, gate delays can be unequal for high and low transitions. Many available logic simulators variously represent these features, although it is apparent that most were really intended to support circuits of SSI and MSI components rather than integrated VLSI systems. For further details on gate-level simulation, the interested reader is referred to [19].

Recently, a new class of logic simulator has emerged specifically for the MOS designer. These switch-level simulators model an MOS system as a network of nodes connected by transistor "switches." This level of simulation can model the wide variety of logic structures used in MOS designs, including logic gates, pass transistor logic, busses, and dynamic memory. Furthermore, such a simulator is fast enough to simulate entire VLSI systems, because behavior is modeled at a logical level rather than at a detailed analog level. An example of such a simulator is MOSSIM II [20].

2.4.6 Register-Transfer Level Simulation

At the highest level in the design hierarchy, the designer is concerned with overall structure and architecture and, as such, is interested in large functional components (adders, memories, etc.) and their interconnections. For this task, logic-level simulation offers too much detail and a higher level of abstraction is required. This leads to the register-transfer level (RTL) at which many hardware description languages are used. At this level, a formal system specification is feasible with functional verification by RTL simulation.

As an example, consider the SABLE/ADLIB/SDL system developed at Stanford [21]. SABLE is a multilevel hierarchical simulation program. The input to the SABLE simulator consists of a block diagram described in SDL, and of the behavior descriptions for each of the blocks in ADLIB. SABLE is a multilevel simulator in several ways: It allows the mixture of logic primitives such as AND, OR with functional blocks such as microprocessors; it also allows the use of multiple values for the signals during the course of a single simulation run (2-valued mixed with 9-valued simulation).

Using the mentioned system, one may formally prove the equivalence of two hardware descriptions. One of these descriptions will represent the specification for a system, while the other represents a proposed design. The hardware designer will supply a simulation relation specifying the desired correspondence between two ADLIB/SDL descriptions. Both descriptions will then be symbolically simulated. The results of two simulations should be consistent with the simulation relation.

2.4.7 Mixed-Level and Mixed-Mode Simulation

A VLSI circuit may be represented at several levels of complexity. It is commonly the case that the performance of a VLSI system cannot be determined adequately without a detailed, circuit-level simulation of some critical cell; however simulation of the complete system at the circuit level would be discouragingly expensive. Considering that a logic-level simulation may be entirely adequate for most of the system, it appears that this conflict might be resolved by modeling different parts of a circuit at different levels simultaneously. Such simulators are labeled mixed level.

Mixed-level simulators also offer the advantage that mixed digital and analog circuits can be modeled. This is a very attractive feature for designers of switched-capacitor filters and data converters enabling the simulation of entire VLSI systems with analog input and digital processing. In mixed-mode simulation, the different programs operate at the same level but the function is different. An interesting combined mixed-level and mixed-mode piece-wise linear modeling approach is described in [22]. SPLICE [12] is an example of a mixed-mode simulator which combines circuit analysis, timing analysis, and logic analysis for MOS integrated circuits.

2.5 Reprints

Computer-aided design systems for VLSI consist of synthesis, analysis, and information management tools. Synthesis tools assist the designer in creating the circuit. Examples of tools in this category include circuit layout editors, silicon compilers for generating datapaths, and PLA generators for generating control. An example of an interactive layout editor for VLSI circuits is the Magic layout system developed at the University of California at Berkeley. It is described in the first reprinted paper by Ousterhout et al. Magic understands layout rules, knows what transistors and contacts are, and knows how to route wires efficiently [23]. It also performs analysis operations, such as design-rule checking and extraction, incrementally, as the circuit is created and modified. Thus, only a small amount of work must be done each time the circuit is changed.

Analysis tools assist the designer in checking that the design is logically correct and behaves as expected (with the desired performance). Examples of these tools include design rule checkers, and various simulators. The second reprinted paper by Baker and Terman, discusses design-rule checking, circuit extraction, and switch-level simulation. A more detailed account of circuit extraction is given in the third reprinted paper by McCormick, which shows how a circuit simulator can be driven from a description of physical layout. An in-depth survey of circuit analysis, logic simulation, and design verification is presented in the fourth reprinted paper by Ruehli and Ditlow. The paper by Hachtel and Sangiovanni-Vincentelli [24] reviews recent work in VLSI circuit simulation. Examples of such simulators are MOTIS, DIANA, and SPLICE.

Hardware accelerators are a recent development in response to the increasing complexity of VLSI designs. Barzilai et al. [25], describe a switch-level simulator that runs on the IBM's Yorktown Simulation Engine (YSE) [26]. Simulation engines, wire routing machines, and design-rule checking processors are surveyed in the fifth reprinted paper by Blank.

The sixth reprinted paper by Ousterhout is concerned with the performance aspect of design validation. Even if a circuit is logically correct and free of layout errors, it will be of little or no use unless it can operate at an acceptable speed. The article describes CRYSTAL, a timing analyzer that has been developed at the University of California, at Berkeley.

Finally, information management tools organize the structure of design data, and form the foundation upon which the other kinds of tools can be built. This set of tools is described in the papers by Katz [27,28].

Over the past few years, the number of silicon foundries and CAD system companies has grown rapidly. The need for a standard format for the exchange of design information between the designer and the foundry has resulted in the introduction of EDIF (the electronic design interchange format). Macro cells, net lists, schematics, layout information, and stimulus/response vectors can be expressed in EDIF [29].

The very high speed integrated circuit (VHSIC) program of the Department of Defense (DoD) has a complementary parallel development program for a hardware description language system (VHDL). This language system is intended to provide a common support environment for various system design tools. It provides a bridge between a design specification and the analysis and design-assistance tools utilized by an engineer to aid the design development and evaluation process [30]. The VHDL system is described in the seventh reprinted paper by Shahdad et al.

References

[1] E.H. Frank and R.F. Sproull, "Testing and Debugging Custom Integrated Circuits," *ACM Computing Surveys*, Vol. 13, No. 4, Dec. 1981, pp. 425-452.

[2] R.F. Lyon, "Simplified Design Rules for VLSI Layouts," *LAMBDA Magazine*, First Quarter, 1981, pp. 54-59.

[3] M.E. Newell and D.T. Fitzpatrick, "Exploitation of Hierarchy in Analyses of Integrated Circuit Artwork," *IEEE Trans. CAD*, Vol. CAD-1, No. 4, Oct. 1982, pp. 192-200.

[4] L. Seiler, "Special Purpose Hardware for Design Rule Checking," *Proc. Second Caltech Conf. on VLSI*, Caltech, Pasedena, CA, 1981, pp. 197-216.

[5] A.R. Neureuther, "IC Process Modeling and Topography Design," *Proc. IEEE*, Vol. 71, No. 1, Jan. 1983, pp. 121-128.

[6] W.L. Engl et al., "Device Modeling," *Proc. IEEE*, Vol. 71, No. 1, Jan. 1983, pp. 10-33.

[7] S. Liu, "Interactive Two-dimensional Design of Barrier-Controlled MOS Transistors," *IEEE Trans. Electron Devices*, Vol. ED-27, No. 8, Aug. 1980, pp. 1550-1558.

[8] L.W. Nagel, "SPICE2: A Computer Program to Simulate Semiconductor Circuits," *Memo No. ERL-M250*, Elect. Res. Lab, U.C. Berkeley, May 1975.

[9] W.T. Weeks et al., "Algorithms for ASTAP—A Network Analysis Program," *IEEE Trans. Circuit Theory*, Vol. CT-20, No. 11, Nov. 1973, pp. 628-634.

[10] E. Lelarasmee et al., "The Waveform Relaxation Method for Time-Domain Analysis of Large-Scale Integrated Circuits," *IEEE Trans. CAD*, Vol. CAD-1, No. 3, July 1982, pp. 131-145.

[11] S.P. Fan et al., "MOTIS-C: A New Circuit Simulator for MOS LSI Circuits," *Proc. 1977 IEEE Int. Sym. on Circuits and Systems*, April 1977, pp. 700-703.

[12] A.R. Newton, "Techniques for the Simulation of Large Scale Integrated Circuits," *IEEE Trans. on Circuits and Systems*, Vol. CAS-26, No. 9, Sept. 1979, pp. 741-749.

[13] G.D. Micheli and A. Sangiovanni-Vincentelli, "Characterization of Integration Algorithms for the Timing Analysis of MOS VLSI Circuits," *Int. J. of Circuit Theory and Appl.*, Vol. 10, 1982, pp. 299-309.

[14] J. Ousterhout, "CRYSTAL: A Timing Analyzer for NMOS VLSI Circuits," *Proc. 3rd Caltech Conf. on VLSI*, March 1983, pp. 57-69.

[15] N. Jouppi, "TV: An NMOS Timing Analyzer," *Proc. 3rd Caltech Conf. on VLSI*, March 1983, pp. 71-85.

[16] C.J. Terman, "Simulation Tools for Digital LSI Design," Ph.D. thesis, MIT, 1983.

[17] J.L. Wyatt et al., "The Waveform Bounding Approach to Timing Analysis of Digital MOS IC's," *Proc. IEEE Int. Conf. on Computer Design*, 1983, pp. 392-395.

[18] M. Horowitz, "Timing Models for MOS Pass Networks," *Proc. IEEE Int. Symp. on Circuits and Systems*, pp. 198-201.

[19] I. Ohkura et al., "VLSI Design Verification and Logic Simulation," *Hardware and Software Concepts in VLSI*, G. Rabbat (Editor), Van Nostrand Reinhold, 1983, pp. 504-551.

[20] R. Bryant, M. Schuster, D. Whiting, "MOSSIM II, A Switch-Level Simulator for MOS LSI: User's Manual," *Tech. Rep. 5033,* Dept. of Comp. Sc., Caltech, March 1982.

[21] D.D. Hill, "Language and Environment for Multi-level Simulation," *Ph.D. thesis,* Stanford, 1980.

[22] W.M.G. Van Bokhoven, "Mixed-Level and Mixed-Mode Simulation by a Piecewise-Linear Approach," *Proc. IEEE Int. Symp. on Circuits and Systems,* May 1982, pp. 1256–1258.

[23] G.T. Hamachi and J.K. Ousterhout, "Magic's Obstacle-Avoiding Global Router," *Proc. 1985 Chapel Hill Conf. on VLSI,* 1985, pp. 145–164.

[24] G.D. Hachtel and A.L. Sangiovanni-Vincentelli, "A Survey of Third-Generation Simulation Techniques," *Proc. IEEE,* Vol. 69, No. 10, Oct. 1981, pp. 1264–1280.

[25] Z. Barzilai et al., "Fast Pass-Transistor Simulation for Custom MOS Circuits," *IEEE Design & Test,* Vol. 1, No. 1, Feb. 1984, pp. 71–81.

[26] M.M. Denneau, "The Yorktown Simulation Engine," *Proc. 19th Design Automation Conf.,* 1982, pp. 55–59.

[27] R.H. Katz, "Managing the Chip Design Database," *IEEE Computer,* Vol. 16, No. 12, Dec. 1983, pp. 26–36.

[28] R.H. Katz, "Computer-Aided Design Databases," *IEEE Design & Test,* Vol. 2, No. 1, Feb. 1985, pp. 70–74.

[29] J.D. Crawford, "EDIF: A Mechanism for the Exchange of Design Information," *IEEE Design & Test,* Vol. 2, No. 1, Feb. 1985, pp. 63–69.

[30] S.G. Shiva and P.F. Klon, "The VHSIC Hardware Description Language," *VLSI Design,* June 1985, pp. 86–106.

The Magic VLSI Layout System

Magic, a smart layout system for large-scale, custom integrated circuits, directly incorporates expertise about design rules and connectivity. The result: powerful new operations.

**John K. Ousterhout,
Gordon T. Hamachi,
Robert N. Mayo,
Walter S. Scott, and
George S. Taylor**

University of California, Berkeley

Magic is an interactive layout editing system for large-scale MOS custom integrated circuits. The system is unusual because it contains knowldge about geometrical layout rules, transistors, connectivity, and routing. Magic uses its knowledge to provide powerful interactive operations that simplify the task of creating layouts. Moreover, Magic makes it easy to modify existing layouts; this encourages designers to fix design errors, experiment with alternative designs, and enhance performance.

Magic provides several new features for its users. It checks design rules continuously and incrementally during editing sessions, to keep up-to-date information about violations —when the layout is finished, so is the design-rule check. An operation called *plowing* allows layouts to be compacted and stretched while observing all the design rules and maintaining circuit structure. The system provides routing tools that can work under and around existing wires in the channels, such as power and ground routing, and still produce results comparable in quality to the best channel routers. A hierarchical circuit extractor works directly from the Magic internal database, to provide fast turnaround during the simulation phase of design.

Two aspects of Magic's implementation make the new operations possible. First, the system is based on a data structure called *corner stitching,* which is both simple and efficient for a variety of geometrical operations.[1] Without corner stitching, operations such as plowing and design-rule checking would be too slow for interactive use. Second, designs in Magic are specified by using abstracted layers called *logs* (after Neil Weste[2]) rather than actual

---

Summary

Magic is a new IC layout system that includes several facilities traditionally contained in separate batch-processing programs. Magic incorporates expertise about design rules, connectivity, and routing directly into the layout editor and uses this information to provide several unusual features. They include a continuous design-rule checker that operates in background and maintains an up-to-date picture of violations; a hierarchical circuit extractor that only re-extracts portions of the circuit that have changed; an operation called *plowing* that permits interactive stretching and compaction; and a suite of routing tools that can work under and around existing connections in the channels. A design style called *logs* and a data structure called *corner stitching* are used to achieve an efficient implementation of the system.

An earlier version of this article appeared in the *ACM/IEEE 21st Design Automation Conference Procedings,* June 1984.

> **Our overall goal for Magic is to increase the power and flexibility of the layout editor, so designs can be entered quickly and modified easily.**

mask layers. The logs design style represents circuit structures, such as contacts and transistors, much in the manner of symbolic layout,[2-5] except that objects appear in their exact sizes and positions. It incurs no density penalty over mask-level layout, but simplifies the designer's view of the system and provides more explicit information about the circuit structure.

This article gives an overview of the Magic system—the specific problems it attempts to solve, its internal data structures and the logs design style, its features and user interface, how it is made technology-independent, and its implementation status. A collection of papers published in the *IEEE ACM 21st Design Automation Conference* provides more detailed information on design-rule checking, plowing, and channel routing.[6-8]

Background and goals

At The University of California, Berkeley, our previous layout editing systems Caesar[9,10] and KIC2,[11] have been used since 1980 for a variety of large and small designs in several MOS technologies. They resemble systems currently in use in industry.

Although Caesar and KIC have proven quite useful, they—and most layout systems—are inadequate in a few areas.

Flexibility is one such area. Once a design has been entered into the layout system, it is almost as hard to change as to re-enter. This makes it difficult to fix bugs found late in the layout process and almost impossible to experiment with alternative designs. If designers cannot try out alternatives, it is hard for them to develop intuitions about what is good and bad.

Routing, for which Caesar and KIC provide almost no support, is the most extreme example of the flexibility problem. We estimate that between 25 and 50 percent of all layout time for our circuits was being used for hand routing global interconnections, a tedious and error-prone task. It takes so long to route a circuit by hand that it is out of the question to reroute a chip to try a new floorplan. Even small cells are difficult to change; modest changes to the topology of a cell might require the entire cell to be reentered. In many industrial settings, layouts are so difficult to enter and modify that designs are completely frozen before layout begins.

Increased power and flexibility. Our overall goal for Magic is to increase the power and flexibility of the layout editor, so designs can be entered quickly and modified easily. In addition to this relatively vague goal, we set three specific ones at the beginning of the project in 1983:

(1) Once a large circuit has been routed, it should be possible to remove the routing and reroute in a few hours. Even the initial routing should not require more than a few days for a large custom circuit. With our old systems, routing requires a few weeks to a few months.

(2) The turnaround time for small bug fixes should be less than 15 minutes. For example, if a bug is found while simulating the circuit extracted from a layout, it should be possible to fix the layout, verify that the new layout meets the design rules, and re-extract the circuit, all in 15 minutes. With the old systems, this process requires several hours of CPU time and at least a half-day of elapsed time.

(3) It should not take more than 30 seconds to one minute to rearrange a cell in order to try out a different topology. With our old systems, this requires anywhere from tens of minutes to several hours.

Magic meets these goals by combining circuit expertise with an interactive editor. It understands layout rules, it knows what transistors and contacts are (and that they must be treated differently than wires), and it knows how to route wires efficiently. Magic uses its circuit knowledge to provide interactive operations that rearrange a circuit as a circuit rather than as a collection of geometrical objects.

It also performs analysis operations, such as design-rule checking and extraction, incrementally, as the circuit is created and modified. Thus, only a small amount of work must be done each time the circuit is changed.

Magic's data structure: corner stitching

In Magic, as in most layout editors, a layout consists of cells. Each cell contains two sorts of things: geometrical shapes and subcells. Magic represents the contents of cells using corner stitching,[1] a geometrical data structure for representing Manhattan shapes (those whose boundaries contain only horizontal and vertical segments). It provides the underlying mechanisms that make possible most of Magic's new features. Corner stitching is simple, provides a variety of efficient search operations, and allows the database to be modified quickly.

Planes and tiles. Planes and tiles are the basic elements in corner stitching. Each cell contains several corner-stitched planes to represent the cell's geometries and subcells; each plane consists of rectangular tiles of different types. A corner-stitched plane has three important properties: coverage, strips, and stitches.

Coverage. Each point in the *x-y* plane is contained in exactly one tile (Figure 1). Empty space is represented, as well as the area covered with material.

Strips. Horizontal strips represent material of the same type (Figure 2). The strip structure provides a canonical form for the database and prevents it from fracturing into a large number of small tiles.

Stitches. The records describing the tile structure are linked in the database through four links per tile, called stitches. The links point to neighboring tiles at two of the tile's four corners (Figure 3).

The stitches permit a variety of search operations to be performed efficiently, including: finding the tile containing a given point, finding all the tiles in an area, finding all the tiles that neighbor a given tile, and traversing a connected region of tiles. The coverage property makes it easy to update the database in response to edits, and the strip property keeps the database representation small. To the best of our knowledge, corner stitching in unique in its ability to provide these efficient two-dimensional searches and yet permit fast updates of the kind needed in an interactive tool. The only disadvantage of corner stitching, in comparison to less powerful data structures, is that it requires more storage space—about three times as much as structures based on linked lists of rectangles. Fortunately, the hierarchical circuit structure results in good paging locality, so the storage requirements do not appear to pose any problems for systems with virtual memory.

Logs

Figure 4 illustrates several ways in which corner-stitched planes might be used to represent the mask geometries in a cell.

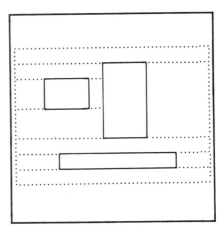

Figure 1. Every point in a corner-stitched plane is contained in exactly one tile. In this case, there are three solid tiles and space tiles (dotted lines) cover the rest of the plane. The space tiles on the sides extend to infinity. In general, a plane can contain many different types of tiles.

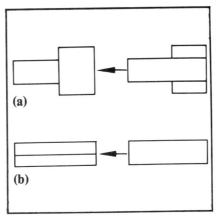

Figure 2. Areas of the same type of material are represented with horizontal strips that are as wide as possible. In each of the figures, the tile structure on the left is illegal and is converted into the tile structure on the right. In (a), it is illegal for two tiles of the same type to share a vertical edge. In (b), the two tiles must be merged, since they have exactly the same horizontal span.

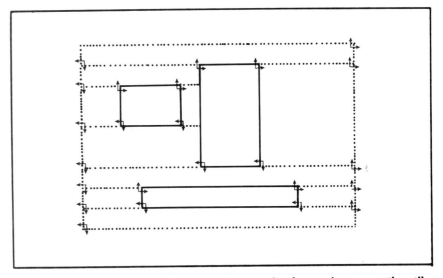

Figure 3. The record describing each tile contains four pointers to other tile records. The pointers are called corner stitches, since they point to neighboring tiles at the lower left and upper right corners. The corner stitches provide a form of two-dimensional sorting.

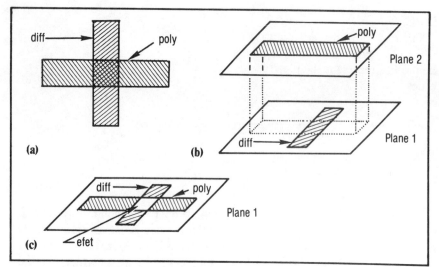

Figure 4. One way of representing the mask pattern drawn in (a) is to use a separate corner-stitched plane for each mask layer, as shown in (b). Another alternative, shown in (c), is to store both layers in a single plane, using a special type of tile (efet) to represent the overlap area.

One alternative is to use a separate plane for each mask layer, so that each plane contains space tiles and tiles of one particular mask type. The disadvantage of this approach is that many operations, such as design-rule checking and circuit extraction, require information about layer interactions, such as polysilicon crossing diffusion to form a transistor, or implants changing the type of a transistor. With a separate plane per mask layer, these operations would spend a substantial amount of time cross-registering the information on different planes.

Another alternative is to place all the mask layers into a single corner-stitched plane. Since only one tile can be at a given point in a given plane, a different tile type must be used for each possible overlap of mask layers. This eliminates the registration problem, but results in a large number of small tiles where several mask layers overlap. Even though many of the layer overlaps, such as metal and implant, are not significant, separate tile types have to be used to represent them. As a result, the database becomes fragmented into a large number of tiles and the overheads for all operations increase.

The Magic alternative. The solution we chose for Magic lies between these two extremes. We decided to use a small number of planes, where each plane contains a set of layers that has design-rule interactions. If layers do not have direct design-rule interactions, such as poly and metal, they are placed in different planes. Contacts between materials in different planes, such as poly-metal contacts, are duplicated in each of the planes. Our single-metal NMOS process has two planes: one for polysilicon, diffusion, transistors, and buried contacts; and one for metal (see Table 1). Our double-metal CMOS process has three planes: one for polysilicon and active areas, one for first-level metal, and one for second-level metal. Technology designers decide the plane structure, which is not normally visible to users.

Abstracted layers. We also decided not to represent every mask layer explicitly. Instead of dealing with actual mask layers, Magic is based on abstracted layers. We use Weste's term *logs* to describe this design style.[2] It resembles symbolic layout or sticks, except that circuit elements appear in their actual sizes and locations. Designers do not draw implants, wells, buried contact windows, or contact vias; they draw only the primary interconnect layers (polysilicon, diffusion, metal). Certain overlaps among the primary layers are drawn in special layers, which represent transistors and contacts of different types. Magic generates the implant, well, and via mask layers automatically when it creates mask files for fabrication.

Table 1 gives the planes and abstract layers used in Magic, and Figure 5 illustrates how they are used in a sample cell. The logs design style changes the way a circuit looks on the screen, but incurs no density penalty.

The Magic design style resembles those of symbolic systems, such as Mulga[2,4] and Vivid,[3] except, again, the geometries appear in their exact sizes and locations (in symbolic systems the sizes and locations are only approximate). As in symbolic design, there are simple operations for stretching and compacting cells.

The advantage of the logs approach is that designers can see the exact size and shape of a cell while it is being edited, so that they work only with a single representation of the circuit. In symbolic design, designers usually go back and forth between the

Table 1.
The corner-stitched planes and tile types used to represent the mask information for an NMOS process with buried contacts and single-level metal.

Plane	Tile types
poly-diff	polysilicon
	diffusion
	enhancement transistor
	depletion transistor
	buried contact
	poly-metal contact
	diffusion-metal contact
	space
metal	metal
	poly-metal contact
	diffusion-metal contact
	overglass via to metal
	space

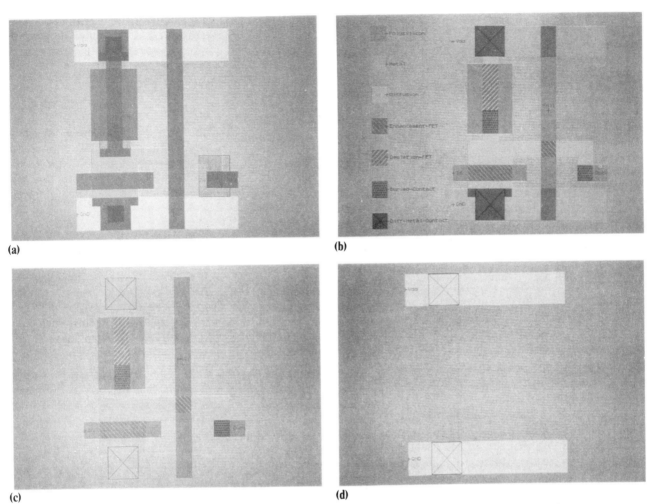

Figure 5. In Magic, transistors and contacts are drawn in an abstract form called logs: (a) a three-transistor shift-register cell, showing actual mask layers; (b) the same cell as it is seen in Magic; (c) the information in Magic's poly-diff plane; (d) the information in Magic's metal plane. Contacts are duplicated in each plane.

symbolic and mask representations; the final size of the cell can be hard to determine until it has been compacted and fleshed out. The following sections show how the abstract layers simplify design-rule checking, plowing, and circuit extraction.

In addition to the planes used to hold mask geometry, each cell contains another plane to hold information about its subcells. Subcells are allowed to overlap in Magic; each distinct subcell area or overlap between subcells is represented with a different tile in the subcell plane. Each tile contains pointers to all the subcells that cover the tile's area. This use of corner-stitching makes it easy to find subcell interactions and determine which, if any, subcells cover a particular area.

Features

Basic commands. The basic set of commands in Magic is similar to the commands in Caesar.[9,10] Mask geometry is edited in a painting style: A rectangle is placed over an area of the layout, and mask layers are painted or erased over the area of the rectangle. Additional operations select the "paint" in a rectangular area and copy it at a different place in the layout. The corner-stitched representation is invisible to users.

Like Caeser, Magic also provides commands for manipulating subcells. Subcells can be placed in a parent, moved, mirrored in x or y, rotated (by multiples of 90 degrees only), arrayed, and deleted. Subcells are handled by reference, modified, the modifications are reflected everywhere that subcell is used.

Incremental design-rule checking. Design-rule checking is an integral part of the Magic system. Our main goal is to make the checker fast, particularly for small changes; the cost of reverifying a layout should be proportional to the amount of the layout that has been changed, not to the total size of the layout.

To achieve this, Magic's design-rule checker runs continuously in the background during editing sessions. When the layout is changed, Magic records the areas that must be reverified. The design-rule checker rechecks these areas while the user is thinking. For small changes, error information appears on the screen in-

> **Plowing provides most of the operations of a symbolic system, yet works with fully fleshed geometry.**

stantly (and disappears instantly when the problem is fixed). For large changes, such as moving one large subcell on top of another, it might take seconds or even a few minutes for the design-rule checker to complete its job. In the meantime, the designer can continue editing.

If reverification is not complete when an editing session ends, the areas still to be reverified are stored with the cell so reverification can be completed the next time the cell is edited. Error information is also stored with cells until the errors are fixed. With this mechanism, there is never a need to check a layout from scratch unless the design rules are changed.

Working from the edges. Magic's basic rule-checker works from the edges of a design. Based on the type of material on either side of an edge, it verifies the presence or absence of certain layers in areas around the edge. Several properties of corner stitching and the logs design style allow edge rules to be checked quickly. First, each corner-stitched plane can be checked independently. Second, all the "interesting" edges are already present in the tile structure, so there is no need to register different mask layers. Third, logs make it unnecessary to check formation rules associated with implants and vias, since Magic generates these layers. Lastly, corner stitching provides efficient algorithms for locating all the edges in an area and for searching the constraint areas.

Handling the hierarchy. In addition to a fast basic checker, the incremental rule checker contains algorithms for handling hierarchy. When a cell in the middle of a hierarchical layout is changed, Magic checks interactions between this cell and its subcells and interactions between this cell and other cells in its parents and grandparents. Interactions are checked by locating interaction areas, combining all the material from the interacting cells into a single set of corner-stitched planes, and then invoking the basic checker. Arrays are treated specially; only four small interaction areas need to be checked for each array, regardless of how many elements the array contains. More details on the basic design-rule checking mechanism and on Magic's hierarchical approach can be found in Taylor and Ousterhout.[8]

The Magic checker is much faster than any system with which we've been able to compare it. We used the Soar microprocessor,[12] which contains about 40,000 transistors in about 200 cells, as a benchmark case, and forced Magic to recheck the entire layout. On a Vax-11/780, this required approximately 18 minutes of CPU time. All of our previous checkers would have required at least four to eight hours of CPU time for the same circuit. In practice, Magic performs better than this comparison suggests, since even large changes do not require rechecking the entire circuit.

Plowing. Plowing is a simple operation for rearranging a layout without changing the electrical circuit it represents. To invoke the plow operation, the user specifies a vertical or horizontal line segment, the plow, and a distance perpendicular to it, the plow distance (Figure 6). Magic sweeps the plow for the specified distance and moves all material out of the area swept by the plow. The edges of this material are likewise treated as plows, pushing other material in front of them. Mask geometry in front of the plow is compacted as it is moved. Jogs are inserted at the ends of the plow. The plow operation maintains design rules and connectivity, so it doesn't change the electrical structure of the circuit. Most material, such as polysilicon, diffusion, and metal, can be stretched or compacted by plowing; transistors and contacts can be moved, but their shape does not change.

Plowing provides most of the operations of a symbolic system, yet works with fully fleshed geometry. If a large plow is placed to one side of a cell and then moved across the cell, the cell will be compacted. If a plow is placed across the middle of the cell and moved, the cell will be stretched at that point. A small plow placed in the middle of a cell can be used to open empty space for new transistors or wiring. Plowing can be used on both low-level cells containing only geometry and high-level cells containing subcells and routing. Plowing moves each subcell as a unit, without affecting its contents.

Implementation. Plowing is implemented by moving edges or pieces of edges. Each edge in the path of the plow must be moved out of the plow's way. When moving an edge, Magic checks the area in front of the edge for additional edges that must be moved to maintain design rules or circuit structure. The same design rules used by the design-rule checker, along with about a half-dozen technology-independent rules that maintain connectivity and circuit structure, guide decisions about which edges to move and how far to move them. The implementation turned out to be surprisingly difficult because of subtle interactions among rules.

The implementation of plowing[7] depends on corner stitching, logs, and the edge-based design rules. Corner stitching provides the fast geometric operations used to search out plow areas, logs tell Magic about materials that cannot be stretched or compacted (such as transistors), and the edge-based design rules indicate what must be moved out of the way when a particular edge of material is moved. By working from the same data structure used for editing and design-rule checking, the plowing operation avoids the overhead of converting between representations.

Circuit extraction and cell overlaps. The Magic database makes circuit extraction much easier than it would be in other representations. Because of logs and corner stitching, the transistor structure and connectivity of individual cells are immediately available. The extractor simply traverses the tile structure and records information about what connects to what. Transistor sizes and parasitic resistances and capacitances are estimated by examining the areas and perimeters of the various circuit elements. There is no need to register layers or infer the structure and type of transistors and contacts; all this information is represented explicitly in the database.

The overlap problem. For hierarchical designs, cell overlaps or abutments complicate the situation. Each cell uses a separate set of corner-stitched planes, so information from the separate planes must be combined in order to find out what connects to what. If we allowed arbitrary overlaps, then transistors might be split between cells or formed or broken by cell overlaps. In such cases, circuits could not be extracted hierarchically, since the structure of a cell might be changed by the way it is used in its parents.

One approach to the overlap problem is to prohibit cell overlaps. This has two drawbacks, however. First, it makes for clumsy designs, since overlap areas must be eliminated by placing all the overlapped material in one of the two cells or in a new cell. This makes it harder to understand designs and harder to reuse cells. Second, it doesn't eliminate the problems in circuit extraction, since information must still be registered along the boundaries of abutting cells. For example, a cell abutment can cause two separate transistors to join.

Instead of prohibiting overlaps, we decided to restrict them. In Magic, cells may abut or overlap as long as this only connects portions of the cells and does not change their transistor structure. Overlaps and abutments may not change the type or number of transistors from what they would be without the ovelap; for example, polysilicon from one cell may not overlap diffusion from another cell, since this would create a new transistor. Magic verifies these restrictions by making additional design-rule checks in the areas where cells overlap. At present, the design-rule chcker does not do this.

The Magic approach still requires information to be registered among subcells, but it allows the extracted circuit to be represented (and extracted) hierarchically. The extracted circuit for any cell consists of the circuits of its subcells, plus the circuit of the cell itself, plus a few connections among the subcells. When subcells overlap, the parent circuit contains adjustments for parasitic resistances and capacitances. For example, if each of two overlapping cells contains polysilicon in a particular area, the parasitic capacitance of the polysilicon is recorded twice, once in each of the cells. The parent contains a compensating negative capacitance value, so when the circuits of all the cells are combined the correct capacitance value results.

Routing. Routing is the single most important area in which we hope Magic will speed up the design process. Most of the Magic routing effort has been spent in making the system flexible enough to be practical for custom VLSI; this has two aspects. First, the routing tools can work around obstacles in the channels, such as hand-routed connections or power and ground routing. Second, the tools do not require designs to be grid-aligned. Although the tools use an internal grid, they can make connections to terminals at arbitrary locations.

There are four phases to routing in Magic.

Net list. The first phase consists of the creation of a net list describing the desired connections. Magic provides an interactive editor for net lists, permitting the designer to view

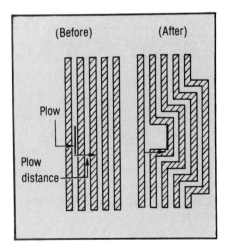

Figure 6. In plowing, a horizontal or vertical line is moved across the circuit, pushing material out of its way. Design rules and connectivity are maintained.

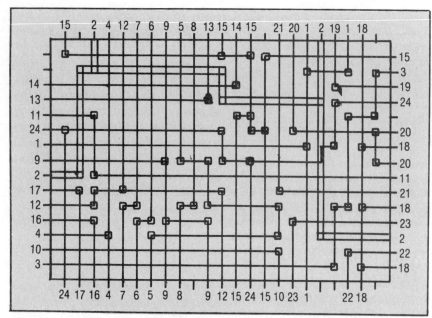

Figure 7. An example of switchbox routing with obstacles. The double-width net was routed by hand, and the Magic router completed the other connections.

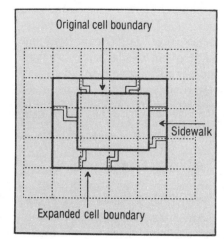

Figure 8. In the sidewalk approach, each cell is enlarged so that its boundary is grid-aligned; then connections on the edge of the original cell are routed to grid points on the outside of the sidewalk.

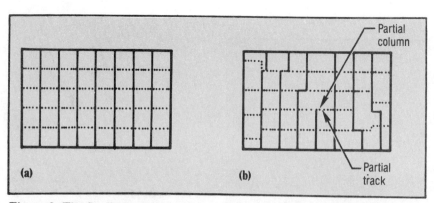

Figure 9. The flexible grid approach modifies the track and column structure of the channel. The channel is grid-based in the center, but the grid lines jog at the edges to meet nongridded connections. The standard orthogonal channel structure is in (a); (b) shows a channel whose grid structure has been flexed.

the describing the desired connections. Magic provides an interactive editor for net lists, permitting the designer to view the terminals in nets and modify their contents. Net lists are stored in textual files, so they can also be created automatically by such programs as high-level simulators and schematic editors.

Channel definition. Magic automatically provides the second, third, and fourth steps in routing. In the second phase, called channel definition, the empty space of the layout is divided into rectangular channels. This is done by examining the space tiles in the subcell plane and rearranging them into more satisfactory channels. The space tiles are long and skinny, because of the horizontal strip property discussed above; we initially used them directly as routing channels, but the results were poor. Our current channel decomposer splits and merges the space tiles to form larger and more square channels.

Global routing. In the third phase, global routing, Magic processes nets sequentially to decide which channels each will cross. The global router does not choose specific paths within channels; that task is left for the last stage of routing. The global router uses a shortest-path algorithm that traces possible paths through channels from one pin of a net to the next. It first explores those paths closest to the destination. Since corner-stitching is used to keep track of the channel structure, there is no need to build a special graph for global routing, and the global router executes quickly. In addition to deciding on a sequence of channels to be crossed by each net, the global router also chooses crossing points—the exact locations where nets cross from one channel to another. A set of rules is used to choose crossings that make it as easy as possible to route the connections within each channel. For example, one of the most important rules is that if a net enters on one side of a channel and leaves on the opposite side, crossings should be chosen so that there is no need to jog the net within the channel.

Channel routing. In the last phase, channel routing,[6] each channel is considered separately and wires are placed to achieve the necessary connections within the channel. The channel router is an extended version of Rivest and Fiduccia's greedy router.[13] It operates by sweeping from left to right across the channel, considering one column at a time. Within each column, a set of rules guides the placement of vertical wiring. For example, one rule is to use vertical wiring to make connections to pins at the top and bottom of the channel. After all the rules for a given column are applied, the tracks are extended to the next column and the process repeats itself.

We have augmented the router with additional rules, so it can handle switchboxes and work around obstacles. For example, one of the new rules is to add vertical wiring to jog nets into tracks they must occupy when they leave the channel at the right side. Another new rule causes vertical wiring to be added to jog a net out of the way of an obstacle that blocks its current track.

Obstacles. To make the routing tools more usable in a custom-design environment, the global router and channel router have been built to work around obstacles in the channels. We believe it is important for designers to be able to wire critical nets by hand and to have the automatic tools route the less critical nets without affecting the hand-routed ones. It is also convenient to run power and ground routing tools as a separate step before signal routing and have the signal router work around the power and ground wires. Magic routes under short obstacles, if possible; it routes around large ones and those that block both routing layers. Where space is insufficient to route around single-layer obstacles, Magic can make interconnections under them by using river-routing.

The channel router and global router must cooperate to avoid obstacles. The channel router contains rules to jog nets around obstacles and/or bridge across them, and the global router contains crossing-placement rules to avoid crossings near obstacles whenever possible. Figure 7 shows an example in which hand-routing and obstacle-avoidance were used to improve channel-routing results. The Magic router was initially unable to route this switchbox; Burstein speculated that it was unroutable.[14] However, after careful hand-placement of one net, shown in the figure with double lines, the router was able to complete the remaining connections while working around the hand-placed wiring.

Flexibility features. The router, combined with plowing and the other editing features, gives designers considerable flexibility. Critical signals and power and ground can be routed by hand or with special-purpose tools; then the router can be invoked to complete the interconnections. If the router is unable to make all connections, the final ones can be placed by hand, or the designer can provide hints to the router by placing a few key connections by hand and letting the router fill in the rest. Yet another alternative is to use plowing to rearrange the placement and then rerun the router. The plowing operation maintains existing connections.

We have also extended the routing tools to handle designs that are not grid-based. Most routers assume a uniform grid determined by the minimum contact-contact spacing; channel dimensions must be an integral number of grid units, and all wires must enter and leave channels on grid points. Unfortunately, custom cells are not usually designed with the router's grid in mind, so the cell boundaries and terminals do not line up on a master grid. We are experimenting with two approaches to this problem, called *sidewalks* and *flexible grid*.

Figure 8 illustrates the sidewalk approach. It involves a prerouting step, in which all cells are expanded so that their boundaries fall on grid lines. This additional cell area is called its *sidewalk*. Wires are added to connect the terminals of the cell to grid points on the outer edge of the sidewalk. In some cases, the sidewalks must be widened to more distant grid lines in order to provide enough space to connect the terminals. After the sidewalk generation stage, everything is grid-aligned, so standard routing tools can be used. Magic currently implements the sidewalk approach. Sidewalks are inefficient because the sidewalk areas cannot be used for channel routing, even though they usually contain little material. Sidewalks typically cause the channels to be reduced in size by two to three tracks and two to three columns.

The flexible grid approach distributes the sidewalks among the channels by jogging the track and column structure at the ends so they match connection points that don't fall on grid lines (Figure 9). In the flexible

The router, combined with plowing and the other editing features, gives designers considerable flexibility.

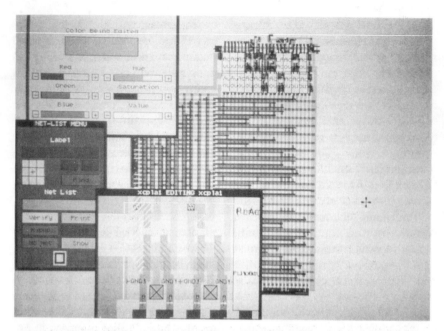

Figure 10. This picture, taken directly from a color display, shows several windows of different types. In the upper left corner is a window used by system maintainers to edit the colors used on the display. Below it is a menu for editing net lists. The main window in the background displays an automatically generated PLA, and the small window at the bottom is a zoomed-in view of a piece of the PLA, with an electrically connected region highlighted.

grid approach, space is wasted within the channel because some columns and tracks cannot extend all the way across the channel. This approach wastes less space than the sidewalk approach, however. In the worst case, the wasted space is equivalent to two tracks and two columns per channel. If connection points are sparse, however (and this appears to occur frequently), the flexible grid approach has almost zero wasted space. We are still in the early stages of exploring this alternative.

User interface. As Figure 10 shows, Magic displays the layout on a color display, and users invoke commands by pointing on the display with a mouse and then clicking mouse buttons or typing keyboard commands. Magic provides multiple overlapping windows on the color display. Each window is a separate rectangular view on a layout. Different windows can refer to different portions of a single cell or to totally different cells. Windows allow designers to see an overall view of the chip while zooming in on one or more pieces of the chip; this permits precise alignments of large objects. Information can be copied from one window to another. In addition to windows on the layout, several other kinds of windows are used for menus and system maintenance functions, such as color map editing and icon design.

Technology independence

Although Magic has considerable knowledge about integrated circuits, the information is not embedded directly in code. All the circuit information is contained in a textual technology file. This file defines the abstract layers for a particular technology, the corner-stitched planes used to represent them, and the assignment of abstract layers to planes. It tells how to display the various layers and defines the semantics of the paint and erase operations (for example, "if poly-metal-contact is painted over diffusion, erase the diffusion and place poly-metal-contact tiles on both the poly-diff and metal planes"). The technology file contains the design rules used in design-rule checking and in plowing and indicates which tile types are electrically connected when they are adjacent. For the extractor, the technology identifies which tile types correspond to transistors and specifies the parasitic resistance and capacitance values for different types of material. For the router, the technology file describes the layers to use for routing, the widths of wires, and the type and size of contacts. Lastly, the technology file tells how to fill in the structural details of transistors and contacts when generating mask patterns for circuit fabrication. The technology file format is general enough to handle a variety of NMOS and CMOS processes. Our technology file for a 4-μm NMOS process with buried contacts and single-level metal contains about 230 noncomment lines. Our technology file for a 1.2-μm P-well CMOS process with two levels of metal contains about 260 noncomment lines.

Performance and size

The implementation of Magic began in February of 1983. By early April 1983, a primitive version of the system was operational, although it did not include any facilities for design-rule checking, plowing, routing, or extracting. The subsystems became operational between the fall of 1983 and the fall of 1984; the plowing system is still undergoing extensive revision. The designers of a 32-bit microprocessor[12] have been using the system since April 1983, and several dozen students used it during the fall of 1983 and the fall of 1984 in introductory VLSI design classes.

Magic is written in C and runs on Vax processors and Sun workstations under the Berkeley 4.2 Unix operating system. The current Vax implementation works only with AED color displays with special Berkeley microcode extensions. In total, Magic contains approximately 90,000 lines of code. Table 2 gives a few sample performance measurements of pieces of the system. All measurements except the Sun redisplay time were made on a Vax-11/780. For most operations, the limiting factor is the redisplay time. Table 3 gives a breakdown of code size. The table reports total lines, of which about half are comments or blank lines.

Table 2.
Sample measurements of Magic system speed.

Operation	Speed
painting tiles into corner-stitched database	1000 tiles/sec.
redisplay (AED display)	100 tiles/sec.
redisplay (Sun workstation)	400 tiles/sec.
design-rule checking (single-cell)	2000 tiles/sec.
design-rule checking (subcell interactions)	600 tiles/sec.
design-rule checking (complete recheck of 40,000 transistors)	18 min.
plowing	150 tiles/sec.
extraction (8000 rectangles, 642 nodes, 153 transistors)	15 sec.
channel routing ("Deutsch's difficult example," 60 nets)	3 sec.
complete routing example (150 nets)	3 min.

Table 3.
Code sizes of elements in Magic.

Component	Size (lines)
database (corner stitching and hierarchy)	14,800
graphics, windows, redisplay	14,600
command interpreter	9700
design-rule checker	3200
plowing	11,000
routing	18,500
circuit extraction	6000
miscellaneous	12,000
total	89,800

Most of Magic's most important features have only recently become operational, so we don't have enough designer experience to evaluate the system thoroughly. However, early feedback has been positive. Designers have extensively and happily used both the continuous design-rule checker and the window package. The design-rule checker is particularly helpful in our classes; the instantaneous feedback makes it easy for students to learn the design rules.

The greatest problem encountered so far has been initial resistance to the logs design style. Designers accustomed to working with actual mask layers found the abstract layers in Magic confusing and irritating at first. This problem was exacerbated in the early versions of the system because the design-rule checker wasn't complete and because there was very little documentation for the design rules. As a result, designers made many mistakes and didn't find out about them until batch design-rule checks could be made, usually late in the design process. Now, in addition to the Magic checker, we have good documentation for the design rules to provide fast feedback when errors are made; and as a consequence, most of our designers prefer the logs style to the old mask-style approach.

Our three stated goals appear well within reach. The first, three-hour rerouting, should be met as soon as we have large circuits to use for benchmarking. For small examples with 150 nets, routing takes about three minutes. The second goal, 15-minute turnaround on bug fixes, depends on the circuit extractor, which has just become operational. Early measurements indicate that the circuit extractor compares in performance to the design-rule checker; this suggests that it will be fast enough to meet the goal. Plowing satisfies the third goal, 30-second cell rearrangement. On typical small- to medium-size cells, a plowing operation requires less than 10 seconds.

The pieces of the Magic system work well together. Corner stitching is a complete success: it provides all the operations needed to implement Magic's advanced features and results in simple and fast algorithms. The design-rule checker's edge-based rule set meshes well with the corner-stitched data, and it is used also for plowing. The logs design style simplifies the design rules, provides information needed for plowing and circuit extraction, and simplifies the designer's view of the layout.

We hope that Magic's flexibility will change the VLSI layout process in two ways. First, it should enable designers to experiment much more than previously. At the cell level, they can use plowing to rearrange cells quickly and easily. Cells can be designed loosely, then compacted. At the chip level, plowing and the routing tools can be used together to rearrange the floorplan, route the connections, compact or stretch, and try again. The ability to experiment means that students will be able to develop better intuitions about how to design chips and that designers will be able to fix bugs and enhance performance more easily.

Second, we hope that Magic will make it easier to reuse pieces of designs. To design a new chip, a designer will select cells from a large library, use plowing and painting to slightly modify their shape or function to suit the new application, and perhaps design a few new cells. Then the routing tools will be used to interconnect the cells. This approach should substantially reduce design time for large circuits. ☐

Acknowledgments

As tool builders, we depend on the Berkeley design community to try out our new programs, tell us what's wrong with them, and be patient while we fix the problems. Without their suggestions, it would be impossible to develop useful programs.

The Soar design team, Joan Pendleton and Shing Kong in particular, have been invaluable in helping us tune Magic; they were willing to use the system even in the dark, early days when Magic was a worse tool than its predecessor. Randy Katz was courageous enough to use Magic in his VLSI design classes. Katz, David Patterson, and Carlo Sequin all commented helpfully on this article, as did the referees.

The Magic work was supported in part by the Defense Advanced Research Projects Agency of the Department of Defense under contract N00039-84-C-0107 and in part by the Semiconductor Research Corporation under grant number SRC-82-11-008.

References

1. J. K. Ousterhout, "Corner Stitching: A Data Structuring Technique for VLSI Layout Tools," *IEEE Trans. CAD/ICAS,* Vol. CAD-3, No. 1, Jan. 1984, pp. 87-99.
2. N. Weste, "MULGA—An Interactive Symbolic Layout System for the Design of Integrated Circuits," *Bell System Technical J.,* Vol. 60, No. 6, July-Aug. 1981, pp. 823-857.
3. J. Rosenberg et al., "A Vertically Integrated VLSI Design Environment," *Proc. 20th Design Automation Conf.,* 1983, pp. 31-36.
4. N. Weste, "Virtual Grid Symbolic Layout," *Proc. 18th Design Automation Conf.,* 1981, pp. 225-233.
5. J. Williams, "STICKS—A Graphical Compiler for High Level LSI Design," *Proc. Nat'l Computer Conf.,* 1978, pp. 289-295.
6. G. T. Hamachi and J. K. Ousterhout, "A Switchbox Router with Obstacle Avoidance," *Proc. 21st Design Automation Conf.,* 1984, pp. 173-179.
7. W. S. Scott and J. K. Ousterhout, "Plowing: Interactive Stretching and Compaction in Magic," *Proc. 21st Design Automation Conf.,* 1984, pp. 166-172.
8. G. S. Taylor and J. K. Ousterhout, "Magic's Incremental Design Rule Checker," *Proc. 21st Design Automation Conf.,* 1984, pp. 160-165.
9. J. K. Ousterhout, "Caesar: An Interactive Editor for VLSI Layouts," *VLSI Design,* Vol. II, No. 4, 1981, pp. 34-38.
10. J. K. Ousterhout, "The User Interface and Implementation of Caesar," *IEEE Trans. CAD/ICAS,* Vol. CAD-3, No. 3, July 1984, pp. 242-249.
11. K. H. Keller and A. R. Newton, "KIC2: A Low-Cost, Interactive Editor for Integrated Circuit Design," *Proc. Compcon Spring,* 1982, pp. 305-306.
12. D. Ungar et al., "Architecture of SOAR: Smalltalk on a RISC," *Proc. 11th Symp. Computer Architecture,* 1984, pp. 188-197.
13. R. L. Rivest and C. M. Fiduccia, "A Greedy Channel Router," *Proc. 19th Design Automation Conf.,* 1982, pp. 418-424.
14. M. Burstein and R. Pelavin, "Hierarchical Wire Routing," *IEEE Trans. CAD/ICAS,* Vol. CAD-2, No. 4, Oct. 1983, pp. 223-234.

John K. Ousterhout received the BS degree in physics from Yale College in 1975 and the PhD in computer science from Carnegie-Mellon University in 1980. Since 1980 he has been assistant professor of electrical engineering and computer sciences at the Berkeley campus of the University of California. His research interests include computer-aided design, VLSI architecture, and operating systems.

Robert N. Mayo is a PhD candidate in the computer science Division of EECS at the University of California, Berkeley. He designed Magic's window package and graphics. His interests include graphical systems to aid expert designers of both hardware and software. Mayo has worked for the Xerox Palo Alto Research Lab (where he is currently a part-time consultant), Tektronix, and Daisy. In 1981 he received the BSCS degree from Washington University in St. Louis, and in 1984 his MSCS degree from the University of California, Berkeley.

The authors may be contacted at the Computer Science Division, Electrical Engineering and Computer Sciences, University of California, Berkeley, CA 94720.

Gordon T. Hamachi is a PhD candidate in computer science at the University of California, Berkeley. He designed and implemented Magic's channel router and global router. His research interests are channel routing, VLSI, and CAD in general. Hamachi has worked for the Xerox Palo Alto Research Lab, Hewlett-Packard, and Alfresco Software. He received both his BS (1977) and his MS (1982) degrees in computer science from the University of California, Berkeley.

Walter S. Scott is a PhD candidate in computer science at the University of California, Berkeley. He is responsible for plowing and Magic's circuit extractor, and also most of the corner-stitching package. He is interested in incremental tools for layout and analysis of integrated circuits. In 1980 Scott received his AB degree from Harvard University in applied mathematics, and in 1984 his MSCS degree from the University of California at Berkeley. He is currently a part-time consultant to Lawrence Livermore National Laboratory.

George Taylor received the BS degree in electrical engineering from Duke University in 1978. Before returning to graduate school at the University of California, Berkeley, he designed the arithmetic processor for the ELXSI 6400 computer. His research interests include VLSI architecture, floating point arithmetic, division algorithms, and computer-aided design tools. He is currently working on the Spur multiprocessor workstation project.

Tools for Verifying Integrated Circuit Designs

Clark M. Baker and Chris Terman, Massachusetts Institute of Technology

So, you've just completed the design and layout of a new architecture that will revolutionize computing, and you'd like to know if it will work before sending it off to be manufactured...

During the past 18 months, many designers at M.I.T. have found themselves in a similar situation. This paper describes a set of tools developed by the authors (Baker 1980) (Terman in preparation) to help designers make the best use of the occasional manufacturing opportunities. Our goal was to provide a variety of feedback on the different aspects of a design, and do it in a timely enough fashion that the designer would be encouraged to repeat the cycle until his design passed through unscathed. Many of the tools evolved from their original conception under pressure from our rather vocal community of users. The intent is to accumulate a repertoire of programs that incorporate all the checks that a designer might wish to do by hand.

Many of the aspects of design verification represented in our tools would be more appropriate in earlier phases of the design process; in the "ounce of prevention" versus "pound of cure" controversy, the tools fall mostly into the latter category because they are used so late in the design cycle. A lot of research is currently being directed toward the construction of ideal design environments. In a couple of years we would like to report that many of our tools have fallen into disuse as their functions were subsumed in new designer support systems. Until that time, however, these tools represent the only opportunity for most M.I.T. designers to run their designs through a gauntlet of tests that catch most ordinary design errors. We are not alone: many of the tools (or their descendants) are in use at Stanford, Carnegie-Mellon, Xerox PARC, and other institutions where small design teams (often one person) are responsible for seeing a design through from conception to final layout.

The organization of our family of verification tools is shown in Figure 1.

During the early stages of design, heavy use is made of checkplots and the design rule checker as subcomponents of the design are laid out and debugged. Once a complete design is ready for testing, the source file (usually stored in Caltech Intermediate Form) is expanded and scrutinized one last time by the design rule checker to ensure that no last-minute errors have crept into the layout. If the layout passes muster, an electrical network is derived from the mask information by the network extractor. This network can be checked for violations of certain electrical rules and then simulated with a variety of inputs to verify correct operation.

The next three sections describe the major components of the verification system in more detail. The last section reports the performance of the system components.

Design-Rule Checks

Due to physical limitations in the manufacturing process, chip designers must obey a set of geometrical constraints (design rules) when laying out their chips. Most of these constraints specify minimum line widths, minimum spacing between lines, and minimum extensions (the minimum amount by which one layer must overlap another). The basic design rules for a given process are usually straightforward, but complications occur when certain deviations from the basic rules are added to allow the designer to save space in special situations. The description of the full set of design rules for a given process is often many pages long. However, if the designer is willing to forego exploiting the details of the current manufacturing processes, it is possible to adopt a simplified set of design rules that is appropriate for a wide variety of current and future processes. In their book on VLSI design, Mead and Conway (Mead and Conway 1980) have specified one such set of simplified design rules for nMOS processes. Their set of lambda-based design rules is checked by the design rule checker described in this section.

Most design rule checkers are actually geometry engines. The input is a list of polygons (each associated with a specific mask layer) and a sequence of commands describing operations on all the objects of one or more layers. Typical commands include the merging of all intersecting polygons, intersections, unions, and differences of layers, along with width and spacing checks. Checking whole chips can involve manipulating quite a large data base of polygons. Care must be used to ensure that not too much time is spent. The design rule checker described below uses a different approach to improve the turnaround time and capacity of the checking process.

All the basic design rules are local, in that they specify only minimum widths and spacings. In the set of design rules we wish to check, the largest minimum width or spacing is three lambda. Therefore, there should be sufficient information in a four-lambda by four-lambda "window" to determine whether any design rules are violated in that window. If a checkplot of a chip is drawn on one-lambda graph paper, then successive windows can be examined by passing a four-

Reprinted with permission from LAMBDA, Fourth Quarter, 1980, pages 22-30. Copyright © 1980 by The Institute of Electrical and Electronics Engineers, Inc.

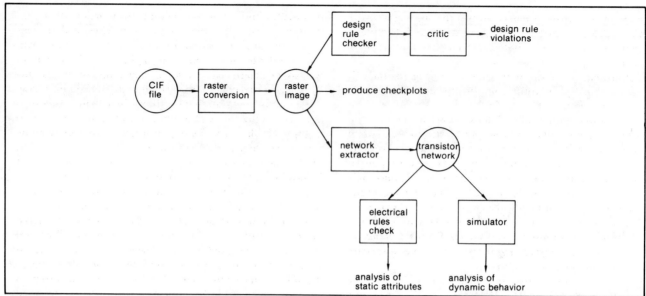

FIGURE 1. Organization of verification tools.

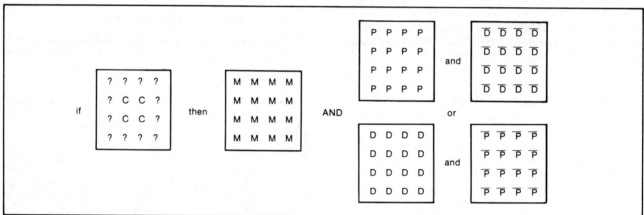

FIGURE 2. Contact cut design rule.

by-four aperture over the plot, moving one square at a time, left to right, top to bottom. If every four-by-four glimpse of the chip obeys the design rules, then the chip as a whole cannot contain any design rule violations.

For example, consider the possible windows for performing a minimum width check for the metal layer (minimum width = 3 lambda). First, look at the problem in one dimension. There are 16 possible four-by-one windows, eight of which are legal (M=metal, W=white space): WWWW, WWWM, WWMM, WMMM, MMMM, MMMW, MMWW, and MWWW. Any other combination represents a width error. It might seem that checking a chip twice, one in the X direction, and then in the Y direction, using a four-by-one window, would be sufficient; however, that method does not check diagonal width or spacing. A four-by-four window is necessary for complete checking.

In addition to the single-layer checks described above, some more complicated multi-layer checks are needed. One typical check pertains to contact cuts. Contact cuts must be surrounded by at least one lambda of metal. If the center four squares of the window are all contact cut, then the whole window must contain metal. If it does not, then an error should be reported. Design rules governing transistors can be checked in a similar way.

The heart of the design rule checker is implemented as a procedure that examines a four-by-four window and reports any errors it discovers. This function is called once for every position on the chip. The errors are accumulated in a file showing which rule was violated, the window containing the error and the coordinates of the window.

The design rules are embodied by the procedure in two ways: tabulation and special purpose code. Two tables are used to perform all width checks (spacing checks are simply width checks of white space). One table indicates which four-by-four windows contain violations of a 3-lambda width rule; the other indicates which three-by-three windows contain violations of a 2-lambda width rule. The first table is for checking metal spacing/width, and the second table is for poly and diffusion checks. Only a single bit (valid/not valid) needs to be stored for each possible window, so the tables are fairly compact. For example, there are 2^{16} possible four-by-four windows, so the corresponding table occupies 2^{16} bits or 8k bytes. Special-purpose code is used to check rules that involve multiple layers because the number of possible windows in these cases prohibits the use of tables.

The initial implementation of this scheme reported

LAMBDA *Fourth Quarter 1980*

spacing errors between mask features that were electrically connected. In the simplest case, these are caused by notches in wires (though some people consider these notches to be real errors); see Figure 3.

Unfortunately, connectivity analysis during design rule checking can become quite complicated, as shown by the example in Figure 4.

In order to eliminate all such false errors, a complete connectivity analysis, like the one performed by the circuit extractor (described below), would be required. A simpler solution that eliminates almost all the false error reports is to include a critic subroutine that performs a connectivity check on a small region (10 lambda by 10 lambda) surrounding any window reported by the checker to contain an error. The additional overhead incurred by the critic is not noticeable, because the number of windows that must be examined by the critic is quite small compared to the total number of windows examined by the checker.

When design rule checks are performed on a whole chip, many of the errors that are reported are actually due to only a small number of errors in symbols that were replicated many times. The checker has a post-processor that attempts to group all errors resulting from the replication of a symbol into a single error message. For each error type (e.g., diffusion spacing), the coordinates of all such reported errors are examined to see if some subset could correspond to replication of a symbol in one or two dimensions. If so, a single error is reported for that subset with information about replication counts, thus reducing the amount of output the designer must examine.

The raster image is generated on the fly and passed to the design rule checker. Since the design rule checker needs only four raster scan lines to perform its checks, older portions of the raster image can be discarded as checking progresses. The current algorithms for raster image generation and design rule checking are reasonably fast, but an increase in speed by a factor of 50 could be gained by implementing the inner loops of both in hardware. The algorithms in both cases are simple enough to be candidates for implementation using only a modest number of MSI chips (or perhaps a single VLSI chip!).

A few limitations of the design rule checker should be mentioned here. The algorithms outlined above are applicable only to dimensional checks. Other design rules, e.g., those involving electrical considerations such as current through a via, cannot be handled. The algorithms are also sensitive to the size of the minimum feature that must be checked. The use of a half-lambda grid would quadruple the number of elements in each window, for example, and cause a corresponding increase in the window processing time. Finally, reducing the mask information to a raster image requires approximating diagonal lines as a "staircase" of raster elements, thus making it impossible to do certain width and spacing checks correctly. Despite these limitations, the programs have been very useful in locating many design rule violations.

Circuit Extraction

The next step is extraction of the actual electrical circuit from the mask descriptions. A major goal is to make this process completely automatic. The designer should not be required to indicate where the transistors are, or where connections exist. The final output is an electrical network consisting of an unordered list of transistors, each one consisting of gate, source, and drain nodes. Other information, such as the length/width ratio for each transistor and the area by layer for each node, is also accumulated during this phase.

The simple extraction algorithm is based on a raster scan approach. The chip is examined in raster scan order (left to right, top to bottom), looking through an L-shaped window containing three raster elements: the current cell, the cell to the left, and the cell above. Using only this information, it is possible to follow connectivity and locate transistors.

There are four layers for which connectivity has to be followed: the original metal layer, M (metal wires); the original polysilicon layer, P (polysilicon wires); a derived layer, D, formed by subtracting polysilicon from diffusion (diffusion wires); and a derived layer, T, formed by intersecting diffusion and polysilicon (transistor gate regions). To follow the connectivity of these layers, the program examines the L-shaped window for each layer in turn and decides between four courses of action:

(1) The current cell is empty. Do nothing.

(2) The layer is present in the current cell, but not in the cells to the left or above. The upper left corner of a new electrical node for the layer has been located and is assigned a new (unique) node number.

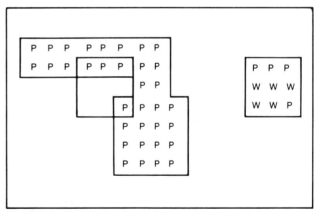

FIGURE 3. Spacing error that should be reported.

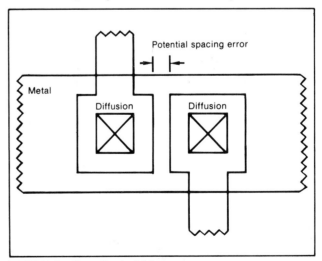

FIGURE 4. Complicated false error.

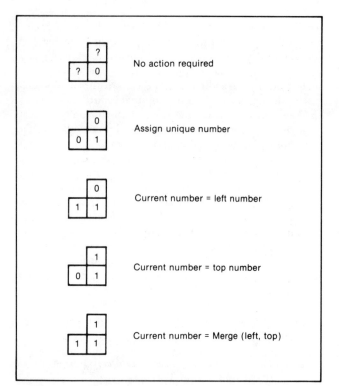

FIGURE 5. Basic raster-scan connectivity following algorithm.

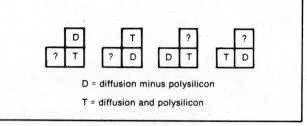

FIGURE 6. Basic nMOS transistor finding algorithm.

(3) The layer is present in the current cell and in one (but not both) of the cells to the left or above. The layer in the current cell is just an extension of the electrical node already found in the neighboring cell, and therefore is given the same node number.

(4) The layer is present in all three cells. If the node number of the layer in the left cell is the same as the node number of the layer in the cell above, then that number is also the number of the current cell. If the two node numbers are different, then two nodes that appeared distinct previously are really part of the same node. In this case, the two node numbers are merged, and the layer in the current node is assigned the merged value.

Merging two nodes requires picking one node as the result and replacing all references to the other node with references to the result. A separate file of obsolete node numbers and the corresponding correct numbers is kept for later use.

In order to account for vias between layers, an additional check of the current cell is made to determine whether contact cut, M, and either P or D is present. If so, the nodes for the M and the P or D layers are merged as in step (4) above. A similar method can be used for buried contacts.

The problem of finding transistors is split into two phases. Transistors are non-local in that the source can be arbitrarily far from the drain, and a whole transistor may not fit into a single window. Therefore, pieces of transistors (i.e., source and drain connections) are located during the first phase, and then all the pieces are combined into whole transistors during the second phase. A transistor piece contains the node numbers of the T, P, and D layers, along with a bit indicating whether the piece was implanted. Any one of four possible window configurations indicates that a transistor piece has been found. The current cell must contain either T or D (it cannot contain both), and either the left or top cell must contain the other (D or T). When this configuration is detected, a transistor record is written out to a file and saved for later processing.

After one pass has been made through the raster image of the chip, two files have been created: a file of merged-out node numbers and a file of transistor pieces. Due to the merging process, some of the node numbers in the transistor pieces may have been subsequently merged into other node numbers. Therefore, an additional pass is made through the transistor pieces, updating any old node numbers to their final value. Next, the pieces are sorted by their T numbers (with duplicate records excluded), bringing all of the pieces with the same transistor gate region together. One pass through the transistor piece file is sufficient to generate a final file of transistors.

Some errors can be detected during circuit extraction. Node numbers that were created but never merged, and which do not appear in the list of transistors, are the result of unnecessary polygons on the chip. These nodes are usually found to be part of the designer's logo, but they may also result from layout errors. If the designer can supply a list of signal names along with X and Y coordinates and the associated layer, the program can perform two additional important checks. First, it can ensure that signal paths with different names are not connected together, and second, it can guarantee that signal paths with the same name are connected together. The signal names are also passed along to subsequent programs so that the designer can refer to nodes using his names instead of the automatically generated node numbers. The extra checking provided by the use of signal names led many designers to include these names in their mask files (e.g., by use of the CIF user extension facility).

A few simple extensions should be noted. It is useful to extract the areas, lengths, and widths of each layer for each node in the circuit, for later use by the static evaluator and the simulator. The static evaluator uses the length and width to calculate pullup/pulldown ratios, and the simulator uses the area and perimeter information to calculate node capacitances (used in charge sharing and timing calculations). It would also be nice if node resistances could be calculated, but the raster approach does not help solve this difficult problem.

An obvious speed improvement in the circuit extraction process would be to intersect the polygon data directly to obtain connectivity information. A simple version (allowing only orthogonal rectangles) of the extractor was constructed that processed designs in this manner. For large designs the speed increase was quite marked (from 4 hours to 15 minutes). However, the implementation has lingering bugs that have proven quite difficult to track down. Moreover, accommodating more exotic mask geometries (diagonal lines, polygons) would complicate an already complex algorithm even further. On the other hand, the raster approach was easily expanded to allow polygons because of the modular nature of the algorithm: geometries need only be handled by the rasterizer, and the remainder of the algorithms use only the resulting raster elements. Choosing to implement circuit extraction using raster elements resulted in a timely implementation, if for no other reason than it kept us out of the quagmire of general-purpose geometry engines.

Like rasterization and design rule checking, circuit extraction is also a prime candidate for hardware implementation. A modest piece of hardware should be able to extract even the largest chips in under an hour.

Static Checks

A variety of checks can be made using the extracted circuit before proceeding to simulation. If the designer has previously entered the circuit schematic, perhaps at an earlier stage in the design process, it is straightforward to compare the extracted circuit with the original. This sort of check is appropriate at development centers where the layout is done (often by a separate team) only after a circuit schematic has been completed by the designers. At M.I.T., most projects are the product of small groups of designers responsible for both design and layout. In this case, layout often proceeds directly from preliminary architectural specifications without creating a complete circuit schematic and entering it into the computer. Thus, with no canonical schematic for comparison, simulation is the only way to verify that the circuit is correct.

However, a collection of "trivial" errors can be discovered without simulation. Because designers in the past often spent much time tracking down errors that could have been detected automatically, a static checker was developed, to be run once over the whole circuit.

These basic checks ensure that each node in the circuit can be potentially pulled up and pulled down. Each depletion-mode transistor is checked to be sure it is used in an appropriate manner. The possibilities are limited (e.g., simple pullup, superbuffer, etc.). A check is also made for threshold drops, and occurrences of two or more threshold drops are flagged as potential errors. Transistors that connect V_{dd} and GND together are located. Using the length and width information, all of the pullup/pulldown ratios are calculated and checked against a range of legal values (taking into account any threshold drops on the gates).

The program that performs these checks is not very complicated, and some checks are dependent on the manufacturing process. It would be reasonable to have a different static checker for each process to detect the specific conditions known to be true and to flag any violations.

Simulation

(Note: The simulator description given here is a shortened version of what was received. Most of what is omitted, including the basic transistor model, is similar to the simulator described in Bryant's article — Ed.)

Nodes are categorized according to how much current they can source or sink, i.e., how they affect neighboring nodes to which they are connected by transistor switches:

input Node is designated input node (e.g., V_{dd} or GND). The value of input nodes can only be changed by explicit simulator commands — the assumption is that they supply enough current to be unaffected by connections (possibly shorts to other inputs) made by transistor switches.

driven Node is connected by closed switches to other driven or input nodes. Driven nodes can affect the value of weak and charged nodes without being affected themselves, but may be forced to an X state if shorted to a driven or input node that has a different logic level.

weak Node is connected to an input node by depletion-mode (implanted) transistor. Weak nodes can affect charged nodes without being affected themselves, but are forced to a driven state when connected to another driven or input node. A weak node returns to the appropriate weak state when completely disconnected from driven or input nodes (i.e., a weak node can never enter the charged state).

charged Node is connected, if at all, only to other charged nodes. Until reconnected to some other part of the network, charged nodes will maintain their current logic state indefinitely (charge storage with no decay). This is the default state of all non-weak nodes.

Table 1 lists the possible node values, each of which has an associated logic state and drive capacity. Note that input and driven nodes are actually assigned similar values; input nodes are recognized by the simulator through a different mechanism.

Four pseudo-values are introduced in this table: DXH, DXL, WXH, and WXL. These values are similar in meaning to DX and WX with regard to the electrical properties of

TABLE 1. Possible node values.

Mnemonic	Logic State	Drive Capacity	Description
DH	1	driven	node is an input, or is connected to an input, specified to be at logic high.
DL	0	driven	node is an input, or is connected to an input, specified to be at logic low.
DX	X	driven	node is part of a possible short circuit path between V_{dd} and GND.
DXH	X	driven	node is connected to DH by a series of transistor switches at least one of which was in an unknown state.
DXL	X	driven	node is connected to DL by a series of transistor switches at least one of which was in an unknown state.
WH	1	weak	node is pulled-up by depletion-mode transistor connected to V_{dd}.
WL	0	weak	node is pulled-down by depletion-mode transistor connected to GND.
WX	X	weak	result of shorting WH and WL, or connecting to either state through a transistor switch in an unknown state.
WXH	X	weak	node is connected to WH by a series of transistor switches at least one of which was in an unknown state.
WXL	X	weak	node is connected to WL by a series of transistor switches at least one of which was in an unknown state.
CH	1	charged	charge nodes maintain state because of associated capacitance.
CL	0	charged	similar to CH.
CX	X	charged	last driven logic state of node was X, or charge sharing failed due to bad capacitance ratio.

a node, and are included only to make the incremental simulation computation easier to implement (see the following section).

What remains to be described is how two or more nodes interact when connected by transistor switches. Rather than treat groups of connected nodes *en masse*, the model specifies interactions of pairs of nodes. The resulting specification is easy to understand, because it is based on only local interactions and can be used to model non-local interactions in a natural way by means of a relaxation computation. For efficiency, the simulator optimizes this calculation when it detects specific simple cases (e.g., when all nodes are connected only by closed switches). The iterative computation that happens during relaxation (when a particular node's value is calculated several times) needs to be done only in complex situations involving the X state. Modelling these situations by relaxation seems far more elegant and simpler to implement than other schemes.

If the transistor switch is closed (gate = 1), then both nodes will eventually have the same electrical potential (the resistance of a closed switch is assumed to be 0), i.e., both nodes will be given the same value by the simulator when it is discovered that they are shorted together. What value that should be depends on the initial value and driving capacity of the connected nodes as shown in Table 2 below:

Because the transistor switch is symmetrical, the table is also symmetrical. If more than two nodes are connected by closed transistor switches, the final value for all the nodes can be computed by looking up the first two nodes in the table, taking the result and looking it up with the third node, etc., until all the nodes have been combined. As one might hope, the final answer does not depend on the order in which the nodes were examined.

If two or more charged nodes in different logic states are connected, then we have a situation called charge-sharing (indicated in Table 2 by **). In this case, the final value of the nodes depends on their relative capacitances. For example, if a large (high-capacitance) node such as a data bus were connected by a pass transistor to a small node such as the input to a register cell, then the small node would "share" the charge of the large node as its final value, regardless of the charge it had initially. The simulator runs through the connected nodes, adding up the capacitances for each logic level. If the ratio of the total capacitance for one level to sum of the others is greater than 3:1, then that level becomes the final state; otherwise, the result is CX. The choice of 3:1 is arbitrary, and was intended to be conservative in situations in which nodes of more or less equal size might become connected.

If two nodes are connected by a transistor switch in an unknown state where the switch may be open, closed, or resistive, then the two nodes may have different final values if they have different driving capacities. This is reflected in Table 3, which specifies the final value of the source node given initial values for the source and drain (source is NODE1 and drain is NODE2).

To determine the final value for the drain node, use the initial value of the drain as NODE1 and the final value of the source node just read from the table as NODE2.

The simulator uses similar tables and computations (with a few optimizations) when performing a basic simulation step. The details are described in the following section.

The Simulation Algorithm

A basic simulation step starts with a list, called the "event list", of nodes whose values have been changed by the designer since the last simulation step. The first node is taken from the event list, its new value is calculated using the model described in the previous section, and any neighboring nodes affected by the new value are added to the end of the list. This simple calculation is repeated until the event list is empty, indicating that the network has settled.

The calculation of a node's new value can be done simply by looking at the values of its neighbors and using the tables above. However, if a group of nodes is connected by closed switches, then eventually they will all reach the same final value. The simulator calculates this final value simultaneously for all the nodes in the group. Even though this optimization requires some extra programming effort, it is advantageous because connected groups of nodes are quite common. The optimized simulation calculation takes place in two phases: (1) determining the new value for the group,

		NODE2											
		DH	DL	DX	DXH	DXL	WH	XL	WX	WXH	WXL	CH	CL	CX
	DH	DH												
	DL	DX	DL											
	DX	DX	DX	DX										
	DXH	DH	DX	DX	DXH									
N	DXL	DX	DL	DX	DX	DXL								
O	WH	DH	DL	DX	DXH	DXL	WH							
D	WL	DH	DL	DX	DXH	DXL	WX	WL						
E	WX	DH	DL	DX	DXH	DXL	WX	WX	WX					
1	WXH	DH	DL	DX	DXH	DXL	WH	WX	WX	WXH				
	WXL	DH	DL	DX	DXH	DXL	WX	WL	WX	WX	WXL			
	CH	DH	DL	DX	DXH	DXL	WH	WL	WX	WXH	WXL	CH		
	CL	DH	DL	DX	DXH	DXL	WH	WL	WX	WXH	WXL	**	CL	
	CX	DH	DL	DX	DXH	DXL	WH	WL	WX	WXH	WXL	**	**	CX

□ more conservative than necessary   **charge sharing (see text)

TABLE 2. Final value of two connected nodes.

NODE1 \ NODE2	DH	DL	DX	DXH	DXL	WH	WL	WX	WXH	WXL	CH	CL	CX
DH	DH	DX	DX	DH	DX	DH	DH	DH	DH	DH	DH	DH	DH
DL	DX	DL	DX	DX	DL	DL	DL	DL	DL	DL	DL	DL	DL
DX	DX	DX	DX	DX	DX	DX	DX	DX	DX	DX	DX	DX	DX
DXH	DXH	DX	DX	DXH	DX	DXH	DXH	DXH	DXH	DXH	DXH	DXH	DXH
DXL	DX	DX	DX	DX	DXL	DXL	DXL	DXL	DXL	DXL	DXL	DXL	DXL
WH	[DXH]	DXL	DX	[DXH]	DXL	WH	WX	WX	WH	WX	WH	WH	WH
WL	DXH	[DXL]	DX	DXH	[DXL]	WX	WL	WX	WX	WL	WL	WL	WL
WX	DXH	DXL	DX	DXH	DXL	WX	WX	WX	WX	WX	WX	WX	WX
WXH	DXH	DXL	DX	DXH	DXL	WXH	WX	WX	WXH	WX	WXH	WXH	WXH
WXL	DXH	DXL	DX	DXH	DXL	WX	WXL	WX	WX	WXL	WXL	WXL	WXL
CH	DXH	DXL	DX	DXH	DXL	[WXH]	WXL	WX	[WXH]	WXL	CH	CX	CX
CL	DXH	DXL	DX	DXH	DXL	WXH	[WXL]	WX	WXH	[WXL]	CX	CL	CX
CX	DXH	DXL	DX	DXH	DXL	WXH	WXL	WX	WXH	WXL	CX	CX	CX

□ more conservative than necessary

TABLE 3. Final value for partially connected nodes.

and (2) adding any nodes affected by the new value to the end of the event list.

Extensions to the Basic Algorithm

Several more recent additions to the basic simulation algorithm deserve brief mention. One of the more ticklish problems associated with logic-level debugging is the initialization of a complex circuit whose state is represented in a distributed fashion. If all the nodes representing state are initialized to X, this usually causes the nodes controlled by the state variables to also be set to X, and so on. Often the "all X" state is self-consistent and the simulator is unable to make any headway. The problem is especially severe in circuits with feedback, which require multiple cycles to initialize and for which arbitrary choices of values for state variables can be mutually inconsistent and as devastating to correct simulation as choosing X. A simple technique, which falls short of a general solution but has proven useful for initializing many circuits, is to locate all nodes whose current value is CX and set them to CL. A simulation step is performed, and the process repeated until no more CX nodes can be found. If this technique is going to result in a successful initialization, only a few iterations seem to be required. More than 3 or 4 iterations usually indicates that the designer will have to initialize certain nodes to a consistent state by hand.

Another extension to the basic algorithm was to accommodate different transition delays when a node changed value. Multiple-delay simulation is quite common in gate-level simulators, and our implementation borrowed many of the techniques they have developed for managing the event list, etc. The novelty of our approach lies in the automatic calculation of these delays based on the electrical properties of the current interconnections in the network. When one or more connected nodes change value, the delay is dynamically calculated using the impedances of the pullup and/or pulldown paths and the capacitances of the nodes. All effects of the value change are then scheduled at the appropriate time in the future. This timing calculation is crude by comparison to circuit analysis techniques, but offers a quick way to do comparative timing of different paths and thus select those paths deserving of more careful scrutiny.

Performance

Simulator peformance was observed to be independent of layout and circuit size; rather it depended on the amount of electrical activity at a particular point in simulated time. One measure of circuit activity is the number of nodes queued on the event list in a single simulation step. The amount of time to process a single node on the list was surprisingly uniform: the simulator processed between 1000 and 2000 events per second. Thus, simulation of one complete clock cycle (four separate phases) of the SCHEME chip required about 2000 events and 1.6 seconds. Simulating execution of a non-trivial program required about 1200 cycles or 30 minutes of CPU time.

Performance

One important aspect of verification tools is their ability to process designs as expeditiously as possible. Considerable effort was expended to maximize the performance of the current tools. Over their lifetimes, many tools were sped up by factors of 2 or more as the algorithms and data representations improved. Most small projects (less than 1000 lambda square) can be processed from CIF to output files suitable for simulation in less than two hours (see Table 4).

Name	X	Y	Area	Rectangles	Nodes	Transistors
Padout	128	167	21376	144	8	10
Fipo	1192	1202	1432784	15887	888	1796
Scheme	3023	2374	7176602	108948	2412	9448

TABLE 4. Project sizes (dimensions in lambda).

For large projects, overnight turnaround is the norm.

As examples of the performance of the various tools, we present some timings for three projects of varying size.

The smallest project is a simple output pad; the FIPO is an 8-deep, 11-bit wide memory and sorting network; and the SCHEME chip (Holloway, et al. 1979) is a 32-bit LISP microprocessor. A variety of timings are given in the Table 5:

Name	T1	T2	T3	T4	T5	T6	T7	T8
Padout	6	1	1	9	15	2	7	9
Fipo	69	35	56	502	2111	92	438	281
Scheme	559	302	297	2527	16756	1380	7834	1951

T1 = Parse CIF and fully instantiate chip
T2 = Sort fully instantiated chip
T3 = Rasterization
T4 = Design rule check
T5 = Node extract
T6 = Rectangle-based node extract (requires no rasterization)
T7 = Post processing (includes additional sorting)
T8 = Make stipple plots (includes additional rasterization)

TABLE 5. Performance of verification tools.

The timings are CPU seconds on a PDP-11/70 running a UNIX timesharing system; all programs are written in the C language.

Conclusions

One of the most important measures of success for design verification tools is their use by the design community. By this standard, our efforts have been well rewarded: almost all M.I.T. designs submitted for inclusion in recent runs of the Xerox-sponsored multi-project chips have been checked using the tools described in this paper. With only one exception, all the designs had fatal flaws that would have passed unnoticed if not for some error message from one of the programs. Designers cheerfully invest additional time in running these tools. If two or three days will greatly increase the likelihood of success for a project in which three months has already been invested, then those days are not begrudged.

There are several observations worth making by way of conclusion. First, the choice to base the mask-specific algorithms on a raster scan data base has proven to be the correct choice. Using the intermediate files has insulated the design rule checker and network extractor from the vagaries of different input formats. It was easy to expand the number of input formats as the need arose, with no modification of the tools themselves. Raster based algorithms have also been straightforward to implement, and offer the prospect of efficient hardware realizations.

Second, as proposed designs become more complex, there is a greater need for simulation to serve as a test bed in which designers may evaluate their ideas. Many errors detected by the simulator are low-level ones that could have been avoided through the use of more comprehensive design systems, but simulation also uncovers conceptual errors with disturbing regularity. This problem is especially noticeable in an academic community, where many would-be architects are not digital designers.

Finally, it is all to easy to bury the designer under a mountain of error printout. The mountain is caused in part by repetitive layouts that lead to repetitive error reports. The problem can be alleviated by using a critic module to telescope these error reports into a manageable size. If designers know that each new line of the report contains new information, they will be much more conscientious in dealing with the errors. False errors can be another factor leading to designer *ennui* when presented with a stack of error messages; we know of no simple cure for false error messages, except, of course, improving the algorithms.

This research was supported by the Advanced Research Projects Agency of the Department of Defense and was monitored by the Office of Naval Research under contract number N00014-75-C-0661.

References

Baker, C. May 1980. *Artwork Analysis Tools for VLSI Circuits*, TR-239, Laboratory for Computer Science, M.I.T.

Holloway, J.; Steele, G.; Sussman, G.; and Bell, A. January 1980. *The SCHEME-79 Chip*, Memo No. 599, Artificial Intelligence Laboratory, M.I.T.

Mead, C.A. and Conway, L.A. 1980. *Introduction to VLSI Systems*, Reading: Addison-Wesley.

Terman, C. (thesis in preparation). *Simulation Tools for VLSI Design*, Laboratory for Computer Science, M.I.T.   λ

About the Authors

Chris Terman received his B.A. (1973) degree in Physics from Wesleyan University, and the S.M., E.E. (1978) degrees in Electrical Engineering and Computer Science from the Massachusetts Institute of Technology, where he is currently a PhD candidate. His research interests include computer-aided design tools for VLSI, system architectures, and compiler technology. Chris is currently working on techniques for rapid simulation of large digital circuits and on the architecture of a low-cost design station for VLSI design, layout, and verification. He is also one of the principal architects of the Nu, the personal computer system for the M.I.T. Laboratory for Computer Science.

Clark M. Baker earned his S.B. (1976), S.M. and E.E. (1980) degrees in Electrical Engineering and Computer Science at the Massachusetts Institute of Technology. Clark is a member of the research staff at the M.I.T. Laboratory for Computer Science where he is working on a variety of tools for verifying VLSI designs. His current interests include tools for artwork analysis (design and electrical rules checks) and network extraction. Clark is also working on the software for a low-cost VLSI design station and on verification algorithms suitable for hardware implementation.

EXCL: A Circuit Extractor for IC Designs

Steven P. McCormick

Research Laboratory of Electronics
Department of Electrical Engineering and Computer Science
Massachusetts Institute of Technology*

Abstract

This paper describes EXCL, an automated circuit extraction program that transforms an IC layout into a circuit representation suitable for detailed circuit simulation. The program has built-in, general extraction algorithms capable of accurate computations of interconnection resistance, inter-nodal capacitance, ground capacitance, and transistor sizes. However, where possible, the general algorithms are replaced with simple techniques, thereby improving execution speed. A basic component of the extractor is a procedure that decomposes regions into domains appropriate for specialized or simple algorithms. The paper describes the decomposition algorithm, the extraction algorithms and discusses how they connect with the rest of EXCL.

1. Introduction

An integrated circuit designer generally wishes to locate circuit design problems before dedicating his design to the expensive IC fabrication line. Inadequate speed, degraded logic levels, and excessive noise are just some of the circuit problems that plague the IC designer. Not too many years ago, active device and connectivity information was sufficient to characterize the IC's circuit behavior. A number of programs were developed to extractor these parameters[1,2,3]. However, as IC structures are scaled down in size, problems associated with IC interconnections become particularly acute, for the effects from parasitic resistances and capacitances begin to dominate over device effects.

Today, a necessary addition to locating potential circuit problems is extracting equivalent circuits for interconnections. The IC designer desires a complete spectrum of circuit parameters, including interconnection resistance, ground capacitance, inter-nodal capacitance (or coupling capacitance), transistor sizes, and transistor areas. General numerical techniques are known for solving all of these extraction problems for arbitrary shapes. However, some of the general techniques—most notably for resistance and inter-nodal capacitance—are limited to very small problems because of their need for vast computing resources. A few automated circuit extractors have been developed which use simpler techniques and do not solve the general problem[4,5]. While the techniques are usually good for long, rectangular field regions, they sacrifice accuracy around irregular regions. This paper describes an automated circuit extractor, EXCL (EXtractor of Circuits from Layout), that uses a range of extraction techniques for each problem—some are special-case and fast, others are general and slow. In the resistance, inter-nodal capacitance and transistor sizing problems, a single and powerful algorithm separates the field regions into *subregions* of three kinds. Where the fields are one-dimensional (as the fields describing conduction in a long, straight wire) one kind of subregion is formed, and the problem is solved using a simple equation. Of the remaining subregions, those with prespecified, commonly-occurring shapes have their solution found in a library. Only those subregions that cannot be solved with the previous two techniques are solved with general techniques. Subdividing the problem in this way enables EXCL to execute with reasonable speed without sacrificing accuracy. This makes EXCL adept at detecting potential circuit hazards in IC designs with "sensitive" circuits.

An important feature of EXCL is its programming modularity supported by the high-level programming language chosen for the project. EXCL exhibits flexibility when being modified to reflect changes in either the IC fabrication technology or the nature of its output analysis.

2. Overview of EXCL

The discussion of EXCL is separated into discussions of each sub-component shown in figure 1. EXCL accepts mask geometric data generated by the IC designer. Initially, all non-orthogonal geometries are converted to a stepped orthogonal equivalent, for internally, EXCL operates only on manhattan geometries. In the geometric processing phase, the extractor locates relevant rectangle intersections and groups together all rectangles that form a single transistor or interconnection. Geometric intersection rules for a specific IC technology are definable by the user and are passed to the geometric processor as program routines.

During the extraction phase, the geometric representation of each transistor and interconnection transforms into its equivalent, lumped-circuit representation. This is accomplished through a set of standard extraction routines. The more complex extraction routines—namely, for resistance, capacitance and transistor size extraction—will be described separately. The extraction phase is user-controllable through routines and parameters passed from the *technology extraction description*.

---

*This work was supported by U. S. Air Force Contract F29620-81-C-0054.

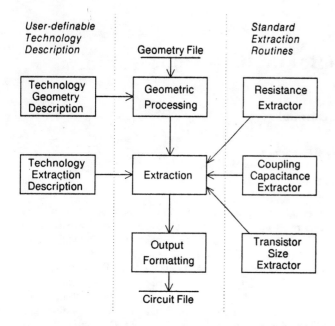

Figure 1: Major program modules of EXCL.

Lastly, EXCL's output formatter generates a complete circuit description in one of several forms. The various output formats of EXCL can feed a circuit or logic simulator directly, or can be used in conjunction with a graphics editor to superimpose circuit information on a layout.

3. Geometric Processing

EXCL *processes* the mask rectangles in a sorted order from maximum y-coordinate to minimum y-coordinate. As the rectangles are sequentially processed, a *scan-line*—defined by the top of the currently processed rectangle—continually moves downward. The connectivity extractor processes a rectangle by locating relevant intersections between it and previously processed rectangles in the *scan-view set*. After a rectangle is processed, it itself is added to the *scan-view set*, but remains there only until the scan-line moves below the bottom of the rectangle. This technique greatly increases the speed of connectivity extraction. Speed is further enhanced by dividing the rectangles into horizontal bins as presented by Bentley[6].

The results of geometric processing are stored in a *geometric network*. An entry in the geometric network contains a list of overlapping rectangles that form either a *switch*, a *path* (or an electrical conducting region on a single IC layer), or a *connection* (which defines an interface region between paths and/or switches). An example *connection* region is a contact cut between layers or a MOSFET gate, source or drain. The rules for creating the geometric network are contained in routines in the *technology geometric description*. These routines itemize all relevant intersections and specifies the appropriate action to take for each. An example entry (stated in English) is: For a **contact cut** rectangle, locate intersections with both **diffusion** and **metal** paths, and add to the geometric network a *connection* between the paths. The user can incorporate IC technology changes in EXCL through modifications of the *technology geometric description*.

During the geometric processing phase, EXCL also reformats switch and path geometric data from the unrestricted format generated by the IC designer into an internal format more easily used by extraction routines. The internal geometric representation contains a set of non-overlapping rectangles that abut only along horizontal edges (see figure 2). Besides simplifying area and perimeter calculations, the internal format has other benefits which will be demonstrated soon. Geometry reformatting involves a scan from maximum y-coordinate to minimum y-coordinate. As the scan moves downward, x-coordinate positions of the original rectangles are tracked, and new rectangles are made to match these positions. At scan lines where x-coordinate positions change, the tops of new rectangles are started and/or the bottoms of new rectangles are finished.

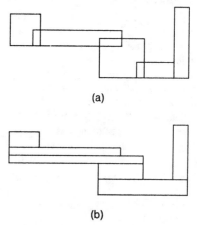

Figure 2: Conversion from random rectangular description (a) to internal format (b).

4. Extraction

Each *path*, *switch*, and *connection* is converted individually from its mask geometry to an equivalent circuit. In EXCL the conversion is controlled by the *technology extraction description*, thus empowering the user to change EXCL to match his needs. The user can trade-off execution speed with circuit detail to define any extraction in the range from rapid, logic-circuit extraction to slow, detailed circuit extraction that includes all parasitics. The *technology extraction description* contains a set of technology "parameters", and a procedure that outlines which standard extraction routines to apply.

When EXCL extracts both the resistance and capacitance of a distributed RC line, it generates an approximate π-ladder network. Often a one-stage π-ladder suffices for an approximation with error less than 3%. But, as the length of the interconnecting line increases, the number of required π-ladder stages needed increases[7]. The number of stages in EXCL is constrained such that all elements are below prespecified maximums, until a maximum of three stages is reached.

5. Resistance Extraction

EXCL uses three techniques to extract interconnection resistance. The general technique is capable of finding the resistance between any set of arbitrarily shaped boundaries through any arbitrarily shaped resistive region. This well-known technique solves Laplace's equation,

$$\nabla^2 V = 0,$$

over the resistive region. Initially, assumed voltages are placed on the *connections* (transistor edges, contact openings, etc.) to the resistive region. Then, the region is divided into a square grid of discrete points and the potential is solved at each point with a discretization of Laplace's Equation,

$$V(x, y) = \tfrac{1}{4}\bigl(V(x-1, y) + V(x+1, y) + V(x, y-1) + V(x, y+1)\bigr).$$

EXCL solves the system of equations with one of two numerical methods—Gaussian Elimination of band-matrices or Successive Over-Relaxation. The method selected for a region depends on the number of discrete points.

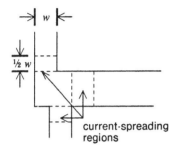

Figure 3: Current-spreading regions.

The second method of resistance extraction is valid only for long, straight resistive regions (which are quite abundant in IC's). The resistance is calculated directly with:

$$R = \rho_{sh} \frac{length}{width},$$

where ρ_{sh} is sheet resistivity. This equation assumes that the field lines are straight and that breaks on both conductor ends fall on equipotential lines. This is not generally true at the ends of straight regions, for the current starts to bend around corners. Thus, a *current-spreading region* is removed from each straight subregions end as shown in figure 3. For a maximum error of 2%, EXCL removes a current-spreading region whose length is one-half the conductor's width.

The third resistance extraction method involves a library lookup scheme for commonly occurring shapes. This method is motivated by the large number of identical corners, T's, and standard contact cut configurations. This extraction method executes very quickly—EXCL needs only to search the library for the geometry, and if found, insert the matching equivalent subnetwork into the complete network. Figure 4 shows a few example library entries and their equivalent resistance subnetworks. Resistor values contained in the library are precomputed with high accuracy using the general methods described in this section or conformal transformations as described by Hall[8]. The library data structures and lookup methods will be discussed soon.

6. Subregion Division

This section examines how EXCL subdivides a complete path region into appropriate subregions for calculation with one of

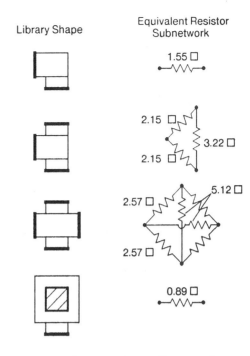

Figure 4: Sample resistance library entries

the three resistance extraction methods. The same subdivision algorithm is also used by the capacitance and transistor extractors, but it is discussed here using resistance as an example.

6.1. Straight Subregions

First, EXCL locates all straight subregions by searching each rectangle of the internal, non-overlapping format for spans that are free of abutments to other rectangles or *connections*. For a given rectangle, the search occurs in one of two possible routines. If the rectangle's width exceeds its height, then one routine searches the rectangle for horizontal straight subregions; if the height exceeds the width, then the other routine searches for vertical straight subregions. Only the routine for rectangles with greater widths is considered here. The other routine is similar; only its use of x and y coordinates is interchanged.

The extractor finds horizontal straight subregions by, first, locating all abutting rectangles and *connections* and storing their x-coordinates as "line segments". Then, the extractor (1) expands each line segment in both directions by an amount equalling one-half the rectangle's height, and (2) combines overlapping line segments. The expanded line segments now cover the current-spreading region, and any portion of the rectangle not covered by a line segment is made into a straight subregion. The process is illustrated in figure 5.

The extractor makes a second pass to isolate straight subregions that are hidden across parallel rectangles of the sort shown in figure 6. These parallel rectangles form only in the horizontal direction as a consequence of the non-overlapping process. When parallel rectangles are found, the extractor repeats the search for straight subregions—treating the parallel rectangles as one.

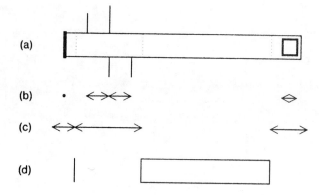

Figure 5: Line egment operations for isolating straight subregions.

(a) Original rectangle and abutments, (b) unexpanded line segments, (c) expanded and combined line segments, and (d) straight subregions (the left region only causes a break in other subregions).

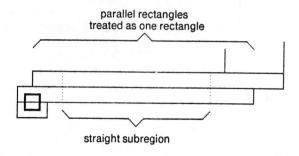

Figure 6: Straight subregion hidden across two parallel rectangles.

6.2. Library Subregions

After all straight subregions are located their areas are removed from the path's geometric area. The remaining rectangles are regrouped into isolated islands of connecting rectangles. Each island represents a non-straight subregion and is checked for inclusion in the library.

A very powerful feature of the internal geometry format is that regardless of how the IC designer forms a region, two equal shapes have identical sets of non-intersected rectangles. It suffices to store only the size and relative placement of the rectangles and boundaries. The library stores its entries in a tree structure. Figure 7 shows the library structure for the resistor library shapes from figure 4. Each leaf node contains a circuit network equivalent; each branch node represents a "*clue*" to the shape description. The branch node contains all valid answers to its *clue*; each valid answer then has a tree path pointing to the next *clue*. The search process traverses the tree structure, checking for valid answers to each *clue*. If the shape has valid answers to all *clues*, the search succeeds, and the leaf node contains the equivalent resistor subnetwork. The tree hierarchy follows the *clues* listed below:

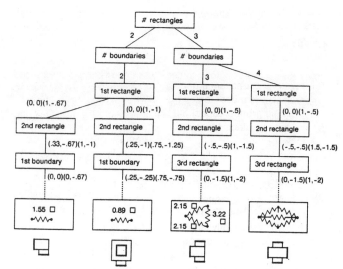

Figure 7: Tree structure for resistance library of figure 4

Resistor shapes are shown below each entry.

1. the number of rectangles. This node immediately drops most complex shapes from the search.
2. the number of network *connections*.
3. the upper left and lower right coordinates of the first rectangle. The origin is always defined as the upper left of the first rectangle. This *clue* is repeated for each rectangle.
4. the upper left and lower right coordinates of the first boundary. This *clue* is repeated for each boundary.

When the relative scale of shapes is unimportant, as with resistance shapes, all dimensions are normalized to a unit defined by the width of the first rectangle. To make the library complete, all orthogonal reflections and rotations of library shapes are stored as a separate entries in the library.

6.3. General Subregions

A subregion failing to match any library shape is rendered "irregular" and must submit to analysis with the general technique.

One often finds that certain irregular shapes are repeated in IC layouts, particularly if cells are replicated. In this case, the library becomes even more useful if it includes the provision for dynamically adding entries. If the extractor stores the results of each time-consuming "general" calculation (memory permitting), then computation time is not wasted by re-extracting the same problem.

7. Capacitance Extraction

Capacitance extraction involves two different problems. One problem, that of finding ground capacitance, involves little more than finding the areas and perimeters of interconnection regions. This is handled easily by EXCL and most other circuit extractors. While the other capacitance problem—finding internodal capacitance—has to date been ignored by many extraction programs, it is becoming increasingly important for both speed and noise analysis of IC's. This is particularly true as process engineers attempt to reduce parasitic resistance by increasing conductor height. Extraction of inter-nodal capacitance is more complex than other circuit extraction problems.

The coupling capacitance on any node is noticeable only when its value approaches the same order of magnitude as (or exceeds) the ground capacitance of the node. Thus, EXCL only solves the problem for nodes whose estimated inter-nodal capacitance, $C_c > \gamma \cdot C_{ground}$, where γ is an accuracy factor typically around 0.05 to 0.3. The inter-nodal capacitance "domain" for any conductor is limited to a small region surrounding the conductor. EXCL eliminates most unimportant inter-nodal capacitances by windowing the coupling capacitance regions. That is, for any given conductor, a window is created that is slightly larger than the conductor region. EXCL locates neighboring conductors with potential coupling capacitance by looking through the window. The window's size is a function of γ and the parallel wire capacitance to be described soon.

7.1. Special Case Inter-Nodal Capacitance Extraction

EXCL uses three special techniques for rapid coupling capacitance calculations along with the general technique. Figure 8 shows a breakdown of where the different methods apply. The first special technique covers the coupling capacitance between two overlapping conductors, and is described by the parallel plate equation:

$$C_{plate} = \frac{\varepsilon_{oxide} \cdot Area}{Separation}$$

An additional "fringe correction", α_{fringe}, is applied to the perimeter of the overlapping areas. For a given combination of overlapping layers, $\frac{\varepsilon_{oxide}}{Separation}$ is a constant denoted by $K_{overlap}$. Thus,

$$C_{plate} = (K_{overlap} \cdot Area) + (\alpha_{fringe} \cdot Perimeter).$$

The second special technique is used between two parallel conductors on either the same layer or different layers. EXCL computes this capacitance with

$$C_{parallel} = (K_{parallel} \cdot Length) + 2\alpha_{end},$$

where $K_{parallel}$ is a line capacitance constant with dimensions of Farads/length, and α_{end} is an "end correction" for each parallel region end. $K_{parallel}$ is a function of three factors: the conductor spacing, width, and layer combination. For typical dimensions, the "width" factor is ignored, since it has a small effect on $K_{parallel}$. (For a discussion of parallel coupling capacitance, see Dang[9].) For each layer combination, EXCL stores a $K_{parallel}(spacing)$ function in tabular form. The functions are determined either with computer simulations[10] or with direct measurements from test circuits.

Like the resistance problem, library shapes can be defined for the coupling capacitance problem. A separate *coupling capacitance library* stores these shapes along with their equivalent capacitance subcircuit. The *coupling capacitance library* tends to be less useful than the *resistance library*

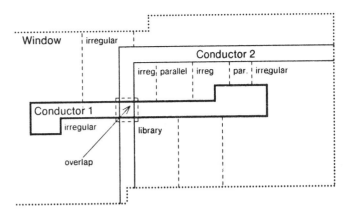

Figure 8: Subregions of coupling capacitance region.

because of the very large number of possible shapes that can exist between two conductors. Figure 8 shows one common library shape that forms near two crossing conductors.

Dividing the capacitive problem into subregions basically follows the same algorithm followed by the resistance extractor. However, the field region is now defined by the oxide gaps between conductors. Initially, all overlapping regions are located and removed. Then, the field region is subdivided into straight (parallel capacitance), library, and irregular subregions as outlined in section 6. The main difference between the two uses of the algorithm is that no *connection* abutments are considered when defining line segments in the capacitance problem.

7.2. General Inter-Nodal Capacitance Extraction

The electrostatic field equations relating charge and voltage are analogous to the conduction equations relating current and voltage. Laplace's Equation holds in a charge-free dielectric region. This might suggest solving the general capacitance problem with the same techniques developed for resistance, and this has indeed been done for careful capacitance calculations of certain geometries. It is, however, not feasible for general capacitance extraction. In electric conduction, field regions have well-defined limits, but, in electrostatics, the dielectric medium extends unbounded in all dimensions.

A faster technique for general coupling capacitance extraction was developed for EXCL. The theory follows from earlier work by Balaban[11] and Benedek[12]. Green's Theorem states that for any point in space, a, its potential is given by

$$V(a) = \int_{all\ charge} G(a; b) \rho(b)\, db. \quad (1)$$

$G(a; b)$ is the appropriate Green's function for the dielectric medium between points a and b. It describes the voltage induced at a by a unit point charge at b. For EXCL's use, it also includes the effects of the substrate or "groundplane".

In the capacitance analysis, conductor edges are divided into N discrete "elements", $e_1, e_2, ..., e_N$, as illustrated in figure 9. EXCL assumes a charge distribution, $\rho(b) = q \cdot f(b)$, where b covers the bottom, edge, and top of each element, q is the total elemental charge, and f is a shape function which describes

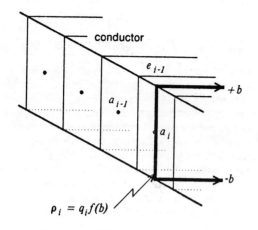

Figure 9: Discrete elements for general coupling capacitance extraction.

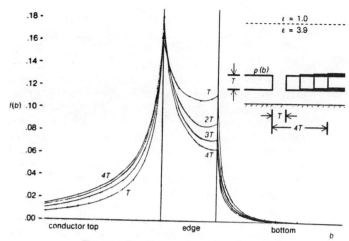

Figure 10: Charge shape functions, $f(b)$.

how a unit charge spreads over the element. Figure 10 shows some sample shape functions. Equation (1) now becomes a summation over all elements:

$$V(a) = \sum_{i=1}^{N} \int_{\text{element } i} G(a; b) \rho(b) \, db = \sum_{i=1}^{N} q_i \int_{\text{element } i} G(a; b) f(b) \, db.$$

Defining the integration quantity as the *Weighted-Green's* function, $H(a; e)$, the potential equation becomes

$$V(a) = \sum_{i=1}^{N} q_i H(a; b).$$

By writing a potential equation for each of the discrete elements, one obtains the matrix equation

$$\mathbf{v} = \mathbf{H} \cdot \mathbf{q}.$$

In this problem, the voltages are known and charge is the unknown. Thus, the extractor sets the voltage values for each conductor and solves for q using a standard Gaussian Elimination procedure.

7.3. Determination of Weighted-Green's Functions

For a given IC process, the Weighted-Green's Functions are computed only and stored as tabular function for use with EXCL. It is difficult to characterize the Weighted-Green's function, $H(a; e)$, by direct analysis or by experimental measurements. Instead, a method presented here applies a series of two-step computer field simulations. The first-step "shape function" simulation finds an approximation to $f(b)$, while the second-step finds an approximation to $H(a; e)$.

For a given pair of layers, the shape function is derived by, first, simulating the parallel conductor charge induced by an arbitrary voltage difference, and then, normalizing the charge distribution to make a total unit charge. (Parallel conductor cross-sections accompany the shape functions of figure 10.) The function is computed for several separation distances; the function that is actually used in extraction is determined by the distance to the "dominant" or nearest points.

The second field simulation computes the voltage field around a unit charge distributed in the shape approximated by the first simulation. The simulation employs spherical coordinates since the effects of a single element away from the conductor are approximately spherical. Figure 11 depicts a simulation along the $\theta = 0$ plane. If point a has a height h, the voltages found along line AB are exactly the values of $H(r, f)$, for shape function f, and lateral distance r between point a and element e.

The accuracy of the general coupling capacitance method depends largely on how accurately the assumed charge distribution follows the actual distribution. The accuracy is particularly good for geometries which have most coupling charge on the conductor edge. These geometries also tend to have the greatest ratio of coupling capacitance to total capacitance; it is for these that coupling capacitance extraction is most important.

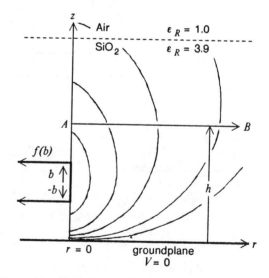

Figure 11: Field simulation for determining $H(a; b)$ for a conductor of height h.

8. Transistor Extraction

Although transistors pose difficulties to simulation modelling, extracting their circuit information is easier. For MOS transistors, for instance, only the transistor's length, width, and gate area need be computed. Most transistor extraction algorithms resemble an algorithm from resistance or capacitance extraction. The notable exception arises when calculating the length and width of an MOS transistor. While the problem is similar to resistance extraction, it differs in that actual linear dimension must be found since many MOS transistor effects are non-linear with channel length.

Rectangular channel regions clearly present no challenge to the transistor sizer. For non-rectangular channel regions, EXCL first decides whether it is "long-and-narrow" or "short-and-wide". The decision is fairly simple and is based the source and drain configurations. Then, the transistor region is subdivided into straight, library, and irregular regions with the same methods as for resistor regions. For each subregion a partial length and width dimension is calculated from: (1) the straight region's rectangular dimensions, (2) length and width dimensions stored in the *transistor library*, or (3) estimates based on sheet-resistance calculations for irregular shapes. In a short-and-wide transistor, the partial lengths of all subregions are checked for coherency, and the partial widths sum to the total width. Conversely, for a long-and-narrow transistor, the widths are checked for coherency, and the partial lengths add. EXCL does not size transistors with non-coherent lengths (or widths) or transistors composed of only "irregular" subregions. Figure 12 shows examples of each transistor sizing case.

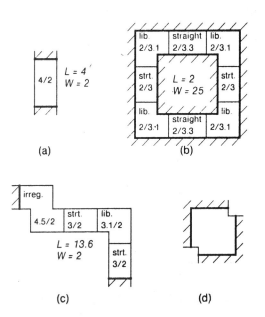

Figure 12: Transistor sizing cases

(a) rectangular, (b) short-and wide, (c) long-and-narrow, and (d) impossible. Subregion dimensions are given as "Length/Width"

9. Output Formatter

After all circuit elements have been calculated, the output formatter joins the circuit information into a complete network, and makes element combinations and reductions where possible. Presently, EXCL can transform the network into one of four output formats. Two formats feed directly into simulators, one to a circuit simulator and the other to a logic simulator. A third format generates display information compatible with a layout editor. When displayed, internal node names and node capacitance information appears at the node's layout position. The fourth format is tabular and lists each node, circuit element, and their coordinates.

10. Sample Problems

Typically, each IC technology has three "technology extraction descriptions" which generate different levels of circuit detail. The simplest level contains only a transistor and connection list and is used for logical verification of the layout. The next level adds rough parasitic resistance and ground capacitance information to the previous level's information. This circuit level is used for preliminary timing analysis. The highest level contains all extractable circuit elements with high precision.

Figure 13(a) shows a very simple NMOS inverter layout. Figures 13(b) and 13(c) show the results from the simplest level, 13(d) and 13(e) show the results from the most accurate extraction level.

The extraction times of a typical CMOS four-bit counter layout were analyzed on a DECsystem 2060. The design contained 92 transistors, in 1421 mask rectangles. A switch-level extraction was completed in 5.1 CPU seconds. A preliminary circuit-level extraction completed in 55.7 seconds. This extraction did not compute coupling capacitance or resistance of irregular shapes. The largest resistance error for any path was around five squares. Extraction of the complete circuit required 167.3 CPU seconds. The run-time requirements apportioned into: 17% for geometric processing, 25% for resistance extraction, 46% for capacitance extraction, and 5% for output formatting. In both resistance and coupling extraction, the general extraction phases occupied most run-time. By dynamically adding all general extraction results to the appropriate library, the extraction time for irregular subregions was roughly halved for the first bit of the counter layout, and over all four bits, the extraction time for irregular regions was reduced by 7.8 times. For this layout, the set of standard library shapes further reduced the resistance extraction time by 39%, and the general coupling capacitance time by 22%.

11. Conclusions

Accurate circuit extraction has become increasingly important in today's technology, particularly extraction of interconnection circuit equivalents. EXCL demonstrates that accurate extraction of large IC layouts is possible without overburdening computer resources. One can integrate a range of extraction techniques into a single, fully-automatic system. It is possible to decompose a geometric region into a set of subregions where a different extraction technique is used for each.

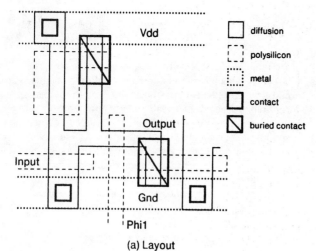

(a) Layout

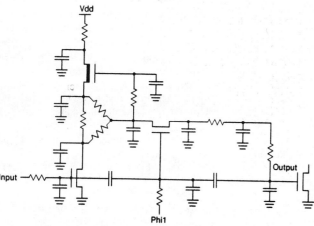

(b) Extracted Circuit Network

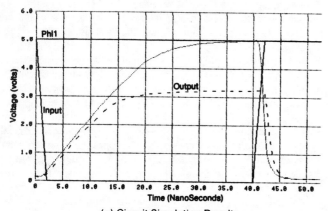

(c) Circuit Simulation Results

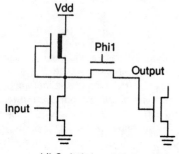

(d) Switch-Level Network

```
input=x   phi1=x   n100=x   output=x
input=0   phi1=0   n100=1   output=x
input=0   phi1=1   n100=1   output=1
input=0   phi1=0   n100=1   output=1
input=1   phi1=0   n100=0   output=1
input=1   phi1=1   n100=0   output=0
input=1   phi1=0   n100=0   output=0
```

(e) Logic Simulation Results

Figure 13: Inverter example

Two extraction problems—resistance and inter-nodal capacitance—involve substantial computation time to solve in the general case. The resistance problem can be solved to a high degree of accuracy. Initially, accurate inter-nodal capacitance extraction was so slow that a considerably faster technique was developed. The method has proven reliable and more accurate, especially for conductor geometries with considerable edge coupling.

12. Acknowledgments

The author wishes to extend his thanks to Professor Jonathan Allen for the initial suggestion and continued support and recommendations for the project; to Donald Baltus, who, as a first user of the system, uncovered many program errors; and to William Evans for his valuable discussions and suggestions on the project.

References

1. B. T. Preas, B. W. Lindsay and C. W. Gwyn, "Automatic Circuit Analysis Based on Mask Information," *Proceedings of the 13th Design Automation Conference*, 1976, pp. 309-317.
2. C. Baker and C. Terman, "Tools for Verifying Integrated Circuit Designs," *Lambda*, Fourth Quarter 1980, pp. Pages 22-30.
3. T. Mitsuhashi, T. Chiba and M. Takashima, "An Integrated Mask Artwork Analysis System," *Proceedings of the 17th Design Automation Conference*, 1980, pp. 277-284.
4. J. D. Bastian, et. al., "Symbolic Parasitic Extractor for Circuit Simulaion (SPECS)," *Proceedings of the 20th Design Automation Conference*, 1983, pp. 346-352.
5. J. Yoshida, T. Ozaki, and Y. Goto, "PANAMAP - B: A Mask Verification System for Bipolar IC," *Proceedings of the 18th Design Automation Conference*, 1981, pp. 690-695.
6. J. L. Bentley, D. Haken and R. W. Hon, "Fast Geometric Algorithms for VLSI Tasks," *Digest of Papers, 20th IEEE Computer Society International Conference (CompCon)*, , San Francisco CA, February 1980, pp. 88-92.
7. Takayasu Sakurai, "Approximation of Wiring Delay in MOSFET LSI," *IEEE Journal of Solid State Circuits*, Vol. SC-18, No. 4, August 1983, pp. 418-426.
8. P. M. Hall, "Resistance Calculations for Thin Film Patterns," *Thin Solid Films*, Vol. 1, 1967/68, pp. 277-295.
9. R. L. M. Dang, and N. Shigyo, "Coupling Capacitances for Two-Dimensional Wires," *IEEE Electron Device Letters*, Vol. EDL-2, No. 8, August 1981, pp. 196-197.
10. W. H. Dierking and J. D. Bastian, "VLSI Parasitic Capacitance Determination by Flux Tubes," *IEEE Circuits and Systems Magazine*, Vol. 4, No. 1, March 1982, pp. 11-18.
11. Philip Balaban, "Calcualtion of the Capacitance Coefficients of Planar Conductors on a Dielectric Surface," *IEEE Transactions on Circuit Theory*, Vol. CT-20, No. 6, November 1973, pp. 725-731.
12. Peter Benedek, "Capacitance of a Planar Multiconductor Configuration on a Dielectric Substrate by a Mixed Order Finite-Element Method," *IEEE Transactions on Circuits and Systems*, Vol. CAS-23, No. 5, May 1976, pp. 279-284.

Circuit Analysis, Logic Simulation, and Design Verification for VLSI

ALBERT E. RUEHLI, SENIOR MEMBER, IEEE, AND GARY S. DITLOW, MEMBER, IEEE

Invited Paper

Abstract—In this paper, we consider computer-aided design techniques for VLSI. Specifically, the areas of circuit analysis, logic simulation and design verification are discussed with an emphasis on time domain techniques. Recently, researchers have concentrated on two general problem areas. One important problem discussed is the efficient, exact-time analysis of large-scale circuits. The other area is the unification of these techniques with logic simulation and design verification technique in so called multimode or multilevel systems.

I. INTRODUCTION

THE FACT that this special issue is dedicated to VLSI design attests to the importance of the subject area. Many talented professionals are presently striving to find solutions to very challenging VLSI problems. The increase in complexity is the one fundamental issue which confronts both chip and Computer-Aided Design (CAD) tools designers [1]. The techniques and tools and algorithms which were successfully applied to LSI are inadequate for VLSI. Mainly, algorithms which are of $O(N^3)$ or worse in time for practical computations may not be extendable to VLSI with a larger number of subcircuits N.

Fundamental design methodologies like structured design [1], hierarchical, and "divide and conquer" strategies [2] are proposed by many authors. The interaction between the design methodology, the designer, and the CAD techniques is of key importance to the ease with which the VLSI chip design is achieved. This paper is mainly concerned with CAD techniques for circuit design and timing-oriented problems. We assess the status of three important areas, circuit analysis, logic simulation, and design verification. Here, we de-emphasize the historical aspects since review papers and texts are available which include extensive references on design tools and techniques pertinent for the electrical chip and package design [1]–[14]. Also, simulation for testing is beyond the scope of this paper.

We will first define the terminology used before giving an introduction to the three main topics—circuit analysis, logic simulation, and verification or checking.

A. Terminology

We start this paper with the definition of a few frequently used terms to be able to give a clear account of the subject. One source of ambiguity is the fusion of the languages of circuit theorists and computer scientists. The subject of this paper is an illustration of a classical "EECS" subject.

Manuscript received May 13, 1982; revised October 15, 1982.
The authors are with the IBM Thomas J. Watson Research Center, Yorktown Heights, NY 10598.

Here, *circuit* is used in two ways. It may be used to describe a collection of gates with possible interconnection models. Unfortunately, this ambiguity is unavoidable since *circuit* can also mean a fraction of a gate or a single gate. We use the term *subcircuit* for a smaller portion of a circuit, especially when we talk about partitioning concepts. Another ambiguity exists in the usage of the term *gate*. We will use gate to mean a logical gate or subcircuit. If we refer to the gate electrode of an MOS transistor, we call it an MOS transistor gate.

Analysis rather than simulation is used to describe techniques which are based on mathematics as opposed to heuristics. The reasons become clear if we consider a dictionary definition for *analysis*: 1) to separate into parts of basic principles so as to determine the nature of the whole; 2) examine methodically; 3) to make mathematical analysis; 4) opposite to synthesize.

The mathematical or physics (basic principles) foundation of analysis is clear from these definitions.

We contrast this definition to the meaning of *simulation* in dictionary terms: 1) to have or take the appearance; 2) assume or imitate a particular appearance or form; 3) counterfeit; 4) to reproduce the conditions of a situation.

The empirical nature of a *simulation model* is apparent from these definitions. Analysis techniques are a subset of simulation approaches since by some lucky insight we may intuitively conceive a model or technique which is mathematically accurate. Simulation techniques span an extremely wide range even if we restrict ourselves to subjects of a technical nature. Here we are interested in topics like *logic* or *circuit simulation* rather than, for example, airplane flight simulation. We use *circuit simulation* to imply that we use approximate, heuristically based models. *Logic simulation* is the conventional process where we replace electrical signals by two or more values, e.g., 0 and 1.

Another specification of a model pertains to its "granularity." Three terms are useful, *micromodel*, *model*, and *macromodel*. A circuit *model* describes the physical situation at an intermediate level. If we desire to analyze at a far more detailed level as, for example, at the level of the transistor physics, we construct very detailed *micromodels* [5], [15], [16]. On the other hand, the need to simulate large VLSI systems has generated much interest in *macromodels* which are designed to trade off internal complexity for speed, e.g., [3], [17], [18]. A macromodel may not necessarily be less accurate than a model, but it may have fewer parameters or exclude unimportant dependencies.

Recently, the usage for the terms *mixed and multilevel (mode) simulation and analysis* techniques has become clear [19], [20]. A *mixed-mode* technique includes two techniques in the same system at the same level [19] while a *mixed-level* technique includes different hierarchical levels which interact with each other. Details are discussed in Section IV.

Another term which we use frequently is *verification* which is the following according to a dictionary: 1) evidence that establishes or confirms the accuracy or truth of; 2) process of research, examination required to establish authenticity.

Verification enters in many aspects of the design processes as will be discussed in Section V. In fact, the major purpose of the techniques in Section II is *timing verification*.

B. Circuit Analysis and Simulation

From the sixties until recently, the emphasis in computer-aided circuit analysis was on general-purpose programs. ECAP [21], which was one of the earliest widely available programs, was replaced by many newer ones. Today, the most used programs are SPICE [22]-[24], ASTAP, [25] and ADVICE [26]. These programs employ techniques which were innovated in the last decade such as 1) sparse tableau analysis method (STA) [27]; 2) modified nodal analysis method (MNA) [28]; 3) implicit integration methods; and 4) sparse matrix techniques.

In the early seventies a few researchers devised techniques suitable for larger circuits [29]-[31]. For a while the field of large-scale circuit simulation and analysis (LSSA) was called macromodeling [31], [17], [32]. Today, the term *macromodeling* is more correctly applied to the *modeling* aspect of large-scale circuits. The field of LSSA has expanded in the last few years due to its importance for *timing verifications* for VLSI circuits.

A multitude of new ideas have been conceived in LSSA in the last few years. MOTIS [33] is one of the first programs where simplified analysis techniques are employed for MOSFET circuits. Other programs include MACRO [34], DIANA [17], and MOTIS-C [35]. The specific techniques employed will be discussed in Section II-A.

We differentiate between *incremental* time techniques which have their origin in circuit analysis and *waveform* techniques which include logic simulation with delay times. In the *incremental* approach the analysis proceeds globally in time steps which are usually smaller than the signal rise times. All the above programs are based on *incremental* time updating techniques. The local circuit waveforms are computed for a sizable time segment in a *waveform* approach [36]-[38]. Two major sources of time saving result from the sparsities present in the space-time relationship of the circuits and from employing *decomposition techniques*. The waveform technique, and *latency*, e.g., [34], [40] are approaches which take advantage of time domain or *temporal* sparsity. Section II-B details these techniques. The other area of *decomposition* pertains to the structure of the logic circuits. The average fan-out for a logic gate is between two and three. Thus the connectivity of most circuits is extremely sparse leading to *structural sparsity*. Usually, for an exact representation of the gates the number of internal nodes of the circuits is much larger than the external nodes. This leads to a *natural* decomposition of the circuits in terms of the gates. Numerous papers referenced give techniques for the exploitation of this structure, e.g., [3], [90], [92]. For MOS devices the MOS transistor gate represents an almost unidirectional node with the exception of the gate-to-drain and -source capacitances. The approach given in [37] exploits this, and the so-called one-way (1-way) scheduling technique is based on this. Here, we discuss mostly graph-based algorithms, while a recent paper [4] concentrates on equivalent sparse matrix techniques. Some of these techniques are detailed in Section II-B.

A source of inaccuracy may be the approximate inclusion of feedback in the solution. Feedback may be *local* or *global*. Further, the strength of the feedback may be *weak* or *strong*. For example, a flip-flop is locally coupled with strong feedback. The Miller feedback in an MOS circuit is local and weak. Further, the feedback extending over several subcircuits in sequential logic is global and strong. Techniques will be discussed below for the inclusion of feedback. For example, the basic waveform 1-way modeling technique [37] works for the flip-flop case. Two recent general methods are the waveform relaxation (WR) technique for a *waveform* analysis [36] and the symmetric displacement [41] for an *incremental* system.

Logic simulation is performed at the functional gate and recently at the transistor level. In this paper we concentrate on simulation for design verification rather than testing. Gate level simulation is an area which evolved over the last two decades. Even the early work was concerned with efficiency improvement techniques, e.g., [42], [43] since the logic circuits of interest were already large and computers were less powerful. Since then a continuous progress has been made towards the simulation of very large circuits, e.g., [7]-[10], [44]-[48]. Both temporal and structural sparsity is exploited in these techniques. Mainly, logic simulation is an area where the need for large-capacity CAD programs is very evident. Many of these simulators include crude timing information.

Transistor level simulation [49], [52] has recently emerged as an important problem due to the emergence of new complex MOS transistor designs which cannot easily be cast into standard logic gates, e.g., [53]. This new field is presently receiving attention from many researchers.

Another area of importance is mixed-mode and multilevel systems. In a mixed system two techniques are used like circuit analysis and simplified macromodels [54] or logic simulation with an electrical WR technique [19]. Many multi-mode and/or multilevel systems include several simulation/analysis levels combining many of the above-mentioned techniques. [19], [20], [54]-[64]. Detailed aspects on multilevel systems are given in Section IV.

C. Verification

There are two fundamental processes in VLSI-CAD. *Design* is the process of construction while *verification* ensures the correctness of the design [65]-[78]. The following are general examples of *verification* steps: timing; layout; mask; electrical connectivity; functional correctness; and design rule.

Some of the verification steps involve mask and layout steps, e.g., [65]-[70] for which we refer the reader to another paper in this issue [70] since we want to concentrate on timing aspects in this paper.

Verification of the functional correctness of the logic design is another area which has intrigued computer scientists for more than two decades [71]-[75], [8], [14], [44]-[48]. This topic is obviously closely related to timing verification since functional correctness is a prerequisite and therefore a subset of timing verification, which is discussed in Section V. Simulation for design verification [48], [75] was successfully applied to truly large-scale problems with more than 0.5×10^6 gates more than a decade ago. However, the chips themselves were LSI rather than VLSI.

Timing verification is perhaps the most evasive aspect of verification, e.g., [76], [78]. Specifically, the physical design may be perfect, but still the electrical performance of the design may be incorrect or insufficient. This is why we concentrate on timing design and timing verification in this paper. The spectrum spans from exact circuit analysis to extremely efficient algorithms designed for timing simulation and verification. A discussion of verification techniques is given in Section V.

II. CIRCUIT ANALYSIS AND SIMULATION

A. General-Purpose Circuit Analysis

The most widely used tools for circuit analysis are the *general-purpose* or *standard* circuit analysis programs like SPICE [22]-[24] or ASTAP [25], based on the modified nodal analysis (MNA) approach and the sparse tableau analysis (STA), respectively. They are designed for the detailed analysis of a large variety of circuits and are characterized by having a multitude of features. For VLSI, these tools have a place in the detailed design of special circuits. Specifically, the following applications are a natural for them.

They can serve as an interface to the usually elaborate models used for semiconductor device analysis, e.g., [15], [16]. In this capacity they are used to verify the validity of the semiconductor circuit models from the detailed usually two-dimensional micromodels.

The individual subcircuits or gates are designed and optimized using these programs. The compute time for this type of work is moderate since usually only a few devices are involved.

The correctness of newly developed large-scale programs is verified using these well-proven exact circuit analysis programs.

Thus even for VLSI, they play an important part in the early design phases where the semiconductor devices and the menu of subcircuits are established. Importantly, the process of circuit analysis is exact within the accuracy of the models used and the mathematical numerical techniques employed.

The circuit equation in both the MNA and the STA approaches lead to a set of coupled equations of the form

$$f_1(x, \dot{x}, y, t) = 0 \quad (1a)$$
$$f_2(x, y, t) = 0 \quad (1b)$$

where x are the differentiated variables and y the nondifferentiated variables. The formulation of the equations in terms of f_1 and f_2 results naturally from both formulations although the STA formulation is a larger, more sparse matrix while the MNA matrix is smaller and more dense.

We are interested in the solution for the analysis time which usually is $t \in [0, T\text{max}]$ where $T\text{max}$ is the final time. Two main techniques are used today to solve (1). The time derivatives are in the most simple case replaced by the backward Euler (BE) formula where

$$\dot{x}_n \cong \frac{x_n - x_{n-1}}{h_n} \quad (2)$$

where n represents the "now" time, $n-1$ the time step before, and $h_n = t_n - t_{n-1}$ is the present time step. Equation (2) is the most simple A-stable formula. However, other integration methods like the second-order backward-differentiation formula [79], [80] and the ACA methods [81] are more desirable for general purpose programs. Numerical damping [82] is severe for the BE formula (2), and misleading results may be obtained especially for oscillatory solutions which are present in high-performance logic. In this discussion, we will proceed using the BE formula for time integration for clarity. The result of applying (2) to the system (1) is

$$\tilde{f}_1(x_n, x_n - x_{n-1}, y_n) = 0 \quad (3a)$$
$$f_2(x_n, y_n, t_n) = 0 \quad (3b)$$

which is a system of nonlinear algebraic equations.

To transform the nonlinear equations into a system of linear equations, we usually apply Newton's method to the system (3) which is written as $g(z) = 0$ for

$$\frac{\partial g}{\partial z} z^i = -g(z)^{i-1} + \frac{\partial g}{\partial z} z^{i-1}. \quad (4)$$

Here $z = \begin{bmatrix} x_n \\ y_n \end{bmatrix}$ and $g = \begin{bmatrix} \tilde{f}_1 \\ f_2 \end{bmatrix}$ and i is the index of the iteration. Usually, (4) is written in the form of a linear system of equations

$$Az = b \quad (5)$$

where A is the Jacobian

$$A = \frac{\partial g}{\partial z} \quad (6)$$

and b is evident from (4) and (5).

The above mathematical procedure for incremental circuit analysis is summarized in ALG. 1. Again, *incremental* refers to the small increments with which time proceeds for all circuits.

ALGORITHM 1: Incremental Circuit Analysis

Inputs:	Circuit	C
	Input Waveforms	IW
	Time Step	h
	Stop Time	$T\text{max}$
Results:	Waveforms of node voltages	$v(t)$
	Waveforms of branch currents	
Note:	$z(n, i)$ Vector of voltages and currents at time n and Newton iteration i	
Procedure	CircuitAnalysis ($C, I, h, T\text{max}$)	
BEGIN		
	Fill in Jacobian Matrix A and right-hand side b with MNA stamps for time invariant circuit elements	
	dc solution at time = 0	
	time = h	
	$n = 1$	
TIMEloop:	for time $\leq T\text{max}$ do	
	BEGIN $i = 1$	
NEWTONloop:	for ABS($z(n,i) - z(n, i-1)$) $<$ EPSILON do	
	BEGIN Fill in A and b with MNA stamps for time variant and nonlinear circuit elements	
	Solve $A z(n,i) = b$	
	$i = i + 1$	
	END	
	time = time + h	
	$n = n + 1$	
	END	
END		

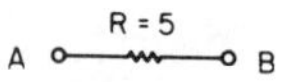

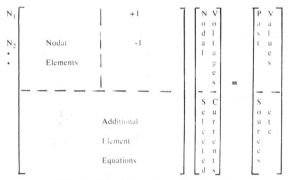

Fig. 1. Processing of resistor R. (a) Graphical representation. (b) Input language statement. (c) Entries into the A-matrix.

Fig. 2. Circuit analysis matrix $Az = b$.

The relevant aspects of the techniques outlined in ALG. 1 are detailed next. The input data may be supplied to the program in one of three forms. Fig. 1(a) shows a circuit element which may be entered with a graphics system. Fig. 1(b) shows the equivalent language statement where A and B are the connecting nodes. In the third way, the circuit element is extracted from the layout with the assistance of a preprocessing program. As shown in Fig. 1(c) the circuit element data are "stamped" directly into the A matrix in the locations which are identified by the connecting nodes. The linear resistor used in this example is a particularly simple case for the way A is assembled inside the NEWTONloop in ALG 1.

Other circuit elements like transistors, capacitors, inductors, current and voltage sources are stamped into the A matrix in a similar way. A schematic representation of A for the MNA approach is shown in Fig. 2.

Many other algorithms are implemented in an efficient program. In the following, a few are enumerated which are not evident from the discussion above: conversion of node-names into program internal numbers; efficient numerical integration techniques for inductances and capacitances; sparse matrix techniques for storing A and solving the system $Az = b$; efficient assignment of variable time steps h; and convergence improvements for the Newton loop (NEWTONloop) which solves the nonlinear equations.

General-purpose analysis programs of the SPICE and ASTAP type may contain many features which add to their flexibility such as the following: statistical analysis; ac sinusoidal steady-state analysis; built-in source time waveforms; changeable parameters and restart facilities; design centering for design improvement, e.g., [83]; and transmission line analysis.

The growth of the solution time for sophisticated circuit analysis programs is $O(N^m)$ where $1.2 \leqslant m \leqslant 1.5$ and N is the number of nodes. These programs are mostly limited to the applications listed in Section I-B since they are exhausted for most practical cases for 50-100 gates. Further, improvements of perhaps factor 3 can be envisioned in the future from the application of techniques like node tearing [84], and programming improvements.

Another effort to improve the speed of circuit analysis programs is based on *vector processing* [85], [86]. Basically, repetitive subcircuits can be analyzed in parallel, which results in time savings. To evaluate the potential gain with this approach, a comparison has been made between SPICE2 and CLASSIE, a vector processing circuit program. For a sufficiently large and regular circuit like an adder, the speed up is about an order of magnitude compared to SPICE2.

B. Large-Scale Circuit Simulation and Analysis

Clearly, substantial improvements must be made for large circuits in the *simulation/analysis* (S/A) techniques above the analysis approaches presented in Section II-A for VLSI. A new technique must lead to at least an order of magnitude improvement in the number of circuits to be analyzed to offer a distinct advantage above the general-purpose circuit analysis programs. Thus we expect to analyze at least 500 subcircuits in several minutes of compute time on a high-performance machine. In the recent past, the following areas of potential gain have been identified: special-purpose programs; repetitive subcircuit structure (modularity); structural sparsity, approximation and simplification techniques; and time domain or temporal sparsity (latency, etc.).

Special-purpose programs can lead to substantial gains in performance. As an example, the analysis of a digital filter subcircuit is efficiently performed by special techniques, e.g., [87] while a general-purpose program lead to an inefficient, time-consuming solution. Another special-purpose program may be involved in the solution of the linear package or interconnection equations [5], [88], [89]. Efficient techniques may involve the symbolic solution [88] of subcircuits like the interconnections and the exploitation of the special sparsity structure [89]. Specifically, if we employ the general solution in ALG. 1 for interconnection circuits, we can completely eliminate the NEWTONloop since f_1 and f_2 in (1) are linear for this case.

Repetitive substructures or modularity lead to gains in several areas of a S/A system. First, we notice that gains can be made in the DATA specifications of ALG. 1. A large amount of repetitive data is present in the input obtained from a VLSI chip. The essential information is usually a very small repertoire of fundamentally different devices and subcircuits. This is in contrast to the general-purpose circuit analysis program where the inherent data structure must accommodate large amounts of data of a different nature. Internally, each element is individually processed. As an example, in a coarse gate oriented MOS circuit simulator, we may specify a few different gates, the interconnections, and perhaps a single capacitance value per interconnection. In this case, we can severely limit the data handling and tailor the algorithms. So-called *modular* methods are based on the regularity of the subcircuits from an analysis point of view [34], [40], [90]–[92]. The major gain in employing modularity is a reduction in storage requirements and convenience in the implementation of other saving concepts rather than time. One of the modular methods will be discussed below in more detail.

Structural sparsity must be exploited beyond the conven-

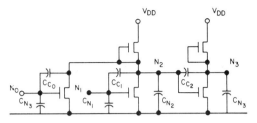

Fig. 3. Example of an MOS circuit.

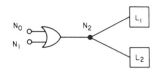

Fig. 4. Illustration of loading.

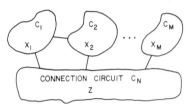

Fig. 5. Circuit with modular subcircuits.

tional sparse matrix techniques for linear systems in circuit analysis programs [93], [94]. Most digital circuits exhibit even more structural sparseness at the global level than analog circuits. The typical fan-out of a logic gate is between 1 and 3 while an analog circuit has about 3 connections per node with other dependent coupled sources. Following this concept, we clearly want to avoid stamping each subcircuit into a single large A matrix as it is done in the general-purpose circuit analysis ALG. 1. Two basic formulations based on the subcircuit structure are widely used. For low- and medium-performance MOS generally connected transistor circuits, a specialized set of nodal equations can be written in the form

$$I(v_N, V_{DD}) + C\frac{dv_N}{dt} = 0 \tag{7}$$

where $I(v_N, V_{DD})$ corresponds to the MOS transistors which are connected between the nodes as exemplified in Fig. 3. C is the matrix of node capacitances, which are of two types, the node-ground C_N and the node–node capacitances C_C. The node capacitance C_N usually lumps several device and wire capacitances into a single value, and the coupling capacitances C_C is C_{gd} the gate-to-drain capacitance in simple cases. This type of model is used in MOTIS [33], MOTIS-C [35], SPLICE [55], MACRO [34], and MEDUSA [92], and other programs with variations. Usually, the voltage sources are treated as having zero internal resistance in this scheme. The time discretization employed is similar to (2) in Section II-A, and the overall time solution approach is *incremental* like ALG. 1.

In the other approach where more elements are present in a subcircuit model, the number of external connections is usually small or sparse compared to the internal connections. For this case, the modular approach is employed at the subcircuit level. Examples of modular techniques are [40], [61], [83], and [90]. Each of the subsystem of equations for the subcircuit is of the form (1) or (7). If we take the discrete time form of the nonlinear equation (3) for the first subcircuit in Fig. 3, it can be written in the form

$$g(v_{n_0}, v_{n_1}, v_{n_2}, v_{n_3}) = 0 \tag{8}$$

where n is the "now" index of time. This model is ideally suited to show how the various approximations employed in the different programs operate.

First, we will exemplify the Gauss–Jacobi and Gauss–Seidel iterative solution methods for the simple example of a system of linear equations $Ax = b$. We decompose the matrix A into $A = L + D + U$ with L being strictly lower triangular, D diagonal, and U strictly upper triangular. Then the system of equations can be written as

$$Lx + Dx + Ux = b. \tag{9}$$

In the Gauss–Jacobi method, we evaluate Dx^k at the present iteration k while Lx^{k-1} and Dx^{k-1} are evaluated at the previous iteration $k-1$. This leads to a very simple solution. In the Gauss–Seidel method, only Ux^{k-1} is evaluated at iteration $k-1$. Thus the Gauss–Seidel case corresponds to a lower triangular system of equations

$$(L + D)x^k = b - Ux^{k-1} \tag{10}$$

which is related to levelizing the circuit graph using 1-way models [37] and can easily be solved by back substitution.

Returning to (8) for the linear case, L corresponds to elements in the inputs or the previously iterated variables v_{n_0} and v_{n_1}, v_{n_2} corresponds to the diagonal or variable to be updated while v_{n_3} corresponds to the not-yet-evaluated voltage, or the Ux contribution. Taking the analogous nonlinear case [95], Gauss–Jordan corresponds to the iterative scheme

$$g(v_{n_0}^{k-1}, v_{n_1}^{k}, v_{n_2}^{k}, v_{n_3}^{k-1}) = 0 \tag{11}$$

while the Gauss–Seidel results in

$$g(v_{n_0}^{k}, v_{n_1}^{k}, v_{n_2}^{k}, v_{n_3}^{k-1}) = 0. \tag{12}$$

The scheme in (12) needs updated input variables before it can update node 2. A source of confusion exists since the Gauss–Seidel and Gauss–Jordan schemes can be applied to the solution of the nonlinear equation as exemplified here, or to the linear equations, (9) and (10). Both approaches are practiced, and the notation employed is usually hard to understand.

Circuit theorists have a similar way of reducing forward coupling among the subcircuits. A circuit motivated approach to reduce the coupling variables in (8) is *loading*. Fig. 4 gives a loading circuit for the analysis of the two input gates in Fig. 3 with the output node $N2$. $L1$ is an approximate circuit for the coupling capacitance C_{C_2}, which is discussed below, while $L2$ approximates the transistor input. Loading models are employed in many programs, e.g., MACRO [34] and RELAX [38].

Many of these approximation techniques apply to both the incremental and waveform approaches. Clearly, the intent is to save computations. In MOTIS [33], transistor characteristics are approximated by secants which results in a linear local equation which does not require multiple Newton iterations. In MOTIS-C [35], table look-up procedures are used for the nonlinear MOS device characteristics to save function evaluations.

MEDUSA [92] is used as an example of a modulator treatment of the circuit matrix A. The circuit is partitioned into "natural" subcircuits with least connections, as shown in Fig. 5.

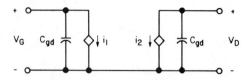

Fig. 6. Equivalent circuit for C_{gd}.

The modular subcircuits C_m with internal variables x_m are embedded in a general interconnection circuit C_N with variables z. The A-matrix (5) can be represented in terms of the modularity as (13)

$$\begin{bmatrix} \frac{\partial f_1}{\partial x_1} & & & & \frac{\partial f_1}{\partial z} \\ & \frac{\partial f_2}{\partial x_2} & & & \frac{\partial f_2}{\partial z} \\ & & \frac{\partial f_m}{\partial x_m} & & \frac{\partial f_m}{\partial z} \\ \frac{\partial f_n}{\partial x_1} & \frac{\partial f_n}{\partial x_2} & \frac{\partial f_n}{\partial x_m} & & \frac{\partial f_n}{\partial z} \end{bmatrix} \begin{bmatrix} \Delta x_1 \\ \Delta x_2 \\ \Delta x_m \\ \Delta z \end{bmatrix} = \begin{bmatrix} -f_1 \\ -f_2 \\ -f_m \\ -f_n \end{bmatrix} \quad (13)$$

where the circuit variables are again combinations of voltages and currents like in ALG. 1.

With $\partial f_m / \partial x_m = A_m$, each row of the matrix is easily solved for Δx_m, or

$$\Delta x_m = -A_m^{-1} f_m - A_m^{-1} \frac{\partial f_m}{\partial z} \Delta z \quad (14)$$

for the mth subcircuit. Δz is found by inserting (14) into the last row of (13). Thus the solution of the structurally bordered block diagonal matrix is formally simple. However, to gain decoupling, the second term on the right-hand side in (14) is ignored in [92]. The omission of these terms results in a lower block triangular system which is used in a Gauss-Seidel iteration scheme.

Approximations for the feedback coupling capacitance C_{c_1} in Fig. 3 are discussed next. It is natural for incremental techniques to attempt to achieve decoupling by replacing unknown values at the "now" time $x(t_n)$ by known values $x(t_{n-1})$ at the previous time steps [105]. This is equivalent to having 1-way properties for t_n.

Explicit forward Euler schemes were used by some authors to eliminate the coupling due to the "floating" capacitances by

$$i_{c,n} = C_{gd} \frac{v_{n-1} - v_{n-2}}{h}. \quad (15)$$

Unfortunately, this scheme severely limits the time step h for the response to be stable. Fig. 6 shows an improved model for a floating capacitor where

$$i_1 = -C_{gd} \frac{dv_2}{dt} \cong -C_{gd} \frac{v_{2_{n-1}} - v_{2_{n-2}}}{h} \quad (16)$$

$$i_2 = -C_{gd} \frac{dv_1}{dt} \cong -C_{gd} \frac{v_{1_n} - v_{1_{n-1}}}{h}. \quad (17)$$

Equations (16) and (17) together with Fig. 6 represent an equivalent circuit interpretation of the IIE method proposed in [96]. Note that this model maintains the 1-way coupling property by restricting the forward Euler integration to the

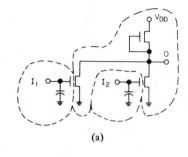

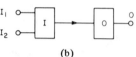

Fig. 7. (a) Two-input gate with strongly connected components. (b) One-way model for gate.

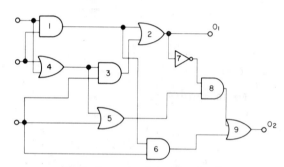

Fig. 8. Example of an adder circuit.

current which represents feedback. A further refinement of this approach is the symmetric displacement method in [41].

In a recent paper a method has been proposed of the form

$$g(v_{n_0}^k, v_{n_1}^k, v_{n_2}^k, v_{n_3}^*) = 0 \quad (18)$$

corresponding to (18) and Fig. 3. Here $v_{n_3}^*$ is computed from

$$v_{n_3}^* = v_{n-1_3} + h v_{n-1_3} \quad (19)$$

which is again a prediction of the feedback variables in the time coordinate.

The "now" time step h_n is more or less synchronized among all subcircuits in the incremental approaches. An attempt has been made at decoupling the time steps by using a master clock with subcircuit running at their own h_n [98]. In a waveform approach, complete decoupling among the subcircuit is achieved and the subcircuits are allowed to run at their own maximum time step [36], [37]. This may result in considerable time saving due to the reduction in the number of computations. This approach is illustrated on the example circuit of Fig. 8, where the model of Fig. 7 is used to represent the 1-way circuits. The interaction graph G of Fig. 9 results if the individual gates or subcircuits are replaced by 1-way models. The key insights gained from this graph are the portions which should be analyzed simultaneously. For example, the subcircuits inside the dashed line B may be strongly connected. Note that the interconnections are conveniently accommodated in this approach. We form a new graph from Fig. 9 which has the strongly connected subcircuits as its nodes. Fig. 10 shows G'' where additionally the nodes have been assigned to levels to further simplify the processing. The levelizing ensures that we analyze the circuit such that all the input information is available for each subcircuit at the processing time. Interpolation must be used in this approach to obtain the input

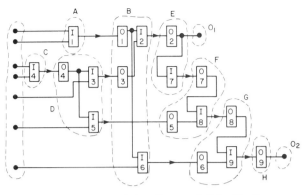

Fig. 9. One-way interaction graph G for Fig. 8.

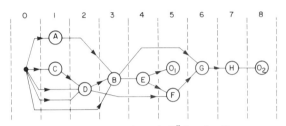

Fig. 10. Levelized graph G'' for circuit.

waveforms at the appropriate time points due to the unequal time steps in the subcircuits [37]. The processing of the subcircuits and the subsequent discarding of this portion of a circuit, is a desirable approach to the time analysis for large-scale circuits. In this process, the waveforms of interest are stored on disk and thus the amount of data present in active store is limited. The accuracy of the solution techniques in the straightforward 1-way technique depends on the size of the gate-to-drain capacitance. Accuracy is always a key issue in the S/A techniques discussed since inaccuracies may lead to faulty chip designs [99]. The WR technique leads to an approach for which the compute time can be reduced at the expense of accuracy and vice versa.

The basic WR approach is easily illustrated for C_{gd} with the assistance of the equivalent circuit Fig. 6. Again, full-time waveforms are involved and we choose

$$i_1(t) = -C_{gd} \frac{dv_2(t)}{dt} \tag{20}$$

$$i_2(t) = -C_{gd} \frac{dv_1(t)}{dt}. \tag{21}$$

The left circuit portion is evaluated first for $t \in [0, T]$ where T is the final time. We iterate between the left and right circuits until convergence, with $i_1(t)$ and $i_2(t)$ being the waveforms at the previous iteration. This process is generalized in ALG. 2.

ALGORITHM 2: Waveform Relaxation for Multiple Circuits

Inputs: Circuit C
         Start time Tmin
         Stop time Tmax
         Input waveforms IW
Result: Waveforms for each node
Note: $S(i, j)$ = subcircuit j for iteration i
         NS = number of subcircuits
         $X(i, j, k, t)$ = voltage of node k in subcircuit j
                         iteration i at time t

Procedure WaveformRelaxation (Tmin, Tmax, C, I)
BEGIN
    Partition C into subcircuits S
    Schedule the subcircuits
    $X(0, *, *, t) = IW$
    $i = 0$
    Repeat
      For $j = 1$ to NS do
      BEGIN
        Solve each subcircuit $S(i, j)$ for voltages
        $X(i, j, k, t)$ at iteration i using the results
        of iteration $i - 1$
        Note: use the Incremental Circuit Analysis
               Algorithm to solve each $S(i, j)$
      END
    until max norm $(x(i, j, k, t) - x(i - 1, j, k, t))$
      $<$ ERROR for all j, k, t
END

Different techniques are employed for choosing the subcircuits in ALG. 2. In [100], [19] the decomposition is by *strongly connected component* while in [36], [38] the decomposition is *by circuit* given by the topology of the logical gate.

In the beginning of this section, the desirable goal of at least an order of magnitude speed enhancement over a general-purpose circuit analysis program was stated and a multitude of methods for improving the solution speed are given. A comparison among the methods may be somewhat misleading due to the different implementation, computers, and accuracy demanded in the solution. An indication of the state of the art may be a factor 50 in speed improvement for RELAX [38] as compared to SPICE for a medium-size circuit.

C. Time Waveform Representation

A key aspect of a circuit S/A program and also a logic simulator is the representation of the time waveforms. This is true for incremental as well as waveform techniques. Some of the factors to take into account are the number of circuit, application, intended accuracy, available computer, and storage space, as well as the compute time to be consumed by a typical run. We easily can think of these considerations as well as the waveform representation as more fundamental than the S/A techniques which lead to the waveforms. Fig. 11 illustrates different time waveform representations. In circuit analysis which has the potential of retaining a better than 1-percent accuracy, all of the details of the waveforms are represented as shown in Fig. 11(a). The high accuracy is obtained at the cost of both high storage and time requirements. For example, at the circuit analysis level, 10^3 time points for a single waveform is not unusually high. If we analyze a problem with $N = 10^4$ nodes we may want to retain 10 percent of the node results for future observation of the resultant waveforms. The storage for these waveforms (time, voltage pairs) will require 2×10^6 words of storage. Usually, this will be a secondary store database. Clearly a form of *data compression* is desirable for realistic large-scale systems. In general, point of the amount of information H in a signal with k levels occurring with probability P_i is

$$H = -\sum_{i=1}^{k} P_i \ln P_i. \tag{22}$$

From this point of view it is clear that the binary simulator (0, 1) employs the minimum waveform representation which

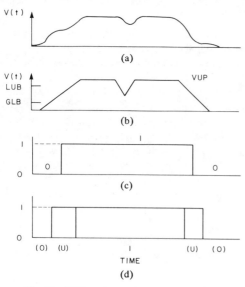

Fig. 11. Different waveform representations.

contains any information at all. However, adding a few additional intermediate levels between the "scaled" levels 0 and 1 is not the most practiced approach to increasing the amount of useful information. Fig. 11 illustrates a few other examples. In (b) a linear rise and fall time has been used as well as a half-signal spike representation. The pure binary signal is given in (c), while in (d), the unknown U- or X-state is added. Both (c) and (d) may ignore the spike shown.

Simulators with 7 or more states are in use, e.g., [102] today attempting to maintain sufficient accuracy while keeping compute time and storage to a minimum. In terms of (22), adding an X-state may increase the useful information content far more than adding an intermediate signal level. It is obvious that the concepts earlier presented unify the techniques employed in analysis and simulation programs. However, new challenges are introduced when different level representations are used simultaneously in a multilevel system. Two fundamentally different types of transitions are necessary: high-to-low information; and low-to-high information.

The first transition is obviously accomplished by discarding information, although challenges remain like the matching of many different waveform representation like the ones with X or high impedance state, etc. The transition from logic variables, Fig. 11(d), to circuit voltages is interesting since information needs to be added. Assume that we want to obtain the constant slope representation of Fig. 11(b). First, the amplitude is scaled by associating a logical 1 with a typical up-level voltage $VUP = 4.5$ V. Further, the slope of the rise time is

$$SL = \frac{V_{LUB} - V_{GLB}}{t_{LUB} - t_{GLB}}. \quad (23)$$

Thus the constant slope in Fig. 11(b) can be constructed. This obviously only works for X-states corresponding to transitions. Most likely, this waveform will be the input to the circuit analysis portion in a multilevel system obtained from the simulation portion.

The concepts given in this section show the similarities between the fundamental aims of the simulation and analysis techniques which is the computation of waveforms. However, it should be clear at this point that the means of obtaining them may be quite different.

D. Temporal Sparsity

The last fundamental technique mentioned in Section II-B which we have not considered so far is temporal sparsity. We choose *temporal sparsity* [90] to describe the basic fact that not all subcircuits are active at the same time. Typically, for sufficiently large circuits only between 0.01 to 10 percent of the gates are active at the same instant.

Numerous clever techniques have been devised to take advantage of temporal sparsity. One of the first techniques reported is the event-driven logic simulation method [43], [44], [103]. In this section we restrict ourselves to the event mechanism while in Section III-B we will discuss the simulation. Usually the events are represented in a time event queue for the actions to be taken. In the scheme in ALG. 3, the actions taken at the present or "now" time t_n are illustrated.

ALGORITHM 3: Time Event Scheduling

Inputs: Circuit C
         Subcircuit Delays $ScD(i)$
         Event Queue EQ
         Initial events IE

Result: Updating of event queue for circuit scheduling

Note: tn is present time

Procedure EventScheduling($C, ScD,$ time)
BEGIN
  $EQ = IE$
  While EQ is not empty do
    BEGIN
    tn is smallest new next event time on EQ
      $P =$ Pop from EQ all events at tn
      For all $P(j)$ do
      BEGIN
        $S =$ Successors($P(j)$)
        For all $S(j)$ whose states have changed do
        SelectiveTracing:
        Push onto EQ the event $[S(i), tn + ScD(i)]$
      END
    END
END

The kernel of ALG. 3 is the *selective tracing* which finds all new events which are caused by the present events at t_n. Then the event queue EQ is updated by these events. A further refinement may be if they actually switch and cause new events before we execute the successors. This scheme leads to an excellent exploitation of temporal sparsity since actions are taken only if they lead to activities, or equivalently computations must be performed. It is noted that the number of computations is directly related to how busy the event queue is. Selective tracing usually assumes the logic circuits models to be simple having 1-way properties. It should be noted that special algorithms must be employed for a circuit with pass-transistors which are 2-way [55].

Temporal sparsity was not taken into account in circuit analysis and large-scale circuit simulation until recently when the latency concepts were introduced, e.g., [34], [84], [90]. Before, all subcircuits were analyzed with the same time-step h, whether the subcircuits were latent or not. In a modular or subcircuit oriented program, the inactive subcircuits can be singled out and no computations are done inside the subcircuit. Then, in an incremental analysis, the internal variables are simply [40]

$$z_n = z_{n-1} \tag{24}$$

which is a zeroth-order integration method with a local truncation error in the variable k is

$$E_{0_k} = \dot{z}_n(t_{n-1})h. \tag{25}$$

Thus a latency detector keeps track of the derivative of all variables in a subcircuit.

The computations for the subcircuit are resumed if changes in one of the subcircuit variables occur which exceed an error criterion. Thus many of the new incremental programs are taking latency into account, e.g., [35], [61], [84], [90]. The waveform programs have independent time steps in the subcircuits. Therefore, if a subcircuit is inactive, its variable time step algorithm will choose a very large time step which is equivalent to the latency concept [36], [37], [19].

Thus it can be concluded that temporal sparsity can be exploited for both general-purpose circuit analysis, as well as for large-scale programs. In logic simulators, the computations are reduced using event scheduling illustrated in ALG. 3.

III. Logic Simulation

By *logic simulation* we mean the simulation of two (0, 1) or more states like H, X, etc. For many years, logic simulation was performed at the *gate type level* only. However, with the advent of custom VLSI pass-transistor designs, logic simulation at the *transistor level* was devised, e.g., [49]. Simply, these designs cannot efficiently be represented by gates. A logical gate is a higher level representation since usually several transistors are mapped into a single gate. Obviously, for a large-scale circuit this reduction in complexity is important if it can be accomplished for at least a portion of the logic at hand.

A. Gate Level Simulation

Today, the majority of simulators are of the gate or subcircuit level type. Logic design is most easily performed in terms of gates which perform logic functions like AND, OR, NAND, etc. This implies that groups of transistors are easily identifiable as one of a few standard gates like an AND or a NOR subcircuit.

Gate level simulation of high-performance machines is perhaps the only discipline which has so far been executed at a large-scale level. In these machines the number of gates per integrated circuit is small by VLSI standards due to the power dissipation required to obtain the high-performance operation. However, simulation runs involving 0.5×10^6 gates have been made already several years ago, e.g., [48]. The simulation methodology does not strongly depend on whether the physical location of the gates is on VLSI chips or on separate chips at least at the gate level for high-performance logic.

The following factors contribute to the convenience and efficiency by which gate-level simulation can be executed.

The gates almost directly correspond to the logical operators OR, AND, NAND, etc., and these operations can efficiently be executed in most computers.

Delay equations for timing verification may yield accurate answers for bipolar transistors. An accuracy of 10 percent is achievable in some cases especially if the interconnections add a large portion to the delay.

The structure of the circuits is extremely sparse since the number of connections per subcircuits is limited for high performance and since the overall size of the circuit is very large.

Temporal sparsity is large since the percentage of active gates usually decreases with increasing circuit size.

Some of the concepts implemented in gate level simulators which were developed in the last two decades are outlined next. Mainly, the design and verification of large, high-performance machines with up to $\sim 10^6$ gates depend heavily on this simulation capability. While simulation is used for both verification and testing, we consider verification only to limit the scope of this paper. Some of the techniques which make the simulation of large-scale circuits possible are these.

Parallel simulation is a technique where for example each bit in a word is used to represent an input variable, e.g., [42]. Thus this process allows for the simulation of 32 input patterns to a circuit in parallel for a 32-bit word machine. Thus this method has the potential of speeding up the simulation time of almost a factor of 32.

Compiled-code simulation is a technique which takes advantage of the high speed of execution of compiled code. The circuit description is programmed in a suitable language like Pascal or Assembler. Either logical macros or the entire circuit is compiled and then run simultaneously. Clearly, scheduling is required which specifies the sequence in which the logical functions are processed similar to G'' in Fig. 10. Hard problems in compiled-code simulators are the methodology for taking advantage of the temporal sparsity and also to efficiently change the logic configuration represented by the compiled code.

Table lookup is a frequently used technique where the truth table and other important information are stored in a table for the logic gates, e.g., [104]. These models are accessed using the input values to determine the resultant output. This approach is quite flexible and is widely used.

As mentioned above, gate level simulation is widely applied to large-scale, high-performance machines where standard gates are employed. The techniques in this discipline are well established and are sophisticated. Timing delay approximations are relatively accurate for high-performance circuits since the delay is mainly in the interconnections [6]. Thus far more accuracy is obtained by this type of timing simulation for high-performance machines than for MOS-VLSI circuits even if standard logical gates are employed.

B. Transistor Level Simulation

Transistor level logic simulators are a relatively new addition to the CAD tools used especially in the design of MOS transistor circuits. Mainly, new circuit designs have been invented which cannot efficiently be treated by a gate simulator. More function per transistor can be obtained by direct transistor designs rather than gate designs [53].

A good example of an MOS logic simulator is MOSSIM [49], [50], which takes the actual transistor circuit into account. MOSSIM is a three-logic level (0, 1, X) simulator. At each *state* of the circuit the steady-state levels (SSL) are established. Note that the transients from one state to the next state are *not* calculated in contrast to the techniques discussed in the last section. However, the SSL (0, 1, X) must be distinguished from the "dc" levels since the SSL may in some case be a dynamic level held by a ratio of capacitances. Dynamic elements like capacitances and inductances are ignored in these simulators unless they are involved in determining the SSL. The fundamental transistor level simulator requires a circuit diagram description where we need to specify the type of each transistor.

Fig. 12. (a) Simplified transistor model. (b) Two resistances in parallel.

The conventional way to solve for the SSL is to apply ALG. 1 with the time-step loop fixed where the time step h is chosen sufficiently large. This large h assures that the companion resistances corresponding to the capacitances are sufficiently large [93], [94]. This approach is both accurate and time consuming. Approximate solutions are of interest since the approximate knowledge of the logic levels suffices in many cases and since we are interested in simulating very large transistor circuits.

The basis of the technique in MOSSIM is what we can call *approximate circuit theory*. The interpretation of the transistor model is best done in terms of resistances of the device. In Fig. 12(a), we show the model for an MOS transistor where R depends on the type of device. For example, a load device may have $R_1 = 10$, while the active devices may have $R_2 = 1$. The switch for an enhancement load is always closed while the active device in an inverter is controlled by the input. Thus the logic output levels v at the inverter are a logical 1 for a 0 input and for a 1 input

$$v = \frac{R_2}{R_1 + R_2} = 0.0909 \tag{26}$$

which is equivalent to a logical zero, since we divide the logical range for v

$$v = 0, \quad \text{for} \quad v < v_L$$
$$v = x, \quad \text{for} \quad v_L \leq v \leq v_H$$
$$v = 1, \quad \text{for} \quad v_H < v. \tag{27}$$

Computations are saved if we use approximate computations. For example, in Fig. 12(b) the parallel resistance is given by

$$R = \min(R_1, R_2) \tag{28}$$

which saves a multiplication, an addition, and a division compared to the exact parallel resistor formula. A similar simplification can be worked out to avoid the computation in (26) for the voltage divider, where one of the conditions is

$$v = 0, \quad \text{if} \quad R_1 > \frac{1 - v_L}{v_L} R_2. \tag{29}$$

A complete set of equations can be worked out for an approximate circuit theory.

A MOSSIM simulator algorithm is outlined below. Here, the nodes are classified as *internal* and *external*. Clock nodes are casually called inputs NE which have a state $E(NE)$. The internal nodes NI have a state $I(NI)$ in ALG. 4.

ALGORITHM 4: Transistor Level Logic Simulator

Inputs: Circuit C
State of each node $E(NE), I(NI)$
State of each transistor $T(NE, NI)$
Result: Updated States (SSL) $I(NI), T(NE, NI)$

Procedure TransistorSimulation (C, S, T, E)
For all MOS-transistor gates connected to NE's update the state $T(NE) = E(NE)$
BEGIN
    For all internal nodes NI do until no changes occur
    BEGIN
    Internal NI connected transistors are held fixed in state $T(NI)$
        Compute new nodal states $I(NI)$ using approximate circuit theory
        Update transistor $T(NI)$ states from new nodal states $I(NI)$
    END
END

All nodal states are fully determined for each input set or time phase. Since the MOS transistor gates form ideal 1-way connections at dc, they control the MOS transistor source and drain nodes. Groups of MOS transistors are formed by the interconnections [50] between the source and drain nodes. These groups suggest another level of hierarchy which is left out of ALG. 4 for simplicity. Specifically, strong components or groups can be formed which are isolated by FET gates while the interconnections join the members of a group. It is interesting to note that for FET circuits, these groups closely resemble the 1-way components in Section II-B.

The relaxation procedure used in MOSSIM given in ALG. 4 starts from one steady-state level defined by all inputs and clock signals $E(NE)$ at the input or external nodes NE. Then a new input excitation $E(NE)$ is applied and the new internal states $I(NI)$ are computed where $E \in [0, 1]$ and $I \in [0, X, 1]$. This basic transistor level MOS simulation procedure is an iterative procedure to avoid the solution of large, sparse matrices in the form of (4). The transistor states $T(NI)$ connected at internal nodes NI are updated simultaneously only once per iteration while the internal computations are performed with fixed transistor states. The nodal states for internal nodes are computed from the simplified computations of the voltage dividers.

MOSSIM [49] has been applied to circuits with 10^4 transistor on a DEC-20 using about 10 s of compute time per input clock cycle. This suggests that the methodology is applicable to about 10^5 transistors corresponding to approximately 3×10^4 logical gates. Presently, many researchers are active in this general area of MOS transistor level simulation.

The question of approximate time-domain simulation at a far coarser level than the techniques reported in Section II-B is of general interest. In [60], [106] an approximate delay simulation is attempted for MOS circuits. The delays are computed using a table lookup and delay equation-type procedure. In [109] the MOSFET problem is formulated in terms of a switching theory in contrast to the resistive model approach discussed here.

IV. MIXED- AND MULTILEVEL SIMULATION/ANALYSIS SYSTEMS

A VLSI circuit may be represented at several levels of complexity depending on the level of the model which may be architectural; functional; gate; device, transistor, circuit; and detailed device macromodel.

In fact, a complete VLSI design of a new technology usually

spans all of these levels. Thus it is very desirable for a design system to encompass as many of these levels as possible.

In the literature a distinction is made between a *multilevel* and a *multimode* system. The multilevel system is based on a hierarchical structure where a common database is employed. Each lower level adds more detail and accuracy to the simulation results like the above list. *Mixed* systems involve two levels and *multi* systems may involve more than two levels.

The first mixed and multilevel systems were different in scope. For example, at the high end logic level simulation was combined with the functional level, e.g., [62]. In the circuit domain, the first macromodels for digital logic were incorporated in a circuit analysis program leading to mixed circuit analysis/simulation systems [32], [55].

The technology of multilevel S/A systems is presently evolving. An early example of such a system is DIANA [87]. Here, the techniques span the digital as well as the analog circuit domain which include analog to digital (A/D), D/A converters as well as filters. With the advent of digital filters, this capability is becoming increasingly important. The internal structure of DIANA is such that even the circuit-analysis-oriented portion can be event driven.

In *mixed-mode* and *multimode* systems, the different programs operate at the same level but the function is different. For example in [19], a logic simulator is employed in the mask design task of the system while the WR time S/A program is used for the timing verification of design. An interesting combined mixed-level and mixed-mode piecewise linear modeling technique is given in [20].

Two recent papers report on CAD systems at Sandia [107] and Hughes [108]. They have a multilevel S/A "engine" as a central part of the system. This approach has many advantages for large-scale VLSI circuits. First, the problems of data conversion from one program to another is avoided in a well-designed system. This is only one reason why the design time is reduced. Other, more important time and storage savings result from the "magnifying glass" technique where, for example, the overall simulation proceeds at a high level while only a few circuits are analyzed accurately at a more detailed level. The potential increase in the number of subcircuits which can be analyzed is substantial. Clearly, a multilevel system provides the environment for these tradeoffs. Further, the danger of losing relevant information by using a too-simple implementation always exists. In a multilevel system the results can always be checked by a lower level technique which ultimately may be the exact circuit analysis level.

V. VERIFICATION

As mentioned above, *design* and *verification* are the two key purposes of a CAD system. The former is the "creation" process while the latter insures the "correctness" of the design. Verification is of concern at all levels in the design. Examples of verification steps are the following: functional design from architectural design; correctness of gate level implementation; electrical connectivity; design rules violations; and timing verification.

Verification is not only of importance in the initial design, but also in the engineering design changes which follow each design phase. The verification process must establish the correspondence among all the models used for the representation of the design [76]. The design and verification processes span the entire spectrum from the system architecture to the implementation of the shapes on silicon wafers. Some of the

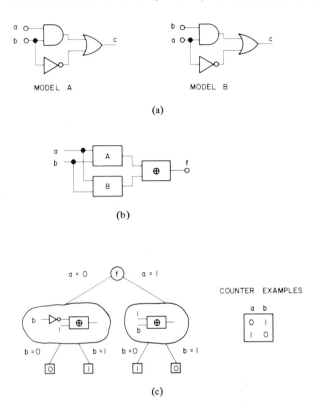

Fig. 13. (a) Example models. (b) Comparison circuit. (c) Expansion procedure.

design procedures at the shapes level involve automatic translation programs [64]–[65], [72] which guarantee correctness. This shifts the burden of verification to the CAD programs. This is a desirable trend since the designs obtained by such a CAD system are automatically correct. Some of the design steps are more amenable to automation, e.g., [71] while others like timing verification are much more difficult to quantize. Here, we give more details on two verification levels, the functional equivalence and timing verification.

A. Functional Equivalence of Logic Circuits

One form of design verification is to determine whether two implementations of the same specification are equivalent. In the IBM 3081 machine, the Static Analysis program [74] was used to find input conditions which would cause two implementations to differ. The two implementations were at the gate level and register transfer level (RTL). Usually the RTL level design is considered correct since a large amount of simulation is possible at this level. However, the 3081 experience demonstrated that errors were found at each level.

The fundamental idea behind this form of design verification is the efficient solution of a set of Boolean equations. The RTL level design is translated to Boolean equations and then each output is exclusive-ored with the corresponding output in the gate level description. All the outputs are ored together to form a single output function f. Solving $f = 1$ is equivalent to finding input patterns which cause the gate level and RTL level designs to differ. If no solutions are found, the two models are declared equivalent. If solutions are found, they represent counter examples which cause at least one of the outputs to be different. Fig. 13 demonstrates the design verification algorithm for comparing two models A and B for equivalence.

ALGORITHM 5 solves for $f = 1$ using symbolic simulation. Inputs are set to 0 or 1, and the effect is propagated through the graph. For example, consider the gate $C = \text{AND}(A, B)$. When $B = 0$, this implies $C = 0$. But for $B = 1$, the output C is equal to A. $C = A$. The input A now becomes the input of a new gate which is one level closer to the output. This procedure is recursively applied by expanding around the other inputs until either $f = 1$ or until $f = 0$. Any path in the expansion process which causes $f = 1$ is a counter example.

ALGORITHM 5: Functional Equivalence

Inputs: Two combinational circuits $C1, C2$
Result: Input patterns which cause output
         values of $C1$ and $C2$ to be different

Procedure BooleanCompare($C1, C2$)
BEGIN
   C = Exclusive-Or($C1, C2$) connections as Fig. 13
   Stack = empty
   Expand(C)
END

Procedure Expand(C)
BEGIN
   Choose input variable X from C
   $CX0 = C(X = 0)$ Symbolic simulate (expand around 0)
   $CX1 = C(X = 1)$ Symbolic simulate (expand around 1)
   Traverse($X, 0, CX0$)
   Traverse($X, 1, CX1$)
END

Procedure Traverse(X, Direction, C)
BEGIN
   Push onto Stack (X, Direction)
   If C not = 0 or 1 then   BEGIN Expand(C)
                                                            RETURN
                                               END
   If $C = 1$ then Print Stack
   Pop Stack(X, Direction)
END

The main advantage of this form of verification is that no simulation is required. The algorithm formally proves the equivalence of two designs. The computational complexity of this algorithm is known to be NP-complete. However, for most practical problems the algorithm exhibits polynomial time behavior. One reason for this is that during the expansion process, two subgraphs may be identical to one another. In this situation, recursively solving for counter examples in two subgraphs reduces to solving for counter examples in one. This verification step clearly precedes the timing verification step given in the next section.

B. Timing Verification

As discussed above, timing verification is an important aspect of the process unless the circuits are designed for a very low performance technology. Here we again will distinguish between high-performance *bipolar* circuits which are packaged on multichip modules and MOSFET circuits where most of the communication among the circuits is on chip.

Again, the *bipolar* transistor delay may be dominated by the interconnection delay [6] and the subcircuits are usually gates which can be described by delay equations. Thus rather accurate delay equation macromodels can be obtained which can be added to obtain path delays. Algorithms for timing verification based on this delay model are applicable for large-scale systems.

Timing verification algorithms using path delays are discussed in [13]. They have the capability of selectively disabling unused paths. This is useful in detecting long functional paths which are hidden by unused paths with worse timing problems. Their implementation propagates a null signal through unused paths which no longer participate in the delay calculation. To approximate statistical delay, they use a min. and a max. range for the rising and falling delays. Furthermore, logic gates fall into one of three categories—positive unate(and,or), negative unate(inverter), and nonunate(xor). A rising delay through a negative unate gate becomes a falling signal. Their implementation also includes the analysis of clock skew, pulsewidth, and both long and short paths. Since the algoritm is block oriented rather than path oriented, the performance is linear with the number of gates in the circuit. The implementation is sufficiently general for large problems of 100 000 logic gates and consumes only 2 ms/gate.

The SCALD timing verifier [76] uses timing assertions based on a calculus of 7 values to verify the design. Here a signal can take the value 0, 1, stable, changing, rising, falling, or unknown. During a block cycle, SCALD determines when a signal is changing and when it is stable. By setting inputs to 0 or 1, certain paths are eliminated from the analysis since changing signals no longer propagate. This is necessary to achieve the proper timing for variable length cycles. To verify the correct functioning of memory chips, a setup and hold check is performed. This guarantees that the data are stable before the clock arrives and that it is held long enough before the clock falls. Finally, a minimum pulsewidth is checked for because of the large variations in circuit rising and falling delays. The algorithm is event driven and is computationally efficient on problems as large as 100 000 gates.

The Timing Analysis (TA) technique [78] is the only algorithm which creates slack diagnostics. The slack for a logic gate is defined as the difference between the expected arrival and the actual arrival time. Slack is a measure of how bad a long or short path is. Another feature of the algorithm is that path lengths are statistical rather than worse case. This allows for process variation rather than worse case. A three sigma design allows for process variation which naturally occurs in the manufacturing cycle. The analysis of paths is block oriented rather than by path enumeration so that an entire IBM 3081 processor of 700 000 gates can be handled efficiently. For problems of this magnitude, a software paging scheme is necessary. Another way of avoiding the data explosion problem for 700 000 gates is to use TA to solve small partitions of the machine and then combine the solutions together. Delay modifiers are also a feature of the TA technique. This allows the user to cut paths that are never active or to adjust delays for multicycle paths. All gates are considered as either inverting or noninverting. The rising and falling delays for each gate use this information to accurately predict delay values as is carefully illustrated in Fig. 14 and (30). The meaning of the labels used is as follows:

a_0, b_0 = rising arrival times
a_1, b_1 = falling arrival times
   d_0 = delay if output is rising
   d_1 = delay if output is falling
   c_0 = rising output arrival
   c_1 = falling output arrival.

Fig. 14. (a) TA model for an inverting gate. (b) Model for a noninverting gate.

For the late mode the following equations hold for Fig. 14(a)

$$c_0 = \text{MAX}(a_1, b_1) + d_0$$
$$c_1 = \text{MAX}(a_0, b_0) + d_1 \quad (30\text{a})$$

while for the early mode

$$c_0 = \text{MIN}(a_1, b_1) + d_0$$
$$c_1 = \text{MIN}(a_0, b_0) + d_1. \quad (30\text{b})$$

Similarly, for a noninverting circuit in Fig. 14(b) the late mode equations are

$$c_0 = \text{MAX}(a_0, b_0) + d_0$$
$$c_1 = \text{MAX}(a_1, b_1) + d_1 \quad (30\text{c})$$

while for the early mode

$$c_0 = \text{MIN}(a_0, b_0) + d_0$$
$$c_1 = \text{MIN}(a_1, b_1) + d_1. \quad (30\text{d})$$

Using this gate or block model, ALG. 6 gives the timing verification for a large-scale clocked circuit.

ALGORITHM 6: Timing Analysis (TA)

Inputs:
   Combinational Circuit C
   Primary input arrival times Tin
   Expected Output times $Tout$
   Circuit delays D
Results: Worst case arrival times $tARR$
   Slack times for each circuit $tSLACK$

Procedure TimingVerification($C,Tin,Tout,D$)
BEGIN
   Levelize C by a topological sort
   Schedule gates in levelized order
   For each gate i do
     $tARR(i) = D(i) + \text{MAX}(tARR(\text{predecessors of } i))$
   For all outputs j do
     $tSLACK(j) = Tout(j) - tARR(j)$

   Schedule gates in reverse levelized order k
   For each connection Edge(m,k) do
   BEGIN
     Edge $tSLACK(m,k) = tARR(k) - D(k)$
     $+ tSLACK(k) - tARR(m)$
   END
END

Worst case arrival times are computed in linear time by making only one sweep through the circuit from primary inputs to primary outputs. Slacks are useful as diagnostics for timing verification. They represent how late (or early) a signal is with respect to the expected arrival time at the primary outputs. The computation of slacks is done in linear time by making one sweep through the circuit in the backward direction from primary outputs to primary inputs. However, it needs the arrival times computed during the forward propagation phase.

The critical path in the circuit is the one with the smallest slack. There are many ways to fix these timing problems. One way is to change low-power subcircuits to high power on nets which drive a high capacitance load. For chip-to-chip timing problems, designers may change to IO assignment or driver and receiver types to prevent electrical reflections. This is another example of how innovative algorithms extend the capabilities of the design procedure to VLSI dimensions.

The area of verification tools is presently evolving [62]-[78] as is illustrated by the above examples. For timing verification at the transistor level, the techniques given in Section II are used. Again, the procedures at the transistor level are more complicated than at the gate level. Mainly MOS transistor responses are highly dependent on the interconnections and the capacitances and thus the delay and rise times include much more variability than the corresponding bipolar transistor gates. This forces the verification to a much more refined level especially for passtransistor circuits.

VI. CONCLUSIONS

At present, the areas reported on in this paper are receiving considerable attention from industry as well as universities since they are of key importance for the design of realistic VLSI circuits. As is evident from the recent references in this paper, much progress has been made in the last few years across the entire spectrum of VLSI problems. The main task that needs to be performed is to find algorithms and implementations which yield acceptable running times at the VLSI level. Programs are in use today in some areas like logic simulation and timing verification which accommodate circuits at the VLSI level without excessive running time and storage requirements.

ACKNOWLEDGMENT

The authors would like to thank G. Almasi for the valuable suggestions and careful reading of the manuscript. They would also like to acknowledge the contributions made by the discussions with other researchers active in the field. Within the space given here they can only mention a few names: C. Carlin, W. Donath, G. Hachtel, I. Hajj, E. Lelarasmee, F. Odeh, G. Rabbat, A. Sangiovanni-Vincentelli, and V. Visvanathan.

REFERENCES

[1] C. H. Sequin, "Managing VLSI complexity: An outlook," this issue, pp. 149-166.
[2] C. Niessen, "Hierarchical design methodologies and tools for VLSI chips," this issue, pp. 66-75.
[3] A. E. Ruehli, N. Rabbat, and H. Y. Hsieh, "Macromodeling—an approach for analyzing large-scale circuits," *Comput. Aided Design*, vol. 10, pp. 121-130, Mar. 1978.
[4] G. D. Hachtel and A. L. Sangiovanni-Vincentelli, "A survey of third generation simulation techniques," *Proc. IEEE*, vol. 69, pp. 1264-1280, Oct. 1981.
[5] A. E. Ruehli, "Survey of computer-aided electrical analysis of integrated circuit interconnections," *IBM J. Res. Develop.*, vol. 23, pp. 627-639, Nov. 1979.
[6] E. E. Davidson, "Electrical design of high speed computer package," *IBM J. Res. Develop.*, vol. 26, pp. 349-361, May 1982.
[7] M. A. Beuer, Ed., *Digital System Design Automation: Languages, Simulation and Data Base*. Woodland Hills, CA: Computer, Sc. Press, 1975.
[8] M. A. Breuer and A. D. Friedman, *Diagnosis & Reliable Design of Digital Systems*. Potomac, MD: Comp. Sc. Press, 1976.
[9] M. A. Breuer, A. D. Friedman, and A. Iosupovicz, "A survey of the state of the art of design automation," *Computer* (IEEE), pp. 58-75, Oct. 1981.
[10] L. Bening, "Developments in computer simulation of gate level physical logic," in *Proc. 16th Design Automation Conf.*, San Diego, CA, June 1979, pp. 561-567.
[11] E. Ulrich and D. Herbert, "Speed and accuracy in digital network simulation based in structural modeling," in *Proc. 19th*

Design Automation Conf., Las Vegas, NV, June 1982, pp. 587-593.
[12] R. B. Hitchcock, Sr., "Timing verification and the timing analysis program," in *Proc. 19th Design Automation Conf.*, Las Vegas, NV, June 1982, pp. 594-604.
[13] L. C. Bening, T. A. Lane, and C. R. Alexander, "Developments in logic network path delay analysis," in *Proc. 19th Design Automation Conf.*, Las Vegas, NV, June 1982, pp. 605-613.
[14] J. P. Roth, *Computer Logic, Testing, and Verification*. Potomac, MD: Comp. Sc. Press, 1980.
[15] D. A. Antoniadis and R. W. Dutton, "Models for computer simulation of complex IC fabrication process," *IEEE J. Solid-State Circuits*, vol. SC-14, pp. 412-422, Apr. 1979.
[16] W. L. Engl, H. K. Dirks, and B. Meinerzhagen, "Device modeling," this issue, pp. 10-33.
[17] G. Arnout and H. De Man, "The use of threshold function and boolean-controlled network elements for macromodelling of LSI circuits," *IEEE J. Solid-State Circuits*, vol. SC-13, pp. 326-332, June 1978.
[18] H. De Man, "Computer aided design for integrated circuits: Trying to bridge the gap," *IEEE J. Solid-State Circuits*, vol. SC-14, pp. 613-621, June 1979.
[19] H. De Man, "Mixed mode simulation for MOS VLSI: Why, Where and How?" in *Proc. IEEE Int. Symp. Circuits System*, Rome, Italy, May 1982, pp. 699-701.
[20] W.M.G. Van Bokhoven, "Mixed-level and mixed-mode simulation by a piecewise-linear approach," in *Proc. IEEE Int. Symp. Circuits System*, Rome, Italy, May 1982, pp. 1256-1258.
[21] R. W. Jensen and M. D. Lieberman, *IBM Electronic Circuit Analysis Program*. Englewood Cliffs, N.J.: Prentice-Hall, 1968.
[22] L. W. Nagel, "SPICE2: a computer program to simulate semiconductor circuits," Univ. of California, Berkeley, ERL Memo ERL-M520, May 1975.
[23] E. Cohen, "Program reference manual for SPICE2," Univ. of California, Berkeley, ERL Memo ERL-M592, June 1976.
[24] A. Vladimirescu, K. Zhang, A. R. Newton, D. O. Pederson, and A. Sangiovanni-Vincentelli, "SPICE Version 2G User's guide," University of California, Berkeley, Tech. Memo., Aug. 10, 1981.
[25] "Advanced statistical analysis program (ASTAP)," Program reference manual, Pub. No. SH20-1118-0, IBM Corp. Data Proc. Div., White Plains, NY 10604.
W. T. Weeks, A. J. Jimenez, G. W. Mahoney, D. Mehta, H. Quassemzadeh, and T. R. Scott, "Algorithms for ASTAP—a network analysis program," *IEEE Trans. Circuit Theory*, vol. CT-20, pp. 628-634, Nov. 1973.
[26] L. W. Nagel, "ADVICE for circuit simulation," in *Proc. IEEE Int. Symp. Circuits and Systems*, Houston, TX, Apr. 1980.
[27] G. D. Hachtel, R. K. Brayton, and F. Gustavson, "The sparse tableau approach to network analysis and design," *IEEE Trans. Circuit Theory*, vol. CT-18, pp. 101-113, Jan. 1971.
[28] C. Ho, A. E. Ruehli, and P. A. Brennan, "The modified nodal approach to network analysis," *IEEE Trans. Circuits Syst.*, vol. CAS-22, pp. 504-509, June 1975.
[29] N. B. Rabbat, W. D. Ryan, and S. Q. Hossain, "Computer modeling of bipolar logic gates," *Electron. Lett.*, vol. 7, pp. 8-10, Jan. 1971.
[30] O. Wing and E. B. Kozemchak, "Computer analysis of digital integrated circuits," in *NERM Conf. REC.*, Boston, MA, Nov. 1971, IEEE Cat. No. 71C51, pp. 189-191.
[31] S. C. Bass and S. C. Peak, "Terminal models of digital gates allowing waveform simulation," in *Proc. IEEE Int. Symp. Circuit Theory*, Apr. 1973, pp. 287-289.
[32] N. Rabbat, A. E. Ruehli, G. W. Mahoney, and J. J. Coleman," A survey of macromodeling," in *Proc. IEEE Int. Symp. Circuits Systems*, Apr. 1975, pp. 139-143.
[33] B. Chawla, H. K. Gummel, and P. Kozah, "MOTIS-a MOS timing simulator," *IEEE Trans. Circuit Syst.*, vol. CAS-22, pp. 301-310, Dec. 1975.
[34] N. Rabbat and H. Y. Hsieh, "A latent macromodular approach to large-scale sparse networks," *IEEE Trans. Circuit Syst.*, vol. CAS-22, pp. 745-752, Dec. 1976.
[35] S. P. Fan, M. Y. Hsueh, A. R. Newton, and D. O. Pederson, "MOTIS-C: a new circuit simulator for MOS LSI circuits," in *Proc. IEEE Int. Symp. Circuits Systems*, 1977, pp. 700-703.
[36] E. Lelarasmee, A. E. Ruehli, and A. L. Sangiovanni-Vincentelli, "The waveform relaxation method for time-domain analysis of large-scale integrated circuits," *IEEE Trans. CAD Integ. Circ. Syst.*, vol. CAD-1, pp. 131-145, Jul. 1982.
[37] A. E. Ruehli, A. L. Sangiovanni-Vincentelli, and N.B.G. Rabbat, "Time analysis of large scale circuits containing one-way macromodels," *IEEE Trans. Circuits Syst.*, vol. CAS-29, pp. 185-189, Mar. 1982.
[38] E. Lelarasmee and A. Sangiovanni-Vincentelli, "RELAX: A new circuit simulator for large scale MOS integrated circuits," Electronic Research Laboratory, Univ. of California, Berkeley, Memo UCB/ERL M82/6, Feb. 1982.
[39] M. Tanabe, H. Nakamura, and K. Kawakita, "MOSTAP: An MOS circuit simulator for LSI circuits," in *Proc. IEEE Intl. Symp. Circuits Systems*, Houston, pp. 1035-1038, Apr. 1980.
[40] N. G. Rabbat, A. L. Sangiovanni-Vincentelli, and H. Y. Hsieh, "A multilevel Newton algorithm with macromodeling and latency for the analysis of large-scale nonlinear circuits in the time domain," *IEEE Trans. Circuits Syst.*, vol. CAS-26, pp. 733-741, Sept. 1979.
[41] G. DeMicheli and A. L. Sangiovanni-Vincentelli, "Numerical properties of algorithms for the timing analysis of MOS VLSI circuits," in *Proc. 1981 Europ. Conf. on Circuit Theory and Design*, Aug. 1981, pp. 387-392.
[42] M. A. Breuer, "Techniques for the simulation of computer logic," *Commun. Ass. Comput. Mach.*, pp. 443-446, Jul. 1964.
[43] E. G. Ulrich, "Time sequenced logical simulation based on circuit delay and selective tracing of active network path," in *Proc. ACM Nat. Conf.*, 1965, pp. 437-448.
[44] E. Ulrich, "Exclusive simulation of activity in digital networks," *Commun. Ass. Comput. Mach.*, vol. 12, no. 2, pp. 102-110, Feb. 1969.
[45] S. A. Szygenda, "TEGAS-Anatomy of a general purpose test generation and simulation at the gate and functional level," in *Proc. 9th Design Automation Conf.*, June 1972, pp. 116-127.
[46] E. G. Ulrich and T. Baker, "The concurrent simulation of nearly identical digital networks," in *Proc. 10th Design Automation Conf.*, June 1973, pp. 145-150.
[47] S. A. Szygenda and E. W. Thompson, "Digital logic simulation in a time-based table-driven environment: part 1, design verification," (IEEE) *Computer*, pp. 24-36, Mar. 1975.
[48] H. E. Krohn, "Design verification of large scientific computers," in *Proc. 14th Design Automation Conf.*, June 1977, pp. 354-361.
[49] R. E. Bryant, "MOSSIM: A switch-level simulator for MOS LSI," in *Proc. 18th Design Automation Conf.*, Jul. 1981, pp. 786-790.
[50] R. E. Bryant, "An algorithm for MOS Logic simulation," *Lamda Mag.*, Fourth Quarter, pp. 46-53, 1980.
[51] J. Watanabe, J. Miura, T. Kurachi, and I. Suetsugu, "Seven value logic simulation for MOS LSI circuits," presented at the IEEE Intl. Conf. Circuits and Computers, Port Chester, NY, Oct. 1980, pp. 941-944.
[52] W. Sherwood, "An MOS modeling technique for 4-state true-value hierarchical logic simulation," in *Proc. 18th Design Automation Conf.*, Nashville, TN, Jul. 1981, pp. 775-785.
[53] C. Mead and L. Conway, *Introduction to VLSI Systems*. Reading, MA: Addison-Wesley, 1980.
[54] N. Rabat, A. Y. Hsieh, and A. E. Ruehli, "Macromodeling for the analysis of large-scale networks," in ELECTRO-76 Professional Program 21, May 1976, pp. 1-8.
[55] A. R. Newton, "Techniques for the simulation of large-scale integrated circuits," *IEEE Trans. Circuits Syst.*, vol. CAS-26, pp. 741-749, Sept. 1979.
[56] D. Hill and W. van Cleemput, "SABLE: a tool for generating structural, multi-level simulation," in *Proc. 16th Design Automation Conf.*, San Diego, CA, June 1979, pp. 403-405.
[57] V. D. Agrawal, A. K. Bose, P. Kozak, H. N. Nham, and E. Pacas-Skewes, "A mixed-mode simulator," in *Proc. 17th Design Automation Conf.*, Minneapolis, MN, June 1980, pp. 1-8.
[58] T. Sasaki, A. Yamada, S. Kato, T. Nakazawa, K. Tomita, and N. Nomizu, "MIXS: a mixed level simulator for large digital system logic verification," in *Proc. 17th Design Automation Conf.*, Minneapolis, MN, June 1980, pp. 626-633.
[59] V. D. Agrawal, A. K. Bose, P. Kozak, H. N. Nham, and E. Pascal-Skewes, "A mixed model simulator," in *Proc. 17th Design Automation Conf.*, Minneapolis, MN, June 1980, pp. 618-625.
[60] H. H. Nham and A. K. Bose, "A multiple delay simulator for MOS LSI circuits," in *Proc. 17th Design Automation Conf.*, Minneapolis, MN, June 1980, pp. 610-611.
[61] P. H. Reynaert, H. De Man, G. Arnout, and J. Cornelissen, "DIANA: a mixed-mode simulator with a hardware description language for hierarchical design of VLSI," in *Proc. IEEE Intl. Conf. Circuits and Computers*, Port Chester, NY, Oct. 1980, pp. 356-360.
[62] D. D. Hill and W. M. Van Cleemput, "SABLE: Multilevel simulation for hierarchical design," in *Proc. IEEE Int. Symp. Circuits and Systems*, Houston, TX, Apr. 1980, pp. 431-434.
[63] W.M.G. van Bokhoven, "Macromodeling and simulation of mixed analog-digital networks by piecewise-linear system approach," in *Proc. IEEE Intl. Conf. Circuits and Computers*, Port Chester, NY, Oct. 1980, pp. 361-365.
[64] M. E. Daniel and C. W. Gwyn, "Hierarchical VLSI circuit design," in *Proc. IEEE Intl. Conf. Circuits and Computers*, Port Chester, NY, Oct. 1980, pp. 92-97.
[65] L. Scheffer and R. Apte, "LSI design verification using topological extraction," in *Proc. 12th Asilomar Conf. Circuits and Systems, and Computers*, Nov. 1978, pp. 149-153.
[66] C. R. McCaw, "Unified shapes checker—A checking tool for LSI," in *Proc. 16th Design Automation Conf.*, June 1979, pp. 81-87.
[67] R. Auerbach, "FLOSS: Macrocell Compaction system," presented at the 1979 IEEE Design Automation Workshop, East

Lansing, MI, 1979.
[68] C. S. Chang, "LSI layout checking using bipolar device recognition technique," in *Proc. 16th Design Automation Conf.*, June 1979, pp. 95-101.
[69] T. Mitsuhashi, T. Chiba, M. Takashima, and K. Yoshoda, "An integrated mask artwork analysis system," in *Proc. 17th Design Automation Conf.*, June 1980, pp. 277-284.
[70] J. P. Avenier, "Digitizing, layout, rule-checking—The everyday tasks of chip designers," this issue, pp. 49-56.
[71] H. C. Godoy, G. B. Franklin, and P. S. Bottorff, "Automatic Checking of Logic design structures for compliance with testability ground rules," in *Proc. 14th Design Automation Conf.*, June 1977, pp. 469-472.
[72] C. M. Baker and C. Terman, "Tools for verifying integrated circuit designs," *Lambda Mag.*, Fourth Quarter, pp. 22-30, 1980.
[73] R. N. Gustafson and F. J. Sparacio, "IBM 3081 processor unit: Design consideration and design process," *IBM J. Res. Develop.*, vol. 26, pp. 12-21, Jan. 1982.
[74] G. L. Smith, R. J. Bahnsen, and H. Halliwell, "Boolean comparison of hardware and flow charts," *IBM J. Res. Develop.*, vol. 26, pp. 106-116, Jan. 1982.
[75] M. Monachino, "Design verification system for large-scale LSE designs," *IBM J. Res. Develop.*, vol. 26, pp. 89-99, Jan. 1982.
[76] T. M. McWilliams, "Verification of timing constraints of large digital systems," in *Proc. 17th Design Automation Conf.*, Minneapolis, MN, June 1980, pp. 139-147.
[77] S. Newberry and P. J. Russell, "A programmable checking tool for LSI," in *Proc. IEE Europ. Conf. Electrical Design Automation*, Brighton, U.K., Sept. 1981, pub. no. 200, pp. 183-187.
[78] R. B. Hitchcock, Sr., G. L. Smith, and D. D. Chang, "Timing analysis of computer hardware," *IBM J. Res. Develop.*, vol. 26, pp. 100-105, Jan. 1982.
[79] C. W. Gear, "The automatic integration of ordinary differential equations," in *Proc. Information Processing 68*, A.F.H. Morrel, Ed. Amsterdam, The Netherlands: North-Holland, 1968, pp. 187-193.
[80] L. O. Chua and P. M. Lin, "Computer-aided analysis of electronic circuits: algorithms and computational techniques." Englewood Cliffs, NJ: Prentice Hall, 1975, chap. 10, pp. 410-431.
[81] F. Odeh and W. Liniger, "On A-Stability of second-order two step methods for uniform and variable steps," in *Proc. IEEE Intl. Conf. Circuits and Computers*, Port Chester, NY, 1980, pp. 123-126.
[82] A. E. Ruehli, P. A. Brennan and W. Liniger, "Control of numerical stability and damping in oscillatory differential equations," in *Proc. IEEE Intl. Conf. Circuits and Computers*, Port Chester, NY, Oct. 1980, pp. 111-114.
[83] V. M. Vidigal and S. W. Director, "A design centering algorithm for non convex regions of acceptability," *IEEE Trans. CAD Integ. Circ. Syst.*, vol. CAD-1, pp. 13-24, Jan. 1982.
[84] P. Yang, I. N. Hajj, and T. N. Trick, "Slate: A circuit simulation program with latency exploritation and node tearing," in *Proc. IEEE Intl. Conf. Circuits and Computers*, Port Chester, NY, Oct. 1980, pp. 353-355.
[85] D. A. Calahan, "Multilevel vectorized sparse solution of LSI circuits," in *Proc. IEEE Intl. Conf. Circuits and Computers*, Port Chester, NY, Oct. 1980, pp. 976-979.
[86] A. Vladimirescu and D. O. Pederson, "Performance limits of the CLASSIE circuit simulation program," in *Proc. Int. Symp. on Circuits Systems* (Rome, Italy, May 1982), pp. 1229-1232.
[87] H. De Man, J. Rabaey, G. Arnout, and J. Vandervalle: "DIANA as a mixed-mode simulator for MOS LSI Sampled-data circuits," in *Proc. IEEE Intl. Symp. on Circuits and Systems*, Houston, TX, Apr. 1980, pp. 435-438.
[88] P. Penfield, Jr. and J. Rubinstein, "Signal delay in RC tree networks," in *Proc. 18th Design Automation Conf.*, June 1982, pp. 613-617.
[89] A. E. Ruehli, N. B. Rabbat, and H. Y. Hsieh, "Macromodular latent solution of digital networks including interconnections," *Proc. IEEE Int. Symp. Circuits and Systems*, New York, NY, Apr. 1978, pp. 515-521.
[90] K. A. Sakallah and S. W. Director, "An activity-directed circuit simulation algorithm," in *Proc. IEEE Intl. Conf. Circuits and Computers*, Port Chester, NY, 1980, pp. 1032-1035.
[91] ——, "An event driven approach for mixed gate and circuit level simulation," in *Proc. IEEE Int. Symp. Circuits and Systems*, Rome, Italy, May 1982, pp. 1194-1197.
[92] W. L. Engl, R. Laur, and H. Dirks, "MEDUSA—A simulator for modular circuits," *IEEE Trans. CAD Integ. Circ. Syst.*, vol. CAD-1, pp. 85-93, Apr. 1982.
[93] L. O. Chua and P-M Lin, *Computer-Aided Analysis of Electronic Circuits*. Englewood Cliffs, NJ: Prentice-Hall, 1975.
[94] J. Vlach and K. Singhal, *Computer Aided Circuit Analysis*. New York: Van Nostrand, 1983.
[95] J. M. Ortega and W. Rheinboldt, *Iterative Solution of Nonlinear Equations in Several Variables*. New York: Academic Press, 1970.
[96] A. R. Newton, "The analysis of floating capacitors for timing simulation," in *Proc. 13th Asilomar Conf. on Circuits Systems and Computers*, Pacific Grove, CA, Nov. 1979.
[97] Y. P. Wei, I. N. Hajj, and T. N. Trick, "A prediction—relaxation based simulator for MOS circuits," in *IEEE Intl. Cir. cuits and Computers* (New York, NY, Sept. 1982), pp. 353-355.
[98] A. L. Sangiovanni-Vincentelli and N. G. Rabbat, "Techniques for the time domain analysis of LSI circuits," *IEE Proc.*, vol. 127, part G, pp. 292-301, Dec. 1980.
[99] R. Bernhard, "Technology '82/82 solid state VLSI/LSI components," *IEEE Spectrum*, pp. 49-63, Jan. 1982.
[100] E. Lelarasmee, A. E. Ruehli and A. S. Sangiovanni-Vincentelli, "Waveform relaxation decoupling (WRD) method," *IBM Techn. Discl. Bulletin*, vol. 24, no. 7B, pp. 3720-3721, Dec. 1981.
[101] P. Goel, H. Lichaa, T. E. Rosser, T. J. Stroh, and E. E. Eichelberger, "LSSD fault simulation using conjunctive combinational and sequential methods," in *Proc. IEEE 1980 Test Conf.*, Nov. 1980, pp. 371-376.
[102] E. M. DaCosta and K. G. Nichols, "MASCOT," *IEE Proc.*, vol. 127, part G, no. 6, pp. 302-307, Dec. 1980.
[103] P. W. Case, H. H. Graff, L. E. Griffith, A. R. LeClercq, W. B. Murley, and T. M. Spence, "Solid logic design automation," *IBM J. Res. Develop.*, vol. 8, pp. 127-140, 1964.
[104] E. Ulrich, "Table lookup techniques for fast and flexible digital logic simulation," in *Proc. 17th Design Automation Conf.*, Minneapolis, MN, June 1980, pp. 560-563.
[105] R. A. Rohrer and H. Nosrati, "Passivity Considerations instability studies of numerical integration algorithms," *IEEE Trans. Circuits Syst.*, vol. CAS-28, pp. 857-866, Sept. 1981.
[106] V. B. Rao, T. Trick, and M. Lightner, "Hazards in a multiple delay logic simulation," *Proc. IEEE Int. Symp. on Circuits and Systems*, Rome, Italy, May 1982, pp. 72-75.
[107] M. E. Daniel and C. W. Gwyn, "CAD system for IC design," *IEEE Trans. CAD Integ. Circ. Syst.*, vol. CAD-1, pp. 2-12, Jan. 1982.
[108] H. W. Daseking, R. I. Gardner and P. B. Weil, "VISIA: A VLSI CAD system," *IEEE Trans. CAD Integ. Circ. Syst.*, vol. CAD-1, pp. 36-51, Jan. 1982.
[109] J. P. Hayes, "A unified switching theory with applications to VLSI design," *Proc. IEEE*, vol. 70, no. 10, pp. 1140-1151, Oct. 1982.

A Survey of Hardware Accelerators Used in Computer-Aided Design

The performance improvements that occur with special-purpose engines permit significant changes in CAD techniques and design size limitations.

Tom Blank, Stanford University

Hardware accelerators, back-end processors, and special-purpose engines are becoming increasingly popular as viable solutions to a wide range of computer-aided design problems, and their use has resulted in notable performance improvements. For example, software logic simulators easily achieve 10^3 gate evaluations per second, and simulation engines have achieved over 10^8 gate evaluations per second.

This increased capability permits significant changes in both design size limitations and design techniques. Superficially, a hundredfold performance improvement in CAD tools allows a similar increase in design size, assuming the same time. More realistically, designs once infeasible because of their size are now becoming possible. Potential design technique changes are also extremely important. Tools previously requiring CPU hours now run in minutes or seconds, allowing interactive design, faster redesign cycles, and more testing before the actual hardware is built. Ideally, special-purpose engines, or SPEs, will eliminate hardware prototypes, making the first hardware implementation the production product.

Trade-offs. Unfortunately, SPEs are not without trade-offs. Figure 1 shows the relationships among hardware flexibility, cost per application, and solution time. The fastest possible solution, an algorithm built in hardware, is also the most expensive and least flexible. At the other extreme, a general-purpose computer and program provides the greatest flexibility and least cost per application, but it is the slowest. A solution can be further complicated if an SPE is not commercially available for a particularly time-consuming problem. Aside from the risks inherent in any hardware development project, the construction of a hardware accelerator is filled with trade-offs between resources for tool development and product development. The optimal solution depends on the problem, design environment, methodology, number of users and uses, etc., and the value associated with each of these factors.

Summary

Hardware accelerators, or special-purpose engines, have been used in computer-aided design applications for nearly 20 years. In this time, roughly 20 machines have been built and tested specifically for such purposes as simulation, design rule checking, placement, and routing. Their uses are increasing, and the machines are becoming commercially available. This survey describes not only the machines but also their problems and limitations. It also gives comparative data on speed-up techniques and performance. Examples include a simulation machine that achieves roughly a million-times speed-up over a conventional 1-MIP mainframe and a very low cost machine for design rule checking that provides a 100-times improvement. These and other examples clearly demonstrate the viability of special-purpose engines.

Besides these trade-offs, there are some fundamental difficulties with SPEs. The accelerator must be connected into the system design environment, and the design tasks must be reasonably partitioned between the host processor and accelerator. Even if the accelerator's data processing time were zero, the problem solution time could still be excessive because of the other steps involved. A typical partition and event sequence is as follows:

- The host contains the design file or database.
- The host converts from the design file format to SPE format.
- SPE initialization data is moved from the host to the SPE.
- The SPE processes the data, usually in a batch mode.
- Results are returned to the host.
- The results are converted from SPE format and placed in the design file.
- The results are analyzed on the host.

An example from IBM[1] demonstrates the total solution dependencies. For a relatively small simulation sequence, a software simulator was over 10 times faster than a high-performance simulation engine. Even if the accelerator parallelism had been doubled for this small example, the software simulator would still have been 10 times faster. The rate-limiting step was the I/O transfer between the host and accelerator. For longer simulation sequences, I/O time no longer dominates, and the simulation engine is faster. In the specific IBM example, the object was to simulate assembly-level instructions of a 500K-gate computer design. A 10^2 instruction sequence required 49 minutes using a special purpose engine and 4.5 minutes using a conventional software simulator. For a 10^6 sequence, however, the results were 66 minutes and 250 hours, respectively; the accelerator was over 200 times faster. For the total problem solution, the cross-over between a conventional software approach and special hardware approach is determined by design size, inputs, user interaction, results, and required computation sequence.

Speed-up techniques. In spite of the trade-offs and difficulties, many different CAD special-purpose engines achieve significant performance improvements, and the speed-up techniques they use fall into similar categories:

- Remove extemporaneous software processes—for example, remove the operating system and run only the algorithm code.
- Use a faster basic processor—for example, a faster system clock.
- Customize a processor for a specific task—for example, task-specific microcode.
- Separate the algorithm parallelism into separate hardware modules.
- Match intraprocessor communication with the algorithm.
- Partition the problem data into separate processors.
- Match interprocessor communication to the problem data.

The rest of this article describes and compares specific special-purpose engines. For each machine, the objective is to provide insight into the system architecture, system limitations, and algorithm-to-hardware mapping demonstrating the various speed-up techniques. Since simulation engines are currently the most prevalent and have the most commonality, they are described together in the next section. Machines for placement, routing, and design rule checking are presented next. In an effort to limit this survey, paper designs and broadly used machines such as graphics engines and floating-point engines are not covered.

Simulation engines

The value of simulation engines has been recognized since the 1960's when Boeing Aircraft started construction of the first simulation engine.[2] Since then, numerous machines have been proposed and built.[3-5] A general-purpose simulation computer, Scientific Machines Corporation's SMC3100,[6] was built

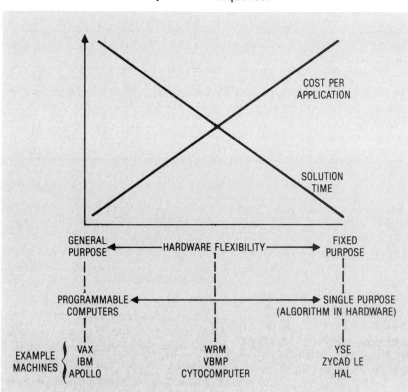

Figure 1. The hardware trade-off.

in the early 1970's specifically for running the S-Lasar simulator. It had 20-bit address/data paths—a cost effective solution at that time. More specialized simulation engines presented in the following sections span the range from microprogrammable machines to machines with embedded hardware algorithms.

All simulation engines use one of two fundamental algorithms: compiled and event driven.

Within that range, all simulation engines use one of two fundamental simulation algorithms: compiled and event driven.[7] A compiled simulator evaluates every element at each time step, whereas an event-driven simulator evaluates only those elements whose inputs have changed. The compiled simulator achieves algorithmic simplicity—that is, no program branches or decisions—at the expense of the time-consuming, brute-force requirement that every gate must be evaluated whether it has changed or not. In contrast, the trade-off for the event-driven algorithm is between the additional bookkeeping to track changing components and the reduced time to evaluate only the gates whose inputs have changed. The architectural differences between the two approaches will be delineated in later examples.

Since the steps of the event-driven technique are common to a number of machines, a further description of the basic technique is in order. Starting with a new time step (a time when some output changes), the first operation removes the most recent event from the event queue. An event states that output X changed to a new value Y at this time. The new value is then assigned to the output as the current value. Using the net list (connectivity description), all components connected to the changed output (X) are scheduled for evaluation (net update phase).

After all events in the current time are processed, the scheduled components are evaluated (evaluation phase). Evaluation takes each scheduled component's input values and calculates a new output value. If the new output value differs from the old one, a new event is stored in the event queue with the appropriate time. Finally, the current time is advanced, and the process is repeated.

After examining the architecture of each machine, we'll specify its relative performance. Unfortunately, no universal performance specification exists for this purpose, any more than one exists for CPU speed comparisons. Wherever possible, we cite the performance increase achieved by the number of logic elements evaluated per second. To further complicate matters, some simulation engines specify speed in terms of two-input-gate equivalents derived from the actual simulation element (for example, a PLA), and the terms "events per second" and "gate evaluations per second" can be confusing. An event is all the activity required to process the change of a gate's output—that is, using the net list to schedule evaluation of all connected elements. A gate evaluation is only the time required to compute a gate's output based on its input values.

Boeing computer simulator. Figure 2 shows the system architecture of the Boeing Computer Simulator.* Its four logic processors operate independently, using an event-driven logic simulation algorithm. Two global communication mechanisms are used: the crossbar switch, which allows the host interface to access internal state information within each processor, and a communication

---

*An interesting historical note: In the early 1960's, Agnus McKay, the Boeing simulator's architect, developed many of the fundamental concepts used in simulation engines today. Ironically, both the patent lawyers and company management decided not to use or pursue the working prototype because it was too expensive and complicated. McKay, whose ideas were too far ahead of the times, was transferred, and the project came to an ignominious end.

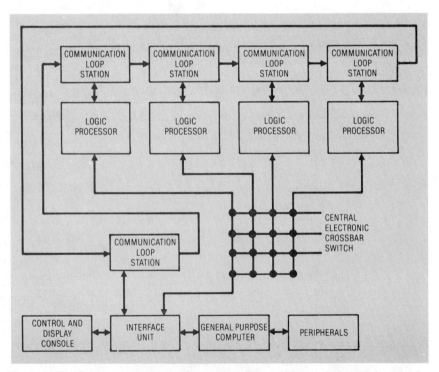

Figure 2. The Boeing computer simulator. (Reproduced by permission of the American Astronautical Society from an article by A. W. Van Ausdal in *Advances in Astronautical Sciences,* Vol. 29, p. 577.[8])

loop, which provides interprocessor communication and communication between the logic processors and the host interface.

The logic processor's architecture is composed of four major parts: paged memory consisting of 16K 48-bit words; a memory switch; three scratch pad memories (equation store (ES), delay (D), and event (E)); and the logic evaluation hardware. Before simulation can start, the host partitions the logic design and stores it in the logic processors. It generates and stores two basic types of information: equations which specify the operations of the primitive logic elements (up to 12 gates can be represented by one equation) and the connectivity information. The basic simulation flow is

(1) The ES (event queue) points to active equations.

(2) For each active equation, the delay value is reduced by the current time step.

(3) Equations whose delay time is reduced to zero are evaluated by propagating the value to all the connected components using the connectivity list, and the ES memory is updated (net update phase).

(4) For equations where the output value changes (evaluation phase),
- the equation output value field is updated and restored to memory;
- the corresponding logic delay value is fetched from the D scratch memory; and
- the new delay and logic value are stored in the E scratch memory.

During the simulation cycle, the smallest delay value is retained and used for the next time step.

The Boeing machine used all the hardware accelerator speed-up techniques except matching the interprocessor communication with the problem data. Instead, Boeing used a fixed communication ring. Constructing the system required roughly 28,000 TTL packages, 40K of fast memory, and 65K of core memory running at a clock speed of 10 MHz.

The system could simulate 36,000 elements (48K two-input gates) including gates, flip-flops, and one-shots specified in equation form. The only specific simulation measurement was stated in terms of a slow-down ratio of 800:1 for simulating a computer with over 4K primitives and a 1-MHz system clock. Using the SDR, a 20-percent gate activity, and the target system clock rate, the estimated performance would be 1M gate evaluations per second.

Tegas accelerator prototype. The Tegas accelerator, developed as an experimental prototype, is similar to the Boeing computer simulator in that it is an event-driven simulator designed for devices such as gates, flip-flops, counters, and memory.[8] The prototype was developed to directly accelerate the Tegas 5 (mode 2) primitives and design environment. Figure 3 shows the system block diagram. Distinguishing features of the system are the 672-bit-wide internal bus and single evaluation processor (many aspects of the machine are adapted from Barto[3]).

The event-driven algorithm is distributed among five hardware modules.

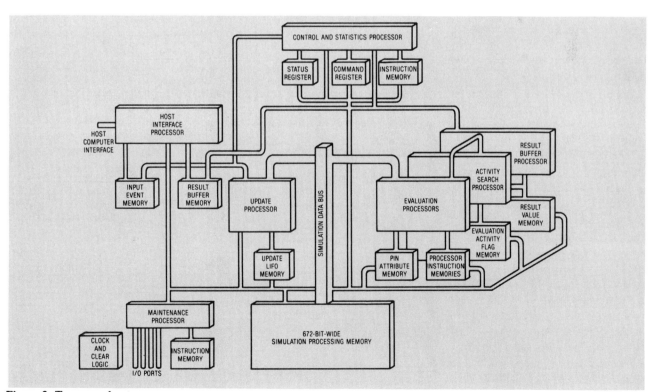

Figure 3. Tegas accelerator.

(1) *Control and statistics* coordinates all processes in the system and regulates simulation time and host/accelerator interaction.

(2) *Update* processes all signal updates and notifies the activity search module of the change.

(3) *Evaluation* is composed of a three-stage data-flow system. Stage one fetches all gate fan-ins. Stage two performs functional evaluation. Stage three schedules the spawned events and stores the results.

(4) *Activity search* maintains a list of active elements and queues descriptors for the evaluation module.

(5) *Result buffer* collects simulation results requested by the host.

A simulation-processing-memory word is composed of three major items: status/data fields, fan-outs (addresses of all driven gates), and fan-ins (addresses of all inputs to the current gate). Each logic element requires one word, but large fan-in/out devices may use up to 32 words. The status/data fields contain information like gate type, signal value (current, previous, and pending), delay before next output, and time the output last changed (allowing inertial delays and spike analysis).

In contrast to the Boeing machine, the Tegas machine speed-up technique uses only one evaluation processor. However, the five distinct hardware modules, used in conjunction with the wide memory, provide significant performance improvements. Physically, the modules are on nineteen 15 × 20-inch boards. The system is limited to 1M primitives or approximately 2.5M gates. Using a 250-ns system clock, performance measurements for a 64-bit ALU with 12-percent average gate activity demonstrated a 370X improvement over a Vax 11/780 running the standard Tegas 5, mode 2 simulator. (The software is usually rated between 1K and 2K primitives per second on the Vax). Other tests over a 10-percent to 20-percent gate activity range demonstrated 200X to 700X improvements. Converting the primitives to two-input gates yield a simulation rate of roughly 1M gates per second.

Hardware logic simulator. HAL, a hardware logic simulator developed at NEC, is an event-driven simulator capable of simulating both memories (RAMs, ROMs, PLAs) and logic blocks where a logic block is defined as a combinational element with 32 inputs and 32 outputs.[10,11] Figure 4 shows the system block diagram, which is composed of an interconnec-

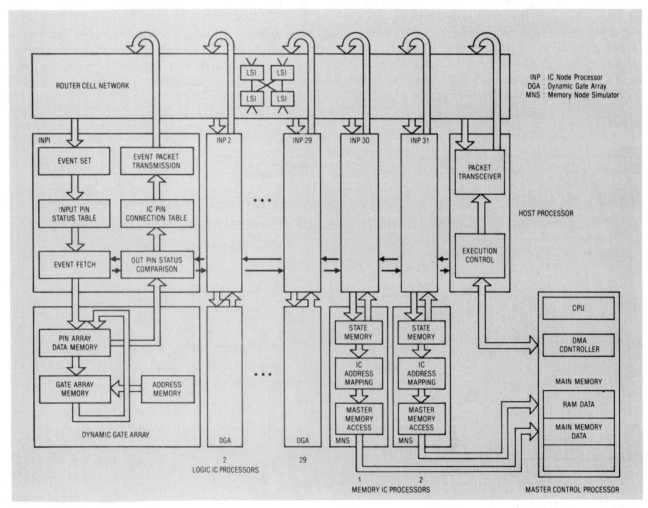

Figure 4. HAL architecture.[10]

August 1984

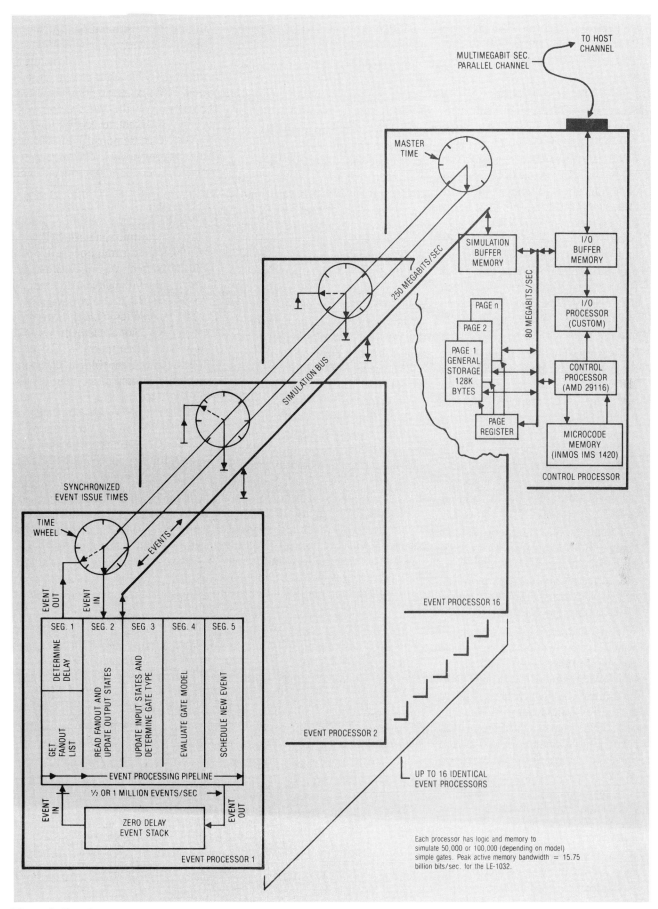

Figure 5. Zycad logic evaluator architecture. (Courtesy of Zycad.)

tion network and four main processor types—IC node, memory node, dynamic gate array, and master controller.

HAL requires both an IC node processor and the dynamic gate array processor to simulate a logic block. The IC node processor is an asynchronous data-flow-type custom processor that regulates all network communication, maintains the event queue, and maintains pin (net) values and connectivity. The dynamic gate array corresponds to the evaluation unit of other simulators: the inputs are the net logic values and block functionality; the output is the new net values. The evaluation is performed using table look-up techniques.

Memory simulation requires an IC node processor and memory node processor. The IC node processor does the event management. The memory node processor simulates a conventional RAM chip in that it simulates both memory control signals (read/write) and data I/O. All physical memory accesses are mapped into the host's address space.

The interconnection network is an omega, store-and-forward type with 32 input and output ports. The internal network nodes are two input/output custom ICs capable of transferring a data packet every 400 ns.

Using all the speed-up techniques except customizing interprocessor communications, HAL achieves its large performance due to its high-level primitives, which are similar to a 32-input and 32-output PLA. The only benchmarks published indicated a performance of six million primitive evaluations per second using 29 evaluation processors and two memory procesors.[10] Using a conversion of 200 two-input gates per 32 I/O primitives, this translates into 1.2 billion gate evaluations per second. If only 30 percent of the gates are active, the rate is 360M gate evaluations per second. System capacity is specified as 15K primitives or 3M gates using the 200X conversion. Memory limits are 2K memory nodes or 2M bytes.

Zycad logic evaluator. Zycad produces a family of event-driven simulation engines, all of which share a common mainframe that provides both expansion capability up to 16 parallel event processors and a control processor. Figure 5 shows the logic evaluator's bus-oriented system architecture.[12] The LE can simulate two basic types of elements: a memory element (RAM, ROM, PLA, etc.) and a three-input, one-output logic element whose function is specified by a truth table using a four-strength, three-value style simulation. The system also includes facilities to simulate pass-gate operation and to separately specify I/O delays for each element.

The control processor is responsible for four primary jobs: controlling the interface with the host processor, keeping the master simulation time used by all event processors, collecting simulation results for the host, and simulating all memory elements. Each event processor is constructed using a five-stage pipeline with a clock rate of 87.5 ns. The stages break the event-driven algorithm into: acquiring the fanout list and delay determination; reading fanout and updating output states; updating input states and retrieving model descriptions; evaluating the model; and scheduling the new events.

Like HAL, the LE uses every speed-up technique except customized interprocessor communication. However, both the interprocessor communication technique and the primitive level are different—omega network versus parallel bus and PLA versus three-input one-output gates, respectively. The LE uses ECL/TTL technology. Since the pipeline clock rate is 87.5 ns, the maximum possible gate evaluation rate is 11.4M per second per processor. For the LE1002, the capacity is 100,000 gates per processor. For the largest system, the 16-processor LE1032, the capacity is 1.6M elements (3.8M two-input gates). At 40M events per second, and assuming a fanout of 2.5 and a normal distribution of interprocessor communication, 60M gate evaluations per second is derived. A benchmark for a 5000-gate test, using one LE1002 evaluation processor, required 1.2 seconds to process 1.4M events for a processing rate of 1.2M events per second.

Megalogician. The Megalogician is a workstation composed of general-purpose computing components (80286, 80287, memory, disk, etc.) and a special-purpose simulation engine.[13,14] Figure 6 shows the Megalogician simulation hardware

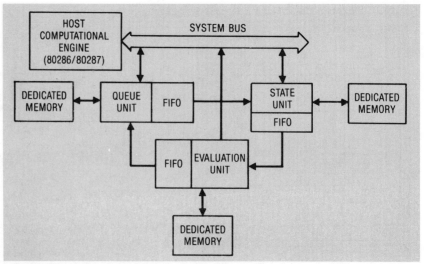

Figure 6. Megalogician architecture. (Reproduced by permission of Hayden Publishing Co. from "CAE Stations' Simulators Tackle One Million Gates," *Electronic Design*, Nov. 1983.[13])

block diagram. Each of three bit-slice engines connects directly to its own dedicated memory and to its two neighboring processors through a hardware queue forming a three-processor ring in a classical data-flow organization.

Using the event-driven simulation technique, the three processors divide the basic simulation tasks. One processor maintains the event queue, taking results from the evaluation processor and providing events in time sequence for the state processor. The state processor collects the net values specified by the host and maintains them, along with the connectivity information. Finally, the evaluation processor takes the logic element's input values and functions and generates the new output value.

Since each logic processor is fully programmable, there is no conceptual limit to the type of simulation the machine can perform. DSL, the Daisy simulation language, a three-value, four-strength simulator includes elements from simple gates to logic gates constructed from multiple input/output Boolean equations.* The models include specifications for separate rise and fall delays.

To speed up simulation, the Megalogician uses faster processors and microcoded processors for specific tasks and partitions the event-driven algorithm into three separate modules. Hardware-accelerated simulations using DSL demonstrate a 100X improvement over software running on an 80286. The hardware simulator is rated at 100,000 gate evaluations per second with a capacity of 64K primitives or 1M gates, but no benchmark was available.

Realfast simulation. Figure 7 shows the architecture of Valid Logic's Realfast simulation machine.[15] Using the event-driven technique, one processor is dedicated to event scheduling and one to evaluation. The inter-

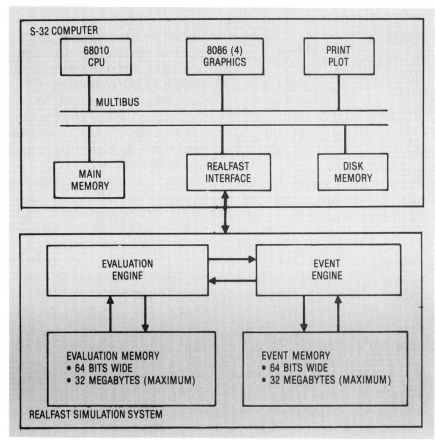

Figure 7. Realfast architecture. (Courtesy of Valid Logic.)

*The Daisy PMX provides additional modeling capability by using a physical chip instead of a software model in a simulation. It can handle both dynamic and static parts.

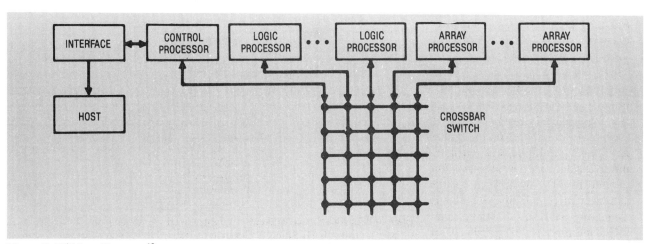

Figure 8. LSM architecture.[18]

nal bus and memory structure is 32 bits wide. Each processor, using the AMD 2900 bit-slice family, has a 250 ns system clock rate.

Since each processor has a fully writable program store, the simulation models in this machine are microprogrammable. Currently, the Scald environment provides models for simple gates, pass gates, registers, multiplexers, memories, etc.,* but this is not a fundamental hardware limitation. As in the other simulation machines, the simulation memory is initially loaded with element functionality and connectivity. For this machine, the simplest gate requires a 32-byte function descriptor and a 32-byte connectivity descriptor per output. More complicated functionality or connectivity requires incrementally more storage. Each element allows separate rise and fall delay specifications.

The Valid simulator, like the Megalogician, relies on three basic acceleration techniques: faster processors, microcoded processors for specific tasks, and two partitions of the event-driven algorithm. Its performance is 500,000 element evaluations per second, roughly a 500X improvement over the 68000-based Valid S32 computer running a software simulation. Using the maximum configuration, 32M bytes per processor, 1M elements or roughly 2.5M gates can be represented.

IBM simulation engines. IBM has built three generations of simulation engines: the logic simulation machine, LSM;[16-20] the Yorktown simulation engine, YSE;[21-23] and the engineering verification engine, EVE.[1] Since all three machines are based on original ideas proposed by John Cocke, Richard Malm and John Schedletsky, only the LSM is discussed in detail but differences between the machines are cited.

In contrast to the other simulation engines discussed, the IBM simulators are not event driven; they evaluate each logic or memory element during each simulation time step. For the LSM, a logic element is defined as a five-input function block capable of 1024 different functions with a separate rise and fall output delay where each input value may be logic 1, logic 0, undefined, or uninitialized. The LSM system architecture, shown in Figure 8, is composed of five primary elements: the host system provides the user interface; the control processor synchronizes the parallel processing elements and collects user output; the logic processor is the evaluation unit for logic elements; the array processor is the evaluation unit for memory elements; and the switch provides interprocessor communication through a 64×64 pipelined crossbar network.

Figure 9 shows a logic processor's architecture where the basic simulation cycle is

(1) the program counter advances;

(2) the 80-bit opcode initiates both operand and function table fetches;

(3) the function unit performs the evaluation;

(4) an intermediate memory stores the result until the delay time specified in the delay memory; and

(5) the loop repeats until the program counter reaches the end of instructions.

Since the simulator is compiled, not event driven, the program counter simply advances through all the operations and memory without any branches providing optimal pipeline operation. All decisions are resolved at compile time, allowing maximum speed-up in the accelerator. All the speed-up techniques are used, including interprocessor communication since the compiler uses its knowledge of the hardware configuration to remove any communication problems before the simulation starts.

Table 1 compares the three IBM hardware simulators. As the chronological progression—LSM, YSE, EVE—indicates, the basic trends are toward larger gate and memory capacities, faster I/O, and increased parallelism. The estimated number of active gate evaluations per second using 15-percent active primitives, 256 evaluation processors, and 80-ns primitive evaluation time per processor provides a meaningful comparison with the event-driven simulators. This yields a primitive evalua-

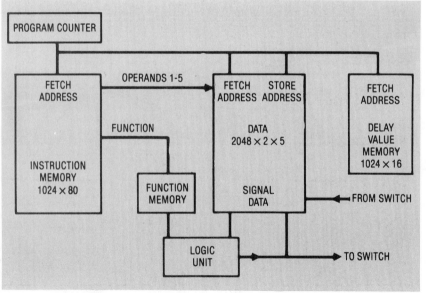

Figure 9. LSM logic processor element.[18]

---

*The Realchip modeling system can use a physical chip instead of a software functional simulation model during system simulation. Thus, designs using microprocessors, custom chips, gate arrays, etc., can be simulated by using the physical chip in the Realchip hardware, while the remaining portions of the system use the software simulator or simulation engine.

August 1984

tion rate of 480M primitives per second or 960M gate evaluations per second. The only specific benchmark is the EVE performance for a 500K-gate design, which required 49 minutes for either one or 10 simulated machine cycles and 66 minutes for 1M simulated machine cycles (the number of evaluation units was not specified).

Simulation engine comparison. Table 2 compares and summarizes the seven simulation engines. The noticeable performance difference between the IBM simulator and the others is due to the extensive parallelism of the 256 evaluation units. Since the IBM simulator uses the compiled approach and the others are event driven, an obvious question is whether an event-driven machine can achieve this level of performance. HAL's performance, roughly 40 percent of the IBM machine's, could clearly be increased by adding processors. However, HAL's performance advantage, measured in gate evaluations, is due primarily to its use of higher level primitives with a resulting loss of circuit detail—for example, timing accuracy. With the Zycad LE, simply using 256 processors would provide an equivalent gate evaluation rate if processor-to-processor communication were free—obviously not the case. Processor-to-processor communication requirements are patently different between the two simulation algorithms. In the compiled case, since all communication decisions are made before simulation begins, no communication "bottlenecks" can occur. But the event-driven tech-

Table 1.
Comparison of IBM hardware simulators.

Criteria	LSM	YSE	EVE
Gate inputs	5	4	4
Simulation Values	3	4	4
Hardware Interactive	NO	YES	YES
Nominal Delay	YES	NO	NO
Capacity (gates)	63K	1.0M	2.0M
Capacity (memory)	512KB	N/A	12.8M
O Speed (bytes/sec)	16.5K	5K	50K
Maximum Evaluation Processors	64	256	256
Processing Cycle Time	80 ns	80 ns	N/A

Table 2.
Summary comparison of simulators.

	Simulation Algorithm	Status	Modeling Level	Maximum Evaluation Units	Maximum Gates*	Active Gate* Evaluations/Second
Daisy Megalogician	Event	Product	RTL/Gate	1	1M	0.1M
Valid Realfast	Event	Laboratory Prototype	RTL/Gate	1	2.5M	0.5M
Tegas Accelerator	Event	Laboatory Prototype	RTL/Gate	1	2.5M	1M
Boeing Simulator	Event	Internal Product Dismantled	Equation	4	48K	1M
Zycad Logic Evaluator	Event	Product	Gate	16	3.8M	60M
NEC HAL	Event	Laboratory Prototype	PLA	31	3M	360M
IBM	Compiled	Internal Product	Gate	256	4M	960M

*A gate is defined as a two-input, one-output logic block.

nique, which determines all communication paths dynamically, has potential communication problems. Clearly, the Zycad bus design (3M events per second) would not be sufficient for system performance in the 1B-gate-per-second range. A difficult question is whether any network design can reasonably handle the event-driven communication requirements at the 1B-gate-per-second rate.

Examining the similarities shows that the simulation engines gain the most performance improvement from three techniques. The first, used by all the examples, divides the simulation algorithm into separate hardware modules. The second, used by four of the systems, divides the simulation task among separate evaluation processors. Finally, all the systems that achieve over 1M gate evaluations per second use a discrete processor design. In general, the performance is proportional to the amount and specialization of the hardware components.

Unfortunately, machine costs were not available. However, using known costs for the two commercial products and estimated costs for the prototypes resulted in an average cost of $.10 per gate evaluation per second. In contrast, using a $200K price for a Vax running at 2K gate evaluations per second, the rate is $100 per gate evaluation per second. The 1000X differential clearly illustrates the significant cost advantage of these engines.

Placement, routing, and design-rule-check engines

There have been a wide variety of approaches to building engines for placement, routing, and design rule checking. The speed-up techniques range from a hardware-implemented algorithm to multipurpose parallel processors. This section describes multipurpose machines used primarily for design rule checking and routing, plus two single-purpose machines—one for design rule checking and one for placement. Even though many routing machine's have been proposed[24,25] and some test routing chips have been made,[26] the discussion deals only with systems that have been prototyped.

Routing is a common application of five machines, and they all use variants of the Lee-Moore algorithm[27,28] illustrated in Figure 10. The objective of the routing algorithm—useful for printed circuit boards and ICs—is to find a path between two points through a maze (Figure 10a). The technique starts at either point and initiates an outward search for the destination using sequentially labeled constant radius circles (Figure 10b). The resulting figure is shaped like a diamond since it has a constant radius using only orthogonal lines. The wave front expands until the destination is hit, then a connecting path is located by following descending wave-front labels back to the starting point.

Distributed array processor. Figure 11 shows the basic system architecture of the ICL distributed array processor.[29] The DAP processor is not connected to the CPU through an I/O channel; instead, it is connected directly to the system memory bus and accessed like conventional memory allowing virtual memory processing. The array is composed of 64×64 one-bit processing elements (Figure 12) connected in a rectangular lattice providing each PE with direct communication to its four orthogonal neighbors. Additionally, each PE has row/column data connections. The array operates in an SIMD fashion where all processors perform the same instruction synchronously at a 200-ns cycle. Within

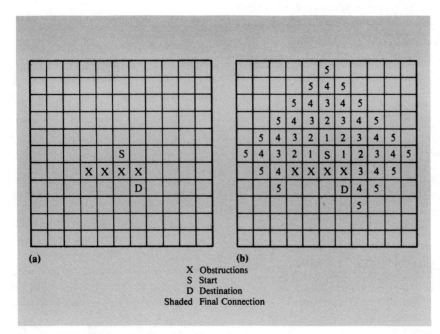

Figure 10. Lee-Moore routing algorithm.

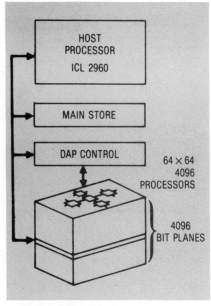

Figure 11. The distributed array processor system architecture.[30]

each processor are its own data store, data multiplexers, and ALU composed of three registers: A, an activity register that provides data-dependent operation; Q, a one-bit accumulator; and C, a carry register.

Adshead[30,31] states that the DAP has been programmed for routing and that algorithms for placement, timing verification, simulation, design rule-checking, and test generation are being developed. (Other non-CAD applictions are described by Gostic.[32]) The routing strategy mapped a two-dimensional plane onto the gridded sheets of the DAP. For example, an 8K CMOS gate array using a 1001×1368 grid was mapped onto 352 64×64 sheets. Then, the Lee-Moore routing algorithm was used for routing. Basically, one processor is responsible for manipulating each point on the 1001×1368 grid plane. The match between the DAP processors connected in a two-dimensional mesh and the two-dimensional wave-front propagation of the Lee-Moore algorithm provides most of the performance advantage. This incorporates two speed-up techniques: partitioning the problem data between multiple processors and optimizing the interprocessor communication.

For the 8K gate array, the routing time on a 1-MIP machine, equivalent to a Vax 11/780, ranged from 8 to 15 hours. Using the DAP, the required time was 30 minutes for loading and 8 to 15 minutes for layout—a 60X improvement over the conventional software approach.

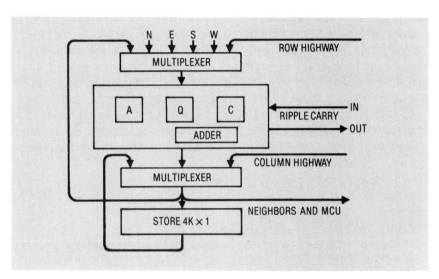

Figure 12. DAP processing element.[30]

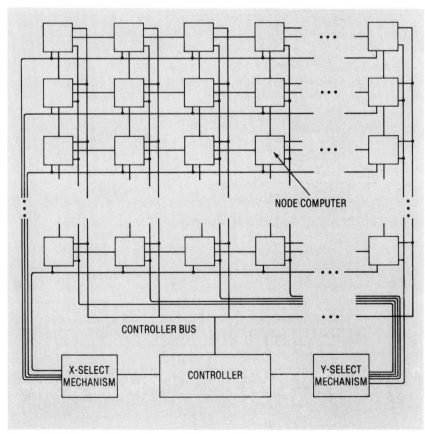

Figure 13. Wire routing machine architecture.[33]

Wire routing machine. Figure 13 shows the system architecture of the wire routing machine developed at IBM Yorktown.[33,34] The WRM's basic system organization is MIMD, where each processor independently executes its own instruction stream on its own data. Figure 14 shows the basic connections of each processing node. Important aspects of the system are

- The system is an 8×8 processor array.
- Each processing node is a 4-MHz Z80 microprocessor connected to 15K bytes of memory.
- All data paths are 8 bits wide.
- Each node is directly connected to its four orthogonal neighbors through a local neighbor bus.
- Processors located on the edge of the array are connected to the processor on the opposite edge of the array.
- The global controller can in-

dividually access any node by using X-Y select lines.

The machine's fundamental architecture is general purpose since it is completely programmable. Nair et al.[33] and Hong et al.[35] describe the techniques for routing a two-layer gate array. They partitioned the task into two parts: global wiring, where the connections are allocated to wiring channels, and exact embedding, where the wires are assigned to a specific track. For both parts, they extended the basic Lee-Moore algorithm to add complex cost functions for occupying a cell. Similar to the DAP, problems larger than the physical number of processor are mapped into the 8×8 array. Also similar to the DAP, the rectangular interconnections provide a significant performance advantage for the Lee-Moore-type algorithms using near-neighbor information. In contrast to the DAP, where all processors operate synchronously, the WRM implemented the wave-front expansion so that active cells located on the frontier of the expanding wave front were processed asynchronously throughout the array, maximizing the number of processing elements doing useful work at each instant.

The system performance for a 19×32 cell problem using the 8×8 prototype was approximately one minute, whereas the same algorithms programmed in PL/I required 45 seconds on a 4.5-MIP 3033 machine or roughly 3.4 minutes on a 1-MIP machine. Using the newer technologies available today and slightly modifying the architecture to provide better communication to the neighbor processors should easily permit a performance increase of over 50 times.[35]

Hardware routing kernel. Damm and Gethoffer[36] describe another approach to hardware-assisted routing using the Lee-Moore algorithm. However, the fundamental approach is different from the DAP or WRM since the routing algorithm is partitioned between a hardware routing kernel and a conventional LSI-11/23 CPU. Figure 15 shows the system block diagram. Kernel operations include calculating the cell cost function and two-dimensional accessing of wave-front values.

In contrast to the other accelerators, this approach uses only two speed-up techniques: separation of the algorithm into separate hardware modules and a custom designed evaluation module. Besides the LSI 11/23 processor, a microprogrammed processor and data path allows efficient access to the 2-D routing cell values and efficient bookkeeping of the Lee-Moore algorithm's active frontier cells. (See Hoel[37] for variations on routing algorithms.)

The routing kernel portion of Figure 15 has been implemented and consists of 160 ICs. Using CAD-Calay's system, performance improved approximately 5.5 times over a software process running on a 250-KIP LSI-11/23, or roughly 1.4 times over a 1-MIP machine. The benchmark is for a printed circuit board. Performance estimates for the microprogram and cell memory additions range from 5 times to 15 times for the complete system over a 1-MIP-machine software solution.

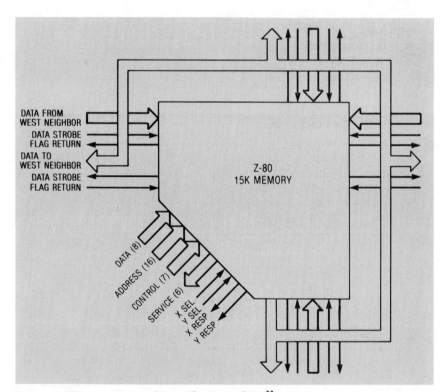

Figure 14. Wire routing machine node connections.[33]

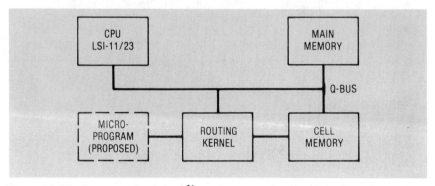

Figure 15. Hardware routing kernel.[36]

Virtual bit-map processor. The VBMP,[38] called a SAM chip in earlier papers,[39] implements a data structure—a three-dimensional array of single binary bits or Booleans—in hardware. The fundamental idea driving the VBMP design is that a bit-map data structure can be used in a wide variety of CAD applications providing performance improvements. Example algorithms using bit data structures include design rule checking, routing, and bit vector manipulations.

Two primary observations underlie the hardware implementation: (1) typical CAD problems require very large data structures, usually over $1K \times 1K \times 16$ bits; (2) with current technology, it is not feasible to assign one processor to each data structure point. Based on these observations, the VBMP system architecture (Figure 16) consists of five primary elements:

(1) *Host.* The host is responsible for all operations not performed within the bit-map data structure. Operations include specifying bit-map operations and data I/O.

(2) *PE array.* The processing element array is a rectangular, interconnected array of one-bit processing elements performing in SIMD fashion. Each PE is directly connected to two large register stores.

(3) *Edge registers.* The edge registers provide a temporary data storage for data communication between the host processor and the PE array.

(4) *R/C generator.* Row/column generation provides row/column enable signals for subregion operations.

(5) *Controller.* The controller performs SIMD control for the PE array, regulates I/O between the host and VBMP, maintains the configuration registers for the virtual problems dimensions, and regulates instruction execution over the virtual address space.

VBMP operations include Boolean operations between bit-map layers, wired OR on both rows and columns, and the neighbor operation, which allows each processor to take the logical OR of any subset of the values of its four orthogonal neighbors and its own value. The neighbor operator, in conjunction with the Boolean operators, allows shift, grow, and shrink operations. Using just the Boolean instructions, bit-serial arithmetic can also be performed.

Although the VBMP and DAP speed-up techniques and architectures are similar, there are two major differences. In the DAP, the algorithm must handle partitioning of the problem data into the 64×64 processor array and commensurate boundary problems. In the VBMP, the hardware automatically handles problems larger than the physical number of processors with no boundary problems. The second difference is that the processor array is independent of the host, allowing simultaneous operation, in contrast to the DAP.

The current VBMP prototype is constructed from standard TTL components implementing a four by four PE array using roughly 420 ICs and an M68000 as the host processor. Memory capacity allows virtual problems ranging from $256 \times 256 \times 8$ to $4 \times 4 \times 32K$. Two algorithms have been implemented on the VBMP: local, grid-based design rule checking[40] and Lee routing. Using the current prototype and a small example not requiring partitioning, design-rule-checking performance improvements ranging between 10 and 100 were observed over conventional software implementations on a Vax 11/780. Estimated performance improvements[38] for one-layer routing are a factor of 10. Changing from the TTL processor implementation to a custom IC allowing construction of a 32×32 array would allow a 64X improvement.

Cytocomputer. The Cytocomputer family, originally developed for image processing,[41,42] has been programmed for both design rule checking and routing problems.[43-45] Based on a three-dimensional bit-map image, Boolean transformations between layers and Boolean operations between the eight nearest orthogonal and diagonal neighbors of all bit positions are possible.

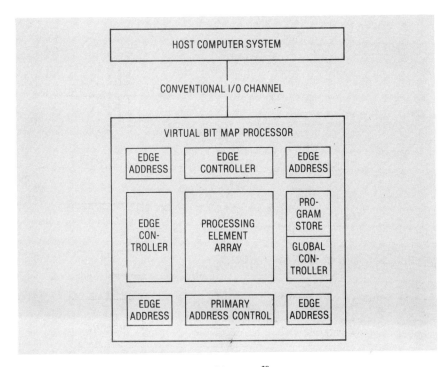

Figure 16. Virtual bit-map processor architecture.[38]

The architecture uses a pipelined approach with identical stages (Figure 17). In operation, the raster line image of problem data flows through a sequence of stages, each of which performs a transformation. The buffers in each stage hold two image lines, allowing two raster lines in Y and one pixel in X to logically separate successive stages. The subarray storage allows the subarray processor access to nine cell values, eight neighbors plus a center. The subarray processor performs all the stage's logical operations.

The Cytocomputer's basic speedup techniques are similar to those of the array processors DAP, WRM, and VBMP. It uses nearest-neighbor information matching the communication paths for algorithms like routing and design rule checking. However, the pipeline versus the rectangular array architecture is a major difference—with all the problems and advantages of each. The pipeline's greatest advantage, as well as the array's potential disadvantage, is the ease of adding additional stages, which, ideally, increase the performance linearly (up to a limit). Conversely, the major pipeline disadvantage is that data-dependent branches cause pipeline breaks.

Using a TTL prototype running at a clock of 0.5 MHz, a small design-rule-checking example demonstrated the approach's viability. Specifically, a local three-grid-width check performed on a 64×4096 pattern required 2.3 seconds of pipeline processing and 2.8 seconds to transfer the bit map using a Vax 11/780 host. Projections for complete design rule checking on a $4K \times 4K$ chip ranged from 737 seconds to 1099 seconds for a 150- and 250-step algorithm, respectively, assuming a 10-stage pipeline. This represents roughly a 10X improvement over a Vax 11/780 software implementation.

The Lee-Moore routing technique was implemented for both its excellent match to the neighbor instruction and path-finding ability. For a 200-net problem on a 512×512 grid, a three-stage Cytocomputer improved performance from 3 times to 10 times, depending on system load, over a Vax 11/780 software solution.

Window processor. The basic technique of the window processor, a design-rule-checking engine developed by Seiler,[46-48] is similar to that of Baker's software checker.[49] It performs in three basic steps: rasterization, local checks, and error processing. The system input format is an edge list that represents rectangles by corner coordinate pairs. The rasterization unit converts the edge-

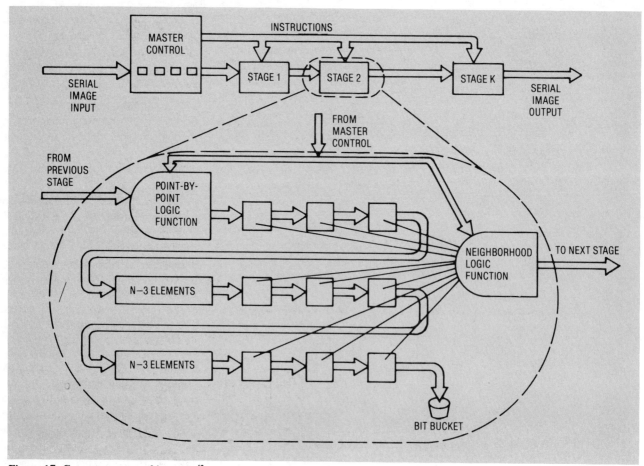

Figure 17. Cytocomputer architecture.[42]

list representation into bit-map representation on a raster line-by-line basis. The raster line data is input to the checking hardware, which builds four by four grid windows and performs the actual checks. Each window check is performed at each grid position in the layout. Rules requiring larger windows are converted into a sequence of smaller window operations. The error-processor run length encodes the output bit stream for transfer back to the host. Using the window processor, 45- and 90-degree checks can be made accurately.

Figure 18 shows the system pipelined architecture matching the algorithm data flow. The additional controller component regulates both I/O with the host and operation of the processing pipeline. Figure 19 shows the internal structure of the local checking hardware. The width hardware has two primary functions: performing width checks and shrinking the input pattern. In conjunction with the derived mask generator, the width hardware can perform arbitrary size, spacing, width, and overlap checks. The edge hardware performs checks between polygon edges from different layers allowing, for example, reflection-rule-style checks.

The window processor, like the hardware routing kernel, uses a speed-up technique that separates the basic algorithm between the host processor and custom accelerator. The most time-consuming portions of the algorithm, the "inner loop" calculations, are implemented in the special hardware module while the host does all the other odd jobs.

The current prototype is a combination of TTL and custom components. Of the 133 parts, 28 are custom in four different types. The controller is an 8086 running at 4 MHz; the rest of the pipeline runs at 2 MHz. Performance measurements show a range of 10X to 245X improvement over Vax 11/780 software approaches. (The slowest software is capable of checking an angle geometry.) The second version, designed with only two types of custom ICs, allows faster cycle times and greatly reduces chip count.

Module interchange placement machine. Figure 20 shows the architecture of a machine designed for accelerating module placement on either printed circuit boards or ICs.[50] For example, suppose the task is to select the "optimal" position for each IC on a printed circuit board

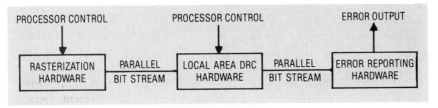

Figure 18. Window processor for DRC.[47]

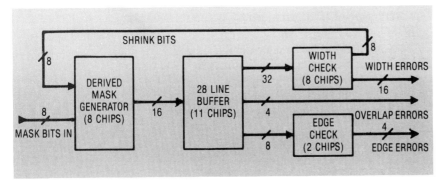

Figure 19. Local checking hardware architecture.[47]

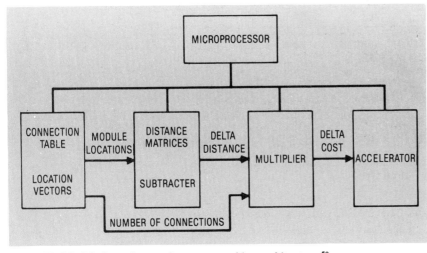

Figure 20. Module interchange placement machine architecture.[50]

where optimal may be defined by many different metrics. After analyzing various placement algorithms, including pairwise interchange, neighborhood interchange, and force-directed interchange, a common time-consuming operation is identified: the evaluation of a particular component placement. In each technique, the modules are given an initial position, then changed or swapped, and the new position is evaluated using various criteria such as minimum weighted wire length. From the quadratic assignment problem, the evaluation requires a subtraction, a multiplication, and an addition repeated over a cost matrix. This type of computational flow is an excellent match to the pipelined architecture shown.

This machine is another example of separating the algorithm into two parts: a time-consuming "inner loop" calculation and the other tasks. A special module is designed for the "inner loop," and the host handles the rest.

The system is currently attached to a HP3000 computer through an M68000 microprocessor, running at a 125-ns clock, using 280 chips. The machine can handle up to 256 modules in the current prototype, which is limited only by the amount of RAM in the various matrices. Using the prototype, various placement results were measured. The average time required to mathematically evaluate a particular swap was approximately 6 μs, a 10X improvement over a 5-MIP machine or roughly a 50X improvement over a 1-MIP machine. Using a pair interchange technique, the average time per interchange was 45 μs, yielding a system overhead of approximately 39 μs per swap. The excessive system overhead could be significantly reduced by using a faster microprocessor interface with the host.[50] Assuming an improved interface, the average time per swap could be 12.5 μs or 80K pair interchange evaluations per second. Thus, a complete pair interchange cycle could be evaluated for a 3000-cell gate array in roughly a minute, plus starting and ending I/O time; no comparative software benchmark is provided.

Non-simulation SPE summary. Table 3 compares the non-simulation special-purpose engines. Although few additional details are available on these primarily lab-prototypes,

Table 3.
Non-simulation, special-purpose engine comparison.

	Application	Status	Basic Architecture	Data Path Width	Number Evaluation Processors	Calculated Vax 11/780 Speed-Up
CAD-Calay	Routing	Product	Pipeline	32	1	1.4
WRM	Routing	Laboratory Prototype	Array	8	8×8	3.4
Cytocomputer	Multiple	Laboratory Prototype	Pipeline	8	3	10
VBMP	Multiple	Laboratory Prototype	Array	1	4×4	50
Placement Machine	Placement	Laboratory Prototype	Pipeline	16	1	50
DAP	Multiple	Product	Array	1	64×64	60
Window Processor	Design Rule Checking	Laboratory Prototype	Pipeline	16	1	100

August 1984

some observations can be made. It seems contradictory that the largest performance improvement, shown in the window processor, used only one evaluation processor. But this was the only system using custom ICs. The architecture matched the algorithm flow, exploiting the maximum parallelism inside the algorithm and intraprocessor communication. Furthermore, the performance improvement doesn't seem to correlate with the number of processors. The difference lies with both the complexity of the individual processors—some are only one bit—and the algorithm implementations. Since these machines are prototypes, the studies are designed to determine approach feasibility, not to generate production solutions. A generic accelerator problem is that all the code is very special purpose and difficult to generate. The accelerators never existed before, so no one is an expert at their programming.

Again, system costs providing a cost-per-performance metric were not available. For example, the CAD-Calay's system, using only 160 ICs and a low-cost host, achieved better performance than a 1-MIP machine. But even without exact figures, the difference ranges from 10 times to 1000 times better for a hardware accelerator than for a conventional software solution. Again, SPEs provide a useful advantage.

Even though simulation, routing, design-rule-checking, and other special-purpose engines provide cost/performance advantages over conventional software solutions, conflicting forces direct their future. Factors favoring new, faster accelerators are increasing design size, increasing design complexity, and decreasing hardware costs. The opposing forces are a combiation of more powerful CPUs, with more MIPs and memory, and general multiprocessor machines aided by better algorithms and design methodologies. The new accelerator commercial offerings, especially in the growing simulation area, represent the current direction.

Besides cost/performance advantages, accelerators also offer practical advantages. For example, mainframe computer cost, maintenance, software support, floorspace, etc., may be unacceptable in a microprocessor-based workstation environment. Adding nodes and partitioning the problem still may not provide adequate performance. Here, an accelerator could be defined as a device (for example, a card for the workstation or a small box) that provides mainframe performance without the related problems. Similarly, designers in a Vax environment may require better performance yet be unwilling to use other, faster mainframes. Again, an accelerator is a possible solution, but an accelerator suitable for the workstation environment will, most likely, be inadequate in the mainframe environment. The performance and capacity needs differ significantly.

A crucial aspect of special-purpose engines is that they are only one link in the total problem solution. Perhaps no commercial routing engines are available because routing alone doesn't solve the physical design problem. The product might need partitioning, placement, and routing to be viable. For simulation engines, the whole solution includes converting the internal design description into simulator elements. Depending on design size and simulation length, model build time might totally dominate. Before deciding that an accelerator is a solution, make sure it solves the real problem—not just the perceived problem. ∎

Acknowledgments

This work was sponsored by the Army Research Office, contract number DAAG29-80-K0046, and by a Tektronix industrial grant.

Many people from numerous companies and universities have helped provide and check this information. The effort by companies releasing new products who were willing to provide architectural details are also appreciated.

References

1. L. N. Dunn, "IBM's Engineering Design System Support for VLSI Design and Verification," *IEEE Design & Test of Computers,* Vol. 1, No. 1, Feb. 1984, pp. 30-40.

2. A. R. McKay, "Comment on 'Computer-Aided Design: Simulation of Digital Design Logic,' " *IEEE Trans. Computers,* Vol. C-18, No. 9, Sept. 1969, pp. 862.

3. R. L. Barto, *A Computer Architecture for Logic Simulation,* PhD dissertation, Univ. of Texas, May 1980.

4. M. Abramovici, Y. H. Levendel, and P. R. Menon, "A Logic Simulation Machine," *IEEE Trans. Computer-Aided Design of Integrated Circuits and Systems,* Vol. CAD-2, No. 2, Apr. 1983, pp. 82-94.

5. W. J. Dally, "The MOSSIM Simulation Engine Architecture and Design," Tech. Report 5123:TR:84, California Institute of Technology, 1984.

6. M. Fatemi, Scientific Machines Corp., private communication, Mar. 1984.

7. B. H. Scheff and S. P. Young, *Gate-Level Logic Simulation,* Prentice-Hall, Englewood Cliffs, N.J., 1972, Ch. 3.

8. A. W, VanAusdal, "Use of the Boeing Computer Simulator for Logic Design Confirmation and Failure Diagnostic Programs," *Advances in the Astronautical Sciences,* J. Vagners, ed., American Astronautical Society, June 1971, pp. 573-594.

9. J. R. Lineback, "Logic Simulation Speeded with New Special Hardware," *Electronics,* June 1982, pp. 45-46.

10. N. Koike et al., "A High Speed Logic Simulation Machine," *Digest of Papers Compcon Spring 83,* IEEE Computer Society, Feb. 1983, pp. 446-451.

11. T. Sasaki et al., "HAL; A Block Level Hardware Logic Simulator," *Proc. 20th ACM/IEEE Design Automation Conf.,* June 1983, pp. 150-156.

12. N. van Brunt, Zycad Corp., private communication, Mar. 1984.

13. "CAE Stations' Simulators Tackle 1 Million Gates," *Electronic Design,* Nov. 1983.

14. B. Paseman, Daisy Systems Corp., private communication, Mar. 1984.

15. B. Harding, Valid Logic, private communication, Mar. 1984.
16. D. Burrier, "The Logic Simulation Machine," Tech. Report LSM-0002, IBM Los Gatos Laboratory, July 1982.
17. J. K. Howard, L. Malm, and L. M. Warren, "Introduction to the IBM Los Gatos Logic Simulation Machine," *Proc. IEEE Int'l Conf. Computer Design: VLSI in Computers,* Oct. 1983, pp. 580-583.
18. T. Burggraff et al., "The IBM Los Gatos Logic Simulation Machine Hardware," *Proc. IEEE Int'l Conf. Computer Designs: VLSI in Computers,* Oct. 1983, pp. 584-587.
19. J. Kohn et al., "The IBM Los Gatos Logic Simulation Machine Software," *Proc. IEEE Int'l Conf. Computer Design: VLSI in Computers,* Oct. 1983, pp. 588-591.
20. J. K. Howard et al., "Using the IBM Los Gatos Logic Simulating Machine," *Proc. IEEE Int'l Conf. Computer Design: VLSI in Computers,* Oct. 1983, pp. 592-594.
21. G. F. Pfister, "The Yorktown Simulation Engine: Introduction," *Proc. 19th ACM/IEEE Design Automation Conf.,* June 1982, pp. 51-54.
22. M. M. Denneau, "The Yorktown Simulation Engine," *Proc. 19th ACM/IEEE Design Automation Conf.,* June 1982, pp. 55-59.
23. E. Kronstadt and G. Pfister, "Software Support for the Yorktown Simulation Engine," *Proc. 19th ACM/IEEE Design Automation Conf.,* June 1982, pp. 60-64.
24. M. A. Breuer and K. Shamsa, "A Hardware Router," *J. Digital Systems,* Vol. 4, No. 4, 1980, pp. 393-408.
25. A. Iosupovicz, "Design of an Iterative Array Maze Router," *Proc. IEEE Int'l Conf. Circuits and Computers,* Oct. 1980, pp. 908-911.
26. C. R. Carroll, "A Smart Memory Array Processor for Two Layer Path Finding," *Proc. Second Caltech Conf. Very Large Scale Integration,* C. Seitz, ed., Caltech Computer Science Dept., Jan. 1981, pp. 165-195.
27. C. Y. Lee, "An Algorithm for Path Connections and Its Applications," *IRE Trans. Electron. Computers,* Vol. EC-10, Sept. 1961, pp. 346-365.
28. E. F. Moore, *Shortest Path through a Maze,* Harvard University Press, 1959, pp. 285-292.
29. P. M. Flanders, D. J. Hunt, and S. F. Reddaway, *High Speed Computer and Algorithm Organization,* Academic Press, New York, 1977.
30. H. G. Adshead, "Employing a Distributed Array Processor In a Dedicated Gate Array Layout System," *IEEE Int'l Conf. Circuits and Computers,* Sept. 1982, pp. 411-414.
31. H. G. Adshead, "Towards VLSI Complexity: The DA Algorithm Scaling Problem: Can Sepcial DA Hardware Help?" *Proc. 19th ACM/IEEE Design Automation Conf.,* June 1982, pp. 339-344.
32. R. W. Gostic, "Software and Algorithms for the Distributed-Array Processors," *ICL Technical J.,* Vol. 1, No. 2, May 1979, pp. 116-135.
33. R. Nair et al., "Global Wiring on a Wire Routing Machine," *Proc. 19th ACM/IEEE Design Automation Conf.,* June 1982, pp. 224-231.
34. S. J. Hong and R. Nair, "Wire-Routing Machines—New Tools for VLSI Design," *Proc. IEEE,* Vol. 71, No. 1, Jan. 1983, pp. 57-65.
35. S. J. Hong, R. Nair, and E. Shapiro, *A Physical Design Machine,* Academic Press, New York, 1981, pp. 257-266.
36. E. Damm and H. Gethoffer, "Hardware Support for Automatic Routing," *Proc. 19th ACM/IEEE Design Automation Conf.,* June 1982, pp. 219-223.
37. J. H. Hoel, "Some Variations of Lee's Algorithm," *IEEE Trans. Computers,* Vol. C-25, No. 1, Jan. 1976, pp. 19-24.
38. W. T. Blank, *A Bit Map Architecture and Algorithms for Design Automation,* PhD dissertation, Stanford, 1982.
39. T. Blank, M. Stefik, and W. van Cleemput, "A Parallel Bit Map Processor Architecture for DA Algorithms," *Proc. 18th ACM/IEEE Design Automation Conf.,* June 1981, pp. 837-845.
40. C. A. Mead and L. A. Conway, *Introduction to VLSI Systems,* Addison-Wesley, Reading, Mass., 1980.
41. R. M. Lougheed and D. L. McCubbrey, "The Cytocomputer: A Practical Pipelined Image Processor," *Proc. Seventh Annual Symp. Computer Architecture,* IEEE/ACM, May 1980, pp. 271-278.
42. R. M. Lougheed, D. L. McCubbrey, and S. R. Sternberg, "Cytocomputer: Architecture for Parallel Image Processing," *Workshop on Picture Data Description and Management,* IEEE, August. 1980.
43. R. A. Rutenbar, T. N. Mudge, and D. E. Atkins, "A Class of Cellular Architectures to Support Physical Design Automation," Tech. Report CRL-TR-10-83, Univ. of Michigan, 1983.
44. R. A. Rutenbar, T. N. Mudge, and D. E. Atkins, "Wire Routing Experiments on a Raster Pipeline Subarray Machine," *IEEE Int'l Conf. Computer-Aided Design,* Sept. 1983, pp. 135-136.
45. R. A. Rutenbar, T. N. Mudge, and D. E. Atkins, "A Class of Cellular Architectures to Support Physical Design Automation," *IEEE Trans. Computer-Aided Design of Integrated Circuits and Systems,* to be published, 1984.
46. L. Seiler "Special Purpose Hardware for Design Rule Checking," *Proc. Caltech Conf. VLSI,* Jan. 1981.
47. L. Seiler, "A Hardware Assisted Design Rule Check Architecture," *Proc. 19th ACM/IEEE Design Automation Conf.,* June 1982, pp. 232-238.
48. L. Seiler, A Hardware Assisted Methodology for VLSI Design Rule Checking, PhD dissertation, MIT, 1984.
49. C. M. Baker, "Artwork Analysis Tools for VLSI Circuits," master's thesis, MIT, May 1980.
50. A. Iosupovici, C. King, and M. A. Breuer, "A Module Interchange Placement Machine," *Proc. 20th ACM/IEEE Design Automation Conf.,* June 1983, pp. 171-174.

Tom Blank is a research associate at Stanford University and has built a prototype hardware accelerator. Between his undergraduate work at the University of Washington and the completion of his PhD in EE from Stanford, he has worked as a production engineer for Hewlett-Packard and as an independent consultant.

His current mailing address is ERL/CISA, Stanford University, Stanford, CA 94305, or Arpanet, TBLANK @ Su-Sierra.

A Switch-Level Timing Verifier for Digital MOS VLSI

John K. Ousterhout

Computer Science Division
Electrical Engineering and Computer Sciences
University of California
Berkeley, CA 94720
415-642-0865

Abstract

Crystal is a timing verification program for digital nMOS and CMOS circuits. Using the circuit extracted from a mask set, the program determines the length of each clock phase and pinpoints the longest paths. Crystal can process circuits with about 40000 transistors in about 20-30 minutes of VAX-11/780 CPU time. The program uses a switch-level approach in which the circuit is decomposed into chains of switches called *stages*. A depth-first search, with pruning, is used to trace out stages and locate the critical paths. Bidirectional pass transistor arrays are handled by having the designer tag such structures with *flow control information*, which is used by Crystal to avoid endless searches. Delays are computed on a stage-by-stage basis, using a simple resistor-switch model based on *rise-time ratios* (a measure of how fully turned-on the transistors in the stage are). The delay modeler executes 10000 times as fast as SPICE, yet produces delay estimates that are typically within 10% of SPICE for digital circuits.

The work described here was supported in part by the Defense Advanced Research Projects Agency (DoD) under Contract No. N00039-84-C-0107

A Switch-Level Timing Verifier   February 20, 1985

1. Introduction

The switch-level approach has been popularized recently by Bryant and others [1,2] as a simulation model for MOS digital systems. It treats a circuit as a collection of bidirectional switches, rather than as a collection of unidirectional Boolean gates. Each transistor is modeled as a perfect switch in series with a resistor, and the circuit operates by opening and closing switches to establish electrical paths of varying strengths between nodes. The switch-level model is conceptually simple, computationally efficient, and flexible enough to handle a wide variety of MOS circuit constructs that are difficult or impossible to model using the Boolean approach. Simulation programs based on this model are widely used in the university and industrial design communities.

This paper describes an application of the switch-level model to timing verification, and presents the implementation of Crystal, a timing verifier that has been used successfully in the design of several large MOS circuits. Crystal's input is an nMOS or CMOS circuit description, usually extracted from mask layout. The circuit is described in terms of transistors of different types and sizes, along with first-order estimates of interconnect resistance and capacitance. Crystal determines the length of each clock phase and identifies the performance-limiting paths. The switch-level approach, in combination with simple delay models, makes the analysis fast: circuits with 40000-50000 transistors can be processed in 20-30 minutes of CPU time (5-10 minutes per clock phase) on a VAX-11/780.

Crystal, like all timing verifiers, uses a value-independent approach. This makes it quite different from simulation; both its strengths and its weaknesses stem from this difference. In simulation, a specific set of input signals is applied to a circuit. The simulator predicts the functional behavior of the circuit so that the designer can see if it matches the desired behavior. In timing analysis, the only goal is to see if the circuit meets its timing specifications; since the function of the circuit is not being tested, specific signal values are largely irrelevant. Timing verification is roughly equivalent to simulating the circuit for a single clock cycle with all possible signal values at the same time. The timing verifier attempts to find a combination of values that results in the worst possible timing behavior, and it reports this to the designer.

Crystal uses a switch-level approach to locate the slowest paths. The circuit is decomposed into *stages*, each of which is a chain of transistors leading from a voltage source (such as *Vdd* or *Ground*) to a transistor gate or output. Then Crystal searches for sequences of stages, each one triggering the next, that result in worst-case behavior. Sections 2 and 3 describe the basic mechanisms.

The value-independent approach provides the main advantage of timing verification over simulation. By considering all possible signal values at once, a timing verifier is guaranteed to locate any performance bottlenecks in a single run. In contrast, the effectiveness of a set of simulations depends on the selection of input data: pathological conditions may be undetected if they aren't triggered by the particular inputs fed to the simulator. Since timing verification only involves simulating a single clock cycle, it can generally be completed much more quickly than an extensive set of simulations.

The value-independent approach is also responsible for the main difficulty in timing verification. When a timing verifier ignores specific signal values, it may report critical paths that can never occur under real operating conditions. These false paths tend to camouflage the real problem areas, and may be so numerous that it is computationally infeasible to process them all. In practice, all timing verifiers include a few mechanisms that the operator can use to restrict the range of values considered by the program, usually by fixing certain nodes at certain values. This process is called *case analysis*; it is used to provide enough information to the timing verifier to eliminate false critical paths. Sections 4 and 5 illustrate the false path problem and describe the mechanisms provided in Crystal for case analysis. In addition to the standard fixed-value approach, Crystal provides a novel mechanism called *flow control*, which is used to handle pass-transistor structures by controlling information flow through switches.

Case analysis must be used with caution. When the user specifies particular values, he restricts the timing verifier from considering certain possibilities; this may cause critical paths to be overlooked. Case analysis generally requires several different runs to be made, with different values each run, in order to make sure that all possible states have been examined (for example, one run might be made with a particular signal forced to 0, and another run with it forced to 1). As the number of cases increases, timing verification becomes more and more like simulation, both in efficiency and in effectiveness. Ideally, the timing verifier should be provided with only the minimum amount of specific information needed to eliminate false paths. This leaves the program as much flexibility as possible to chase out critical paths that might not have been foreseen by the operator.

An important part of a timing verifier is the portion of the program that estimates delays through the circuit. The traditional approach to delay modeling is the one used in circuit simulation programs such as SPICE [8]: circuits are modeled with collections of differential equations, and the equations are solved to predict the delays. Although this approach is accurate, it is too inefficient for timing verification of a large circuit, where hundreds of thousands of delay calculations may have to be made. Crystal uses a simple switch-level approach to delay modeling: each transistor is characterized as a perfect switch in series with a resistor. Instead of solving differential equations, Crystal uses tables to compute the value of the series resistance. For each transistor, the input waveform, transistor type and size, and output load are combined into a single number called the *rise-time ratio*; this ratio is used to interpolate into a small table for the series resistance. Using rise-time ratios, Crystal can usually predict delays to within 10% of SPICE. The simplicity of the models allows Crystal to calculate delays in a few hundred microseconds per transistor, instead of a few seconds per transistor in SPICE. Section 8 describes Crystal's delay models.

2. The Basic Approach: Stages and Depth-First Search

In the simplest approach to timing verification, the user indicates when a particular input node rises or falls. Crystal finds all other nodes that may possibly change as a consequence of the change in the input, computes the worst-case delays to those nodes, and presents the designer with information about the last nodes to settle and the worst-case paths leading from the input to those nodes. Typically, the input is a clock signal, and the designer compares the Crystal output with the desired length of the clock phase. If the circuit is taking too long to settle, the designer uses information about the critical paths to optimize the design.

This simple approach is is not sufficient by itself for complex, state-of-the-art circuits, but it forms the core of the timing verifier. This section describes the basic mechanisms, and in later sections the approach will be embellished with additional techniques to obtain better results for real circuits.

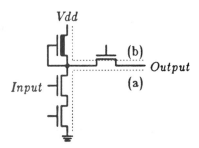

Figure 1. If *Input* changes from 0 to 1, it could activate an electrical path (a) from *Ground* to *Output*. If *Input* changes from 1 to 0, it will disable path (a), and could thereby permit the weaker path (b) to drive *Output* from *Vdd*. Each of the paths (a) and (b) is a stage.

Given that one node has changed value, the first step is for Crystal to determine what other nodes may change as a consequence. It uses a switch-level approach for this. If a change at one node is to cause another node to change value, one of two things must happen: either (a) the change must cause a transistor to turn on and establish an electrical path through a sequence of switches from a strong signal source to the other node, or (b) the change must cause a transistor to turn off, breaking one electrical connection to the other node and thereby permitting a weaker opposing path to drive the node. Figure 1 illustrates the two cases. Crystal considers only these two possibilities; it does not deal with capacitive or inductive coupling.

Such an electrical path, triggered by a change in one node and causing another node to change value, is represented in Crystal with a structure called a *stage*. A stage is a chain of transistors and nodes between a signal source and a place where the signal is used. A signal source is any strong source of a logic 0 or 1. Anything labelled by the user as a chip input (such as Vdd or Ground) is considered to be a signal source. Some sources have known values (for example, *Vdd* always has the value 1), while others have unknown values. Signals are "used" at nodes that attach to transistor gates, and at nodes the user has explicitly labelled as outputs, such as pads. In this paper, the term "output" refers to the end of a stage, whether it is a gate or an explicit chip outputs. One of the transistors in the stage is identified as the *trigger*. This is the one whose gate is controlled by the input. A stage generally corresponds to a path through a logic gate plus any pass transistors following the logic gate.

Stages form the basis for tracing out critical paths. When told that an input node changes, Crystal first finds all stages that may possibly be triggered by the change (this process is described in Section 3). Each of the stages is then passed to a delay modeler that computes how long it will take for the output of the stage to change value. The delay modeler uses an RC approach: an effective resistance and capacitance is computed for each transistor,

A Switch-Level Timing Verifier February 20, 1985

parasitic resistances and capacitances are computed for the nodes, and the RC product is used as the delay. See Section 8 for details on the delay calculation.

Once stages have been found and the delay has been calculated through each stage, the whole algorithm is repeated recursively, treating the output of each stage as an input, finding stages it may trigger, evaluating their delays, and so on. If the signal source for a stage is known to be 0 or 1, then only a 1-to-0 or 0-to-1 transition (respectively) is considered in the recursive analysis of its output. If the value of the source is not known, then both transitions are considered. This automatically accounts for the inverting nature of most digital logic. The recursive search continues until all paths have been found to all nodes that could possibly change state as a consequence of the original input change. Path tracing thus takes the form of a recursive depth-first graph search (see Figure 2). During path tracing, Crystal records information about the slowest nodes found, including the times when they change and the sequences of stages leading from the input to the nodes. At the end of path

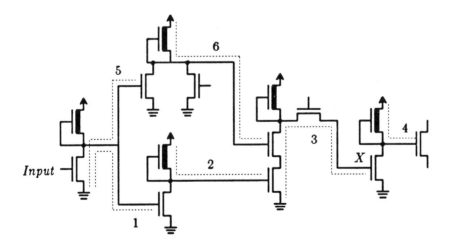

Figure 2. If *Input* changes from 0 to 1, Crystal will trace out the numbered stages in a recursive search. If stage 1 is processed before stage 5 (the choice is arbitrary), then the order will be 1, 2, 3, 4, 5, 6, 3. Stage 3 is processed twice with different triggers. Stage 4 will only be processed once unless the delay to X through 5-6-3 is greater than the delay through 1-2-3. Inversions are handled automatically in the switch-level approach: when *Input* rises, for example, it can only trigger stages leading to ground.

tracing, this information is presented to the designer.

The total number of paths through any real circuit is so large (possibly exponential in the size of the circuit) that it is impractical to examine every possible sequence of stages. Fortunately, it is also unnecessary to examine every possibility when there are reconvergent paths. Instead, Crystal stores in each node the slowest time that has been seen so far for that node. When a node appears as the output of a stage, the new delay time from the stage is compared to the time stored in the node. If the new time is less than the node's time, there is no need to examine the node recursively, since any critical paths involving that node have already been traced using the later time. On the other hand, if the new time is greater than the node's time, then the node must be examined recursively, and the node's time is set to the new time. This simple form of search pruning can still result in exponential search times in the worst case, but in practice the running time is approximately linear in the size of the circuit.

Static memory elements, such as cross-coupled NOR gates, result in circular sequences of stages (see Figure 3). A simple depth-first search will loop infinitely inside such circular paths, computing ever-greater delays. To avoid this problem, Crystal marks all the nodes in its current path (at any

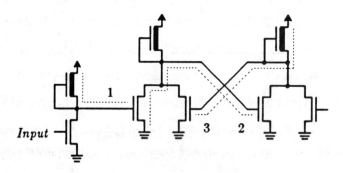

Figure 3. Static memory formed from cross-coupled gates results in circular chains of stages. In this example where *Input* falls, then the output of stage 2 will trigger stage 3 and the output of stage 3 will trigger stage 2.

given time in path tracing, there is a single sequence of stages pending from the current node back to the input). A stage is ignored if it ends in a node that is already marked (Crystal assumes that all feedback is positive, so the last stage will not change the value of its output). In this way Crystal can process memory elements, and will compute the time required for the memory to switch states without looping infinitely.

In depth-first search, delays are propagated from a node as soon as a new worst time is found for the node; this can result in the same node being processed several times with new worst times. An alternative to depth-first search is breadth-first search, such as the one used in Jouppi's TV program [4]. In breadth-first search, a node is not processed until all nodes that can drive it have been completely processed. As a result, each node is only processed once, and the search algorithm is guaranteed to have a linear running time. The disadvantage of the breadth-first approach is that it can only be used on circuits without cycles. This means that static memory elements cannot be represented in their switch-level form. They must be replaced by some higher-level construct without circularity, and must be handled specially. The general problem of finding critical paths in graphs with cycles is NP-complete, so no guaranteed-linear-time solution exists or is likely to be found in the near future.

Using stages as the basis for timing verification has worked out well, because both normal gates and pass transistors can be handled in a uniform fashion. Most previous timing analyzers, such as SCALD [6], have been organized around logic gates. Such programs have difficulty tracing paths through bidirectional arrays of pass transistors such as barrel shifters. The gate-level approach also has difficulty making accurate delay estimates where there are pass transistors, since the non-linear pass transistor effects cannot be separated cleanly from the preceding gate. The stage approach handles gates with or without pass transistor structures uniformly as chains of switches. It also accomodates complex gates such as AND-OR-INVERT.

3. Stage Extraction and Value Independence

The core of Crystal is the code that finds all stages triggered by a given node change. One approach is to extract the stages once and for all when the circuit is read in, and represent the circuit in terms of them. However, a single transistor can participate in many different stages during a single timing verification run so it appeared that this approach would result in a bulky circuit representation. In addition, the stage structure of the circuit can change during timing verification. Because of this, Crystal's approach is to extract stages from the circuit dynamically during timing verification and discard them as soon as they have been used.

The circuit is represented in Crystal as a collection of records linked together in a graph structure. Each transistor and each node is represented as a separate record. A transistor record contains pointers to the nodes of its source, drain, and gate, in addition to information about the size and type of the transistor. A node record contains pointers to each transistor connecting to the node, plus other information such as its parasitic capacitance and resistance. Altogether, the circuit structure requires about 100 bytes of storage per transistor (see Table III in Section 10).

The stage extraction algorithm embodies Crystal's value-independent approach. In general, very little will be known about specific values; in these cases, a stage must be considered if it could possibly be triggered by the change at the input node. For example, if the input node could be either rising or falling, both possibilities must be considered. However, there will be some cases where more specific information is known, either because the user provided it or because Crystal deduced it. For example, the user may indicate that a particular node is precharged, so it can only fall. Or, if the source of a stage is Ground, then this stage can only cause its output to fall, so in extracting later stages triggered by the output, there is no need to consider those that are triggered when the node rises (of course, some other stage may connect the node to Vdd; when this stage is found, the consequences of a rise at the node will be considered).

To find all the stages triggered by a change at a node, Crystal first locates the transistor gates attached to the node and determines whether the change causes each transistor to turn on or off. A table based on transistor type indicates whether a transistor turns on when its gate rises or when its gate falls. Some transistor types are always turned on (depletion loads, for example, or p-channel devices with their gates grounded); since the input will not change their on-off state, they are not considered at this point in the algorithm.

If a transistor is turned on by the change, then Crystal searches for paths from a signal source through the transistor to a gate or output (see Figure 4). First, Crystal searches out from one side of the trigger transistor to see if a signal source can be reached through a series of transistor channels. Then for each source found, Crystal searches out from the other side of the trigger transistor to see if a transistor gate or output node can be reached through a series of transistor channels. A stage is created for each path from a source to a gate or output. Because transistors are bidirectional, the entire process is repeated starting from the other side of the transistor to find additional

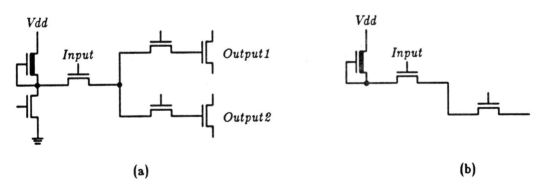

Figure 4. If *Input* turns on in (a), any of four paths could be triggered, leading from one or both of the output gates to either *Vdd* or *Ground*. One stage will be generated for each of the four paths. One of the stages is shown in (b).

sources and outputs. In searching for stages, Crystal assumes that any transistor it encounters could conceivably be turned on (Sections 4 and 5 will discuss exceptions to this rule).

If the trigger transistor is turned off by the input change, then this may disable a strong path and permit some weaker opposing path to drive a stage. In MOS, the only case where two opposing paths to the same node are enabled at the same time is where one of the paths is a depletion load or weak device, which may be overridden by a stronger enhancement device. Crystal handles this case by searching out stages as before, but also looks for loads attached to the stage. A load is any transistor that is always turned on and that has source or drain connected to a supply rail; in order for it to drive the stage once the trigger turns off, the load must connect to the stage between the trigger transistor and the output. For each such load found, a stage is created leading from the supply rail through the load to the output or gate. Except for this special case for loads, Crystal requires all stages to be activated by a transistor in the stage turning on.

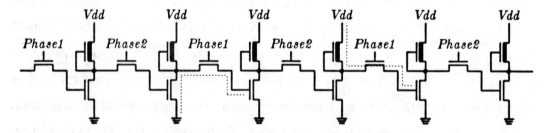

Figure 5. When tracing critical paths from *Phase1*, it is important to consider that *Phase2* is 0. Otherwise, a single critical path will be traced through a sequence of stages from the left side of the shift register to the right side. If Crystal knows that *Phase2* is 0, it will not consider stages containing transistors controlled by *Phase2*. As a result, only short paths like the dotted ones will be found.

4. Case Analysis: Fixed Values

The stage extraction algorithm described above assumes that any transistor could conceivably be turned on at any time. This causes the program to examine the greatest possible number of stages. In practice, though, the actual node values in the circuit at any given time prevent certain transistors from being turned on. For example, Figure 5 shows a two-phase dynamic shift register. When *Phase1* is high, *Phase2* will always be low, thereby disabling alternate pass transistors. If Crystal does not use this information, then when it is tracing paths starting from *Phase1* it will examine a sequence of stages leading from one end of the shift register to the other. This path will never occur in the actual circuit, and will result in an overly pessimistic worst-case time for *Phase1*.

Crystal provides two mechanisms that designers can use to exclude such false paths from consideration. This section describes a mechanism based on fixing node values, and the next section describes another mechanism based on restricting the flow of information through transistors.

The fixed-value mechanism is similar to the case analysis method used in other timing verifiers such as SCALD. Before invoking the path tracer, the user tells Crystal that certain nodes are fixed at certain values. For example, in the case of Figure 5 the user will fix *Phase2* to 0 before invoking path tracing on *Phase1*. The fixed-value approach is also used to set *Vdd* to 1 and *Ground* to zero. When a node is fixed in value, Crystal executes a switch-level simulation algorithm to see if this may also fix other nodes. For example, if any input of a NOR gate is fixed at 1, then the output must be fixed at 0. If a node is fixed in value, then any transistor whose gate is attached to that node is forced to be either turned on or turned off, depending on the type of the transistor and the value of the fixed node.

The fixed-value information is used by the path tracer in two ways. First, if a transistor is forced off because its gate is fixed, then no stages involving that transistor are considered. This will eliminate the problem in Figure 5: the path tracer will consider only short paths leading from one *Phase2*-triggered pass transistor up to the next *Phase2*-triggered pass transistor. The second use of fixed-value information concerns nodes. No stage is considered if it contains a fixed node between the trigger transistor and the stage's output: the fixed value "blocks" any new value from from reaching the output.

The primary use of fixed values is to handle multi-phase clocks and to disable diagnostic devices such as scan-in-scan-out loops. A typical timing verification consists of one run for each clock phase, with all the other clock phases disabled.

The simulation algorithm uses a mechanism similar to the one described by Bryant in [2]. A node becomes fixed in value under either of two circumstances. First, the user may explicitly fix it. Second, a node is fixed if there is an unopposed path from it to a signal source. To determine whether this is true for a node, Crystal locates all stages leading from the node through transistor channels to signal sources. Each stage is assigned a strength based on the strengths of its transistors. Strength is a rough estimate of the driving

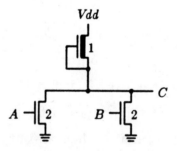

Figure 6. A simulation example. The number next to each transistor is its strength for simulation purposes. If *A* is fixed at 1, then there is a definite path from *Ground* to *C* of strength 2. Since the only opposing path (to *Vdd* through the load) is weaker in strength, *C* is fixed at 0. If *A* is fixed at 0, then it is possible that the pullup will drive *C* to 1; however, if *B*'s value isn't known, then there may possibly be a stronger opposing path to *Ground*, so *C* cannot be fixed. If both *A* and *B* are fixed at 0, then the only possible path to C is through the load to *Vdd*, so *C* is fixed at 1.

power of a transistor, and is determined by the type of the transistor and by whether it is driving a logic 1 or logic 0 (the size of the transistor is not considered). In the standard nMOS process used at Berkeley, for example, enhancement transistors have a greater strength pulling to 0 than depletion loads. The strength of a stage is the strength of the weakest transistor in the stage.

The stages are divided up into two classes: those with every transistor in the stage forced on (called *definite stages*), and those for which every transistor could possibly be turned on (called *possible stages*). If a stage contains a transistor that is forced off, then the stage is not considered. In addition, if a stage contains a fixed node whose value is different from the value of the signal source, then the path is not considered (this corresponds to Bryant's notion of path blocking). Crystal finds the strongest definite path to either a logic 1 or logic 0, and compares its strength to the strongest possible path to the opposite logic level. If the definite path is stronger, then the node is fixed at that value. See Figure 6 for examples.

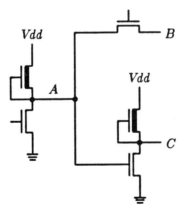

Figure 7. If A is determined to have a fixed value, Crystal's simulator checks B and C to see if they are now fixed in value too. B is part of the vicinity of A because it is in a stage containing A (the same conditions that fixed A may also fix B). C is part of the vicinity of A because it is in a stage triggered by A.

When one node becomes fixed in value, this may cause other nearby nodes to become fixed in value also. Thus the simulation algorithm is recursive: each time a node becomes fixed, Crystal locates all the nodes in its vicinity and checks to see if they are now fixed as well. The vicinity of a node includes all nodes in stages containing the fixed node, and all nodes in stages triggered by the fixed node. See Figure 7 for examples.

The switch-level approach to simulation, based on definite and possible paths, is simple, efficient, and powerful enough to handle a variety of structures. It deals equally well with logic gates, simple pass transistors, and arrays of pass transistors used for charge steering.

5. Flow Control

The bidirectional nature of MOS transistors causes special problems for the path tracer, as illustrated in Figures 8 and 9. There are two related problems. The first problem is that Crystal may examine false paths. For example, in Figure 8 Crystal will consider a path passing both forwards and

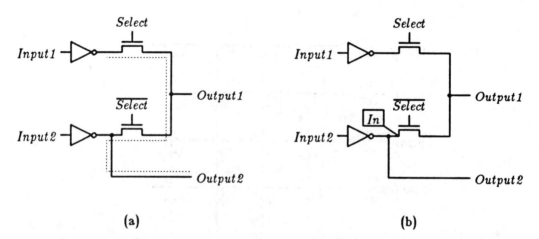

Figure 8. Without any information about how information flows through pass transistors, Crystal will consider the dotted path in (a), which can never occur in practice. In (b) the user has tagged a transistor to show that signals must always flow into it from the left side; Crystal uses this information to eliminate the false path.

backwards through the multiplexor, since it doesn't know that only one of the two transistors can be enabled at any given time. This will result in a pessimistic delay estimate through the multiplexor. Figure 9 illustrates a more severe problem: Crystal will consider contorted paths through barrel shifters and other pass transistor arrays. The number of possible paths is so large (exponential in the size of the array) that the program cannot examine them all in a reasonable amount of time.

One approach to the problem is to use fixed values to limit the possible paths. For example, in the case of Figure 8, two separate runs could be made, one with *Select* fixed at 0 and $\overline{Select}$ fixed at 1, and another with the values reversed. This is the approach taken by the SCALD system. However, it has two disadvantages. First, it is tedious and expensive to make separate runs with different fixed values. The situation of Figure 9, for example, has many different valid combinations of gate inputs on the pass transistors. The second problem is that no delays are calculated to fixed nodes (as explained in Section 4). If *Select* and $\overline{Select}$ are fixed in Figure 8 and one of those nodes is on the critical path, Crystal will not discover that fact (it assumes any fixed node

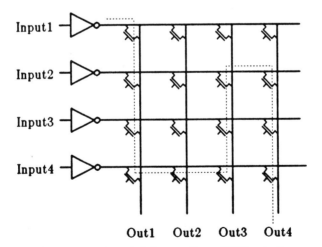

Figure 9. Another example of false paths. Without flow control information, Crystal will examine a large number of paths like the dotted one, which will never occur in the actual circuit.

settles before the clock cycle begins).

Crystal's solution to the problem is a new mechanism called *flow control* whereby the user indicates how information flows through pass transistors. Flow control information can be provided in two ways. The simplest and most common way is when the transistor is really unidirectional: one side of the transistor is tagged with a *flow attribute* indicating that the source of the 0 or 1 signal is on that side of the transistor. Figure 8 illustrates how this eliminates the false path from the multiplexor case. When extracting stages, Crystal skips over any stage that would result in a signal flowing against an "In" attribute.

A similar but more powerful technique is used for bidirectional structures such as the one in Figure 10. Instead of placing an "In" flow attribute on pass

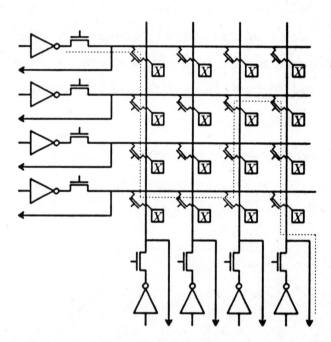

Figure 10. An example of a bidirectional pass transistor array. The flow attributes X allow information to flow in either direction across the pass transistor array as long as signal flow in any path is in the same direction with respect to the X tag. The dotted path is prohibited, since it requires information to flow in through X in one place and out through X in another.

transistors, the designer uses some other name in the flow attribute. When a stage contains more than one transistor tagged with the same flow attribute, the signal flow through each transistor must be the same with respect to the flow attribute. If the signal passes into one transistor on the tagged side and out another on the tagged side, then the stage is considered to be impossible and is discarded. This technique permits paths leading in either direction across a structure, but ignores paths that go back and forth. Each different structure in the circuit can use a separate tag for its flow control so Crystal handles them independently.

In the Berkeley design environment, flow tags are entered into the mask layout and passed along by the circuit extractor using a general-purpose attribute facility. It is relatively easy to tag large arrays: the designer tags the master cell, and the tag is replicated along with the rest of the array.

In addition to eliminating false paths, information about flow is used to speed up the basic path tracing and simulation algorithms. At the beginning of a run, Crystal scans the entire circuit and sets flags in each transistor indicating whether signals can flow through the transistor from source to drain and/or vice versa. This is done by searching outward from every transistor gate and output node for a path to a strong signal source. For each path found, flow is marked as permissible along the path from the source to the gate or output (if there are no contradictory flow tags). For almost all transistors this will eliminate one of the two directions of flow. The designer can find out about any transistors that still appear to be bidirectional: these are likely candidates for flow tags. During simulation and path tracing, Crystal can immediately discard any stage as soon as a transistor is found that cannot flow in the right direction. Since most transistors are unidirectional, this provides almost a factor of two speedup in simulation and path tracing.

The idea of a pre-pass to mark flow originated in Jouppi's TV program [4]. In TV, all transistors are required to be unidirectional; the program makes several passes over the circuit using a number of techniques to deduce the flow of each transistor. This prohibits bidirectional structures, but eliminates the need for the designer to enter flow tags.

6. Precharging

In addition to fixed values and flow control, Crystal allows the designer to indicate that certain nodes are precharged or predischarged. If a node is marked as precharged, Crystal assumes that the node is set to 1 before the beginning of the clock phase, and can only be discharged during the phase. Thus it ignores any stage that would drive the node to 1. Predischarging works in a complementary fashion.

7. Clocking and Memory Nodes

In the previous sections, timing verification was treated in a simplified fashion: the only goal was to compute how long it takes a circuit to settle after a clock rises or falls. In practice, this approach is too conservative, since many signals will settle across multiple clock phases. For example, the input to an arithmetic unit might be latched during *Phase1* and the output might be latched during *Phase2*. In this circuit, the outputs may or may not settle during *Phase1*; the only thing that really matters is that they settle in time to be latched during *Phase2*. Unfortunately, if Crystal uses the simple approach it will "bill" the entire delay to *Phase1*: it assumes that if a node can change during a clock phase, then it must change during that clock phase.

The key aspect to be verified is not that all nodes settle within a particular clock phase, but that all memory elements are latched safely (their setup and hold times are met). In the SCALD system [6], this is done by collecting a large amount of waveform information for each node during path tracing. The waveform information provides worst-case estimates for the time periods when the node is rising, falling, stable, unknown, etc. For each memory element, the waveform information for its clock is then compared to the information for its data input to be sure that the setup and hold times are met.

The SCALD approach works well for edge-triggered circuits, but is much more difficult to apply in circuits with level-sensitive latches like those used commonly in MOS designs. The reason for the difficulty is that level-sensitive latches can change value at any point in an interval, whereas an edge-triggered latch changes value at a single well-defined time. Level-sensitive latches allow time to be "borrowed" between functional units: one unit can take longer to settle if its predecessor takes less time than expected. Borrowing can also occur between clock phases. Borrowing is typically handled using an iterative approach that repeatedly solves for critical paths, using information from the previous solution to compute how much borrowing can occur. Unfortunately, the iterative solution appears to converge very slowly in some cases, resulting in very long running times.

It appeared that the SCALD approach would substantially increase both the complexity of Crystal's code and its time and memory requirements, so I decided to use a simpler approach instead. MOS chips, particularly those in the Mead-Conway style, tend to use a small number of non-overlapping clock phases. This results in much simpler clocking disciplines than the ECL circuits for which SCALD was designed. Crystal was designed for such simple clocking disciplines, and uses a simple mechanism for dealing with memory: instead of requiring all nodes to settle in each clock phase, it only requires memory nodes to settle.

Certain nodes in the circuit are known to Crystal as memory nodes. Memory nodes are identified in two ways, depending on whether they are static (cross-coupled gates) or dynamic (nodes that store charge capacitively). The first mechanism occurs automatically during path tracing. Whenever a circularity is found (i.e. a node is reached that is already on its own critical path), the node is marked as a memory node. This situation corresponds to static memory elements such as cross-coupled NOR gates. The second mechanism for identifying memory nodes is invoked before path tracing. All the clock lines are turned off, using the simulation mechanism described in Section 4, and then any node that is electrically isolated (i.e. cannot be driven) is marked as a memory node. This situation corresponds to dynamic memory elements where charge is stored on a node.

When Crystal records the slowest paths during path tracing, it keeps separate records for the slowest paths overall, and the slowest paths that lead to memory nodes. Users can examine these two groups of slow paths separately. In general, the clock length is determined by the slowest memory node that is modified by that clock.

This simple mechanism for dealing with memory nodes still processes each clock phase separately, which means that it does not handle borrowing correctly. If a node starts to settle during one clock phase and finishes settling during the next clock phase, the time required during the second phase will not be considered during the processing for that phase, since each clock phase is handled as a separate verification run. Crystal will not report critical paths that cross clock phases. This could result in an underestimation of the time required for the second phase. Designers can use the "slowest path overall" information for the first clock phase to get a rough idea where such problems may occur, but the manual nature of this step may result in errors. As long as simple clocking disciplines are used, Crystal's simple approach will probably be adequate. However, if MOS clocking disciplines become as complex as those for ECL, then it will become necessary to develop automatic tools for checking interactions between clock phases. This appears to be a difficult problem and is an important area for future research.

8. Calculating Delays

Crystal uses an RC approach to calculate the delay through each stage. It computes a resistance value and capacitance value for each transistor, a parasitic resistance and capacitance for each node, and then uses the RC product as the delay through the stage. This section describes three different models that use different mechanisms for the summation and transistor resistance calculation to produce more and more accurate results. The models were evaluated using two large circuits, a microprocessor [5] and an instruction cache [12]. Crystal was run on the circuits to extract 12 critical paths containing a total of 157 stages. Delays through the critical paths were calculated with each of Crystal's models, and the critical paths were also processed by SPICE for comparison purposes. Table I summarizes the results. The best of Crystal's models, the distributed slope model, produces results that are typically within 10% of SPICE.

Crystal's delay models achieve their accuracy and speed by capitalizing on the uniform design style used in large circuits. For example, digital VLSI circuits tend to have only a few different sizes of transistor and a few pullup-pulldown ratios, used over and over. Since the pieces of the circuit have about the same structure, they also have about the same delay properties. Even where the structures vary (such as different loading characteristics), they tend to vary in simple ways that can be characterized by one or a few parameters

Model	Overall Error	Av. Stage Error	Av. Path Error	Std. Deviation in Path Error
Lumped RC	-22%	45%	24%	10%
Lumped Slope	4%	23%	8%	9%
Distributed Slope	2%	20%	6%	7%

Table I. A comparison of the delay models. "Overall Error" compares the sum of all the delays computed by Crystal with the sum of all delays computed by SPICE, to point out consistent underestimates or overestimates. For example, the lumped RC model's delay estimates are 22% less than SPICE's, on average. "Av. Stage Error" is the average of the absolute value of errors for individual stages (the points in Figures 12 and 15). "Av. Path Error" is the average absolute error in estimating total delays in the critical paths up to each stage (the points in Figures 13 and 16). The rightmost column gives the standard deviation in the errors from the "Av. Path Error" column.

(such as load capacitance). The delay properties of the basic constructs are measured by running SPICE on small examples and distilling the results down to a few tables. When analyzing large circuits, delay estimates are computed quickly using the tables. If there isn't much variation in the structures used in the circuit, small tables will produce accurate results. The simple models tend not to work as well for sensitive analog components or circuits with large variation in design style.

8.1. The Lumped RC Model

The RC model is the simplest and least accurate of Crystal's models, and is a slight generalization of the "tau" models of Mead and Conway [7]. Each type of transistor is characterized by two resistance values, each expressed in ohms per square. The first resistance value is used if the transistor is transmitting a logic one, and the second is used if the transistor is transmitting a logic zero. The table value is multiplied by the transistor's length/width ratio to obtain its effective resistance. The path tracer provides the delay modeler with information about whether the signal source for the stage is a zero or one; the zero-one distinction results in more accurate modeling of pass transistors. The values currently used in Crystal are given in Table II.

The table values are generated by running SPICE simulations of simple stages. In the SPICE simulations a step function is applied to the gate of the trigger transistor; the effective resistance is computed by measuring the time for the output to reach the logic threshold voltage and then dividing this time by the load capacitance. This generally results in an underestimation of effective resistance (see below).

In the lumped RC model, the delay through a stage is computed by lumping all of the resistances and capacitances together and using the product as the delay:

Transistor Type	Ohms/square (transmitting 1)	Ohms/square (transmitting 0)
Enhancement	30000	15000
Enhancement driven by pass transistor	---	48000
Depletion Load	22000	---
Super-buffer (depletion, gate 1)	5500	5500
Depletion (gate 0 or unknown)	50000	7000

Table II. The effective resistances used by the RC model for a 4-micron nMOS process. The missing entries are for situations that do not ever occur (for example, depletion loads are used only to transmit 1's).

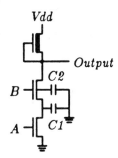

Figure 11. A switch-level approach to delay analysis automatically accounts for different delay characteristics at different inputs of a NAND gate. Capacitances $C1$ and $C2$ will be included in the delay calculation if A is the trigger, but not if B is the trigger.

$$delay = \left[\sum R\right]\left[\sum C\right]$$

The resistances include those for each transistor and the parasitics for each node along the stage. The capacitances include only those between the trigger transistor and the output of the stage: since the trigger transistor is the last to turn on, all the capacitance between it and the signal source is assumed to have discharged already. This means that different delays will be computed from each input of a NAND gate (see Figure 11). If a gate-level model were used instead of a switch-level approach, such asymmetries would be difficult to capture.

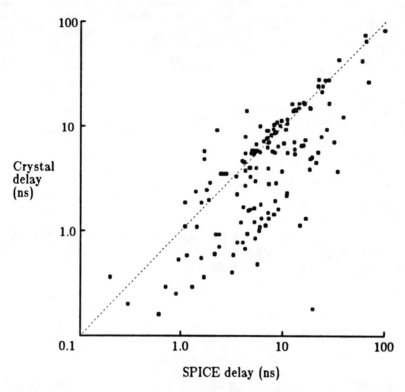

Figure 12. A comparison between the lumped RC model and SPICE, using 12 critical paths (157 stages) extracted by Crystal from two large circuits. Each point compares Crystal's delay estimate for a stage with the corresponding SPICE time, measured from a simulation of the critical path. Ideally, all points should fall along the diagonal line. The RC model consistently underestimates delays by about 20%.

Figures 12 and 13 compare the lumped RC model to SPICE. Figure 12 makes a stage-by-stage comparison and Figure 13 compares total delays through the critical paths. Although Crystal often erred by a factor of 2 or more on individual stages, the errors of successive stages tend to cancel. On average, the lumped RC model can usually estimate the delay through a path to within 25% of SPICE (see Table I). Most of the errors are due to underestimations of delays.

There are two sources of error in the RC model. One is the lumping of resistances and capacitances. This tends to overestimate the delays since it assumes that all capacitance must discharge through all resistance. Fortunately, most stages contain only a single transistor and small parasitic

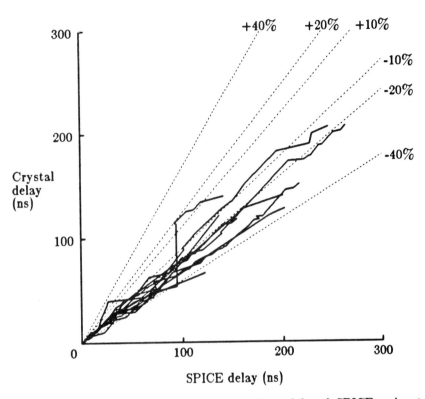

Figure 13. A comparison between the lumped RC model and SPICE, using total delays through critical paths. Each line corresponds to one critical path, and each point in the line corresponds to a stage in the path. The point compares Crystal's and SPICE's estimates for the total delay in the critical path up through that stage.

resistances, so lumping introduces only a small error and is not the major problem with the RC model. Section 8.3 improves on the lumped approach by applying the Penfield-Rubinstein models for distributed capacitance [13,15].

The second and most significant source of error in the lumped RC model comes from its inability to deal with waveform shape. In practice, the effective resistance of a transistor depends on the waveform on its gate. If the trigger transistor turns on instantaneously, then its full driving power is used to drain the output capacitance and the transistor has a relatively low effective resistance. If the trigger turns slowly, then it may do much or all of its work while only partially turned-on. In this case its effective resistance will be higher.

If all waveforms in a circuit have the same shape, then the effective resistances of transistors can be characterized using that waveform and the RC model will produce accurate results. Unfortunately, this is not the case in actual VLSI circuits. Although almost all waveforms have an exponential shape, they vary by more than three orders of magnitude in their slopes. As a result, the effective resistance of the transistors varies by more than a factor of ten and the RC model produces only a rough estimate for delays.

8.2. The Lumped Slope Model

The lumped slope model incorporates information about waveform shape in order to make more accurate delay estimates. It assumes that all waveforms are exponential in overall shape but vary in their slopes. Each waveform is represented by its inversion time (the time when it reaches the logic threshold) and its rise-time, in ns/volt at the logic threshold. A rise-time of zero corresponds to a step function, and a large rise-time corresponds to a slowly rising or falling signal.

Unfortunately, the effective resistance of a transistor depends not just on the rise-time of its gate, but also on the load being driven by the stage and on the sizes of the transistors in the stage. If a stage is driving a large load, or has small transistors, then only very slow input rise-times will affect the stage's delay (it will take the stage so long to drive its output that the trigger transistor will be completely turned on long before the output settles). If a stage is driving a small load or has very large transistors, its delay will be more sensitive to the rise-time of its input. The important issue is whether or not the trigger transistor is fully turned-on when it does most of its work, and this depends on the input rise-time, the output load, and the transistor's size.

The key to implementing the lumped slope model was the discovery that all of these factors can be combined into a single ratio, which alone determines the transistor's effective resistance. First, the output load and transistor size are combined into a value called the *intrinsic rise-time* of the stage; this is the rise-time that would occur at the output if the input were driven by a step function (i.e. the value that would be computed under the RC model). The input rise-time of a stage is then divided by the intrinsic rise-time of the stage's output to produce the *rise-time ratio* for the stage. The rise-time ratio gives an estimate of how fully turned-on the trigger transistor is when it is doing its work. SPICE simulations showed that the rise-time ratio is an accurate predictor of the effective resistance of a transistor, independent of the specific input rise-time, transistor size, or output load. Pilling and Skalnik were the first to suggest the ratio approach [14].

In the slope model, each transistor type is characterized by two resistance tables, one used when the transistor is transmitting a logic 0 and one used when it is transmitting a logic 1. Each table gives effective resistance values as a function of rise-time ratio. During timing analysis, the effective resistance of the trigger transistor is computed by interpolating in the appropriate table. When a stage contains several transistors, the slope model is applied only to the trigger transistor. The other transistors are assumed to be fully turned-on, so the RC model is used to compute their resistances. Once a resistance has been computed for each transistor, the separate values are summed as before, and multiplied by the total capacitance to compute the delay.

The parameter tables for the slope model were generated in much the same way as for the RC model: SPICE simulations were run on simple stages and parameters were extracted from the output. Each table contains six to ten values. Figure 14 plots the contents of a few of the tables for our 4-micron nMOS process.

The ratio approach is important because it allows transistors to be parameterized with one-dimensional tables, instead of three-dimensional tables based on input rise-time, load, and transistor size. The three-dimensional approach would require large amounts of CPU time to generate the tables and would also make the delay analysis slower by requiring three-dimensional

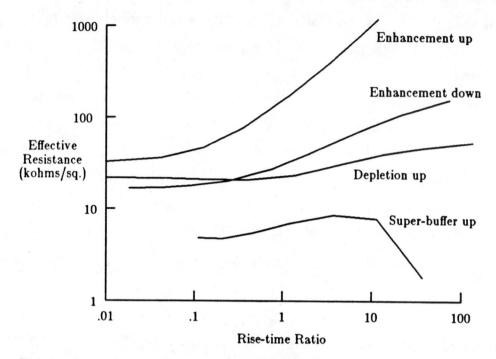

Figure 14. A plot of the tables that characterize the resistance of different types of transistors as a function of the rise-time ratio of the stage. A rise-time ratio of zero means the stage's input rises very quickly in comparison to the output. Different tables are used when transmitting zero (e.g. the "enhancement down" curve) and transmitting one (e.g. the "enhancement up" curve). The curves represent a few of the transistor types for a 4-micron nMOS process. The curves were generated using SPICE simulations on simple circuits. Different rise-time ratios were produced by varying the load on a standard inverter and using the inverter's output to drive the test circuit.

interpolation.

The slope model requires the actual rise-time of each stage's input to be known in order to calculate the delay through the stage. The current implementation of the slope model uses the intrinsic rise-time of the stage driving the input as an estimate of the actual input rise-time. Thus the rise-time ratio is just the ratio of the intrinsic rise-time of the previous stage to the intrinsic rise-time of the current stage. This approximation means that a very slow stage only affects the delays through stages it drives directly. It cannot affect second- or third-level stages (Section 11.3 discusses this limitation). Intrinsic rise-times are computed under the assumption that inputs are step

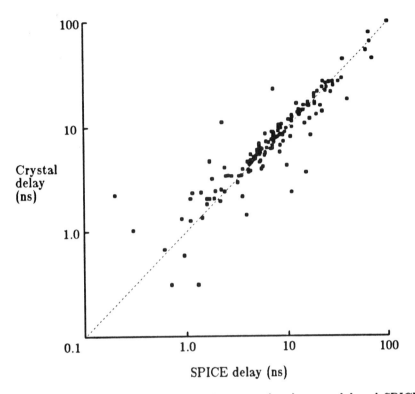

Figure 15. A stage-by-stage comparison between the slope model and SPICE, for the same critical paths as in Figures 12 and 13.

functions and outputs have exponential waveforms, which means that the intrinsic rise-time for a stage is proportional to its delay under the RC model.

Figures 15 and 16 compare the slope model to SPICE with the same data as in Figures 12 and 13. They show that the slope model is substantially more accurate than the RC model. The average error for individual stages was reduced from 45% to 23%, and the average overall error over critical paths dropped from 24% to 8%. Only rarely does the estimate for a critical path differ from SPICE by more than 20%.

8.3. The Distributed Slope Model

Although most stages in MOS circuits contain only a single transistor, the delay modeler must still be able to make reasonable delay estimates for more

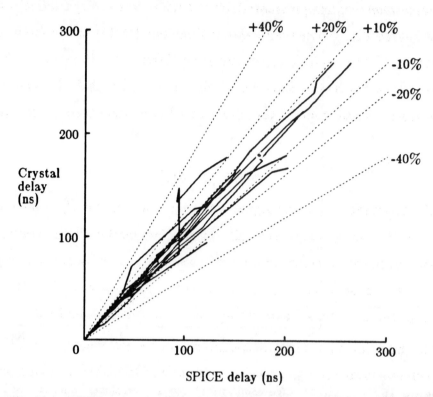

Figure 16. A total-delay comparison between the slope model and SPICE, using the same critical paths as in Figures 12 and 13. The two largest vertical jumps are due to lumping errors, and were eliminated by the distributed slope model.

complex stages. In the RC and slope models, complex stages are handled by lumping resistances and capacitances. The lumped approach is pessimistic for complex paths with distributed capacitance, since it assumes that all the capacitance must be discharged through all the resistance (see Figure 17 for an example).

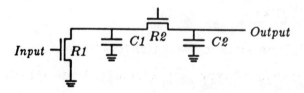

Figure 17. The lumped models will compute a delay of $(R1+R2) \times (C1+C2)$, which is pessimistic since it treats $C1$ as if it were actually in the same location as $C2$. The distributed slope model computes a delay of $(R1 \times C1) + (R1+R2) \times C2$.

In order to handle such situations more accurately, a third model called *distributed slope* was implemented. It is similar to the slope model except that it uses the results of Penfield and Rubinstein [13,15] to avoid lumping the capacitance. [13] provides upper and lower bounds for the delay; Crystal uses the average of the two. Since each stage is a linear chain of transistors with no side trees, the average of the upper and lower bounds simplifies to Elmore's delay:

$$delay = \sum_i R_i C_i$$

R_i is the resistance from the signal source to point i, and C_i is the capacitance at point i. Instead of weighting all the capacitance by all the resistance, each separate capacitor is weighted only by the resistance between it and the signal source; this model is slightly optimistic. Aside from the difference in weighting, the distributed and lumped slope models are identical.

The addition of the Penfield-Rubinstein models made a small additional improvement in accuracy, which is summarized in Table I. The average error for single-stage estimates dropped from 23% to 20%, and the average error over paths containing several stages dropped from 8% to 6%. The small improvement supports the conclusion that complex stages are rare in large circuits. However, the PR-slope model is almost as efficient as the basic slope model, and can avoid gross over-estimates that will occasionally happen in the basic slope model (for example, the two largest vertical jumps in Figure 16 were due to lumping errors, and disappeared with the PR-slope models).

9. Transistor Types

In Crystal, information about transistors in contained in a table with one record for each transistor type. By keeping all the transistor information in tables (instead of hardwiring it into the code), and by allowing users to modify and extend the tables, Crystal can handle both nMOS and CMOS, and can also deal gracefully with some analog subcircuits. The table entry for each transistor type contains the following information:

Strengths. Two integer strength values are provided, one for when the transistor is transmitting a 0, the other for when it is transmitting a 1. The strengths are used during simulation to see which of several paths is stronger.

Activation. This value is one of "gate 1", "gate 0", or "always". It indicates when the transistor is turned on, and is used during path tracing to see which paths might be triggered by a node change.

Resistance. For the RC model, this consists of two resistance values, as in Table II. For the slope and distributed slope models, this consists of two tables of values, as in Figure 14. The values are used by the delay modeler to compute delays.

Capacitance. Two values are provided, which are used to compute the gate-channel and gate-source/gate-drain capacitances, based on the geometry of the transistor. The capacitance values are used during delay calculations.

Users can add new transistor types to the tables and label transistors in the circuit with the new types. This provides a simple facility for dealing with special circuit constructs such as bootstrap drivers. As yet we have only limited experience using this feature, but the initial results look quite promising. SPICE is used to pre-characterize the analog circuits, and these results can be fed into Crystal with the transistor type facility.

In the normal case where a transistor isn't labelled with an explicit type, Crystal chooses a default type based on the physical structure of the transistor (enhancement, depletion, p-channel, etc.), and also based on how the transistor is used in the circuit. For example, enhancement transistors driven through pass transistors have different characteristics than enhancement transistors driven directly by depletion loads; Crystal distinguishes these two cases and uses a different transistor type for each.

10. Experiences with Crystal

The initial implementation of Crystal was developed in the summer and fall of 1982, using only the RC delay model and with nMOS hard-coded into the program. It was almost completely re-implemented in the summer and fall of 1983 to use transistor tables and to add the slope-based delay models. The two test circuits were developed at the same time as the first version of Crystal. Crystal was used for analysis of both those circuits, although only the lumped RC model was available at the time. See Table III for a summary of the experience with the circuits. In one case (the RISC II cache) Crystal located about a half dozen performance bugs that would have slowed the cycle time by nearly a factor of two. In the other case (the RISC II CPU), Crystal located several performance bugs and one functional error. Both of the chips have been fabricated and tested. Both are fully functional, and both run at approximately the predicted speeds. Overall, Crystal required about 10 milliseconds of processing time per stage, of which most was spent in the simulation and path-tracing phases. Only a few hundred microseconds per stage were required for delay calculations.

	CPU [5]	Cache [12]
Size (transistors)	41000	46500
Clock phases	4	4
Crystal processing time (CPU minutes, VAX 11/780)	35	15
Stages processed	154000	72000
Memory required (bytes)	4100000	4800000
Estimated cycle time (ns, RC model)	480	450
Actual cycle time (ns)	500	600

Table III. Experiences with Crystal on two nMOS circuits. The processing time includes circuit pre-processing and searching functions as well as delay modeling; delay modeling accounts for only a small fraction of the total time.

A Switch-Level Timing Verifier

11. Crystal's Limitations

Although Crystal has been used successfully on several circuits, there are several weaknesses remaining in the program. This section discusses the four most significant of these.

11.1. Clocking

The independent treatment of clock phases, described in Section 7, is probably Crystal's most important limitation. So far, the design projects at U.C. Berkeley have not had troubles with this limitation, because they use very simple clocking disciplines. However, if more complex clocking disciplines begin to be used, this loophole may result in undetected timing problems. Level-sensitive latches make this a very difficult problem for which I know of no good solution. In my view, this is the single greatest problem yet to be solved in timing verification.

11.2. Linear Stages

Crystal requires each stage to be a linear chain of transistors and nodes. All side transistors connecting to the stage are assumed to be turned off: their gate-source capacitance is included in the parasitic capacitance, but

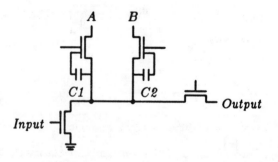

Figure 18. In computing delays, information from side paths is not included. For example, in this figure the capacitance at A and B is not included. However, the gate-source overlap capacitance from the side transistors ($C1$ and $C2$) is included in the parasitic capacitance of the stage.

information on the far side of transistors is ignored (see Figure 18). This approach is used in Crystal because the program does not have specific information about whether side transistors are turned on or off; it it automatically included all side capacitance, its delay estimates would be unrealistically high. Busses and pass transistor arrays account for most of the situations with many side paths, and in these cases only a single path is usually active through the structure at once. However, in cases where there are active side paths, such as PLA ground switches and some carry-chain prechargers, Crystal will underestimate delays. Experience so far suggests that these situations are rare.

11.3. Rise-time Estimation

Crystal estimates rise-times by using intrinsic rise-times, which is only a first-order approximation. If the input to a stage has a large rise-time, then the actual rise-time of the output will be substantially larger than the intrinsic rise-time. On average, Crystal's rise-time estimates differed from SPICE's by 30% over the test data. To find out the maximum improvement one can expect with better rise-time estimates, I ran an experiment on the test data using both Crystal and SPICE. Crystal's distributed slope model was used to compute delays over the 12 critical paths, but SPICE provided all the rise-time estimates. There was no significant improvement in Crystal's delay estimation with "perfect" rise-time estimation. This suggests that rise-time estimation is not a major source of errors.

11.4. Complex Stages

Although the distributed slope model makes a first-order attempt to deal with complex stages, there are still several situations that can cause it to produce inaccurate results. One source of error in complex stages is the assumption that only a single transistor is turning on or off at once. If two transistors turn on simultaneously in a NOR gate, the gate's delay will be less than predicted; if two transistors turn on simultaneously (and slowly) in a NAND gate, its delay will be greater than predicted. Another source of error in complex stages is the additive treatment of different transistors and resistors: total resistances and rise-times for stages are computed by summing the contributions of each device along the stage. This is an accurate approximation for resistors, but it is less accurate for non-linear devices such as pass transistors.

12. Related Work

Timing verification was popularized by McWilliams' SCALD system [6]. Although Crystal's overall function is similar to that of SCALD, it differs in its mechanism for dealing with clocks (as described in Section 7), and also in its approach to path tracing and delay calculation. SCALD uses a gate-level approach, which is appropriate for TTL and ECL circuits, whereas Crystal's switch-level approach appears to have significant advantages for MOS.

Norman Jouppi developed the TV MOS timing verifier [4] at about the same time that Crystal was being written, and the two programs are quite similar. The main differences are that TV uses a breadth-first search instead of depth-first search (this results in faster execution but prohibits circular structures such as static memory), and that TV does all flow control automatically, instead of with designer-placed tags (this results in less work for the designer, but prohibits bidirectional structures).

The importance of using waveform information in computing delays has been recognized for some time [3,9,16,17]. Tokuda et al. developed analytic models for the effects of waveforms and those models were validated against circuit simulation, but only over a relatively small range of rise-time ratios. As a consequence, they concluded that load transistors were insensitive to waveform. Tamura et al. have also developed an analytical model for waveform effects [16], but their model appears to apply only to simple gates without pass transistors. A recent thesis by Mark Horowitz also explores simplified analytical models for MOS circuits [3]. The ratio approach was suggested in 1972 by Pilling and Skalnik [14], although they also did not deal with pass transistors.

13. Conclusions

There are three overall aspects of Crystal that, together, have resulted in an efficient and useful tool for locating timing problems. The first of these is the use of a switch-level approach rather than a gate-level approach. This allows Crystal to deal simply and accurately with a large variety of MOS circuit constructs that would be difficult or impossible to accomodate using a gate-level approach. The second important aspect of the program is its value-independent approach, which is common to all timing verifiers. By ignoring specific signal values wherever possible, the program can deal with a large number of potential cases in a small amount of computing time. Fixed-value and flow-control facilities provide the user with the ability to eliminate false paths. The third important aspect of Crystal is its use of rise-time ratios for delay calculation. This results in small parameter tables, fast interpolation, and accurate delay estimates.

14. Acknowledgements

Many people have contributed to the development of Crystal. Most important of these are the circuit designers in the U.C. Berkeley EECS community, who provided the initial motivation for building Crystal, used the early unstable versions, and provided additional advice to help tune the program into a useful tool. Dimitris Lioupis, Robert Sherburne, and Gaetano Borriello were among the earliest users. Dan Fitzpatrick modified our circuit extractor to provide information about parasitics and flow tags. Mark Horowitz gave me several valuable pieces of technical advice to fill gaps in my knowledge of Electrical Engineering. Randy Katz provided helpful comments on an early draft of this paper.

15. References

[1] Baker, C. and Terman, C. "Tools for Verifying Integrated Circuit Designs." *Lambda Magazine*, 4th Quarter 1980, pp. 22-30.

[2] Bryant, R.E. "A Switch-Level Model and Simulator for MOS Digital Systems." *IEEE Transactions on Computers*, Vol. C-33, No. 2, February 1984, pp. 160-177.

[3] Horowitz, M. "Timing Models for MOS Circuits." Technical Report SEL83-3, Stanford University, December 1983.

[4] Jouppi, N.P. "Timing Analysis for nMOS VLSI." *Proc. 20th Design Automation Conference*, 1983, pp. 411-418.

[5] Katevenis, M., Sherburne, R. Patterson, D., and Séquin, C.S. "The RISC II Micro-Architecture." *Proc. IFIP TC10/WG10.5, International Conference on VLSI*, North Holland, 1983, pp. 349-359.

[6] McWilliams, T.M. "Verification of Timing Constraints on Large Digital Systems." *Proc. 17th Design Automation Conference*, 1980, pp. 139-147.

[7] Mead, C. and Conway, L. *Introduction to VLSI Systems*. Addison-Wesley, 1980.

[8] Nagel, L.S. "SPICE2: A Computer Program to Simulate Semiconductor Circuits." ERL Memo ERL-M520, University of California, Berkeley, May 1975.

[9] Okazaki, K., Moriya, T. and Yahara, T. "A Multiple Media Delay Simulator for MOS LSI Circuits." *Proc. 20th Design Automation Conference*, 1984. pp. 279-285.

[10] Ousterhout, J.K. "Crystal: A Timing Analyzer for nMOS VLSI Circuits", *Proc. 3rd Caltech Conference on VLSI*, March 1983. Reprinted as technical report UCB/CSD 83/119, Computer Science Division (EECS), University of California, Berkeley, January 1983.

[11] Ousterhout, J.K. "Switch-Level Delay Models for Digital MOS VLSI," *Proc. 21st Design Automation Conference*, June 1984, pp. 542-548.

[12] Patterson, D.A., et al. "Architecture of a VLSI Instruction Cache for a RISC." *Proc. 10th International Symposium on Computer Architecture*, 1983 (*SIGARCH Newsletter*, Vol. 11, No. 3), pp. 108-116.

[13] Penfield, P. Jr. and Rubinstein, J. "Signal Delay in RC Tree Networks." *Proc. 18th Design Automation Conference*, 1981, pp. 613-617.

[14] Pilling, D.J. and Skalnik, J.G. "A Circuit Model for Predicting Transient Delays in LSI Logic Systems." *Proc. 6th Asilomar Conference on Circuits and Systems*, 1972, pp. 424-428.

[15] Rubinstein, J., Penfield, P. Jr., and Horowitz, M.A. "Signal Delay in RC Tree Networks." *IEEE Transactions on CAD/ICAS*, Vol. CAD-2, No. 3, July 1983, pp. 202-211.

[16] Tamura, E., Ogawa, K. and Nakano, T. "Path Delay Analysis for Hierarchical Building Block Layout System." *Proc. 20th Design Automation Conference*, 1984, pp. 411-418.

[17] Tokuda, T., et al. "Delay-Time Modeling for ED MOS Logic LSI." *IEEE Transactions on CAD/ICAS*, Vol. CAD-2, No. 3, July 1983, pp. 129-134.

VHSIC Hardware Description Language

Moe Shahdad, Roger Lipsett, Erich Marschner, Kellye Sheehan,
and Howard Cohen, Intermetrics
Ron Waxman, IBM
Dave Ackley, Texas Instruments

Reprinted from *Computer*, February 1985, pages 94-103. Copyright
© 1985 by The Institute of Electrical and Electronics Engineers, Inc.

VHDL incorporates a versatility that enables it to support design, documentation, and simulation of many circuit technologies while committing to none. This capability makes it a suitable standard language for describing defense systems.

In March 1980, the Department of Defense launched the Very High Speed Integrated Circuits program to advance the state of the art in high-speed integrated circuit technology,[1] specifically for defense systems. Early in the VHSIC program, it became apparent that a standard hardware description language was needed to communicate design data, and in summer of 1981, the Institute for Defense Analyses arranged a workshop to define the requirements for such a standard.

The DoD used the final report of the IDA workshop[2] as a basis for defining a set of language requirements for the VHSIC Hardware Description Language, issuing a request for proposal for a two-phase procurement of VHDL and its support environment.[3] The team of Intermetrics, IBM, and Texas Instruments was awarded a contract to design and implement VHDL and its support environment. The program, which started on July 31, 1983, called for a 12-month design phase, followed by two months of design reviews and a 14-month implementation period. One major requirement was that VHDL use Ada[4] constructs wherever possible.

VHDL supports the design, documentation, and efficient simulation of hardware from the digital system level to the gate level. While designed to be independent of any underlying technology, design methodology, or environment tool, the language is also extendable toward various hardware technologies, design methodologies, and the varying information needs of design automation tools.

The primary objective of VHDL is to support technology insertion—a term given to the use of the latest technology in the development of new systems and the upgrading of existing ones to relax environmental requirements or to improve performance. VHDL is expected to reduce both the time lag and the cost involved in technology insertion.

We begin our discussion of VHDL by describing the concept of *design entity*, the language's primary abstraction mechanism. We then present a time-based execution model and describe VHDL's features, using a coded four-bit adder to illustrate the use of the most significant ones. Figures 1 and 2 contain the block diagrams and Figures 3 through 8 contain the code for this example.

Design entities

A design entity, the primary way a hardware component is represented in a language, is composed of an interface and one or more bodies. The interface of an entity defines that entity's external characteristics, and each body represents an alternative design approach consistent with those characteristics. Figure 4 contains the interface of a four-bit adder, and Figures 5 and 6 contain bodies corresponding to that interface.

Interface. Certain items of information in a design entity *interface* are externally visible; these include the ports and generic parameters. Other items are not visible externally but are common to all alternative bodies; these, also specified in the interface, include declarations and assertions. Each body of a design entity conforms to the information specified in its interface. Figures 4 and 7 are examples of interfaces.

Interface declarations include type, subtype, and attribute declarations and assertion specifications. Assertions defined in the interface constrain all alternative bodies, while those defined within a particular body apply only to that body.

Ports define the channels of communication between the entity and the outside world. A port declaration involves specification of its mode and type. The mode of a port specifies the direction of information flow, which may be "in," "out," or "inout." The type name specifies the kind of values that can pass through the port, such as integer or bit.

A design entity interface may also contain generic parameters, which allow the interface to define a class of components. For example, the four-bit adder interface in Figure 4 defines a class of adders that have some delay. When a generic design entity is used to describe a more complex component, the values of its generic parameters are specified. For instance, when the four-bit adder in Figure 4 is used, the delay might be specified as 30 ns. This specification makes the generic entity unique and thereby allows a certain member of the class of components defined by the entity to be selected. Generic parameters also allow a description to be reused more often. Uses of generic parameters include the specification of technological dependencies such as the timing characteristics, size, or fanout of a component.

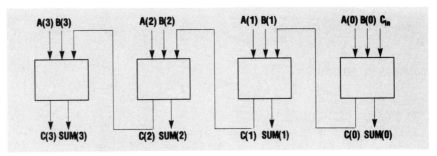

Figure 1. A four-bit adder composed of four full adders.

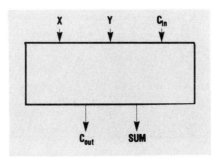

Figure 2. A full adder.

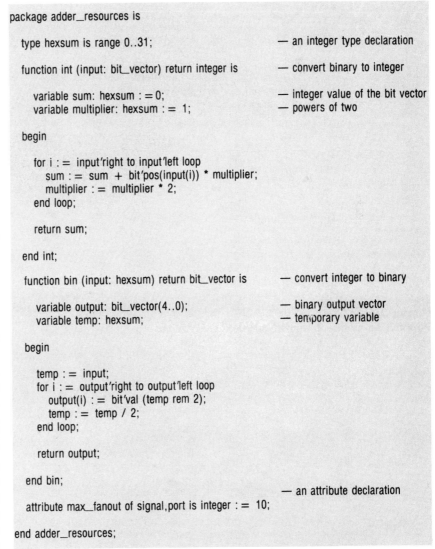

Figure 3. Package defining resources needed by the four-bit adder.

```
with adder_resources;                  — make use of the conversion function
   — interface specification
entity FOUR_BIT_ADDER
   — ports
   (A,B: in BIT_VECTOR(3..0);
    Cin: in BIT;                       — carry input
    Cout: out BIT;                     — carry output
    SUM: out BIT_VECTOR(3..0)) is      — output sum
   — parameter
generic
   (DELAY: time := 36ns);
   —alternative bodies conform to the following assertion
assertion
   DELAY > 3ns;
   SUM'FANOUT <= max_fanout;
end FOUR_BIT_ADDER;
```

Figure 4. Interface of the four-bit adder.

```
                                       — structural decomposition
architectural body PURE_STRUCTURE of FOUR_BIT_ADDER is
   signal C : BIT_VECTOR (3..0);       — internal carries
                                       — local declaration of subcomponents
   component FULL_ADDER (Cin, I1, I2: in BIT; Cout,RES: out BIT);
begin
   for i in 3..0 generate
      if i = 0 generate                — connect first stage
            FULL_ADDER (Cin, A(i), B(i), C(i), SUM(i));
      end generate;

      if i > 0 generate                — connect other stages
            FULL_ADDER (C(i − 1), A(i), B(i), C(i), SUM(i));
      end generate;

   end generate;

   Cout <= C(3);
end PURE_STRUCTURE;
```

Figure 5. An architectural body of the four-bit adder.

```
behavioral body ADDER4B of FOUR_BIT_ADDER is
   variable t, ta, tb: hexsum          — temporary variables
begin
   ta := int(A);                       — convert A from binary to integer
   tb := int(B);                       — convert B from binary to integer
   t := ta + tb;                       — compute sum
   Cout & SUM <= bin (t) after DELAY;  — convert from integer to binary
                                       —    and return the sum
end ADDER4;
```

Figure 6. A behavioral body of the four-bit adder.

Bodies. A body describes an alternative implementation of a design entity and is either behavioral (Figure 6) or architectural (Figure 5 and 8). A behavioral body is a control flow description of the behavior of a component. The designer may declare data structures and specify algorithms that operate on such data structures to determine the values of output signals. Data structures are specified by declaring variables, and algorithms are described using assignment, loop, and selection statements. A behavioral body contains no information on structural decomposition of the design entity.

An architectural body is a data flow description of the architecture of a design entity. Unlike the statements of behavioral bodies, which execute sequentially, the statements of architectural bodies execute in parallel. The execution of statements in an architectural body may be controlled by a guard—a Boolean condition associated with one or more statements in an architectural description. A statement is executed when either (1) its guard becomes true, or (2) its guard has been true, and one or more of its inputs change. For all guards that evaluate to True, the associated statements are considered to execute simultaneously.

The designer can use either a behavioral or architectural body to describe the behavior of a design entity. However, if any structural information is to be specified, an architectural body must be used.

A design entity body must contain local declarations of any design entities or functions used within that body. Because this feature essentially isolates one level of design from another, either the top-down or bottom-up design methodology can be used. In a top-down approach, the designer can specify in a local component declaration what the ports and parameters of a lower level component should look like; such a component may then be fully described at a later time. In a bottom-up approach, the local component declaration specifies the portion of the interface from an existing design entity that will be used in a new design.

Depending on the level of detail present, an architectural body may imply varying degrees of structural information about the design entity. Figure 8 fully specifies the structure of the full adder in terms of its subcomponents.

A VHDL design entity is a template to be used in creating specific instances of a component via the component instantiation statement. The component instantiation statement provides a list of generic parameters that make a design entity unique. These statements express the structural decomposition of a component and may be used only in an architectural body.

Regular structures. A regular structure is a pattern that is repeated many times. Because VLSI designs often contain regular structures, VHDL includes the *generate statement* to facilitate the description of such structures. This statement essentially provides a macro expansion capability within an architectural body. Figure 5 shows how the generate statement is used to highlight the regularity of the structure of the full adder.

Often, the boundaries of a regular structure exhibit slightly different connectivity characteristics than the rest of the structure. In Figure 5, note that the full adder stage corresponding to the least significant bit uses the input carry to the whole component. For any other stage, the carry is internally generated. The conditional variation of the generate statement is used to describe this irregularity in the structure of the four-bit adder.

Time-based execution

Time is one of the most important aspects of a hardware description language because the timing characteristics of hardware are perhaps the most difficult to represent in text. In part, this difficulty is due to the massive parallelism that may exist in a hardware description. Not only must the language represent the dynamic behavior of one hardware component over time, but it must also describe the temporal interactions among all components involved in the design.

The model of time in VHDL is similar to that in Conlan,[5] having both a macro-time scale and micro-time scale (Figure 9). The macro-time scale represents real time (nanoseconds, microseconds, etc.) and is measured in discrete units of type time (Figure 10). The micro-time scale

```
entity FULL_ADDER
   — ports
   (Cin, X, Y: in BIT;
    Cout, SUM: out BIT)

end FULL_ADDER;
```

Figure 7. Interface of the full-adder.

```
architectural body GATE_LEVEL_STRUCTURE of FULL_ADDER is
   component AND_GATE (A,B: in BIT; C: out BIT);
   component XOR_GATE (A,B: in BIT; C: out BIT);
   component OR_GATE (A,B: in BIT; C: out BIT);
   signal S1,S2,S3: BIT;

begin

   X1: XOR_GATE (X,Y,S1);          — exclusive or gate
   X2: XOR_GATE (S1,Cin,SUM);      — exclusive or gate
   A1: AND_GATE (Cin,S1,S2);       — and gate
   A2: AND_GATE (X,Y,S3);          — and gate
   OR1; OR_GATE (S2,S3,Cout);      — or gate

end GATE_LEVEL_STRUCTURE;
```

Figure 8. An architectural body of the full-adder.

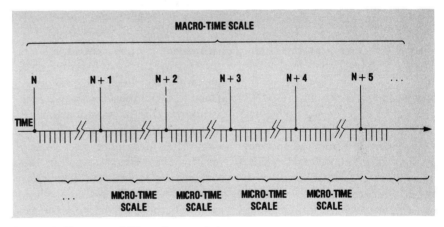

Figure 9. Macro- and Micro-time scales.

```
type time is range 0..1E20
   units
      fs;                    — femtosecond
      ps  = 1000 fs;         — picosecond
      ns  = 1000 ps;         — nanosecond
      us  = 1000 ns;         — microsecond
      ms  = 1000 us;         — millisecond
      s   = 1000 ms;         — second
      min = 60 s;            — minute
      hr  = 60 min;          — hour
end units;
```

Figure 10. Definition of physical type time.

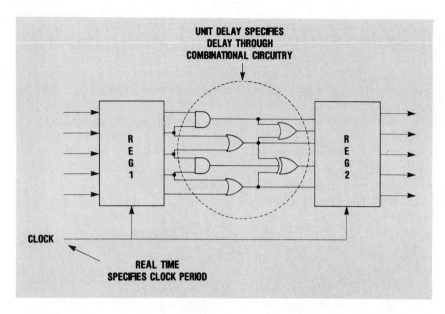

Figure 11. Mixing real time and unit delay.

represents unit-delay and is essentially not measurable. Any number of micro-units of time may exist between any two macro-units of time.

With two time scales, the designer can perform unit-delay simulations, simulate with real timing data, or even mix the two. For instance, in the design illustrated in Figure 11, in which clocked registers alternate with combinational logic, the designer might ignore the propagation delays through the combinational logic but still consider the overall timing of the design. This approach can be realized by specifying the clock cycle in real time units while simulating the combinational logic with unit delay.

In a VHDL description of a component, the instantaneous mapping from the past and present values of the component's inputs to the future values of its outputs is specified. That is, given a set of inputs, a VHDL description predicts the expected outputs at future points in time. As time advances, projected outputs become actual outputs, at which point they are propagated along data paths to become inputs to other components.

Time is involved in this description in the form of delays on the outputs of components. Such delays are always relative to the current time, the value of which is invisible to the description. The inability of a description to determine the current time forces it to behave in a time-invariant fashion, which is appropriate for hardware description. Since delays are specified by values of physical type time, which are effectively fixed point quantities, the timing accuracy essential to synchronization is maintained throughout the system.

A VHDL description is evaluated when an event occurs at one of the component's inputs. The evaluation yields a new set of projected values for the outputs of the component. This stimulus response approach to computation in VHDL is a fairly natural way to describe the behavior of hardware from the digital system to logic gate level and permits efficient simulation.

The event-driven semantics of VHDL are based on the assumptions that all signals in a design propagate in well-defined directions and that signal propagation always involves some delay. These assumptions, however, are not valid for electrical level descriptions of hardware, in which the behavior of a system can be accurately determined only by simultaneous solution. Rather than force higher level descriptions to be simulated less efficiently, VHDL provides an escape from the event-driven model in the form of purely structural descriptions. Because these descriptions have no behavioral semantics, they cannot be simulated within the VHDL system. They can, however, effectively describe an electrical network whose behavior can be determined by simulation with existing low-level tools.

VHDL features

To support the wide range of descriptive capabilities required to model hardware from the logic gate to digital systems level, VHDL incorporates user-defined data types, attributes, and assertions. In addition, the language supports both algorithmic and data-flow-oriented descriptions of hardware, and provides for user definition of functions and packages. The collection of these facilities allows a variety of descriptive styles and supports descriptions at many levels of abstraction.

Types and subtypes. A type is a collection of values and a set of operations on those values. Hardware description languages typically include a data type representing bit values 0 and 1, as well as the standard operations on bits and bit vectors. Such a data type is sufficient for describing hardware at the level of logic gates, where truth tables are usually expressed in those terms (with the possible addition of "ambiguous," "unknown," and "high-Z" values). However, to describe systems at higher levels of abstraction, designers would need more abstract data types, such as "integer," "address," and "instruction." VHDL does not attempt to predefine all abstract data types a designer might require; rather, it provides a mechanism that allows the designer to define the needed data types.

The two categories of types in VHDL are scalar types and composite types. Scalar types include numeric types (integer and floating point), enumeration types (lists of values), and physical types that represent physical measurements such as length and voltage.

Numeric types include both integer and floating point. The types Integer and Real are predefined as are the standard numeric operations; the designer may define additional numeric types by specifying the range of values to be included in the type. Integer types may be declared with arbitrary size, so the designer can control the characteristics of integer arithmetic in the hardware being described. Floating point types, on the other hand, are limited to the capabilities of the machine that supports the VHDL system.

Enumeration types include values that are either character literals, such as the character *A*, or identifiers. Enumeration types representing logic values (True and False) and bit values (0 and 1) are predefined; the standard logical operations are defined on both of these types. Designers may easily define their own enumeration types to represent multi-valued logic, for instance (0, 1, Z, A, U), or processor opcodes (add, sub, mul, div, loa, sto). Character literals and identifiers may be mixed in a single enumeration type, such as in the predefined type Character which contains the ASCII character set represented by mnemonics for nonprinting characters and character literals for printing characters.

Physical types allow the expression of quantities that carry a unit of measurement. A physical type declaration specifies a set of such units, all defined in terms of some base unit and all measuring the same quantity. Since all units are integral multiples of the base unit, the underlying representation of such physical quantities is an integer value, and therefore calculations involving physical types exhibit absolute accuracy. The different units, however, allow the designer to include scale factors so that values can be expressed in units appropriate for a given situation.

Figure 10 illustrates the definition for type Time. In this definition, the base unit is femtoseconds, and all other units are defined as multiples (directly or indirectly) of that base unit. Consequently, the designer can use nanoseconds or minutes as appropriate in different parts of a design, yet maintain absolute accuracy when nanoseconds are added to minutes. Such accuracy is necessary for accurate description of synchronization. The range specified in a physical type definition indicates the quantities of base units that can be measured with the type. As defined here, type Time includes values from

VHDL uses signals to support nonprocedural, data-flow-oriented descriptions of hardware.

0 to 100 quintillion femtoseconds (100,000 seconds or about 2.77 hours).

The second VHDL type category consists of composite types, which are built from elements that are any of the scalar types just described or other composite types. There are two classes of composite types: array types and record types. An array type consists of elements that are all alike, while elements of a record type may be different. Predefined array types include Bit_Vector and String, which represent arrays of bits and arrays of characters, respectively. The logical operations defined on Bits are defined on Bit_Vectors as well.

A type definition may be constrained by the definition of a subtype. A subtype of a type contains a restricted set of (contiguous) values of the type to facilitate the detection of errors that might arise during simulation. For instance, if a record that represents a machine instruction contains an element that represents one of 16 registers, a designer can use predefined type Integer as the type of that element. However, a more exact specification (and one that can be checked by the VHDL system) is a subtype of Integer constrained to have only the values 0 through 15.

Signals and variables. VHDL supports nonprocedural, data-flow-oriented descriptions of hardware through the use of signals. Signals can be used to represent wires or buses in a structural description or to represnt data paths in an architectural description. Signals may retain state and as such may be used to represent memory elements such as flip-flops and registers.

A signal has historic values that are visible but unalterable and future values that are alterable but invisible. Assignment to a signal may change the projected (future) values of the signal but not its current or past values. A signal may be assigned a single value, delayed in time; the time specification is the time relative to "now," when the new value is to take effect for that signal. A signal may also be assigned a waveform, which is expressed as a sequence of values and associated times.

Signals driven by multiple sources are called buses. To resolve the values supplied by multiple sources into a single value for the bus, VHDL allows the designer to specify a bus resolution function to be associated with each bus in a description. Such functions can be used, for instance, to model a wired-OR connection. This approach, as opposed to building bus resolution logic into the language, allows the designer to define a bus that carries information of arbitrary type and a bus resolution function appropriate for that type.

VHDL also provides *variables* for use in abstract computations. Because variables have no relationship to time, they have only a single value that is both visible and alterable. Assignment to a variable changes its value immediately, and therefore variables can be used in algorithmic descriptions. Variables may be either static or dynamic; static variables retain their values from invocation to invocation, while dynamic variables are reinitialized each time.

Attributes. One of the goals of VHDL is to provide a means of expressing the information required to drive various design tools. The minimum required to satisfy this goal is that all information contained in a data sheet for a component be expressible in VHDL. However, since new technologies are continuing to emerge, each with different characteristics, it is difficult to provide specific descriptive facilities that can satisfy all future needs. Instead, VHDL provides a general-purpose mechanism that allows the designer to define attributes of objects. The designer can thus "decorate" a description with extra information about a component or its parts.

Attributes are intended to be processed by arbitrary design tools and therefore must represent arbitrarily complex information. For example, in ISPS[6] this is accomplished by using untyped attributes. In VHDL, however, the combination of data abstraction facilities and typed attributes enables both representation of arbitrarily complex information and verification that an attribute has the appropriate type of information.

In contrast, attributes in VHDL are strongly typed, being either predefined or user-defined. They may contain arbitrarily complex information, but because they are typed, the language processor can determine if the information associated with an attribute is appropriate for that attribute. An additional consequence is that attributes may be referenced within expressions in the language, thus enabling the information contained in a design as attributes to be used to express behavior or structure.

Attribute definition involves two steps. The first is attribute declaration, which defines an attribute name and specifies the classes of entites (signals, ports, functions, etc.) that have this attribute. The declaration may also specify a default value for the attribute. For instance, the attribute declaration in Figure 3 defines the identifier Fanout to designate an attribute of ports; this declaration implies that any port within its scope has an attribute called Fanout.

The second step in defining an attribute is attribute specification, which takes into account that each entity with a given attribute may have a different value assigned to that attribute. The value supplied in an attribute specification overrides any default value supplied as part of the corresponding attribute declaration.

Assertions. Assertions allow a designer to specify conditions that are expected to be true during design execution. They allow the designer to

---

VHDL accommodates emerging technologies by incorporating a general-purpose mechanism that allows the designer to define attributes of objects.

---

specify information about the intent of a design so that errors can be detected closer to their source. Assertions may involve expressions concerning set-up and hold times, operating temperature, radiation level, electrical characteristics, and other operating conditions and characteristics.

An assertion definition consists of a Boolean expression that specifies the condition being asserted, optionally followed by a severity level and an error message. An example of an assertion definition is:

($S1'$fanout < 5)
   severity : = fatal
   report "Signal S1 is driving too
      many lines";

At simulation time, if the fanout attribute of signal $S1$ (the number of sinks being driven by $S1$) is not less than five, an assertion violation is reported. At that time, the message, "Signal $S1$ is driving too many lines" is printed. The severity level provides the ability to specify which levels of assertion violation are to be reported during a given simulation. The simulator can be directed to report only those violated assertions above a particular level, or to terminate execution when an assertion violation is detected at or above a given severity level.

Functions. A function is an algorithmic description for computation of a value. It is invoked in an expression and may be used for an abstract computation, type conversion, or bus resolution.

A function is declared by specifying the name of the function, its input parameters, and the type of its return value. (Figure 3 shows sample function declarations.) The body of a function consists of a sequence of statements that implements an algorithm. A function may not retain state or cause side-effects.

Unlike functions that perform mathematical computations, functions that perform type conversion or bus resolution are not explicitly invoked. They are implicitly inserted at the appropriate places in the description by the system and invoked when necessary.

Packages. A package contains a group of declarations that are related in some way. A package may be created to share declarations among many design units or to collect declarations that relate to a particular abstraction. The package mechanism in VHDL provides a general-purpose facility that may be used in a number of ways.

Several different kinds of declarations may appear within a package, including type and subtype declarations, attribute declarations and specifications, and function declarations. Notably absent from packages are signals and variables; this restriction forces a description to explicitly specify (as ports in the interface) its communication channels with other components.

Once a package is defined, it may be referenced by any other description. A context clause at the beginning of a description specifies which packages constitute the context in which that unit is to be compiled. All declarations that appear in each package mentioned in the context clause are then visible within the description. Different descriptions may reference the same package in their context clauses to share the declarations contained therein.

Packages are also a convenient way to encapsulate all declarations relating to some abstraction. For instance, a package might contain declarations relating to 1's-complement arithmetic, including a bit vector type and integer type, conversion functions between the two, and the standard arithmetic functions. Another package might contain RS-232 declarations, including an enumeration type specifying signal values, a record type representing the set of signals involved, and perhaps attribute declarations that specify the pin numbers of each signal in the standard DB-25 connector.

A set of standard packages is supplied with every installation of VHDL. Package Standard contains all predefined types, attributes, and functions—a package implicitly referenced by every description and not alterable by the designer. Package Measure contains a number of commonly used physical type declarations (distance, voltage, capacitance, temperature, frequency, etc.). Packages Ones_Comp, Twos_Comp, Signed_Magnitude, Unsigned_Magnitude, and Biased contain type and function declarations for the corresponding representations of integers in digital hardware.

The design process

The components that make up the VHDL hardware support environment (Figure 12) are the design library and the analyzer, profiler, and simulator. The environment is an open-ended system designed to allow additional tools to be added easily. Figure 13 shows the flow of data in the design process.

The design library contains intermediate representations of VHDL descriptions. A single, installation-wide database shared by all users of the support environment, it contains facilities for recording the status of each intermediate form representa-

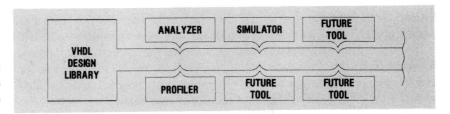

Figure 12. VHDL hardware support environment.

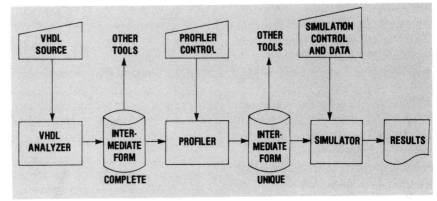

Figure 13. Flow of design data.

tion and the relationships among these representations.

The analyzer accepts the VHDL description of a hardware design, translates it into the intermediate form, and stores it in the design library. It enforces the syntax and static semantic rules of the language and alerts the designer to any violations of these rules.

The VHDL description of a hardware design may be contained in one or more design files. Each design file is a partial or full description of the design of a hardware entity. A complete design may be described in one design file; alternatively, several design files may be used to describe the design incrementally. Because several designers may contribute to a design at different times, each design file in the hardware design contains one or more blocks of VHDL code, each of which is called a design unit.

VHDL provides four kinds of design units: design entity interfaces, design entity bodies, functions, and packages. Two rules govern how VHDL design units are analyzed: (1) a body cannot be analyzed before its interface and (2) a unit that references a package cannot be analyzed before the package it references.

The profiler configures a cross-section of a design hierarchy by pulling all necessary design entity interfaces, bodies, functions, and packages from the library. This configuration may be used by the simulator and other design tools. Any generic parameters referenced within the units are assigned values at profile time. Generic design entities are instantiated, and additional consistency checks are also performed at profile time.

The simulator simulates the configuration produced by the profiler, providing mixed-mode and mixed-level simulation. The simulator will record and report signal histories and dynamic errors, including any assertion violations.

The future of VHDL

A number of language-related documents are now being or will be finalized as the language is implemented. *VHDL Reference Manual*[7] specifies the syntax and semantics of the language. *VHDL Design Analysis and Justification*[8] discusses the rationale for the design of VHDL and *VHDL Language Requirements*[9] justifies the language against its requirements. Finally, *VHDL User's Manual*[10] describes how features of the language should be used by design engineers, describes scenarios in which the language may be used, and contains eight benchmarks coded in VHDL.

VHDL forms a part of the ongoing VHSIC Phase 2 program, primarily a technology development and insertion effort for submicron chips. As part of that effort, each contractor would describe its chips in VHDL for delivery to the DoD, and verify those designs by simulating the VHDL descriptions with the VHDL simulator. Further, to verify the capability of the contractor to support VHSIC technology insertion efforts, contractors would accept a VHDL description of a chip, together with other documentation, and produce and test the chips. This demonstration will be proof that the range of description provided by VHDL is sufficient to support a wide range of design methodologies and applications.

The government intends to make VHDL a standard language, making mandatory the use of VHDL as the design and description mechanism in DoD hardware design efforts. To maximize the usefulness of VHDL in this capacity, a set of design automation tools centered around VHDL and around the VHDL support environment definition must be built—the overall objective of the Integrated Design Automation System program, now underway in the DoD. The result of the program will be a sophisticated design automation system, including tools at all levels of design, centered around a database used for the storage and query of design information. The language used to enter information into the database (to describe the components in the systems) will be VHDL; the database will be built around the VHDL design library.

We have shown that VHDL supports the design, documentation, and efficient simulation of hardware. The overall organization of the language reflects the hierarchical structure of hardware designs, and the design entity concept provides an appropriate abstraction for describing hardware components.

The language allows all hardware levels—from digital systems to logic gates—to be described independent of technology, methodology, and design tools. The ability of the designer to define new data types and bus resolution functions means that VHDL can support many technologies while committing to none. Finally, because descriptions at different levels may be processed in an arbitrary order, both top-down and bottom-up design approaches can be supported. ☐

Acknowledgments

The work described in this article is supported by Air Force Contract F33615-83-C-1003, administered by Wright-Patterson AFB, Ohio.

References

1. D. F. Barbe, "VHSIC Systems and Technology," *Computer,* Vol. 14, No. 2, Feb. 1981, pp. 13-22.
2. G. W. Preston, *Report of IDA Summer Study on Hardware Description Language,* Institute for Defense Analyses, Arlington, Va., Oct. 1981.
3. A. Dewey, "VHSIC Hardware Description (VHDL) Development Program," *Proc. 20th Design Automation Conference,* IEEE Press, New York, 1983.
4. *Military Standard Ada Programming Language,* US Department of Defense, Jan. 1983.
5. R. Piloty et al., *CONLAN Report,* Springer-Verlag, Berlin, 1983.
6. M. R. Barbacci, "Instruction Set Processor Specification (ISPS): The Notation and Its Applications," *IEEE Trans. Computers,* Vol. C-30, No. 1, Jan. 1981, pp. 24-40.
7. *VHDL Language Reference Manual: Version 5.0,* tech. report IR-MD-025-1, Intermetrics, Bethesda, Md., July 30, 1984.
8. *VHDL Design Analysis and Justification,* tech. report IR-MD-018-1, Intermetrics, Bethesda, Md., July 30, 1984.
9. *VHDL Language Requirements,* tech. report IR-MD-020-1, Intermetrics, Bethesda, Md., July 30, 1984.
10. *VHDL User's Manual: Vol. I, II, and III,* tech. report IR-MD-029, Intermetrics, Bethesda, Md., July 30, 1984.

Moe Shahdad is the task leader for the design of the VHSIC hardware description language. Prior to joining Intermetrics in 1981, he was director of the Information Systems Division at Inter Systems, Inc., in Virginia. From 1974 to 1979 he was assistant professor of computer science in the Department of Mathematics and Computer Science at the Tehran University of Technology.

Shahdad received a BS in electrical engineering from the University of Tehran in 1967, an MS in mathematics and computer science from the Queen's University of Belfast in 1971, and a PhD in computer science from the University of London in 1974. He is a member of the ACM and IEEE-CS.

Erich Marschner is a software engineer at Intermetrics, Inc. Prior to that, he was employed by Advanced Computer Techniques Corporation. He has been involved with the design and implementation of languages and language processors since 1980. His interests include both natural and computer languages, integrated programming environments, and CAD systems.

Marschner holds a BS in computer science and a BA in medieval languages and literature from the University of Maryland. He is a member of the ACM and IEEE-CS.

Kellye Sheehan is a software engineer with Intermetrics and a member of the VHDL design team. Graduated with honors from Texas A&M University with a BS in computing science in 1981, she has since been active in compiler and language work. Sheehan is a member of the ACM.

Ronald Waxman, a senior engineer at IBM's Federal Systems Division in Manassas, Virginia, is lead engineer for IBM in the VHDL effort. Recently he was lead engineer for the CAD aspect of the Phase I VHSIC program at IBM. In 29 years with IBM, he has been involved with circuit, logic, and system design; CAD; and hardware design languages. He holds patents in circuit and logic design and has authored papers on design automation and on the use of hardware design languages in the design process.

Waxman received a BSEE from New Jersey Institute of Technology and an MSEE from Syracuse University. He is a senior member of the IEEE, a member of the ACM and IEEE-CS, and chairman of a subcommittee of the IEEE-CS Design Automation TC, which is establishing design automation standard interfaces.

Roger Lipsett is Intermetrics' program manager for the VHSIC hardware description language. He has worked at Intermetrics since 1978 as a compiler and language designer. Prior to 1978, he was employed by the Large Systems Group of Digital Equipment Corporation, where he was responsible for APL and Basic development on the PDP-10 and PDP-20 computers.

Lipsett received his BS and MS in mathematics from Brandeis University in 1971, and his PhD in mathematics in 1974 from the Massachusetts Institute of Technology. He is a member of the ACM and IEEE-CS.

Howard Cohen is a member of the VHSIC hardware description language design team for Intermetrics. Prior to joining Intermetrics in 1982, he was a member of the technical staff at GTE Laboratories. His interests include language and compiler design, software development methodologies, and design of software engineering tools.

Cohen received an MS in computer science from the State University of New York in 1978. He is a member of the ACM and IEEE-CS.

David A. Ackley is the deputy program manager of Texas Instruments' part of the VHDL contract. He joined TI in 1959 and has spent the last 25 years developing digital systems and components. Over the last several years his work has included CAD tool and hardware design language development in support of the VHSIC program. He received a BS in physics and an MS in electrical engineering from the California Institute of Technology in 1958 and 1959.

Questions concerning this article can be addressed to Shahdad at Intermetrics, Inc., 4733 Bethesda Ave., Suite 415, Bethesda, MD 20814.

Chapter Three: Failures, Fault Models, and Testing

The failure of a VLSI chip may be due to design or fabrication flaws, environmental factors, or a combination of the two. The resulting physical defects consist mainly of breaks in lines, shorts between lines at the same interconnection level (metallization, diffusion, and polysilicon), shorts through the insulator separating different levels, shorts to the substrate, point defects, and large imperfections such as scratches across the chip. Other possible defects are incorrect dosage of ion implants, contact windows that fail to open, and misplaced or defective bonds. During the operation of the chip, faults may also be caused by power supply fluctuations, and ionizing or electromagnetic radiation.

In this chapter we are mainly concerned with failures that can occur during the manufacture or use of the integrated circuit (IC) chip.

3.1 Flaws and Testing

During the complicated stages of IC processing, several factors can introduce flaws, resulting in imperfect circuits. How these flaws are introduced and how they affect the final circuit depends primarily on the particular circuit topology and process type used. Such flaws are tested for at various stages during the fabrication and lifetime of the circuits. It is important to differentiate which physical and process mechanisms cause what kinds of deviation from the expected behavior. How well these mechanisms are understood and modeled determines the success of detecting failures. This affects both the testing and final yield of the ICs.

3.2 Failures

Ideally, in a properly fabricated wafer of VLSI circuits, we would expect all of the circuits on the wafer to be good, functional circuits. In practice, the number of good circuits may range from nearly 100% to only one (or even none) good circuit per wafer. Usually, the causes for a less than perfect yield fall into three basic categories: parametric processing problems, circuit design problems, or random point defects in the circuit.

If one looks at a map or photograph of a tested wafer, one of the most obvious features is that the wafer may be divided into regions with a very high proportion of good chips and other regions where the yield of good chips is very low or even zero.

3.2.1 Processing Effects

During processing, it is not uncommon for wafer size to vary in excess of 20 ppm. Thus, a 125-mm wafer changes size by 2.5 μm. If such variations in wafer size are not compensated for, some areas of the wafer may have inoperative circuits because of misalignment. In addition, we may find areas of the wafer where improper cleaning has left a chemical residue that can lead to excessive formation of oxidation-induced stacking faults. These stacking faults may result in excessive leakage currents and subsequent circuit failure.

Poor starting materials as well as contaminated chemicals and gases used in wafer fabrication can cause a host of problems including conducting "spikes" in pn junctions. Phosphorus-contaminated furnaces can also affect device parameters.

Accurate environmental control of the processing area is absolutely important. The performance of organic resists used to define device structures depends on the humidity and temperature of the processing area, as do most other fabrication steps.

The source of many problems, such as defective contacts, can be found in incorrect etching and misalignment of masks, which are examples of poorly processed wafers.

3.2.2 Circuit Sensitivities

Often two circuits of nominally identical size and complexity, processed in the same technology, have vastly different yields. The low yield in the one case is due to a lack of understanding of the sensitivity of the circuit to device parameters. Here, higher yield requires a cooperative effort between the circuit designer, who identifies the specific device parameters to which the circuit is sensitive, and the process engineer, who optimizes both the value and range of variation of those parameters. Once the circuit sensitivities to specific process parameters have been determined, redesign of the circuit to reduce these sensitivities produces high yield and low cost, with minimal attention from the process engineer.

Threshold voltage (V_T) and channel length (L) of the MOS transistors are two of the most important parameters in MOS-circuit design. Variations in substrate doping, ion implantation dose, and gate oxide thickness will cause variations in the threshold voltage. Variations in gate length, and source and drain junction depth cause the channel length to vary. The threshold voltage and channel length variations are generally not correlated with each other. However, the speed of a circuit generally increases as the threshold voltage and channel length decrease. Circuit performance is often simulated under high-speed (small V_T and L) and low-speed (large V_T and L) conditions. It is important that circuit performance should also be simulated for small V_T and large L as well as for large V_T and small L. Circuit design must also consider variations in other device parameters, such as the resistance of implanted regions, capacitance of conductors to substrate, contact resistance, and leakage currents.

3.2.3 Point Defects

A point defect is a region of the wafer where the processing is imperfect, and the size of this region is small compared to the size of the chip. Many types of processing defects are considered to be point defects. One of the most common defects is dust or other particulates from the environment. Isolated oxidation-induced stacking faults that cause excessive junction leakage and circuit failure can also be considered point defects, as can isolated spikes in an expitaxial film or pinholes in a dielectric film.

Point defects can occur on lithographic masks as well as on silicon wafers. Dust or particles on the mask substrate during the mask-generation process can cause permanent defects in the mask. Particles that adhere to the mask during use cause a gradual increase in the density of point defects that are reproduced on the wafers. Periodic cleaning of the mask will remove these defects.

Successful IC fabrication requires continued monitoring of the density of point defects. Monitoring may be done visually or by scanning electron microscopy to observe circuit quality during all steps of fabrication. When any variation in the defect density of a particular operation is observed, the appropriate corrective action can be taken. Special operations may also be performed to monitor defects. Wafers can be etched to remove all films, and the silicon substrate can be treated to reveal stacking faults, whose density is then monitored. Other wafers may be patterned by using special masks that allow electrical measurement of the incidence of dielectric breakdown defects. As new types of defects that produce circuit failure are identified, methods for monitoring the density of these defects must be instituted.

The circuit yield and the cause of circuit failure must also be monitored. The use of a circuit, such as a memory chip, in which a failure can be related back to a specific area of the chip (e.g., one of the six transistors of a static memory cell), has been most useful. The successful operation of an IC facility requires that the density of point defects be continuously monitored, controlled, and reduced.

3.3 Testing

Testing is intended to detect the presence of permanent physical failures. Thorough testing will ensure that defective chips or subsystems can be removed from the overall system and, thus, prevent these defects from affecting the operation of the system. During the evolution of a manufacturing process, it is also necessary to locate the failure site so that any deficiency in the process responsible for the failures can be rectified. Since there is a very large number of possible physical failure sites in a complex system, it is practically impossible to attempt to test for each individual failure.

Traditionally, testing has been done at the gate level (where physical failures are modeled by permanent "stuck" faults on the lines of the equivalent gate model). Empirical evidence indicates that detecting failures at the gate level will, in fact, detect actual physical failures in the circuit. However, certain physical failures cannot be detected by gate-level tests. Recent work, therefore, has studied failures at the transistor level [1,2]. Because of the possible large number of transistors, these studies have been limited to relatively small cells. Although a test for gate-level stuck faults would detect many physical failures, a transistor-level approach would be necessary to more accurately locate failures, since a fault at the gate level may not correspond to a physical failure, and vice versa. A transistor-level approach is compatible with a design methodology that uses a library of predesigned cells.

Another approach is to try to describe failures at a level higher than the gate level. The advantage here is that the number of primitives in the overall system will be much smaller, since each primitive is a much more complex circuit [3,4,5]. However, this approach is still in its infancy and more work is needed to see whether simple descriptions of failures are possible for these higher level cells. These higher-level fault models could be developed for commonly used standard cells and macros.

3.4 Test Generation

Automatic test generation has been applied very successfully at the gate level for combinational circuits. Design methodologies that allow controllability and observability of each of the combinational blocks, by appropriate scan-path or other techniques, have been extremely successful in achieving high testability of the overall circuit. However, these approaches are somewhat restrictive to the designer and may require too high an overhead for full custom designs. In addition, failures such as stuck-open faults in CMOS cause sequential behavior in combinational circuits and, thus, cannot be tested for by the gate-level combinational test generation techniques. More work, therefore, is required to automate the generation of test patterns that will detect failures at the transistor level.

3.5 Fault Simulation

When a circuit is designed with a mixture of combinational and sequential blocks with limited controllability and observability, one approach to deriving tests is to start with a set of test patterns compiled by the designer or test engineer. These test patterns are then graded for their fault coverage by using a fault simulator. Again, existing fault simulators are primarily at the gate level, but some recent work has addressed the question of transistor-level fault simulation. More work must be done in this area, as well as in the area of simulating very large circuits in a short time. Hardware simulation engines [6] are seen as a possible solution to the problem of simulating very large circuits. For a survey of hardware simulation engines, refer to the paper by Blank that is reprinted in Chapter Two.

3.6 Reprints

The expanding role of testing in semiconductor manufacturing is emphasized in the reprinted paper by Mahoney. In this paper the use of lasers and electron beams for testing and making repairs is emphasized. The reprinted paper by Menzel and Kubalek, and the one by Henley, have detailed discussions of electron-beam and laser-beam testing, respectively.

Test chips are of tremendous value in process evaluation [7], and process development [8]. Parametric test structures and measurement methods are discussed in the reprinted paper by Buehler. Such tests focus on the extraction of device and process parameters, the evaluation of design rules and process limits, and the impact of all of these elements on process yield.

Transmission electron microscopy (TEM) is a useful tool for solving problems in VLSI technology that require high spatial resolution. Information is obtained by an analysis of electron diffraction patterns and by an analysis of the microscope image. In [9], sample preparation methods are discussed, and micrographs are presented which demonstrate some of the TEM's capabilities. The instrumental methods found useful to solving problems arising in VLSI technology development efforts are discussed in the reprinted paper by Marcus.

While the stuck-at fault model can represent the effects of a significant percentage of the physical defects that occur in NMOS and CMOS VLSI circuits, it is inadequate [1]. The reprinted papers by Galiay et al. and by Hayes each propose a more comprehensive fault model for VLSI circuits. For a description of a fault-simulator based on the Hayes' model, see [10].

Testing combinational circuits using the D-algorithm is summarized in the Abadir and Reghbati paper reprinted in chapter one. The paper by Goel [11] and those by Fujiwara and Shimono [12] and Motohara and Fujiwara [13] describe the PODEM and FAN algorithms, respectively, which are improvements to the original D-algorithm.

The classical D-algorithm is not directly applicable to MOS circuits. Jain and Agrawal describe in their paper reprinted here how the D-algorithm can be extended to generate tests for both the stuck type faults and the transistor faults (open and short). The MOS circuits considered may contain transmission gates and busses. The paper by Banerjee and Abraham [14] illustrates various types of non-stuck-at behavior resulting from physical failures. It is shown that for detecting certain types of failures, it may be necessary to apply test vectors in a certain sequence.

Stuck-open faults in CMOS combinational logic circuits create memory and require test sequences of length two. It has been shown that tests for stuck-open faults in CMOS combinational logic circuits could potentially be invalidated because of delays in the circuit being tested and/or because of timing skews in the input changes. The method given in the reprinted paper by Reddy et al. derives tests, when they exist, that will remain valid in the presence of arbitrary delays in the circuit being tested and/or in the presence of timing skews in the input changes.

Many current logic-simulation techniques for digital systems use gate-level circuit models and, if they allow fault simulation, employ the single-stuck line fault model. Some major difficulties are encountered when these simulators are applied to VLSI circuits, especially circuits based on MOS technology. The papers by Kawai and Hayes [10] and Schuster and Bryant [15] describe fault simulators for MOS circuits.

Complexity of a fault simulator in terms of the CPU time and the memory requirements is known to grow at least as the square of the number of gates in the circuit. For VLSI circuits this is a serious limitation [16]. Abramovici et al. [17] and Jain and Agrawal in the reprinted paper describe recent attempts to combat fault simulation complexity.

Even with proper design, VLSI systems can have timing problems due to physical faults or to variation of parameters. Malaiya and Narayanaswamy [18] describe a fault model that takes into account timing-related failures in both the combinational logic and storage elements.

References

[1] T.E. Mangir, "Sources of Failures and Yield Improvement for VLSI and Restructurable Interconnects for RVLSI and WSI: Part I—Sources of Failures and Yield Improvement for VLSI," *Proc. of IEEE,* Vol. 72, No. 6, June 1984, pp. 690–708.

[2] J.P. Hayes, "Fault Modeling," *IEEE Design & Test,* Vol. 2, No. 2, April 1985, pp. 88–95.

[3] M.S. Abadir and H.K. Reghbati, "Test Generation for LSI: A Case Study," *Proc. of ACM-IEEE Design Automation Conf.,* 1984, pp. 180–195.

[4] M.S. Abadir and H.K. Reghbati, "Functional Test Generation for LSI Circuits Described by Binary Decision Diagrams," *Proc. of IEEE International Test Conference,* 1985 (to appear).

[5] M.S. Abadir and H.K. Reghbati, "Functional Specification and Testing of Logic Circuits," *International Journal of Computers and Mathematics*, 1985 (to appear).

[6] D.M. Lewis, "A Hardware Engine for Analog Mode Simulation of MOS Digital Circuits," *Proc. 22nd Design Automation Conf.*, 1985, pp. 345-351.

[7] M.G. Buehler and L.W. Linholm, "Toward a Standard Test Chip Methodology for Reliable, Custom Integrated Circuits," *Proc. 1981 Custom Integrated Circuits Conference*, 1981, pp. 142-146.

[8] R.T. Jerdonek et al., "The Use of a Silicon-Gate CMOS/SOS Test Vehicle to Evaluate Technology Maturity," *IEEE Trans. Electron Devices*, Vol. ED-25, 1978, pp. 873-878.

[9] R.B. Marcos and T.T. Sheng, "TEM Studies of VLSI Circuits," *Proc. of 1982 MIT Conf. on VLSI*, 1982, pp. 78-83.

[10] M. Kawai and J.P. Hayes, "An Experimental MOS Fault Simulation Program: CSASIM," *Proc. of ACM-IEEE 21st Design Automation Conf.*, 1984, pp. 2-9.

[11] P. Goel, "An Implicit Enumeration Algorithm to Generate Tests for Combinational Logic Circuits," *IEEE Trans. Comput.*, Vol. C-30, No. 3, March 1981, pp. 215-222.

[12] H. Fujiwara and T. Shimono, "On the Acceleration of Test Generation Algorithms," *IEEE Trans. Comput.*, Vol. C-32, No. 12, Dec. 1983, pp. 1137-1144.

[13] A. Motohara and H. Fujiwara, "Design for Testability for Complete Test Coverage," *IEEE Design & Test*, Vol. 1, No. 4, Nov. 1984, pp. 25-32.

[14] P. Banerjee and J.A. Abraham, "Characterization and Testing of Physical Failures in MOS Logic Circuits," *IEEE Design & Test*, Vol. 1, No. 3, Aug. 1984, pp. 76-86.

[15] M.D. Schuster and R.E. Bryant, "Concurrent Fault Simulation for MOS Digital Circuits," *Proc. of MIT VLSI Conf.*, 1984, pp. 129-138.

[16] S.K. Jain and V.D. Agrawal, "Statistical Fault Analysis," *IEEE Design & Test*, Vol. 2, No. 1, Feb. 1985, pp. 38-44.

[17] M. Abramovici et al., "Critical Path Tracing: An Alternative to Fault Simulation," *IEEE Design & Test*, Vol. 1, No. 1, Feb. 1984, pp. 83-93.

[18] Y.K. Malaiya and R. Narayanaswamy, "Modeling and Testing for Timing Faults in Synchronous Sequential Circuits," *IEEE Design & Test*, Vol. 1, No. 4, Nov. 1984, pp. 62-74.

CLOSING THE LOOP

An Expanding Role for ATE in Semiconductor Manufacturing

Matthew Mahoney
Senior Technical Consultant
LTX Corporation
Westwood, MA 02090

ABSTRACT

Automatic Test Equipment is generally used to screen out bad units and sort the remainder into useable categories, but in itself does not guide the fundamental processes in which device failures take place. Until recently, the best that ATE could do to improve the process was to provide information to management on the results of its tests.

Increasingly, however, ATE is cast into a new role whereby it acts directly on the process steps to prevent failures and to alter the mix of product grades--ultimately a far more valuable role for productivity than its old one. This paper provides an overview of direct process control by ATE, and presents a few examples of present and future applications in semiconductor manufacturing.

INTRODUCTION

The theme of this year's conference is quality, productivity and profit. As users or manufacturers of Automatic Test Equipment, we like to believe that ATE already contributes heavily to the areas. But as I step back to take a broader view of industry, especially to see how electronics and computers fit into other aspects of manufacturing, I am surprised to find how little ATE is really doing. There are several valuable roles that ATE might be playing, but we seem to have asked it to play just one. From the stand-point of productivity, it is probably the wrong one at that.

There was a story told in management seminars some years ago about the potato farmer who hires a helper to grade potatoes by size and quality. At the end of the day, the farmer returns to find that nothing has been done. The bins are empty, the original pile of unsorted potatoes still stands, and the helper is still holding potato number 1. The farmer, quite angry, berates the new helper for his laziness. "Hey, I'm not afraid of hard work," the helper replies, "it's just all these blasted decisions I can't stand."

This is certainly a problem that ATE has solved most effectively. Our machines make sorting decisions tirelessly all day long. With each new year they sort ever faster, more thoroughly and more precisely.

ATE has been so successful in this particular role, though, that it seems to have blinded us to the fact that the problem we have solved is not the major problem of production--whether potatoes or integrated circuits. What good is sorting if there is nothing to sort--if there has been a crop failure? Or if the percentage of good items after sorting and screening is so low that the manufacturer goes out of business?

I think the problem that the manufacturer really wants to solve is how to ensure consistently uniform, high quality production--not only to eliminate wasted labor and material, but also to control the ratio or mix of products to meet the market demand. It is ironic that, in this situation, there would be little need for conventional ATE. Yet the manufacturer would by definition have high quality, and obviously be productive and profitable.

A psychiatrist might conclude that ATE, far from being a tool of efficient manufacturing, is a symptom of an unhealthy manufacturing process. The more variable and uncertain the quality at each stage of manufacturing, the greater the need for screening and sorting. We are so used to using ATE in this role, however, that it is hard to envision it doing something different. When I ask users or makers of

ATE how new equipment might be used to improve the fundamental yield of a factory, the answer is all too often to do more of the same: to grade more potatoes, grade them faster, and throw them into higher resolution bins.,

Oh yes, a sophisticated system should remember how many potatoes it has thrown into each bin. And a really sophisticated system will today communicate this information by wire. But equipping the farmer's helper with a hand tally and a telephone has not changed the fundamental nature of the job. Screening, sorting and reporting occur only after the fact; they do not act directly to prevent the problems in the first place.

What is the alternative? To answer this, let us step backwards in time a quarter-century, and visit a factory making plastic film. The thickness of the film is controlled mostly by the spacing of two heated rollers. It is influenced, however, by a number of factors like temperature, speed, and plastic formulation, that will vary in the course of production. Today, the manufacturer wants 2 mil film, nothing else. How does he do it?

Figure 1 shows the answer in highly simplified form. A lamp or low-level radiation source is placed on one side of the film, and a sensor is placed on the other. Should the film thickness begin to deviate from its desired amount, the detector output will begin to rise or fall relative to some pre-set reference. The error signal then starts a motor which adjusts the roller spacing in a way to reduce the error.

This is a simple example of a servomechanism at work. The information involved in this action flows not in a network, but in a loop--a servo loop. What is important is that the manufacturing process is self-adjusting on the run, not after the fact. The error it takes to activate the loop is always within the acceptable thickness tolerance, so the process continuously produces the right thickness, all the time. The only out-of-spec material, or waste, is the "tailings"--the beginning and end of a roll. This is a vanishingly small percentage in a good process.

Incredible! There is no ATE system farther down the line, sorting the film into sheets of different thickness, or reporting at the end of the day that everything we made was wrong. Instead, this instrumentation is used directly to ensure that the output is what the manager wants--nothing more or less. The measurement system is used to control the process, not to judge it after it is too late.

Can we do the same for semiconductor manufacturing? The answer is yes. Modern computer-based ATE has the diagnostic and control ability to participate in, or direct, a number of self-correcting process steps. Dozens of applications exist today and the use is growing rapidly. While new technology is partly responsible, the biggest factor is probably a change in viewpoint, a re-thinking of the process and how to control it.

Re-defining the Process

To see what this means, let us start by defining the "process" in its broadest sense--everything from materials in, to shipments out (Figure 2). The intent of this is a bit Zen-like. We clear the mind, so that we may next ask where-- anywhere and in any way--the sensing and analytical abilities of computer-based test equipment can be put to use. No preconceived notions.

The old approach was to divide the process into the obvious, visible sub-processes, then to use automatic or manual measurement for screening and sorting. Military electronics provides an extreme example where, as in Figure 3, screening occurs after every step.

Screening is one way to achieve output quality, but it is a very expensive way. It is essentially a procedure of progressively diminishing returns. Each stage might have an acceptance ratio of, say, 50%, but with 10 such stages, the overall yield is a dismal 0.1%!

To see the alternative, it helps to use an analogy from optics. In Figure 4, we see an optical train made up of divergent lenses--each analogous to one stage of an ordinary process. In maintaining quality by screening, each succeeding stage captures only some fraction of the light coming from the previous one. The yield of the process is the product of all these fractions--usually a small number. A lens made this way would be very dim indeed.

This is an example of a totally divergent process--one where each element is divergent. Each divergent element increases the scatter or spread of output parameters, or is said to disperse them. If it is not the intent of the whole process to have significant scatter, screening is required along the way to constrain the total scattering. The process is inherently one of high loss.

Now suppose we replace some of the lenses with convergent elements, as in Figure 5. The dispersion of light rays is constrained this time by occasional reversals in the process. Very little light is lost this way, and the rays may be focussed at the desired target. The optical process is now one of inherently low loss, or high yield. It can also provide an extremely wide choice of systems--far greater than a divergent system with screening.

Are there manufacturing equivalents of convergent lenses? There are several types, in fact, two of which we will put to work today. In doing so, we should observe an important consideration evident in Figure 5: _it is not necessary to make every element in a manufacturing process convergent in order to have a convergent process_. There indeed may be steps in semiconductor processing that cannot be perfectly controlled, and are thus divergent; but that is okay as long as we follow that stage with a convergent sub-process.

Convergent Processes

There are a number of schemes to make processes converge. The two most common ones involve feedback loop mechanisms or feedforward error correction mechanisms. Both are important to us here because they are "programmable" or easily altered when desired.

The plastic film example shown earlier employs a feedback loop. The ouput is continually sampled or sensed, and this information compared with a preset requirement called the set point (Figure 6). If it does not match, an error signal is produced which causes the process to adjust itself in the proper direction. It is the error signal, not the set point, which actually adjusts the process.

A poorly designed feedback loop can be unstable, causing the output to exceed the acceptance limits of the next stage either occasionally, or in a periodic "hunting" action. One of the several requirements to ensure that the output always stay within its performance limits is to make the time delay in the feedback loop small compared to the response time of the forward path to sudden changes in the set point, or in the disturbance ("noise") effects. In this way, any unwanted trend in the output can be offset by corrective action before it becomes large enough to be significant.

For many processes, however, there is no feedback correction possible. Typically, these involve batch or transient inputs, as in Figure 7. Here, let us say the forward lag is one hour, perhaps just the propogation delay to a group of wafers travelling through an oven. If there is only one group to be handled today, which is loaded or unloaded in one batch, feedback is meaningless. The knowledge that too much diffusion has occurred comes too late to prevent it. Feedback is thus limited primarily to "steady state" processes-- at least, to those in which the process is sustained for a much longer time than the process lag.

A useful alternative for transient processing is feedforward correction-- actually an older process than feedback, but one almost forgotten until recently. Figure 8 shows a rearrangement of the earlier control system for feedforward. As before, the output is sensed and compared with the set point, and an error signal is produced. But this time, the error is sent forward, amplified or converted to the same form as the output, then subtracted from the output. This arrangement does not tend to hunt or oscillate. There is a closed path in the diagram, but it is not a feedback loop.

Both methods produce convergence, however: they accept a wide range of material input parameters and tolerate a wide range of disturbance while producing outputs that are consistent about the set point. The two immediate effects in production are to produce high yield and quality, and (by virtue of the set point input) to allow the manager a reliable means of adjusting the output to whatever is required.

There is a third, far less obvious advantage to convergent elements, concerning the way information is used. Consider again the old screening concept. By definition, the nature of each device reaching a screening station is uncertain. Information is thus generated, continually, and in large amounts (Figure 9).

Also by definition, this screening equipment is helpless to act directly to control its surrounding processes. There is only one place that its information can go--to a human, who must decide what everything means, and what to do about it.

Networking is a popular topic these days, particularly over high-speed lines. But I think the kind of network I have just shown you is the wrong kind--a terribly inefficient way to run a factory. Imagine that those ten stages in the diagram deliver their data by nothing more elegant than simple telephone-grade lines, e.g., RS232. Even those lines, if kept busy 8 hours, can collectively transmit 48 thousand pages of dense data--a pile of printed paper over twenty feet high each shift!

By contrast, the information flow with covergence is far more sane. Figure 10 shows that the bulk of ongoing information flows backwards (or forwards) in a simple loop, and does not involve a human. In fact, we want very much to keep people out of these loops. Even with the best intentions, they would add to the loop delay and create instabilities.

Really, what information does the test manager want to receive or send? He wants to adjust the set points, and he wants assurance that the system is following his commands. Since the process is self adjusting, and therefore less variable, less information is generated. And of this, only the important information reaches--or is sent by--the human. From a system viewpoint, the human acts as a feedback link only for very slow aspects of the forward process, where the lag in thinking is small by comparison, and where real business judgement is required. I believe that ATE networks a few years from now are going to take a very different form than what we tend to popularize today. Part of the reason is that ATE itself is going to be used quite differently.

Let us now turn to some of these applications.

Convergance by Trimming

An historical application of "convergence" is alignment--or calibration, as we tend to call it today. Originally, a technician received uncalibrated assemblies which exhibited a wide spread of characteristics, then, by trimming, brought them to a uniform behavior. Three things are noteworthy about this:

1. Trimming can be done by feedback-- e.g., by watching a meter and turning a trimpot until the parameter in question converges on the desired value.

2. Trimming can be done by feedforward-- e.g., by measuring the initial error, then selecting a fixed value of trim capacitor to be inserted.

3. The assembly has to be designed to allow testing and trimming. The designer has to understand the manufacturing process, and that he is part of it. This is a hurdle that we have not yet overcome, especially in university engineering programs.

In using this approach, the manufacturer acknowledges the impracticality of making the assemblies all uniform in some previous manufacturing stage. He compensates by adding a "convergent" stage afterwards to restore uniformity. The optical analogy is the simple picture of Figure 11.

In a modern factory, the technician is replaced by two coordinated machines: the ATE system, which measures the response and decides what action is to be taken, and to replace the technician's hands. Figure 12 shows a not-uncommon version of this for FV chassis alignment, combining the ATE system with a robot to do the trimming, and a mechanical handler other to move the chassis along.

On-line laser trimming is today a very common closed-loop calibration process (Figure 13). Between laser pulses, the ATE system reads the affected parameter via a sample-hold technique, then decides if and where additional cutting is needed. In A/D or D/A trimming, the procedure is usually a feedback procedure.

Many semiconductors are trimmed using the laser to cut links. Often this is a feedforward procedure using a single set of initial measurements to calculate which links should be cut. In principle, it is possible to calculate in advance how to trim a resistor ladder in a one-pass, feedforward technique as well.

Memory Repair

It is difficult to make large memory circuits without any defective cells. ATE/Laser combinations are used to repair faulty memory chips today by disabling bad cells and, by vaporizing other links, activating circuits that switch in good, spare cells.

This is another instance of feedforward correction. The computer must first memorize the failure pattern, then decide how to change (by laser) the circuit connections to make the memory right.

Camera Chip Normalization

A new and more dramatic example of convergence by feedforward trimming occurs with camera chips. These chips have tens or hundreds of thousands of cells (picture elements or "pixels"), each of which has a grey or continuous output. Each chip is physically large; there are not very many on one wafer, so yield is particulary important. Yet because or their large size they are subject to variations from one region to another in mask registration diffusion, etc., and this variability is easily seen in the analog output. A chip illuminated with uniform light will not, for example, produce a uniform field on a TV monitor. It is entirely possible that no chip on a given wafer is acceptable without correction.

How do we add a convergent process to make these chips uniform? One technique just now coming into use is to test the chip for gamma (light-to-voltage gain) and bias--for every individual cell, or at least for many regions of cells. The cells or regions may number in the thousands. The chip is designed with on-board sequential correction circuitry that allows the gamma and offset of each cell to be normalized by an arbitrary pattern stored in memory or programmed into a logic arrary. The ATE system must first remember all the individual cell parameters, calculate the correction pattern, and finally direct the programming (Figure 14).

Controlling the Wafer Process

Looking ahead, we see a more direct assault on wafer processing itself. A modern computer-based system is capable of diagnosing--from parameter variations of the wafer it is probing--the nature of certain errors in mask alignment, implantation, diffusion, etc. It is easy to envision this diagnostic information being used in a feedback loop for E-beam lithography, which has long been computer-driven. By extension, we picture the loop being closed around more common forms of lithographic equipment as they become more heavily based on computer control (Figure 15).

With feedback, it is desirable to reduce the loop delays where possible. Part of the likely evolution of ATE-driven lithography is thus the consolidation, of the ATE and lithographic computers into a single high speed unit l(Figure 16). One may speculate on the nature of the corporation capable of designing such a system.

Another saving in loop time comes from sensing the desired parameters as early as possible in each sub-process. In laser trimming, this means testing during the trim procedure--actually, between pulses. But what can be done at the level of wafer manufacturing itself? A year ago at this conference, G. V. Lukianoff of IBM presented a paper which described how electrical measurements were made on VLSI circuits directly by electron beam. It is fascinating to envision an E-beam lithographic system probing, monitoring, and correcting its own performance without physically contacting the wafer.

Along with its increased diagnostic and control duties, future ATE must thus develop new sensing abilities as well.

Non-contact measurement such as suggested above, is inevitable. Already, production laser trimmers are using pattern recognition to position, automatically, wafers in X, Y, and rotary alignment. But this is not nearly as daring as making circuit parameter measurement by non-contact methods. The need for the latter is obvious, as VLSI features grow smaller. A memory cell with, say, 30 femtofarads of nodal capacity cannot drive any existing metal probe. Methods involving charged beams or polarized radiation may do the job.

SUMMARY

It is time we stopped limiting ATE to its stereotyped role. There are many more productive and profitable roles it can play. Let us make the 1980's the decade in which we promote the farmer's helper from his old job as sorter to a more responsible and rewarding one: a full partner in a really productive farm.

REFERENCES

[1] H. Seidel, "Feedforward Technology;" Proc. 1973 IEEE Int. Symposium on Circuit Theory (April 1973).

[2] H. S. Black, "Translating System;" U. S. Patent 1686792 (1928).

[3] J. M. Morrisey, et al, "An Approach to Memory Testing, Diagnostics and Analysis;" Digest of Papers, 1981 IEEE International Test Conference (October 1981).

[4] G. V. Lukianoff, J. Wolcott, J. Morrissey, "Electron-Beam Testing of VLSI Dynamic RAMS;" Ibid.

[5] G. Crichton, P. Fazekas, and E. Wolfgang, "Electron Beam Testing of Microprocessors;" Digest, 1980 IEEE Test Conference (Nov. 1980).

[6] "Feedback Control Systems," Chapter 16, Reference Data For Radio Engineers, Sixth Edition; Howard W. Sams & Co., Inc.

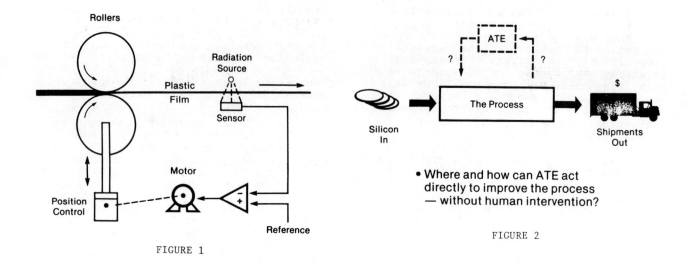

FIGURE 1

FIGURE 2

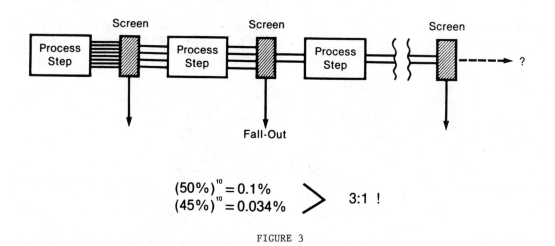

$$(50\%)^{10} = 0.1\%$$
$$(45\%)^{10} = 0.034\%$$
$$> 3:1 \ !$$

FIGURE 3

An Optical Analogy to Screening

FIGURE 4

- Not every element need be convergent to make a convergent **process!**

FIGURE 5

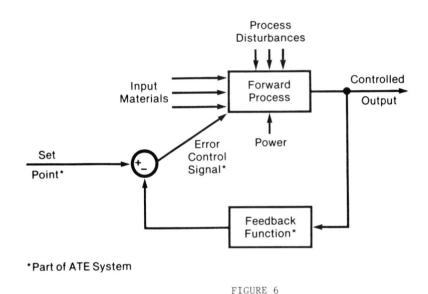

*Part of ATE System

FIGURE 6

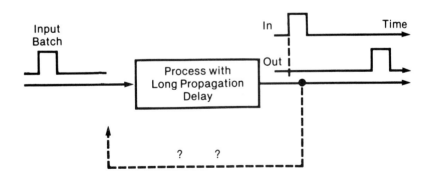

Feedback Does Not Work in Pipeline Processes

FIGURE 7

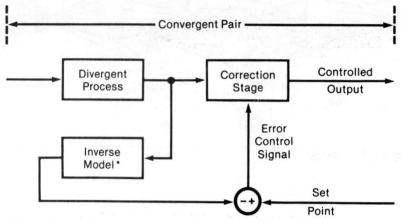

The Second Type of Convergent Process: Feedforward Correction

FIGURE 8

Information Flow with Conventional ATE Application

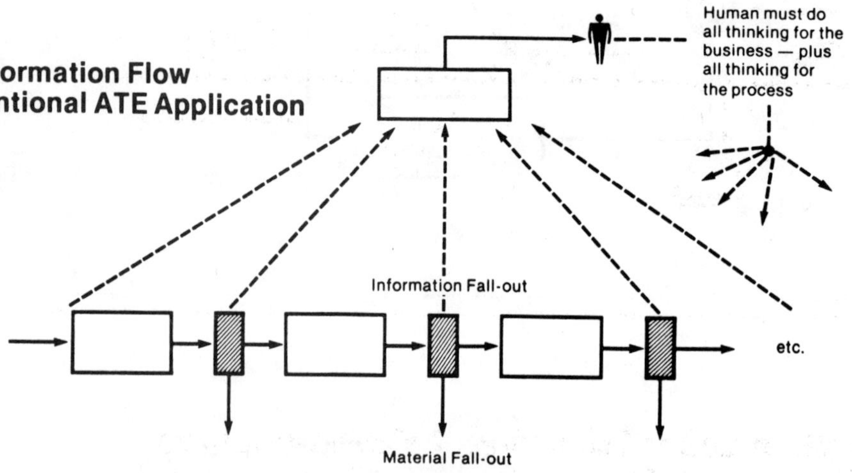

FIGURE 9

Information Flow with Convergent Loops

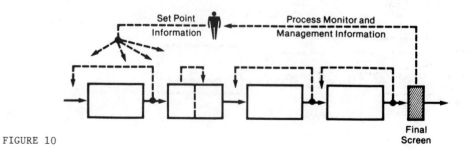

FIGURE 10

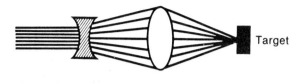

FIGURE 11

Alignment Revisited:

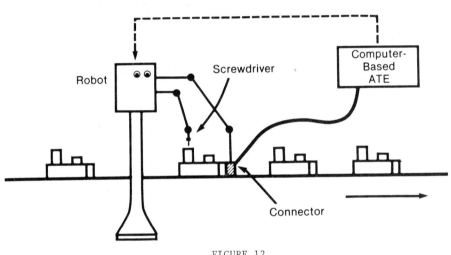

FIGURE 12

Active Laser Trimming: A Commonplace Example of Calibration by ATE Computer

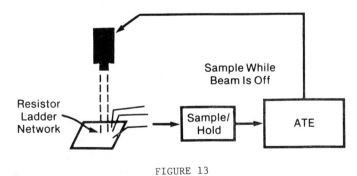

FIGURE 13

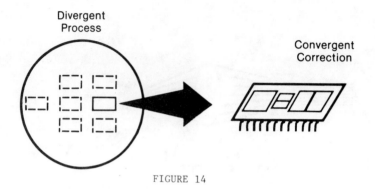

FIGURE 14

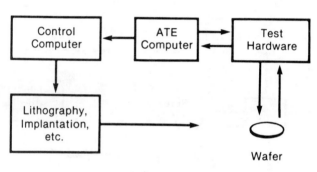

FIGURE 15

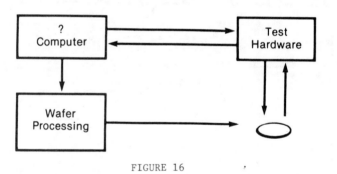

FIGURE 16

Next...

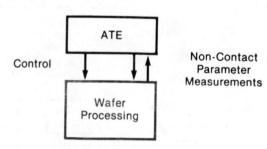

FIGURE 17

Review

Fundamentals of Electron Beam Testing of Integrated Circuits

E. Menzel and E. Kubalek

Universität Duisburg, Fachbereich Elektrotechnik, Werkstoffe der Elektrotechnik, Kommandantenstr. 60, D-4100 Duisburg, FRG

1. Introduction

The complexity of integrated circuits (IC) is steadily growing, because it allows the speed and reliability of ICs to be increased and the costs of single gates to be reduced. Increasing the complexity is accomplished by reducing the device sizes (e.g. micron or submicron wide conductor lines), enlarging the chip size (up to 100 mm^2) and improving IC-technology and design.

With growing complexity the testing of an IC is rendered more difficult and more and more determines IC costs. "Classical" test methods have been improved and new testing strategies are currently being developed aiming at keeping development times of ICs at a justifiable level and at assuring and improving the quality and reliability of ICs.

One of the new test methods that have received a tremendous impetus in the past few years is electron beam testing. The attractive properties of an electron beam lie in its capability of being used as a finely focussed, easily aligneable contactless probe, which, under certain conditions, can be operated in a "passive" mode to measure IC-internal signals in a non-loading and nondestructive way. Thus the e-beam represents the only alternative to the mechanical probe used so far for electrical signal measurements in medium scale (MSI) and large scale integrated circuits (LSI) and which completely avoids its disadvantages.

In this "Voltage Measurement Mode" the e-beam acts as a tool, which measures time dependent voltages on IC-nodes or which displays voltage distributions of large IC areas at different operating states by utilizing the dependence of the secondary electron (SE) spectrum on the potential of the emission point and on electrical fields above the IC (1-20).

Besides the well established Voltage Measurement Mode, were the e-beam is required to have as little influence on the device operation as possible, the e-beam can also take over an "active" role in IC-testing. In the "Current Impressing-Mode" the e-beam serves as a micron scaled current source taking advantage of absorbed currents flowing into the irradiated area (19, 21, 22), or of carrier multiplication in thin insulating films (23, 24) or in pn-junctions (25). By means of specially designed e-beam sensors on the chip logical states within a device can be changed allowing e.g. the IC operation mode (normal-selftesting) to be e-beam controlled (19, 21, 22) or IC subsystems to be connected to a common data bus (25).

While the Current Impressing Mode provides volatile switching operations, the "Restructuring Mode" makes non-volatile links between IC subsystems possible by utilizing e-beam induced irradiation damage in MOS-devices or e-beam induced charges in floating gate field effect transistors (25).

The full capability of the e-beam is obtained when it is operated in a combination of these three modes and when special test aids on the chip are provided (e.g. selftest) (19, 21, 22). This offers the possibility for on-wafer tests without the necessity for applying waferprobers (19, 21, 22) or would allow waferscale integrated systems on which defective subsystems could be easily localized and the remaining subsystems be interconnected (25).

The importance of e-beam testing will increase further because its field of application will enlarge by the use of additional modes and by the use of e-beam testing of interconnections. Thus, for example, the "*Magnetic Contrast Mode*" for the investigation of magnetic devices and the "*Cathodoluminescence Mode*" for testing of integrated optics devices are discussed. Testing of interconnections in multichip packages has already been used (26, 27).

In the following the fundamentals of the Voltage Measurement, Current Impressing and Restructuring Mode are treated with stress on the up till now most established Voltage Measurement Mode.

2. Voltage Measurement

Voltages on electronic devices operated in a SEM modulate the detected secondary (SE) signal. This so called "*Voltage Contrast*" is caused by the influence of the SE emission point potential and of the electrical field distribution between device and detector on the SE.

The generated contrast is non-definite and non-linear. Therefore on the one hand only a logic state analysis, limited to devices with logical swings of 1 V is possible providing the localization of soft- and hardware failures (6). On the other hand, a coarse function control and failure analysis by estimation of IC-internal voltages with typical voltage resolutions of 1 V can be achieved (7, 11, 16).

By aid of the stroboscopic or sampling mode (46) this is possible with high time resolution of the order of several ps (28–36) up to frequencies in the GHz range (28, 29, 32, 34).

The voltage information then is displayed either as a micrograph in the stroboscopic mode showing the voltage distribution of an IC area at a fixed operating state (scanned e-beam, fixed delay, e-beam pulses synchronously with IC signal), or as a waveform in the sampling mode (fixed e-beam, scanned delay, e-beam pulses synchronously with IC-signal).

By properly controlling the SE trajectories, thus rendering the contrast definite, semi-quantitative measurements become possible that allow an improved distinction between logical states enabling real time logic state analysis up to frequencies of 4 MHz at voltage swings as low as 700 mV (20).

Truly quantitative voltage measurements are obtained if a SE energy analyzing scheme is added and disturbing influences such as topography and material contrast are excluded. Provided these preconditions are fulfilled, an evaluation of voltages down to less than 1 mV (37) is possible with high time resolution in the low ps regime (29, 33, 36) up to device operation frequencies of about 10 GHz (29) and with a high spatial resolution that is adequate to cope with submicron VLSI (very large scale integration) dimensions. Voltages can be measured with typical accuracies of 5% or less depending on device geometries and applied voltages (37–39). Depending on the e-beam parameters, namely the primary electron (PE) energy, measurements can be carried out without introducing any load to the device and with negligible damage (40).

Operating in this quantitative Voltage Measurement Mode the e-beam provides informations on rise and fall times, amplitudes and delays of IC internal signals (7, 9, 11, 12, 13, 14, 15, 18, 30, 37) and in this way allows the control of IC internal signal timing, failure localization and classification (13) and even the control of computer simulation programs (41, 42).

2.1 Qualitative voltage contrast

The qualitative voltage contrast observable in any unmodified commercial scanning electron microscope on operating electronic devices is caused by the specimen voltage dependent detection of the SE. By convention, SE are those electrons generated by a primary electron (PE) beam having kinetic energies between 0 and 50 eV.

These electrons show a maximum in their energy distribution at several eV (metals : 1 to 3 eV, insulating materials : 0,5 to 2 eV (43, 44) and a full width at half maximum (FWHM) of some eV (metals : 4 to 6 eV, insulators : 2 to 3 eV (43, 44).

A typical SE spectrum of a metal (aluminium) widely used in ICs is illustrated in Fig. 1. This theoretically determined spectrum (45) which is given analytically by

$$N(E) = K \frac{E - E_F - \Phi}{(E - E_F)^4} \qquad (1)$$

where K is a constant, E the electron energy, E_F the Fermi energy and Φ the work function, shows a good agreement with experimentally obtained spectra (43).

The low energy of the SE, 75% of which having energies smaller than 5 eV, is the reason for their high sensitivity to local fields on an operating electronic device.

Different local potentials yield different signals at a detector. This contrast appears superimposed to the normal topography and material contrasts. An example is shown in Figs. 2 and 3, Fig. 2 showing an unbiased analog bipolar IC and Fig. 3 the same device with +10 V working voltage applied.

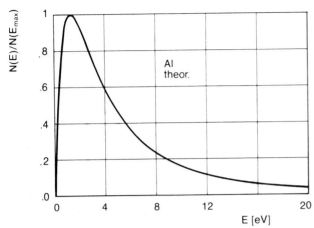

Fig. 1 SE spectrum of aluminum, normalized to E_{max} (after /45/)

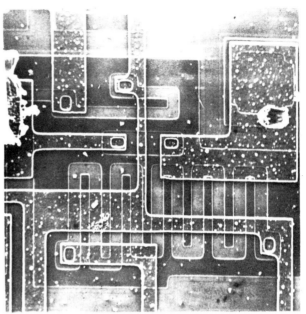

Fig. 2 SE micrograph of a section of an unbiased bipolar IC

Those parts of the device to which a positive voltage is applied, show up dark, parts at zero voltage usually remain the same, and parts at negative voltage show up brighter.

The reason for this contrast is the existence of retarding electrical fields above positively biased parts of the IC. This is explained by aid of Fig. 4 which shows the lines of equipotentials of a two-dimensional IC model in scale with the real SEM specimen chamber dimensions and a detector (Everhart-Thornley) with +300 V extraction voltage.

If the electron beam impinges on a negatively biased or grounded line of the IC, SE are continuously accelerated towards the detector. The detector integrates over the SE energy distribution and this gives rise to a signal V_s at the specimen potential V_{sp} shown as measuring values "a" for the negative and "b" for zero bias (right column of Fig. 4). If the electron beam is positioned on a positive biased line the emitted SE have to pass a retarding electrical field of some volts, so that only electrons with higher energies reach the detector causing a lower signal "c". Note the nonlinearity between the SE-signals V_s and the specimen potentials V_{sp}, giving large signal variations near the maximum of the SE spectrum and small variations at high positive specimen voltage V_{sp}.

The correlation between V_s and V_{sp} strongly depends on the barrier height of the retarding fields, on the microfields above the IC surface, and on the fields between IC-surface, polepiece, and detector. These fields between the IC surface and the detector, which vary with the IC operation influence the SE trajectories and render contrast non definite.

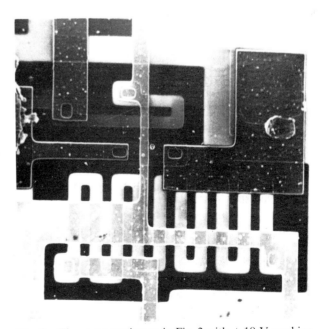

Fig. 3 The same section as in Fig. 2 with +10 V working voltage applied

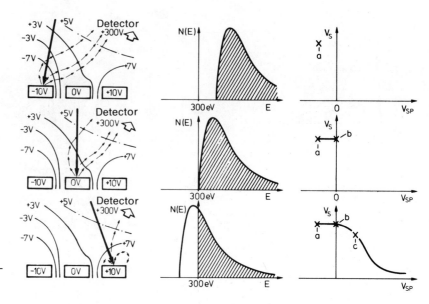

Fig. 4 Origin of qualitative voltage contrast

2.2 Quantitative voltage measurements

Specimen voltage dependent signals are obtained only by detection of Auger and secondary electrons. When SE are used for voltage measurements a number of problems (local fields, topography, material contrast, work function variations) have to be solved, which will be treated in detail in the following. These problems may be avoided by measuring the linear shift of Auger-peaks with applied specimen voltage. Using the 270 eV carbon Auger peak present on any e-beam irradiated surface allowed voltage differences of less than 1 V to be determined with a cylindrical mirror analyzer (62). However, the disadvantages of this technique, the very low Auger electron currents leading to long measurement times and the need for clean vacuum condition (ultra high vacuum or gas jet), make it impractical for voltage measurements on ICs. Therefore only the SE are of practical usefulness.

The gain or loss of the SE kinetic energy with applied specimen voltage is the basis of the quantitative voltage measurement. The whole energy spectrum is linearly shifted with applied specimen voltage as shown in Fig. 5. The linear shift of the maximum of the spectrum then is a direct measure of the applied specimen voltage. Obviously this can only be true, if the shape of the SE spectrum is in no way influenced as for example is the case, if retarding fields as mentioned in 2.1. are present on an IC surface. From these considerations the two main preconditions for a quantitative measurement can be deduced:
1. Energy analysis of the SE
2. Suppression of local electrical fields above the IC surface.

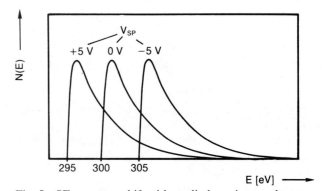

Fig. 5 SE spectrum shift with applied specimen voltage

If the voltage measurements have to be carried out on different materials (aluminium or silicon dioxide) or if quantitative voltage distribution images of the IC surface are to be made, the change of the SE-spectrum due to topography, material and local variations of the work function has to be taken into account. So the third precondition for quantitative measurements has to be:
3. Elimination of topography, material contrast, and work function variations.

2.2.1 Energy analysis of SE

Analyzing the energy of the SE may be done by analyzing the shift of the peak of the differential SE energy spectrum with applied specimen voltage (Fig. 6a) using cylindrical mirror analyzers (1) or 63° cylindrical analyzer (5).

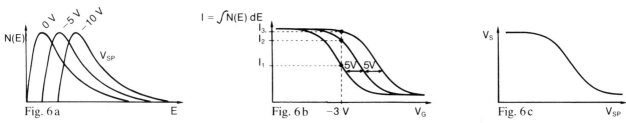

Fig. 6 Shift of differential (a) and integral (b) SE spectrum and nonlinear dependence of measured signal V_s on applied voltage V_{sp} (c)

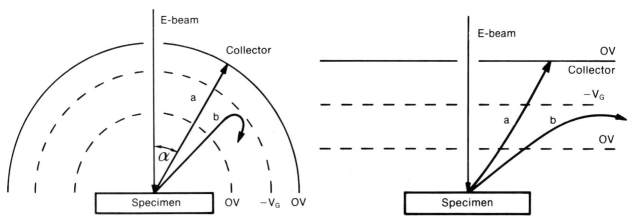

Fig. 7 Hemispherical grid retarding field spectrometer. Emitted SE enter retarding field region perpendicular to grid. Passing condition (electron a): $E_{SE} \geq e V_G$

Fig. 8 Planar grid retarding field spectrometer. Emitted SE enter retarding field region with angle α. Passing condition: $E_{SE} \geq e V_G/\cos^2 \alpha$.

The second and more widely used technique is by analyzing the integral form of the SE-spectrum as shown in Fig. 6b. The integral form of the spectrum is obtained by sending the SE through retarding fields established by either hemispherical grids (49) (Fig. 7) or planar grids (2, 3, 4, 8, 14) (Fig. 8).

As can be seen in Fig. 6b the integral SE distribution is shifting linearly as does the differential SE distribution. However, as the current I is measured at a detector the same energy shifts generate different current differences resulting in a nonlinear dependence between measured signal V_s and applied specimen voltage V_{sp} as shown in Fig. 6c.

This disadvantage of the retarding field method has been overcome by additional installation of a feedback loop (2, 4, 8, 14), which keeps the current I constant by varying the retarding field voltage V_G. The necessary variation of V_G is equal to the applied specimen voltage V_{sp} and thus provides a direct measure of V_{sp}.

2.2.2 Suppression of local electrical fields

Local electrical fields at the surface of IC, which are generated by different voltages at neighbouring lines during IC operation on the one hand act as retarding fields above positively biased IC lines and influence the SE spectrum by cutting off the low energy tail of the SE distribution. On the other hand these local fields influence the flight direction of the SE and alter their angular distribution.

Both effects can cause voltage measurements errors of up to several volts (37–39). Whereas the influence of local retarding fields can be suppressed, the dependence of the angular SE distribution on the local electrical surface fields could only be diminished, and a complete elimination has not been realized so far.

2.2.2.1 Local retarding fields

Retarding electrical fields are, for example, formed above positively biased conductor lines surrounded by

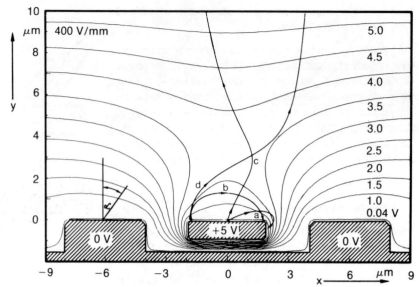

Fig. 9 Retarding field above positively (+5 V) biased conductor line with an extraction field of 400 V/mm and SE trajectories (37).
a: $E_{SE} = 1.5$ eV, $\alpha = 60°$; b: $E_{SE} = 1.5$ eV, $\alpha = 20°$; c: $E_{SE} = 1.5$ eV, $\alpha = 90°$; d: $E_{SE} = 2$ eV, $\alpha = 90°$.

grounded lines. This is illustrated in Fig. 9, which shows a computer simulation (37) of the equipotential lines and some SE trajectories at an IC surface with three neighbouring conductor lines with geometries and applied voltages typical for LSI devices.

SE generated at the center line only contribute to the signal if their energy is sufficiently high to overcome the barrier of the retarding field (electrons c, d in Fig. 9). Low energy SE are reflected back to the conductor (electrons a, b in Fig. 9), resulting in an altered SE spectrum and causing measurement errors (37, 50).

A reduction of the influence of these local fields on the SE spectrum can be realized by installation of strong extraction fields above the specimen (3, 5, 8, 10, 14, 37, 38, 39) either by placing a planar grid at high positive potential (3, 14) or an immersion objective close to the specimen surface (5, 8). It was shown (14, 37, 38, 39) that with increasing extraction field strength the barrier above positively biased IC lines decreases and vanishes at certain field strengths that depend on the device geometry and the voltage applied to the IC lines. For example, with 4 μm wide IC-lines and +5 V applied to one line an extraction field of ≃ 1 kV/mm is necessary to remove the retarding field at least in the center region above the positive line (37). However, the influence of the fields on the angular SE distribution is still present.

2.2.2.2 SE angular distribution changes by local fields

The angular distribution of the SE is approximately a cosine distribution (57) if the e-beam impinges normal to the surface:

$$N(\alpha) = N_O \cos \alpha \qquad (2)$$

with α the angle between normal and emission direction, $N(\alpha)$ the number of electrons emitted with an angle α and N_O the number of electrons emitted into the direction of the plane normal.

If the SE pass through the inhomogeneous local fields above the IC surface (compare Fig. 9) the original cosine distribution is altered (37, 38, 39, 50). SE emitted from positive lines neighboured by grounded lines are focussed, while SE from grounded or negatively biased lines neighboured by positively biased lines are defocussed (37, 38, 39, 50).

These effects decrease with increasing extraction field strength, but are still existent even if the retarding field barrier above positive lines is eliminated. Therefore the influence on the angular SE-distribution depends on the amount and sign of the voltages at the neighbouring lines (37, 38, 39, 50).

If the SE-analyzing system is sensitive to SE angular distribution, which is the case for planar retarding field spectrometers (compare Fig. 8), the obtainable voltage measurement accuracy also is dependent on the amount and sign of voltages on neighbouring lines. For the IC structure of Fig. 9 with the same applied line voltages and an extraction field strength increased to 1 kV/mm a measurement error of about 10 % was calculated (37). A reduction of this error is expected from a SE analyzing system combining an extraction field with a SE spectrometer insensitive to the SE angular distribution. This could be realized by a planar extraction field and a curved retarding field electrode, the curvature of which has to assure that most SE trajectories are perpendicular to the retarding grid. This curvature has to be determined by computer simulation (37).

2.2.3 Elimination of topography, material contrast and work function variations

Quantitative voltage measurements and quantitative voltage micrographs are possible only if the shape of the SE spectrum remains the same for different measuring points on the IC surface. However, remaining influences of the angle of incidence of the e-beam (topography), the material of the irradiated measuring spot and its surface condition (variations of work function) on the SE spectrum have to be eliminated.

2.2.3.1 Influence of topography

When the PE hit the surface with an angle of incidence Θ with respect to the normal of the plane the SE yield δ varies approximately proportional to $1/\cos\Theta$ (47, 51). This effect forms the basis of the topography contrast in the SEM, but causes measurement errors in the Voltage Measurement Mode. The dependence of the SE spectrum on topography is illustrated schematically in Fig. 10 for both SE analyzing methods, showing the SE spectra for two measurement spots, one at the edge of an aluminum line (2) and the second at the center of the line (1). With increasing Θ the SE yield increases and the SE spectrum is shifted towards lower energies (47). If Θ varies from 0° to 45° the SE yield increases by a factor of about 1.4 and the SE peak shifts by several tenths of an eV (47). This gives rise to measurement errors of several 100 mV in the differential SE analyzing technique, where only the peak position and not its height is analyzed. As the integral SE analyzing technique is also sensitive to the SE yield, much higher measurement errors of typically 1.5 V are present (compare Fig. 10).

2.2.3.2 Influence of material contrast

The SE yield δ is roughly correlated with the densitiy of the irradiated material (metals) (44). Also, the peak of the SE spectrum shifts when different materials are irradiated (compare (45)). For example if measurements are to be made on gold bond wires and aluminum lines the yield at 1.4 keV is $\delta_{Al} = 0.63$ and $\delta_{Au} = 1.1$ with the SE peak positioned at $E_{max\,Al} \simeq 1,33$ eV and $E_{max\,Au} \simeq 1.6$ eV. This results in measurement errors of 0.27 V in the differential spectrum analysis and about 3 V in the integral spectrum analysis.

Even if performed on the same material, voltage measurements are strongly influenced by contamination, especially in spot measurements. With increasing contamination layer thickness the SE yield varies from that of the irradiated material to that of the contamination layer (e.g. at 20 keV $\delta_{Al} = 0.175$ drops to $\delta_{cont} = 0.105$ (52)) causing measurement errors during the measurement time.

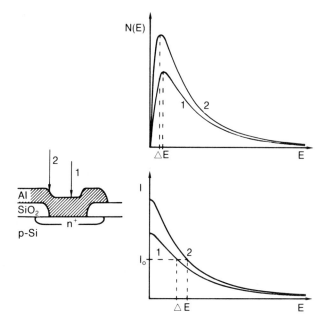

Fig. 10 Dependence of the SE spectrum on topography

2.2.3.3 Influence of work function variations

On any material surface that was exposed to air prior to voltage measurements, physisorbed and chemisorbed atoms and molecules are present, causing local variations of the material work functions and of the surface potential. Work function changes generate differences of the SE yield (45) and cause shifts of the SE spectrum (59) ("random potentials" (66)). In this way measurement errors between measuring spots on the same material of the order of several 100 mV are introduced (53, 54, 59).

These local work function variations have until the present prevented the absolute determination of voltages with the e-beam.

2.2.3.4 Elimination of these influences

The influences of all these three effects can be eliminated by electronic techniques that isolate the voltage information from the total signal. By a subtraction of the SE signal bearing voltage information (e.g. + 5 V) from that SE signal of the same measuring point with no voltage information (0V). These subtractions have been realized by various techniques (2, 8, 14, 55, 56, 57, 58), requiring highly sophisticated setups. Complete elimination of these influences, however, has not been achieved so far. The question arises as to whether a complete voltage contrast isolation as is necessary for quantitative voltage micrographs is also really necessary for quantitative voltage spot measurements. Most of the quantitative measurements are carried out today as spot measurements on known material, such as IC aluminum lines, so that the material contrast can be neglected. Aluminum lines normally do not have a flat surface due to elevations and voids giving rise to topographic contrast. The influence of this measurement error source may be reduced by scanning along a line during the measurement and thus integrating over the topographic effects to a certain extent.

Finally the influence of work function variations can be reduced by using dry vacuum systems (61) and by investigating carefully cleaned surfaces (18). In this way the necessary experimental equipment can be reduced to a minimum, still allowing quantitative voltage spot measurement to be performed.

2.2.4 Voltage resolution

The trend in IC technologies towards low voltage swings (voltage difference between logical "0" and "1"), which at the moment is of the order of 300 mV in ECL (emitter coupled logic) devices, and the low sense signals in MOS (Metal Oxide Semiconductor) memory devices necessitate voltages resolutions of better than about 10 mV for quantitative voltage measurements.

The voltage resolution is determined by noise in the process of generating and amplifying the voltage signal. Here, the noise contributions from the scintillator, photomultiplier and amplifiers can be neglected, because the main noise source is that of the SE current caused by PE current fluctuations and fluctuations in the generation of SE.

2.2.4.1 Noise in the SE current

Due to the random process of electron emission at the cathode the PE current I_{PE} fluctuates. These fluctuations show a frequency dependent power spectral density $S(f)$ given by (60):

$$S(f) = e\, I_{PE} + \text{const}\, \frac{I_{PE}^2}{f^m} \quad (3)$$

with e = electron charge, $1 < m < 2$, f = frequency.

The first term represents the frequency independent shot noise (white noise) for which the mean square current fluctuations are given by the Schottky formula (63):

$$\overline{i_{PE}^2} = 2e\, \Delta f\, I_{PE} \quad (4)$$

where Δf is the bandwidth of the detection system.

The second term in eq. (3) represents flicker noise generated by work function variations at the cathode caused e.g. by molecules hitting the cathode surface. For a tungsten cathode operated in the saturation mode m is about 2 (60). With decreasing frequency the second term increases proportional to f^{-2} which e.g. leads to a doubling of $S(f)$ at frequencies of typically 0.1 Hz (60). For low emission currents and a clean vacuum, where the impingement rate of molecules at the cathode is low, the onset of flicker noise contributions shifts towards lower frequencies (10^{-2} to 10^{-3} Hz).

As total measurement times in the SEM usually are in the range of less than 100 s the contributions of flicker noise can be neglected here.

The PE current fluctuations (eq. 4) give rise to SE current fluctuations, which are amplified or reduced by the average yield, :

$$\overline{i_{SE}^{2'}} = \delta^2\, \overline{i_{PE}^2} \quad (5)$$

However, there will also be fluctuations in δ appearing as an additional noise source

$$\overline{i_{SE}^{2''}} = 2e\, \delta\, I_{PE}\, \Delta f \quad (6)$$

The toal noise in the SE current then is given by (64):

$$\overline{i_{SE}^2} = \overline{i_{SE}^{2'}} + \overline{i_{SE}^{2''}} = 2e\, \Delta f\, I_{PE}\, \delta(1 + \delta) \quad (7)$$

If the SEs through grids (spectrometer, detector) interceptions of electrons at the grid bars occurs, leading to a reduced transmitted SE current:

$$\overline{i_{SE}^2} = 2e\, \Delta f\, I_{PE}\, T\, \delta(1 + \delta) \quad (8)$$

where $T \leq 1$ is the transmission of the grids given by the ratio of the transmitted current to the incident current.

The signal to noise ratio S/N in the detected SE current is given by

$$\frac{S}{N} = \frac{I_{SE}}{\sqrt{\overline{i_{SE}^2}}} = \frac{I_{PE}\, \delta\, T}{\sqrt{2e\, \Delta f\, I_{PE}\, T\, \delta(1 + \delta)}} \quad (9)$$

and thus decreases with decreasing T proportional to $T^{1/2}$. An increase of the average yield δ increases the S/N ratio proportional to $(\delta/\delta+1)^{1/2}$. For the sake of a high signal-to-noise ratio grids with high transmissons T and PE energies that allow high SE yields δ to be obtained should be used.

2.2.4.2 Minimum detectable voltage

For a retarding field spectrometer the minimum detectable voltage can be calculated (compare (40, 65)) on the following assumptions (compare Fig. 12). The spectrometer operates at a fixed retarding field

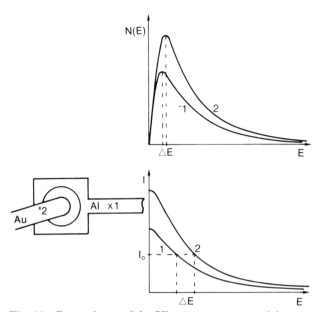
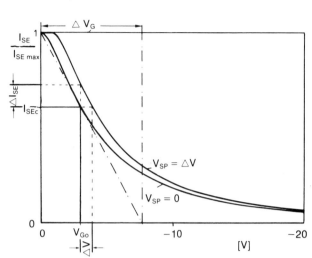

Fig. 11 Dependence of the SE spectrum on material

Fig. 12 Shift of integral SE spectrum with applied specimen voltage $V_{sp} = \Delta V$

voltage V_{Go} which gives rise to a SE current at the detector I_{SEo}. A small voltage $-\Delta V$ applied to the specimen shifts the spectrum by ΔV which results in an increase of I_{SEo} by ΔI_{SE} (Fig. 12). If this increase of the SE current is to be detected it has to be n times larger than the noise current i_{SE}^2 (normally n = 3). With $a = \Delta I_{SE}/\Delta V = I_{SEmax}/\Delta V_G = \delta\, T\, I_{PE}/\Delta V_G$ the integral spectrum curve slope at the working point, ΔV is given by

$$\Delta V = \frac{3\sqrt{i_{SE}^2}}{a}. \quad (10)$$

As only a fraction $\gamma = I_{SEo}/I_{SEmax}$ contributes to the signal, ΔV then is

$$V_{min} = \Delta V = 3\,\Delta V_G \underbrace{\sqrt{\frac{2e\,\delta\,(1+\delta)\,\gamma}{T\,\delta}}\sqrt{\frac{\Delta f}{I_{PE}}}}_{c} \quad (11)$$

The obtainable voltage resolution is therefore strongly dependent on the measuring time and the PE beam current. As stated before, the transmission and the yield should be as high as possible to obtain the optimum voltage resolution.

Due to the counteracting dependence of γ and ΔV_G on the retarding field voltage V_{Go} (γ decreases and ΔV_G increases with increasing V_{Go}) an optimum retarding field voltage exists at $V_{Go} \simeq -2.5$ to 3 V for metals (37, 50, 65).

If in eq. (11) $\Delta V_G = 6$ V (spectrum of Al, comp. Fig. 12) $\delta = 0.8$ (SE yield of Al at $E_{PE} = 1$ keV), $T = 0.64$ (transmission of two grids in series, each with $T = 0.8$), $I_{PE} = 5 \times 10^{-8}$ A, $\Delta f = 2$ Hz, corresponding to an integration time of 500 ms, $\gamma = 0.6$ at the optimum retarding field voltage of -2.5 V, then V_{min} becomes 0.08 mV. Under these conditions a minimum detectable voltage of 0.5 mV was determined experimentally using a planar retarding field spectrometer (37). The difference between theoretical and experimental resolutions (factor of $\simeq 6$) is most probably due to the influence of backscattered electrons (BSE) and SE generated at the grids that contribute to background signals, which have not been taken into account in the theoretical considerations.

2.2.5 Time resolution

The fastest voltage variations on an electronic device that can be measured with the electron beam is determined by the time resolution of the voltage measurement system which is determined by the generation time of the SE by the beam, the SE flight time, the bandwidth of the detection system, the required voltage resolution and the shortest e-beam pulse width in the stroboscopic or sampling mode.

2.2.5.1 Generation time of SE

The e-beam with PE-energy E_{PE} travels a certain distance, the penetration depth, into the solid. As the velocity v_{el} of electrons of energy E is given by

$$v_{el} = \sqrt{\frac{2E}{m_e}} = 0.579 \sqrt{E} \qquad (12)$$

with v_{el} in µm/ps and E in eV, where m_e is the electron mass, e.g. a 1 keV primary electron travels down to its penetration depth of $\simeq 0.1$ µm in an aluminum conductor line within about 0.01 ps. An 1 eV– SE generated at an escape depth of 20 nm below the surface leaves the surface after about 0.03 ps.

From this estimation it can be concluded, that the generation time of SE is of the order of 10^{-14} to 10^{-13} s, which is sufficiently short to allow voltage measurements up to frequencies in the high GHz range (100 GHz $\triangleq 10^{-11}$s period time).

2.2.5.2 SE flight time

Due to the velocity spread of the SE, low energy SE will arrive at the positively biased detector later than high energy SE.

Assuming a homogeneous accelerating field between the IC surface and the detector gives a flight time t_f of

$$t_f = \frac{2d}{v_{el}} \qquad (13)$$

where d is the distance between emission point and v_{el} is the electron velocity at the detector, held at a voltage v_D.

For $V_D = 300$ V and d = 40 mm a 0.5 eV SE has a flight time of 7.69 ns compared to 7.57 ns for a 10 eV SE. The flight time differences of 120 ps would limit the time resolution to about 4 GHz, as the time dispersion of SE should be smaller than half the signal period time. This example shows that the geometries of detection system and the extraction fields have to be optimized with respect to short SE flight times, because otherwise the SE flight time will limit the obtainable time resolution.

The time dispersion is reduced and the time resolution is increased by keeping d as small as possible, by applying high accelerating voltages (extraction fields) and by analyzing only the high energy SE. The effect of time dispersion has to be taken into account only in the stroboscopic or sampling mode (see section 2.2.5.5) because real time measurements have been limited to the MHz range until the present (20).

2.2.5.3 Detection system bandwidth

The voltage measuring system bandwidth is characterized by the performance of scintillator, photomultiplier, amplifier and the linearization feedback loop. Scintillators like e.g. P 47 or plastic scintillators have rise times of $\simeq 80$ ns and $\simeq 1$ ns, respectively, corresponding to $\simeq 12$ MHz and $\simeq 1$ GHz bandwidth. Photomultipliers show rise times between $\simeq 1$ ns and $\simeq 30$ ns corresponding to $\simeq 1$ GHz and $\simeq 30$ MHz bandwidth. The transit time of signals through photomultipliers is of the order of 15 ns to 80 ns equivalent to 12 to 70 MHz. Amplifiers allow rise times of 1 ns to 1 µs corresponding to $\simeq 1$ GHz and $\simeq 1$ MHz.

All these factors which influence bandwidth are negligible compared with that of the linearization feedback loop. Because of instability problems its maximum operating frequency is about 300 kHz (14) which represents the upper frequency limit of voltages that can be quantitatively measured with the e-beam in real time.

2.2.5.4 Required voltage resolution

According to eq. (7) the minimum detectable voltage V_{min} is proportional to $f^{1/2}$. Increasing the bandwidth is, therefore, possible only at the expense of a reduced voltage resolution.

2.2.5.5 Electron beam pulse width

The problems of a low bandwidth of the linearization feedback loop and the drop of voltage resolution with increasing time resolution can be solved by the stroboscopic or sampling mode (3, 7, 8, 9, 11 to 19, 28 to 37, 41, 42).

In both modes short e-beam pulses generated by fast beam blanking system (14, 28 to 37) with the same repetition rate as the periodic signal on the electronic device impinge on the device causing SE emission only at a specific phase intervals of the device signal.

In the stroboscopic mode the phase relation between e-beam pulse and device signal is kept constant while the e-beam scans the surface. In this way stroboscopic voltage distribution images are obtained. As an example, Fig. 13 shows different logical states of a nonpassivated 4 bit MOS divider. Fig. 14 shows the transition of logical states at a bipolar passivated NAND gate.

In the sampling mode the phase relation (delay) between e-beam pulse and device signals is varying while the beam is fixed at an IC-node of interest. In this way the waveform of the device signal is measured. As an example Fig. 15 shows four different waveforms obtained on a non passivated I^2 L circuit.

As in both modes the generated SE signal is integrated over many samples, Δf can be chosen to be very small (typically several Hz), which improves the voltage resolution at high frequencies. The time resolution now is determined by the shortest e-beam pulse and the flight time of the SE (see section 2.2.5.2) limits the maximum device frequency.

Time resolutions of 0.2 ps (33) at a fixed frequency of 1 GHz, of 5 ps at a fixed PE energy and at a frequency of 9 GHz (29 to 31) and of 10 ps at variable energies and frequencies (36, 37) have been demonstrated.

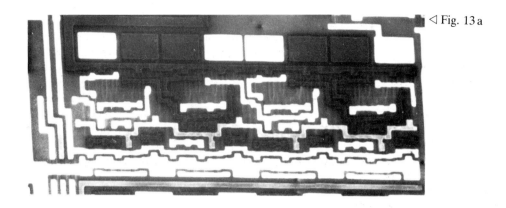

Fig. 13a

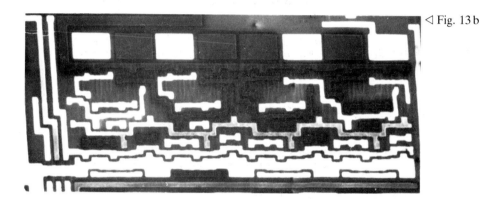

Fig. 13b

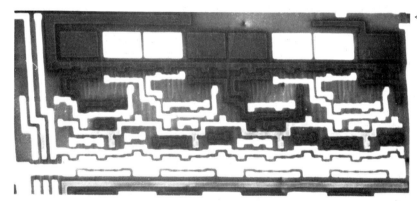

Fig. 13c

Fig. 13 Stroboscopic voltage contrast micrograph of an unpassivated 4 bit MOS divider $E_{PE} = 1.2$ keV, $I_{PE} = 10^{-10}$ A (average), pulse duration $t_p = 200$ ns, repetition rate 16 μs, a: delay time $t_D = 3.5$ μs, b: $t_D = 6.5$ μs, c: $t_D = 10.5$ μs.

2.2.6 Spatial resolution

The spatial resolution in the Voltage Measurement Mode is not only determined by the e-beam current I_{PE}, the PE energy E_{PE} and lens aberrations as in normal SEM modes, but is also affected by the chopping degradation, an elliptical deformation of the spot diameter due to the chopping action, and by astigmatism or defocussing introduced by the extraction field above the device surface.

2.2.6.1 Spot diameter

According to (66) the e-beam spot diameter d_{tot} at the specimen surface is given by

$$d_{tot}^2 = \frac{I_{PE}}{2.47 \, B \, \alpha_o^2} + (0.61 \frac{\lambda}{\alpha_o})^2 + (0.3 \, C_s \alpha_o^3)^2 + (C_c \, \alpha_o \frac{\Delta E_{PE}}{E_{PE}})^2 \quad (14)$$

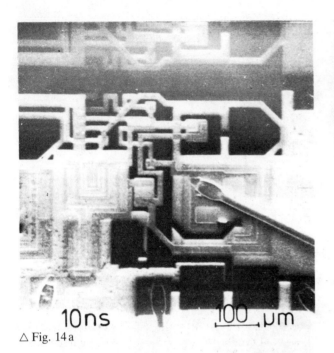

△ Fig. 14a

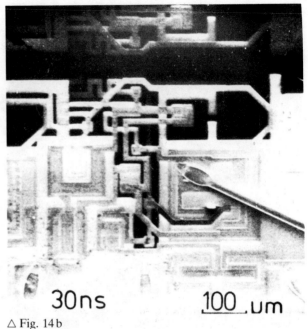

△ Fig. 14b

Fig. 14c ▽

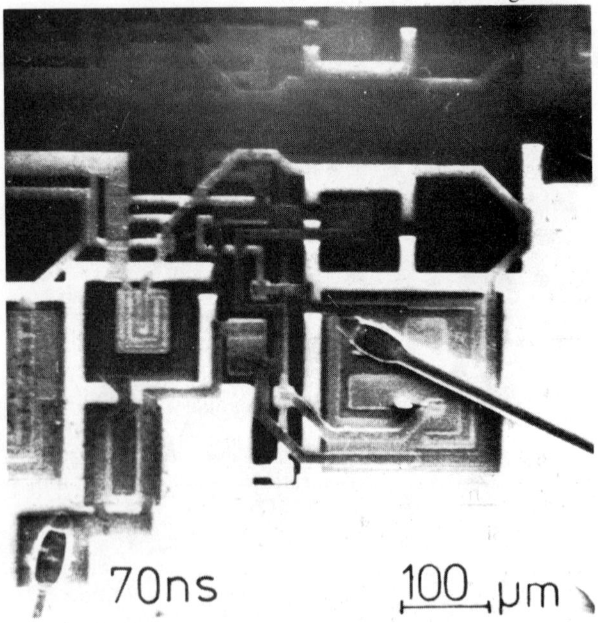

Fig. 14 Transition of logical states at a passivated bipolar NAND gate $E_{PE} = 10$ keV, $t_p = 10$ ns, a : $t_D = 10$ ns, b : $t_D = 30$ ns, c : $t_T = 70$ ns.

where B = brightness of the gun (A/cm² sr), C_s = spherical aberration coefficient of the final lens (cm), C_c = chromatic aberration coefficient of the final lens (cm), ΔE_{PE} = energy spread of PE, α_o = aperture semiangle. For high PE currents ($\geq 10^{-8}$ A) the first term in eq. (14) mainly determines d_{tot}.

Combining eq. (11) and (14) under this condition shows the interdependence of voltage and spatial resolution:

$$V_{min} = C \sqrt{\frac{\Delta f}{2.47 \, B \, \alpha_o^2}} \frac{1}{d_{tot}} \quad (15)$$

For a given specimen material, spectrometer, bandwidth, electron gun, and primary energy, a particular compromise between spatial and voltage resolution has always to be found.

2.2.6.2 Chopping degradation

Chopping the electron beam over an aperture (28 to 37) results in an apparent movement of the source, which may lead to an elliptically deformed spot diameter at the specimen surface (astigmatism). This astigmatism can be corrected by the stigmator control of the SEM (36, 46), but at the expense of an enlarged spot diameter. By properly designing the chopping structure and also taking advantage of the electron optics, the elliptical deformations are reduced to negligible values (36, 37).

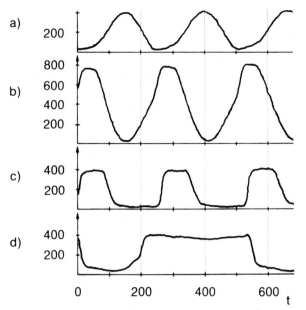

Fig. 15 Sampling voltage measurement at an I²L device. $E_{PE} = 2$ keV, $I_{PE} = 10^{-9}$ A (average), $t_p = 5$ ns, repetition rate 1 µs.
a : base and b : collector of oscillator stage, c : before and d : behind first frequency divider stage

2.2.6.3 Extraction and deflection fields above the device

As shown in section 2.2.2, high extraction fields above the specimen surface are a precondition for quantitative voltage measurements.

If these fields are planar (3, 14) there is only a defocussing effect that can be compensated by readjusting the final lens of the SEM.

In the case of nonplanar fields (immersion objective (8, 37), tubular extraction field electrode (5, 10)) astigmatic spot deformation is introduced which cannot be compensated at that level. Also large deflection fields installed for bending the SE trajectories towards the detector (8, 37) also cause deterioration of the spot diameter.

It is therefore essential to use planar extraction fields and a deflection field that is small enough not to influence the PE beam diameter (14).

2.2.7 Electron beam influence on device operation

For the testing of ICs with the e-beam under real IC operation conditions it is important that the device is in no way influenced by the beam, i.e. the e-beam should be as passive as possible. However, due to the fact that the e-beam penetrates a certain distance R_G, the projected penetration depth (67)

$$R_G = 4 E_{PE}^{1.75} \quad (16)$$

with R_G in µg/cm² and E_{PE} in keV, into the solid, losing its energy primarily through ionization processes, irradiation damage by trapped charges and increased number of interface states (68, 69) may occur leading to a failure of the device.

Furthermore, in order to balance the sum of SE and BSE currents with the PE current, an absorbed or impressed current flows into the irradiated IC area, which may momentarily alter the device operation.

2.2.7.1 Irradation damage

As the e-beam penetrates into the IC surface it generates electron-hole pairs of which the holes may become trapped in silicon dioxide (SiO_2) layers because holes have small mobilities in SiO_2 compared to electrons (68 to 70). Additionally, the number of interface states at the Al–SiO_2 and SiO_2–Si transitions increases (69).

In MOS devices, which are most sensitive to radiation these effects alter device parameters like the threshold voltage V_{th}, gain, and leakage currents. Bipolar devices show decreased common emitter gain, and increases in leakage current and noise (72). These parameter changes may be annealed by a heat treatment of the devices (several hours at temperatures > 150° C (70)).

The amount of device parameter changes generally increases as a function of the energy deposited in SiO_2 layers.

In an MOS transistor as illustrated in Fig. 16 the energy deposited in the gate oxide with a layer thickness d covered by an aluminum line of thickness b is a function of E_{PE}, I_{PE} and the irradiation time t.

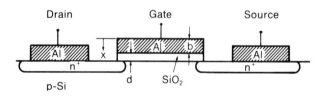

Fig. 16 Schematical cross section of a MOS transistor

According to (67) the energy deposition in keV/µg/cm² versus the penetration depth of the PE in µg/cm² is shown in Fig. 17. The absorbed energy or dose D in the SiO_2 layer is given by (72):

$$D(rad) = \frac{I_{PE} E_{PE} t(1-\eta)}{e A_s d \varrho} \int_b^{b+d} \frac{dE}{dx} dx \bigg/ \int_0^{R_G} \frac{dE}{dx} dx \quad (17)$$

where t = irradiation time (s), A_s = scanned area (cm²), ϱ = material density (g/cm³), η = backscattering coefficient, e = electron charge.

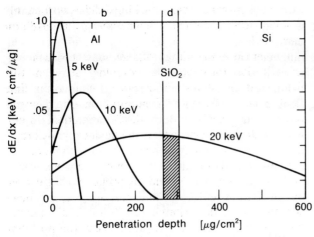

Fig. 17 Deposited energy versus penetration depth (after (67))

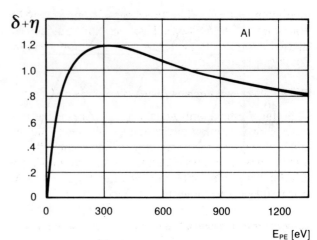

Fig. 18 PE energy dependence of $\delta + \eta$ of aluminum according to the universal yield curve (75) and assuming a constant backscattering coefficient $\eta = 0.2$.

If, for example, $E_{PE} = 20$ keV, $b = 1$ μm, $d = 0.1$ μm, $I_{PE} = 10^{-9}$ A, $A_s = 0.01$ cm^2, $\eta = 0.1$, $t = 10$ s the absorbed radiation dose (hatched area in Fig. 17) is $D = 1.6 \times 10^6$ rad. Such doses lead to typical threshold voltage changes in MOS transistors of several volts (73), the change in threshold voltage ΔV_{th} being proportional to the dose D (73).

The change of the threshold voltage V_{th} has been shown to be dependent on the electrical field within the gate oxide (70 to 74). A positive bias $+V_B$ applied to the gate electrode causes the positive charges to accumulate near the SiO_2–Si transition where the influence on V_{th} is large. Negative bias attracts the positive charges towards the Al-gate electrode, where the influence on the device characteristics is smaller. The threshold voltage change ΔV_{th} is proportional to $V_B^{1/2}$ and is smallest with grounded gate electrode (74).

Irradiation damage can be drastically reduced by using low energy primary electrons which deposit their energy within the aluminum gate electrode above the critical oxide. For example at 1 keV beam energy threshold, voltage changes as little as 0.1 V are obtained after irradiation times of some 1000 s with beam currents of 10^{-7} A (73).

2.2.7.2 Impressed currents

The incident e-beam current I_{PE} is divided into three portions, the SE current I_{SE}, the BSE current I_{BSE}, and the absorbed current I_{ab}:

$$I_{PE} = I_{SE} + I_{BSE} + I_{ab}. \qquad (18)$$

This may be rewritten for the absorbed current using δ and η to

$$I_{ab} = (1 - \delta - \eta) I_{PE} \qquad (19)$$

As δ and η are a function of the primary electron energy E_{PE}, the absorbed or impressed current is also a function of E_{PE}. However, for most metals the variation of η is only small in the E_{PE}-range of 0.5 to 10 keV (75) (typically several % per keV), so that the main factor of the PE energy dependence of I_{ab} is that of the SE yield δ.

The sum of δ and η is shown in Fig. 18 for aluminum, assuming $\eta = 0.2$ and the ratio of absorbed current I_{ab} to the primary electron current I_{PE} constant over the range shown. In the range 120 eV $< E_{PE} <$ 700 eV positive currents are impressed, whereas for $E_{PE} >$ 700 V and $E_{PE} <$ 120 V negative currents are impressed. The exact location of the crossover points I and II is strongly dependent on the surface conditions (oxide layers, contamination layers).

This unwanted effect of impressed currents can influence the device operation if the impressed current is of the same order as the switching currents within the device. However, by properly adjusting the PE energy, the impressed currents may be reduced to zero (crossover II in Fig. 18). As the IC conductor materials vary (Al is often used with several % of Si or Cu dopant) this crossover has to be determined for every IC.

The first crossover (I) is of no practical use because the spatial resolution in this case would be too low. The second crossover is already situated at a PE energy much lower than those used for normal SEM operation (10 to 30 keV). The resultant reduction in spatial resolution has to be taken into account if a non-loading e-beam is to be used.

2.2.8 Voltage measurements on passivated devices

So far, we have only considered nonpassivated devices. However, most devices to be subjected to failure analysis are covered with a glass passivation layer. Furthermore, with growing application of multi-layer ICs the need to be able to measure voltages on oxide covered layers becomes more stringent.

2.2.8.1 Electron beam induced conductivity in insulators

When a thin insulating film on a metal base plate (like the passivation layer of ICs above the Al-lines) is bombarded with a high energy e-beam, the PE generate electron-hole pairs in the energy dissipation volume. With ionization energies of 10 to 20 eV a single 10 keV primary electron is able to produce about 500 to 1000 free electron-hole pairs. If the penetration depth of the beam is large enough to reach the conducting aluminum layer underneath the passivation layer ($\geq$ 10 keV for 1 µm passivation (67)), a conducting channel is formed, so that the insulator surface assumes the same potential as the metallization (24). Thus the SE emitted from the insulator surface also show shifts of the energy spectrum. The obtainable voltage measurement accuracy compared to that on unglassed ICs has, however, not yet been investigated in detail, but voltage accuracies of about 200 mV have been reported (76).

The voltage resolution (eq. 11) should be somewhat better than that for pure metals as the SE spectrum of insulators show a steeper peak (section 2.1) so that the slope of the integral SE spectrum increases.

The thickness of the passivation layer is not uniform on an IC, so that the electron beam may travel deeper into the device at places where the layer is thinner. This then results in a possible irradation effect which makes this technique rather dangerous, especially for MOS devices; bipolar devices being not very critical (see section 2.2.7).

2.2.8.2 Capacitive coupling voltage contrast

Irradation effects (section 2.2.7.1) may be reduced by choosing very low primary electron energies. Most insulator materials used for passivating IC show a SE yield of $\delta > 1$ at 30 eV $< E_{PE} <$ 2 to 3 keV. Operating at a primary electron energy within this range allows the surface potential of the insulator to be stabilized. Positive charges accumulated in the insulator surface build up a retarding field above the surface that reduces the SE yield δ to unity, so that no charging is observed provided the retarding field is sufficiently homogeneous (scanned e-beam) above the insulator surface.

Under these conditions voltage contrast may be observed due to capacitive coupling between the irradiated insulator surface and the conducting line underneath (16). With this method only the changes of states on an IC are imaged, DC potentials do not contribute to the contrast which may be an advantage if only alternating signals have to be investigated.

However, the origin of this contrast mechanism is not yet well understood and the influence of the kind of passivation and layer thickness on voltage measurement accuracy, spatial resolution and repeatability needs further investigation. Also, the problem of charging of the passivation layer observed in the spot mode has to be solved.

3. Current Impressing Mode

In section 2.2.7.2 currents impressed to the IC-node under e-beam irradiation have been shown to have a negative effect on the device operation. On the other hand this effect may be taken advantage of, thus converting the role of the e-beam from a passive to an active one. The effect of impressed currents offers the possibility of using the e-beam as a contactless probe that applies input signals to the device or subsystem under test.

3.1 Primary energy dependence of impressed currents

Due to the PE energy dependence of the SE yield the absorbed or impressed current is also PE energy dependent.

This current I_{ab} flowing into an aluminum surface is shown in Fig. 19. Although a clean high vacuum (10^{-7} Pa residual pressure of hydrocarbons) was used (61), a drift of the impressed current of several seconds depending on I_{PE} was observed. This drift was only observed on Al surfaces. It could be stabilized after completely bombarding the IC surface with the e-beam for several minutes. It follows from Fig. 19, that positive and negative currents may be impressed to aluminum layers of an IC.

The I_{ab} versus E_{PE} curve strongly depends on the material (Al, possible doping material, surface conditions) and has to be determined prior to e-beam application. The magnitude of impressed currents I_{ab} is of the order of the PE beam current.

PE beam currents of 10^{-7} A at spot sizes of 1 µm are obtainable using a LaB$_6$ electron gun at e.g. 2 keV (37).

Switching currents in integrated circuits vary from 10^{-3} A to 10^{-8} A depending on the device technology. Especially in MOS devices low switching currents are used so that the e-beam may be used directly for

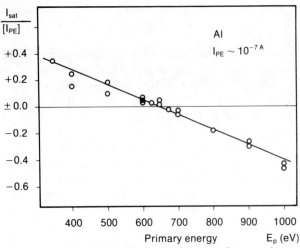

Fig. 19 Dependence of impressed current I_{sat} on primary electron energy

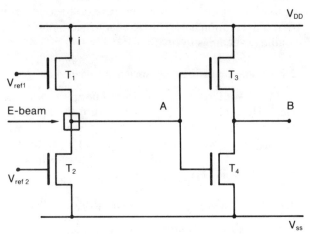

Fig. 20 Possible layout for an e-beam receiver

switching operations by compensating IC-internal current sources (19, 21, 22, 37). Fig. 20 shows a possible lay-out in a C-MOS device. The current i through transistors T_1 and T_2 is determined by the reference voltages V_{ref1} and V_{ref2}.

If the e-beam is directed toward the e-beam receiver A and compensates the current i, the logical state of the CMOS gate (T_3, T_4) changes from "0" to "1". This ability of the e-beam was used for initiating a built-in self-test on a 1 kbit ROM device, which was designed to be functionally convertable from the normal mode to a testing mode (19, 21, 22, 37).

A test controller is initialized by an electron beam pulse of some 100 μs compensating the built-in current source of 50 nA. With the 1024 pulses of the built-in clock the self-test is carried out.

Fig. 21 shows two different e-beam operations modes: on the left the impressing mode where the beam current compensates the internal current source. Then the beam is deflected toward the clock circuitry where it measures the clock pulses. The two top curves show the x- and y-beam deflection signals; the x-deflection being zero and the y-deflection being about 300 μm. After the 1024 pulses of the clock the compressed test data is read out at the signature register by the beam in the Voltage Measurement Mode.

Each chip on a wafer equipped with a wafer interconnections for the supply voltage could be automatically e-beam tested without application of wafer probers using the electron beam as a measuring probe and a current source.

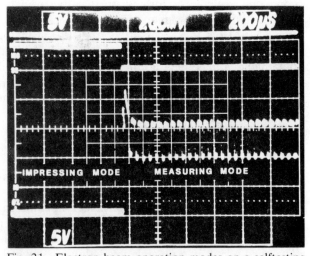

Fig. 21 Electron beam operation modes on a selftesting ROM device (19, 21, 22).

3.2 Current multiplication in thin dielectric layers

If the e-beam completely or partly penetrates a dielectric layer of an IC (e.g. the passivation layer, usually 1 μm thick) it generates electron-hole pairs in the dissipation volume and causes a current I to flow into the metal electrode underneath the dielectric (23, 24, 77).

In SiO_2, these currents may be 10 to 20 times larger than the primary beam current if a voltage of about 10 V is present between the insulator surface and the metal electrode underneath. The maximum current multiplications with a 1.1 μm insulator thickness occur at $E_{PE} \simeq 5$ keV (10 fold multiplication of electron flow), and at $E_{PE} \simeq 15$ keV (20 fold multiplication of hole flow) (77).

This effect of current multiplication may be taken advantage of in the Current Impressing Mode of the e-beam if beam and switching currents have to be larger than those described in 3.1. However, possible irradation damage has to be taken into account with the large primary electron energies that have to be used (see section 2.2.7.1).

3.3 Current multiplication in pn-junctions

If electron-hole pairs are generated by an e-beam in the space charge region of a pn-junction, they are separated by the electrical field of the junction, causing an electron beam induced current (EBIC) to flow in an external circuit (78 to 81). The induced current is determined by the ionization energy of electron-hole pairs, the primary electron beam energy and current. In silicon a 2 keV electron produces about 500 electron-hole pairs corresponding to a carrier multiplication of 500. It is necessary that the pn-junctions be acessable to the e-beam. Fig. 22 shows a possible lay-out, comparable to that of Fig. 20 except for the fact that T_2 has been replaced by a diode. The e-beam has to be positioned toward the pn-junction of the diode D_1. Transistor T_1 serves as resistor. If D_1 is not electron beam bombarded it is blocking, so that point A is at positive voltage corresponding to "0" at point B. E-beam bombardment will drive D_1 towards forward bias causing the voltage V_B to drop corresponding to a logical "1" at B. A similar arrangement was used by Shaver (25) to connect different testpoints of an IC to a common test bus and to read their logical states on an IC external node. On the basis of two e-beam irradiated pn-junctions set-reset flip-flops forming e-beam controlled latches were realized (25) to provide e-beam controlled input or clocks signals or programmable links during the test procedure.

4. Restructuring mode

In the Current Impressing Mode the e-beam is used for initiation of switching operations within the IC. These switching operations are reversible either by blanking off the beam (inverter operation of e-beam receiver) or by directing it towards a second "reset" e-beam receiver (electron beam switched latch ESL).

It may be useful for an e-beam testing procedure not only to control the proper function of devices (chip or complete wafers) but also to provide possibilities for connecting operating subsystems at the chip or wafer scale with an e-beam or to disconnect faulty subsystems from the data bus. These e-beam modified interconnects should hold their state indefinitely.

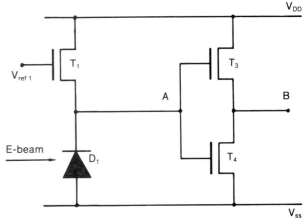

Fig. 22 Possible layout of an e-beam receiver using a pn-junction

Electron beam activated interconnections can be realized by taking advantage of the irradiation damage effect. As described in 2.2.7.1 the threshold voltage of MOS transistors can be altered by e-beam irradiation and can thus be turned on or off dependent on the applied gate voltage. However, at high temperatures ($\simeq 150°$ C), irradiation influences of the e-beam anneal within a certain time (typically several hours) so that these e-beam activated switches have to be regarded as non-volatile if the system is exposed to high temperatures. The successful switching of FET with an e-beam was demonstrated by Shaver (25) using parasitic field oxide FETs (programming time $\simeq$ 1 μs).

The disadvantage of erasure can be avoided by use of floating gate FET used for erasable EPROMS. Electrons delivered from the e-beam charge the floating gate thus altering the state of the transistor (25). These FET store the acquired charge for more than 10 years and therefore can be regarded non-volatile links.

Conclusions

Testing of LSI and VLSI devices would be unthinkable without e-beam testing. This is because the e-beam on the one hand represents the only possibility of measuring IC-internal signals in a nondestructive and nonloading manner and on the other hand can actuate IC-internal volatile and non volatile switching operations.

This paper summarizes the fundamentals of the different e-beam operations modes for IC-testing: the Voltage Measurement, Current Impressing, and Restructuring Mode.

The physics of the latter two modes are straightforward and therefore further research and development has to be concentrated on circuit design, taking advantage of these two modes preferably in combination with the Voltage Measurement Mode in order to speed up VLSI testing. The problems with the Voltage Measurement Mode are much more severe because of the large number of factors influencing the voltage measurements. Although many of these problems have already been recognized and at least partly been solved, some of them require further detailed investigations. The voltage measurement accuracy has to be improved, presumably by computer aided designs for SE spectrometers. Research work in the field of capacitive coupling voltage contrast is necessary in order to determine those device parameters that limit the voltage, spatial, and time resolutions and hopefully also make this method quantitative, as it is the only one to determine voltages on oxide covered conductors without causing irradiation damage.

In this context it should be pointed out that in order to avoid charging, irradation damage and current load within the devices, low PE energies have been favoured in the past two years. The consequent loss in spatial resolutions requires optimization of the electron optics for low PE energies.

Acknowledgement

Thanks are due to the Bundesminsterium für Forschung und Technologie (BMFT) of Germany and to the Ministerium für Wissenschaft und Forschung (MWF) of North-Rhine-Westfalia for financial support of parts of the work presented here.

References

(1) Wells O C, Bremer C G: Voltage measurement in the scanning electron microscope. J Phys E 1, 902–906 (1968)

(2) Fleming J P, Ward E W: A technique for accurate measurement and display of applied potential distributions using the SEM. Scanning Electron Microscopy 1970, IITRI, Chicago, pp 465–470

(3) Plows G S: Stroboscopic scanning electron microscopy and the observation of microcircuit surface voltage. Ph D Thesis Cambridge University Library, Cambridge, Great Britain, 1969

(4) Gopinath A, Sangen C C: A technique for linearization of voltage contrast in the scanning electron microscope. J Phys E 4, 334–336 (1971)

(5) Hannah J M: Scanning electron microscope applications to integrated circuits testing. Ph D Thesis, University of Edinburgh Library, Edinburgh, Scotland, 1974

(6) Lukianoff G V, Touw T R: Voltage Coding: Temporal versus spatial frequencies. Scanning Electron Microscopy 1975, IITRI, Chicago, pp 465–471

(7) Gonzales A J, Powell M W: Internal waveform measurements of the MOS three transistor, dynamic RAM using SEM stroboscopic techniques. Technical Digest of IEDM, IEEE, New York, USA, 1975, pp 119–122

(8) Balk L J, Feuerbaum H P, Kubalek E, Menzel E: Quantitative voltage contrast at high frequencies in the SEM. Scanning Electron Microscopy 1976/I, IITRI, Chicago, pp 615–624

(9) Thomas P R, Gopinathan K G, Gopinath A, Owens A R: The observation of fast voltage waveforms in the SEM using sampling techniques. Scanning Electron Microscopy 1976/I, IITRI, Chicago, pp 609–614

(10) Dyukov V, Kolemeytsev M, Nepijko S: A simple technique of potential distribution mapping in a scanning electron microscope. Micr Acta 80, 5, 367–374 (1978)

(11) Fujioka H, Hosokawa T, Kanda Y, Ura K: Submicron electron beam probe to measure signal waveform at arbitrarily specified positions on MHz IC, Scanning Electron Microscopy 1978, SEM Inc, AMF O'Hare, p 755

(12) Feuerbaum H P, Kantz D, Kubalek E: Quantitative measurement with high time resolution of internal waveforms on MOS-RAMs using a modified scanning electron electron microscope IEEE, SC-13, 319–325 (1978)

(13) Fazekas P, Lindner H, Lindner R, Otto J, Wolfgang E: On-wafer defect classification of LSI-circuits using a modified SEM. Scanning Electron Microscopy 1978/I, SEM Inc, AMF O'Hare, pp 801–806

(14) Feuerbaum H P: VLSI testing using the electron probe. Scanning Electron Microscopy 1979/I, SEM Inc, AMF O'Hare, pp 285–296

(15) Wolfgang E, Lindner R, Fazekas P, Feuerbaum H P: Electron-beam testing of VLSI circuits. IEEE J Sol State Circuits, SC-14, (2), 471–481 (1979)

(16) Kotorman L: Non-charging electron beam pulse prober on FET wafers. Scanning Electron Microscopy 1980/IV, SEM Inc, AMF O'Hare, pp 77–84

(17) Crichton G, Fazekas P, Wolfgang E: Electron beam testing of microprocessors. IEEE Test Conference 1980, pp 444–449

(18) Menzel E, Kubalek E: Electron beam test techniques for integrated circuits. Scanning Electron Microscopy 1981, SEM Inc, AMF O'Hare, pp 305–322

(19) Kubalek E, Menzel E: Microcircuit inspection techniques using an electron beam probe. Microcirc Engineering, Lausanne 1981, 495–506

(20) Ostrow M, Menzel E, Kubalek E: Real time logic state analysis of IC-internal signals with an electron beam probe-logic state analyzer combination. Microcirc Engineering, Lausanne 1981, 563–572

(21) Könemann B, Mucha J, Zwiehoff G: Built-in test for complex digital integrated circuits. IEEE J Solid-State Circuits, SC-15, (3), 315–319 (1980)

(22) Theus U, Leutigen H: Gleichzeitiger Selbsttest an der unzersägten Si-Scheibe. NTG Fachberichte „Großintegration" 77, 89–90 (1981)

(23) Pensak L: Conductivity induced by electron bombardment in thin insulating films. Phys Rev, 75 472–478 (1949)

(24) Taylor D M: The effect of passivation on the observation of voltage contrast in the scanning electron microscope. J Phys D Appl Phys, 11, 2443–2454 (1978)

(25) Shaver D C: Electron beam testing and restructuring of integrated circuits. J Vac Sci Technol 19 (4), 1010–1013 (1981)

(26) Chang T H P, Hohn F J, Coane P J, Kern D P: Electron beam testing of packaging modules for VLSI chip arrays. 16th Symp Electron Ion and Photon Beam Techn Dallas 1981, Abstract Q 2

(27) Pfeiffer H C, Langner G O, Stickel W, Scinpson R A: Contactless electrical testing of large area specimens using electron beams. J Vac Sci Technol 19 (4), 1014–1018 (1981)
(28) Robinson G Y: Stroboscopic scanning electron microscopy at gigahertz frequencies Rev Sci Instr, 42, 251–255 (1971)
(29) Gopinath A, Hill M S: A technique for the study of gun devices at 9.1 GHz using a scanning electron microscope. IEEE Trans Electron Dev ED-20, 610–612 (1973)
(30) Gopinath A, Hill M S: Some aspects of the stroboscopic mode: a review. Scanning Electron Microscopy 1974, IITRI, Chicago, pp 235–242
(31) Gopinath A, Hill M S: Deflection beam-chopping in the SEM. J Phys E 229–236 (1977)
(32) Hosokawa T, Fujioka H, Ura K: Two-dimensional observation of gunn domains at 1 GHz by picosecond pulse stroboscopic SEM. Appl Phys Lett 31, 340–505 (1977)
(33) Hosokawa T, Fujioka H, Ura K: Generation and measurement of subpicosecond electron beam pulses. Rev Sci Instr 49, 624–628 (1978)
(34) Hosokawa T, Fujioka J, Ura K: Gigahertz stroboscopy with the scanning electron microscope. Rev Sci Instr 49, 1293–1299 (1978)
(35) Feuerbaum H P, Otto J: Beam chopper for subnanosecond pulses in scanning electron microscopy. J Phys E 11, 529–532 (1978)
(36) Menzel E, Kubalek E: Electron beam chopping systems in the SEM. Scanning Electron Microscopy 1979/I, SEM Inc, AMF O'Hare, pp 305–318
(37) Menzel E: Elektronenstrahltestsystem für die Funktionskontrolle und Fehleranalyse höchstintegrierter Schaltkreise. Ph D Thesis University Duisburg, Duisburg 1981
(38) Ura K, Fujioka H, Yokobayashi T: Calculation of local field on voltage contrast of SEM. Proc 7th European Cong on Electron Microscopy, vol 1, 1980, pp 330–331
(39) Fujioka H, Nakamae K, Ura K: Local field effects on voltage measurement using a retarding field analyser in the SEM. Scanning Electron Microscopy 1981, SEM Inc, AMF O'Hare, pp 323–332
(40) Gopinath A: Estimate of minimum measurable voltage in the SEM. J Phys E 10, 911–913 (1977)
(41) Feuerbaum H P, Hernaut K: Application of electron beam measuring techniques for verification of computer simulations for large-scale integrated circuits. Scanning Electron Microscopy 1978/I, SEM Inc, AMF O'Hare, pp 795–799
(42) Höfflinger B, Sibbert H, Zimmer G, Kubalek E, Menzel E: Model and performance of hot-electron MOS transistors for high-speed low power LSI. IEEE/IEDM Technical Digest, 463–467 (1978)
(43) Kollath R: Zur Energieverteilung der Sekundärelektronen. Ann der Physik 1, 357–380 (1947)
(44) McKay K G: Secondary electron emission. Adv Electronics 1, 65–130 (1948)
(45) Chung M S, Everhart T E: Simple calculation of energy distribution of low-energy secondary electrons emitted from metals under electron bombardment. Appl Phys 45, 2, 707–710 (1974)
(46) Fujioka H, Ura K: Electron beam blanking systems. Scanning, 3–13 (1983)
(47) Koshikawa T, Shimizu R: Secondary electron and backscattering measurements for polycristalline copper with a spherical retarding-field analyser. J Phys D 6, 1369–1380 (1973)
(48) Muratal K, Matsukawa T, Shimizu R: Monte Carlo calculations on electron scattering in solid target. Jap J Appl Phys 10, 678–686 (1971)
(49) Tee W J, Gopinath A: Improved voltage measurement system using the scanning electron microscope. Rev Sci Inst 48, 350–355 (1977)
(50) Menzel E, Kubalek E: Secondary electron detection systems for quantitative voltage measurements. Scanning 1983 (to be published)
(51) Kanaya K, Kawakatsu H: Secondary electron emission due to primary and backscattered electrons. J Phys D 5, 1727–1742 (1972)
(52) Seiler H, Stärk M: Bestimmung der mittleren Laufwege von Sekundäelektronen in einer Polymerisatschicht. Physik 183, 527–531 (1965)
(53) Parker J H, Warren R W: Kelvin device to scan large areas for variations in contact potential. Rev Sci Inst 33, 948–950 (1962)
(54) Petit-Clerc Y, Carette J D: Effect of temperature on surface charges caused by an incident electron beam on a metallic surface. App Phys Lett 12, 227–228 (1968)
(55) Oatley C W: Isolation of potential contrast in the SEM. J Phys E 2, 742–744 (1969)
(56) Gopinath A, Sanger C C: A Method of isolating voltage contrast in the scanning electron microscope. J Phys E 4, 610–611 (1971)
(57) Piwczyk B, Siu W: Specialized SEM voltage contrast techniques for LSI failure analysis. 12th Ann Proc IEEE Reliability Physics Symp, 1974, pp 49–53
(58) Rau E I, Spivak G V: On the visualization and measurements of surface potentials with SEM. Scanning Electron Microscopy 1979/I, SEM Inc, AMF O'Hare, pp 325–332
(59) Janssen A P, Akhter P, Harland C J, Venables J A: High spatial resolution surface potential measurements using secondary electrons. Surf Science 93, 453–470 (1980)
(60) Bittel H, Storm L: Rauschen. Springer, Berlin–Heidelberg–New York 1971
(61) Menzel E, Kubalek E: Electron beam test system for VLSI circuit inspection. Scanning Electron Microscopy 1979/I, SEM Inc, AMF O'Hare, pp 297–304
(62) MacDonald N C: Potential mapping using Auger electron spectroscopy. Scanning Electron Microscopy 1970, IITRI, Chicago, pp 481–485
(63) Schottky W: Spontaneous current fluctuations in various conductors. Ann Physik 57, 541–567 (1918)
(64) Birdsall C K, Bridges W B: Electron dynamics of diode regions, Academic Press, New York 1966
(65) Lin Y C, Everhart T E: Study on voltage contrast in SEM. J Vac Sci Technol 16, 1856–1860 (1979)
(66) Wells O C: Scanning electron microscopy. McGraw-Hill, New York 1974, pp 184–185
(67) Everhart T E, Hoff P H: Determination of kilovolt electron energy dissipation vs. penetration distance in solid materials. J Appl Phys 42, 5837–5846 (1971)
(68) Johnson W C: Mechanismus of charge build-up in MOS insulators. IEEE Trans Nucl Sci NS-22, 6, 2144–2150 (1975)
(69) Ma T P, Scoggan G, Keone R: Comparison of interface-state generation by 25 keV electron beam irradiation in p-type and n-type MOS capacitors. Appl Phys Lett 27, 61–63 (1975)
(70) Mitchell J P, Wilson D K: Surface effects on radiation on semiconductor devices. Bell Syst Techn J 46, 1–80 (1967)
(71) MacDonald N C, Everhart T E: Selective electron-beam irradiation of metal-oxide-semiconductor structures. J Appl Phys 39, 2433–2447 (1968)
(72) Keery W J, Leedy K O, Galloway K F: Electron beam effects on microelectronic devices. Scanning Electron Microscopy 1976/IV, IITRI, Chicago, pp 507–514
(73) Nakamae K, Fujioka H, Ura K: Measurements of deep penetration of low-energy electrons into metal-oxide-semiconductor structure. J Appl Phys 52 1306–1308 (1981)
(74) Holmes-Siedle A G, Zaininger K H: The physics of failure of MOS devices under radiation. IEEE Trans Rel R-17, 1, 34–44 (1968)
(75) Dekker A J: Secondary electron emission. Solid State Physics, Prentice-Hall 1957, pp 251–311

(76) Touw T R, Hermann P A, Lukianoff G V: Practical techniques for application of voltage contrast to diagnosis of integrated circuits. Scanning Electron Microscopy 1977/I, IITRI, Chicago, pp 177–182

(77) Taylor D M: Electron-beam-induced conductivity and related processes in insulating films. IEE Proc 128, A, 174–182 (1981)

(78) Holt D B, Chase B D, Censlive M: SEM studies of electroluminescent diodes of GaAs and GaP. Phys stat sol (a) 19, 467–478 (1973), (a) 20, 135–144 (1973), (a) 20, 459–467 (1973)

(79) Bresse J F, Lafeuille D: SEM beam-induced current in planar p-n junctions, diffusions lengths and generation factor measurements. Proc 25th Aniv Meeting of Emag Inst Phys, London–Bristol 1971, pp 220–223

(80) Balk L J, Kubalek E, Menzel E: Microcharacterization of electroluminescent diodes with the SEM. IEEE Trans El Devices ED-22, 707–712 (1975)

(81) Balk L J, Kubalek E, Menzel E: Time resolved and temperature dependent measurement of electron beam inducent current (EBIC) and voltage (EBIV) and cathodoluminescence with SEM. Scanning Electron Microscopy 1975/I, IITRI, Chicago, pp 447–455

AN AUTOMATED LASER PROBER TO DETERMINE VLSI INTERNAL NODE LOGIC STATES

F. J. HENLEY

Dataprobe Corporation*
3050 Oakmead Village Dr.
Santa Clara, CA 95051
(408) 727-2863

Anritsu Electric Co., Ltd.
1800 Onna, Atsugi City
Kanagawa Pref., 243 Japan
(0462) 23-1111

Mitsui & Co., Ltd.
Electronic Industries Project Dept.
Transportation & Communications Div.
2-1 Ohtemachi 1-Chome
Chiyoda-ku, Tokyo, Japan
03-285-4640

ABSTRACT

A laser probing system for VLSI internal functional test and debug is introduced and described. Advances in the detection circuitry allows maximum clock speed testing and high sensitivity. Computer control of beam position permits automated operation suitable for a fast turnaround environment.

I. INTRODUCTION

The evolution of digital systems from vacuum tubes in the 1940's to VLSI technology today has prompted different solutions for the testing or characterization of these systems. Fig. 1 explains the major driving force for probing technology innovation as being the changing medium within which the measurement(s) must be extracted. For discrete digital systems (1940's to 1960's), probing was easily accomplished with "oscilloscope" probes, where all circuit points could be directly accessed by hooking a wire to it. As digital integrated circuits were first designed and commercially produced in the early 1960's, mechanical needle probes were developed to gain access to their planar metal contacts and interconnections. These probing systems basically consisted of an optical microscope fitted onto a mechanically secure base where needles could be visually placed on the desired integrated circuit metal or poly line.

*correspondence

With the emergence of VLSI densities and the need for rapid turnaround custom/gate array designs, the traditional mechanical needle probing technology for internal fault detection and functional testing is rapidly becoming obsolete. The major difficulties are the slow and strictly manual process of positioning these fine needles on small ($\sim$2um metal traces), the finite capacitance of the probes (sometimes seriously affecting circuit performance), passivation layer removal, and the chance of destroying the circuit due to probe slip.

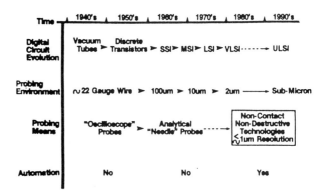

Figure 1: Evolution of digital probing technologies.

Non-contact methods such as e-beam and laser probing have been proposed as a substitute for the cumbersome mechanical probe [1,2]. Ideally, such a method should be automated (for high testing speeds and ease of use), non-destructive, have high system resolution/probe positioning accuracy (~1um), and be non-loading to the circuit under test.

Until recently, laser probing for internal VLSI logic state analysis suffered from serious speed and signal detection problems which offset its inherent simplicity (no vacuums or feedthroughs) and completely non-destructive nature. The laser probing technique presented in this paper consists of irradiating a transistor whose logic state analysis is desirable with a relatively weak laser spot (on the order of tens of microwatts) and monitoring the photo-electric interaction via a novel detection scheme* which allows high speed (tens of MHz) timing diagrams to be extracted and can reliably detect photocurrents down to the sub-microamp range [5]. It has been shown that, depending on the state of the transistor being illuminated, the coupling of the induced photocurrents to the power bus will be different [2,3,4]. By concatenating this photocurrent coupling of succeeding states, computer processing of the data can yield the corresponding timing diagram (logic polarity vs time). Fig. 2 shows an internal logic state analyzer's basic function of extracting the logic diagram of specific nodes. Fig. 3 presents a diagram of the multiplicity of uses such an automated probing system could have to aid the design engineer in modifying/characterizing a faulty or new IC design.

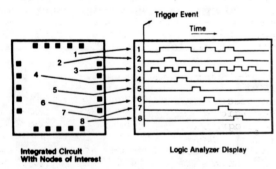

Figure 2: Laser prober function of extracting logic states from specific IC nodes.

The block diagram of a recently developed instrument (called DATA*PROBE) which can perform automated, high speed logic state analysis of VLSI internal nodes in a non-contact and completely non-destructive manner is shown in section II. Section III describes the

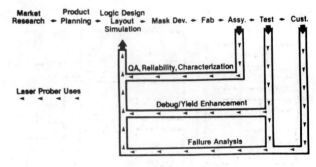

Figure 3: Laser probe uses in IC fabrication.

architecture, exterior detail, and demonstrates the dynamic performance of the detection scheme with an example logic analysis on an LSI CMOS microprocessor. This section also includes a brief discussion of the practical implementation of automating this contactless technique.

II. DATA*PROBE SYSTEM BLOCK DIAGRAM

The DATA*PROBE system block diagram is shown in Fig. 4. The device under test (DUT) rests on a computer controlled X-Y translation stage (1) incorporating laser interferometer encoders for precise positioning (~1um accuracy and ~0.3um repeatability over 2.5 cm travel). The DUT is biased using up to 3 programmable power supplies (+15, +12, and -12 Volts) and clocked via an external clock or an internally synthesized frequency source (100KHz to 40MHz in 100KHz increments). The DUT is at the focus of an optics assembly serving as a laser spot demagnifier and node location finder. A 5mW He-Ne laser (3) is mounted inside the optics assembly and is used as the state detection source. A computer adjustable analog acousto-optic modulator with associated D/A and driver electronics serves as the laser shutter. An auto-focus mechanism (4) mounted between the optics column and the microscope objective (2) assures focus of the laser beam on the DUT during stage movements. The flooding illuminator (5), CCD camera (6), and the monitor (9) allow the user to view the DUT's area of interest (surface viewed by the camera objective) with the laser spot's reflection superimposed. The laser spot size of DATA*PROBE is ~1.8um (2 sigma).

The laser induced photocurrents are

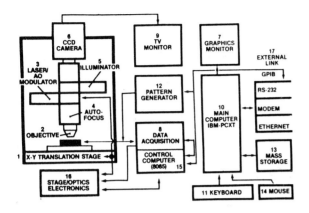

Figure 4: Block diagram of DATA*PROBE IC test system.

detected by a special measurement scheme* described in detail elsewhere [5] The data gathered by the data acquisition circuitry (8) is sent to the control computer (15) to be pre-processed before transmission to the main computer (10) for final signal processing and display (7). The DUT's electrical stimulus is generated by a pattern generator (12) (such as a Pulse Instruments PI-5800 or Interface Technology RS-4000) under GPIB control from the main computer. If an external link (17) is used to acquire complex test vectors from a tester, the main computer can reformat the vectors suitable to the pattern generator, thus avoiding "tying up" an expensive tester or engineering a specialized test head interface to DATA*PROBE.

The main computer (10) (an IBM-PCXT) performs the human and machine interface functions. It incorporates mass storage (13) in the form of a 10 Mbyte hard disk, a 360 Kbyte floppy disk and a streamer back-up tape for the hard disk. The system can be driven through a manual mouse (14)/keyboard (11) and menu mode, through command files from mass storage, or as a slave to a host system via the external link (17).

III. DATA*PROBE FUNCTIONAL DESCRIPTION

Exterior Detail

The exterior detail of the DATA*PROBE model 500 is shown in Fig. 5. The "optics table" is on the left and supports the complete optics/stage assembly. The control computer and the data

*Patent Pending

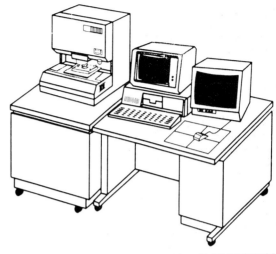

Figure 5: Exterior detail of DATA*PROBE model 500 IC test system.

acquisition/control circuitry are located inside the cabinet. The "user table" on the right is for supporting the main computer, the TV monitor and the mouse.

The GPIB interface between the control/main computer and a 50 ohm BNC connection for the TV monitor are the only cabling necessary between the two tables. Both tables are supported by lockable wheels.

The DUT stimulus signals are available through an array of 16 connectors (40 pin with alternate ground lines) giving a maximum DUT pin count of 320. The connector plate appears on the bottom right side of the optics assembly. The length of the signal lines are approximately 12 inches from the connector plate to the DUT.

Functional Description

The function of DATA*PROBE is to extract the logic state diagram from a number of pre-programmed internal integrated circuit nodes. Before such a data acquisition run can be initiated, the user must load a DUT and, using the mouse, find four reference marks on the die surface, preferably near the corners. If the intended test is new, the user must enter every target node's X-Y coordinate which will be stored in a "node coordinate file" (see Fig. 6). Such a node coordinate file can also be loaded directly from a CAD tool which can access its chip database and send the node X-Y coordonates relative to arbituary reference marks. If the user desires to redo a previous test on a new chip, the main computer will access the desired node coordinate file and, by comparing the stored vs entered reference marks, recompute the node X-Y stage coordinates from a tilt and rotation correcting algorithm (see Fig. 7).

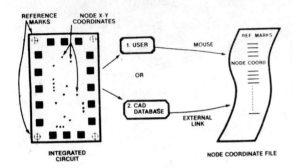

Figure 6: Generating a "node coordinate file" manually (through a mouse) or from a CAD database link.

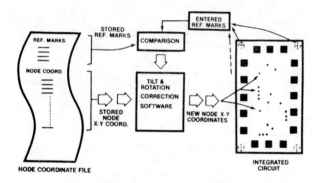

Figure 7. Using a "node coordinate file" by correcting the IC's tilt & rotation from a comparison of stored vs entered reference marks.

The DUT's I/O vectors must also be programmed into the pattern generator before a test. Fig. 8 shows the two different ways in which this can be accomplished: manually via a pattern generation software module on the main computer or from a tester via a reformatting software module.

The last step before execution is entering the test configuration options (such as auto-trigger, state delay, and clock polarity) and the DUT parameter values (bias voltages, test frequency) manually or through the execution of a configuration file.

The user can now initiate a test which will cause the system to acquire the timing diagram of each node in rapid succession. When the test is finished,

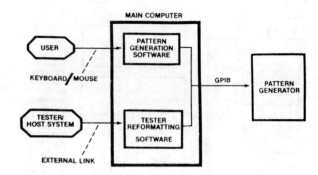

Figure 8: Entering the DUT test vectors by using resident pattern generation software on the main computer or from a tester/host system after reformatting.

the resulting timing diagram is stored in a "data file" which can be displayed on the high resolution graphics screen, compared to other files to highlight any state discrepancies, or sent via an external link to a host system.

A schmoo function is available to help characterization of a DUT by automatically performing tests at different DUT biases and clock frequencies, comparing the resulting timing diagrams against a "standard", and graphing the corresponding schmoo plot. A "single-shot" node is also available allowing the user to manually move the stage via mouse control and, when a mouse button is depressed, perform a single test at the selected location.

DATA*PROBE Architecture and Performance

The architecture of the DATA*PROBE system is shown in Fig. 9 and incorporates 2 computers and one finite-state controller for maximum system throughput.

The finite-state controller is an 8 input/16 output finite-state machine (FSM) having 1K (x 8 bit) vector word depth and operating at 2.5MHz. The FSM has the responsibility of efficiently controlling the data acquisition, DUT control circuitry, and managing the high speed transfer of raw data between the A/D converter and the control computer memory via a DMA channel. It is slaved to the control computer and only requires test parameters and a "start test" command to initiate a complex series of data acquisition actions.

An 8085-based micro-computer (having 16K ROM, 48K RAM and GPIB/DMA capability) is used as a system controller and data pre-processor. This control computer accepts high level commands from the main computer through its GPIB interface and initiates the required sctions such as moving the stage, changing a DUT parameter, or starting a test.

The main computer performs the higher level processing and display, directs overall system operation, and interfaces between the mass storage, the external links, and the user.

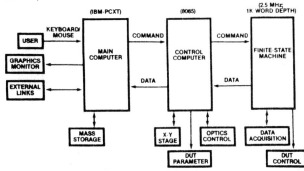

Figure 9: DATA*PROBE system architecture showing each subsystem's I/O and control responsibility.

The pipelined operation of each subsystem during a data acquisition run is shown in Fig. 10. A complete test cycle, in which the logic analysis of a single node is performed, occurs during time intervals t1 to t4. During t1, the main computer sends a stage move command which is accepted and immediately processed by the control computer. During t2 (stage movement) and t3 (node data acquisition), the main computer calculates the next stage move command and signal processes the data acquired from the previous node test cycle. The FSM is only active during t3, when the raw data (from a high performance 12 bit A/D converter with 1uSec conversion time) is directly transmitted to the control computer memory via a dedicated DMA channel. During t4, the pre-processed data taken at time interval t3 is sent from the control computer to the main computer via GPIB, after which the t1-t4 cycle repeats itself if other nodes are to be tested.

The overall test time per logic node is shown in Fig. 10 as being approximately 320 mSec, calculated for a 1024 state timing diagram, an average stage movement of 100um, and a DUT test clock frequency of 10MHz. The average test rate is therefore 1/320 mSec, or greater than 3 nodes/second.

	SEND MOVE COMMAND t_1	STAGE MOVE t_2	DATA ACQUISITION t_3	DATA TRANSMISSION t_4	SEND MOVE COMMAND t_1 ----
MAIN COMPUTER	SEND MOVE COMMAND TO NODE 'n'	PROCESS NODE 'n-1' DATA	PROCESS NODE 'n-1' DATA	ACCEPT NODE 'n' DATA	SEND MOVE COMMAND TO NODE 'n+1'
CONTROL COMPUTER	ACCEPT STAGE COMMAND	MOVE STAGE TO NODE 'n'	PRE-PROCESS NODE 'n' DATA	SEND NODE 'n' DATA	ACCEPT STAGE COMMAND
FINITE-STATE MACHINE	IDLE	IDLE	ACQUIRE NODE 'n' DATA	IDLE	IDLE
TIME TO ACCOMPLISH (TYPICAL)	50uSec	200mSec	100mSec	20mSec	50uSec

Figure 10: DATA*PROBE pipelined operation of the 3 subsystems during a data acquisition run.

Example Logic Analysis

The dynamic operation of the logic analysis detection scheme is shown using a manual prototype built by the author and described in detail elsewhere [5]. The example shown here will demonstrate the laser probe's capability of capturing logic information at high clock rates by presenting normal and high speed failure logic diagrams from a working CMOS microprocessor.

Fig. 11 shows an example logic analysis of a CMOS microprocessor (COSMAC 1802) register circuitry while the uP is executing a small program at a clock rate of 4MHz. Fig. 11a) shows a schematic of a small portion of the uP's register circuitry. The circled numbers are circuit locations where the laser probe was used. Fig. 11b) shows the laser probe generated timing diagram of 11 channels/256 states. The first 4 channels are the least significant bits of the program counter keeping track of the memory address while the other 7 channels measure various register control and internal data bus signals.

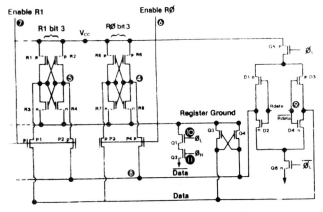

a) Section of the CMOS uP register circuitry showing laser probing points (circled numbers).

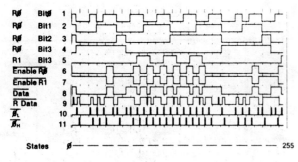

b) Laser probe generated timing diagram of correct uP operation (fc=4MHz). Channels 1-4 are program counter bits ∅-3 respectively. Channels 4-11 represent the probe points shown in a) as circled numbers.

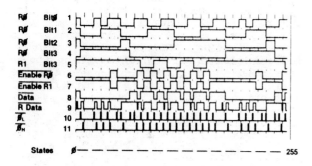

c) High speed data bus failure taken at fc 4MHz.

Figure 11: Example laser probe logic analysis of a CMOS microprocessor (COSMAC 1802) register circuit while the uP is executing a small program loop.

At slightly higher clock frequency, Fig. 11c) shows the uP presenting incorrect data on its internal data bus ($\overline{DATA}$ and $\overline{RDATA}$) causing register errors.

DATA*PROBE/CAD Interface Requirements

A laser probing system will gain significantly in usefulness and effectiveness if a CAD link is established. Fig. 12 shows a block diagram of how a CAD tool and a pattern generator/tester can be linked to DATA*PROBE.

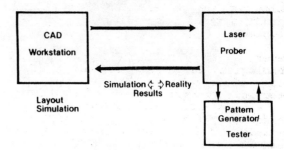

Figure 12: Implementation of test automation showing CAD workstation downloading X-Y information, test patterns, and expected node results. Laser prober performs actual probe tests at specified operating conditions and flags any discrepancies.

The transmission of three different file types are required for the successful implementation of this link. The first is a complete "node coordinate file" (see section III) to be transmitted from the CAD database to DATA*PROBE. The existence of this file would allow the laser prober to access the nodes of interest without prior manual coordinate entries. The second file type is a "data file" where the expected node signature for a particular test is present. The prober can use this file in test result comparisons to assess the correct operation of the DUT. The last file type can be sent by the CAD tool (to recreate the DUT I/O patterns used in the simulations) or from a tester to exercise the DUT with the same test vectors. In each case, the main computer would reformat the data vectors for programming the pattern generator. If test feedback is desired, the prober can transmit a file containing a concise report of the test parameters, test results, and any discrepancies between the simulated node signatures and the prober extracted data.

With these interface capabilities, the time for determining an integrated circuit's full or partial functionality and subsequent characterization can be substantially shortened. For fast turnaround environments, this can be a significant factor in productivity.

System Limitation Due to Multiple-level Metalization Coverage

A Limitation of the laser probe that is shared by all other probing techniques in some form is the "screening" effect of top layer(s) of metalization from desired information within the die. In single-level metal IC's, this potential problem is not an issue because of the general optical accessability of the diffusions.

However, in Multiple-layer metal technologies, large metalization coverage could ultimately limit access to appreciable numbers of important devices. For e-beam and mechanical probing techniques, the existence of these layers generally limits the probing to the uppermost metal layer without the design of inter-layer via "stubs". This requirement can be prevalent because of the usual design methodology of having the inner level of metal carry the bulk of the signals. For the laser probe, the only probing requirement is the optical accessability of a diffusion connected to the node of interest. This could be met in the design phase by designing optical "windows" for nodes having 100% metal coverage.

IV. CONCLUSIONS

An improved technique of extracting logic state information from internal VLSI nodes using a focused laser beam was introduced and described. The technique relies on the measurement of power bus coupled laser injected photocurrents which are strongly modulated by the digital state of the illuminated device. This measurement was accomplished using a novel AC transient scheme which is first order insensitive to test clock frequency and power bus switching transients.

A new laser probe IC test system incorporating these recent advances (the DATA*PROBE model 500) was introduced and briefly described. It is a dedicated logic analysis system with substantial automation capability. Local automation of the system is made possible with a computer controlled laser interferometer encoded X-Y translation stage. The DATA*PROBE system architecture and pipelined operation during a data acquisition run was discussed and shown to increase system test speed through the use of two computers and one controller which are often working in parallel. The remote automation requirements of interfacing DATA*PROBE to CAD design tools and testers was also discussed as being a viable solution for fast turnaround IC design environments.

V. ACKNOWLEDGEMENTS

The author wishes to thank the following persons for their enthusiasm and support of this work: Mr. Weeks of Dataprobe Corp., Mr. Utsumi, Mr. Urao, Mr. Yoshida, Mr. Tachikawa, and Mr. Lynch of Mitsui & Co. Tokyo, Mr. Langston of Mitsui & Co. Semicon (New Jersey, USA), and Dr. Tsunoda, Mr. Shiyomi, Mr. Takabayashi, Mr. Hiratsuka, and Mr. Nakamura of Anritsu Electric Company. The author would also like to extend a big Domo Arigato Gozaimas to Patricia Cornwell for her unfailing support.

REFERENCES

[1] M. Macari, K. Thangamathu, and S. Cohen, "Automated Contacless SEM Testing for VLSI Development and Failure Analysis", Proceedings of the IEEE 1982 International Reliability Physics Symposium, April 1982.

[2] M.T. Pronobis and D.J. Burns, "Laser Die Probing for Complex CMOS", Proc. of the 1982 ISTFA, pp. 178-181, Ocotber 1982.

[3] R.E. McMahon, "A Laser Scanner for Integrated Circuit Testing", Proc. 1972 International Reliability Physics Symposium, pp. 23-25.

[4] F.J. Henley, "Functional Testing and Failure Analysis of VLSI using a Laser Probe", Proceedings of the 1984 Custom Integrated Circuits Conference (C.I.C.C.), May 1984.

[5] F.J. Henley, "Logic Failure Analysis of CMOS VLSI Using a Laser Probe", Proceedings of the IEEE 1984 International Reliability Physics Symposium, April 1984.

Microelectronic Test Chips for VLSI Electronics

MARTIN G. BUEHLER

Jet Propulsion Laboratory
California Institute of Technology
Pasadena, California

I. Historical Perspective
II. Introduction
   A. Test-Chip Measurement Environment
   B. Objections to the Use of Test Chips
III. Types of Test Structures
   A. Test Structures for Device Parameter Extraction
   B. Test Structures for Layout Rule Checking
   C. Test Structures for Process Parameter Extraction
   D. Test Structures for Random Fault Analysis
   E. Test Structures for Reliability Analysis
   F. Test Structures for Circuit Parameter Extraction
IV. Test-Chip Organization and Test-Structure Design
   A. Test-Chip Organization
   B. Test-Structure Design Rules
V. Test-Chip Testers and Advanced Test Structures
VI. Future Directions
   A. Data Acquisition
   B. Data Reduction
   C. Vendor Transactions
   Appendix
   References

I. HISTORICAL PERSPECTIVE

The microelectronic test chip is an ancillary test device that is manufactured along with product circuits on wafers. It is composed of numerous devicelike test structures that are measured by a variety of means to obtain information that is difficult, if not impossible, to obtain from product circuits. The test structures are designed to provide a rapid analysis of a specific portion of the wafer fabrication process. Traditional failure analysis techniques, when applied to a product circuit, are time consuming and hence expensive tasks. The use of test chips is expected to enhance the reliability of the final product and reduce its cost.

Test chips have been used by the integrated circuit industry since its beginnings in the early 1960s. In general, test chips have been used for component characterization, manufacturing process control, equipment and operator performance evaluation, look-ahead indicators of circuit yield, and new wafer fabrication process evaluation. In the early days test chips were rarely used in the purchase or rejection of circuits or manufacturing equipment. As explained later, this situation is changing.

An early test chip was described in the 1968 work of Barone and Myers [1]. They developed a test chip (1.3 mm × 3.0 mm) with 44 probe pads to aid in the evaluation of a bipolar 8-bit adder circuit that contained 448 components. The test chip consisted of the following eleven test structures: a bipolar transistor, base resistor, base-under-emitter resistor, step-coverage resistor, metal-sheet resistor, base-collector diode with buried layer, buried-layer resistor, metal–semiconductor contact resistor array, metal–metal contact resistor array, multiemitter transistor, and MOS capacitor. These test structures were used in device characterization, process control, and circuit reliability evaluation. The number of elements (100) in the contact resistor array structures was remarkably large compared with other contemporary test structures where the number was generally much less.

In this same time period, two other test chips are noteworthy. In 1969 Schnable and Keen [2] describe a test chip for monitoring LSI reliability life aging. Their chip was designed to allow the measurement of first-level to second-level metal contact resistance, dielectric pinhole density and breakdown, metal step coverage, metal sheet resistance, and the resistance of an array of diffused resistors. In 1970 Sahni [3] described a chip for evaluating the reliability of bipolar integrated circuits. His structures examined the leakage current of transistors, integrity in the metallization, moisture resistance of the passivation layer, integrity of the bonds, and resistance of the metallization. Both test chips [2, 3] were packaged and subjected to thermal stress tests.

A test chip that is representative of this era is shown in Fig. 1. It was described in 1972 by Penney and Lau [4] and was developed for aluminum-gate PMOS integrated circuits having about 6000 individual transistors. The test chip (~0.9 mm × 1.3 mm) had 200 probe pads and was designed as a reliability evaluation device. It contains seven electrically testable test structures as listed in Table I.

Back then Penney and Lau [4] were rather optimistic about the use of test chips.

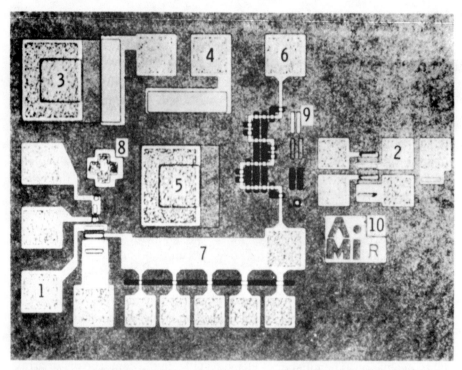

Fig. 1. An early test chip (0.9 mm × 1.3 mm) for use in evaluating a PMOS circuit wafer fabrication process [4]. Test structures are listed in Table I. (Used with permission.)

TABLE I

Test Structures for the Test Chip Shown in Fig. 1

Number	Test structure	Comment
1	Inverter	
2	Input transistor and field threshold transistor	
3	Large P–N junction diode with enhancement gate	
4	MOS capacitor	Identical in size to the enhancement gate of structure 3
5	Large P–N junction diode	Identical in size to structure 3 but without enhancement gate
6	Metal step-coverage resistor	Number of steps = 40
7	Metal resistors with different linewidths	Used to evaluate metal thickness
8	Alignment marker	
9	Etch control structure	
10	Logo	

If such a device (test chip) is included with production circuits, it is valid to conclude that good performance of the test chip implies good working circuits. There is merit to being able to evaluate a standard circuit (test chip) on a daily basis rather than to randomly evaluate different types of circuits where data comparison is meaningless.

These views are just now beginning to be accepted by the semiconductor industry.

The above test chips were intended for internal use by chip manufacturers. The first test chip used to accept or reject wafers was described in 1974 by Reynolds *et al.* [5]. Their chip (2.54 mm × 2.54 mm) contained 35 test structures and 35 probe pads. It was intended to validate the layout rules, wafer-fabrication process, and reliability of aluminum-gate PMOS integrated circuits. The purpose of the test chip was not to control fabrication practices but merely to assure that the process was under control.

Early test chips had limited usefulness, for they were not comprehensive and were not designed for automatic wafer probing. In addition, they were usually designed as an afterthought to a circuit design effort. The situation is summed up nicely by Tingley and Johnson [6].

> Process development (test) chips usually result from hurried efforts by process development engineers with limited experience in actual circuit design or layout. The resulting designs tend to be highly idiosyncratic, with numerous omissions which any experienced circuit or layout designer, MOS physicist, or reliability specialist would readily notice. Usually poorly designed for automatic testing, process development chips often tie up one or more highly paid technical people for weeks doing manual probing, testing, and data reduction.

Recently a comprehensive test structure was developed by Ham [7] for characterizing SOS technology. His chip (6.6 mm × 6.6 mm) contained 175 test structures and 1250 probe pads. This chip has a very large number of probe pads and requires 21 probe cards to access all the electrical test structures.

Another present day test chip developed by Mitchell and Linholm [8] is shown in Fig. 2. This test chip (5.1 mm × 5.1 mm) has 216 probe pads and 40 test structures, which can be accessed by one probe card. These structures are listed in Table II. A comparison of this list with the structures listed in Table I provides a measure for the progress that has been made in the past 10 years. For instance, the structures shown in Fig. 2 have been arranged to allow for automatic data acquisition. The pattern is probe-pad intensive so that test structures are electrically isolated from each other. Finally, the chip area is 25 times larger so that a statistically significant number of steps can be included in the step-coverage resistors (structures 1–3). The number of steps (115,200) included in the structures on the NBS-28A test chip is con-

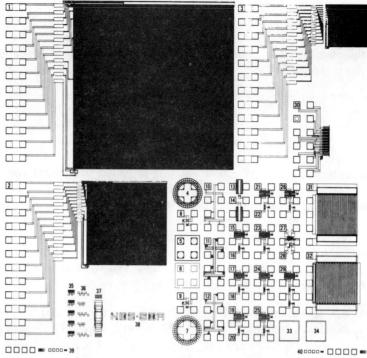

Fig. 2. Comprehensive test chip (5.1 mm × 5.1 mm) NBS-28A for analyzing either an NMOS or PMOS self-aligned polysilicon gate, junction isolated process [8]. Test structures are listed in Table II.

siderably more than the 40 steps seen in structure number 6 of Fig. 1. Such large test structures are necessary in order to provide more than a go/no-go test and to properly emulate today's complex circuits which can have one-half million transistors [9].

In recent years there has been a growing awareness of the vital role test chips can play in the manufacture of electronic devices. This awareness can be seen in the commercial availability of multifunction parametric testers [10], which greatly facilitate the acquisition of test-chip data. Professional societies have devoted special sessions to this topic such as the process/device monitors session at the Semiconductor Silicon 1981 Symposium [11]. Finally, since 1974 [12] the National Bureau of Standards has made and is making significant contributions to test-chip methodology [8].

To date, test chips have been most extensively used within circuit fabrication facilities. Typically a wafer found in a high-volume circuit manufacturing facility will appear as seen in Fig. 3 where 8 test chips are interspersed among 550 identical circuits.

Test chips are now being used more frequently in the acceptance or rejec-

TABLE II

Test Structures for Test Chip NBS-28A [8] Shown in Fig. 2[a]

Number (#)	Test structure	Comments
1, 2, 3	Metal step-coverage serpentine resistor over polysilicon grid	Linewidth and spacing: #1 = 4 μm, #2 = 2 μm, #3 = 1 μm. Number of steps = 115,200.
4, 5	Polysilicon-gate oxide-silicon capacitor	Area = 829 μm². #4 = round. #5 = square.
6, 7	Metal-field oxide-silicon capacitor	Area = 829 μm². #6 = square. #7 = round.
8, 9	Contact resistor	Contact area = 64 μm² (square). #8 metal-implant, #9 = metal-polysilicon.
10, 11, 12	Cross-bridge resistors	Linewidth = 16 μm. #10 = metal, #11 = polysilicon, #12 = implant.

13–22, 27–28	Field effect transistors (3-terminals) #14 = metal gate, field oxide. All others = polysilicon gate, gate oxide. #27 and 28 have all dimensions reduced, whereas #17–#22 have only W and L reduced relative to #15 and #16.	#	W (μm)	L (μm)	Reduction
		13	160	28	—
		14	160	26	—
		15	120	8	—
		16	12	8	—
		17	60	4	$\frac{1}{2}$
		18	6	4	$\frac{1}{2}$
		19	30	2	$\frac{1}{4}$
		20	3	2	$\frac{1}{4}$
		21	15	1	$\frac{1}{8}$
		22	1.5	1	$\frac{1}{8}$
		27	15	1	$\frac{1}{8}$
		28	1.5	1	$\frac{1}{8}$

Number (#)	Test structure	Comments
23–26	Inverter	#23 = #15 + #16 ($L = 8$ μm). #24 = #17 + #18 ($L = 4$ μm). #25 = #19 + #20 ($L = 2$ μm). #26 = #21 + #22 ($L = 1$ μm).
29	NAND gate	#29 = #15 + 2(#16) ($L = 8$ μm).
30	Multigate field-effect transistor (polysilicon gate)	L = 1, 1.5, 2, 2.5, 3, 3.5, 4, 6, 8 μm.
31, 32	Comb field-effect transistor	W = 14.5 mm, L = 8 μm, #31 + metal gate, #32 = polysilicon gate.
33, 34	Secondary ion mass spectrometry (SIMS) analysis area	#33 = implant, #34 = polysilicon. Area is a square 240 μm on a side.
35	Resolution structure for each photomask	Linewidth = 1, 2, 4, 8, 16 μm.
36	Etch control structure for each photomask.	Spacing = 0, 1, 2, 4, μm. Also 8 μm for window, contact, and passivation levels.
37	Surface profilometer structure	
38	Chip identification number	
39, 40	Photoresist mask alignment marks	#39 for positive photoresist and #40 for negative photoresist

[a] Photomask levels are implant, polysilicon, contact, metal, and passivation.

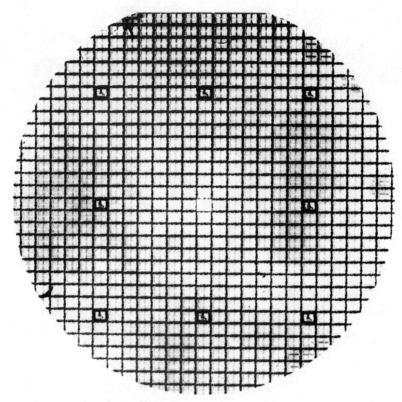

Fig. 3. High-volume integrated circuit wafer.

tion of wafer lots of custom-integrated circuits produced by "silicon foundries" [13] that specialize in the manufacture of custom circuits. This is not a new role for test chips, for the pioneering work of Reynolds *et al.* [5] used the same principle. What is new is the number of circuits found on a wafer. This is illustrated in Fig. 4 by a wafer that consists of 50 different circuits and 5 different test chips. In this multicircuit environment, it is no longer possible to accept or reject wafer lots based on circuit performance. Instead, lot acceptance is based on results obtained from test chips. In this environment it is essential that a test-chip methodology be perfected that will facilitate the buying and selling of custom circuits. This chapter describes the current state of the test-chip art and indicates where improvements are needed.

II. INTRODUCTION

Over the years a number of alternate names have been given to the test chip. Many of the names are indigenous to a particular company. The names

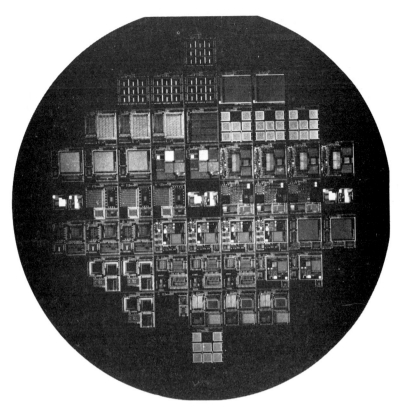

Fig. 4. Custom multicircuit wafer.

vary from test pattern, test coupon, test pellet, reliability test site, reliability evaluation device, process control bar, PCM (process control monitor), to a most descriptive name, "panic." The name panic is derived from the use of the test chip as a diagnostic in emergency situations. The name test pattern is used quite frequently, but this name conflicts with the meaning assigned by a test engineer who refers to a test pattern as a train of electrical pulses used to test a circuit. In recent years test chip seems to be the most universally accepted term.

The test structures discussed in this chapter are those that are measured by electrical methods at the end of the wafer-fabrication process. Test chips usually contain other large-area test structures that are measured by a variety of techniques. For example, this text does not discuss in-process tests where mechanical probes are used to measure bipolar transistor characteristics on partially fabricated wafers. Nor do we discuss the use of four-point probes to measure the sheet resistance of implanted layers, or optical techniques to measure the thickness of dielectric layers. In addition, various physical analysis techniques such as Auger, SEM, SIMS [14], or angle lap and stain are not discussed. Many of these techniques are described in a special issue of *IEEE Transactions on Electron Devices* [15] and in the other chapters of this book.

A. Test-Chip Measurement Environment

The electrical test methods described in this chapter are restricted to those methods where test structures are measured in the dark in wafer form with an automatic multifunction tester in a near-room ambient. This restriction eliminates very low-current and high-frequency measurements; it eliminates measurements made far from room temperature and humidity; and it restricts the use of light to photon flooding where a burst of light is used to quickly pull a device out of deep depletion. The restriction is imposed in order to focus attention on the need to make rapid and statistically meaningful measurements in a production environment. Instead of having a stand-alone measurement capability dedicated to the measurement of a single quantity, the thrust of this approach is to measure as many quantities as quickly as possible with a single set of computer-controlled instrumentation.

The goal is to gather as much information as economically as possible while preserving the integrity of the data. The restrictions mentioned above have the following important implications for the test technology:

(1) *Measurements must be rapid.* The measurement time to evaluate a single parameter must be less than a few seconds. This means that simple procedures are needed for both data acquisition and on-line data reduction. Results from these stripped-down measurements must be carefully correlated with results from more accurate measurements to ensure their validity.

(2) *Test chips must be measurable in wafer form with the use of a multifunction tester in a near-room ambient.* The measurement environment is rather noisy, especially for low-level signals. To aid the measurement process it may be necessary to build in buffer amplifiers on-chip to increase signal levels.

(3) *Test chips must be organized so that all test structures can be accessed by a single probe array.* As will be shown, a scheme has been developed that uses a 2-by-N probe array to access all test structures.

(4) *Test structures must be designed so they are modular,* can be stored in a computer cell library, and can be rapidly assembled into a new test chip. This goal is achieved by designing test structures so they can be probed by a 2-by-N probe pad array.

(5) *Test structures must be designed to allow the characterization of very small regions,* since many of the electrical properties of VLSI devices are dominated by peripheral rather than bulk transport. Because microelectronic test structures are fabricated from the very elements that compose the

integrated circuits, they are uniquely qualified to be used as analytical tools in evaluating VLSI circuits.

B. Objections to the Use of Test Chips

As described in Section I, test chips have been used for a number of years; however, their utility is often questioned. An analysis of the objections to their use reveals their strengths and weaknesses. Some of the objections follow.

(1) *Test chips take up the space of functional circuits* and so are an automatic yield loss. The space taken up by a test chip must be justified by the return received in improved process control and diagnostic capability. These are difficult to justify on an absolute basis.

(2) *Test results from test chips reveal proprietary information* about the wafer fabrication process. Test chips greatly assist the "reverse engineering" of a wafer fabrication process. When test chips are used to transact business, the results often must be kept proprietary.

(3) *Test chips divert attention* away from making circuits perform properly to making the test chips perform. This objection is not relevant in the "silicon foundry" business where fabricated wafer lots are accepted based on test-chip results rather than on circuit performance.

(4) *The measurement of test chips requires an excessive amount of engineering effort* to develop test software. The impact of this objection can be minimized through careful planning aimed at minimizing software costs. Currently no standards exist for test software that will allow the transfer of test programs between different test equipment.

(5) *Advanced test chips cannot be measured with a digital integrated circuit tester*. To take full advantage of the diagnostic capabilities of a test chip, a special tester, a multifunction parametric tester, must be used.

The use of test chips in the semiconductor industry is becoming increasingly important. As Potter has observed [16], test chips are essential in many commercial transactions. Scott *et al.* [17] indicate that test chips offer an effective tool to speed up the LSI fabrication, assembly, and test processes and to enhance the reliability of the final product—all at a substantial cost savings. Currently test chips are supplied by the wafer fabricator, but user-supplied test chips are destined to be used in the procurement process.

III. TYPES OF TEST STRUCTURES

Test structures are used for a variety of purposes in the fabrication of integrated circuits. The six categories of their use are to extract device parame-

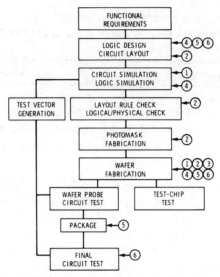

Fig. 5. Simplified integrated circuit production sequence illustrating where results from the six types of test structures are used. Test structures are used for (1) device parameter extraction, (2) layout rule checking, (3) process parameter extraction, (4) random fault analysis, (5) reliability analysis, and (6) circuit parameter extraction.

ters, check layout rules, extract wafer-fabrication parameters, analyze random faults, analyze reliability, and extract circuit parameters. Results from the test structures are used at a number of steps in the production of integrated circuits as illustrated in Fig. 5. Here a typical manufacturing sequence is shown, and it is seen that the most heavy utilization of test structures is found in wafer fabrication. Of course, one must go through the process at least once and collect results from test structures before the results can be used to influence the fabrication process.

In this chapter, test structures are discussed according to their use. The relationship between test structure uses and their description as circuit elements is shown in Table III, where it is seen that a particular circuit element has a variety of uses. For example, a discrete resistor may be used to extract process doping information, such as sheet resistance; it may be used to provide wire resistance data needed in modeling circuit propagation delay. Large-area capacitors may provide a host of process parameters, such as oxide thickness, interface density, or bulk dopant density; they may be used to evaluate random fault densities, such as shorts between conducting layers; they may be used in reliability analysis, such as time-dependent oxide breakdown studies; or they may provide circuit parameters, such as wire capacitance data.

Each of the six types of test structures is described below. At the beginning of each section a number of examples is listed, but only one is described in detail.

TABLE III

Relationship between Test Structures as Circuit Elements and Test Circuits and Their Use in Data Acquisition

Test structure use	Resistor		Capacitor large area	Diode discrete	Transistor		Test circuit
	Discrete	Array			Discrete	Array	
Device parameter extraction	X		X	X	X		
Layout rule checking		X					
Process parameter extraction	X		X	X	X		X
Random fault analysis		X	X			X	X
Reliability analysis		X	X		X		X
Circuit parameter extraction	X		X				X

A. Test Structures for Device Parameter Extraction

These structures are used to extract the parameters for circuit simulation and wafer fabrication control. Examples are

(1) test transistor for threshold voltage, conduction factor, transconductance, breakdown voltage, and leakage current measurements, and

(2) resistors and capacitors for interconnection parameter measurements such as wire resistance and capacitance.

As an example consider the extraction of MOS transistor parameters in both the linear and saturation portion of the transistor current–voltage ($I-V$) characteristics. The method follows from Ham [7] and serves to illustrate the strategy used in measuring transistors in wafer form. The method is illustrated by the $I-V$ characteristics shown in Fig. 6 where seven measurements of the drain–source current I_{DS} establish seven transistor parameters. This is a very small number of measurements to characterize a complex device such as a transistor.

Before device parameters are measured, an untested device is given a qualification test. The strategy calls for an initial stress test to ensure that the transistor has no major faults. In essence one approaches the untested device as a skeptic, requiring that the device prove that it is worthy of further detailed characterization.

The parameter extraction procedure [7] is outlined in Table IV. The initial screening tests consist of measurements that evaluate threshold voltage stability, gate-leakage current, current handling capability, and low-current lin-

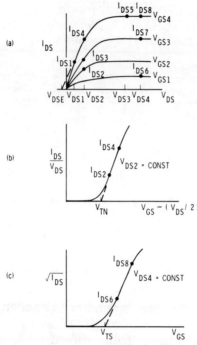

Fig. 6. NMOS transistor characteristics. (a) Common source characteristics, (b) nonsaturation region conductance curve, and (c) saturation region $\sqrt{I_{DS}}$ curve.

earity. The screening parameters must be within previously established parameter bounds or the device is rejected. This saves overall measurement time in that defective devices are not characterized further.

Referring to Table IV, it is seen that the extraction procedure requires the user to define eleven quantities: I_{DS0} is the drain–source current and V_{DS2} is the drain–source voltage; they are used to determine the threshold voltage in the linear region. V_{Tmax} is the maximum allowable threshold voltage. V_{GS0} is the gate–source leakage current, and ΔV_{Tmax} is the maximum allowable shift in the threshold voltage. V_{GS} is an estimate of the upper bound for the gate–source voltage. V_{DS4} is the maximum drain–source voltage used to characterize the device in the saturation region. I_{DSmin} is the minimum allowable drain–source current that is used to evaluate the current handling capability of the device. V_{DSmax} is the maximum allowable extrapolated drain–source voltage V_{DSE} that is used to verify the linearity of the $I-V$ curve in the linear region; see Fig. 6a. The measured quantity in most cases is a current that is measured with an electrometer. The V_{GS} measurement requires the use of an operational amplifier [7].

The parameter extraction process assumes a device model for the transistor in both the nonsaturation (linear) and saturation regions. In the

TABLE IV
MOS Transistor Parameter Extraction Procedure

Test	Stimulus	Measured quantity	Derived parameter	Fail test	User-defined quantity[a]
Initial threshold	V_{DS2}, I_{DS0} $+V_{GS0}$, t	V_{GS01} I_{GS1}	$V_{T1} = V_{GS01}$	$V_{T1} > V_{Tmax}$ $I_{GS1} > I_{GSmax}$	V_{DS2}, I_{DS0}, V_{Tmax} V_{GS0}, t, I_{GSmax}
Gate leakage (positive stress)					
Threshold stability	V_{DS2}, I_{DS0} $-V_{GS0}$, t	V_{GS02} I_{GS2}	$V_{T2} = V_{GS02}$, $\Delta V_{T1} = V_{T2} - V_{T1}$	$|\Delta V_{T1}| > \Delta V_{Tmax}$ $|I_{GS2}| > I_{GSmax}$	ΔV_{Tmax}
Gate leakage (negative stress)					
Threshold stability	V_{DS2}, I_{DS0}	V_{GS03}	$V_{T3} = V_{GS03}$, $\Delta V_{T2} = V_{T3} - V_{T1}$ $V_{GS4} = V_{T1} + V_{GS}$ $V_{GS3} = V_{T1} + 0.9 V_{GS}$ $V_{GS2} = V_{T1} + 0.5 V_{GS}$ $V_{GS1} = V_{T1} + 0.25 V_{GS}$ $V_{DS1} = 0.5 V_{DS2}$ $V_{DS4} = 0.9 V_{DS4}$	$|\Delta V_{T2}| > V_{Tmax}$	V_{GS}
Current handling ability	V_{DS2}, V_{GS4}	I_{DS4}		$I_{DS4} < I_{DSmin}$	V_{DS4} I_{DSmin}
Linearity check	V_{DS1}, V_{GS4}	I_{DS1}	$V_{DSE}(V_{GS4}) = \dfrac{(V_{DS1}I_{DS4} - V_{DS2}I_{DS1})}{(I_{DS4} - I_{DS1})}$	$|V_{DSE}| > V_{DSmax}$	V_{DSmax}
Threshold (nonsaturation)	V_{DS2}, V_{GS2}	I_{DS2}	$V_{TN}(V_{DS2}) = \dfrac{[I_{DS4}(V_{GS2} - (V_{DS2}/2)) - I_{DS2}(V_{GS4} - (V_{DS2}/2))]}{(I_{DS4} - I_{DS2})}$		
Conduction factor (nonsaturation)	V_{DS2}, V_{GS3}	I_{DS3}	$k_N(V_{DS2}) = \dfrac{(I_{DS4} - I_{DS2})}{[2 V_{DS2}(V_{GS4} - V_{GS2})]}$		
Transconductance (nonsaturation)	V_{DS2}, V_{GS3}	I_{DS3}	$g_{mN}(V_{DS2}) = \dfrac{(I_{DS4} - I_{DS3})}{(V_{GS4} - V_{GS3})}$		
Threshold (saturation)	V_{DS3}, V_{GS4}	I_{DS5}	$V_{TS}(V_{DS4}) = \dfrac{(V_{GS1}\sqrt{I_{DS8}} - V_{GS4}\sqrt{I_{DS6}})}{(\sqrt{I_{DS8}} - \sqrt{I_{DS6}})}$		
Conduction factor (saturation)	V_{DS4}, V_{GS1}	I_{DS6}	$k_S(V_{DS4}) = \left[\dfrac{(\sqrt{I_{DS8}} - \sqrt{I_{DS6}})}{(V_{GS4} - V_{GS1})}\right]^2$		
Transconductance (saturation)	V_{DS4}, V_{GS3}	I_{DS7}	$g_{mS}(V_{DS4}) = \dfrac{(I_{DS8} - I_{DS7})}{(V_{GS4} - V_{GS3})}$		
Output conductance (saturation)	V_{DS4}, V_{GS4}	I_{DS8}	$G_{oS}(V_{GS4}) = \dfrac{(I_{DS8} - I_{DS5})}{(V_{DS4} - V_{DS3})}$		

[a] User-defined quantities are noted on the first test that requires them. The parameter t is stress time.

nonsaturation (linear) region where $(V_{GS} - V_{TN}) > V_{DS}$, the Sah equation [18] is used to describe the drain–source current:

$$I_{DS} = k_N[2(V_{GS} - V_{TN})V_{DS} - V_{DS}^2], \tag{1}$$

where V_{GS} is the gate–source voltage, V_{DS} is the drain–source voltage, V_{TN} is the nonsaturation (linear) threshold voltage, and k_N is the nonsaturation (linear) conduction factor. The analysis for the device parameters follows by rearranging the above equation

$$I_{DS}/V_{DS} = 2k_N[V_{GS} - (V_{DS}/2) - V_{TN}]. \tag{2}$$

A plot of this equation is shown as the linear portion of the curve in Fig. 6b, where the intercept at $I_{DS} = 0$ denotes the threshold voltage, V_{TN}. From the above equation it is seen that the conduction factor follows from

$$k_N = (1/2V_{DS})(\partial I_{DS}/\partial V_{GS})|V_{DS} = \text{const}. \tag{3}$$

The transconductance is defined by

$$g_m = (\partial I_{DS}/\partial V_{GS})|V_{DS} = \text{const}. \tag{4}$$

This relation is used to evaluate g_m in both the nonsaturation (linear) and saturation regions.

In the saturation region where $(V_{GS} - V_{TS}) \le V_{DS}$, the drain–source current is described by

$$I_{DS} = k_S(V_{GS} - V_{TS})^2, \tag{5}$$

where k_S is the saturation conduction factor and V_{TS} is the saturation threshold voltage. A plot of the square root of this equation is represented by the linear portion of the curve shown in Fig. 6c. The threshold voltage V_{TS} is found by extrapolating the straight-line portion of the curve to $I_{DS} = 0$. The conduction factor follows from

$$k_S = [(\partial \sqrt{I_{DS}}/\partial V_{GS})|V_{DS} = \text{const}]^2. \tag{6}$$

The output conductance is determined from

$$G_{oS} = (\partial I_{DS}/\partial V_{DS})|V_{GS} = \text{const} \tag{7}$$

and provides an evaluation of the degree of flatness of the I–V characteristics in saturation. In the device parameter extraction procedure presented here the body effect was ignored for the sake of simplicity.

This procedure is noteworthy in that numerous qualification tests are made to ensure the integrity of the device parameters that are determined from seven current measurements. The approach wisely requires that k and V_T be evaluated in both the nonsaturation (linear) and saturation regions, for significant differences have been observed between these parameters. The procedure for obtaining the k and V_T values utilizes a two-point method based on predetermined V_{GS} and V_{DS} values in the nonsaturation (linear) and saturation regions. Other more detailed approaches have been suggested. For example, k and V_T values can be determined at the maximum slope of the curves shown in Figs. 6b and c. Tradeoffs must be made between detailed measurements that take considerable time and simple measurements that can lead to erroneous results.

B. Test Structures for Layout Rule Checking

These structures are used to evaluate those geometrical circuit layout features that form the layout rules. Examples are

1. cross-bridge sheet resistor for linewidth measurements [19] and
2. alignment resistor [20, 21] or a comb resistor [22] to evaluate feature-to-feature spacing.

As an example of the use of this type of test structure, consider establishing the optimum contact window opening from the yield of good contacts in a serpentine array of contacts. The metal mask for such a structure, shown in Fig. 7 [23], consists of eight subarrays where the number of contacts progresses from a few (200) contacts to many (19,200) contacts. A cross-section of the structure is shown in Fig. 8 where the metal strap represents one of the small elements in Fig. 7. The structure is repeated many times (about 100) across a wafer. Each subarray is tested for open-circuit fault condition, and results from each string are used to construct a yield curve as seen in Fig. 9. That is, for contacts with a dimension of 3.6 μm on a side and for the subarray with 4800 contacts, the percentage of good subar-

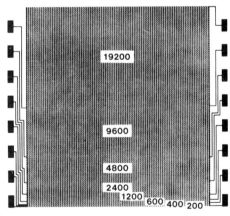

Fig. 7. Metal mask used to fabricate metal-to-silicon contact arrays with various contact window dimensions. The numerical value indicates the number of contacts in each string [23].

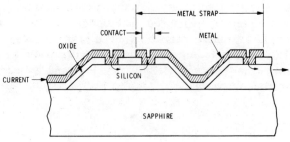

Fig. 8. Cross section of an element from the metal-to-silicon contact array shown in Fig. 7.

rays was 33. The results shown in Fig. 9 were obtained from four different wafers each fabricated with a different contact window opening dimension as indicated in the figure.

From the data shown in Fig. 9 optimum layout rules can be determined by trading off the higher yield of a larger contact window against fewer chips per wafer [23].

The analysis of serpentine structures requires that one assume that *the detected fault is the intended fault*. For the serpentine contact resistor structure, the intended fault is a failure of the metal to make contact with the silicon. Other faults can cause an open circuit and corrupt the data. For example, an open circuit can be caused by a step-coverage fault, a photolithographic or etching fault that omits a silicon island, or a probe fault where a probe fails to make contact with a probe pad. Also, spurious leakage currents between strings in an array can cause unintended faults.

To minimize the occurrence of unintended faults, test structures must be designed with oversize layout rules except for the feature under study, and the structure should be as simple as possible. In addition, the structures should contain double-probe pads to minimize probing errors; see Fig. 3 in [24]. Structures must be inspected visually to verify that the detected fault is the intended fault.

The expression used in the analysis of the data seen in Fig. 9 follows an exponential law. This indicates that faults are uniformly distributed across

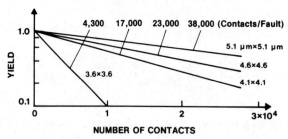

Fig. 9. The yield (number good) of metal-to-silicon contacts with different contact window dimensions [23].

the wafer and occur one at a time in any subarray. But data from other types of arrays have been measured where the yield curves are nonexponential [25]. In such cases multiple faults may be present [22], faults may be clustered [26], or faults may have a radial dependence due to increased misalignment errors toward the periphery of wafers.

Results from this kind of test structure need to be treated cautiously. Their use should be limited to evaluating layout rules and to serving as a guide in wafer processing [7]. Attempts to use the results from different arrays to predict the yield of complex circuits have not met with success. In fact, such attempts detract from the important use of these structures in establishing layout rules.

C. Test Structures for Process Parameter Extraction

These structures are used to evaluate the uniformity of the semiconductor doping processes, the quality of the interfaces between the various semiconductor materials, and the quality of the etching processes used to define the various features in the semiconductor materials. Examples are

(1) cross-bridge sheet resistor for sheet resistance and linewidth measurements [19];

(2) contact resistors for metal-to-silicon or metal-to-polysilicon contact resistance measurements;

(3) MOS capacitors for oxide thickness, interface-state measurements, flat-band voltage, and dopant density measurements [27];

(4) diodes for leakage current measurements;

(5) alignment resistors for evaluating the registration of photomask generated features;

(6) MOSFET dopant profiler for profiling the dopant profiles of various layers [28].

Also included in this type of test structure are nonelectrical structures such as alignment markers, photomask layer designators, surface topology structures, and critical dimension structures.

As an example of the use of this type of test structure, consider the evaluation of the linewidth of the metallization process. The cross-bridge sheet resistor [19] that was used is shown in Fig. 10. The linewidth is determined after measuring the sheet resistance R_s, which is determined from the van der Pauw relation [29]:

$$R_s = (\pi/\ln 2)(\Delta V/I), \tag{8}$$

where I is the current forced between I_1 and I_2, and ΔV is the voltage difference between V_1 and V_2. The linewidth W is determined from

$$W = R_s L I^*/\Delta V^*, \tag{9}$$

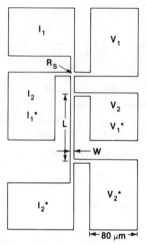

Fig. 10. Cross-bridge sheet resistor used to measure the sheet resistance and linewidth of a metal layer.

where L is the distance between the voltage tapes shown in Fig. 10, I is the current forced between I_1^* and I_2^*, and ΔV^* is the voltage difference measured between V_1^* and V_2^*.

Using the test structure shown in Fig. 10, a single photomask test chip was prepared on a 10× master reticle. The test chip was composed of a 12-by-20 array of identical test structures with a design linewidth of 6 μm. The final photomask was prepared from the master reticle by a step-and-repeat process. The photomask was used in conjunction with a contact printer and a photolithographic process to etch the test-chip pattern into an 800-nm-thick aluminum layer. The aluminum had been electron-gun evaporated and deposited on an oxide film thermally grown on a 2-in. (50.8-mm) diameter silicon wafer.

The linewidth variations shown in Fig. 11 are from a single row of test structures measured across the diameter of the wafer. The plot shown in Fig. 11 indicates that the linewidth varies periodically with the chip dimension of 250 mils (6.35 mm). This periodic or *intrachip* variation is superimposed on a nonperiodic or *interchip* linewidth variation due to those factors that affect the contact between the photomask and the photoresist-covered wafer. The periodic or intrachip linewidth variation is due to aberrations in the optics of the image repeater used to step and repeat the 10× reticle. Similar results have been reported for a 15-μm line [30]. The absolute magnitude of the variations shown in Fig. 11 is independent of the magnitude of the linewidth. Thus, the impact of such linewidth variations on device characteristics is quite dramatic especially for small devices. The linewidth variations for the lines shown in Fig. 11 is about 13%. For 1-μm lines, the variation would be 70%.

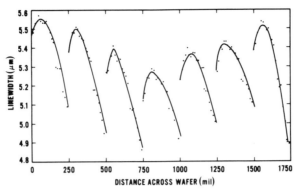

Fig. 11. The variation of the linewidth as determined from an array of identical cross-bridge resistors as shown in Fig. 10.

These results raise some important issues in developing a test-chip methodology. For instance, the circuit simulator SPICE [31] calls for one value of each parameter to be entered into the computer program. But as shown in Fig. 11, parameters can have a distribution of values that varies with position in the test chip as well as with position on the wafer. One usually does not have the luxury of detailed parametric data like that shown in Fig. 11, so guidelines must be established to indicate where test structures should be placed in the test chip and where test chips should be placed on the wafer to facilitate the acquisition of the most representative data. Such guidelines will be important in providing the circuit designer with the best data for the circuit simulators and in establishing the criteria for the positioning of test structures used in the purchase of fabricated wafers.

D. Test Structures for Random Fault Analysis

These structures are used to evaluate the physical faults in the semiconductor material system before it is subjected to significant stressing. A knowledge of the faults is necessary for logic design, logic simulation, and test-vector generation. To detect these faults, test structure arrays are constructed out of series, parallel, or addressable arrays of elements. Examples are

1. serpentine resistor for metal step coverage analysis,
2. comb resistor for measuring the quality of the etching process in separating conducting lines,
3. MOS capacitor for oxide integrity (pinhole) measurements, and
4. addressable MOSFET for identifying the exact nature of a defect and for accumulating fault statistics.

The serpentine and comb-type structures are designed to evaluate the occurrence of certain physical failures. For example, the metal step-coverage resistor is used to evaluate breaks in the metallization at oxide steps. These failures are known as "intended failures" for the test structure. But other unintended failures can masquerade as intended failures. For the example of the metal step-coverage resistor, an open circuit can occur due to the failure of a probe to touch a probe pad or the failure of the photomasking process to properly define the metal lines. In order to identify unintended failures, the test structure must be examined visually. This is a time-consuming process when looking for a failure in a large number of elements. In order to reduce the search time, an addressable MOSFET array was developed to pinpoint the location of physical failures so that the site can be examined visually.

The MOSFET array shown in Fig. 12 is composed of 100 MOSFETs where the gate is connected to the drain. This structure appears on test chip NBS-16 [32], which includes two p-channel and two n-channel MOSFET arrays. On a 76.2-mm-diameter wafer, 380 arrays containing 38,000 MOSFETs were tested. The results shown in Table V are from 26,760 MOSFETs located in the interior portion of the wafer. Both the fault location and the relative density of different fault types, both clustered and nonclustered, can be determined from the electrical data. A fault is considered to be clustered when two or more adjacent MOSFETs containing the same fault type are detected in an array. As seen in the table, the most frequent fault was number 8, a combination of excessive leakage current and low breakdown voltage.

The advantage of the MOSFET array test structure is that it pinpoints the location of faults so that one can visually examine the site to determine the exact nature of the failure. The disadvantage of this test structure compared with a serpentine or comb structure is twofold. First, the time to test the structure is much longer. Second, the number of elements tested is much smaller. The number of elements tested could be increased by using a memorylike structure with addressable registers. But in this kind of structure one would have to sort out faults in the address registers from faults in

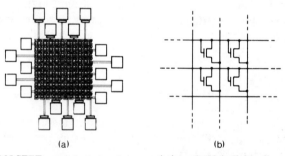

Fig. 12. (a) MOSFET array test structure consisting of 100 individually addressable transistors. (b) A schematic diagram. (Developed by L. W. Linholm.)

TABLE V

MOSFET Array Test Results

Number	Fault			Number of faults	
				Nonclustered	Clustered
1	Polycrystalline void/break			4	0
2	Epitaxial void			0	1
3	Metal void/break			0	0
4	Metal bridge			1	0
5	Gate short			4	0
	$V_T{}^a$	I_L	V_B		
6	—	—	L	5	0
7	—	H	—	25	0
8	—	H	L	62	0
9	L	—	—	1	1
10	L	H	—	0	1
11	L	H	L	0	1
12	H	—	—	2	2
13	H	H	L	0	1

^{a} V_T = threshold voltage; I_L = leakage current; V_B = breakdown voltage. H = parameter too high; L = parameter too low.

the array. An ideal structure is one where a large number of elements can be tested quickly and where failure sites can be pinpointed.

E. Test Structures for Reliability Analysis

These structures are used to identify failure mechanisms that result from atomic motion or changes in ionic charge states. The approach is to measure the change in a parameter after the semiconductor material system is given an environmental stress such as overvoltage, temperature, humidity, and radiation. Both the reliability of circuit chips and the package are addressed by these structures.

Examples of the kinds of structures used to evaluate the reliability of silicon devices are listed in Table VI. As seen in the table, the structures vary from single devices to arrays of devices. This table is similar to the one presented by Peck [33] and includes in a similar manner the stress conditions for each failure mechanism.

Test structures for measuring package-induced stress are discussed first. A strain gauge test structure [34] was fabricated by diffusing or implanting a serpentine resistor. With this structure Spencer and his co-workers showed

TABLE VI

Test Structure for Evaluating Some Time-Dependent Failure Mechanisms in Silicon Devices[a,b]

Device association	Failure mechanism	Test structure	Stress condition[b]	Reference
Package	Chip stress	Strain gauge resistor	P, T	[34], [35], [36]
Oxide	Dielectric breakdown	Large-area MOS capacitor	T, J	[37], [38]
Oxide	Charge injection	MOSFET	E, T	[39]
Oxide	Surface charge spreading	MOSFET	T	[40]
Oxide	Ion migration	MOSFET	E, T	[41]
Metallization	Electromigration	Serpentine resistor	T, J	[42]
Metallization	Corrosion	Comb resistor	H, V, T	[43], [44]

[a] See Peck [33].
[b] E = electric field, H = humidity, J = current density, P = pressure, T = temperature, V = voltage.

that package-induced stresses can be quite large, approaching 60% of the breaking strength of silicon. In addition, they indicate that these stresses can produce significant shifts in the k' of MOSFETs. Other workers [35] observed that the package-induced stresses caused significant shifts in resistors used in monolithic D/A converters. Their strain gauge, which consisted of three diffused silicon resistors arranged at 60° with respect to each other, was used to map the strain in chips with different crystal orientations. They discovered the best crystal orientation for the chip and the correct crystal direction along which to fabricate their resistors so as to minimize stress-induced resistance changes. Still other workers [36] have used commercially available Wheatstone-bridge-type strain sensors in their studies.

Test structures for measuring oxide-related failure mechanisms [37–41] are listed in Table VI. These structures are usually large-area devices or special transistor structures. Schroen and his co-workers [40] show a combination of test structures for measuring local stress, ionic accumulation, leakage current, and metal corrosion.

The test structures for measuring metallization-related failure mechanisms [42–44] are the serpentine and comb resistors. The aluminum electromigration studies [42] used a structure that had a serpentine resistor 3-cm long. This is a rather long structure when compared to most electromigration test structures, but, as the authors pointed out, the total length of the wires on present-day VLSI chips is about 100 cm. Structures with long wires are complicated by the increasing probability that a fatal structural defect will occur. This means that the structures must be prescreened for photolithographic defects before electromigration studies can begin or that outliers from the main distribution be discarded. Fortunately in this study outliers due to photolithographic defects were not present. The structures were fabricated with linewidths and spacings that varied from 1 to 8 μm. Accelerated life testing was done at temperatures ranging from 150 to 250°C and at current densities of 1×10^5 to 2×10^6 A/cm². The failure rates were observed to gradually increase as the linewidth decreased, although for the smallest width of 1 μm the failure rates were lowest in certain large-grain aluminum films.

A summary of comb and serpentine test structures is shown in Figure 13. The comb resistor shown in Fig. 13a is used to measure the corrosion currents flowing between the metal fingers [44] and to determine layout rules by measuring the frequency of bridging faults [23]. The comb MOSFET shown in Fig. 13b is used to measure surface leakage currents and appears in the test chip shown in Fig. 2 (structures 31 and 32). By measuring the connectivity of the serpentine resistor shown in Fig. 13c, layout rules or the effect of metal corrosion can be evaluated by analyzing the frequency of opens. The step coverage resistor shown in Fig. 13d appears in the test chip shown in Fig. 1 (structures 1-3). The combined serpentine-comb resistor shown in Fig. 13e was used by Bartlett and Schoenberg [22] to evaluate linewidth and spacing layout rules for printed circuit boards. Finally, the combined serpentine-comb resistor shown in Fig. 13f was used by Vaidya *et al.* [42] to study electromigration and by Skar and Kozakiewicz [43] to study metal corrosion.

The combined serpentine-comb resistors are the recommended struc-

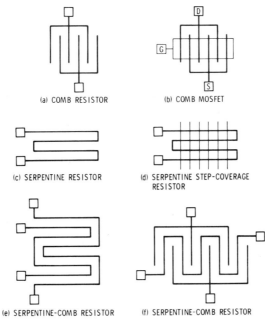

Fig. 13. Schematic diagrams of various comb and serpentine resistors used to evaluate electromigration, metal corrosion, layout rules, and leakage currents.

tures because both electrical connectivity and isolation can be measured, which allows an efficient use of test structure area. As illustrated above, these structures are used not only for reliability analysis but also for layout rule checking and device parameter extraction (leakage current).

F. Test Structures for Circuit Parameter Extraction

These structures are used to extract circuit parameters that characterize ac and dc circuit performance and to verify that wafer fabrication process can produce functional circuits. Examples are

1. inverters for measuring the inverter threshold, gain, and noise immunity and
2. ring oscillators for measuring the oscillation frequency and stage delay.

The full characterization of circuit performance involves evaluating such parameters as power dissipation, maximum clock frequency, fan-out capability, and the propagation delay.

The ring oscillator is a popular test structure for characterizing the capabilities of a wafer fabrication process. The oscillation frequency and power are measured, and the stage delay and gate power dissipation are calculated. Relating these results to processing details is difficult. As expressed by Ham [7] the frequency of oscillation is of secondary interest, since it is very difficult to relate the frequency to a specific parameter or processing step. Operation in a reasonable frequency range guarantees that the process is capable of producing working circuits.

A simplified expression for the pair delay of a ring oscillator was derived as an aid in their design and analysis; the derivation is presented in the appendix. The pair delay was derived for the NMOS inverter pair shown in Fig. 14 where the load transistors Q_{L1} and Q_{L2} are depletion-mode devices, the input transistors Q_{I1} and Q_{I2} are enhancement-mode devices, and C ($C = C_1 = C_2$) is the equivalent load capacitance. As depicted in the lower portion of the figure, the analysis assumes the input to stage 1 is a falling step voltage and the output of stage 2 is a rising ramp [45]. The output of stage 2 is shown in detail in Fig. 15 for both a simplified and a more complete analysis. The Mead–Conway NMOS design parameters [46] were used in the analysis. The pair delay is the time for the output of stage 2 to fall to the inverter threshold V_{TINV} for a falling step input to stage 1. The pair delay is

$$t_D = t_{on} + \tau_r \sqrt[3]{(V_{DD} - V_{TINV})/V^*}, \tag{10}$$

where

$$t_{on} = \tau_r (V_{TI} - V_L)/(V_{DD} - V_L), \tag{10a}$$

$$\tau_r = C/k_L |V_{TL}|, \tag{10b}$$

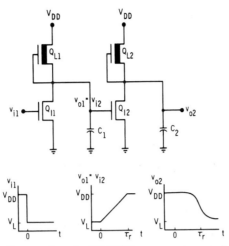

Fig. 14. Schematic diagram of a pair of NMOS inverters where the input transistor Q_I is an enhancement device and the load transistor Q_L is a depletion device.

$$V^* = (k_I/k_L)(V_{DD} - V_L)^2/(3|V_{TL}|). \tag{10c}$$

The important parameters are V_D, the power supply voltage; V_{TI}, the input transistor threshold voltage; V_{TL}, the load transistor voltage; V_L, the inverter low-state voltage; k_I, the input transistor conduction factor; k_L, the load transistor conduction factor; and C, the equivalent load capacitance. The oscillation frequency for the ring oscillator is $1/Nt_D$, where N is the number of inverters in the ring.

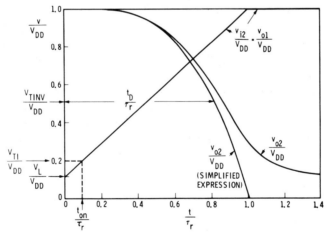

Fig. 15. Input v_{i2} and output v_{o2} waveforms for inverter stage 2 for the inverter chain shown in Fig. 14 where the input to inverter stage 1 is a falling voltage step at $t = 0$. The simplified expression is derived in the appendix and labeled Eq. (A7).

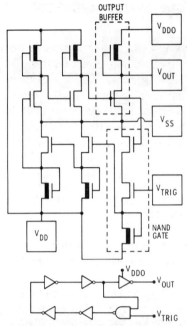

Fig. 16. Five-stage, NMOS, high-speed ring oscillator with a fan-in and fan-out of one, a NAND gate trigger, and output buffer inverter separately powered.

In designing ring oscillators, additional design factors must be considered. The number of stages in the oscillator must be an odd number (usually a prime number) that is large enough to allow each inverter to reach the high or low state after switching state. Ring oscillators with a few numbers of stages oscillate at abnormally high frequencies, because each stage does not have enough time to reach a final value. For the oscillator shown in Fig. 16, the output buffer inverter is a minimum geometry transistor that adds a minimum capacitive load to the oscillator. In addition, the output inverter is separately powered so that the power drawn by the oscillator can be accurately measured [47]. The circuit also includes a NAND gate trigger [48] that allows one to control the onset of oscillation preventing the propagation of multiple pulses.

The overall architecture of the ring oscillator can take many forms. The traditional design is the so-called high-frequency configuration where inverters are arranged as shown in Fig. 16. Here the path length and hence the capacitive load are minimized. Alternative designs involve the use of NOR gates [7] or programmable logic array-type architectures [49]. These configurations are intended to more closely simulate circuit environments by using fan-in, fan-out, and wiring loads that are typical of actual circuits. Yu *et al.* [49] show that the circuit delay is about 100 psec for the high-speed ring oscillators whereas for heavily loaded arrays the delay is about 2 nsec.

IV. TEST-CHIP ORGANIZATION AND TEST-STRUCTURE DESIGN

As mentioned in Section II, the test chip must be organized so that all test structures can be accessed by a single probe array. This restriction has two important economical considerations. First, only one common probe array need be purchased for all test chips, and second, time is saved in not having to change probe cards. Test-structure design is guided by two factors. First, the design must minimize the possibility that faults will degrade the results. Second, the design must allow for modular structures and test procedures that can be easily transferred to new test designs. This feature is particularly important in maintaining a consistent data base between various test chips.

A. Test-Chip Organization

The major issue in organizing a test chip concerns the placement of the probe pads. Over the years several philosophies have developed that are noteworthy.

Early test chips generally were designed with the probe pads located at the periphery of the chip [50]. For the test chip shown in Fig. 17, the location of the peripheral probe pads coincides with the location of the probe pads on the integrated circuit. Such a test chip, when used as a drop-in in a high-volume integrated circuit wafer as seen in Fig. 3, can be probed along with the circuits. The test chip is encoded (a short between two pads, for example) to alert the tester that the unit under test is a test chip and not a circuit. The tester then changes the test sequence from the circuit test to the test-chip test. Another example of the use of the peripheral probe approach can be seen in the work of Reynolds [5].

The peripheral probe pad approach is severely pad limited. The number of probe pads located at the periphery is limited by the dimension of the chip or the number of probes in the probe array. For example, if the test chip shown in Fig. 2 were redesigned with 100-μm square peripheral probe pads, only 88 pads could be accommodated, whereas the chip has 216 probe pads.

The peripheral probe pad approach encourages the use of common bussing where parts of several test structures are connected to the same probe pad. This practice can lead to interferences between structures and places an additional burden on the test engineer to demonstrate for all test conditions that no interference exists. If an interference can occur, then it must be shown that the interference is easily recognized and is not of a subtle nature that will lead to results that will be taken as acceptable. The transistor structure shown in Fig. 18 contains all gates and sources in common. A fault in one gate can cause all the other transistors to be inoperable, thus preventing one to isolate which device is failing. If all devices work perfectly, no difficulty is encountered. It should be noted that the test of the first

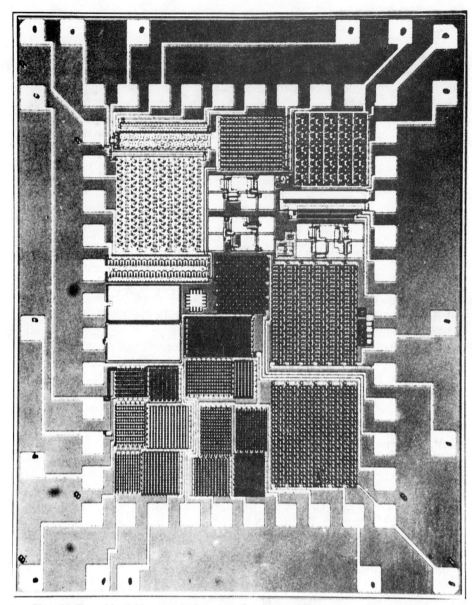

Fig. 17. Test chip (2.36 mm × 2.92 mm) designed with peripheral probe pads and internal probe pads. The outer set of probe pads is located at the same site as the probe pads on the product chips [50].

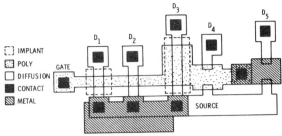

Fig. 18. Five test transistors, where all the gates are connected together as well as all the sources, illustrating the common bussing concept.

Device number	1	2	3	4	5
Transistor	Min. geom. depletion	Min. geom. enhancement	Pull-up depletion	Min. geom. field	Min. geom. enhancement
Gate	Poly	Poly	Poly	Poly	Metal
Oxide	Gate	Gate	Field	Field	Field

transistor prestresses all other transistors, which can be deleterious to the characterization of subsequent transistors, especially if oxides are unstable. In this case an additional 8 pads are required in order to isolate all transistors.

Another objection is that the peripheral probe pad approach does not enable the use of modular test structures that can be called from a computer cell library and placed anywhere on the test chip. With pads on the periphery, each test-chip design may be a unique design experience that does not build on previous design experience. Because of the uniqueness of a new design, it may not be possible to relate results from the new chip with previous results. In this case the data in the old data base may be useless.

Various modular probe arrays are found in the literature. Initial attempts consisted of subdividing test chips into minichips with peripheral probe pads. In the work of Jerdonek et al. [51], 20 probe pads were located on three sides of a 1.5-mm × 1.5-mm minichip. Tingley and Johnson [6] located 21 probe pads on three sides of a 0.71-mm × 2.03-mm minichip. The objective of both sets of workers was to provide for wafer probe and package measurements. Zucca et al. [52] placed 14 probe pads on all four sides of a 0.22-mm × 0.44-mm minichip. The minichip allows for a degree of modularity, but the level of modularity is above test structure level.

To achieve modularity at the test structure level where structures can be called from a cell library requires the use of the 2-by-N probe pad array [53]. The parameter N is an arbitrary positive integer limited by the number of probes allowed in the measurement system. In many test chips N is 10. The array is illustrated in Fig. 19. The layout requires a square pad 80 μm on a side with 80-μm spacing between pads. In order to conserve space, the

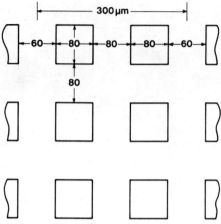

Fig. 19. The 2-by-N probe pad array [53].

pad-to-pad spacing between arrays is 60 μm. The overall width of the array is 300 μm.

The probe pad size of 80 μm was chosen after a study of the probeability of various pad sizes, including 40, 60, 80, and 100 μm [54]. A special test chip was designed, and probing failures were measured for each probe pad size. Excellent results were achieved for the 60-, 80-, and 100-μm pads. To be on the safe side, the pad size was chosen as 80 μm. It should be noted that the test chip of Zucca *et al.* [52] used 60-μm pads spaced 10 μm apart. Their test chip was fabricated in a 14-mm (0.55-in.) GaAs wafer. With a small wafer, proper alignment and run-out across the wafer is less severe than with larger 75- and 100-mm silicon wafers. Thus, a smaller probe pad size is probably acceptable for a smaller wafer.

At NBS, over 30 test chips have been designed using the 2-by-N probe pad array. As an example consider the test chip shown in Fig. 20. This chip contains 640 probe pads and 140 test structures that are electrically isolated from each other [55]. The ability to design variations of test structures for study purposes is greatly enhanced by this pad arrangement. Other investigators [56] have used this pad layout in their test chips. The flexibility of the 2-by-N probe concept was demonstrated recently by Wetterling [57]. He confined all his test structures to a 2-by-2 array. This allowed test chips to be assembled in a variety of configurations that matched the size of the circuit chips. For large chips the structures are arranged in a test strip that accompanies each circuit.

The test strip, in which test structures are accessed by a single 2-by-N pad array, is becoming an important vehicle in the purchase of wafers from silicon foundries. It may prove feasible to place a test strip on each circuit that is produced and to develop chip acceptance criteria based on test of the test strip. At this time it is not clear which structures should be included on the test strip and how they should be tested.

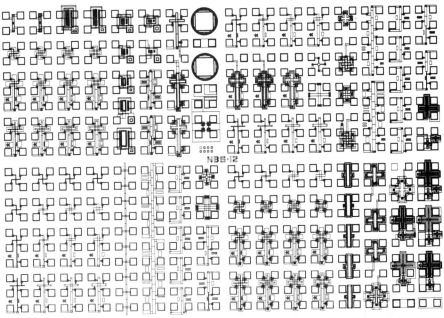

Fig. 20. Test chip NBS-12 (3.2 mm × 4.8 mm) illustrating the use of modular test structures probeable with a 2-by-10 probe array [55].

Future organizational schemes must allow for the bonding and packaging of test chips. The 2-by-N array is unsuitable for this application. A floor plan for such a test chip is shown in Fig. 21 where peripheral probe pads access selected structures that will be stressed. The central portion of the chip is organized using the 2-by-N array where structures are measured in a near-room environment. Since packaging test chips will greatly increase the cost of the measurements, alternatives should be sought. One possible alternative is to stress the test chips while they are still in wafer form. Some of this technology has been developed [58]. However, packaging test chips cannot be avoided entirely, for it will be necessary to evaluate the reliability of the package using test chips [40].

B. Test-Structure Design Rules

From the previous discussion it is clear that an appropriate choice for the probe pad arrangement is the 2-by-N probe pad array. With this as a design constraint, certain other restrictions must be followed in order to properly emulate the functional circuits and to obtain highly accurate measured values. These restrictions follow.

(1) *Use the 2-by-N probe pad array approach.* This allows (a) the test structures to be assembled in a machine-readable cell library, (b) the probe

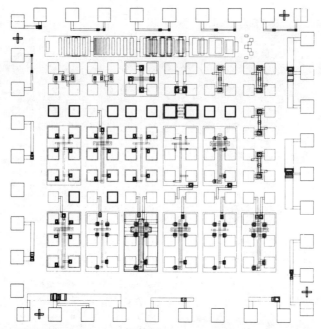

Fig. 21. Test chip organized to accommodate both wafer probe measurements and packaged measurements. The larger pads around the periphery are intended to be used for wire bonding the chip to a package.

pads to be an integral part of the test structure, (c) the electrical isolation of test structures thus minimizing structure-to-structure interference, and (d) the "standardization" of the geometries of the test structures so that many users can use the same design.

(2) *Use the same layout-rule set to design both the test structures and the functional circuits.* This is perhaps the oldest rule in test-chip design [1]. The reason for this rule is that the test structures must emulate the actual circuit geometries as much as possible. There is a notable exception to this rule. When engaging in process evaluation, it is appropriate to shrink layout rules to discover the limits of the process.

(3) *Design test structures so that only critical parts are sensitive to the layout rules.* Many parts of a structure can be designed with oversize features that will lessen the occurrence of a faulty structure. For example, contact size, metal overlap at contacts, and linewidths and spacings can be increased in noncritical areas. Design structures so that they are tolerant to photomask misalignment. Also be aware of pinholes in insulating layers as a potential source of faults; excessively large conducting features can be faulted with pinholes. Only the critical dimensions of the structure need be specified, and these can be tailored to each manufacturing facility. Recently a computer program was developed that allows one to specify certain features of a test structure while holding others constant [59]. This is illustrated

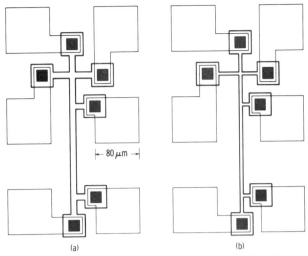

Fig. 22. Test structures drawn by a layout program where critical features are parameterized [59]. The cross-bridge test structures are shown with two linewidths for the conducting channel (outlined with a heavy line). The linewidth is 10 μm in (a) and 5 μm in (b).

in Fig. 22, which shows two cross-bridge sheet resistors with different linewidths but the same probe pad dimensions. With this approach, the critical features of a test structure can be tailored to the layout rules of a particular chip manufacturer. This saves having to redesign the entire test chip.

(4) *Avoid common bussing between test structures.* Structures that are connected together are said to share a common bus. The danger in such connections is that a fault in one structure may influence values measured in the other structure. The objective in good test structure design is to use the "separation of variables" concept to design structures that have a minimum of interferences. The 2-by-N array provides a pad-rich environment in which to lay out test structures where interferences can be avoided. Sometimes it is desirable to design test structures with common connections. For example, when the results of one structure are to be correlated or used with another, it is highly desirable to place the structures as close together as possible. A successful combination of structures requires a rigorous fault analysis to assure that a fault in one structure will not affect another. An example of a combined structure is the cross-bridge sheet resistor; see Figs. 10 and 22. This structure requires 6 probe pads. If the structures are separated into the cross and bridge structures, then 8 pads are required.

(5) *Use four-terminal resistors for sheet resistance and linewidth measurements.* The Kelvin contact scheme for measuring resistance uses a four-terminal resistor where current is passed between two terminals, and the voltage is measured between the other two terminals. The cross-bridge sheet resistor is an example of such a test structure. The advantage to this approach is that contact resistance encountered in injecting current into the

structure is eliminated from the measured resistance. Do not use two-terminal or "dog bone" resistors.

(6) *Use channel stops wherever possible.* A channel stop serves to shut off surface currents that result from a surface inversion channel. There are several ways to eliminate these channels. The surface can be doped more heavily than the region that is inverted or a field plate can be used to accumulate the surface. Results taken on sheet resistors where surface inversion was present revealed that sheet resistance values can be five times larger than the true value [60]. In addition, the values were in the wrong direction from an intuitive explanation.

When designing test structures it is useful to keep in mind the numerous interferences that can plague the measurements. Many failure mechanisms are listed in Table VI. Additional interferences and their possible causes are listed in Table VII. These interferences were identified in a study of the measurements of junction sheet resistance [60] and can be detected by measuring the various factors listed in Table VII.

TABLE VII

Interferences Affecting the Measurement of Sheet Resistance[a]

Factor	Interference	Possible cause
Asymmetry factor large	Nonuniform sheet resistance across structure	Nonuniform lateral doping and/or junction depth variations
	Nonuniform structure boundaries	Poor photomasks and/or poor wafer fabrication
	Bulk leakage current	Low junction breakdown voltage
Zero offset factor large	Photovoltaic effect	Too much light
	Thermoelectric effect	Thermal voltages at relays and/or probe contacts
	DVM zero offset	Instrumentation problem
Linearity factor large	Joule heating	Excessive current; poor design (structure too narrow)
	Nonlinear resistivity	Grain boundaries and/or junctions
	Oxide–silicon interface currents	Oxide charge causes channels
Sheet resistance value incorrect	Part of structure missing	Poor fabrication and/or structure near edge of wafer
	High-contact resistance	Poor fabrication
	Metal-to-substrate shorts	Pinholes in oxide
	Surface currents across oxide	Incomplete metal removal
	Oxide–silicon interface currents through inversion layer	Excessive oxide charge and/or interface states
	Poor junction isolation	Low breakdown voltage or too much light

[a] See [60].

V. TEST-CHIP TESTERS AND ADVANCED TEST STRUCTURES

The commercial availability of several multifunction parametric testers [10, 61] has greatly facilitated the use of test chips in the manufacturing environment for such applications as wafer-fabrication control or equipment evaluation. These testers generally provide the capability of sequentially forcing dc currents or applying dc voltages at selected contacts of a test structure. Three quantities are measured on processed wafers: voltage, current, and capacitance. Voltage is measured in the 1-mV to 100-V range, current is measured in the 1-nA to 1-A range, and capacitance is measured in the 10 to 1000-pF range. The time to acquire one measured value can vary from 1 msec to 10 sec depending on the measurement impedances, electrical filtering requirements, and settling time. Measurements are usually made in the dark in a room ambient although a hot–cold chuck can be used for measurements at temperatures other than room temperature.

The configuration of a multifunction parametric tester is shown in Fig. 23. The essential feature of this architecture is the connection of the stimulus/measure instruments to the structures on the wafer through a mechanical switch matrix. (Capacitance measurements have not been shown because they require a special probe card in order to make precision mea-

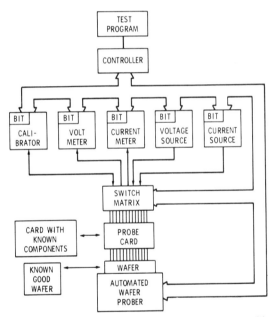

Fig. 23. Multifunction parametric tester used in characterizing test chips. The cards with known components and the known-good wafer are used in system calibration.

surements and disrupt the flow of a fully automatic test system.) The entire system is supervised by a minicomputer that controls the system through a digital data bus. The system is generally connected to a large main-frame computer that performs such higher-level tasks as data analysis and storage.

The first part of system validation concerns hardware verification to ensure that the system instrumentation is within prescribed precision and accuracy limits. Calibration techniques are similar to those used by the automatic test equipment community and consist of instrument and system level approaches. The instrument level calibration consists of

(a) external calibration where an instrument is removed from the system and calibrated in a calibration laboratory,

(b) self-calibration where the system is reconfigured and self-tests itself using the switch matrix and built-in calibrator, and

(c) built-in test (BIT) where the controller executes a diagnostic program that activates the BIT circuitry in each instrument.

Because the calibration of a system is always a compromise, some shortcomings of the above approaches are mentioned. The external calibration approach has the disadvantage that instruments are not calibrated in the environment in which they are used. The self-calibration approach relies on the accuracy and precision of the calibrator, which can be very good. The BIT approach relies on generally ill-defined self-test routines within each instrument so that the confidence level in system performance is not easily stated.

Once each instrument is calibrated, the system performance must be validated. System level calibration consists of testing the system after various measurement artifacts have been inserted into the front end of the system.

(a) *Known-good-component card.* Such a card is inserted in place of the probe card and allows a check of the system through the switch matrix.

(b) *Known-good wafer.* This wafer allows a system check that best simulates the actual wafer measurement environment.

The known-good wafer can take several forms. For instance, if the wafer is made of silicon it should be extra thick to reduce the chance of breakage, it should have a metal that can be probed many times, and it should be coated with an insulator that prevents ionic contamination of test structures and drift of their characteristics. Because of the difficulty in producing such a wafer, alternative approaches are being sought. At this time it is not clear what technique will be used in the future for the known-good wafer, but the need is real.

The second part of system validation concerns software validation to ensure that the test programs for each test structure are correct. This involves validating the data acquisition sequences and the data reduction routines. The data acquisition sequences must allow for the proper sequencing of the

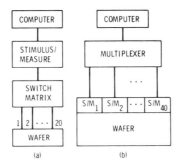

Fig. 24. System architectures of a multifunction parametric tester where (a) there is the existing, commercially available switch matrix approach and (b) there is the proposed pin-electronics approach.

forcing voltages and currents with no glitches between values, and settling times must be set correctly in order to achieve the desired measurement precision and accuracy. This task is simplified when identical systems are used where test programs can be transported between systems in a machine-readable format. The task is very difficult when systems differ in both instrumentation and test languages. A good solution to this probem seems to be the use of well-documented English-language tests that included a flow diagram [7]. Data reduction routines are more easily validated through the use of bench-mark test data and results.

A final note concerns the architecture of multifunction parametric testers. A block diagram of typical commercially available testers is shown in Fig. 24a. The disadvantage of the switch matrix-based testers is due to (a) the mechanical switch matrix, which generates thermal voltage noise and reduces the speed of the measurements (relay settling time is in the 1–10-msec range) and (b) cable noise introduced by the often long cables (10 m in some cases) that connect the stimulus/measure instruments to the probe card. An alternative is shown in Fig. 24b where the stimulus/measure (S/M) instruments are placed as close to the wafer as physically possible. This requires the replication of the S/M instruments by the number of probe pads, 40 times in Fig. 24b. With this architecture, digital signals are switched electronically (multiplexed), whereas in the current approach shown in Fig. 24a analog signals are switched mechanically. The advantages of the tester architecture shown in Fig. 24b are reduced measurement noise and improved measurement speed. For example, if the tester shown in Fig. 24b is used to test four-terminal test structures with a 40-pin probe array, the data acquisition time will be reduced by a factor of 10 since the 10 structures can be measured simultaneously. In addition, the acquisition time will improve, since the system is no longer restricted to the 10-msec matrix switch settling time.

When developing new test structures, one must decide if there is a measurement advantage to be gained by incorporating a portion of the tester into

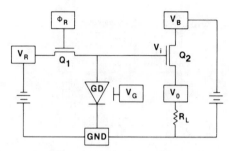

Fig. 25. Schematic diagram of an integrated gated-diode electrometer. The boxes represent probe pads. The off-chip devices are the two dc supplies V_R and V_B and the resistor R_L.

the test structure. Such an advantage has been found with the integrated gated-diode electrometer [62] shown schematically in Fig. 25. Low-level diode leakage currents can be determined by measuring the time decay of the output voltage V_o resulting from the momentary application of reverse-bias voltage V_R to the gated diode GD through the MOSFET switch Q_1. The internal gated-diode current I is determined from the expression:

$$I = (C/\beta)(-dV_o/dt), \tag{11}$$

where C is the diode capacitance and β is the incremental gain of MOSFET Q_2. The diode capacitance can be determined from $C = \epsilon A/W$ where ϵ is the permittivity of silicon, A is the area of the diode, and W is the width of the depletion region. For a one-sided step junction, $W = [2\epsilon(V_i + V_b)/(qN)]^{1/2}$ where V_i is the diode voltage, V_b is the built-in voltage, q is the electronic charge, and N is the dopant density. The dc gain of MOSFET Q_2 is determined from $\beta = \Delta V_o/\Delta V_i$, which is evaluated by closing the MOSFET switch Q_1 ($V_i = V_R$) and measuring V_o at two different values of V_R. The expression above assumes that the capacitance of the gated-diode C is large compared to the gate–source capacitance of MOSFET Q_2. Examples of this test structure are shown in the lower right-hand corner of Fig. 20 in [55, 63].

Another integrated test structure has recently been reported by Iwai and Kohyama [64]. This structure is shown schematically in Fig. 26. Here, the unknown capacitance

$$C_X = C_R[\beta(v_i/v_{oa}) - 1] \tag{12}$$

where C_R is a known reference capacitor, β is the ac gain of the output MOSFET, v_i is the rms value of the ac input signal, and v_{oa} is the output at v_o for v_i connected to point a. To measure C_X, v_i is applied to point a and the MOSFET switch Q_1 is opened by properly biasing ϕ. The ac gain β is determined from $\beta = v_{ob}/v_i$ where v_{ob} is the output at v_o for v_i connected to point b. In this measurement the MOSFET switch Q_1 is closed by properly biasing ϕ. This structure is useful in determining the value of the small capacitances typical of VLSI device geometries by connecting many capacitors into an array.

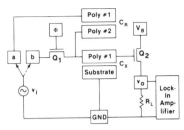

Fig. 26. Schematic diagram of an integrated precision capacitance meter. The off-chip devices are v_i, R_L, and the lock-in amplifier.

VI. FUTURE DIRECTIONS

Test chips have been used for many years, but little effort has gone into their development compared to the massive effort that has gone into the development of other facets of the integrated circuit business. The state of affairs can be judged by the modest amount of literature that exists concerning test chips. Articles often appear in the literature that describe a test chip used in the development of a new device, but very little space is devoted to explaining the test-chip measurements. On a more positive note, the appearance of the multifunction testers has greatly enhanced the usage of test chips. They enable the chip manufacturers to concentrate on the business of producing circuits by removing the burden of having to design and build test equipment.

The major goal for test-chip metrology is to provide diagnostic information in a manufacturing environment. Thus the kind of measurements and the time allowed to acquire the diagnostic data are governed by the return on investment. The goal is met by providing the highest quality data at the lowest possible price on a timely basis. Providing high quality data means that the data are correct and that correct decisions result from the use of the data. The characteristics of test-chip metrology are as follows.

(1) Measurements must be made quickly in order for the results to be of value. This is true in both a manufacturing environment and in the purchase of fabricated wafers. An electrical parameter that takes longer than a few seconds to measure is probably not worth acquiring. For such parameters alternative measurement schemes that allow rapid data acquisition must be sought.

(2) Data must be reduced quickly to a usable form. Data that is analyzed a week or a month from the time it was taken is of little use in a manufacturing environment.

(3) Parameters must be measured at more than a few sites on a wafer in order to properly characterize the parameters. Because parameters vary sig-

nificantly across wafers and within chips, parameter correlations must be done on a site-by-site basis. The use of "average values" in logic simulations must be viewed with caution.

(4) Test structures must be viewed as "virgin" untested devices. The testing strategy must include a qualification step that an untested device must pass before further tests are undertaken in order to save time and to keep from corrupting the data set.

Improvements that are needed in test chip metrology cover a wide spectrum of activities. For discussion purposes the activities are grouped into data acquisition, data reduction, and vendor transactions.

A. Data Acquisition

Some future needs have already been mentioned, such as the need for improved tester architecture using the pin electronic approach, advanced test circuitry using on-chip test circuitry, and system validation using known-good wafers. Other needs include a portable test language and stress testing for reliability evaluations.

Portable test languages would facilitate the rapid, accurate, and economical transfer of tests between groups. One impediment to the implementation of test-chip metrology is the cost of developing software. A common test language should have a high-level description that can be compiled by machine into the test languages compatible with existing multifunction testers.

Stress testing for reliability evaluation requires the use of special wafer probers equipped with hot/cold wafer chucks and controlled environmental chambers [58]. Thermal stressing in wafer form is more economical than stressing in packages. As mentioned previously, stressing in packages is essential to verify the reliability of the entire chip-package system.

B. Data Reduction

Improvements in data reduction call for developing the fundamental knowledge necessary to more fully understand the devices and the semiconductor material systems. In addition, the application of existing practices in statistical engineering will greatly enhance the data handling problems.

The data reduction algorithms used to convert the measured values to more useful values depend on physical models. For VLSI-type geometries, current transport is better modeled in two dimensions. Simplified expressions are needed that embody the essence of the complex phenomena.

The nature of test-chip metrology is characterized by rapid data acquisition and rapid data reduction using simplified algorithms. In such a situation it is essential that the test-chip measurements be constantly verified against

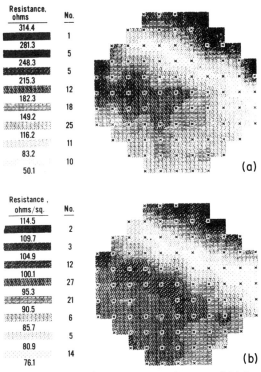

Fig. 27. Wafer maps of (a) the metal-to-n^+ contact resistance and (b) the n^+ sheet resistance. The high correlation between these two parameters leads to the conclusion that excessively high contact resistance was due to the lack of adequate control of a phosphorus implant process step [66].

measurements that allow an in-depth analysis by such analytical measurements as SIMS, AUGER, XPS, SEM, [14] spreading resistance, and ellipsometry.

Information handling is a difficult problem in test-chip metrology because parameters depend on the location on the wafer where they were measured. This causes difficulties in averaging results, in identifying outliers, and in correlating parameters. A number of data reduction and display techniques have been developed and are summarized by Cave and Smith [65]. Recently Linholm [66] presented a technique for averaging results. Also, Ham [7] has described the difficulty of the problem. An example of the nature of the problem is shown in Fig. 27. A contact resistance problem was identified and correlated to variations in the sheet resistance as shown in the figure [66]. The correlation was identified by observing a high correlation of these parameters from a host of other parameters. The maufacturer of these wafers could not have identified this problem because his production wafers contained only two test chips.

C. Vendor Transactions

Test chips are presently being used in the buying and selling of custom-integrated circuit wafers produced by the "silicon foundries." The current usage is limited to very simple parameters usually measured on various transistors. However, a more sophisticated usage can be envisioned where the reliability of the silicon material system is evaluated. Also one could expect that test-chip measurements would take the place of visual inspection performed on the final circuits. In VLSI-type circuits it is no longer economical to perform a detailed inspection for cosmetic defects. As pointed out some years ago [1], visual inspection is limited to the topmost features, for the lower layers are often hidden beneath opaque layers.

APPENDIX

The simplified expression for the pair delay of two NMOS inverters as seen in Fig. 14 was derived with the use of certain assumptions:

(1) The transistors have no storage of transit time effects [4].

(2) The circuit capacitances are combined into an equivalent load capacitance C that is voltage independent [4].

(3) The rise time of the output of inverter stage 1 is independent of the input waveform of stage 1 provided the delay time is computed from the point where the falling input crosses the inverter threshold voltage, V_{TINV} [45].

(4) The load transistor current i_L is approximated by a linear expression derived from the expression for the load transistor operating in the linear region:

$$i_L = k_L |V_{TL}|(V_{DD} - v_o), \tag{A1}$$

where k_L is the load transistor conduction factor, V_{TL} is the load transistor threshold voltage, and v_o is the output voltage.

(5) The two inverter stages are identical so $C_1 = C_2 = C$, $k_{L1} = k_{L2} = k_L$, etc.

Assumption (3) allows the rise time of the output of stage 1 to be calculated for a falling voltage step applied to the input of stage 1. This step input cuts off Q_{I1} allowing Q_{L1} to charge C_1. For these conditios, $i_L = i_{C1}$ where the capacitor current is $i_{C1} = C_1 \, dv_{o1}/dt$.

The solution to the resulting differential equation for the initial and final conditions $v_{o1}(0) = V_L$ and $v_{o1}(\infty) = V_{DD}$, is

$$v_{o1} = V_{DD} - (V_{DD} - V_L) \exp(-t/\tau_r), \tag{A2}$$

where the inverter rise-time time constant is

$$\tau_r = C/k_L|V_{TL}|. \tag{A3}$$

The analysis for the simplified pair-delay expression uses the following linear expression derived from the initial slope of Eq. (A2):

$$v_{o1} = ((V_{DD} - V_L)/\tau_r)t + V_L \tag{A4}$$

for $0 \leq t \leq \tau_r$. For $t > \tau_r$, $v_{o1} = V_{DD}$.

The fall-time analysis for inverter stage 2 assumes that only Q_{I2} redischarges C_2 so $i_{L2} = 0$. Initially Q_{I2} is cut off but after v_{i2} exceeds V_{TI}, Q_{I2} operates in the saturation region with a current

$$i_{I2} = k_I(v_{i2} - V_{TI})^2, \tag{A5}$$

where k_I is the input transistor conduction factor and V_{TI} is the input transistor threshold voltage. The output of stage 2 follows from the solution to the differential equation derived from $i_{I2} = -i_{C2}$ where $i_{C2} = C_2\,dv_{o2}/dt$. The equation uses the rising ramp described by Eq. (A4) where $v_{o1} = v_{i2}$ and the initial condition that $v_{o2} = V_{DD}$ for $t \leq t_{on}$. The time when Q_{I2} turns on, t_{on}, is calculated from Eq. (A4) for $v_{i2} = v_{o1} = V_{TI}$:

$$t_{on} = \tau_r(V_{TI} - V_L)/(V_{DD} - V_L). \tag{A6}$$

The output voltage for stage 2 for $t \geq t_{on}$ is:

$$v_{o2} = V_{DD} - V^*(t - t_{on})^3/\tau_r^3, \tag{A7}$$

where

$$V^* = (k_I/k_L)(V_{DD} - V_L)^2/(3|V_{TL}|). \tag{A8}$$

The pair delay t_D is defined as the time for v_{o2}, as given in Eq. (A6), to fall to the inverter threshold, V_{TINV}. This time is

$$t_D = t_{on} + \tau_r\sqrt[3]{(V_{DD} - V_{TINV})/V^*}. \tag{A9}$$

An expression for the inverter threshold V_{TINV} is derived from $i_L = i_I$ for $v_o = v_i$. For $v_o = v_i$, i_I is in saturation and described by

$$i_I = k_I(v_i - V_{TI})^2. \tag{A10}$$

The combination of this equation with the expression for i_L as given by Eq. (A1) yields the inverter threshold voltage

$$V_{TINV} = \tfrac{1}{2}[2V_{TI} - (|V_{TL}|/k_R) + \sqrt{(2V_{TI} - (|V_{TL}|/k_R))^2 + 4((|V_{TL}|V_{DD}/k_R) - V_{TI}^2)}] \tag{A11}$$

where $k_R = k_I/k_L$. The inverter output low-state voltage is calculated from $i_L = i_I$ for $v_i = V_{DD}$. For this condition i_I is in the linear range and described by

$$i_1 = k_1(2(v_i - V_{TI})v_o - v_o^2). \quad (A12)$$

The combination of this equation with the expression for i_L as given by Eq. (A1) yields the inverter output low-state voltage:

$$V_L = \tfrac{1}{2}[(2(V_{DD} - V_{TI}) + (|V_{TL}|/k_R)) \\ - \sqrt{(2(V_{DD} - V_{TI}) + (|V_{TL}|/k_R))^2 - 4(|V_{TL}|V_{DD}/V_R)}]. \quad (A13)$$

The pair delay is shown in Fig. 15, which illustrates the input and output waveforms for inverter stage 2. In the computation of these waveforms, the Mead and Conway [46] design parameters were used. That is, $V_{TI} = 0.2V_{DD}$, $V_{TL} = -0.8V_{DD}$, and $k_R = k_I/k_L = 4$. For these values, $V_L = 0.119V_{DD}$, $V_{TINV} = 0.513V_{DD}$, and $t_{on} = 0.092\tau_r$. In Fig. 15, Eq. (A7) is plotted using the above design parameters. For the sake of comparison, a more complete solution for the output voltage of stage 2 is also shown in Fig. 15. In computing the waveform, the rising ramp given by Eq. (A4) was used, but the differential equation was derived from $i_{L2} = i_{I2} + i_{C2}$. Transistor Q_{I2} was allowed to operate in all three modes starting in cutoff, then saturation, and terminating in the linear region. A comparison of the curves indicates that the pair delay as given by Eq. (A9) is a good approximation for the case considered here.

ACKNOWLEDGMENTS

This work was performed under the Product Assurance Technology Program jointly sponsored by the National Aeronautics and Space Administration and the Defense Advanced Research Projects Agency. The author wishes to acknowledge the efforts of the staff engineers and scientists at NBS who participated with the author in developing a number of test structures during the eight years the author was with NBS.

REFERENCES

1. F. J. Barone and C. F. Myers, *Electronics* **41**, 84 (July 1968).
2. G. L. Schnable and R. S. Keen, *IEEE Trans. Electron Devices* **ED-16**, 322 (1969).
3. R. J. Sahni, *Reliability Phys. Symp. Proc.* **8**, 226 (1970).
4. W. M. Penney and L. Lau (eds.), "MOS Integrated Circuits," Van Nostrand-Reinhold, New York, 1972.
5. F. H. Reynolds, R. W. Lawson, and P. J. T. Mellor, *Proc. IEEE* **62**, 223 (1974).
6. R. G. Tingley and D. W. Johnson, *Circuits Manufact.* **17**, 30 (April 1977).
7. W. E. Ham, Comprehensive Test Pattern and Approach for Characterizing SOS Technology. National Bureau of Standards Special Publ. 400-56 (January 1980).
8. M. A. Mitchell and L. W. Linholm, Test Pattern NBS-28 and NBS-28A: Random Fault Interconnect Step Coverage and Other Structures. National Bureau of Standards Special Publ. 400-65 (March 1981).
9. J. M. Mikkelson, L. A. Hall, Aron K. Malhotra, S. D. Seccombe, and M. S. Wilson, *IEEE Int. Solid-State Circuits Conf.* 106 (1981).
10. C. Chrones, Semicond, Int. **3**, 113 (October 1980).

11. H. R. Huff, R. J. Kriegler, and Y. Takeishi (eds.), "Semiconductor Silicon 1981." Chapter 12. Electrochemical Society, Princeton, New Jersey, 1981.
12. M. G. Buehler, Microelectronic Test Patterns: An Overview. National Bureau of Standards Special Publ. 400-6 (August 1974).
13. W. D. Jansen and D. G. Fairbairn, *VLSI Design* (formerly *Lambda*) **2**, 16 (First Quarter 1981).
14. R. Kossowsky, *Reliabil. Phys. Symp. Proc.* **16**, 112 (1978).
15. M. G. Buehler and W. M. Bullis (eds.), Special issue on characterization techniques for semiconductor materials, processes, and devices, *IEEE Trans. Electron Devices* **ED-27** (December 1980).
16. G. Potter, *VLSI Design* (formerly *Lambda*) **1**, 12 (Fourth Quarter 1980).
17. W. R. Scott, L. M. Hess, and F. R. Ault, *IEEE Test Conf.* 494 (1980).
18. C. T. Sah, *IEEE Trans. Electron Devices* **ED-11**, 324 (1964).
19. M. G. Buehler and W. R. Thurber, *J. Electrochem. Soc.* **125**, 650 (1978).
20. T. J. Russell and D. A. Maxwell, A Production-Compatible Microelectronic Test Pattern for Evaluating Photomask Misalignment. National Bureau of Standards Special Publ. 400-51 (April 1979).
21. T. F. Hasan, S. U. Katzman, and D. S. Perloff, *IEEE Trans. Electron Devices* **ED-27**, 2304 (1980).
22. C. J. Bartlett and L. N. Schoenberg, *Circuits Manufact.* **21**, 54 (April 1981).
23. A. C. Ipri and J. C. Sarace, *RCA Rev.* **38**, 323 (1977).
24. M. G. Buehler, *J. Electrochem. Soc.* **127**, 2284 (1980).
25. A. C. Ipri, *RCA Rev.* **41**, 537 (1980).
26. O. Paz and T. R. Lawson, *IEEE J. Solid-State Circuit* **SC-12**, 540 (1977).
27. K. H. Zaininger and F. P. Heiman, *Solid State Technology* **13**, 49 (May 1970).
28. M. G. Buehler, *J. Electrochem. Soc.* **127**, 701 (1980).
29. L. J. vander Pauw, *Philips Res. Rep.* **13**, 1 (1958); *Philips Tech. Rev.* **20**, 220 (1958).
30. L. W. Linholm and M. G. Buehler, Electrochem. Soc. Ext. Abstr. Abstract No. 191 (May 1979).
31. L. W. Nagel, SPICE2: A Computer Program to Simulate Semiconductor Circuits, Memorandum No. ERL-M510, Electronics Research Laboratory, Univ. of California, Berkeley, California (May 9, 1975).
32. L. W. Linholm, The Design, Testing, and Analysis of a Comprehensive Test Pattern for Measuring CMOS/SOS Process Performance and Control, National Bureau of Standards Special Publ. 400-66 (August 1981).
33. D. S. Peck, *Reliabil. Phys. Symp. Proc.* **13**, 253 (1975).
34. J. L. Spencer et al., *Reliabil. Phys. Symp. Proc.* **19**, 74 (1981).
35. S. Komatsu, K. Suzuki, N. Iida, T. Aoki, T. Ito, and T. Sawazaki, *Int. Electron Devices Meeting Proc.* 144 (1980).
36. R. J. Usell, Jr., and S. A. Smiley, *Reliabil. Phys. Symp. Proc.* **19**, 65 (1981).
37. S. P. Li, S. Prussin, and J. Maserjian, *Reliabil. Phys. Symp. Proc.* **16**, 132 (1978).
38. D. L. Crook, *Reliabil. Phys. Symp. Proc.* **17**, 1 (1979).
39. S. Rosenberg, D. Crook, and B. Euzent, *Reliabil. Phys. Symp. Proc.* **16**, 19 (1978); A. K. Sinha, W. S. Lindenberger, W. D. Powell, and E. I. Povilonis, *J. Electrochem. Soc.* **127**, 2046 (1980).
40. W. H. Schroen, J. L. Spencer, J. A. Bryan, R. D. Cleveland, T. D. Metzgar, and D. R. Edwards, *Reliabil. Phys. Symp. Proc.* **19**, 81 (1981).
41. W. H. Schroen, *Reliabil. Phys. Symp. Proc.* **16**, 81 (1978).
42. S. Vaidya, D. B. Fraser, and A. K. Sinha, *Reliabil. Phys. Symp. Proc.* **18**, 165 (1980).
43. N. L. Sbar and R. P. Kozakiewicz, *IEEE Trans. Electron Devices* **ED-26**, 56 (1979).
44. R. B. Comizzoli, L. K. White, W. Kern, G. L. Schnable, D. A. Peters, C. E. Tracy, and R. D. Vibronek, *Reliabil. Phys. Symp. Proc.* **18**, 282 (1980).

45. I. Ohkura, K. Okazaki, and Y. Horiba, *Proceedings IEEE Int. Conf. Circuits Comput.* 953 (1980).
46. C. A. Mead and L. A. Conway, "Introduction to VLSI Systems," Chapter 1. Addison-Wesley, Reading, Massachusetts, 1980.
47. J. Shott and T. Walker, Private communication, (1981).
48. F. L. Schuermeyer, H. P. Singh, Russell L. Scherer, and D. L. Mays, *Int. Electron Devices Meeting Proc.* 441 (1980).
49. H.-N. Yu, A. Reisman, C. M. Osburn, and D. L. Critchlow, *IEEE Trans. Electron Devices* **ED-26**, 318 (1979).
50. W. Murakami, *in* ARPA/NBS Workshop III. Test Patterns for Integrated Circuits, (H. A. Schafft Ed.), National Bureau of Standards Special Publ. 400-15 (January 1976).
51. R. T. Jerdonek, H. F. Bare, Jr., and G. J. Fromen, *IEEE Trans. Electron Devices* **ED-25**, 873 (1978).
52. R. Zucca, B. M. Welch, C.-P. Lee, R. C. Eden, and S. I. Long, *IEEE Trans. on Electron Devices* **ED-27**, 2292 (1980).
53. M. G. Buehler, *Solid State Technol.* **22**, 89 (October 1979).
54. R. L. Mattis and M. R. Doggett, *Solid State Technol.* **21**, 76 (November 1978).
55. G. P. Carver, R. L. Mattis, and M. G. Buehler, Microelectronic Test Patterns NBS-12 and NBS-24, NBSIR 81-2234 (May 1981).
56. D. S. Perloff, F. E. Wahl, C. L. Mallory, and S. W. Mylroie, *Solid State Technol.* **24**, 75 (September 1981).
57. S. Wetterling, *in* "Semiconductor Silicon 1981," (H. R. Huff, R. J. Kriegler, and Y. Takeishi, eds.), p. 896. Electrochemical Society, Princeton, New Jersey, 1981.
58. R. Y. Koyama and M. G. Buehler, A Wafer Chuck for Use Between −196 and 350°C. National Bureau of Standards Special Publ. 400-55 (January 1979).
59. C. A. Pina, Private communication, (1981).
60. M. G. Buehler and W. R. Thurber, *J. Electrochem. Soc.* **125**, 645 (1978).
61. J. S. Howard and J. Nahourai, *Solid State Technol.* **21**, 48 (July 1978).
62. G. P. Carver and M. G. Buehler, *IEEE Trans. Electron Devices* **ED-27**, 2245 (1980).
63. G. P. Carver, L. W. Linholm, and T. J. Russell, *Solid State Technol.* **23**, 85 (September 1980).
64. H. Iwai and S. Kohyama, *Int. Electron Devices Meeting Proc.* 235 (1980).
65. T. Cave and D. Smith, *Comput. Design* **17**, 161 (May 1978).
66. L. W. Linholm, R. L. Mattis, and R. C. Frisch, *in* "Semiconductor Silicon 1981" (H. R. Huff, R. J. Kriegler, and Y. Takeishi, eds.), p. 906. Electrochemical Society, Princeton, New Jersey, 1981.

DIAGNOSTIC TECHNIQUES

R. B. MARCUS

12.1 INTRODUCTION

This chapter summarizes instrumental methods found to be useful for solving problems that arise in VLSI technology development efforts, and explains their application to the problems. Some of the instrumental methods described in this chapter are most applicable to the analysis of circuit structures; others are more applicable to problems generated during experiments on the preparation of new materials for VLSI processing programs. Of the large class of diagnostic techniques available, only those are discussed which are presently useful or which show high potential for becoming useful in the immediate future.

Four areas of application of instrumental methods to VLSI problems can be described: the determination of morphology, chemical analysis, the determination of crystallographic structure and mechanical properties, and electrical mapping of sites of device leakage and breakdown. The application of instrumental methods to these problems is summarized in Table 1. Crosses indicate cases where the instrumental methods are primary sources of information for the indicated areas of application; the cross marks in parenthesis (×) are cases where special accessory equipment is needed for the indicated application. The four areas of application constitute four main sections of this chapter, and instrumental methods constitute subsections.

In a number of the analytical procedures described in Table 1 the sample is bombarded with a beam of x-rays or electrons, and analysis requires a measure of the resulting radiation. The procedures based on these interactions and typical energy ranges of incident and secondary radiations are given in Table 2.

12.2 MORPHOLOGY DETERMINATION

One of the first steps in most diagnostic efforts is examination of the shapes of relevant features: edge acuity of patterned lines, proximity between features, misalignment, and so on. These features are examined by optical microscopy, scanning electron microscopy, and transmission electron microscopy. The maximum useful magnification of these three methods is approximately $1,000\times$, $50,000\times$, and $500,000\times$, respectively. Since the magnification ranges overlap, few questions on morphology of device features can escape scrutiny.

Table 1 Application of instrumental methods to VLSI problems

Instrumental method	Acronym(s)	Morphology determination	Chemical analysis	Crystallographic structure and mechanical properties	Electrical mapping
Auger electron spectroscopy	AES		×		
Electron beam induced current microscopy	EBIC				×
Laser reflectance	LR			×	
Neutron activation analysis	NAA		×		
Normarski interference contrast optical microscopy	...	×			
Rutherford backscattering spectroscopy	RBS		×	(×)	
Scanning electron microscopy	SEM	×	(×)		(×)
Secondary ion mass spectroscopy	SIMS		×		
Transmission electron diffraction	TED			×	
Transmission electron microscopy	TEM	×	(×)	×	(×)
Voltage contrast microscopy	VC				×
X-ray diffraction	XRD			×	
X-ray emission spectroscopy	XES		×		
X-ray fluorescence	XRF		×		
X-ray photoelectron spectroscopy	XPS, ESCA		×		

Table 2 Types of radiation resulting from electron or x-ray bombardment of a sample surface and instrumental methods of analysis based on these interactions.

Incident beam		Secondary radiation				
		Electron			X-ray	
Radiation	Energy E_0 (keV)	Type	Energy (eV)	Analytical Procedure	Energy	Analytical Procedure
Electron	2–10	Auger	20–2000	AES		
	2–40	Secondary	<10	SEM(VC)		
	2–40	Back-scattered	$<E_0$	SEM(BS)		
	20–200				$<E_0$	XES
X-Ray	<2	Primary ionized	20–2000	XPS		
	<50				$<E_0$	XRF

12.2.1 Nomarski Interference Contrast Optical Microscopy

Nomarski interference contrast microscopy[1,2] is the most generally useful form of optical microscopy for solving VLSI processing problems. With this method, surface features of different elevations appear as different colors or as different shades of gray. This contrast is achieved by splitting the illuminating beam into two beams displaced by a short distance on the sample surface, followed by reflection and reconstitution of the reflected beams. The optical path length changes because of the presence of a step or a change in the index of refraction (caused by a phase boundary). These changes in optical path length produce a contrast change in the reconstituted beam, which appears in the microscope image.

The resolution of an optical microscope is governed by the wavelength of illuminating light λ and the numerical aperture (NA) of the objective lens:

$$r = \frac{0.61\lambda}{NA} \quad (1)$$

The resolution limit of the optical microscope is approximately 0.25 μm. If visual resolution with the unaided eye is assumed to be 0.1 mm, then loss of image sharpness can begin to be detected above 400× magnification, and the upper limit to useful magnification (when images become too blurred to be "acceptable") is between 1000× and 2000×. For many problems requiring an analysis of the morphology of VLSI features, a lateral resolution of 0.25 μm or even 1.0 μm is quite acceptable. A level of vertical (depth) resolution is also desirable, however, which permits the clear identification of features that are about 200 Å thick (for some MOSFET gate oxides) or thicker. This depth resolution is obtained by using Nomarski interference contrast optics.

Figure 1 shows the essential features of a Nomarski interference contrast microscope operating in the reflectance mode. Light passes through a polarizer and is reflected downward toward birefringent crystals that together make up a Wollaston

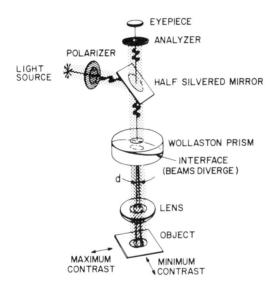

Fig. 1 Features of a Nomarski interference contrast microscope operating in the reflectance mode. The divergence of the two beams that illuminate the sample is given by d.

prism, that is, a prism in which the light is split into two mutually perpendicular polarized components that move at different velocities with an angular divergence d. After emerging from the prism and reflecting off the sample, the two beams recombine by passing once again through the Wollaston prism in the opposite direction. The reconstituted beam then passes through an analyzer, where its intensity changes are observed.

The microscope image contains contrast effects that depend on differences in optical path length caused by changes in the geometrical profile of the surface, and differences in phase changes resulting from variations in index of refraction, as may occur across a phase boundary. Figure 2 illustrates intensity variations seen in the microscope when monochromatic light illuminates a substrate. Figure 2a shows a cross section of the sample. The sample consists of two phases with different refractive indices; B represents their phase boundary. Figure 2b shows the wave fronts of the reflected beams after they pass through the Wollaston prism. The optical path differences between the two reflected light beams give rise to the intensity variation shown in Fig. 2c. Both the polarizer and analyzer settings and the position of the prism can be varied. The resulting shapes of the emerging wave fronts and intensity profiles reflect these settings as well as the optical properties of the phases. Figure 2b and c, therefore, describes the situation for one specific setting of the three adjustments for a given sample orientation.

Interference contrast is maximized in a direction parallel to the maximum displacement of the two beams and is essentially zero in the orthogonal direction. Figure 3 shows two micrographs of a sample consisting of oxide features. The micrographs are taken under identical polarizer, analyzer, and Wollaston prism settings, but the sample in Fig. 3b has been rotated 90° compared with Fig. 3a. The oxide pattern contains a slight offset, which gives rise to an approximately 100-Å step along C‒‒C

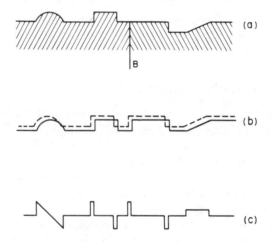

Fig. 2 (a) Representation of a cross section of a sample at a surface. (b) The wave fronts of the reflected beams after emerging from the prism, and (c) an intensity distribution in the image plane in a Nomarski interference contrast microscope. The sample is assumed to consist of two phases (with different refractive indices) that meet at the boundary B.

(Fig. 3a). The interference contrast image in Fig. 3a clearly shows the offset, since the polarizer, analyzer, and prism are adjusted for maximum contrast for the sample position used for that photograph; the contrast disappears when the sample is rotated 90° and the offset becomes extremely hard to see (Fig. 3b). The contrast features in Fig. 3b are essentially the same as those that would be seen in a microscope with ordinary optics. Figure 3 excellently illustrates the use of Nomarski interference microscopy in revealing subtle changes in morphology.

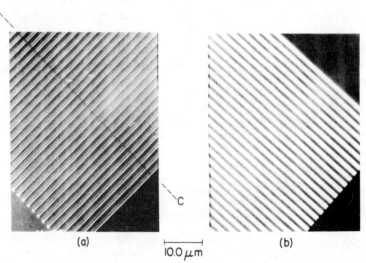

Fig. 3 (a) Nomarski interference contrast reflection images of 1-μm features showing contrast due to slight pattern error at C‒‒C. (b) This contrast feature is absent when the sample is rotated 90°.

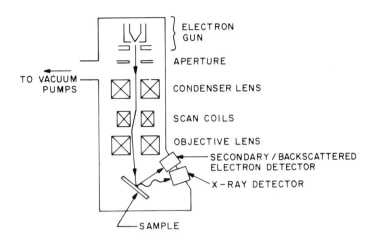

Fig. 4 Schematic drawing of a scanning electron microscope. The incident electron beam of energy 2 to 40 keV produces low-energy secondary and higher-energy backscattered electrons as well as x-radiation (see Table 1), all of which can be analyzed to provide useful information.

12.2.2 Scanning Electron Microscopy (SEM)

Scanning electron microscopy is a standard analytical method in VLSI laboratories, mainly because it provides increased spatial resolution and depth of field compared with optical microscopy, and because chemical information can be obtained from the x-ray spectra generated by electron bombardment (see Section 12.3.7). Resolution better than about 100 Å can be achieved under optimum conditions, and typical depths of field are 2 to 4 μm at 10,000× magnification and 0.2 to 0.4 mm at 100× magnification.

Figure 4 gives a schematic drawing of a scanning electron microscope. An electron gun, usually consisting of a tungsten or LaB_6 filament, generates electrons, which are accelerated to an energy of 2 to 40 keV. A combination of magnetic lenses and scan coils produces a small-diameter beam which is rastered across the sample surface. The electron bombardment produces three useful types of radiation: x-rays, secondary electrons, and backscattered electrons (Table 2).

Figure 5 shows the energy spectrum of electrons emitted from a sample that has been bombarded with an electron beam. A large fraction of the spectrum consists of electrons with energy less than 50 eV peaking at less than 5 eV. These electrons are referred to as secondary electrons. The other electrons peak at an energy close to E_0 and are referred to as backscattered electrons.

The secondary or backscattered electron current is used to modulate the intensity of an electron beam in a cathode ray tube (CRT). Since the CRT electron beam moves in synchronism with the rastering incident beam of the scanning electron microscope, the CRT beam produces an image of the sample surface whose contrast is determined by variations in the secondary or backscattered electron flux. The x-ray signal (Table 2) is useful for chemical analysis, and a discussion of this diagnostic procedure appears in Section 12.3.7.

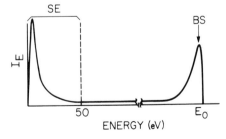

Fig. 5 Energy distribution of electrons emitted from sample bombarded by electron beam of energy E_0. A peak of low-energy secondary electrons (SE) occurs near 5 eV, and the backscattered electron (BS) peak occurs near E_0. Auger electron peaks (not shown) occur between these two peaks.

The incident electron beam undergoes multiple collisions as it penetrates a sample; incident electrons that are not backscattered finally come to rest after traversing a range R that can be calculated or measured. Figure 6 gives the electron ranges calculated from one set of expressions[3] for silicon, aluminum, and gold. The electron range increases with decreasing atomic number and with increasing incident beam energy E_0. The electron trajectory changes with each collision, causing a narrow incident beam to spread as it penetrates into a sample. Figure 7 shows the maximum penetration depth of incident electrons at four different beam energies. The penetration is described by a series of pear-shaped envelopes, which increase in depth and width with increasing energy.

For normal incidence on a bulk sample, the surface area contributing backscattered electrons is a disc with a diameter approximately equivalent to the electron

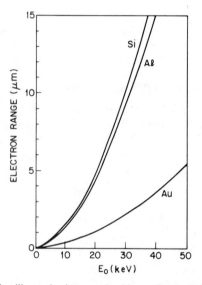

Fig. 6 The electron range in silicon, aluminum, and gold as a function of incident beam energy. *(After Everhart and Hoff, Ref. 3.)*

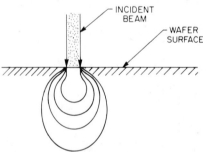

Fig. 7 Illustration of volume spread of incident electrons within sample after sufficient multiple scattering to reduce electron energy to nearly zero. Envelopes of increasing size represent maximum electron penetration at increasing beam energy.

range. The resolution of a backscattered electron image improves as the sample thickness decreases; thus, the resolution of a backscattered electron image from a thin metal film on silicon or SiO_2 is better than the resolution from the same metal in bulk form. Resolution of images formed from secondary electrons is partly determined by the lateral size of material within which most secondaries are generated at a depth less than the escape depth. The escape depth[4] of electrons in metals reaches a minimum of 4 Å at 70 eV, and increases with decreasing energy to a value of 25 Å at 10 eV. The escape depth is more than 50 Å for insulators. The lateral resolution of secondary electron images from a flat surface is therefore given by the diameter of the incident beam plus a lateral increment that is due to the electron mean free path.[5] Note that secondary electrons generated by emerging backscattered electrons within an escape depth of the surface come from a larger area and they therefore contribute to resolution degradation.

The contrast of both backscattered and secondary electron images depends on variations in the flux of electrons arriving at the detector. The yield of backscattered electrons increases with increasing atomic number Z and is 10 times higher for gold than for carbon (Ref. 6, pp. 75 to 77). Because the yield depends on the atomic number, contrast is produced in the backscattered electron image between regions with different atomic numbers. The contrast between adjacent pairs of elements in the periodic table decreases with increasing Z, and for aluminum and silicon the contrast is 6.7% (Ref. 6, p. 159). Thus, backscattered electron detection can be used to distinguish aluminum particles from a silicon background.

The secondary electron yield is a weaker function of Z than is the yield of backscattered electrons; the secondary electron yield only increases by about a factor of 2 from carbon to gold. Yield is a stronger function of the work function of the material,[7] and is significantly higher for oxides and other wide-band gap materials than for silicon.[8] This source of contrast makes the use of secondary electron SEM imaging in VLSI studies a major advantage, since metallization, oxide, and silicon regions are easily distinguishable from each other. A second source of contrast in secondary electron images is due to the dependence of secondary electron yield on surface curvature. The secondary electron flux from a surface of changing slope varies with the secant of the tilt angle. Therefore, surfaces that differ significantly in slope can be clearly distinguished. The secondary electron flux that is detected is also a strong function of the orientation of the emitting surface with respect to the detector; surface regions that face the detector appear significantly brighter than other surface regions.

Spatial resolution depends on the size of the sample surface contributing to secondary or backscattered electrons, and on local changes in phase, composition, and sample orientation which influence the secondary or backscattered electron flux as discussed above. Resolution also depends on the condition of the scanning electron microscope. These factors interact intimately and cannot be separated. For example, the diameter of the electron probe decreases with decreasing beam current and increasing beam energy, and is smaller for a LaB_6 source than for a tungsten filament. A 30-keV electron beam with a current of 10^{-11} A has probe diameters of about 40 and 90 Å for a LaB_6 and tungsten filament, respectively; these values change to 60 and 130 Å, respectively, at 10 keV. The actual resolution achieved in SEM study, however, can be considerably poorer than 40 to 130 Å. The minimum beam current I_{min} needed to produce a detectable contrast C between adjacent regions is given by (Ref. 6, p. 174)

$$I_{min} = \frac{4 \times 10^{-12}}{\epsilon C^2 t_f} \qquad (2)$$

where ϵ is the efficiency of signal collection and t_f is the time needed to scan an SEM frame. "Difficult" samples have low contrast between adjacent regions, say 1 to 5%. For a thermionic tungsten filament the minimum beam diameter needed to provide sufficient current (I_{min}) to detect this brightness difference is 2300 Å for $C = 1\%$ and 460 Å for $C = 5\%$ (Ref. 6, p. 177). Thus, a major limitation on spatial resolution is the need to have sufficient beam current for the experimenter to be able to distinguish regions that have similar brightness.

Three common difficulties which occur during SEM study of VLSI circuits are sample contamination, electron beam induced damage, and surface charging during examination. The major type of sample contamination is hydrocarbon polymerization, which occurs as the electron beam strikes the surface. Although modern microscopes are well pumped and maintain a vacuum (at the specimen chamber) of about 10^{-6} torr, contamination cannot be avoided. A related problem is damage to oxides by electron beams and the effect of this damage on device performance. Electron irradiation produces positive oxide charges and interface traps which can be avoided by maintaining a sufficiently low beam energy to prevent penetration to an active area (such as a gate oxide). Once they have formed, defects can be annealed out[9] at temperatures between 400 and 550°C.

The third problem frequently encountered in SEM studies of insulating surfaces is surface charging. This problem sometimes occurs when beam energies are above the second secondary electron yield crossover point (the incident beam energy above which the secondary electron yield is less than 1). The surface becomes negatively charged, disturbing the trajectory of the incident electron beam and degrading the image. A method for avoiding this problem uses a low-energy incident beam. A field emission source is the only electron source that seems capable of producing high-resolution images at low-beam energies,[10] and some commercially available SEM units have these sources. Applying grounded metallic coatings approximately 100 Å thick to sample surfaces is another method for avoiding charge buildup during study. The metallic coatings provide a conducting path to ground. Since the escape depth of secondary electrons is much smaller in metals ($\sim$5 Å) than in insulators, a thin metal surface coating also greatly improves spatial resolution in secondary electron imaging. Unfortunately, such coatings make the samples unsuitable for further processing.

12.2.3 Transmission Electron Microscopy (TEM)

Transmission electron microscopy (TEM) is a useful tool for solving problems in VLSI technology that require high spatial resolution. TEM offers a resolution of about 2 Å. In a transmission electron microscope an electron beam passes through the thin-film sample, and forms an image that displays morphological and crystallographic features of the film components. Commercial TEM instruments use electron beams with energy between 60 and 350 keV. Higher beam energy permits greater sample penetration; the maximum thickness of silicon which permits TEM image formation is 1.5 μm for a 200-keV beam, but only 0.5 μm for an 80-keV beam.[11]

VLSI specialists are usually concerned with the morphology of features whose phase boundaries extend to both surfaces of the TEM sample. These boundaries limit further the maximum sample thickness. For example, a well-oriented thin-film TEM sample which contains a 1-μm-wide polysilicon runner on oxide produces an image of two lines, one corresponding to each interface (Fig. 8a). In Fig. 8 it is assumed

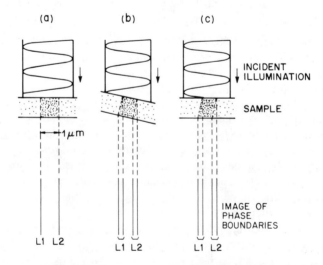

Fig. 8 Illustration of problems of angular misorientation of a sample in TEM study of VLSI circuits. (a) TEM study of a well-oriented vertical cross section of a two-phase region (such as 1-μm-wide polysilicon runner over oxide) produces a clean image of two lines, corresponding to the two interfaces. (b) A misoriented sample or (c) a misoriented feature in an oriented sample produces line doubling.

that the sample is a vertical cross section through a chip (e.g., the plane of the page is perpendicular to the original surface of the chip). If the sample is tilted slightly during study, the image of the interfaces L1 and L2 doubles (Fig. 8b). Such doubling becomes harder to avoid as the sample becomes thicker, and places a severe practical limit on permissible sample thickness: Although a 0.5° misorientation of the sample causes a doublet spacing of 10 Å from a 1000-Å-thick film, the same misorientation produces a doublet spacing of 50 Å from a 5000-Å-thick film.[12] Figure 8c illustrates another related problem caused either by misorientation of the sample initially cut from the wafer (or chip), or by texture at an interface. Texture often results at the polysilicon-oxide interface (after the polysilicon has thermally oxidized) and at the edges of patterned features.

Contrast in a TEM image can be described for two situations. One is TEM study of crystalline materials (such as silicon, aluminum, polysilicon, and various silicides), and the other is the study of amorphous materials.

In crystalline materials, the incident electron beam is diffracted by the material, and local variations in diffraction intensity produce contrast in an image from the undiffracted beam (bright field image) or from one or more diffracted beams (dark field image). The intensity of the emergent beam is periodic with sample thickness. The sample thickness corresponding to one period in silicon is 602 Å for a (111) reflection and 757 Å for a (220) reflection.[13] Thus, a thicker region of the sample does not necessarily look lighter in the negative; a wedge-shaped crystalline sample produces a TEM image that has alternate light and dark bands called thickness extinction contours. Dark bands are also caused by a bent sample and are then called bend extinction contours. Abrupt changes in thickness, phase structure, or crystallographic orientation cause corresponding abrupt changes in contrast, and these crystallographic features can be easily imaged at high resolution.

In the case of nearly amorphous material, contrast is determined by local changes in electron scattering which result from differences in sample thickness or from differences in chemical or phase composition. A sample region whose thickness varies continuously produces a corresponding continuous variation in image intensity, unlike the case of diffraction contrast. TEM images obtained from oxides, nitrides, and other amorphous materials are therefore somewhat easier to interpret intuitively than images obtained from crystalline samples.

Sample preparation Difficulties in sample preparation have been the main factor limiting the application of TEM methods to VLSI programs. One difficulty is preparing a sample that is sufficiently thin for TEM study. Another difficulty is that, after a thin-film sample is prepared, the morphological feature of interest to the diagnostician must be present in the thinned region. The solutions of these problems are briefly discussed in the following paragraphs.

Thin-film sections most useful to VLSI diagnosticians are those made orthogonal to the wafer surface (vertical cross section). TEM study of such samples provides information on the relationship among multiple layers as well as information on the shapes of steps created by edges and contacts. Figure 9 describes the main features of vertical cross section sample preparation. Samples can be prepared in less than 2 days, particularly when ion milling machines are used with hole detection capability which obviates the need for constant attention. More details on this procedure as well as methods for the preparation of horizontal sections (parallel to the wafer surface) are described in Refs. 12 and 14.

The second problem, assuring that the morphological feature(s) of interest is (are) in the thin part of the TEM preparation, has been solved by incorporating special TEM test chips on each processed wafer; the test chip includes every morphological feature relevant to the particular VLSI technology (see Fig. 10). Each feature appears within 1 to 3 μm in one dimension, and extends about 2 mm in the orthogonal direction; in the figure all the features are contained within a 23-μm distance. This "repeat unit" is replicated a number of times over a distance of 2 mm. Using the

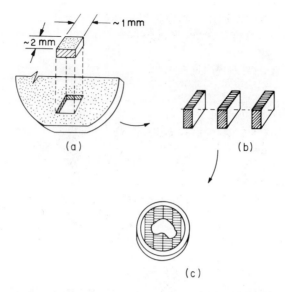

Fig. 9 Preparation of a vertical cross section for TEM study. (a) A small piece (~1 mm × ~2 mm) is cleaved from the wafer. (b) A number of such pieces are epoxy bonded face to back. (c) After lapping and ion milling a hole is produced. A region sufficiently thin to permit TEM study is usually produced within a 50- to 100-μm area around the hole perimeter.

method of sample preparation described in Fig. 9, at least one complete set of morphological features in one test-chip repeat unit can be produced within a region of sample that is sufficiently thin for TEM study.

Figure 11 shows an application of TEM to a VLSI problem. The photograph is of a vertical section through a charge-coupled device test structure and shows the presence of very thin oxide at the tips of "horns" at the polysilicon corners. These thin oxide regions are sites of potential failure; after a second layer of metallization is deposited over the oxide, the oxide breakdown voltage is significantly lowered. This failure mode would have been extremely difficult to detect with any other diagnostic technique.

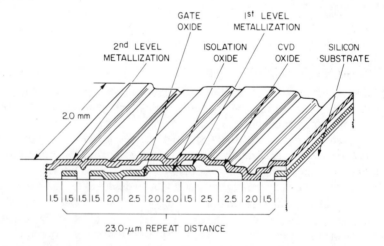

Fig. 10 Schematic drawing of a TEM test pattern. All morphological features essential to the technology appear within a short (1.5- to 2.5-μm) distance, comprising a "repeat" unit of 23.0 μm for the case shown. This unit is repeated 87 times to cover a distance of about 2 mm; each feature is extended 2 mm in the orthogonal direction.

12.3 CHEMICAL ANALYSIS

Many different methods are required for the chemical analysis of materials used in

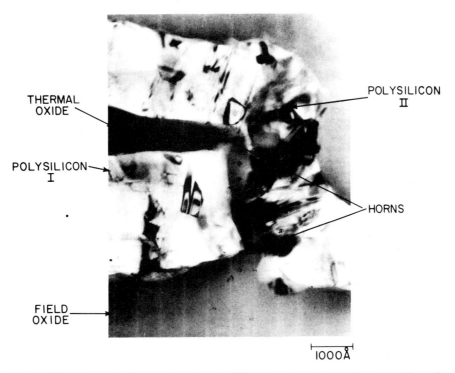

Fig. 11 TEM photograph of cross section through a CCD test structure showing "horns" which constitute a failure mode. This defect would have been extremely difficult to detect by any other technique.

VLSI technology. Spatial resolution requirements vary from atomic dimensions, as in the depth profiling of intermetallics or dopants, to essentially macroscopic dimensions, as in the bulk analysis of large area films or substrates. Vertical (depth) and lateral spatial resolution requirements for these studies are often quite different. Sensitivity requirements range from 10^{11} atoms/cm^3 to 10^{21} atoms/cm^3. The chemicals usually sought in these studies are silicon dopants (arsenic, phosphorus, boron), oxygen, carbon, resist residue, various components of metallizations, and metallic impurities. Thus, the chemicals run the gamut from light elements to heavy elements, such as gold, platinum, and tungsten.

12.3.1 Auger Electron Spectroscopy (AES)

Auger electron spectroscopy (AES) analyzes a certain class of electrons, called Auger electrons, that are generated when either an electron beam or a photon beam strikes

Table 3 Escape depth[†] of Auger electrons from a silicon matrix

Element	Transition energy (eV)	Escape depth (Å)
Phosphorus	120	5
	1859	32
Boron	179	6
Oxygen	507	12
Arsenic	1228	23
Aluminum	1396	26
Silicon	92	4
	1619	29

[†]Escape depths are interpolated from data presented in Table 2 of Ref. 17.

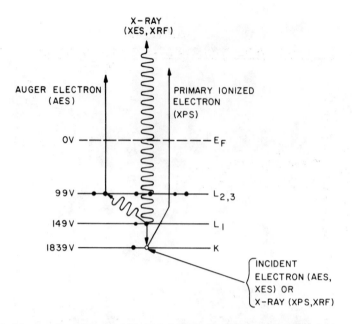

Fig. 12 Partial energy level diagram of the singly ionized silicon atom. An incident electron or x-ray photon is assumed to eject an electron from the K level into the vacuum. An electron transition from the L_1 level to the empty position in the K level releases energy, which may be detected directly as x-radiation, or may eject a second electron into the vacuum; this latter electron is called an Auger electron. This diagram illustrates the types of interactions detected and analyzed by four types of spectroscopy: AES, XES, XRF, and XPS. The transitions shown are only one set of a number of possible transitions.

the surface of the sample; in its usual mode, an incident electron beam is used (Table 2). Auger electrons are ejected from an atom in response to core-level ionization.[15,16] As shown in Fig. 12, an incident electron or photon of sufficient energy can eject an electron in the K shell of a target atom (the material assumed in Fig. 12 is silicon). An electron transition from the L_1 to the empty position in the K level releases energy, which ejects an Auger electron from the $L_{2,3}$ level.

The incident electron beam usually falls between 2 and 10 keV and penetrates only a short distance into the sample (Fig. 6). Most Auger electron energies fall within 20 to 2000 eV, and appear in the spectral range between the low-energy (secondary electron) and high-energy (primary backscattered electron) peaks shown in Fig. 5. The escape depth of Auger electrons is generally less than 50 Å and is considerably less for the lower-energy transitions (see Table 3).[17] Thus, a chemical analysis of surface regions can be made from AES data.

Many diagnostic problems require depth analysis beyond the escape depth of Auger electrons, and the sample must be ion milled in order to continuously create a new surface which essentially moves through the sample during AES analysis. Data are obtained by interrupting the milling process at regular intervals and obtaining Auger spectra, or by continuously recording spectra during ion milling. The Auger peak heights can then be plotted as a function of milling time or milling depth, and a depth profile can be obtained as shown in Fig. 13.

For quantitative analysis, the concentration C_i of element i in the host material (matrix) is calculated from the expression

$$C_i = \frac{\alpha_i I_i}{\sum_j \alpha_j I_j} \qquad (3)$$

where I_i is the intensity of the Auger peak of element i and I_j is the Auger peak intensity from an element in the matrix. The proportionality constants α are most easily determined from known standards. Methods for quantitative analysis are discussed in Refs. 4, 15, and 16. Table 4 gives the approximate sensitivity of AES detection of some species important to VLSI technology.[18] Sensitivity is given as the detection limit for an element homogeneously distributed within a silicon matrix.

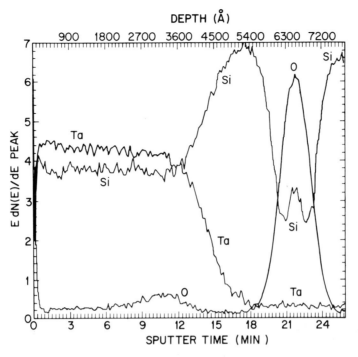

Fig. 13 AES depth profile from a tantalum silicide film on polysilicon on an oxidized silicon wafer showing the changing distribution with depth. *(Courtesy C. C. Chang, Bell Laboratories.)*

Lateral spatial resolution is governed mostly by the size of the incident beam which, in turn, is related to the minimum current necessary to generate useful data (about 5 nA). In modern commercial instruments the resolution can be higher than 0.1 μm. Depth resolution in nonprofiling AES study is limited by the Auger electron

Table 4 Detectability limits for AES analysis of various species in silicon[18]

Element	C_{min} (atoms/cm^3)
Phosphorus	1×10^{19}
Arsenic	5×10^{18}
Oxygen	5×10^{17}
Carbon	5×10^{17}

escape depth and is essentially one monolayer for lower-energy Auger transitions. The greater escape depth of higher-energy transitions permits the thickness of very thin surface films to be measured. The relative heights of a 1619-eV Si Auger peak from elemental silicon in the substrate and the 1607-eV chemically shifted silicon peak from a surface oxide film can be used to measure the oxide thickness.[19] The method is useful for measuring films with oxide thickness less than about four times the escape depth of the Auger electrons.

12.3.2 Neutron Activation Analysis (NAA)

Neutron activation analysis (NAA) is the most sensitive analytical method for many elements. This method uses nuclear irradiation of a sample to produce radioisotopes that can then be analyzed. NAA is most useful for solving problems involving gettering, or problems involving surface or bulk appearance of trace impurities on whole wafer samples. It is also useful in measuring contamination introduced by processing furnaces. Only light elements (such as boron, oxygen, nitrogen, and carbon) do not produce radioactive isotopes suitable for analysis.

Irradiation is performed with a thermal neutron flux of 10^{13} to $10^{14}/cm^2$-s over a period of 0.5 to 12 h. During irradiation of silicon wafers a number of radioactive species are produced, including ^{31}Si. The half-life of ^{31}Si is 2.6 h. After irradiation the sample is allowed to "cool" for 24 to 48 h to permit the radiation level from the silicon to fall below the levels from the other elements.

The species most commonly monitored during these experiments is γ radiation, with energies between 0.1 and 2.5 MeV. This radiation is detected by a lithium-drifted germanium detector and analyzed by a multichannel analyzer.

Both the energy of γ emission and the measured half-life are used to identify the isotope giving rise to the radiation. To measure the amount of the element present, several factors must be known: the radiation count over a time interval, the detector and emitter efficiencies for a particular peak, the thermal neutron flux, the time lapse since irradiation, the irradiation time, and other factors listed in Table 5. The number of atoms of the particular element in the counted sample can be determined from known formulas,[20] and the corresponding volume concentration can be computed from the original sample shape.

The minimum detectable limits for a number of species are listed[21] in the last column of Table 5. Although the sensitivity varies with sample size, irradiation time, flux, and other factors, the last column indicates the extremely high sensitivity of this analytical method.

By counting before and after removal of layers of controlled thickness, the depth distribution of elements within a sample can be determined. Thus, by repeatedly etching and counting, it can be shown that phosphorus gettering is successful in reducing bulk gold impurity atoms by a factor of 1/50 from an initial concentration of 3×10^{14} atoms/cm^3, and that gettering confines gold atoms to within 2 μm of the back surface.[22]

Table 5 Radioisotope parameters and sensitivity limits

Element	Atomic mass	$\tau_{1/2}$	f	$\sigma(b)$	γ Energy (MeV)	C_{min} (atoms/cm^3)
Arsenic	75	26.4 h	100	4.3	0.560	7.1×10^{11}
Copper	63	12.75 h	69.17	4.5	0.511	2.3×10^{12}
Gold	197	2.69 d	100	98.8	1.34 0.411	1.1×10^9
Sodium	23	15.0 h	100	0.53	1.37	6.3×10^{12}
Tantalum	181	115 d	100	21.0	1.121 1.221	1.1×10^{12}
Tungsten	186	23.9 h	28.41	40.0	0.686	4.9×10^{11}

Notes:
$\tau_{1/2}$ = the half-life.
f = the percent abundance of the radioisotope.
σ = the thermal neutron cross section in barns.
C_{min} = the minimum detectable concentration assuming analysis was performed on an entire 7.3-cm-diameter silicon wafer 510 μm thick. Detection limit values also assume a 10-h irradiation with a neutron flux of $1 \times 10^{13}/cm^2$-s and a 40-h delay before counting.

12.3.3 Rutherford Backscattering Spectroscopy (RBS)

RBS uses an incident 1- to 3-MeV ion beam (usually He^+) directed on the surface of a sample. The beam diameter is typically 10 μm to 1 mm. Elastic collisions between the incident ion and a target atom cause the ion to lose energy. The kinematic factor K relates the incident ion energy E_0 to the energy of the ion after backscattering E_0' as:

$$E_0' = KE_0 \qquad (4)$$

The scattered ions are detected by an energy-dispersive silicon barrier detector, and the signal is fed into a multichannel analyzer (see Fig. 14). Since values of K are known for each element,[23] the chemical composition at the surface of a sample can be determined by measuring the energies of the backscattered ions.

Incident ions suffer energy loss as they penetrate a sample and scatter. Ions scattered at a depth ΔZ have to return through the same path before they can escape and be detected, and additional energy is lost on this return trip. The total energy difference between ions scattered at the surface and ions emerging from the sample after scattering at a depth ΔZ is

$$\Delta E = KE_0 - E_1 = [\epsilon] N \Delta Z \qquad (5)$$

where $[\epsilon]$ is the stopping cross-section factor, and N is the atomic density of the sample. A depth profile is obtained by monitoring the number of backscattered ions as a function of backscattered ion energy E_1, as illustrated by the RBS spectrum[24] in Fig.

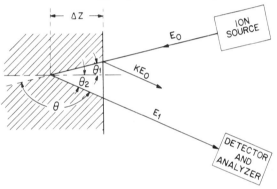

Fig. 14 Features of a Rutherford backscattering spectrometer. The incident beam energy E_0 becomes KE_0 after scattering; the emergent beam from scattering at a depth ΔZ has an energy E_1 because of the loss of energy of the incident beam during penetration to ΔZ, and loss of energy of the scattered beam during the return trip. The scattering angle is θ.

15. Note that the abscissa is also a depth axis (for a particular phase) by the relationship given in Eq. 5.

The sample used to generate the RBS spectrum of Fig. 15 is shown in the upper part of the figure. An aluminum film was used as the top layer of the sample, and the energy position of the leading (high-energy) edge of the Al peak is at $K_{Al}E_0 = 1.1$ MeV, while the energy position of the trailing edge is less by ΔE and appears at 0.950 MeV. The leading edge of the Ti peak in TiN is not given by $K_{Ti}E_0$, since the incident beam energy at that depth has been attenuated while passing through the aluminum layer, and the scattered beam undergoes further attenuation as it passes back through the aluminum. The leading edges of the titanium, silicon, and platinum peaks are all lower than their positions would have been in the absence of surface films; these positions are indicated by the vertical arrows in the figure.

RBS is one of the few chemical analysis methods that can provide quantitative information without the use of standards. The total number of counts detected in an RBS spectrum is a product of the differential scattering cross section of the scattered species $d\sigma/d\Omega$, the number of scattering centers per cm^2 ($N\Delta Z$), the acceptance angle of the detector ($\Delta \Omega$), and the beam current Q:

$$H = \frac{d\sigma}{d\Omega} N \Delta Z Q \Delta \Omega \qquad (6)$$

Values for $d\sigma/d\Omega$ have been tabulated[23] for all elements as a function of scattering angle θ for ^{4}He$^+$ and ^{1}H$^+$ incident ions. For the special but frequently occurring case of the analysis of a homogeneous film that contains only one compound of unknown composition $A_m B_n$, the ratio of the peak heights H_A and H_B can be found from Eq. 6:

$$\frac{H_A}{H_B} = \frac{\frac{d\sigma_A}{d\Omega}}{\frac{d\sigma_B}{d\Omega}} \frac{m}{n} \frac{[\epsilon]_B}{[\epsilon]_A] \tag{7}$$

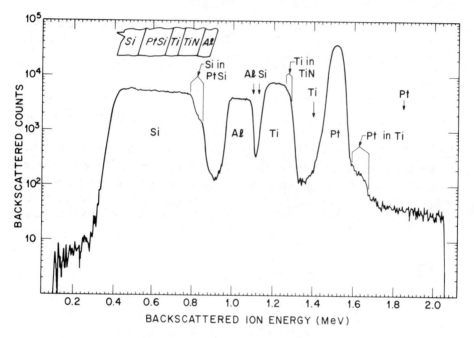

Fig. 15 RBS data from the structure diagramed at the top of the figure. The positions of the leading edges of each peak are shown by arrows for the case of no absorption. Only the leading edge of the aluminum peak coincides with its arrow; the other edges are shifted to lower energies because of energy loss by absorption by surface layer(s). *(Courtesy R. Schutz, Bell Laboratories.)*

Since $d\sigma_i / d\Omega = k Z_i^2$ where Z_i is the atomic number of element i, Eq. 7 can be written as

$$\frac{m}{n} = \frac{H_A}{H_B} \left[\frac{Z_B}{Z_A} \right]^2 \frac{[\epsilon]_A}{[\epsilon]_B} \tag{8}$$

By measuring H_A / H_B and finding the appropriate values for $[\epsilon]$ in a table, we can determine the quantity m/n.

The energy resolution of modern detectors is about 15 keV, which translates to a depth resolution of ~300 Å for silicon, and a depth resolution of about 100 Å for heavier metals such as those present in silicides. Unfortunately, the large incident ion-beam diameter (typically 10 μm to 1 mm) precludes the use of RBS for the analysis of most VLSI features. The sensitivity of RBS is determined by ion straggling, mass separation of peaks, and beam current. Table 6 gives estimates of the sensitivity limits for the detection of elements relevant to VLSI technology. The sensitivity for phosphorus is poor because of the close proximity of the phosphorus and silicon peaks.

12.3.4 Scanning Electron Microscopy (SEM)

The scanning electron microscope provides information on chemical composition by

Table 6 Detectability limits for RBS analysis of various species in silicon

Species	C_{min} (atoms/cm^3)	Reference
Arsenic	9×10^{18}	Calculated
Oxygen	5×10^{21}	25
Antimony	4×10^{18}	Calculated
Phosphorus	Poor	

use of x-ray spectrometer attachments; an SEM-like instrument specifically designed for quantitative chemical analysis has been called an electron probe. A description of x-ray emission spectroscopy is found in Section 12.3.7.

12.3.5 Secondary Ion Mass Spectroscopy (SIMS)

In the SIMS method, an ion beam sputters material off the surface of a sample, and the ionic component is mass analyzed and detected (Fig. 16). Sputtered ions are extracted and mass analyzed with a magnetic prism or quadrupole analyzer. In a system that uses a magnetic prism, a two-dimensional image of the distribution of an ionic species across the surface can be obtained by directing the secondary ion beam onto a channel plate. In quadrupole instruments the image is formed by recording the changing secondary ion-beam current as the primary beam is rastered across the sample surface. The intensity of the detected signal is related to the mass concentration.

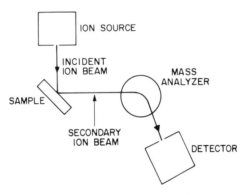

Fig. 16 Schematic diagram of a secondary ion mass spectrometer. An ion source creates a beam which rasters across a sample surface and sputters material off that surface. The ion fraction of sputtered material is mass analyzed, and displayed as a current intensity for a particular mass, or as a two-dimensional image of the distribution of that mass species.

The detected signal can be displayed as a mass spectrum during ion milling, giving a depth profile of the chemical species.

Both positive and negative incident ions are used with a beam energy typically between 5 and 15 keV. Since only the ionic fraction of sputtered material produces a SIMS signal, ion beams are chosen that produce the highest ion yields of the species under study. Positive cesium ion beams are generally useful for producing high negative ion yields of electronegative species from a target,[26] and O_2^+ ion beams are usually used for generating high positive ion yields from electropositive species.[27]

The incident beam is rastered across a small area of the surface to create a crater with a nearly flat bottom. Mass analysis is performed on the ionic fraction of sputtered material only from a central portion of the crater. When very low primary ion currents are used, sputtering rates are lowered to the point where data can be collected from a few monolayers, and surface analysis can be performed. Depth profiles are obtained by using higher primary ion currents. Lateral spatial resolution depends on the type of ion optics used in the instrument. Increased resolution must be traded off with sensitivity. SIMS achieves lateral resolutions of about 0.5 μm which is useful for the analysis of some problems with patterned VLSI chips. Vertical (depth) resolution is controlled by many factors, such as texture at the bottom of a crater, the contribution of signals from the crater wall, and impurity redistribution during ion milling.

Sensitivity limits are set by the factors described above and by problems of mass interference. The mass resolution of an experiment is defined by $M/\Delta M$, where ΔM is the minimum mass difference that can be detected at a mass level M. Typical values of $M/\Delta M$ are 250 to 5000. An example of the influence of mass resolution on sensitivity is the difficulty in detecting phosphorus in silicon with background water vapor present in the system. The $^{31}P^+$ peak is very close to the peak from $^{31}SiH^+$ (formed from the interaction of water with silicon). At the mass resolution of >3500 required to distinguish the two masses, the sensitivity limit becomes 1×10^{19} atoms/cm^3. Table 7 lists sensitivity for the detection of a number of species, plus it identifies the most useful primary ion(s) and the detected ion species. Note that because of the complexity of the SIMS process, standards are required to apply SIMS to a problem requiring quantitative analysis.

Table 7 SIMS parameters for detection of some elements relevant to VLSI problems[28]

Element[†]	Primary beam	Detected species	C_{min} (atoms/cm^3)
Arsenic	Cs$^+$	^{75}As$^-$	5×10^{14}
Phosphorus	Cs$^+$	^{31}P$^\pm$	5×10^{15}
Boron	O$_2^+$, O$^-$	^{11}B$^+$	1×10^{13}
Oxygen	Cs$^+$	^{16}O$^-$	1×10^{17}
Hydrogen	Cs$^+$	^{1}H$^-$	5×10^{18}

[†] In all cases a silicon matrix is assumed.

12.3.6 Transmission Electron Microscopy (TEM)

The transmission electron microscope provides information on chemical composition by use of an x-ray spectrometer attachment. Information on this instrumental method is found in the following section.

12.3.7 X-Ray Emission Spectroscopy (XES)

Both SEM and TEM instruments can generate x-rays by bombarding a sample with electrons. Electron bombardment of a sample produces an x-ray continuum, as well as x-ray peaks which are characteristic of the material. The intensity of background continuum at a particular wavelength λ is given by[29]

$$I_{\lambda,c} \sim \frac{iZ(E_0 - E_\lambda)}{E_\lambda} \tag{9}$$

where i is the electron beam current, Z is the atomic number of the material under bombardment, E_0 is the beam energy, and E_λ is the x-ray energy at wavelength λ. For many elements the intensity of an x-ray peak occurring at an energy E_λ is[30]

$$I_{\lambda,p} \sim i \left[\frac{E_0 - E_\lambda}{E_\lambda}\right]^n \tag{10}$$

where n is a constant, typically 1.7. The peak-to-background ratio becomes

$$\frac{I_{\lambda,p}}{I_{\lambda,c}} \sim Z \left[\frac{E_0 - E_\lambda}{E_\lambda}\right]^{0.7} \tag{11}$$

Thus, a higher beam energy produces stronger signals. A high beam energy is also used to stimulate x-ray emission of higher-energy peaks. Offsetting these arguments for higher beam energy, however, is the need to maintain a sufficiently low primary beam energy in studies on multilayer structures, where the stimulation of x-ray signals from lower layers is to be avoided. Typical beam energies fall between 15 and 40 keV.

Spatial resolution depends on the sample volume contributing x-ray spectra. X-rays are generated during scattering of the incident beam and backscattered electrons; they originate from a sample volume determined by the electron range (Fig. 7). This volume is quite large for bombardment of silicon, as can be determined from the electron range values shown in Fig. 6.

The loss in lateral spatial resolution caused by the size of the irradiated volume is not a major problem for qualitative analysis of patterned features or small particles on wafers, as long as the feature's chemical composition and "background" (i.e., the region beneath or adjacent to the feature) are known to be different. For example, identifying a gold particle on a boron nitride x-ray mask is easy, but identifying a particle in a 1.0-μm-wide space between two aluminum lines may be difficult if the particle contains aluminum.

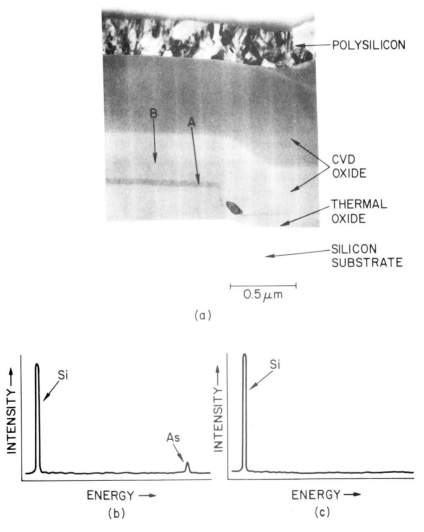

Fig. 17 (a) TEM cross section through an NMOS structure. (b) XES analysis of the 300-Å-wide dark band [region A in (a)] shows the presence of arsenic. (c) Analysis of background region B shows the absence of arsenic.

For thin (~ 0.1-μm) films studied by TEM, lateral spatial resolution is determined by the diameter of the electron beam; small-diameter (<100-Å) probes are available in modern TEM and STEM instruments. Figure 17a shows a TEM image of a cross section through part of an NMOS TEM test pattern structure made with arsenic source-drain implantation through a thin thermal oxide.[31] The top edge of the dark band correlates with the top edge of the thermal oxide. XES analysis of the band, region A, using an electron probe with diameter about 100 Å, and analysis of the background region B, produced the spectra shown in Fig. 17b and c, respectively. The figure shows that the band is a region of oxide which contains arsenic.

The sensitivity and accuracy of quantitative analyses are determined by the efficiency of x-ray generation, interference from other peaks and from background, and by detector and other instrumental parameters. When all conditions are optimal, the sensitivity of XES is often better than 10^{-15} g on bulk samples (Ref. 30, p. 355) which corresponds to a gold particle with diameter about 100 Å or a silicon particle with diameter about 950 Å. On a thin-film TEM sample the sensitivity is described by the expression[32]

$$M_{min} = \frac{1}{P_A \tau J} \quad (12)$$

where M_{min} is the minimum detectable mass, P_A is a constant for a given element and detector geometry, τ is the counting time, and J is the current density of the electron beam. With a thermionic electron source operating at 100 keV, a beam current density of 20 A/cm^2, and a counting time of 100 s, M_{min} for silicon is calculated to be 2×10^{-20} g, which corresponds to a silicon particle with diameter 20 Å.

Quantitative XES analysis generally requires the use of standards. The main problem in such analysis is the determination of the proportionality constant k, which relates the peak height (or integrated counts) H to the concentration C of element i in matrix j:

$$C_{ij} = k_{ij} H \quad (13)$$

Refer to Refs. 6 and 30 for an excellent discussion of these topics for the case of XES during SEM analysis, and to Ref. 33 for the case of XES during TEM and STEM analysis.

Both energy-dispersive and wavelength-dispersive detectors are used for XES studies. Energy-dispersive detectors are lithium-drifted silicon diodes, which convert an x-ray photon to a voltage pulse. The detector is kept at the temperature of liquid nitrogen to reduce noise, and must be kept in a vacuum. A thin window of beryllium or mylar (or other low-Z material) is usually used to provide vacuum isolation from the SEM. The energy distribution of the voltage pulses is proportional to the incident photon energy, and is displayed on a screen or read into a computer for storage and analysis. The main advantages of an energy-dispersive detector are its ability to detect x-ray peaks over a broad spectral range simultaneously and its ease of operation. A major limitation is the absorption of low-energy x-rays by the thin window, which effectively prevents x-ray peaks from low-Z elements (fluorine and below) from reaching the detector. A second disadvantage of energy-dispersive detectors is their limited ability to discriminate between two adjacent x-ray peaks. Typical detector linewidths (full-width at half-maximum) are 150 to 170 eV. Although this energy resolution is acceptable in many analyses some cases require greater resolution. For example, using an energy-dispersive detector to determine the stoichiometry of a tantalum silicide film is difficult, because the silicon K_α peak is at 1.74 keV and the strong tantalum M_α peak is at 1.71 keV; the separation is only 31 eV.

Wavelength-dispersive detection is achieved by using a crystal analyzer which intercepts the emitted x-radiation. The analyzer is adjusted until maximum Bragg reflection occurs. The resulting signal is then detected by a gas proportional counter. Such an analyzer can only be used to detect one peak at a time and is, therefore, more tedious to operate than an energy-dispersive analyzer. However, this type of detector is attractive because of its high-energy resolution (~5 to 10 eV) and ability to detect low-Z elements (the analyzing crystal is directly exposed to the x-radiation and need not be isolated by a vacuum).

12.3.8 X-Ray Fluorescence (XRF)

A primary x-ray beam of sufficient energy can stimulate the emission of x-radiation from a bombarded sample (Fig. 12), and both qualitative and quantitative analysis may be obtained from the spectra. Methods for detection of x-ray spectra are similar to the case of x-rays generated by electron bombardment. Refer to Section 12.3.7 for details.

Three features of XRF and XES, however, are quite different. Insulating samples, such as oxides and polymers used in packaging, are often difficult to examine by XES, because they charge (negatively) or decompose during electron bombardment. These types of samples may be easily examined by XRF. The large width of the primary x-ray beam makes XRF useless for analyzing small features on a VLSI chip, but quite useful for large-area analyses. Finally, in XRF, the characteristic x-radiation originates from a sample volume deeper than the volume which emits x-rays during electron stimulation (XES), because of the greater attenuation of electrons. This deeper penetration must be considered when applying XRF to the study of multilayered structures.

12.3.9 X-Ray Photoelectron Spectroscopy (XPS, ESCA)

X-ray bombardment of a sample can stimulate the emission of core-level electrons if the incident x-ray energy is sufficiently high (Fig. 12). Letting E_0 be the energy of the incident x-radiation, E_b the binding energy of the emitted electron, $\Delta \phi$ the work function difference between the sample and spectrometer surfaces, and E_{xp} the energy of the emitted electron (photoelectron kinetic energy),

$$E_{xp} = E_0 - E_b - \Delta \phi \qquad (14)$$

with constant $\Delta \phi$ and E_0 in a given experiment, electrons of different binding energies give rise to separate peaks in the photoelectron spectrum. The application of this phenomenon to chemical analysis of the bombarded surface is called x-ray photoelectron spectroscopy (XPS); this procedure is also called electron spectroscopy for chemical analysis (ESCA).

The incident x-ray beam is usually generated by low-energy electron bombardment of an aluminum or magnesium anode. K_α radiation from magnesium has an energy of 1253.9 eV and a linewidth of 0.7 eV, and K_α radiation from aluminum has an energy of 1487.0 eV and a linewidth of 0.85 eV. Since $\Delta \phi$ is about 1 eV, photoelectron energies from aluminum or magnesium sources are sufficiently low so that escape depths are less than 50 Å. The two methods, XPS and AES, are therefore similar in providing chemical information from a region within a few monolayers of the surface.

The electron detection and analysis instrumentation used in XPS is similar and, in some cases, nearly identical to that used for AES. As in AES, ion milling is used to obtain chemical depth profiles, and the depth resolutions of XPS and AES are similar. Lateral spatial resolution is quite poor in XPS, however, because x-ray beams used have cross sections 1 to 2 mm in diameter.

XPS is often used as a complement or back-up to AES, because of the following three advantages. First, radiation sensitive material can be nondestructively studied, because the scattering cross sections for x-ray induced desorption and dissociation are significantly lower than the corresponding cross sections for electron bombardment. Second, insulators can be studied with less surface charging, since a neutral incident beam is used, and third, information on chemical bonding can be obtained from XPS data. The energy levels of core electrons are affected by the valence state and type of chemical bonding. The energy resolution of XPS peaks is typically about 0.5 eV, and since different chemical bonds often produce shifts in the binding energy by larger amounts, these shifts can be detected and the bond identified.[34]

12.4 CRYSTALLOGRAPHIC STRUCTURE AND MECHANICAL PROPERTIES

An important aspect of device materials and process development programs is the analysis of crystallographic and mechanical properties of films and substrates. These analyses include the determination of substrate orientations and the determination of the degree of preferred orientation and crystallite size in grown and deposited films, the identification of phases and determination of unit cell parameters, the identification of amorphous regions and characterization of crystallographic defects, and the measurement of film stress. Five types of analyses are considered: laser-reflectance measurements of wafer curvature used to compute film stress, Rutherford backscattering spectroscopy (channeling), x-ray diffraction (camera and diffractometer methods), transmission electron diffraction, and transmission electron microscopy. A summary of the degree of success of each method in solving problems relating to structure is given in Table 8.

12.4.1 Laser Reflectance (LR)

The stress of a film on a thick substrate, assuming uniform film thickness and otherwise isotropic film stress, is given as[35]

$$\sigma = \frac{E}{6(1-\nu)} \frac{D^2}{Rt} \qquad (15)$$

where E and ν are Young's modulus and Poisson's ratio, respectively, for the substrate, D and t are the substrate and film thicknesses, respectively, and R is the radius of curvature of the composite. By convention R is negative for a convex wafer surface (compressive film stress) and positive for a concave surface (tensile film stress). The quantity $E/(1-\nu)$ has the value 1.8×10^{12} dyn/cm² for {100} silicon.[36]

Table 8 Applicability of various analytical methods to solution of structure related problems

	XRD					
Type of problem	Camera	Diffracto-meter	TED	RBS (channeling)	LR	TEM
Phase identification	Good	Good	Good	N.A.	N.A.	N.A.
Preferred orientation	Good	Poor	Good	N.A.	N.A.	N.A.
Unit cell parameter	Fair	Good	Fair	N.A.	N.A.	N.A.
Presence of amorphous regions	Good	Good	Good	Good	N.A.	Good
Lattice location of impurity atoms	N.A.	N.A.	N.A.	Good	N.A.	N.A.
Substrate orientation	Good	Fair	N.A.	N.A.	N.A.	Poor
Crystallographic defect analysis	N.A.	N.A.	N.A.	N.A.	N.A.	Good
Film stress	Poor	Good	Poor	N.A.	Good	Poor

Note: N.A. means either that the method would be a very poor choice or that it cannot be used.

The radius R can be determined by measuring the deflection of a light beam reflected off the wafer surface as the wafer is moved a fixed distance. This method uses a collimated laser beam which reflects from a wafer surface and projects on a screen over 10 m away. The wafer is moved a known and fixed distance while being illuminated by the incident laser beam. Wafer curvature causes the image of the beam to appear shifted on the screen; the shift is magnified by the optical leverage of the system. A measure of the wafer translation x, the corresponding translation of the position of the reflected beam d, and the reflected beam path length L gives the radius of curvature R

$$R = 2L\frac{x}{d} \qquad (16)$$

The minimum detectable curvature for the case $L = 10$ m, $x = 7$ cm, and $d = 0.1$ cm is 1400 m, assuming simple wafer curvature. Unprocessed wafer surfaces are sometimes not sections of spheres but saddle shapes and other complex shapes that make detection of such large radii of curvature difficult. Because of this problem, film stress is best determined by measuring wafer curvature before and after film deposition (or removal), using approximately the same endpoints for wafer translation. The curvature used for film stress calculation is computed from these measurements by

$$\frac{1}{R_f} = \frac{1}{R_T} - \frac{1}{R_S} \qquad (17)$$

where R_f, R_T, and R_S are curvatures due to the film alone, composite, and substrate alone, respectively.

Coefficients of thermal expansion can be obtained by curvature measurements of the wafer at different temperatures. Film stress σ is usually resolvable into two components, σ_i and σ_{th}, which are the intrinsic stress and the stress introduced by differential thermal expansion of the film and substrate:

$$\sigma = \sigma_i + \sigma_{th} \qquad (18)$$

In Eq. 18,

$$\sigma_{th} = \frac{E_f}{1 - v_f} \int_{T_1}^{T_2} (\alpha_s - \alpha_f)\, dT \qquad (19)$$

E_f and v_f are Young's modulus and Poisson's ratio, respectively, for the film; α_s and α_f are the coefficients of thermal expansion of the substrate and the film, respectively. Assuming both coefficients of thermal expansion are constant over the temperature range,

$$\frac{d\sigma}{dT} = \frac{E_f}{1 - v_f}\Delta\alpha \qquad (20)$$

where $\Delta\alpha = \alpha_s - \alpha_f$. When α_s is known, α_f can be determined. This method has been used for determining values of α for various silicides.[37]

12.4.2 Rutherford Backscattering Spectroscopy (Channeling)

Rutherford scattering intensity is reduced by a factor of about 100 when the incident ion beam is coincident with a low index crystallographic axis or with a plane of high symmetry. In such a configuration the beam is "steered" down crystallographic channels by low-angle repulsive interaction with the cores of lattice atoms, and penetrates to a depth about ten times greater than depths achieved with off-axis (random) orientations. If the incident beam angle is changed slightly from the value for optimum channeling, the backscattered yield increases sharply as shown in Fig. 18a. Yield versus energy is plotted in Fig. 18b, and shows the strong attenuation of the yield when channeling conditions are exactly met. The ratio H_A/H, measured just to the left of the leading edge of the substrate signal in a spectrum such as shown in Fig. 15, is known as the minimum yield. Disturbances to crystalline perfection, such as those caused by interstitial atoms or linear defects, cause an increase in the yield for an aligned orientation. In the extreme case of an amorphous layer, the yield is identical to that from a random orientation of a perfect lattice.

The distribution of dopant atoms between interstitial and substitutional sites can be determined by taking random and aligned RBS spectra and comparing the minimum yield H_A/H for dopant and silicon signals. Dopant atoms positioned substitutionally are shielded by silicon atoms from the channeled beam, and H_A/H is the same for both silicon and for the dopant. As the dopant atoms begin to occupy interstitial positions, H_A/H rises proportionally, and the fractional interstitial component can be computed (see Ref. 23, p. 269). RBS (channeling) has been used to compute the distribution of arsenic implanted into silicon after various anneals.[38, 39]

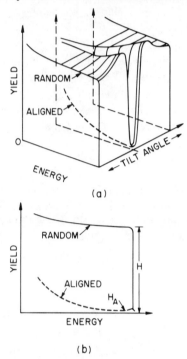

Fig. 18 (a) Three-dimensional representation of RBS yield as a function of tilt angle. (b) Typical yield plots for channeled (aligned) and random orientation. *(After Chu, Mayer, and Nicolet, Ref. 23.)*

12.4.3 X-Ray Diffraction (XRD)

An analysis of the angular position and intensity of x-ray beams diffracted by crystalline material gives information on the crystal structure (phase) of the material. Four types of x-ray diffraction methods have been found to be useful: the back-reflection Laue technique for determining wafer orientation, the Read camera technique, the Huber-Seemann-Bohlin camera technique, and diffractometer methods. All x-ray methods depend on establishing conditions that satisfy the Bragg requirement for diffraction of x-rays by a crystal lattice:

$$n\lambda = 2d \sin\theta \qquad (21)$$

where λ is the x-ray wavelength, d is the interplanar spacing, θ is the Bragg diffraction angle, and n is an integer giving the order of the diffraction. Diffraction occurs only when Eq. 21 is satisfied.

Substrate orientation can be quickly determined from a back-reflection Laue photograph. An incident beam of "white" radiation (the continuous unfiltered output of an x-ray tube) is collimated and strikes a wafer placed parallel to a photographic film. The incident beam passes through a hole in the film, and the interaction with the wafer produces a large number of diffracted beams. Diffraction occurs simultaneously from a large number of crystallographic planes and leaves an image of spots on the film. The spots lie on hyperbolae, and the distribution and symmetry of spots directly relate to the crystallographic orientation of the substrate. Back-reflection Laue patterns from {100}, {111}, and {110} silicon wafer surfaces are clearly distinguishable. A pattern from a wafer of unknown orientation can be quickly compared with standard patterns to determine the orientation.

A version of the classical x-ray powder diffraction camera that is particularly suited for analysis of thin films is the Read camera, illustrated in Fig. 19. The camera is a cylinder with a 5-cm radius and 13 cm high and has an entrance hole for the x-ray beam. Photographic film is laid against the inside wall of the cylinder. During sample irradiation by a monochromatic x-ray beam, a cone of diffraction is created for each set of diffracting planes in a polycrystalline film sample. The intersection of each cone with the film produces a curved line of intensity.[40] The cone axis is coincident with the incident beam direction, and the angle at the cone apex is four times the Bragg angle. A template that matches the camera dimensions can be laid over the developed photographic film to identify the lines. A line has an associated "d-value," which is the interplanar spacing which produced the diffraction. These "d-values" plus a semiquantitative measure of their intensity are usually sufficient to either identify the material by referring to standard x-ray crystallographic tables, or establish the spacings of a newly discovered phase. Interplanar spacings measured by this method are usually accurate to 0.01 Å.

The two features of the Read camera that make it particularly useful for thin-film analysis are the shallow angle of incidence, and the large photographic area. The shallow angle of incidence permits greater beam penetration into the sample film material than would be the case with normal incidence. The large photographic film

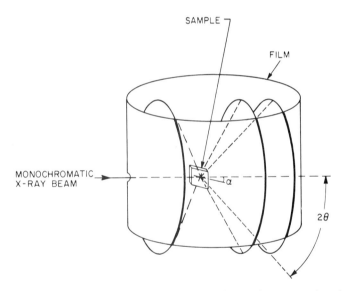

Fig. 19 Read x-ray diffraction camera method. A monochromatic x-ray beam passes through a hole in the camera and strikes a sample surface at an angle α, typically 15°. Diffraction cones which intersect the film produce lines of intensity (dark lines); the angle 2θ is twice the Bragg angle.

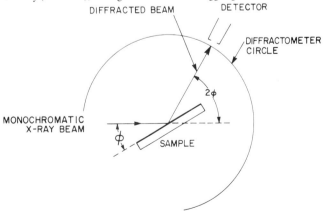

Fig. 20 X-ray diffractometer configuration. As the sample rotates over an angle ϕ the detector moves along a circumference an angular distance 2ϕ; a diffraction maximum occurs when ϕ is equivalent to a Bragg angle.

width permits information to be recorded on the degree of preferred orientation of the sample film. Preferred orientation has the effect of producing nonuniform cone intensities. Instead of curved lines of approximately constant intensity from a polycrystalline sample, preferred orientation introduces elongated spots and arcs. Most of the information on preferred orientation would be lost with a Huber-Seemann-Bohlin camera or with diffractometer methods.

The Huber camera with Seemann-Bohlin geometry is more sensitive to very thin films than is the Read camera. This improvement in sensitivity is due to decreasing the angle of incidence to a few degrees which permits more of the material to diffract. The sample is located on the wall of a cylinder, and a convergent monochromatic x-ray beam becomes incident over an elongated path on the sample surface. The camera is designed so that beams diffracted from the same d-spacing at the near and far edge of the sample both converge at the same point on the film.

More precise measurements of lattice parameters result from using a diffractometer geometry, shown in Fig. 20. A monochromatic x-ray beam strikes the film surface at an angle ϕ. The sample is slowly rotated and a detector simultaneously moves, at twice the rate, along the circumference of a circle with the same center as the sample. Diffraction maxima appear wherever ϕ coincides with a Bragg angle. A measure of 2ϕ is used to identify the interplanar spacing (d-values) of the crystalline material giving rise to those peaks. Data are usually fed into a strip chart recorder where the peak positions, intensities, and half-widths can be easily read. The identification of a phase on the basis of calculated lattice parameters and the determination of lattice parameters of a new phase both depend on the precision of the measurement of θ. The highest precision results from diffractometer measurements near $2\phi = 180°$ (see Fig. 20). Most diffractometer measurements can give precision better than 0.01 Å.

12.4.4 Transmission Electron Diffraction (TED)

Equation 21 establishes the basic condition for obtaining the diffraction of electrons as well as of x-rays. A typical transmission electron diffraction (TED) pattern is a set of concentric rings from polycrystalline material (see Fig. 28) or spots from a single-crystal sample (Fig. 21a). The rings are formed by diffraction cones intersecting the screen, analogous to the formation of diffraction arcs in a Read camera as previously described. TED patterns can be generated with a transmission electron microscope. In electron microscopes that permit diffraction patterns to be obtained from regions no smaller than about 1 μm in diameter, diffraction rings are continuous for crystalline sizes less than 100 Å, and become increasingly spotty as grain size increases. In the limiting case where only one crystallite contributes to the diffraction pattern, a single set of diffraction spots appear. Other TEM instruments, which include scanning transmission electron microscope (STEM) capability, can probe regions of a film smaller than 100 Å diameter; the shape of the diffraction pattern in this case is partly determined by the size, number, and shape of the crystallites within this small region, and by beam spot size. Microdiffraction studies of small areas are particularly useful in the study of individual layers within TEM samples of VLSI circuits.

The most common structural problem that occurs during TEM study of metallization layers or crystallographic inclusions is identifying unknown phases. These phases are identified by measuring the diameters of a number of diffraction rings of a pattern, and computing the corresponding interplanar spacings by using a pattern of a known standard taken under identical conditions. The phase is identified with techniques similar to those used to identify unknown phases by x-ray analysis.

Figure 21 shows how TED can be used to determine the orientation of a small silicon crystallite. One of the features produced during thermal oxidation of polysilicon is inclusions of silicon within the polysilicon oxide. Figure 21b shows the TED pattern from the inclusion shown in the micrograph of Fig. 21c. The TED pattern from the silicon substrate (Fig. 21a) is compared with the pattern from the inclusion. The comparison shows that the inclusion is oriented with a $\langle 110 \rangle$ axis normal to the wafer surface. This information was used to develop a model of inclusion formation.[41]

12.4.5 Transmission Electron Microscopy (TEM)

Information on structure is obtained from a transmission electron microscope by an analysis of electron diffraction patterns (see Section 12.4.4) and by an analysis of the microscope image (see Section 12.2.3).

12.5 ELECTRICAL MAPPING

In electrical mapping, an electron beam is used to locate regions in a device structure that differ in electrical activity, and in some cases, to measure the difference. Two mapping methods are used. In one, the energy distribution of secondary electrons, produced when an electron beam strikes a device sample, is influenced by the local

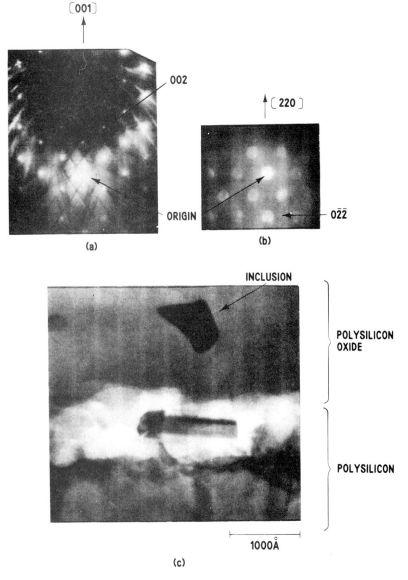

Fig. 21 Electron diffraction patterns (a) from silicon substrate and (b) from silicon inclusion within polysilicon oxide. (c) A TEM photo of the polysilicon, polysilicon oxide, and inclusion. The orientation of the inclusion was determined from the diffraction patterns.

surface potential, and the secondary electron flux reaching the detector reflects this potential. This phenomenon, essential to voltage contrast imaging, can be used to determine the potential of an element on a device surface. The other involves generation of charges in the device by the electron beam, and the collection of the charges

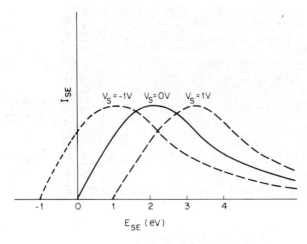

Fig. 22 Secondary electron current (I_{SE}) as a function of secondary electron energy (E_{SE}) for cases where the surface potential V_s is -1, 0, and $+1$ V.

via a capacitor or junction. Local changes in morphology, material properties, and junction electric field cause corresponding local modulations of the collected current.

This procedure is called EBIC (electron beam induced current) or charge-collection microscopy. Both types of electrical mapping can be performed with a scanning electron microscope; a transmission electron microscope can be used for EBIC studies.

12.5.1 Voltage Contrast Microscopy

The energy distribution of secondary electrons, produced when an electron beam strikes a surface, shows a peak at an energy near 5 eV (Fig. 5). This energy is sufficiently low that changes in surface potential on the order of 1 V have a significant effect on the energy distribution as illustrated in Fig. 22. In Fig. 22, a secondary electron distribution from a surface at zero potential is assumed to peak at 2.1 eV. A change of surface potential V_s of ± 1 V shifts the distribution right or left by an approximately proportional amount. Such shifts can be qualitatively observed for voltage contrast analyses of images, or quantitatively measured for voltage contrast analyses of surface potentials.

The secondary electron image in most SEM instruments is created by an extraction field of a few hundred volts that brings the electrons to a detector where a strong field attracts them to a scintillator surface. A light pipe transfers the signal to a photomultiplier tube. Not all electrons reach the scintillator. Local fields at the sample surface affect the trajectories of the secondary electrons and permit only those above a threshold energy, about 1 eV, to reach the detector. The total flux of electrons reaching the detector and, therefore, the intensity of the signal depend on the number of secondary electrons that exceed the threshold energy. This number is strongly affected by the surface potential as shown in Fig. 22. The minimum intensity difference that can be noted on a screen corresponds to a surface potential difference of about 1 V, which can be considered a practical limit.

Because of this sensitivity of the secondary electron image intensity to surface potential, an SEM can be applied to detect electrical discontinuities in conducting regions of VLSI circuits. Electrical discontinuity can be easily seen by applying a dc bias to a metallization, as shown in Fig. 23. The bias can be applied by using a movable mechanical probe inside the SEM chamber or by fixed contact to input pads in a packaged device.

A bias may also be applied as a pulse,[42] i.e., the device can be operated dynamically inside the SEM and the voltage contrast image obtained by stroboscopically pulsing the incident electron beam in synchronism with the circuit.[43] Time resolutions of 0.2 ps have been achieved in the stroboscopic voltage contrast mode.[44]

Circuit testing, analysis, and design verification often require the characterization

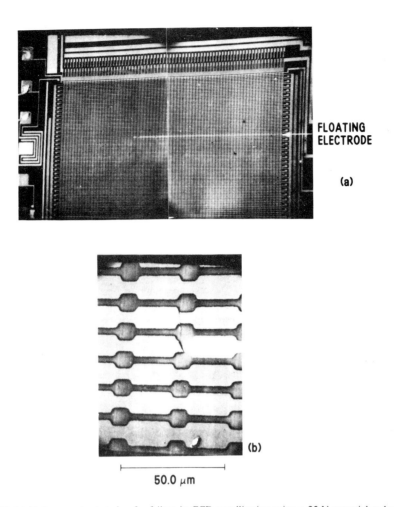

Fig. 23 (a) Voltage contrast study of a failure in CCD metallization using a 25-V potential to locate the failure. (b) Higher-magnification study shows the failure to be due to a loss in metallization caused by a mask error.

of waveforms at selected internal nodes of the circuit. Such measurements are usually made by placing a fine mechanical probe at these modes or at pads connected to them. As metallization linewidths approach 1 μm, placing a probe at a selected site and avoiding metallization damage during probing becomes increasingly difficult. A second problem with mechanical probes is that probe capacitance, generally larger than 0.1 pF, can interfere with the measurement by placing a parasitic capacitance on the node under test. These problems can be solved by using an electron beam as a small-diameter (~0.1-μm) probe; the beam does not capacitively load the circuit and can be easily positioned to small circuit elements.[43]

12.5.2 Electron Beam Induced Current (EBIC) Microscopy

An electron beam penetrating into a device can generate carriers which, in turn, can be collected as current to modulate the intensity of a CRT image. This mode of electron beam induced current (EBIC) or charge-collection scanning electron microscopy is useful in spatially locating device failure sites within a junction or capacitor region. Figure 24 shows the arrangement for EBIC imaging using a Schottky barrier, a pn junction, and a capacitor. A mechanical probe inside the SEM specimen chamber applies the bias. An electron beam current of 10 nA is usually sufficient. The choice of beam energy is a compromise between the need to penetrate to a depth that maximizes EBIC signal and the loss of spatial resolution with increasing sample penetration. Excessive beam energy can damage a device although this damage is annealable (see Section 12.3.1); such a consideration may be an additional factor in limiting

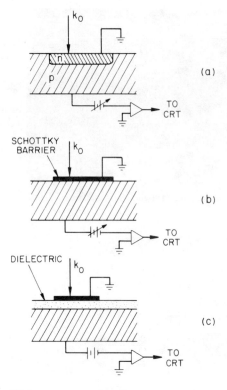

Fig. 24 Arrangement for EBIC measurement (a) on a np (or pn) junction, (b) on a Schottky-barrier junction, and (c) on a capacitor. The incident electron beam is k_0.

beam energy. For pn-junction or Schottky-barrier analysis, the beam energy is usually set high enough so that the electron range exceeds the thickness of the top electrode (or layers). This technique generates the highest electron-hole pair within or below the space-charge region. Capacitor analysis is somewhat more complicated. Useful EBIC information can be obtained by generating carriers in the field plate, oxide layer, or substrate, depending on the specimen geometry, applied bias, and beam energy. These points are discussed in the following paragraphs.

Useful information on junction leakage and breakdown sites is obtained by EBIC studies of Schottky barriers and pn junctions under zero or reverse bias conditions. The incident beam generates charges within silicon that diffuse to the space-charge layer where they are swept across by the field to contribute a current pulse. The current is given by

$$I_{EBIC} = \frac{I_0(E_0 - E_{BS})}{E_{eh}} \eta \qquad (22)$$

where I_0 and E_0 are the incident beam current and energy, respectively, E_{BS} is the energy lost due to backscattering, E_{eh} is the energy needed to create an electron-hole pair in silicon (3.6 eV), and η is the charge collection efficiency.[45] At an incident beam energy of 10 keV and with $E_0 - E_{BS}$ approximately $0.9 E_0$,

$$\frac{I_{EBIC}}{I_0} \sim 2500 \eta \qquad (23)$$

Since η is essentially unity in the space-charge charge region, EBIC currents can be large multiples of the incident beam current.

The efficiency of charge collection falls off with increasing electron beam penetration, due partly to recombinations that limit the lifetimes of minority carriers, and partly to loss of carriers that diffuse away from the junction. Recombination centers in the path of carriers diffusing to the space-charge layer decrease the current collected locally by the junction. These recombination centers are seen as regions of decreased EBIC signal. The magnitude of recombination efficiency of defects is characterized by the "defect strength"; the EBIC contrast, created by the defect, is a function of the defect strength, as well as the depth of the defect.[46] Although the defect strength of a dislocation is a function of the orientation of the dislocation,[47] defect strength appears to be dominated by the amount of impurity decoration at the crystalline defect.[48,49]

Figure 25a and c shows Nomarski interference contrast micrographs of epitaxial stacking faults on silicon substrates of two different orientations, and Fig. 25b and d shows the corresponding EBIC images; gold Schottky barriers were used to collect the EBIC current. The most obvious contrast features in these EBIC images are at the dislocations at the junction of stacking faults. The defect strengths of two opposite dislocations in the stacking fault tetrahedron on the {100} surface (Fig. 25b) are clearly larger than the defect strengths of the remaining two dislocations. The difference in defect strengths of the dislocations in the pyramidal stacking faults on the {111} surface can be clearly seen with the aid of the EBIC line scans (Fig. 25d).

EBIC studies of Schottky barriers and pn junctions have been used to reveal silicon defects, especially stacking faults, dislocations, and inhomogeneities segregated during crystal growth. Except for the case of analysis of defects resulting from pn-junction processing, Schottky-barrier junctions are preferred for EBIC defect analysis because of the ease in forming such junctions. Thin films of gold and aluminum are generally employed for forming Schottky barriers on p-type and n-type silicon, respectively.

Defects responsible for capacitor leakage and breakdown can also be spatially located with EBIC methods.[50] EBIC current in this case is controlled by four mechanisms: secondary electron emission from the field plate or substrate across an oxide interfacial barrier, current generation within the oxide, electron tunneling across an interfacial barrier, and displacement currents induced in a space-charge layer. Fowler-Nordheim tunneling begins to be important as a current generation mechanism at higher fields and as oxide films become very thin. EBIC microscopy can be used

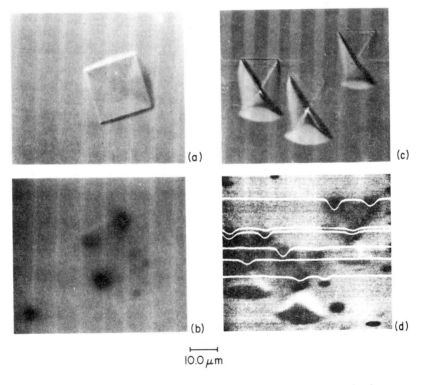

Fig. 25 Nomarski interference contrast micrographs of epitaxial stacking faults (a) on {100} silicon, and (c) on {111} silicon. EBIC photographs of the same regions made with gold Schottky barriers and a beam energy of (b) 15 keV, and (d) 12 keV. The line scans in (d) give a quantitative measure of the EBIC "defect strength" of the partial dislocations. The EBIC images were made at 0 V applied bias.

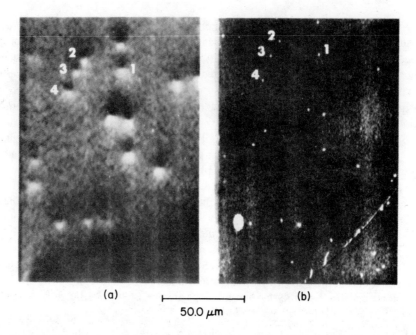

Fig. 26 (a) EBIC image of a capacitor, and (b) EBIC image of a Schottky barrier at the same site showing EBIC defect sites. The appearance of EBIC signals at the same locations (see, for example, sites 1–4) shows that the defect is within the substrate.

to reveal defects in thin oxide capacitors because of locally enhanced tunneling due to higher fields created at these sites.[51] In all cases defects create local disturbances of current, which create contrast in the image.

The displacement current mechanism is operative only for capacitors under reverse bias or very weak forward bias conditions which maintain a space-charge layer. Figure 26a shows an EBIC image taken with a beam energy of 8 keV of a capacitor which was biased at nearly 0 V. The capacitor had an oxide thickness of 250 Å and a 1000-Å phosphorus doped polysilicon field plate. The white-on-black features are EBIC signals due to uncanceled displacement currents at the defects.[52] After removing the field plate and oxide and depositing a thin aluminum film over this region, Schottky-barrier EBIC imaging shows the same sites, indicating that the defects are in the substrate. Subsequent TEM analysis showed the defects to be heavily decorated stacking faults.[52]

12.6 SUMMARY AND FUTURE TRENDS

The diagnostic procedures described are useful in solving problems occurring in a VLSI technology development program. Future developments will undoubtedly improve the capabilities of these methods as well as produce new analytical methods.

SEM stages are now available that can handle wafers 125 mm and larger in diameter; full translation coverage of large-diameter wafers including eucentric tilts with short working distances (for improved resolution) are expected to become available. Scanning electron microscopes will thereby become more generally useful for a variety of VLSI diagnostic studies.

As methods for preparing TEM samples and the use of TEM test chips become better known, it is expected that TEM methods will become an integral part of VLSI programs, and XES studies using TEM methods will be more widely used. EBIC studies will also gain in importance since an electrically active defect spatially located by EBIC is (often) most easily analyzed by TEM.

An increased ability to perform quantitative measurements through the use of more and better standards will be a major improvement in AES, XPS, and SIMS.

The application of these methods to VLSI problems is still in its early stages.

A major growth in the field of electrical mapping is expected to be in the increased use of electron beams for waveform analysis and voltage contrast imaging. Once the barrier to accepting this qualitatively different method of circuit diagnostics and testing is overcome, the advantages of having an automated, nondestructive, high-speed, low-capacitance probe method for circuit analysis should result in further growth.

REFERENCES

[1] G. Nomarski and A. R. Weill, "Application á la Métallographie des Méthodes interférentielles á Deaux Ondes Polarisées, *Rev. Metall.*, **52**, 121 (1955).

[2] D. C. Miller and G. A. Rozgonyi, "Defect Characterization by Etching, Optical Microscopy and X-Ray Topography," in S. P. Keller, Ed., *Handbook on Semiconductors*, North-Holland, New York, 1980, p. 217.

[3] T. E. Everhart and P. H. Hoff, "Determination of Kilovolt Electron Energy Dissipation vs. Penetration Distance in Solid Materials," *J. Appl. Phys.*, **42**, 5837 (1971).

[4] A. Joshi, I. E. Davis, and P. W. Palmberg, "Auger Electron Spectroscopy," in A. W. Czanderna, Ed., *Methods of Surface Analysis*, Elsevier, New York, 1975, p. 164.

[5] T. E. Everhart, O. C. Wells, and C. W. Oatley, "Factors Affecting Contrast and Resolution in the Scanning Electron Microscope," *J. Electron. Control*, **7**, 97 (1959).

[6] J. I. Goldstein, D. E. Newbury, P. Echlin, D. C. Joy, C. Fiori, and P. Lifshin, *Scanning Electron Microscopy and X-ray Microanalysis*, Plenum, New York, 1981.

[7] K. G. McKay, "Secondary Electron Emission," *Adv. Electron.*, **1**, 65 (1948).

[8] P. R. Thornton, *Scanning Electron Microscopy*, Chapman and Hall, London, 1968, p. 105.

[9] J. M. Aitken, "1 µm MOSFET VLSI Technology: Part VIII. Radiation Effects," *IEEE Trans. Electron Devices*, **ED-26**, 372 (1979).

[10] L. M. Welter and V. J. Coates, "High Resolution Scanning Electron Microscopy at Low Accelerating Voltages," in O. Johari, Ed., *Scanning Electron Microscopy/1974*, IITRI, Chicago, 1974, p. 59.

[11] Gareth Thomas, "Electron Microscopy at High Voltages," *Philos. Mag.*, **17**, 1097 (1968).

[12] R. B. Marcus and T. T. Sheng, *Transmission Electron Microscopy of Silicon VLSI Devices and Structures*, Wiley, New York, 1983

[13] J. M. Oblak and B. H. Kear, "Analysis of Microstructures in Nickel-Base Alloys: Implications for Strength and Alloy Design," in G. Thomas, R. M. Fulrath, and R. M. Fisher, Eds., *Electron Microscopy and Structure of Materials*, University of California Press, Berkeley, 1972, p. 566.

[14] T. T. Sheng and R. B. Marcus, "Advances in Transmission Electron Microscope Techniques Applied to Device Failure Analysis," *J. Electrochem. Soc.*, **127**, 737 (1980).

[15] C. C. Chang, "Analytical Auger Electron Spectroscopy," in P. F. Kane and G. B. Larrabee, Eds., *Characterization of Solid Surfaces*, Plenum, New York, 1974, Chap. 20.

[16] J. M. Morabito, "A First-Order Approximation to Quantitative Auger Analysis in the Range 100–1000 eV Using the CMA Analyzer," *Surf Sci.*, **49**, 318 (1975).

[17] D. R. Penn, "Electron Attenuation Lengths for Free-Electron-Like Metals," *J. Vac. Sci. Technol.*, **13**, 221 (1976).

[18] C. C. Chang. Bell Laboratories, unpublished data.

[19] C. C. Chang and D. M. Boulin, "Oxide Thickness Measurements up to 120 Å on Silicon and Aluminum Using the Chemical Shifted Auger Spectra," *Surf. Sci.*, **69**, 385 (1977).

[20] Paul Kruger, *Principles of Activation Analysis*, Wiley, New York, 1971, pp. 44–47.

[21] S. P. Murarka, Bell Laboratories, unpublished data.

[22] S. P. Murarka, "A Study of the Phosphorus Gettering of Gold in Silicon by Use of Neutron Activation Analysis," *J. Electrochem. Soc.*, **123**, 765 (1976).

[23] Wei-Kan Chu, James W. Mayer, and Marc-A. Nicolet, *Backscattering Spectroscopy*, Academic, New York, 1978.

[24] R. Schutz, Bell Laboratories, unpublished data.

[25] G. Mezey, E. Kotai, T. Nagy, L. Lohner, A. Manuba, J. Gyulai, V. R. Deline, C. A. Evans, Jr., and R. J. Blattner, "A Comparison of Techniques for Depth Profiling Oxygen in Silicon," *Nucl. Instrum. Methods* **167**, 279 (1979).

[26] Peter Williams, R. K. Lewis, Charles A. Evans, Jr., and P. R. Hanley, "Evaluation of a Cesium Primary Ion Source on an Ion Microprobe Mass Spectrometer," *Anal. Chem.*, **49**, 1399 (1977).

[27] J. A. McHugh, "Secondary Ion Mass Spectrometry," in A. W. Czanderna, Ed., *Methods of Surface Analysis*, Elsevier, New York, 1975, p. 223.

[28] Vaughn Deline, Chas. Evans. Associates, San Mateo, California, private communication.

[29] H. A. Kramers, "On the Theory of X-ray Absorption and of the Continuous X-Ray Spectrum," *Philos. Mag.*, **46**, 836 (1923).

[30] E. P. Bertin, *Principles and Practice of X-ray Spectrometric Analysis*, Plenum, New York, 1970, p. 27.

[31] E. K. Kinsbron, W. Fichtner, T. T. Sheng, and R. B. Marcus, Bell Laboratories, to be published.
[32] D. C. Joy and D. M. Maher, "Sensitivity Limits for Thin Specimen X-ray Analysis," in O. Johari, Ed., *Scanning Electron Microscopy/1977*, IITRI, Chicago, 1977, Vol. 1, p. 325.
[33] J. I. Goldstein, "Principles of Thin Film X-ray Microanalysis," in J. J. Hren, J. I. Goldstein, and D. C. Joy, Eds., *Introduction to Analytical Electron Microscopy*, Plenum, New York, 1979, Chap. 3.
[34] C. R. Brundle and A. D. Baker, *Electron Spectroscopy*, Academic, New York, 1977, Vols. 1–4.
[35] R. W. Hoffman, "The Mechanical Properties of Thin Condensed Films," in G. Haas and R. E. Thun, Eds., *Physics of Thin Films*, Academic, New York, 1966, Vol. 3, p. 211.
[36] W. A. Brantley, "Calculated Elastic Constants for Stress Problems Associated with Semiconductor Devices," *J. Appl. Phys.*, **44**, 534 (1973).
[37] T. F. Retajczyk and A. K. Sinha, "Elastic Stiffness and Thermal Expansion Coefficients of Various Refractory Silicides and Silicon Nitride Films," *Thin Solid Films*, **70**, 241 (1980).
[38] C. E. Christodoulides, R. A. Baragiola, D. Chivers, W. A. Grant, and J. S. Williams, "The Recrystallization of Ion Implanted Silicon Layers. II. Implant Species Effect," *Radiat. Eff.*, **36**, 73 (1978).
[39] D. G. Beanland and J. S. Williams, "The Damage Dependence of the Epitaxial Regrowth Rate During the Annealing of Amorphous Silicon Formed by Ion Implantation," *Radiat. Eff.*, **36**, 15 (1978).
[40] B. D. Cullity, *Elements of X-Ray Diffraction*, Addison-Wesley, Reading, Mass., 1978, pp. 96–99.
[41] R. B. Marcus, T. T. Sheng, and P. Lin, "Polysilicon/SiO_2 Interface Microtexture and Dielectric Breakdown," *J. Electrochem. Soc.*, **129**, 1282 (1982).
[42] J. K. Rodman and J. T. Boyd, "Examination of CCD, Using Voltage Contrast with a Scanning Electron Microscope," *Solid State Electron.*, **23**, 1029 (1980).
[43] Peter Fazekas, Hans-Peter Feuerbaum, and Eckhard Wolfgang, "Scanning Electrom Beam Probes VLSI Chips," *Electron.*, **54**, 14, July 14, 1981, p. 105.
[44] T. Hoskowa, H. Fujioka, and K. Ura, "Generation and Measurement of Subpicosecond Electron Beam Pulses," *Rev. Sci. Instrum.*, **49**, 624, (1978).
[45] H. J. Leamy, L. C. Kimmerling, and S. D. Ferris, "Silicon Single Crystal Characterization by SEM," in O. Johari, E., *Scanning Electron Microscopy/1976*, IITRI, Chicago, 1976, Vol. 2, p. 529.
[46] C. Donolato, "Contrast Formation in SEM Charge-Collection Images of Semiconductor Defects," in O. Johari, Ed., *Scanning Electron Microscopy/1979*, IITRI, Chicago, 1979, Vol. 2, 257.
[47] M. Kittler and W. Seifert, "On the Characterization of Individual Defects in Silicon by EBIC," *Crystal Res. Technol.*, **16**, 157 (1981).
[48] R. B. Marcus, M. Robinson, T. T. Sheng, S. E. Haszko, and S. P. Murarka, "Electrical Activity of Epitaxial Stacking Faults," *J. Electrochem. Soc.*, **124**, 425 (1977).
[49] H. Blumtritt, R. Gleichmann, J. Heydenreich, and H. Johansen, "Combined Scanning (EBIC) and Transmission Electron Microscopic Investigations of Dislocations in Semiconductors," *Phys. Status Solidi*, A, **55**, 611 (1979).
[50] W. R. Bottoms, P. Roitman, and D. C. Guterman, "Electron Beam Imaging of the Semiconductor-Insulator Interface," *CRC Critical Reviews in Solid State Sciences*, **5**, 297 (1975).
[51] P. Lin and H. Leamy, *Appl. Phys. Lett.*, to be published.
[52] P. S. D. Lin, R. B. Marcus, and T. T. Sheng, "Gate Oxide Leakage and Breakdown Caused by Oxidation Induced Decorated Stacking Faults," *J. Electrochem. Soc.*, to be published.

Physical Versus Logical Fault Model in MOS LSI Circuits: Impact on Their Testability

J. GALIAY, Y. CROUZET, AND M. VERGNIAULT

Abstract—At the end of an IC production line, integrated circuits are generally submitted to three kinds of tests: 1) parametric tests to check electrical characteristics (voltage, current, power consumption), 2) dynamic tests to check response times under nominal operating conditions, and 3) functional tests to check its logical behavior.

This paper, dealing with functional testing of MOS monochannel large-scale integrated circuits, is divided into three parts. The first part is devoted to the analysis of the failure mechanisms for such circuits, the second to the generation of test sequences for these observed failures, and the third to the definition of a set of layout rules enabling to improve the testability of such circuits.

Index Terms—Failure characterization, fault models, testability improvement, testing procedures, test sequences generation.

INTRODUCTION

At the present time, testing techniques are generally based on the assumption that all failures may be modeled by stuck-at-0, stuck-at-1 logical faults associated with the logic diagram of the circuit to be tested. As soon as the integration density increases, it appears that this hypothesis is less and less sound. This paper tries to solve this problem and is divided into three parts.

The first part is devoted to the definition of fault assumptions motivated by the physical origin of failures. This has been done by direct inspection of 4-bit microprocessor chips. It appears that a great majority of physical failures are shorts and opens. Furthermore, the commonly used gate level model does not allow one to account for the effect of these common physical failures.

The second part deals with the generation of test sequences for these observed failures.

The third part describes layout rules which will facilitate testing procedures. These rules are intended to decrease the number of possible kinds of failures and to avoid the occurrence of those which are not easily testable. If the design adheres to these rules, then gate level models again become usable and are able to accurately represent the effects of formerly encountered physical failures.

I. FAILURE MODE ANALYSIS FOR MOS LSI [1]

The problem of fault test generation can be formulated as follows. Given a description of the circuit and a list of faults, one must derive a sequence of input vectors, as short as possible, enabling the detection of an eventual presence in the circuit of some faults in the list. This detection must be ensured by observing only the primary outputs of the circuit. It then appears that the nature of the considered list of faults is of great influence with respect to test sequence generation. The more these faults are related to the physical nature of the circuit, the higher the quality of the test, but as a general consequence, the more laborious is the generation of a test sequence. So the list of faults must be carefully selected in order to satisfy the contradictory requirements of sufficient failure coverage and easy test generation.

In practice, rather than considering individually each kind of physical failure of the circuit, it is customary to treat a more general fault model which can represent all of these failures. In the work which has been devoted to test sequence generation, the most often considered fault model is that of stuck-at-0 or stuck-at-1 of any connection of the logic diagram representing the circuit to be tested. Even if that model is relatively satisfactory for small-scale integration, one can doubt its validity for large-scale integrated circuits. To answer such a question, we have first tried to carry out a characterization of LSI failure modes by analyzing a set of failed circuits. The application circuit is a 4-bit microprocessor designed by EFCIS and described in [2].

Due to the size of this circuit, to detect failures by direct observation of the chip is a very complex task. Consequently, to reduce the region of investigation, an initial step aimed at a prelocalization of the failure was introduced. The specially designed test sequence is hierarchically organized using a "start small" approach.

The second step of the circuit analysis consists of a direct observation of the chip in the region determined by the prelocalization sequence. The different techniques which were used are

1) parametric measurements which give information about processing quality,
2) research of the schmoo plot domain (i.e., the domain of correct operation) for different parameters such as temperature, frequency, supply voltage,
3) visual inspection with an optical microscope,
4) potential cartography with a scanning electron microscope,
5) electrical analysis of the circuits nodes by placing probes directly onto the chip.

This method has been applied to a set of 43 failed circuits; the two main results obtained from this study are: 1) it appears that the failures are randomly distributed and no block is more susceptible than any other, and 2) concerning the nature of the failures, Fig. 1 indicates the observed failure modes. They consist mainly of shorts and opens at the level of the metallizations and at the level of the diffusions. It should be particularly noted here that no short was observed between metallization and diffusion.

For 10 percent of the cases, a logical fault was clearly established, but no physical failure could be observed.

For 15 percent of the cases, the circuit presented a very large imperfection (e.g., a scratch from one side to the other of the chip) which can be considered as "insignificant" for test purposes because such faults can be easily detected by any test sequence.

Short between metallizations	39%
Open metallization	14%
Short between diffusions	14%
Open diffusion	6%
Short between metallization and substrate	2%
Inobservable	10%
Insignificant	15%

Fig. 1. Observed failure modes.

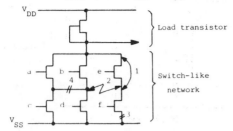

Fig. 2. Failure examples in an MOS gate.

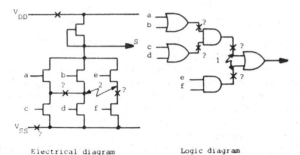

Fig. 3. Relations between electrical and logic diagrams.

II. BASIC CONSEQUENCES OF THE FAILURE MODES ON THE CIRCUIT MODEL: GENERATION OF TEST SEQUENCES FOR SHORTS AND OPENS

A. Basic Consequences

Concerning test sequence generation and fault simulation, the results of the failure mode analysis have two very important consequences.

1) First, all failures cannot be modeled by stuck-at faults. This can be clearly illustrated by the following example. Fig. 2 represents the electrical diagram of a MOS gate on which two possible shorts (numbered 1 and 2) and two possible opens (numbered 3 and 4) are indicated. Short number 1 and open number 3 can, respectively, be modeled by a stuck-at-1 at input e and by a stuck-at-0 at input e (or input f or both). On the other hand, short number 2 and open number 4 cannot be modeled by any stuck-at-fault because they involve a modification of the function realized by the gate. For the same reason, a short between the outputs of two gates cannot be modeled by any stuck-at faults.

2) The second consequence concerns the representation of the circuit. Taking into account physical failures such as shorts and opens implies consideration of the actual topology of the circuit. This leads to the rejection of the representation of the circuit by a logic diagram in favor of an electrical one because the former does not constitute a real model of the physical circuit. Some connections of the real circuit are indeed not represented on the logic diagram and, inversely, some connections appearing on the logic diagram do not exist in the physical circuit. As an example, Fig. 3 represents the logic and electrical diagrams of the same gate; the indicated faults in each diagram are those which cannot be represented on the other or even cannot occur. For instance, short number 2, which is physically possible, cannot be represented on the logic diagram, and short number 1 in the logic diagram has no physical meaning.

Consequently, all methods for test sequence generation and fault simulation based on a stuck-at fault model at the logic diagram level are not well adapted. A possible new approach for fault simulation may be to introduce short failures or, better, to work directly with the transistor diagram. For test sequence generation, it is necessary to use a new method accounting directly for the failures at the gate and blocks levels.

B. Test Sequence Generation at the Gate Level

For failure detection at the level of a gate, we first need to define the notion of a *conduction path* that we will use later. MOS technology enables the realization of complex gates including several cascaded AND/OR basic functions (Fig. 2). Schematically, such a gate can be divided into two parts: a load transistor and a set of "command" transistors which can be considered as a switch-like network.

The switch-like network constitutes the active part of the gate. It permits, by applying convenient input patterns, the realization of a set of conduction paths between the output node and the V_{SS} power supply node. We will say that a conduction path is activated when all of its command transistors are on; inversely, a conduction path is blocked when at least one of its command transistors is off. At the level of the whole gate, when one or more conduction paths between the output node and the V_{SS} node is activated, the output of the gate is at the V_{SS} potential, i.e., logical state 0; inversely, when all conduction paths between the output node and the V_{SS} node are blocked, the output of the gate is at the potential V_{DD}, i.e., logical state 1.

	a	b	c	d	e	f	activated path
T1	1	0	1	0	0	–	a c
T2	1	0	0	1	0	–	a d
T3	0	1	1	0	0	–	b c
T4	0	1	0	1	0	–	b d
T5	0	0	–	–	1	1	e f

Fig. 4. Test sequence for opens in gate of Fig. 2.

1) Opens: An open in the switch-like network, depending on its location, involves the removal of one or more conduction paths. In order to detect such an open, two conditions are required: 1) activation of at least one of the conduction paths between the output node and the V_{SS} node which pass through the open equipotential, and 2) blocking of all the conduction paths between the output node and the V_{SS} node which do not pass through the open equipotential.

When these requirements are fulfilled, if the considered open is not present, at least one conduction path is really activated and the output of the gate is at logical state 0, but if this open is present, no conduction path is really activated and the output of the gate is at logical state 1. The eventual presence of the open is indeed observed at the output of the gate.

In order to obtain a test sequence which detects all possible opens in the switch-like network, a *systematic* procedure deduced from the general graph theory is given in [3]. It consists of first listing all the conduction paths between the output node and the V_{SS} node and then successively activating one of these conduction paths and simultaneously blocking all the others. As an example, Fig. 4 gives a sequence of five vectors obtained for the gate of Fig. 2.

The sequence obtained with such a systematic procedure is generally redundant, but minimizing its length is only feasible if we have information about the actual layout of the command transistors. For instance, if the layout is exactly the one of Fig. 2, only the three tests $T2$, $T3$, and $T5$ are required to detect all opens of the switch-like network.

2) Shorts: A short between any two nodes in the switch-like network involves the creation of one or more conduction paths. Two conditions are required in order to detect a short between nodes i and j: 1) activation of at least one conduction path between the output node and the i (respectively, j) node and at least one conduction path between the j (respectively, i) node and the V_{SS} node, and 2) blocking of all other conduction paths of the network.

When these two conditions are realized, if nodes i and j are shorted, the output node is at logical state 0, but if there is no short, the output node is at logical state 1: the eventual presence of the short is indeed observed at the output of the gate. For a given short, there generally exist several test vectors enabling its detection; Fig. 5 gives, for instance, the set of six tests which enable detection of short number 2 of Fig. 3. In order to obtain a complete test sequence for all shorts of the switch-like network, it is first necessary to determine in this way the set of test vectors for each of them, and then to research a minimal cover enabling detection of all these shorts. If n is the number of nodes of the network, this minimal cover presents a maximum number of $\binom{2}{n}$ vectors.

a	b	c	d	e	f
1	0	0	0	0	1
0	1	0	0	0	1
1	1	0	0	0	1
0	0	1	0	1	0
0	0	0	1	1	0
0	0	1	1	1	0

Fig. 5. Set of test vectors detecting short number 2 of Fig. 3.

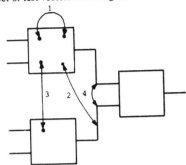

Fig. 6. Shorts in a block structure.

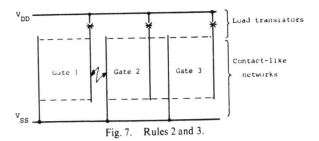

Fig. 7. Rules 2 and 3.

C. Test Sequence Generation at the Block Level

After considering gate level tests, we will now approach testing at the level of a block of several interconnected gates. Two specific problems appear.

1) The first problem is related to controllability and observability. Assuming that a complete test sequence has been determined for each gate of the block, we must now apply to each gate its test sequence even if all its inputs are not easily controllable and/or all its outputs are not easily observable. This can be done by using a path sensitization method based on propagation to primary outputs of the block and consistency to primary inputs [4].

2) The second problem is related to the additional failures introduced by the block structure itself.

   a) In addition to the opens in the gates, the block structure introduces opens in interconnections between gates. Such an interconnection always connects the output of a gate to the gates of one or more command transistors of other gates. The open of such a connection thus leads to a floating gate potential of the transistors which are located after the open.

In static operation, the leakage current is sufficient to set that potential to V_{SS}, and the transistors concerned are therefore always off. The open will indeed be detected by the test sequence of the gates including these transistors if a low testing rate is assumed.

   b) The problem concerning shorts is more difficult. Fig. 6 represents all types of shorts occuring in the block structure. In addition to the shorts inside a gate (1), we find shorts between a connection inside a gate and the output of this or another gate (2), shorts between two internal connections belonging to two different gates (3), and shorts between the outputs of two different gates (4).

Each of these shorts involves a specific malfunction (introduction of asynchronous sequential loop, modification of the realized function, introduction of analog behavior, . . .). In order to determine test sequences, they must be individualy analyzed and, as the possibilities of shorts are very numerous in LSI circuits, this requires a very large amount of work.

Consequence: Thus, it seems more realistic to simplify the problem by reducing the number of potential shorts and preferentially removing those which are the most difficult to test, even if this leads to a chip area increase. Such an approach (which will be presented in the next section) is based on restricting the circuit layout by a set of rules concerning relative arrangement of the gates at the block level and relative arrangement of nodes at the gate level.

III. Improvement of Circuit Testability

The layout rules which will improve circuit testability are based on a set of failure hypotheses that we will first justify by an analysis of the manufacturing process.

A. Manufacturing Process Analysis

1) Schematically, a basic MOS integrated chip consists of three levels of interconnection (assuming an aluminum gate): 1) a lower level of connections made by diffusions in an insulating substrate, 2) an upper level of connections made by metallizations, and 3) a medium level of oxide insulating the two previous levels and presenting two kinds of discontinuities: holes which enable contact between a metallization of the upper level and a diffusion of the lower level and thindowns which correspond to transistor gates.

The realization of the diffusions and then of the metallizations each requires selective masking of very precise regions on the surface of the chip. The inherent failure mode of such a process consists of diffusing (or etching) regions which do not have to be diffused (or etched) and, inversely, of not diffusing (or etching) regions which have to be diffused (or etched). On a manufactured chip, such failures involve only shorts between diffusions or metallizations and opens of diffusions or metallizations.

The growth of thick and thin oxide levels uniformly over the whole surface of the chip implies homogeneous levels with relatively few defects. Due to its thickness, the thick oxide constitutes a very good insulator between the diffusion level and the metallization level and, as a consequence, shorts between a metallization and a diffusion are very unlikely to occur. Concerning the thin oxide, two kinds of failures can occur: local thindowns enabling breakdown by electrostatic discharge, and local contamination involving threshold voltage drift for the corresponding transistor. Electrostatic breakdown affects mainly input and output buffers where it can be detected by parametric testing, and more rarely affects transistors in the middle of the chip. Threshold voltage drift is a gradual phenomenon appearing during aging and implying, at the logical level, that the concerned transistor is either off or on. This has the same effect as an open of the drain or source diffusion or a short between these two diffusions.

Lastly, for pin holes in the oxide, the only possible failure consists of a bad contact between the diffusion and the metallization and acts as an open of any one of these connections.

2) According to these results, an incomplete but satisfactory fault coverage will be ensured for monochannel MOS integrated circuits if the two following failure assumptions are retained: 1) all possible failures consist of opens of diffusions or metallizations and shorts between two neighbouring diffusions or metallizations, and 2) there is no short between a metallization and a diffusion.

As a general remark, it should be noted that this evaluation qualitatively agrees with the experimental results described in Section I.

B. Rules for Improving Testability

1) Block Level: Using the failure assumptions defined above, we will first define layout rules governing the relative arrangement of gates.

Rule 1 is based on failure hypothesis 2; it is intended to avoid shorts between an internal connection between the gates (short of type 2 in Fig. 6).

Rule 1: Make all internal gate connections entirely with diffusions and all interconnections between gates entirely with metallization.

Rules 2 and 3 are intended to control the short possibilities between two internal connections of two different gates (short of type 3 in Fig. 6).

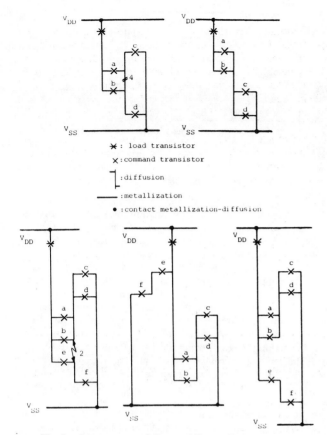

Fig. 8. Improvement of the testability at the gate level.

Rule 2: Arrange all internal gate connections (which are made by diffusions according to rule 1) inside a domain whose external limits are either the output diffusion or a V_{SS} diffusion of this gate (the latter diffusion can be commonly used by two neighboring gates).

Rule 3: Arrange any two neighboring gates to ensure that, if the output diffusion of the one constitutes one of its external limits, then it adjoins a V_{SS} diffusion of the other gate.

Fig. 7 illustrates rules 2 and 3. All connections of each gate are enclosed in a domain bounded by the vertical lines representing diffusions and the horizontal dotted lines. With such a layout, the only possible short between two gates, for instance, gates numbered 1 and 2 on the figure, always involves the output diffusion of the first and a V_{SS} diffusion of the second. This short can be modeled by a stuck-at-0 fault of the output of the former.

Rules 4 and 5 are intended to control the short possibilities between the outputs of two different gates (short of type 4 in Fig. 6).

Rule 4: Arrange the gates of the block along a given direction in increasing level order according to their logic level in the block.

Rule 5: Arrange the interconnections between the gates (which are realized by metallizations according to rule 1) in order to avoid the shorts which can introduce asynchronous sequential loops. For instance, given three interconnections *A, B,* and *C,* if a short between *A* and *B* leads to a loop and shorts between *A* and *C* on one hand and *B* and *C* on the other hand do not lead to a loop, then *A* and *B* can be isolated by placing *C* between them. As a practical matter, this rule is not very systematic, and therefore is not as easy to apply as the previous one.

2) Gate Level: In the same way, it is possible to reduce the probability of failures which cannot be modeled by stuck-at faults by taking some precautions during the implementation of a gate, e.g., open number 4 in Fig. 8(a) which cannot be modeled by a stuck-at fault is physically impossible for the implementation given in Fig. 8(b).

A similar reasoning can be applied to the only short that cannot be modeled by a stuck-at fault (i.e., short number 2 for the gate of Fig. 2). Among the last three layouts represented in Fig. 8, the first [Fig. 8(c)] permits the short to occur, but the other two prevent its occurrence either because the diffusions concerned are separated by the output diffusion [Fig. 8(d)] or simply because they are farther apart [Fig. 8(e)].

C. Test Strategies

According to all the rules that we have just defined, two different test strategies can be considered.

S1): The first consists of applying only the layout rules related to the relative arrangement of the gates. With such a topology, the possible failures of a block are: 1) shorts and opens inside the gates which cannot all be modeled by stuck-at faults, 2) opens of gate interconnections which can all be modeled by stuck-at faults, and 3) shorts between two gate interconnections which cannot induce asynchronous sequential loops and can be modeled by OR wiring of the two shorted connections.

The determination of a test sequence for the whole block can thus be divided into three steps.

1) Determine a complete test sequence for each gate of the block with the method described in Section II using the transistor diagram of this gate.

2) Determine a test sequence at the level of the whole block, applying to each gate its own test sequence. This can be done by using a path sensitization method on the logic diagram of the block. Notice that this sequence also detects all interconnection opens.

3) The test sequence obtained after step 2) will generally also detect some shorts between interconnections. For each still undetected short, a specific test vector must be added to that sequence. Using the logic diagram of the block, this test vector can be derived by first setting any one of the shorted interconnections at logical state 0, the other at logical state 1, and then propagating the logical state of the latter interconnection to an observable output of the block.

S2): The second test strategy consists of simultaneously applying all the rules concerning the relative arrangement of the gates and all the rules concerning the relative arrangement of the connections inside the gates. At the failure level, this strategy differs from the previous one by the fact that all shorts and opens inside the gates can be modeled by stuck-at faults. For test sequence generation, this offers the advantage of saving step 1) of the previously mentioned procedure, thus reducing computation time and avoiding the use of the transistor diagram. Test sequence generation can therefore be realized in two steps.

1) Determination on the logic diagram of a test sequence for stuck-at faults of the connections of that diagram.

2) Identical to step 3) above.

D. Application

The application of these layout rules does, of course, involve, for a given function, an increase of the chip area. To evaluate this approach, these layout rules were applied to a set of basic building blocks of the application circuit. As a general remark, there is a large difference between blocks with a relatively regular layout such as programmable logic arrays (PLA) and the blocks with a random layout.

For PLA's, due to the array nature, the relative arrangement of the gates is fixed and agrees with all the layout rules that we have defined. For such blocks, it is therefore easy to determine test sequences with satisfactory fault coverage.

The situation is more difficult for random logic. Although it is always feasible to apply all the rules, the relative freedom offered by some of them means that, for a given function, several possible layouts are possible. Each, however, has very different areas, and determining the minimal one is a complex task. In the present state of the art, no design tool exists that enables an automatic solution to such an optimization problem, and the search for a solution is essentially based on the designer's ability. So in order to reduce the design delays required for a circuit, two methods can be retained, depending on the nature of the block. For the specific blocks appearing only on one particular chip (customer design), one can systematically apply all layout rules without optimization. Inversely, for the basic blocks appearing on many chips (flip-flops, registers, counters, RAM, etc.), a greater optimization can lead to a block library which will then be used by the designer for future circuits.

These general remarks can be illustrated by a particular block in the application circuit: a master-slave flip-flop used in the realization of registers or counters. Because conventional layout rules do not consider testing requirements, no preestablished test sequence can be formed for 30 percent of the shorts between two metallizations and for 28 percent of the shorts between two diffusions. This situation arises from the fact that the above failures introduce analog behavior. Systematic application of all layout rules leads to a layout for which all shorts and opens can be detected by a very simple test sequence (Set-Reset-Set), but with an area increase of 30 percent after optimization.

This area increase does not, however, apply to the whole circuit because at a chip level we have to account for the following factors: 1) the I/O buffers and the interconnections take up a nonnegligible area which is not influenced by the application of these rules, and 2) certain blocks (e.g., PLA) have an arrangement such that the respect of the layout rules leads to a very slight or even negligible increase.

IV. Conclusion

It is recognized that the results presented in this paper are specific because they refer to one of the presently available IC technologies. However, disregarding this particular technology, one can retain the procedure which has been followed and which can be reproduced for any circuit realized with any technology. In this area, notice that [5] deals with fault modeling for CMOS technology.

To test a circuit, the first step must consist of an analysis of the failure mechanisms of this circuit to obtain information about their nature and their probability. Then in order to facilitate test sequence generation, it is essential to derive a general model rather than to individually consider all types of individual failures. However, as manufacturing processes become more and more sophisticated, it appears that the stuck-at faults model, very often used because of its practical interest, will cover a more and more reduced part of the failure modes. One can now thus adopt two different approaches: the first consists of defining a specific test generation method taking directly into account the failures of the circuit, and the second consists of submitting the layout of the circuit to a set of rules in order that all the failures be covered by the stuck-at fault model. As the first solution will generally lead to very great complexity, the second one seems more realistic for most cases, although it implies layout constraints and an increase in chip area. The study we have carried out shows that this second approach appears to be quite efficient.

Acknowledgment

The authors would like to thank Prof. A. Costes, Dr. M. Diaz, and Dr. C. Landrault of LAAS, and also P. Rousseau and X. Messonnier of EFCIS for their helpful comments, suggestions, and assistance.

References

[1] Y. Crouzet, J. Galiay, C. Landrault, P. Rousseau, and M. Vergniault, "Definition and design of easily testable or self-testing LSI circuits" (in French), Contract Rep. DRET 77/008, LAAS Publ. 1787, July 1978.

[2] Y. Crouzet and C. Landrault, "Design of self-checking MOS-LSI circuits: Application to a four-bit microprocessor," this issue, pp. 532-537.

[3] J. Galiay, "Design of easily testable LSI circuits" (in French), Docteur de Spécialité thesis, Univ. Paul Sabatier, Toulouse, France, Nov. 15, 1978, available from LAAS.

[4] J. P. Roth, W. G. Bouricius, and P. R. Schneider, "Programmed algorithms to compute tests to detect and distinguish between failures in logic circuits," *IEEE Trans. Comput.*, vol. C-16, pp. 567-580, Oct. 1978.

[5] R. L. Wadsack, "Fault modeling and logic simulation of CMOS and MOS integrated circuits," *Bell Syst. Tech. J.*, vol. 57, pp. 1449-1473, May-June 1978.

Fault Modeling for Digital MOS Integrated Circuits

JOHN P. HAYES, SENIOR MEMBER, IEEE

Abstract— A new fault modeling technique aimed at efficient simulation and test generation for complex digital MOS IC's is described. It is based on connector-switch-attenuator (CSA) analysis, which employs purely digital models of switching transistors, resistive/capacitive elements, and their associated signals. The use of CSA networks to model the digital behavior, both static and dynamic, of MOS circuits is reviewed. It is shown that most physical failure modes in such circuits, including short-circuit, open-circuit, and delay faults, can be modeled more efficiently by CSA models than by conventional approaches. A generalized single stuck-line (GSSL) fault model is suggested as a uniform and practical method for fault representation.

I. INTRODUCTION

THE TRADITIONAL approach to fault analysis for digital integrated circuits employs logic gates, latches, flip-flops, etc., as the basic circuit components. The digital signals being processed are represented primarily by the logical zero and one values. Faults are almost always modeled by the *single-stuck line* (SSL) model [1] which allows any connector to be stuck at logical zero (s-a-0) or stuck at logical one (s-a-1). This approach has some major drawbacks when applied to such CAD tasks as fault simulation and test generation for MOS VLSI circuits [2], [3]. For example, electrical short circuits and open circuits account for most physical failures in IC's [4]-[6]. The classical SSL fault model provides, at best, only a rough approximation to these physical faults. Moreover, the SSL model cannot deal with delay faults that produce changes in a circuit's timing characteristics. The number of logic values available is often insufficient when fault conditions must be modeled. At least four logical values: 0, 1, Z (high-impedance state), and U (indeterminate or unknown state) are encountered in MOS circuits [7].

The foregoing problems can be attributed to the fact that they involve analog electrical considerations like bidirectional current flow and resistive/capacitive effects, which are incompatible with conventional digital logic concepts. These issues are typically addressed by a heuristic combination of analog and digital methods, and by the construction of special "workaround" subcircuits to replace circuit elements or faults whose behavior cannot be modeled directly [7]. For example, a VLSI chip design may be verified by using an analog simulation program like SPICE in conjunction with an digital simulator like TEGAS. This mixed analog-digital approach, which underlies the well-known Mead-Conway VLSI design methodology [8], tends to be costly and inefficient.

Recently, a number of simulators have been developed which, although digital in the sense that all signals are represented by a small set of discrete values, nevertheless, capture some of the relevant properties of analog simulators [9]-[12]. The logic simulator LOGIS [10], for example, in addition to handling conventional undirectional logic elements, allows bidirectional transistor switches and pull-up elements (termed resistive gates) to be modeled directly. It also recognizes up to nine logic values, including special values that represent a charge packet stored at a signal node. A related class of switch-level digital simulators such as MOSSIM [11], which analyze MOS circuits directly at the transistor level, are also widely used in VLSI design. None of these simulators address all the fault modeling issues mentioned above, however.

In this paper, a switch-level modeling technique called connector-switch-attenuator (CSA) theory [2], [3] is applied to the fault analysis of digital MOS circuits. The basic components of CSA models are connectors, representing both logical and electrical (power and ground) conductors, switches representing switching transistors, and attenuators representing resistive load elements. Logic behavior is defined by the four-valued set $V_4 = \{0, 1, Z, U\}$, which can be extended systematically to increase modeling accuracy to essentially any desired level. Timing analysis is based on a capacitive component termed a well which can represent the charge-storage behavior of MOS devices. CSA networks thus incorporate such analog parameters as electrical resistance and capacitance within a rigorous digital framework. Furthermore, since the elements and connectors of a CSA network can be placed in one-to-one correspondence with those of the physical circuit it represents, IC layout information can also be represented.

Section II summarizes combinational and sequential CSA theory. Fault modeling is examined in Section III. It is shown that two important fault types, (adjacent) short-circuit faults and stuck-open faults, can be modeled far more efficiently by using the CSA approach instead of conventional approaches. A generalized single-stuck-line (GSSL) fault model is proposed which allows diverse fault types to be simulated in a uniform manner.

II. CSA THEORY

The basic logic elements of static or combinational CSA circuits and the physical devices in MOS circuits that they typically represent are illustrated in Fig. 1. CSA connectors, which are inherently bidirectional, play a central role in the logical behavior of CSA circuit models, since the joining of two or more connectors performs a wired-logic operation of

---

Manuscript received June 28, 1983; revised January 9, 1984. This research was sponsored by the National Science Foundation under Grant MCS-8213884. A preliminary version of this work was presented under the title "A Fault Simulation Methodology for VLSI" at the 19th Design Automation Conference, Las Vegas, NV, June 1982.

The author is with the Department of Electrical Engineering and Computer Science, University of Michigan, Ann Arbor, MI 48109.

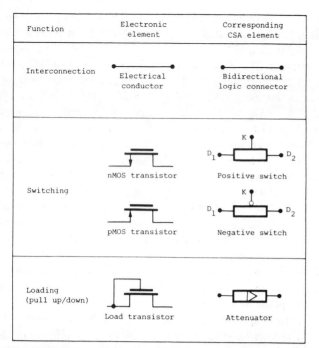

Fig. 1. Some basic components of MOS circuits and their CSA models.

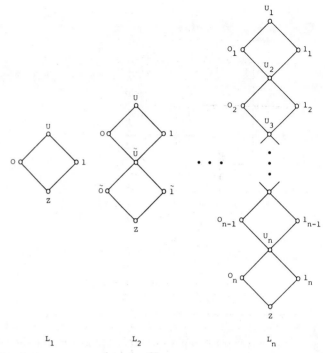

Fig. 2. Lattices $L_n = \{V_{3n+1}, \leqq\}$ showing the logic values and their strength relationship.

the AND or OR type. A switch S is a three-terminal device with a control terminal K and two symmetric data terminals D_1 and D_2. When S is turned on by applying an appropriate logic value to K, D_1 and D_2 are connected together. When S is turned off by applying some other value to K, D_1 and D_2 are disconnected from each other. S represents the idealized digital behavior of a switching transistor. The third CSA element, dubbed an attenuator, is primarily intended to represent the pull-up or pull-down load elements encountered in pMOS and nMOS circuits. An attenuator can be interpreted as a digital resistor.

To describe the behavior of a CSA network, a set V_N of digital or logic values is used that corresponds to a subset of the analog voltage–current signals in the underlying electronic circuit. These logic values, whose number N may be increased to improve modeling accuracy, are of four types:

1) one (1) values denoting signals in the high voltage V_H range,
2) zero (0) values denoting signals in the low voltage V_L range,
3) unknown (U) values denoting indeterminate voltages that can lie between V_H and V_L,
4) the Z value denoting the high-impedance or disconnected state.

The smallest set of logic values for describing CSA network behavior is $V_4 = \{0, 1, Z, U\}$. Suppose that a set of k signals $x_1, x_2, \cdots, x_k$ defined on V_4 is applied to a connector c. c assumes an (output) value $z_c = \#(x_1, x_2, \cdots, x_k) \in V_4$, where # denotes the connection operation performed by c. The behavior of # may be understood by interpreting the x_i's and z_c as voltages and c as an equipotential body. If all the x_i's are Z, which is equivalent to $k = 0$, then $z_c = Z$. If, say, x_1 is set to 0, then x_1 overrides the previous high-impedance state,

changing z_c to 0. Thus we can say that 0 is logically "stronger" than Z, denoted $0 > Z$. If $x_1 = 0$ and $x_2 = 1$, then z_c assumes the indeterminate value U; hence $U > 0$ and $U > 1$. In general, c assumes the value $z_c = \#(x_1, x_2, \cdots, x_k)$, where z_c is the logically weakest member of V_4 such that $x_i \leqq z_c$ for $i = 1, 2, \cdots, k$. The strength relationship among the members of V_4 is represented graphically in Fig. 2. The set V_4 with the ordering relation $\leqq$ forms an algebraic structure $L_1 = \{V_4, \leqq\}$ called a lattice [13]. The connection operation # defined above is identical to the lattice operator called least upper bound or join.

The behavior of a switch, which is essentially a two-state controlled connector, can readily be defined with respect to V_4. In the on state, the data terminals D_1 and D_2 of the switch (see Fig. 1) form part of a single connector. The values assumed by D_1 and D_2 are therefore the same, and are determined by the connection function and the signals applied to D_1 and D_2. In the off state, D_1 and D_2 are disconnected. The signal applied to the control terminal K determines the state of the switch. $K = 1$ turns a positive switch on, while $K = 0$ turns it off. A negative switch is defined similarly, with the roles of 0 and 1 interchanged. The effect of $K = U$ or Z on the switch state may be defined in various ways to reflect the underlying device technology. For example, $K = Z$ may be defined to leave the switch in its previous state.

Some useful circuit types, including relay and CMOS circuits, can be modeled using connectors and switches alone; these are called CS networks. Fig. 3(a) shows a standard CMOS NOR gate, while Fig. 3(b) shows the equivalent CS network. Note the one-to-one correspondence between the connectors and switches of the CS model, and the connectors and transistors of the electronic circuit. If the input signals x_1 and x_2 are confined to the set $\{0, 1\}$, then it is easily seen that the output

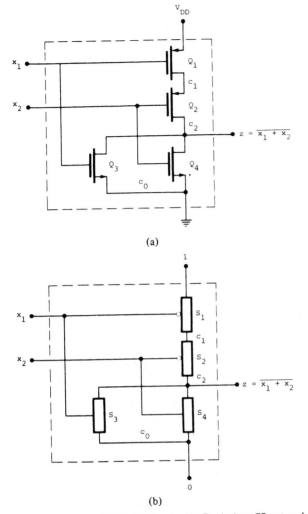

Fig. 3. (a) Two-input CMOS NOR gate. (b) Equivalent CS network.

z of the CS network is the usual NOR function $\overline{x_1 + x_2}$ defined on $\{0, 1\}$.

To model the structure and behavior of other MOS circuit types, additional logical devices and signal values are required. These values are obtained by subdividing the 0, 1, and U values into n subvalues corresponding to n different logical strength levels thus

$$1 = 1_1, 1_2, \cdots, 1_n = \tilde{1}$$
$$0 = 0_1, 0_2, \cdots, 0_n = \tilde{0} \qquad (1)$$
$$U = U_1, U_2, \cdots, U_n = \tilde{U}.$$

where $d_i > d_{i+1}$ for $d \in \{0, 1, U\}$. Each set V_N of $N = 3n + 1$ values obtained by adding Z to (1) forms a lattice $L_n = \{V_{3n+1}, \leqq\}$. Fig. 2 shows the standard representations (Hasse diagrams [13]) of these lattices, in which d_i is (logically) stronger than d_j, denoted $d_i > d_j$, if d_i appears above d_j. Note that only $L_1 = \{V_4, \leqq\}$ is a Boolean algebra. If $d_i, d_j \in V_N$ are applied simultaneously to a connector, the resulting output signal always assumes the least upper bound value $\#(d_i, d_j)$. For example, $\#(0_1, 0_2, U_2) = 0_1$ and $\#(0_1, 1_1, U_3) = U_1$.

The logical strength concept is closely related to the electronic concept of current drive capability. A logically strong signal is derived from a low-resistance source such as a power or ground line, whereas a logically weak signal is derived from a high-resistance source such as a load resistor. The logical element in CSA theory that provides weak signals is termed an attenuator [2]. The number of possible attenuator sizes, corresponding to possible resistance values, increases with the number of logical values used, i.e., with the number of strength levels n that are recognized. The seven-valued lattice $L_2 = \{V_7, \leqq\}$ shown in Fig. 2 allows each $d \in \{0, 1, U\}$ to have two strength levels which are denoted by d (strong) and $\tilde{d}$ (weak). Only one attenuator size A is required for CSA circuits with values defined on V_7; when d is applied to one terminal of A, $\tilde{d}$ is produced at the other terminal. An attenuator may be viewed as the inverse of an amplifier, which converts a weak signal $\tilde{d}$ to the corresponding strong signal d. With the enlarged set of logic values V_7, various new types of switches can be defined. The positive switch of Fig. 1 is assumed to be turned on by 1 and turned off by 0, but is unaffected by $\tilde{0}$ and $\tilde{1}$. If an amplifier is inserted in the control line K, the result is a (positive) amplifying switch which is turned on by 1 and $\tilde{1}$, and is turned off by 0 and $\tilde{0}$. Amplifying switches of this sort can model the gain inherent in electronic switches.

Many nMOS and pMOS combinational circuits can be adequately modeled by CSA networks composed of connectors, amplifying or nonamplifying switches, and attenuators, whose behavior is defined with respect to V_7. Typically, the attenuators denote the pull-up load elements found at the outputs of MOS gates. For example, Fig. 4(a) shows an nMOS inverter, while Fig. 4(b) shows its CSA equivalent. The latter's behavior is summarized in the truth table of Fig. 4(c). When the input signal x is 0 or $\tilde{0}$, the positive amplifying switch S is turned off causing its connection b to the primary output z to float. The attenuator A produces the weak signal $\tilde{1}$ on line a, hence z assumes the value given by

$$z = \#(a, b) = \#(\tilde{1}, Z) = \tilde{1}.$$

Thus A "pulls up" z from Z to 1. When $x = 1$ or $\tilde{1}$, S is switched on causing the strong signal 0 representing ground to be applied to b changing z to $\#(\tilde{1}, 0) = 0$. Hence the output signal z only assumes values on the set $\{0, \tilde{1}\}$. This asymmetry in output signal strength reflects the asymmetry in the current drive capability of the underlying electronic circuit.

To represent the dynamic or sequential aspects of MOS circuits we must account for the way in which time-varying signals are delayed by the circuit elements and interconnection lines. The major physical cause of delay phenomena in integrated logic circuits is stray electrical capacitance. Such capacitance, in conjunction with load elements and other resistance sources, causes digital signals to rise and fall at a finite rate. Consider for example, the stray capacitance C between the output line z of the inverter of Fig. 4(a) and ground, which results in the dynamic behavior depicted in Fig. 5. Here the ideal digital pulse of Fig. 5(a) is applied to the inverter's input x. The finite charge and discharge times of C yield the distorted analog output pulse of Fig. 5(b). This analog pulse can be approximated by the digital waveform of Fig. 5(c), which shows that the falling and rising output signals are subject to discrete delays

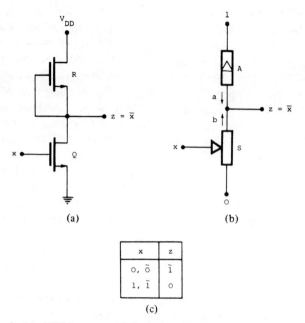

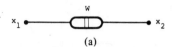

Fig. 6. (a) Well W. (b) Behavior of W with respect to $\{0, \tilde{0}, 1, \tilde{1}\}$ with $x_2 = 0$.

Fig. 4. (a) nMOS inverter. (b) Equivalent CSA network. (c) Truth table with respect to $\{0, \tilde{0}, 1, \tilde{1}\}$.

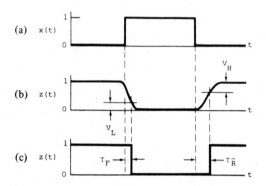

Fig. 5. (a) Input pulse to the nMOS inverter. (b) Analog output pulse. (c) Corresponding digital output pulse.

of τ_F and $\tau_{\tilde{R}}$ s, respectively. The values of these delay parameters depend primarily on the size of the inverter's load resistance R and the capacitance C.

Traditional logic simulators model signal propagation delays, like those illustrated in Fig. 5(c), by means of unidirectional lumped delay elements [1]. Such delay elements are artificial constructs from a simulation viewpoint, since in most cases they do not correspond to a well-defined physical element of the circuit under consideration. Simulators such as MOTIS [9] and all analog simulators include detailed electrical models of capacitance from which dynamic behavior is computed. The CSA approach also considers capacitive effects explicity, but uses a relatively simple capacitor model termed a well [3], and represented by the symbol of Fig. 6(a). A well W is a two-terminal device capable of charge storage but, unlike a capacitor, its input/output signals are restricted to a finite set of logic values V_N. The storage capacity or size of the well corresponds to that of the underlying analog capacitance C. W may be characterized by four stable states, namely a discharged state s_Z, and three charged states s_0, s_1, s_U corresponding to the storage of logical one, zero, and unknown signals, respectively.

The state of W is determined by the signals x_1 and x_2 applied to its terminals. Suppose that the well is in the discharged state s_Z. If x_1 is now set to Z, which corresponds to disconnecting a terminal of W and allowing it to float, no charging can occur, so the state of W does not change. W also remains in state s_Z if identical signals such as $x_1 x_2 = 11$ or $x_1 x_2 = 00$ are applied to its terminals. If, however, $x_1 x_2 = 01$, W charges to the s_1 state. If U is applied to one terminal of W, and any value except Z is applied to the other terminal, W enters the indeterminate state s_U, which typically corresponds to a partially charged condition.

Time is introduced into CSA modeling by making the rate at which a well W changes state depend on the logical strength of the signals $x_1 x_2$ applied to its terminals. In general, strong signals cause W to charge or discharge rapidly, while weaker signals cause it to charge or discharge more slowly. Hence the dynamic behavior of W can be defined by a set of state transition delays, whose numerical values can be estimated by standard electrical RC analysis. Suppose, for instance, that one terminal x_2 of W is held at the strong 0 level, denoting the ground or IC substrate potential. Then if x_1 assumes any of the values $\{0, \tilde{0}, 1, \tilde{1}\}$, four distinct state transition delays can be obtained as indicated in Fig. 5(b). Additional delay values can be obtained by allowing x_1 to assume other values in V_N, and allowing W to assume other sizes.

Fig. 7(a) shows the use of the well concept to model the dynamic behavior depicted in Fig. 5 for the NMOS inverter of Fig. 4(a). A well W is connected between the output line z and the strong 0 source to represent the inverter's output capacitance C. Suppose that W is initially in the discharged state s_Z. As long as $x = 1$, which turns switch S on and makes $z = 0$, W remains discharged. When x changes to 0, S switches off and z becomes $\tilde{1}$. W then charges slowly via the attenuator A until it reaches state s_1, a process taking time $\tau_{\tilde{R}}$. When z returns to 0, W discharges rapidly through S in time τ_F. Thus when the digital input pulse of Fig. 5(a) is applied to the CSA model of the inverter, the digital output pulse of Fig. 5(c) is obtained directly.

It is instructive to compare this CSA model of an inverter to the conventional logic model of Fig. 7(b) which employs a delay element D. If the delay parameters $\tau_{\tilde{R}}$ and τ_F used in the CSA circuit are assigned to D, then both circuits exhibit

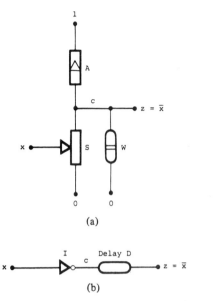

Fig. 7. (a) Dynamic CSA model of nMOS inverter. (b) Gate-level model with delay element.

the same normal behavior. However, there are significant differences between the two circuits when faults are present. The well W can store a one, i.e., remain in the state s_1, for an indefinite period. The delay D on the other hand, has a finite memory. Thus if line c in Fig. 7(a) is cut, due to an open-circuit fault, W can maintain the output line z at the one value indefinitely. If an equivalent cut is made in line c of Fig. 7(b), the output signal z becomes indeterminate after some finite delay period when the last value provided by the inverter has been passed through D.

The accuracy with which a CSA circuit reflects the behavior of the underlying electronic circuit can be increased by expanding the set of recognized logic values V_N. This, also increases the number of switch, attenuator, amplifier, and well types that must be dealt with, thereby increasing the model's computational complexity. It is desirable from a computational viewpoint to use the smallest value of N and the smallest set of component types that allow the structural and behavioral parameters of interest to be identified and measured. The logic value set V_7 is of particular interest in the simulation of MOS logic circuits. As observed earlier, only one type of attenuator A is needed with V_7. In the case of pMOS and nMOS circuits, A is used to model the dominant resistance sources present in the circuit, namely output loads. Other resistances, such as the source-to-drain resistance of a transistor, are modeled by either open circuits (infinite resistance) or short circuits (zero resistance). In many cases it suffices to use single well size W to represent the dominant capacitance, which is the gate-to-substrate capacitance of a transistor switch. The behavior of W can be defined by four delay parameters of the kind given in Fig. 6(b).

III. Fault Analysis

It is well known that many physical faults exist in MOS VLSI circuits which are not functionally equivalent to SSL faults in gate-level circuit models [4]-[7]. These include short circuits between logic signals in all types of logic circuits, and open-circuit faults occurring in certain logic families including CMOS. Such faults are generally modeled indirectly by workaround circuits that introduce numerous artificial elements into the simulation process. In this section, we show that most faults affecting logical behavior can be directly and efficiently modeled using the CSA approach.

Various studies [4], [5] of the actual fault mechanisms in IC's suggest that up to a half of all failures are due to short circuits between interconnection lines. Some short-circuit faults, such as a logic signal shorted to a power or ground line, can be adequately modeled by the classical SSL model. Most faults involving short signals between lines carrying logic signals cannot be so modeled, however. There are several reasons for this.

1) Connections between certain pairs of lines, such as the outputs of two gates, may result in a signal that is required to be both 0 and 1 simultaneously.

2) Lines in the gate-level model may not exist in the underlying physical circuit, and vice versa.

3) The number of possible short circuits may be too large.

The CSA modeling technique alleviates each of the foregoing difficulties as follows:

1) The connection between any pair of lines carrying the values $x_i, x_j \in V_N$ always produces a unique value $\#(x_i, x_j) \in V_N$.

2) As Figs. 3 and 4 demonstrate, a one-to-one correspondence between lines in the physical circuit and connectors in the CSA model can easily be established.

3) Unlike traditional logic circuits, a CSA circuit can provide layout information which may be used to reduce the number of short circuits that must be considered.

It seems useful to restrict attention to short-circuit faults occuring between physically adjacent lines. We define an *adjacent short-circuit* (ASC) fault to exist between two connectors c_1 and c_2 in an IC or a structurally equivalent CSA circuit, if a line L can be drawn from c_1 to c_2 that does not cross any other connector or circuit, and does not pass outside the perimeter of the circuit. (A restriction may also be placed on the maximum length of L.) In a complex m-line circuit, the number of unrestricted short-circuit faults is proportional to m^2, whereas in large circuits the number of ASC faults, like the number of SSL faults, is proportional to m.

Consider again the NOR circuit of Fig. 3(a). The connectors c_1 and c_2 are adjacent, whereas c_0 and c_1 are nonadjacent, assuming the broken line denotes the circuit's perimeter. An ASC fault between c_1 and c_2, which could represent transistor Q_2 in the stuck-on condition, changes the function z realized by the circuit from $\overline{x_1 + x_2}$ to $z' = \#(z_1, z_2)$, where z_1 and z_2 denote the normal functions appearing on c_1 and c_2, respectively. Fig. 8 specifies the behavior of the circuit with this ASC fault present. Note that short-circuit faults can generate the U value which is not present in well-behaved fault-free circuits.

Open-circuit faults can affect the logical behavior of MOS circuits in ways that are very hard to model via conventional approaches. One such fault, which is of special interest, is the *stuck-open fault* [7] occurring when a logic connection becomes improperly isolated from all sources of zero and one signals. Stuck-open faults can cause complex and unexpected

x_1	x_2	z_1	z_2	$z' = \#(z_1, z_2)$
0	0	1	1	1
0	1	1	0	U
1	0	0	0	0
1	1	Z	0	0

Fig. 8. Effect of an adjacent short-circuit (ASC) fault between connectors c_1 and c_2 of the two-input CMOS NOR gate.

behavior in dynamic MOS and CMOS circuits by preventing capacitors (wells) from charging or discharging. An example of this is the *parasitic flip-flop* (PFF) fault [14], which can convert a combinational to a sequential circuit. A PFF fault occurs when a line L becomes isolated from all signal sources except stray or parasitic capacitance that is capable of maintaining a previous value on L for a relatively long period of time.

To illustrate the foregoing fault modeling issues, we consider the task of representing the faults encountered in a CMOS NOR circuit of the kind shown in Fig. 3(a). Among the possible failure modes occurring in this circuit are short circuits between logic, power or ground lines, stuck-open faults including PFF faults, and transistors that are stuck in the on or off state. The simplest conventional logical model for this circuit, which is a single two-input NOR gate G_0, appears in Fig. 9(a). This allows only the six SSL faults associated with the three input/output lines x_1, x_2, and z to be modeled via normal fault injection techniques.

Fig. 9(b) shows a workaround model of the same CMOS circuit developed by Wadsack [7] which covers three major PFF faults, in addition to the six SSL faults associated with G_0. A clocked D-type latch GL simulates the charge–storage property of the CMOS circuit's output line z. Under fault-free conditions, the clock input CK of GL is 1, causing GL to behave like a (unit) delay element. Thus the logic gate G_0 and the latch GL with $CK = 1$ model a fault-free NOR gate having unit delay; cf., Fig. 7(b). Stuck-at-0/1 faults can be injected into G_0 in the standard fashion. The remainder of Fig. 9(b) is used to model the three main PFF conditions which are:

F_1 transistor Q_3 stuck open or missing,
F_2 transistor Q_4 stuck open or missing,
F_3 an open circuit in the path V_{DD} to z.

For example, to model F_1, the output of gate G_1 is forced to 1. When the input combination $x_1 x_2 = 01$ occurs, G_1 forces CK to 0, thereby causing GL to latch, and store the preceding output value $z(t - 1)$ indefinitely. Fig. 9(c) shows another model of a different type of two-input CMOS NOR gate [14], which covers five PFF faults and three short-circuit faults, in addition to the six basic SSL faults. Again SSL faults can be injected into certain gates in Fig. 9(c) in order to simulate the behavior of the nonstandard faults of interest.

Although the circuits of Fig. 9(b) and (c) are quite efficient in the sense that they use the minimum amount of logic needed to represent the faults of interest using a typical 0/1 fault simulator, they introduce many redundant components and connectors. The original electronic circuit of Fig. 3(a) contains four components, (the transistors Q_1-Q_4) and six interconnections. The logical models of Fig. 9(b) and (c) contain 12 and 26 components, respectively, with a corresponding increase in the number of interconnections. This complexity increase limits the use of this type of fault modeling in VLSI circuits. Another drawback of the above approach is that it creates an undesirable distinction between elements that may be faulted (gates G_0-G_8 in Fig. 9(b), for instance) and elements which must remain fault-free (gates G_4-G_{10} and latch GL in Fig. 9(b)) since they are artifacts of the simulation model.

If the CSA approach is used, all the fault types mentioned so far can be modeled directly with essentially no artificial added elements. Fig. 10 shows a CSA network for modeling the NOR-gate faults discussed above. Only the wells W_p and W_n need to be added to the original four-switch CSA model of Fig. 3(b) in order to simulate the storage effect of PFF faults. W_p and W_n represent the gate-to-substrate capacitance of the pMOS and nMOS transistors, respectively, that are driven by z. This circuit can simulate directly not only the faults covered by the circuits of Fig. 9, but can also model all internal short-circuit and open-circuit faults. Consider, for example, the stuck-open fault F_1 defined earlier. This is represented in Fig. 10 by line c_0 s-a-Z. Fault injection merely requires breaking c_0 as indicated; the logic value Z is then applied to the broken ends. Now, whenever $x_1 x_2 = 10$, S_8 applies Z instead of 0 to the output line z. If the previous value of z was $z(t - 1) = 0$, then W_n is discharged at time t, while W_p is charged. W_p, therefore holds z at the zero level indefinitely. Similarly, if $z(t - 1) = 1$ immediately before F_1 occurs, W_n holds z at the one level.

The SSL fault model has the practical advantage of being easy to incorporate into the logical framework of conventional CAD programs. A particular fault, say line L s-a-1, is injected into a circuit by cutting L and applying the constant 1 signal to the input side of the cut. Thus the basic SSL model is implicitly restricted to undirectional circuits, where all lines have well-defined input and output ends. CSA networks require a more general fault model that is applicable to both unidirectional and bidirectional lines, and accommodates a larger set of stuck values V_N. These considerations motivate the *general SSL* (GSSL) fault model described in Fig. 11 with V_4 as the set of possible stuck values. As in the SSL case, the affected line is cut, but constant signals from V_N are applied to both cut ends. This permits various faults, including classical SSL faults, short-circuit faults, and stuck-open faults to be modeled uniformly.

Another characteristic of SSL faults is that they can only represent the total failure of a conducting or an insulating medium, i.e., SSL faults are of the all-or-nothing variety. Many physical failures are of a less extreme variety, e.g., line L develops excessive leakage to ground rather than the complete short to ground represented by L s-a-0. Partial failures of this type can be directly represented by GSSL faults involving weak stuck values. If line L normally carries a signal d_i of logical strength i, then inserting a s-a-d_j fault into L effectively increases the resistance (attenuation) of L if $j > i$, and decreases its resistance if $j < i$. Such changes in resistance can also alter a circuit's dynamic behavior. Hence weak GSSL faults can be employed to model changes in a circuit's timing characteristics, i.e., delay faults.

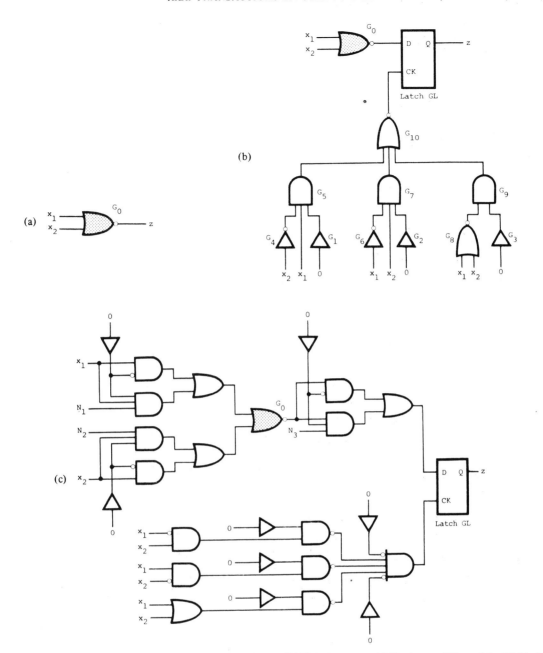

Fig. 9. Conventional fault analysis models for the two-input CMOS NOR gate. (a) Single-gate SSL model. (b) Model covering three PFF faults [7]. (c) Model covering five PFF and three short-circuit faults [14].

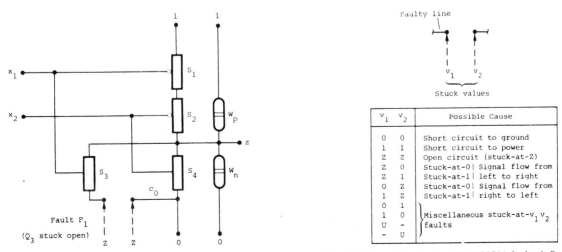

Fig. 10. CSA fault analysis model for the two-input CMOS NOR gate.

Fig. 11. General single stuck-line (GSSL) faults defined on V_4.

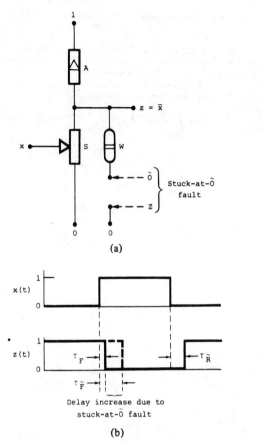

Fig. 12. (a) The nMOS inverter with a weak GSSL fault present. (b) Resultant circuit behavior.

To illustrate this important aspect of GSSL faults, consider again the nMOS inverter of Fig. 7(a). Suppose that a weak s-a-$\tilde{0}$ GSSL fault is injected into the lower connector of the well W as illustrated in Fig. 12(a). This is equivalent to inserting another attenuator between W and ground. Thus the well's normally fast discharge time τ_F through S changes to the slow discharge time $\tau_{\tilde{F}}$ as indicated in Fig. 12(b). Hence the s-a-$\tilde{0}$ fault has the effect of increasing the signal fall delay from τ_F to $\tau_{\tilde{F}}$ at the inverter's output. By injecting similar weak GSSL faults of the s-a-$0_i/1_i$ variety into any circuit, its dynamic behavior can be modified in various ways that realistically represent the physical failure modes present.

IV. DISCUSSION

CSA methodology provides an alternative to conventional fault modeling methods for digital circuits. CSA circuits can be viewed as simplified analog circuits that employ only the relatively small number of signal values V_N needed for digital simulation. These values, beginning with the four-valued set $V_4 = \{0, 1, U, Z\}$, can be extended systematically to increase the modeling accuracy by using additional strength levels. The logical strength concept introduces the notions of signal amplification and attenuation into logical analysis, and allows digital models of resistance (the attenuator) and capacitance (the well) to be defined. Thus various design verification and fault analysis tasks currently performed by analog simulators can be performed by a CSA-based simulator. A CSA circuit can also be regarded as a refinement of a standard logic circuit which reveals some hidden digital components.

CSA models have several advantages from a fault modeling viewpoint. Many standard and nonstandard failure modes, including stuck-at faults, short circuits, open circuits, and delay faults, can be modeled directly at the CSA level, eliminating the need for complex and costly workaround circuits. Partial failures that are intermediate between short circuits and open circuits can be represented by GSSL faults of the weak variety. Unlike gate-level circuits, a CSA circuit can represent the geometric structure of an integrated circuit. This fact can be used to identify quickly layout-dependent faults like adjacent short-circuit faults. An experimental fault simulation program called CSASIM that implements the CSA and GSSL concepts, and demonstrates their feasibility, has recently been implemented [15].

REFERENCES

[1] M. A. Breuer and A. D. Friedman, *Diagnosis and Reliable Design of Digital Systems.* Woodland Hills, CA: Computer Science Press, 1976.
[2] J. P. Hayes, "A logic design theory for VLSI," in *Proc. Second CalTech Conf. on VLSI* (Pasadena, CA), Jan. 1981, pp. 455-476.
[3] —, "A unified switching theory with applications to VLSI design," *Proc. IEEE*, vol. 70, pp. 1140-1151, Oct. 1982.
[4] D. B. Nicholls, "Digital evaluation and generic failure analysis data," Reliability Analysis Center, Rome Air Development Center, NY, Rep. MDR-10, Jan. 1979.
[5] J. Galiay, Y. Crouzet, and M. Vergniault, "Physical versus logical fault models in MOS LSI circuits: Impact on their testability," *IEEE Trans. Computers*, vol. C-29, pp. 527-531, June 1980.
[6] B. Courtois, "Failure mechanisms, fault hypotheses, and analytical testing of LSI-NMOS (HMOS) circuits," in *VLSI 81*, J. P. Gray, Ed. London, England: Academic Press, 1981, pp. 341-350.
[7] R. L. Wadsack, "Fault modeling and logic simulation of CMOS and MOS integrated circuits," *Bell Syst. Tech. J.*, vol. 57, pp. 1449-1474, May-June 1978.
[8] C. Mead and L. Conway, *Introduction to VLSI Systems.* Reading, MA: Addison-Wesley, 1980.
[9] B. R. Chawla, H. K. Gummel, and P. Kozak: "MOTIS—An MOS timing simulator," *IEEE Trans. Circuits Syst.*, vol. CAS-22, pp. 901-910, Dec. 1975.
[10] ISD Inc.: *Logis User's Manual*, Santa Clara, CA, 1978.
[11] R. E. Bryant, "An algorithm for MOS logic simulation," *Lambda*, vol. 1, no. 3, pp. 46-53, fourth quarter 1980.
[12] W. Johnson et al., "Mixed level simulation from a hierarchical language," *J. Digital Syst.*, vol. IV, pp. 305-335, 1980.
[13] F. P. Preparata and R. T. Yeh, *Introduction to Discrete Structures.* Reading, MA: Addison-Wesley, 1973.
[14] M. W. Sievers and A. Avizienis: "Analysis of a class of totally self-checking functions implemented in a MOS LSI general logic structure," in *Dig. 11th Symp. on Fault-Tolerant Computing*, June, 1981, pp. 256-261.
[15] M. Kawai and J. P. Hayes: "An experimental MOS fault simulation program CSASIM," in *Proc. 21st Design Automation Conf.*, (Albuquerque, NM), June 1984, to appear.

★

John P. Hayes (S'67-M'70-SM'81) received the B.E. degree from the National University of Ireland, Dublin, in 1965, and the M.S. and Ph.D. degrees from the University of Illinois, Urbana, in 1967 and 1970, respectively, all in electrical engineering.

While at the University of Illinois he participated in the design of the ILLIAC 3 computer, and carried out research in the area of fault diagnosis of digital systems. In 1970 he joined the Operations Research Group at the Shell Benelux Computing Center of the Royal Dutch Shell Company in The Hague, The Netherlands, where he was involved in mathematical programming and software development. From 1972 to 1982 he was a faculty member of the Departments of Electrical Engineering and Computer Science of the University of Southern California, Los Angeles. He is currently a Professor in the Electrical Engineering and Computer Science Department of the University of Michigan, Ann Arbor. He was Technical Program Chairman of the 1977 International Conference on Fault-Tolerant Computing. He is the author of the books *Computer Architecture and Organization* (New York: McGraw-Hill, 1978), and *Digital System Design and Microprocessors* (New York: McGraw-Hill, 1984). He served as Editor of the Computer Architecture and Systems Department of *Communications of the ACM* from 1978 to 1981, and was Guest Editor of the June 1984 Special Issue of IEEE TRANSACTIONS ON COMPUTERS. His current research interests include fault-tolerant computing, computer architecture, VLSI design, and micro-processor-based systems.

Dr. Hayes is a member of the Association for Computing Machinery and Sigma Xi.

TEST GENERATION FOR MOS CIRCUITS USING D-ALGORITHM

Sunil K. Jain
Vishwani D. Agrawal

Bell Laboratories
Murray Hill, New Jersey 07974

ABSTRACT — An application of the D-algorithm in generating tests for MOS circuit faults is described. The MOS circuits considered are combinational and acyclic but may contain transmission gates and buses. Tests are generated for both, the stuck type faults and the transistor faults (open and short). A logic model is derived for the MOS circuits. In addition to the conventional logic gates, a new type of modeling block is used to represent the "memory" state caused by the "open" transistors. Every fault, whether a stuck type fault or a transistor fault, is represented in the model as a stuck fault at a certain gate input. For generating tests, however, the D-algorithm needs modification. The singular cover and the D-cubes for the new gate include some memory states. To handle the memory state, an initialization procedure has been added to the consistency part of the D-algorithm. The procedure of modeling and test generation is finally extended to transmission gates and buses.

INTRODUCTION

MOS circuits contain transistors interconnected through wires. A simple fault model should, therefore, include stuck faults (stuck-at-1 and stuck-at-0) on wires and switch faults (stuck-open and stuck-short) in transistors. Some recent workers [1-3] have reported test generation algorithms for transistor faults from the MOS device level description of the combinational circuit. Most of the conventional test generation procedures, however, require a gate-level description of the circuit. Notable among these is the D-algorithm [4].

This paper describes a procedure to convert a transistor structure into an equivalent logic gate structure. In order to model the memory state of a transistor, one new gate is defined. Every fault of the transistor circuit (whether it is a fault in wire or in transistor) is equivalent to a stuck fault in the logic structure. If the circuit is combinational, then the D-algorithm can be used for generating a test for any fault. Certain transistor faults, however, produce memory states. Handling of these faults requires some modifications of the D-algorithm. Basically, an initialization algorithm has been added to the consistency part of the D-algorithm.

An additional motivation for the present work is derived from some recent developments in fault simulation [5]. It is now possible to consider the transistor faults in the fault coverage analysis of tests. Particularly, for CMOS circuits, the coverage of transistor faults has been found to be significantly lower than that of the stuck type faults. One reason for this may be that the tests are often aimed at detecting stuck type faults only while not every transistor fault (e.g., stuck-open fault) can be mapped into a stuck-type fault. A test generation capability for transistor faults is, therefore, of significant value for CMOS technology.

HOW DOES AN MOS GATE FUNCTION?

Figure 1 shows a simple CMOS gate. It contains two clusters of transistors — one nMOS cluster and the other pMOS cluster. All input signals are connected to the "gate" terminals of transistors. Depending upon the state (logical 0 or 1) of an input signal, the corresponding transistor behaves like an "open" or a "short." If the output is to be set "low" then a conducting path is created between the output and ground while all the paths between the output and VDD are opened. Similarly, the output is set "high" by creating a path to VDD, while all the paths to ground are blocked. This CMOS gate, does not pass on any of its input values but, in fact, regenerates a 1 from VDD or a 0 from ground. This function is represented by the switch model of Fig. 2. The operation of the switches S_0 and S_1 is a function of the inputs A, B and C. Either switch can be open or short. In a CMOS circuit, since S_0 and S_1 are complementary signals, only one switch is short at a time. The open switch, presents a floating (or high impedance) state to the output node. The output node behaves as a bus, on which the high impedance value is overridden by any other logic value. If, however, both inputs are in high impedance state, then the bus will retain its 'past' value. This state results from the charge retention capability of an isolated node in an MOS circuit. If both switches, S_0 and S_1, are shorted then the output will be a 0 or a 1 depending upon the relative resistances of the load and the driver transistors in their conducting states.

A MODEL FOR TEST GENERATION

Most test generation algorithms work on a logical description of the circuit. In these cases, the circuit should be modeled using the building blocks whose input/output behavior can be described logically (e.g., by truth tables). We will discuss a logic model for MOS transistor structures in terms of conventional logic gates, AND, OR, NOT, NAND, NOR, and one additional modeling block. The additional block is necessary for handling the high impedance (or memory) state and for eliminating VDD and ground from the logic model [6]. This modeling block is referred to as block 'B'.

The logic gate model for the MOS circuit of Fig. 1 is shown in Fig. 3. Simple rules, given in Table 1, explain how to first replace the transistors of each type (nMOS and pMOS) by logic gates to produce the switch functions S_1 and S_0. Each of these functions is logic-1 if the switch is shorted and it is logic-0 if the switch is open. The output is then generated by combining S_1 and S_0 in the block B given in Table 1. Notice that the behavior of block B is not symmetric. The signal S_1 which is the output of the paths from VDD should be connected to the terminal marked as 1 while the output of the

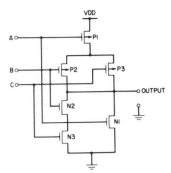

Fig. 1. A CMOS gate.

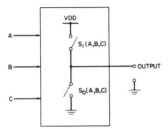

Fig. 2. A switch model for CMOS gate.

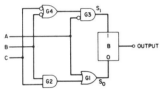

Fig. 3. A logic gate model for the CMOS circuit of Fig. 1.

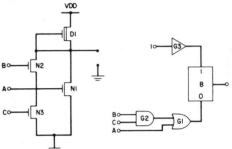

Fig. 4. An nMOS gate and its logic model.

TABLE 1
MOS TO LOGIC TRANSFORMATION

MOS		LOGIC	
GATE TERMINAL INPUT TO NMOS TRANSISTOR		INPUT TO A LOGIC GATE	
GATE TERMINAL INPUT TO PMOS TRANSISTOR		INVERTED INPUT TO A LOGIC GATE	
DEPLETION LOAD		LOGIC 1	
SERIES ELEMENTS		AND GATE	
PARALLEL ELEMENTS		OR GATE	
MOS GATE OUTPUT		S_1 S_0 Y / 1 1 0 / 1 0 1 / 0 1 0 / 0 0 M	

TABLE 2
EQUIVALENT FAULTS

Fault in MOS circuit (Fig. 1)	Fault in Logic circuit (Fig. 3)
A stuck at j*	A stuck at j
B stuck at j	B stuck at j
C stuck at j	C stuck at j
output stuck at j	output stuck at j
P1 short	G3, input A stuck at 0
P1 open	G3, input A stuck at 1
P2 short	G4, input B stuck at 0
P2 open	G4, input B stuck at 1
P3 short	G4, input C stuck at 0
P3 open	G4, input C stuck at 1
N1 short	G1, input A stuck at 1
N1 open	G1, input A stuck at 0
N2 short	G2, input B stuck at 1
N2 open	G2, input B stuck at 0
N3 short	G2, input C stuck at 1
N3 open	G2, input C stuck at 0

* j can be 0 or 1.

ground paths, S_0, should be connected to the 0 terminal of block B. When both the inputs are 0, the output is shown as M. Here M refers to the memory or the previous state (before S_1 and S_0 changed to 0) of the output line. M can be any value among 0, 1 or "unknown." When both the inputs are 1, the output is shown as 0. It is assumed that the path from ground dominates over the path from VDD and hence, the output is pulled to the logic value of 0. If the assumption is not true for any particular technology, this value can be set to 1 or "unknown," as appropriate.

It is easy to verify that the logical behavior of the equivalent logic circuit in Fig. 3 is identical to that of the CMOS circuit in Fig. 1. A similar transformation of a nMOS circuit is shown in Fig. 4. Here the depletion mode transistor D1 is always "on" as it would require a negative voltage at its gate to be in the "off" state.

REPRESENTATION OF FAULTS

We will consider two types of faults in the MOS circuits. These are the stuck faults (s-a-0 and s-a-1) at the inputs and outputs of the MOS gates and the transistor (short or open) faults. Table 2 gives a list of faults in the MOS circuit of Fig. 1 and their corresponding equivalent faults in logic circuit of Fig. 3. Every fault in the MOS circuit has an equivalent stuck type fault in the logic representation. The input/output stuck faults in MOS circuit correspond to identical stuck faults in the logic model. A transistor fault in the MOS circuit corresponds to a stuck fault on that input lead of the logic gate which is the replacement for the transistor (see Table 1). Thus transistor P1 short corresponds to input A of gate G3 stuck-at-0, or transistor P1-OPEN corresponds to the input A of gate G3 stuck-at-1.

The faults in the depletion mode transistors used in nMOS circuits need special mention. In Fig. 4, the fault "transistor D1-SHORT" corresponds to the input of G3 stuck-at-1. Since in our logic model the input of G3 is permanently set to a 1, this fault is undetectable. Indeed this fault is also undetectable in the MOS gate. The fault D1-OPEN, on the other hand, corresponds to the input of G3 stuck-at-0 and this is detectable in the MOS circuit as well as in our logic model. In general, tests need to be generated for only one fault among those that collapse together.

APPLICATION OF D-ALGORITHM

The basic D-algorithm as described in the literature [4] requires every gate to be described by its singular covers and the D-cubes. The only additional gate that has been introduced here is the MOS output modeling block B, as shown in Table 1. The singular cover and D-cubes for this block are given in Fig. 5. Singular cover is used for determining the inputs (during consistency operation) or determining the output (during forward implication) when either the output or some of the inputs of a gate are 0 or 1. D-cubes are used for representing the faults at the site of the faulty gate and for propagating the 'D' (during the D-drive).

A D-cube is obtained by combining two rows of the truth table. For example, the first two rows of the truth table of the MOS output gate produce the following D-cube:

$$S_1 = 1, \quad S_0 = D, \quad OUTPUT = \bar{D}$$

By substituting $D = 1$, in this D-cube we get the first row of the truth table and $D = 0$ leads to the second row. If we remember that the memory state M is the previous state (0 or 1) of the output line, then its value for this D-cube is "don't care." This is shown as the first D-cube in Fig. 5. The first four D-cubes do not involve any M state. The remaining D-cubes have the values of M specified. For example, consider the fifth D-cube which is obtained by combining the following rows of the truth table:

S_1	S_0	OUTPUT
0	1	0
0	0	M

If the previous state M of the output was 0 then the output will be 0 for any value of S_0. This state will, therefore, lead to the following cube:

S_1	S_0	OUTPUT
0	D	0

which does not propagate the D to the output. If, however, $M = 1$, then the D on S_0 is propagated to the output as $\bar{D}$. This is shown as the fifth D-cube in Fig. 5 where the corresponding value of M is shown in the parenthesis. This assumes that if the output of the gate can be *initialized* to a 1 in both the faulty and the fault-free circuits then a 0 on S_1 and a D on S_0 will produce a $\bar{D}$ at the output. Other D-cubes in Fig. 5 were obtained in a similar fashion. In our notation, D(X) or $\bar{D}$(X), (where X in parenthesis means that initialization at the output is not needed) are written simply as D or $\bar{D}$.

The test generation for any given fault can now proceed in the conventional manner. First a D or a $\bar{D}$ is placed at the fault site. D or $\bar{D}$ assume the logic values 1 or 0, respectively, in the fault-free circuit and the complementary values in the faulty circuit. Consistency, implications and D-drive procedures [4] are repeatedly applied until a D or a $\bar{D}$ arrives at a primary output. In addition, a back-up (back tracking) procedure in case of an inconsistency or a premature disappearance of all D and $\bar{D}$ guarantees a test if one is possible. Whenever a D-drive is carried through an MOS output gate (B), the associated memory state of the particular D-cube is checked. If this memory state is 0 or 1 then the following consistency operation includes an additional procedure of *initialization*.

SINGULAR COVER		
S_1	S_0	OUTPUT
X	1	0
1	0	1
0	0	M

D-CUBES		
S_1	S_0	OUTPUT
1	D	$\bar{D}$
1	$\bar{D}$	D
D	$\bar{D}$	D
$\bar{D}$	D	$\bar{D}$
0	D	$\bar{D}$(1)
0	$\bar{D}$	D(1)
D	0	D(0)
$\bar{D}$	0	$\bar{D}$(0)
D	D	$\bar{D}$(1)
$\bar{D}$	$\bar{D}$	D(1)

Fig. 5. Singular cover and D-cubes for the MOS output gate B.

THE INITIALIZATION ALGORITHM

The initialization algorithm described here has been derived from the D-algorithm [4]. It differs from the D-algorithm in two aspects. First, D-drive is not used and second, a modified singular cover table is used for the faulty gate. To illustrate this modification in the singular cover table, consider an AND gate with one of the inputs (B), stuck-at-1 as shown in Fig. 6. Since the initialization requires that both, the faulty and the fault-free circuits behave identically, the faulty line should not be set to a logic value 0 during the initialization. All singular cubes that attempt to set the input B to a 0 are, therefore, deleted from the singular cover table. In this case, the modified table, as shown in Fig. 6, contains two cubes.

For the initialization, a second copy of the circuit is used so that the states of the circuit on which the D-algorithm is running are not disturbed. The initialization procedure begins by placing unknown or "don't care" values on all lines in this copy of the circuit. The required initialization value (0 or 1) is then placed at the line to be initialized. Consistency and implication procedures are repeatedly applied to all gates until no more changes in line values are necessary. These procedures involve intersection, as defined in [4], of the singular cubes with the existing line-values. Singular cubes are the rows of the singular cover. In the case of an inconsistency, the usual procedure of back-up to the last available choice of an unused singular cube is employed. If the procedure backs up to the very line which needs initialization and no more singular cubes are available, then this initialization is considered impossible and the back up is carried through to the previous D-drive where the memory state was generated. Hence, for a transistor fault, one test pattern is sufficient for initialization and a second test pattern for detecting the fault. These two pattern tests are sufficient to test any transistor fault in a circuit, which does not have memory in the fault-free state. For stuck faults on wires, only one test pattern is sufficient.

EXAMPLE 1

The circuit shown in Fig. 7 will be used to illustrate the test generation algorithm. This circuit consists of three CMOS gates. It is converted into an equivalent logic gate model by using the rules given in Table 1. The logic gate model for the circuit is shown in Fig. 8. Consider the fault, "transistor T5-OPEN" in the MOS circuit. This fault is equivalent to the fault "line 1b stuck-at-0" in the logic circuit of Fig. 8. The step-by-step procedure of generating a test for this fault is as follows:

SINGULAR COVER

A	B	C
0	X	0
X	0	0
1	1	1

SINGULAR COVER WITH FAULT, B S-A-1

A	B	C
0	X	0
1	1	1

Fig. 6. Modified singular cover of an AND gate used for initialization.

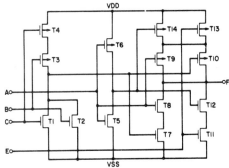

Fig. 7. CMOS circuit for Example 1.

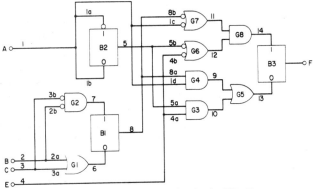

Fig. 8. Logic model for the circuit of Fig. 7.

Step 1: Since the fault is "line 1b stuck-at-0," this line is set to 1. This sets the primary input A = 1 and line 1a = 1. To represent the fault we rewrite, line 1b = D. No further implications are necessary. All other lines have unknown values represented by X.

Step 2: D-drive through the output gate B2. Using the fifth D-cube from Fig. 5, $\overline{1a} = 0$, 1b = D, 5 = $\overline{D}(1)$, and line 5 is to be initialized to 1.

Step 3: Initialization. Assume all lines at value X. Place a 1 on line 5. For justifying this value, we need the modified singular cover for gate B2. Due to the fault (stuck at 0) on line 1b, this singular cover consists of only the last two rows of the singular cover given in Fig. 5. Using the second row, we get, $\overline{1a} = 1$, 1b = 0, 5 = 1. No further implications are necessary. Save initialization pattern, A = 0, B = C = E = X and restore the logic values on various lines to those that existed at the end of Step 2.

Step 4: D-drive through gate G3. 5a = $\overline{D}$, set 4a = 1, 10 = $\overline{D}$. This sets primary input line E to 1.

Step 5: D-drive through gate G6. Since $\overline{4b}$ = 0, $\overline{5b}$ = D, 12 = D.

Step 6: D-drive through gate G5. Set line 9 to 0 so that 13 = $\overline{D}$.

Step 7: Consistency check for gate G4. 9 = 0 and 1C = 1 requires 8a = 0. This further requires 6 = 1, 7 = 0, B = 1, 11 = 1.

Step 8: D-drive through G8. Since 12 = D, 11 = 1, 14 = D.

Step 9: D-drive through B3. Using the third D-cube in Fig. 5, we get 14 = D, 13 = $\overline{D}$, 15 = D. Thus the test is A = 1, B = 1, C = X, E = 1. This test should be preceded by the initialization pattern found in Step 3.

The logic gate model for the CMOS circuit can also be used to generate tests for stuck faults at the inputs and outputs of the gates. As an illustration, we generate a test to detect the fault, "input B stuck-at-1." This fault is equivalent to the "line 2 stuck-at-1" in the logic model of Fig. 8. We proceed as follows:

Step 1: The initial test cube for the fault line 2 stuck at 1 is: line 2 = 0, line 3 = 0 and line 6 = $\overline{D}$.

Step 2: To justify the logic value 0 on line 2 and logic value 0 on line 3 requires: B = 0 and C = 0.

Step 3: The D-drive of $\overline{D}$ from line 6 to the output line 15 requires the following values on the inputs: A = 1 and E = X.

Hence, the test requires the logic values 100X on the primary inputs A, B, C and E, respectively.

EXAMPLE 2

Tests were generated for all stuck and transistor faults in a two-bit adder. A CMOS circuit implementing this two-bit adder is shown in Fig. 9 and the corresponding logic gate model is shown in Fig. 10. A test set for all stuck faults and transistor faults was manually written. Transistor faults were first considered. For the first fault, T17-OPEN, the test consisted of two patterns:

A0	B0	C0	A1	B1
1	1	0	X	X
0	1	0	X	X

Upon fault simulation [5] of the circuit of Fig. 9, it was found that these patterns also detected the faults T13-SHORT, T16-OPEN, T1-SHORT, T9-OPEN and T8-OPEN. The next fault for which a test was then generated was T3-OPEN. In this way 38 patterns were generated to cover the 52 transistor faults. These patterns are shown in Table 3. From the fault simulation of the MOS circuit it was found that all of the 48 stuck type faults were already covered by these patterns. Therefore, no more patterns were generated for them. It should be noted that the number (38) of test patterns is greater than the total number (2^5 = 32) of possible input combinations. This is due to the sequential behavior of the circuit in the presence of certain transistor faults.

MODEL FOR A TRANSMISSION GATE

A transmission gate functions as a switch with a charge retention (memory) capability in its OFF state. Even though the transmission gates are bidirectional devices they are often embedded between unidirectional gates and thus used as unidirectional elements whose function can be represented by a truth table (Fig. 11). Functionally this transmission gate can be modeled using the conventional logic gates and the new output gate "B", introduced in Table 1. The logic model for the transmission gate is shown in Fig. 12. It is easy to verify that the logical behavior of the circuit in Fig. 12 is identical to that of the transmission gate represented in Fig. 11.

In a transmission gate, the transistor -SHORT or -OPEN faults produce the same effects as the CLOCK input stuck-at-1 or stuck-at-0, respectively. Hence, modeling faults at

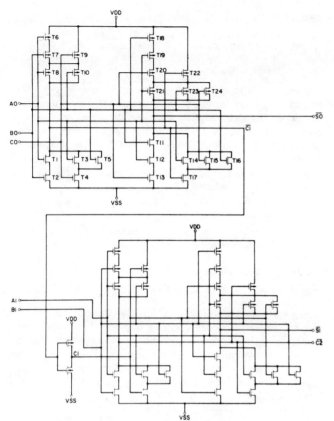

Fig. 9. CMOS implementation of the two-bit adder used in Example 2.

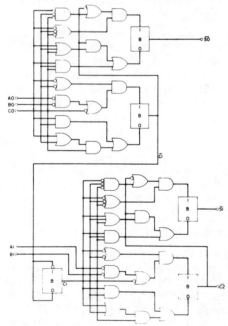

Fig. 10. Logic gate model for the circuit of Fig. 9.

TABLE 3
Test Patterns for two-bit adder

Pattern No.	A0	B0	C0	A1	B1
1	1	1	0	X	X
2	0	1	0	X	X
3	1	1	1	X	X
4	1	0	0	X	X
5	1	0	1	X	X
6	0	0	1	X	X
7	0	1	0	X	X
8	1	1	0	X	X
9	0	0	1	X	X
10	0	1	1	X	X
11	0	0	1	X	X
12	0	0	0	X	X
13	0	0	0	X	X
14	0	1	0	X	X
15	1	1	0	X	X
16	1	1	1	X	X
17	0	0	1	X	X
18	0	1	1	1	0
19	0	1	1	X	X
20	0	0	1	1	0
21	1	1	1	1	1
22	1	0	0	0	1
23	1	1	X	0	1
24	1	1	X	0	0
25	0	X	0	0	1
26	0	X	0	1	1
27	1	1	X	1	0
28	0	0	X	0	0
29	X	0	0	1	0
30	1	X	1	1	0
31	0	0	X	0	0
32	X	0	0	1	0
33	X	1	1	0	1
34	0	X	0	0	1
35	0	X	0	1	1
36	1	1	X	0	0
37	1	1	X	1	0
38	1	1	X	1	1

nMOS TRANSMISSION GATE

TRUTH TABLE		
DATA	CLOCK	Q
0	1	0
1	1	1
0	0	M
1	0	M

Fig. 11. Transmission gate and its truth table representation as an unidirectional gate.

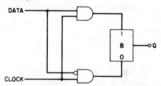

Fig. 12. Logic model for tansmission gate.

transmission gate. Let us generate a test to detect the fault "transistor T3-SHORT" (or input line C stuck-at-1). This fault is equivalent to the fault "line 2 stuck-at-1" in the logic model shown in Fig. 14. We proceed as follows:

Step 1: The initial test cube for fault line 2 stuck-at-1 is: line 2 = 0, line 4b = 1, line 5 = $\bar{D}$.

Step 2: To justify the above values on lines 2 and 4b requires: C = 0 and A = 0

Step 3: The D-drive of $\bar{D}$ through output gate B2 is accomplished using the eighth D-cube from Fig. 5.
line 5 = $\bar{D}$ line 6 = 0   line 7 = $\bar{D}$ (0)

Step 4: Line 7 has to be initialized to 0 in presence of the fault. The initialization requires following values on the inputs: A = 1, C = 1 and B = X.

transmission gate inputs and output pins is sufficient. Stuck faults on input and output lines of the transmission gate correspond to stuck faults on the inputs and output lines in the model of Fig. 12. The test generation for any given fault can be done in a manner discussed in the previous section.

Example:

The MOS circuit shown in Fig. 13 is used to illustrate the test-generation procedure for a circuit containing a

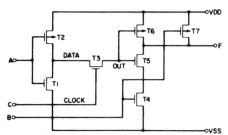

Fig. 13. Circuit used as example of test generation for transmission gate faults.

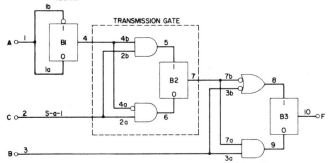

Fig. 14. Logic model for the circuit of Fig. 13.

Step 5: Return to the states at the end of Step 3. Driving $\bar{D}$ from line 7 to the output F only requires setting the input B to 1.

Hence, the test requires two vectors. The initialization vector sets the values 1X1 on the primary inputs A, B and C, respectively. The second test vector then changes the input values to 010.

A MODEL FOR BUS

The model for a "bus" structure is dependent upon the technology. The bus structure in logic circuits is either modeled as Tied-AND or Tied-OR. We will discuss the Tied-AND structure and the reader can easily adapt the model for the Tied-OR structure.

A bus structure is a physical connection of outputs of two or more gates. Hence, all the gates that are tied to a bus should be considered as a family. In order to ascertain the logical value of the bus all the members of this family should be examined. If this is done, then there will be no need to consider the bus separately. The main problem with this approach is that one will have to write D-cubes separately for each bus connection. For simplicity, therefore, we will model the bus as a separate gate.

The inputs to the bus, i.e., the outputs of gates feeding the bus, in a fault-free MOS circuit, can have three states (logical 0, 1 or high impedance). When all the inputs to the bus are in high impedance state, then the bus retains its past value. In addition, we will consider the bus as an unidirectional gate. The high impedance state at an input is represented as "M." Just like logic values 0 and 1 this state can also be justified in certain cases. The truth-table for a two input AND-bus is given in Fig. 15. Singular cover and the D-cubes for the AND-bus are given in Figs. 15 and 16, respectively. Similarly, singular cover and D-cubes can be written for the AND-bus having more than two inputs.

A few things, that are special to the handling of buses, should be noted here. First, if a gate, whose output feeds into a bus, is to a be initialized to a certain logic value, then initialization should be done after propagating the D-Drive through the bus. For example, consider the fourth D-cube in

I1	I2	OUTPUT
I	0	0
I	I	I
I	M	I
0	0	0
0	I	0
0	M	0
M	0	0
M	I	I
M	M	M

SINGULAR COVER		
I1	I2	OUTPUT
0	X	0
X	0	0
I	I	I
I	M	I
M	I	I
M	M	M

Fig. 15. Truth table model of a two-input AND-BUS and the corresponding singular cover.

I1	I2	OUTPUT
I	D	D
I	$\bar{D}$	$\bar{D}$
I	$\bar{D}(0)$	$\bar{D}$
I	$\bar{D}(1)$	$\bar{D}$
M	D	D
M	$\bar{D}$	$\bar{D}$
M	D(0)	D(0)
M	D(1)	D(1)
M	$\bar{D}(0)$	D(0)
M	$\bar{D}(0)$	D(1)
D	I	D
$\bar{D}$	I	$\bar{D}$
$\bar{D}$	M	$\bar{D}$
D(0)	M	D(0)
D(1)	M	D(1)
$\bar{D}(0)$	I	$\bar{D}$
$\bar{D}(1)$	I	$\bar{D}$
$\bar{D}(0)$	M	$\bar{D}(0)$
$\bar{D}(1)$	M	$\bar{D}(1)$
D	D	D
$\bar{D}$	$\bar{D}$	$\bar{D}$

Fig. 16. D-cubes for the two-input AND-BUS of Fig. 15.

Fig. 16. Here, the gate whose output is I2 is to be initialized to the value of 1 during the test. Instead of performing the initialization, if we use this D-cube for D-drive, the output of the bus will be set to a $\bar{D}$ and no initialization will be needed, provided a 1 can be justified for I1. If, however, a 1 is not possible on I1, then the tenth D-cube will be used and the output of the bus will have to be initialized to a 1. Second, some D-cubes are not considered because they will never be needed under the single fault assumption. For example, it is not possible that two gates whose outputs are bused, would require a simultaneous initialization during the test generation.

Example:

Consider the simple MOS circuit shown in Fig. 17. The logic gate model for test-generation is shown in Fig. 18. Let us generate a test to detect the fault, "transistor T2-OPEN (or control input, B of transmission gate T2 stuck-at-0)." This fault is equivalent to the fault "line 2 stuck-at-0" in the logic gate model of Fig. 18. Test generation proceeds as follows:

Step 1: Primary input B is set to 1; this produces a D on 2a and 2b. By considering the D-cubes for logic gate G2, we get: 2b = D, 1b = 0 and 6 = D. This also sets A = 0.

Step 2: Forward implication at G1: 1a = 0 implies line 5 = 0.

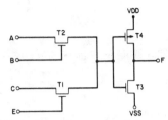

Fig. 17. MOS circuit used in test generation example involving a bus.

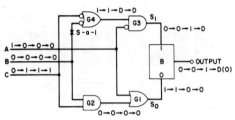

Fig. 19. Propagation of transitions of a two-pattern test.

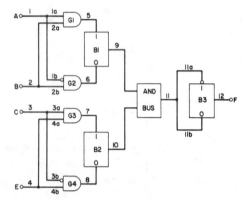

Fig. 18. Logic model for the circuit of Fig. 17.

Step 3: D-drive through B1: line 5 = 0, line 6 = D, line 9 = $\bar{D}(1)$. Since line 9 is an input to bus, the initialization is to be done after D-drive through the bus.

Step 4: D-drive through AND-bus: line 9 = $\bar{D}(1)$, line 10 = 1 and line 11 = $\bar{D}$. Thus, no initialization is needed. [Note: If we had chosen a different D-cube, for example, line 9 = $\bar{D}(1)$, line 10 = M, and line 11 = $\bar{D}$, then, the initialization at the bus would be necessary.]

Step 5: Consistency: To justify the value 1 on line 10 we set C = 1 and E = 1. Hence, the test requires the values 0111 on the primary inputs A, B, C and E, respectively.

EFFECT OF GATE-DELAYS ON TESTS

The test generation procedures described above treat MOS devices as ideal gates, which are described by their logical behavior alone. This description is sufficient for one-pattern, purely combinational, tests. In general, however, transistor faults may require two-pattern tests, where the first pattern is used for initialization. At the transition from the first pattern to the second pattern, the test generator assumes that all logical values on various lines in the circuit change simultaneously. In practice, the delays in devices and lines will cause the line values to change sequentially. This can result in multiple transitions which, in turn, can change the initialization. As an example, consider the fault P2-OPEN in the circuit of Fig. 1. In the logic model of Fig. 3, this fault is represented as "input B of G4 stuck-at-1." Consider the test A = 0, B = 0, C = 1 with the initialization pattern 100. Let us assume that each gate, with the exception of the "B" gate, in the model of Fig. 3 has one unit of delay. Then a transition at the circuit input will propagate to the output in two units of time. In Fig. 19 these transitions are shown for each line as $i \rightarrow j \rightarrow k \rightarrow \ell$, where the first transition $i \rightarrow j$ takes place at time zero, second transition $j \rightarrow k$ occurs at time = 1 unit and final transition $k \rightarrow \ell$ occurs at time = 2 units. Notice that at the output of gate G3, two transitions are occurring. The intermediate transition $0 \rightarrow 1$ causes the output initialization to change to 1 from the original 0 that was required by D(0). Thus the fault will remain undetected. This phenomenon can be traced in the CMOS circuit of Fig. 1 also. The initialization pattern shorts N1 to set OUTPUT to 0. It also opens P1 and shorts P2 (only in fault-free circuit) and P3. Transition $1 \rightarrow 0$ at A opens N1 and shorts P1. If the transition $0 \rightarrow 1$ at C is slightly delayed, OUTPUT will be changed to 1 before P3 is opened. Now whether P2 is short (fault-free) or open (faulty), OUTPUT will be 1. However, if transition $0 \rightarrow 1$ at C had occurred before the transition $1 \rightarrow 0$ at A, then OUTPUT remains at 0 and the fault will be detected. It can be easily verified that an initialization pattern 011 would have detected the fault irrespective of the circuit delays. In this case only one input of G3 changes and therefore, the hazard-like transition at its output is avoided. During the generation of a test pattern, the possibilities of such hazards can be reduced by so choosing the D-cubes and singular cubes that the Hamming distance between the individual gate inputs for two-pattern tests is minimized.

CONCLUSION

A logic model which allows test generation for faults in MOS circuits is described. The algorithms have been illustrated by examples. An implementation of these algorithms should not be much more complex than the conventional D-algorithm as the added initialization phase requires only the basic features of the D-algorithm. Past experience has demonstrated the usefulness of the D-algorithm in generating tests and in detecting redundancies. Also the technology dependent faults (particularly in CMOS circuits) are beginning to cause testing-related concerns. It is believed that the present work will serve a useful purpose.

ACKNOWLEDGMENT

Authors acknowledge cooperation from A. K. Bose and M. R. Mercer.

REFERENCES

[1] Y. M. El-ziq and R. J. Cloutier, "Functional-Level Test Generation for Stuck-Open Faults in CMOS VLSI," IEEE International Test Conference, Philadelphia, PA, October 27-29, 1981, *Digest of Papers*, pp. 536-546.

[2] K. W. Chiang and Z. G. Vranesic, "Test Generation for MOS Complex Gate Networks," 12th International Symposium on Fault-Tolerant Computing, Santa Monica, CA, June 22-24, 1982, *Digest of Papers*, pp. 149-157.

[3] Y. H. Levendel and P. R. Menon, private communication.

[4] J. P. Roth, W. G. Bouricius, and P. R. Schneider, "Programmed Algorithms to Compute Tests to Detect and Distinguish Between Failures in Logic Circuits," *IEEE Transactions on Electronic Computers*, Vol. EC-16, October 1967, pp. 547-580.

[5] A. K. Bose, P. Kozak, C-Y Lo, H. N. Nham, E. Pacas-Skewes, and K. Wu, "A Fault Simulator for MOS LSI Circuits," *Proceedings of 19th Design Automation Conference*, Las Vegas, Nevada, June 14-16, 1982, pp. 400-409.

[6] Y. H. Levendel, P. R. Menon and C. E. Miller, "Accurate Logic Simulation Models for TTL Totempole and MOS Gates and Tristate Devices," *Bell System Technical Journal*, Vol. 60, September 1981, pp. 1271-1287.

ROBUST TESTS FOR STUCK-OPEN FAULTS IN CMOS COMBINATIONAL LOGIC CIRCUITS

Sudhakar M. Reddy
Madhukar K. Reddy
Department of Electrical and Computer Engineering
University of Iowa, Iowa City, Iowa 52242

Vishwani D. Agrawal
AT&T Bell Laboratories
Murray Hill, New Jersey 07974

ABSTRACT — Recently it has been shown that tests for stuck-open faults in CMOS combinational logic circuits could potentially be invalidated due to delays in the circuit under test and/or due to timing skews in the input changes. A method is given to derive tests, when they exist, that will remain valid in the presence of arbitrary delays in the circuit under test and/or timing skews in the input changes.

I. INTRODUCTION

It has been shown by several researchers [1-6] that stuck-open faults in CMOS combinational logic circuits create memory and require test-sequences of length two. These test-sequences are called two-pattern tests [6]. A two-pattern test for detecting a stuck-open fault consists of an initializing input pattern T_1 followed by a test input pattern T_2; T_1 to initialize the output of the faulty CMOS gate to the logic value which is the complement of the expected output of the gate when the circuit is fault-free, and T_2 to cause a high impedance state at the output of the faulty gate and to sensitize one of the observable outputs of the circuit to the output of the faulty gate. Spurious logic values may occur in the circuit when the transition from T_1 to T_2 is made, due to time-skews in input changes and/or unequal delays along different paths in the circuit. These spurious logic values may change the initialized values (which are necessary for the detection of the stuck-open fault) resulting in the invalidation of the two-pattern test [6,7]. It is possible to prevent such test invalidation by carefully choosing T_1 and T_2 [6,7]. In this paper, we present a procedure to generate robust two-pattern tests (i.e., two-pattern tests which are not invalidated by arbitrary skews in input changes and/or circuit delays). Our procedure will generate a test, if it exists, for a stuck-open fault in a combinational logic circuit constructed from "full" CMOS gates only.

II. PRELIMINARIES

Several types of CMOS gates are used in designing logic circuits. These gates may be referred to as "full" CMOS gates, tristate CMOS gates, transmission gates, and domino CMOS gates [8]. In this paper, we consider combinational logic circuits constructed from full CMOS gates only. A block diagram of a full CMOS gate is shown in Figure 1 and an example is given in Figure 2. Every input to the gate is connected to a PFET and a NFET, and for every input combination, when the gate is fault-free, there is a path from the output node to either V_{DD} or V_{SS}, but not to both, through conducting FETs. When there is a path from the output to $V_{DD}(V_{SS})$ through conducting FETs and there is no path to $V_{SS}(V_{DD})$, the output is at logic 1(0). (Some gates may have the property that for some inputs, conducting paths from the output node to V_{DD} and V_{SS} occur simultaneously. Designers normally warn users to avoid such inputs).

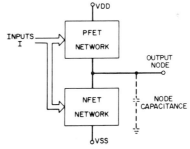

Fig. 1   Block diagram of a CMOS gate.

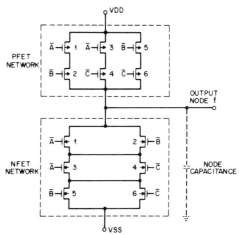

Fig. 2   A full CMOS gate realizing $f(A, B, C) = AB + AC + BC$.

It was shown in [1,9] that the behavior of complex CMOS combinational circuits, both fault-free and faulty, cannot be modeled by logic circuits made up of classical gates (e.g., AND, OR, NOT, NAND, NOR gates). For this reason several researchers have proposed switch-capacitance models [10], graph models [5], and logic models using gates and memory type elements [1,6]. There are relative advantages and disadvantages of the models proposed. In this paper we use the gate plus memory model proposed by Jain and Agrawal [6]. The main reason for our choice is that most currently existing fault simulation and test pattern generator programs use gate level models and these programs can be readily augmented to implement the new test procedure. A second reason for our choice is that the model allows a simple and uniform method to treat most types of CMOS gates.

The derivation of the model of a CMOS combinational logic circuit using the procedure given by Jain and Agrawal [6] requires that:

(i) each CMOS gate in the circuit be replaced by a network consisting of a B block (the memory element shown in Figure 3) driven by two networks constructed from AND, OR, NOT gates, and

(ii) the gate networks driving the S_1 (S_0) input of the B block are obtained by iteratively replacing series connections of PFETs (NFETs) by AND gates and parallel connections of PFETs (NFETs) by OR gates and finally applying the inputs to PFETs (NFETs) as inputs to the gate network driving S_1 (S_0) input through inverters (straight connections).

This procedure is applied to the circuit of Figure 2 to obtain the circuit of Figure 4. Notice that the S_1 (S_0) input of the B block takes the value of 1 whenever the combination of inputs is such that a conducting path from V_{DD} (V_{SS}) to the output node occurs. The truth table for the B block then indicates the logic operation occurring at the output node of the gate.

Next we will define stuck-open faults and review the problem of detecting them in CMOS combinational logic circuits.

Definition 1: A *stuck-open fault* in a CMOS logic gate G is any failure that permanently leaves at most one serially connected set of FETs in G in a non-conducting state.

Definition 2: A *single stuck-open fault* in a CMOS logic circuit is a failure that causes a stuck-open fault in at most one gate in the circuit.

Several authors have proposed procedures to detect all single stuck-open faults in CMOS logic circuits [1-6,11]. We illustrate the current approach to the detection of stuck-open faults in CMOS combinational logic circuits by an example. Consider a stuck-open fault that leaves NFET 1 in Figure 2 open. To detect this fault, a "test input" should be applied to the gate such that the output of the gate is "affected" in the presence of the fault (i.e. the output should be different from the expected value in a fault-free gate) and if the gate is part of a larger circuit then the circuit inputs must also propagate this affected value to an observable output. Since the fault is a stuck-open fault, affecting an NFET, the only effect it can have is to put the output node, for some input(s), in a high impedance state instead of at logic value 0. Since the stray capacitance and gate capacitance of the driven gates at the output node will retain the previous output value, the change due to the effect of fault can be noticed at the output node only if the previous output value was 1. This essentially implies that the "test" for a stuck-open fault has two input patterns:

(i) an initializing input, T_1, to place the output of the faulty gate at 1 (0) to detect NFET (PFET) stuck-open fault, and

(ii) a test input, T_2, to sensitize the effect of the fault to the output node of the gate as well as to propagate this effect to an observable output.

For the NFET 1 stuck-open fault in Figure 2, an appropriate initializing and test input pair $<T_1, T_2>$ could be $<101, 010>$.

III. PROBLEM FORMULATION

As illustrated in the last section, a test to detect a stuck-open fault in a CMOS logic circuit requires a pair of inputs $<T_1, T_2>$. However, due to circuit delays and/or timing skews

Fig. 3   Truth table of B block. M represents the high impedance state. The last 0 output indicates 0 as the dominating logic when conducting paths from the output node to VDD and VSS are present, simultaneously.

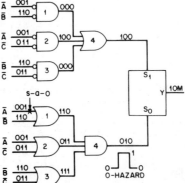

Fig. 4   Equivalent circuit of the CMOS gate of Figure 2.

in the changes of input values the tests may be invalidated (i.e. the tests may not detect the faults they were meant to detect) [6,7]. As a matter of fact, there may be stuck-open faults in some circuits for which no tests that will not be potentially invalidated, exist [7]. As an example of this invalidation process, consider the test $<101, 010>$ that was expected to detect NET 1 stuck-open fault in the circuit of Figure 2. If in changing from input 101 to 010, input C changed earlier than inputs A and B, the inputs to the gate under test could be 100. This input would make NFETs 2, 4 and 6 to conduct momentarily, driving the output node to 0 and hence discharging the capacitor at the output. When the inputs ultimately change to 010, the output node does go to a high impedance state if NFET 1 was faulty, however, the output logic value under test input 010 would be zero (having become zero due to earlier discharge of the capacitor at the output) which is the logic value expected at the output of the fault-free gate under input 010.

To better understand the mechanisms causing the test invalidation and to motivate the methods used in deriving the test procedure described in the next section, consider the gate equivalent circuit of Figure 2, given in Figure 4. On this equivalent circuit the inputs $<101, 010>$ as well as an intermediate input value (100), potentially entered during the change from 101 to 010 are shown. At this point, we need to digress to review the model for faults and test generation mechanism for the equivalent circuit and some ideas on hazards in logic circuits.

In the gate equivalent circuit, derived by the procedures given by Jain and Agrawal [6], for a CMOS combinational circuit the tests are derived for appropriate stuck-at-0 and stuck-at-1 faults. The correspondence of the faults in the equivalent circuit to the stuck-open faults in the modeled CMOS circuit is:

a stuck-open fault on a PFET (NFET) corresponds to a stuck-at-1 (stuck-at-0) fault on the corresponding input to the gate network driving S_1 (S_0) input of the B block.

For example, the faults in the equivalent circuit shown in Figure 4 for PFET 2 and NFET 1 stuck-open are $\bar{B}$ input to

AND gate 1 stuck-at-1 and $\bar{A}$ input to OR gate 1 stuck-at-0, respectively.

Definition 3: A static 0-hazard (static 1-hazard) is said to exist on a line ℓ, in a logic circuit, for input change from T_1 to T_2, if and only if:

(i) under inputs T_1 and T_2 the expected logic value on line ℓ is 0 (1), and

(ii) during the change in inputs from T_1 to T_2 there exists a possibility (due to delays in the circuit and/or skews in input changes) of line ℓ momentarily taking the logic value 1 (0). (In the literature some times static hazards have been associated with input changes that require only one input variable to change. The definition given here allows T_1 and T_2 to be different in more than one position and is essentially the one used in [14]. The word "static" is to be associated with the fact that the value on line ℓ is to remain at a fixed value for inputs T_1, T_2 and during the change of inputs from T_1 to T_2).

We can now articulate the cause for test invalidation for the stuck-open fault on NFET 2 of Figure 2. Referring to the equivalent circuit of Figure 4, we see that when inputs change from 101 to 010 a static 0-hazard exists on the S_0 input of the B block, which causes the output of the B block change to zero before entering the M (high impedance) state. Note that the hazard is present in the *faulty circuit*. A test for NFET 1 stuck-open which will remain valid in the presence of arbitrary delays in the circuit under test (CUT) can be obtained by choosing the initializing input T_1 to be 011 and the test input T_2 to be 010 (same as the earlier test input). This can be readily verified.

IV. TEST PROCEDURE

In this section, we give a procedure to derive robust tests to detect single stuck-open faults in full CMOS combinational logic circuits. Clearly, one can take three approaches. One is to assume that the test input T_2 of a two-pattern test is given and then determine an appropriate initializing input T_1. Second approach is to assume that an initializing input T_1 of a two-pattern test is known and then determine an appropriate test input T_2. Lastly, one can attempt to derive both T_1 and T_2 simultaneously. In this paper we take the first approach, as it is conceptually simplest to derive.

We make the following assumptions:

A1: T_1, the initializing input of a two-pattern test is applied at time t_1,

A2: T_2, the test input of a two-pattern test is applied at time t_2,

A3: For the purposes of two-pattern generation the output of a CMOS gate G is to be set to 0 only by attempting to create a conducting path through its NFET network while making sure that no conducting path is created through the PFET network (i.e. the inputs to the B block of the gate equivalent circuit modeling G should be $S_1=0$ and $S_0=1$. This is a prudent assumption to insure that the tests do not depend on the dominance of logic 0 over logic 1), and

A4: T_2 is known and hence the state of all lines in the faulty CUT at t_2 are known or are don't care (i.e., left unspecified). (Even the value of the output of the faulty gate is known to be zero (one) if the fault being considered is a PFET (NFET) stuck-open. Of course to assure this value at time t_2, the initializing input T_1 should properly initialize the gate output. A procedure based on the D-algorithm [12] was given in [6] to derive T_2.)

At t_2 the S_1 and S_0 inputs of the B block modeling the output node of the faulty gate will be both zero (indicating that the faulty gate output is in high impedance state). The desired values on these S_1 and S_0 lines at t_1 are 1 and 0 (0 and 1), respectively, if the fault under consideration is a NFET (PFET) stuck-open. Theorem 1, given next, is a necessary and sufficient condition on the S_1 and S_0 lines of the faulty gate.

Theorem 1: Let $<T_1, T_2>$ be a two pattern test to detect, in the presence of arbitrary delays in the CUT and/or timing skews in input changes, a NFET (PFET) stuck-open fault in gate G of a CMOS combinational logic circuit. It is necessary and sufficient that S_0 (S_1) input of the B block modeling the output of the faulty gate G be free of static 0-hazard.

Proof: Assume that the fault under consideration is PFET stuck-open.

Necessity: Assume that the S_1 line of the B block corresponding to the faulty gate has a static 0-hazard under input sequence $<T_1, T_2>$. Then the S_1 line may momentarily take the value of 1 after the S_0 line has changed to 0, thus changing the output of the gate to 1. Afterwards, when both S_0 and S_1 take the value 0, the output of the gate reaches high impedance and retains the logic value 1. This contradicts that the two-pattern test $<T_1, T_2>$ detects the fault under consideration.

Sufficiency: Assume that the S_1 line of the B block is free of static 0-hazards. In this case, independent of the timing of the changes on S_0 line of the B block, the output will remain at 0 since the only way the output will go to 1 is for S_1 to be 1 and S_0 to be 0, which is not possible. All gates, other than G, are combinational even in the presence of the fault, hence, their steady state response is not affected by hazards.

The theorem can be similarly proved for NFET stuck-open faults.   Q.E.D.

Theorem 1, proved above, provides a method to derive T_1 given T_2. Essentially T_1 should be chosen so as to avoid static 0-hazards on S_1 or S_0 line of the faulty gate. Earlier, Breuer and Harrison [13] have studied a similar problem related to the elimination of static and dynamic hazards in generating tests for sequential circuits. The procedures given in [13] are applicable to sequential circuits constructed from gates (AND, OR, NAND, etc.) and flip-flops. The problem at hand requires two modifications to the procedures developed in [13]. One is that we need to avoid only static hazards as opposed to both static and dynamic hazards for sequential circuits, and the other enchancement needed is to accommodate the B block.

The basic idea is to start from the faulty gate and proceed towards the primary inputs by specifying the state, at time T_1, of all lines driving the faulty gate to avoid static 0-hazards on the S_0 or S_1 line of the faulty gate. To facilitate this process we will use the following notation.

Each line in the faulty circuit has associated with it a sequence of two values from (0, 1, d) representing its state at time t_1 and t_2 (d is don't care). Also associated with a line is the hazard status of the line; hp for hazard present, hu for hazard status unknown and hf for hazard-free. We also assume that a list names *HFR* is maintained while deriving T_1. *HFR* contains the names or labels of lines which are required to be hazard-free. A member of *HFR* is deleted when its hazard status becomes known to be hazard free. A member is added to *HFR* when a new hazard-free requirement is created.

The procedures to derive T_1 given T_2 are similar to the line justification step of the D-algorithm [12]. Hence we will not attempt to give a formal procedure. An intuitive version is given below. Since we need to determine the state, at time t_1, of only those lines that drive the gate under test it is simpler to derive a subcircuit of the CUT that includes only these lines and then process this circuit to derive T_1. We call this circuit CUT' (note that for our purposes CUT is the gate equivalent circuit of the given CMOS logic circuit). In CUT' the faulty line is isolated and permanently assigned its stuck-at-value, together with the hazard status of hf. For convenience, the primary inputs and their fanout branches are not assigned the hazard status hf before hand, instead, a hazard-free requirement is assumed to be automatically satisfied when it arrives at a primary input.

Procedure Robust_Test:

1. Set *HFR* to null. Set the state of the faulty line to its stuck-at value. Set the state of all lines to d at time t_1. Set hazard status of the faulty line to hf. Set the hazard status of all other lines to hu.

2. Add S_0 (S_1) line of the B block of the faulty gate to *HFR*, if the stuck-open fault is a NFET (PFET) fault. Depending upon the fault under consideration, set, the inputs and outputs of the B block of the faulty gate as shown in Figures 5(a) and 5(b).

3. If *HFR* is not empty, pick an element, say x, of *HFR* and attempt to specify the values on the lines driving the gate or B block whose output is x such that the hazard status of x becomes hf. If no choice to set the hazard status of x to hf exists, back track to the last choice made and make an alternate choice, if possible. If a choice existed, then delete x from *HFR*. Add lines to *HFR* when new hazard-free requirements are created. Perform line justification needed to satisfy any new line values specified.

4. Perform implication of the choices made at Step 3. If in the process of implication the hazard status of a line in *HFR* becomes hf, then delete this line from *HFR*. If a conflict occurs in the implication due to hazard conditions or logic values, then backtrack to the last choice made and make an alternative choice, if possible. (In this step, the hazard status of a line could change to hf or hp from hu, but never from hp to hf or vice versa.)

5. If *HFR* is empty and no line justifications are to be done, exit declaring T_1. Otherwise, go back to step 3.

It can be seen that Step 3 is similar to D-drive and line justification portion of the D-algorithm (in our case we are driving towards primary inputs). What remains to be done, to complete the implementation of Procedure Robust_Test is to specify how to compute the hazard status of AND, OR, NOT gates and B block from the status of the lines driving them and how to specify the inputs to these gates to satisfy a hazard-free requirement specified on their outputs. The requirements to

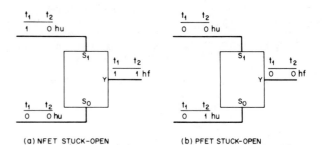

Fig. 5 Initial setting of the B block of the faulty gate for stuck-open fault. The output at t_2 is due to the memory state (M).

propagate (backwards) hazard-free conditions are given in Table I. The methods to calculate hazard status of outputs of gates, given the status of their inputs, are given in Figures 6-8 and Table II. Note that the values 01hp, 10hf, 01hp, and 10hp never appear on any lines as dynamic hazards are of no significance here.

To illustrate the procedure to derive input T_1, we apply it to the circuit of Figure 9. The equivalent circuit for this CMOS logic circuit is given in Figure 10. Consider the fault NFET 2 (refer to Figure 9) stuck-open. This fault is equivalent to the B input to OR gate 3 of Figure 10 stuck-at-0. A test T_2 to detect this fault can be obtained by using the test procedure given in [6]. One such T_2 is A=B=D=1 and C=0. As the first step to the derivation of T_1 for the given T_2 we determine that part of the CUT which drives block B numbered 5 (see Figure 10). This part of the circuit is shown in Figure 11, in which the faulty line (line 9 in Figure 11) is isolated and is shown to be permanently at 0. In Table III steps of the procedure to determine T_1 for given T_2=1101 are shown. The '__' entries in this table indicate that the previous assignment to the line holds. In the last step, we obtain T_1 = 0d0d.

V. SUMMARY

In this paper, a procedure to derive robust two-pattern tests to detect stuck-open faults in combinational logic circuits made of full CMO gates is given. The distinguishing feature of the tests derived is that they will not be invalidated by arbitrary delays in the circuit under test.

Our approach was to assume that the test T_2 in a two pattern test $<T_1, T_2>$ is given and then provide a procedure to determine an appropriate initializing input T_1. If no appropriate T_1 is found, another test T_2 should be obtained and the procedure iterated until one $<T_1, T_2>$ pair is determined. If our procedure fails to provide a T_1 for all possible T_2s this would imply that no robust test exists for the corresponding fault. Under these conditions one may have to redesign the circuit under test, to render it testable [7].

The procedure Robust_Test can be extended to generate tests for stuck-open faults in CMOS combinational logic circuits which also include tristate gates and transmission gates. In the general case, the initial hazard-free requirements may not only be generated by the faulty CMOS gate, but also by some fault-free tristate gates or transmission gates. An extra logic value will be needed to represent the "unknown" Steady-state signal which may be generated at the outputs of tristate gates and transmission gates. The tables for determination of the output and hazard state of AND, OR, NOT gates and the B block should also be expanded to accommodate this new value, but this is simple.

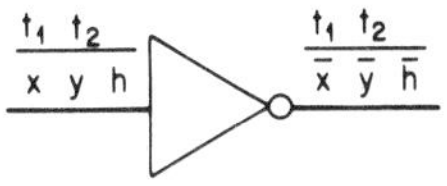

$x \in \{0, 1, d\}$, $y \in \{0, 1, d\}$ and

$h \in \{hp, hu, hf\}$

note that $\bar{d} = d$.

Figure 6: Calculation of the hazard status of NOT gate

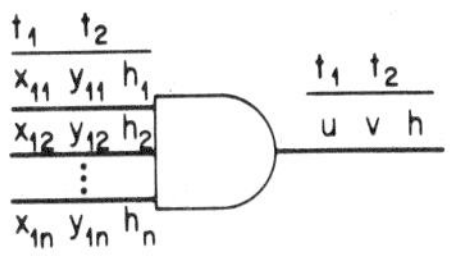

$x_{lj}, y_{lk}, u, v \in \{0, 1, d\}$

$h, h_j \in \{hp, hu, hf\}$

$u = x_{11} \wedge x_{12} \wedge ... \wedge x_{ln}$

$v = y_{11} \wedge y_{12} \wedge ... \wedge y_{ln}$

If $u=v=1$ then $h=hf$, iff $h_1=h_2=...=h_n=hf$

$=hp$, iff at least one of $\{h_1, h_2,...,h_n\}$ is hp

$=hu$, otherwise

If $u=v=0$ then $h=hf$, iff at least one input is $00hf$

$=hu$, iff no input is $oohf$ and at least one input

is $00hu$ or $0dhu$ or $d0hu$ or $ddhu$

$=hp$, otherwise

If $u=v=d$, or $u \neq v$ then $h = hu$

Figure 7: Calculation of the hazard status of AND gate

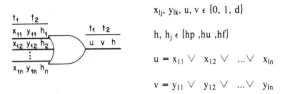

$x_{lj}, y_{lk}, u, v \in \{0, 1, d\}$

$h, h_j \in \{hp, hu, hf\}$

$u = x_{11} \vee x_{12} \vee ... \vee x_{ln}$

$v = y_{11} \vee y_{12} \vee ... \vee y_{ln}$

If $u=v=0$ then $h=hf$, iff $h_1=h_2=...=h_n=hf$

$=hp$, iff at least one of $\{h_1, h_2,...h_n\}$ is hp

$= hu$, otherwise

If $u=v=1$ then $h=hf$, iff at least one of the inputs is $11hf$

$=hu$, iff no input is $11hf$ and at least one input is
$11hu$ pr $1dhu$ or $d1hu$ or $ddhu$

$= hp$, otherwise

If $u=v=d$ or $u \neq v$ then $h = hu$

Figure 8: Calculation of the hazard status of OR gate

Detection of stuck-open faults in logic circuits constructed from domino CMOS gates has been studied in [15]. It was shown that the test invalidation, due to delays in the CUT, does not occur in combinational circuits using domino CMOS gates.

It is important to point out that gate model for CMOS logic circuits given in [6] has certain limitations. For example, the model, (i) is applicable only to CMOS gates realizing series-parallel circuits, (ii) uses a memory module (B Block) and separate equivalent circuits for NFET and PFET circuits, and (iii) may not accurately represent some circuits using transmission gates. The first two problems are addressed in [16] where the model is generalized to CMOS gates realizing non-series parallel circuits. It is further shown that for full CMOS gates the B block is not needed in the gate level model and also only either the NFET or the PFET network need be considered in deriving the gate level model. Combining the results of [16] and those presented here will give a simpler and complete procedure to derive robust tests for all detectable single line stuck-at, stuck-open and stuck-on faults in logic circuits constructed from full CMOS gates.

Acknowledgment - The research of S. M. Reddy and M. K. Reddy was supported in part by Army Research Office Contact - No. DAAG29 - 84-K-0044. S. M. Reddy also acknowledges the support from the AT&T Bell Laboratories.

Table I: **Propagation of hazard free requirements**

Gate Type	Hazard free Requirement on the output	Required hazard-free requirements on the inputs to gates
NOT	0-hazard free	1-hazard free
NOT	1-hazard free	0-hazard free
AND	0-hazard free	at least one input 0-hazard free
AND	1-hazard free	all inputs 1-hazard free
OR	0-hazard free	all inputs 0-hazard free
OR	1-hazard free	at least one input 1-hazard free
B block	0-hazard free	S_1 input 0-hazard free, S_0 input 1 in t_1
B block	1-hazard free	S_0 input 0-hazard free, S_1 input 1 in t_1

Table II: Output of the B block of a fault-free "full" CMOS gate in t_1, t_2 and the hazard status

S_0	00hf	00hp	00hu	01hu	0dhu	10hu	11hf	11Lp	11hu	1dhu	d0hu	d1hu	ddhu
S_1	11hf	11hp	11hu	10hu	1dhu	01hu	00hf	00hp	00hu	0dhu	d1hu	d0hu	ddhu
Output of B block	11hf	11hp	11hu	10hu	1dhu	01hu	00hf	00hp	00hu	0dhu	d1hu	d0hu	ddhu

Note: At any given time, the S_0 and S_1 inputs to the B block of a fault-free "full" CMOS gate are complementary and hence the entries in the table are not given for situations like S_0:01 hu, S_1:11hf, etc. Furthermore the hazard status of S_1 and S_0 are always identical for full CMOS gates and hence the table does not have entries for situations like S_0:00hf, S_1:11hp etc.

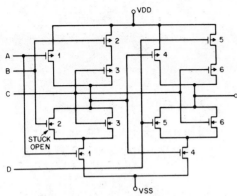

Fig. 9   A CMOS circuit realizing the function $A(B+C) + \overline{CD}$

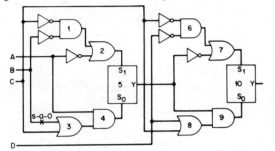

Fig. 10   The gate equivalent circuit for the CMOS circuit of Figure 9.

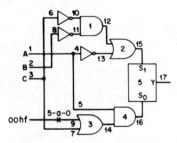

Fig. 11   CUT' derived from the CUT given in Figure 10.

REFERENCES

[1] R. L. Wadsack, "Fault Modeling and Logic Simulation of CMOS and MOS Integrated Circuits," *Bell Syst. Tech. J.*, vol. 57, pp. 1449-1474, May-June 1978.

[2] Y. M. El-Ziq, "Automatic Test Generation for Stuck-Open Faults in CMOS VLSI," *Proceedings of the Eighteenth Design Automation Conference*, Nashville, Tennessee, pp. 347-354, June 1981.

[3] Y. M. El-Ziq and R. J. Cloutier, "Functional-Level Test Generation for Stuck-Open Faults in CMOS VLSI," *Digest of Papers, 1981 International Test Conference*, Philadelphia, IEEE CH1693-1, pp. 536-546, October 1981.

[4] R. Chandramouli, "On Testing Stuck-Open Faults," *Proceedings of the 1983 International Symposium on Fault-Tolerant Computing*, Milano, Italy, June 28-30, 1983.

[5] K. W. Chiang and Z. G. Vranesic, "On Fault Detection in CMOS Logic Networks," *Proceedings of the Twentieth Design Automation Conference*, Miami Beach, FL, pp. 50-56, June 1983.

[6] S. K. Jain and V. D. Agrawal, "Test Generation for MOS Circuits Using D-Algorithm," *Proceedings of the Twentieth Design Automation Conference*, Miami Beach, FL pp. 64-70, June 1983.

[7] S. M. Reddy, M. K. Reddy, and J. G. Kuhl, "On Testable Design for CMOS Logic Circuits," *Proceedings of the 1983 International Test Conference*, Philadelphia, PA, pp. 435-445, October 1983.

[8] R. H. Krambeck, C. M. Lee, and H. S. Law, "High Speed Compact Circuits with CMOS," *IEEE Journal of Solid-State Circuits*, Vol. SC-17, pp. 614-619, June 1982.

[9] J. Galiay, Y. Crouzet and M. Vergniault, "Physical Versus Logical Fault Models in MOS LSI Circuits: Impact on their Testability," *IEEE Trans. Comp.*, vol. C-29, pp. 527-531, June 1980.

[10] R. E. Bryant, "A Switch-Level Simulation Model for Integrated Logic Circuits," Ph.D. thesis, Dept. of Electrical Engineering and Computer Science, MIT, March 1981.

[11] M. W. Levi, "CMOS is Most Testable," *Proceedings of the 1981 International Test Conference*, Philadelphia, PA, pp. 217-220, October 1981.

[12] J. P. Roth, "Diagnosis of Automata Failures: A Calculus and a Method," *IBM Journal of Research and Development*, vol. 10, pp. 278-291, July 1966.

[13] M. A. Breuer and L. Harrison, "Procedures for Eliminating Static and Dynamic Hazards in Test Generation," *IEEE Trans. on Comp.*, vol. C-23, pp. 1069-1078, October 1974.

[14] M. A. Breuer and A. D. Friedman, *Diagnosis and Reliable Design of Digital Systems*, Rockville, MD: Computer Science Press, 1976.

[15] S. R. Manthani, "On CMOS Self-Checking Checkers," Master's thesis, Dept. of Electrical and Computer Engineering, University of Iowa, Iowa City, 1984.

[16] S. M. Reddy, V. D. Agrawal and S. K. Jain, "A Gate Level Model for CMOS Combinational Logic Circuits with Applications to Fault Detection," *Proceedings of the Twenty-First Design Automation Conference*, Albuquerque, NM, June 25-27, 1984.

Table III: Generation of T_1 for given $T_2 = 1101$ for the circuit of Figure 11 with line 9 s-a-0 (equivalent to NFET 2 stuck-open in Figure 9)

Step	\multicolumn{17}{c}{Line Number}	HFR
	1	2	3	4	5	6	7	8	9	10	11	12	13	14	15	16	17	
Initial Value	d1hu	d1hu	d0hu	d1hu	d1hu	d0hu	d0hu	d1hu	00hf	d1hu	d0hu	d0hu	d0hu	d0hu	d0hu	d0hu	d1hu	null
1	--	--	--	--	--	--	--	--	--	--	--	--	--	--	10hu	00hu	11hf	{16}
2	--	--	--	--	--	--	--	--	--	--	--	--	10hu	00hu	--	00hf	--	{14}
3	--	--	--	01hu	--	--	00hu	--	--	--	--	--	--	00hf	--	--	--	{7}
4	01hu	--	00hf	--	01hu	00hf	00hf	--	--	11hf	--	--	--	--	--	--	--	null

$T_1 = 0d0d$

STAFAN: AN ALTERNATIVE TO FAULT SIMULATION

Sunil K. Jain
Vishwani D. Agrawal

AT&T Bell Laboratories
Murray Hill, New Jersey 07974

ABSTRACT — STAtistical Fault ANalysis (STAFAN) is proposed as an alternative to fault simulation of digital circuits. In this analysis, controllabilities and observabilities of circuit nodes are defined as probabilities which are estimated from signal statistics obtained from fault-free simulation. Special procedures are developed for dealing with these quantities at fanout nodes and at feedback nodes. The computed probabilities are used to derive unbiased estimates of fault detection probabilities and overall fault coverage for the given set of input vectors. Fault coverage and the undetected fault data obtained from STAFAN for actual circuits are shown to agree favorably with the fault simulator results. The computational complexity added to a fault-free simulator by STAFAN grows only linearly with the number of circuit nodes.

1.0 INTRODUCTION

Fault simulation is an essential part of VLSI design. Complexity of a fault simulator in terms of the CPU time and the memory requirements is known to grow at least as the square of the number of gates in the circuit [1]. For VLSI circuits, this is a serious limitation. One approach that has been used to combat fault simulation complexity is the random sampling of faults [2]. This requires simulation of a randomly selected subset of faults and gives an estimate of the fault coverage. The accuracy of this estimate depends only upon the number of simulated faults and not on the total number of faults in the circuit. The disadvantages of fault sampling technique are that 1) it assumes the availability (or development) of a complete fault simulation program, and 2) it gives absolutely no information about the detectability of the faults that were not sampled. A list of undetected faults is desirable for test vector development and for checking redundancies.

In this paper we develop a STAtistical Fault ANalysis (STAFAN) technique which requires only the fault-free simulation. Fault-free (or true-value) simulation is essential for design verification and is considerably less complex than the fault simulation. The presented analysis calculates the probability of detection for each stuck fault in the circuit from the signal statistics. In a recent work [3] fault detection probabilities were computed for random stimuli. In general, upper and lower bounds for the detection probabilities could be obtained. These bounds could be narrowed down at the cost of greater amount of computation time. However, the method could handle only combinational circuits. In contrast, the present method computes the detection probabilities for the given stimuli (vector set) with which the fault-free simulation is carried out. No elaborate software development is needed. The calculation adds only a small overhead to the fault-free simulation. The method is, in no way, limited to combinational circuits. It produces an estimate of fault coverage by the vector stimuli used in simulation. Since the estimate is statistical, higher accuracy could be expected for larger circuits. Another technique, known as critical path tracing [4], has been proposed as an alternative to fault simulation but it too can analyze only the circuits that are basically combinational.

STAFAN makes use of the concepts of *controllability* and *observability* [5], [6]. These quantities are redefined as probabilities of controlling and observing the lines. Controllability of a line is estimated by collecting the statistics of activity on that line. Observability is then computed from the estimated controllabilities. The product of the appropriate controllability and observability gives the detection probability of a fault. Faults can be graded in terms of their detection probabilities and the faults with lower detection probabilities are the likely ones to be left undetected by the vector set. A simple calculation leads to fault coverage.

2.0 THEORY

Consider the logic model of the circuit as is often used by simulators [7]. In this model the circuit is described in terms of single-output boolean (AND, OR, NAND, NOR and NOT) gates. The gates are connected by paths which are referred to as *lines*. The primary input line values are specified by *input vectors*. Simulator then computes the logical values (0 or 1) for all other lines. We will consider the stuck-type ($s-a-1$ and $s-a-0$) faults on input and output lines of each gate.

Let us define the following quantities for each line:

$C1(\ell)$ or *one-controllability* of line ℓ is the probability of line ℓ having a value 1 on a randomly selected vector.

$C0(\ell)$ or *zero-controllability* of line ℓ is the probability of line ℓ having a value 0 on a randomly selected vector.

$B1(\ell)$ or *one-observability* of line ℓ is the probability of observing the line ℓ at a primary output when the value of line ℓ is 1. This is the conditional probability of sensitizing a path from line ℓ to a primary output, given that the value of line ℓ is 1. Similarly,

$B0(\ell)$ or *zero-observability* of line ℓ is the path sensitization probability from line ℓ to a primary output when the value of line ℓ is zero.

2.1 Statistical Estimation of Controllabilities

Suppose we simulate N input vectors and for every line in the circuit keep two counters in the fault-free simulator. These counters will be called the one-counter and the zero-counter, respectively. Whenever a line assumes a value one or zero, its one-counter or the zero-counter is incremented. Most simulators use three values, 0, 1 and unknown (X). When the line value is X, none of the counters is incremented. If the simulator also generates the high-impedance state on a line then the same counter that was incremented on the previous vector is incremented again. After simulating N vectors, the one and zero controllabilities are estimated as

$$C1(\ell) = \frac{one\text{-}count}{N},$$

and

$$C0(\ell) = \frac{zero\text{-}count}{N}.$$

2.2 Computation of Observabilities

For observability computation the simulator keep an additional sensitization counter for each input line of a gate. This counter is incremented on a vector only if a path from the corresponding line to the gate output is sensitized. Figure 1 gives two examples of input states for which the sensitization counter of the line ℓ will be incremented. After simulating N vectors the one-level sensitization probability for line ℓ is estimated as

$$S(\ell) = \frac{sensitization\text{-}count}{N}.$$

For observability computation, first the one and zero observabilities $B1$ and $B0$ for all primary output lines are set to 1.0. Next the following procedure of propagating observabilities from a gate output to inputs is used until the observabilities of all lines have been obtained.

Consider the four input AND gate shown in Fig. 2. Lines i, j, k and ℓ are the input lines and m is the output line. Since

$$C1(m) = Prob\{i=1, j=1, k=1, \ell=1\}$$

$$= Prob\{i=1, j=1, k=1 | \ell=1\} \cdot C1(\ell),$$

therefore $Prob\{i=1, j=1, k=1, | \ell=1\} = C1(m)/C1(\ell)$
Since $B1(\ell)$ is the probability of observing line ℓ, when the value of line ℓ is 1, we have

$$B1(\ell) = B1(m) \cdot Prob\{i=1, j=1, k=1 | \ell=1\}$$

$$= B1(m) \cdot C1(m)/C1(\ell).$$

Further,

$$S(\ell) = Prob\{i=1, j=1, k=1\}$$

$$= Prob\{i=1, j=1, k=1, \ell=1\} + Prob\{i=1, j=1, k=1, \ell=0\}$$

$$= C1(m) + Prob\{i=1, j=1, k=1 | \ell=0\} \cdot C0(\ell)$$

and therefore

$$Prob\{i=1, j=1, k=1 | \ell=0\} = \frac{S(\ell) - C1(m)}{C0(\ell)}.$$

Thus,

$$B0(\ell) = B0(m) \cdot Prob\{i=1, j=1, k=1 | \ell=0\}$$

$$= B0(m) \cdot \left[\frac{S(\ell) - C1(m)}{C0(\ell)}\right].$$

Similarly, for OR gate,

$$B1(\ell) = B1(m) \left[\frac{S(\ell) - C0(m)}{C1(\ell)}\right]$$

$$B0(\ell) = B(0) \cdot C0(m)/C0(\ell),$$

for NAND gate,

$$B1(\ell) = B0(m) \cdot C0(m)/C1(\ell)$$

Fig. 1 Sensitazation of path from line ℓ to gate output.

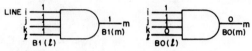

Fig. 2 Propagation of observabilities from output to input of an AND gate.

$$B0(\ell) = B1(m) \left[\frac{S(\ell) - C1(m)}{C0(\ell)}\right],$$

for NOR gate,

$$B1(\ell) = B0(m) \left[\frac{S(\ell) - C1(m)}{C1(\ell)}\right]$$

$$B0(\ell) = B1(m) \cdot C1(m)/C0(\ell),$$

and for NOT gate,

$$B1(\ell) = B0(m)$$

and

$$B0(\ell) = B1(m).$$

The above formulas allow propagation of observabilities through logic gates. Two other situations that frequently occur are discussed below.

2.3 Fanouts

Consider the line ℓ which fans out into two lines i and j as shown in Fig. 3. We will analyze one-observabilities recognizing

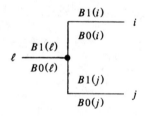

Fig. 3. Observabilities at a fanout.

that similar arguments apply to zero-observabilities. $B1(i)$ and $B1(j)$ are the observabilities of lines i and j, respectively. The value of line ℓ can be observed through either line i or line j. Therefore

$$B1(\ell) \geq B1(i),$$

and

$$B1(\ell) \geq B1(j),$$

or

$$B1(\ell) \geq \max[B1(i), B1(j)].$$

If the observation paths through i and j are completely independent then

$$B1(\ell) = B1(i) \cup B1(j)$$

$$= B1(i) + B1(j) - B1(i) \cdot B1(j).$$

This independent path expression represents the upper bound for $B1(\ell)$, Thus

$$\max[B1(i), B1(j)] \leq B1(\ell) \leq \bigcup_{k=i,j} B1(k),$$

where $\cup$ represents the joint probability of the events (assumed independent) whose probabilities are $B1(i)$ and $B1(j)$ [8]. Similarly, if line ℓ fans out to m branches, then,

$$\max_{1 \leq k \leq m} [B1(i_k)] \leq B1(\ell) \leq \bigcup_{k=1}^{m} B1(i_k).$$

Actual computation of these bounds would require a detailed analysis of the circuit topology. These bounds would depend upon the degree of reconvergence of the fanouts. However, advantage gained from such an analysis would depend upon how wide this range is. For simplicity, we will introduce an arbitrary constant α such that

$$B1(\ell) = (1-\alpha) \max_{1 \leq k \leq m} [B1(i_k)] + \alpha \left\{ \bigcup_{k=1}^{m} B1(i_k) \right\}.$$

When $\alpha = 0$, $B1(\ell)$ is minimum and $\alpha = 1$ corresponds to the upper bound. In the general case, the value of α may differ for different fanout branches. Later we will experimentally study the sensitivity of the fault analysis to the value of α and determine its suitable value for actual circuits.

2.4 Feedback

We will demonstrate the calculation of observability in a feedback situation by the NAND flip-flop circuit of Fig. 4. Suppose the observabilities of lines 7 and 8 are known and the controllabilities of all lines have been computed. The observabilities of lines 1, 3, 4, and 6 are to be determined. Let us assume that the signals on lines 1 and 6 are independent. We will establish an iterative procedure for determining the observabilities of lines 3 and 4. Suppose $B1_0(4) = 0$ is the zero-order value of one-observability of line 4. Then

$$B1_0(2) = B1(7)$$

$$B0_0(3) = B1(7) \left[\frac{S(3) - C0(2)}{C0(3)} \right].$$

$B0_0(5)$ is obtained by combining $B0_0(3)$ and $B0(8)$ according to the fanout formula. Now

$$B0_0(5) = (1-\alpha) \max \{B0_0(3), B0(8)\}$$
$$\quad\quad + \alpha [B0_0(3) + B0(8) - B0_0(3) B0(8)]$$

$$B1_1(4) = B0_0(5) \cdot C0(5)/C1(4).$$

$B1_1(2)$ will be obtained by combining $B1_1(4)$ and $B1(7)$ in the fanout formula and we proceed further in the above manner to calculate $B0_1(3)$, $B0_1(5)$, $B1_2(4)$, $\cdots$ until the successive iterations produce identical values. It can be verified that such iterations will converge to a unique value. Thus $B1(4)$ and $B0(3)$ are computed. Similarly, starting with $B0_0(4) = 0$, the values of $B0(4)$ and $B1(3)$ can be calculated. Once these are available, observabilities of lines 1 and 6 follow easily.

The above iterative procedure can be applied to feedback structures containing any number of gates. The iterations, in fact, correspond to expanding the structure of Fig. 4 by cutting the feedback loop at line 4 and tracing the circuits along the lines 2, 3 and 5 until we again arrive at 4 at which point the structure consisting of lines 4, 2, 3, 5, 4, is duplicated as shown in Fig. 5. In some simple cases the observability of the signals on the feedback path can be analytically computed. In such cases the iterative procedure is not necessary. Suppose after n-th iteration $B1_n(4) = x$. If n is sufficiently large such that

$$B1_{n+1}(4) = B1_n(4) = x,$$

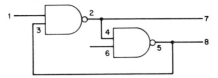

Fig. 4 NAND flip-flop.

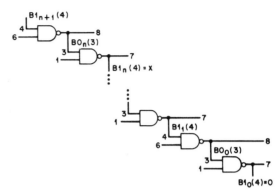

Fig. 5 Iterative analysis

then we can write an expression for $B1_{n+1}(4)$ in terms of x and solve the above equation for x.

In the above analysis of feedback, we assumed that the signals 1 and 6, which sensitize the feedback loop, are independent. Sensitization of the loop corresponds to storing the values of signals in the feedback (provided the number of inversions around the loop is even; for odd number of inversions the loop will give rise to oscillations and the fault analysis will not be valid). In a clocked circuit, the signals 1 and 6, which are derived from the same clock signal, can not be assumed to be independent. In order to adequately analyze such a circuit, we must calculate the *loop sensitization probability*, L. This can be done by identifying all the input signals of the gates forming a loop and removing from them the output signals of these gates which form the feedback. For example, in Fig. 4, the input signals are 1, 3, 4 and 6. If we remove the output signals, 3 and 4, we are left with 1 and 6. In order to sensitize the loop, 1 and 6 should be simultaneously set to "one". In the fault-free simulator if we set up a loop-counter for each loop to count the occurrences of loop sensitization states, then the loop sensitization probability can be estimated after simulation of N vectors as

$$L = \frac{loop - count}{N}.$$

Now L is the probability with which all inverting gates in the loop functionally become inverters and all non-inverting gates become simple buffers. This is shown for the NAND flip-flop in Fig. 6. The iterative structure in Fig. 6(b) shows that line 4, after n vectors, can be observed either through line 7 or through line 8 with one inversion or through the previous (n-1) vectors. This observability, which is shown in Fig. 6(c), will be extended to next vector, $(n+1)th$, if the path through the cross-hatched AND gate is sensitized. Node 4• represents the observability after $(n+1)$ vectors, if the loop is sensitized. For this hatched gate the probability of sensitization is L. Again the observability x of line 4 can be obtained by solving

$$B1_{n+1}(4) = L.B1_{n+1}(4•) = x.$$

where $B1_{n+1}(4•)$ is the observability due to three fan out observabilities, $B1(7)$, $B0(8)$ and x.

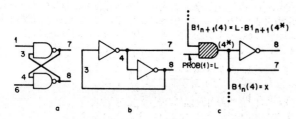

Fig. 6 Model for NAND flip-flop with sensitized loop.

The value of x can be analytically computed as follows:

$$L \cdot B1_{n+1}(4\bullet) = x$$

$$L \cdot [K + x - Kx] = x$$

where,

$$K = (1-\alpha)\max[B1(7), B0(8)]$$
$$+ \alpha [B1(7) + B0(8) - B1(7)B0(8)],$$

therefore

$$x = \frac{L \cdot K}{L \cdot K + (1 - L)}.$$

2.5 Computation of Fault Detection Probability

Consider the fault "line ℓ stuck-at-1". To detect this fault the line ℓ must be set to a value 0 and then its value should be observed at a primary output. Thus the probability of detection of this fault is given by

$$D1(\ell) = Prob\{line\ \ell\ set\ to\ 0\ and\ observed\}$$

$$= Prob\{\ell\ is\ set\ to\ 0\} \cdot Prob\{line\ \ell\ observed\ |\ \ell\ is\ set\ to\ 0\}$$

$$= B0(\ell) \cdot C0(\ell).$$

Similarly, the probability of detecting the fault "line ℓ stuck-at-0" can be calculated as

$$D0(\ell) = B1(\ell) \cdot C1(\ell).$$

If for a given fault the detection probability is x, then the probability of detecting this fault by a set of N vectors is $X(N) = 1 - (1-x)^N$.

2.6 Unbiasing Fault Detection Probability

Suppose we simulate N vectors and from the statistical data collected during simulation obtain the detection probabilities of all the faults. The probabilities thus computed are statistical estimates and can be regarded as random variables whose values converge to their exact values as N tends to infinity. Let us assume that $p(x)$ is the probability density function of the detection probability of certain fault. Also assume that the mean value, x_0, of this random variable is the exact value of the detection probability. Since we do not know the exact probability distribution function, for simplicity, we assume that x is uniformly distributed in the domain $[x_0-\Delta, x_0+\Delta]$. The estimated probability of detecting the fault by at least one among N vectors is given by the random variable,

$$1 - (1-x)^N.$$

In order to find the actual probability, $1 - (1-x_0)^N$, we proceed by computing the expected value,

$$E(1-(1-x)^N) = 1 - E(1-x)^N = 1 - \frac{1}{2\Delta}\int_{x_0-\Delta}^{x_0+\Delta}(1-x)^N dx$$

$$E(1-(1-x)^N) = 1 - \frac{1}{2\Delta(N+1)}[(1-x_0+\Delta)^{N+1} - (1-x_0-\Delta)^{N+1}]$$

$$= 1 - (1-x_0)^N \left[1 + \frac{N(N-1)}{3!}(1-x_0)^{-2}\Delta^2 + \cdots\right]$$

The above expression shows that the mean value of our estimation is biased away from the required value $1 - (1-x_0)^N$. In order to remove this bias, we modify the estimation as follows:

$$X(N) = 1 - \frac{(1-x)^N}{W(x_0)}$$

where

$$W(x_0) = 1 + \frac{N(N-1)}{3!}(1-x_0)^{-2}\Delta^2 + \cdots$$

If the detection probability was determined through N Bernoulli trials ([8], pages 495-496) then its standard deviation would be $\sqrt{x_0(1-x_0)}/\sqrt{N}$. Let us assume that

$$\Delta = \beta \sqrt{\frac{x_0(1-x_0)}{N}}$$

where β is a constant of proportionality which should account for the determining process not being Bernoulli trials and for the assumed uniform density function for x. Thus

$$W(x_0) = 1 + \frac{N-1}{6}\beta^2 \frac{x_0}{1-x_0} + \cdots$$

Finally, since the true mean x_0 is not known in practice we substitute the sample mean x in its place to obtain

$$X(N) \simeq 1 - \frac{(1-x)^N}{W(x)},$$

leaving the value of β to be determined empirically.

2.7 Fault Coverage

Consider a circuit with K faults and V test vectors. We divide the set of vectors in k subsets, each containing $V/k = N$ vectors. We will calculate the fault coverages $F(N), F(2N), \cdots, F(kN)$. For the i-th fault, we compute the detection probabilities from the N-vector statistics in each subset as $x_{i1}, x_{i2}, \ldots, x_{ij}, \ldots, x_{ik}$. The cumulative detection probability of this fault is given by

$$f_i(jN) = f_i\{(j-1)N\} + [1-f_i\{(j-1)N\}]\left[1 - \frac{(1-x_{ij})^N}{W(x_{ij})}\right]$$

$$= 1 - [1-f_i\{(j-1)N\}]\frac{(1-x_{ij})^N}{W(x_{ij})}.$$

Since $f_i(0) = 0$, iteratively writing $f_i(N), f_i(2N), \cdots$, we get

$$f_i(jN) = 1 - \prod_{m=1}^{j}\frac{(1-x_{im})^N}{W(x_{im})}.$$

Now the fault coverage for the whole circuit is obtained as

$$F(jN) = \frac{1}{K}\sum_{i=1}^{K} f_i(jN).$$

2.8 Complexity of STAFAN

STAFAN requires two types of operations to be included in a fault-free simulator. The first type, involving the updating of counters, must be performed on every vector. The other type, which include computation of controllabilities, observabilities, etc., are performed once every N vectors, where N can be chosen arbitrarily. In general, N can be large. In the extreme case, the second type of operations may be performed only once after simulating through all the vectors. Also the operations like loop identification and the initialization of counters are performed only once during preprocessing. Thus the main overhead due to STAFAN in the fault-free simulator is due to the first type of operations. These operations are listed in Table 1. Each counter is updated after examining certain line values. The number, G is the total number of gates in the circuit. Every gate output line has a 0-counter and a 1-counter. Each counter requires two operations: 1) examine line value and 2) update counter. Thus there are $2G$ operations for each type of counter. In practice, some of these operations may overlap. If n_f is the average number of inputs per gate then the number of sensitization counters is $n_f G$. In order to update this counter for a gate-input line, all other inputs of this gate must be examined. The total numbers of operations connected with these counters is $n_f^2 G$. Updating of a loop counter requires examining whether or not certain sensitization counter of each of the gates in the loop was updated. In general, the number of loops in a circuit will be less than G. If n_ℓ is the average number of gates in a loop, the number of operations associated with these counters will be less than $(n_\ell+1)G$. The total overhead in the fault-free simulator due to all the four types of counters is

$$\text{overhead} \propto G(4+n_f^2 + an_\ell + a)$$

where $a < 1$.

Table 1. Operations performed in STAFAN

Counter	Required for	Number of values examined	Number of counters	Number of operations
0-counter	every gate	1	G	$2G$
1-counter	every gate	1	G	$2G$
sensitization-counter	every gate-input line	$n_f - 1$	$n_f G$	$n_f^2 G$
loop-counter	every loop	n_ℓ	$< G$	$< (n_\ell+1)G$

3.0 IMPLEMENTATION

In an experimental study STAFAN algorithms were implemented in an existing MOS simulator [9]. Since this simulator is also capable of simulating circuits at gate level with $s-a-0$ and $s-a-1$ faults, it was possible to compare its fault simulation results with STAFAN fault coverage estimation.

In the implementation it was necessary to set the values of the empirical constants α and β. For calculation of observabilities at fanouts, suitable value of constant α was determined by an experiment conducted on a large circuit. A value $\alpha = 1$ showed good agreement with the fault simulation result. Notice that this corresponds to the assumption that the fanout branches are independent. In fact the results were rather insensitive to the actual value of α. The other empirical constant β is used for unbiasing the fault detection probabilities. The following approximation was used:

$$X(N) \simeq 1 - \frac{(1-x)^N}{W(x)}$$

where $W(x) \simeq 1 + \frac{N-1}{6} \beta^2 \frac{x}{1-x}$.

Again by experimenting with an actual circuit it was found that $\beta^2/6 = 5.0$ produced a good match with fault simulation result. The values of α and β which were thus set in the analysis, upon trials on several other circuits, produced consistently good results. In this implementation the feedback in the cross-coupled gates was handled using the loop-sensitization probability. But the larger feedback loops were processed by the iterative procedure.

4.0 RESULTS

Results obtained by STAFAN on several circuits varying in size from 102 gates to 2,723 gates are summarized in Table 2. The plots of fault coverage versus number of vectors for two of the circuits are shown in Figs. 7 and 8. Figure 7 shows the fault coverage plot for a 64-bit ALU (Arithmetic-Logic Unit). Figure 8 shows the fault coverage for the sequential circuit with 1983 gates. From the graphs it can be seen that the STAFAN fault coverage estimation closely follows the actual fault coverage obtained from fault simulator. Different periods (N) of analysis were taken for various circuits. It was found that this period of analysis has very small effects on the final fault coverage estimates. From the data in the Table 2, it is evident that STAFAN is a very effective technique for small to large circuits. Average deviation of STAFAN fault coverage from the actual fault simulator result varies from 0.29% to 3.65%.

TABLE 2 — STAFAN RESULTS

CIRCUIT	#GATES	PRIM. INPUTS/ OUTPUTS	FLIP-FLOPS	Number of Vectors	Number of Faults	Period of estimation in STAFAN (N)	FINAL COVERAGE FAULT SIM. %	FINAL COVERAGE STAFAN %	Average Difference
4-bit ALU	102	14/8	0	52	263	10	96.57	96.25	0.47
64-bit ALU	1827	134/69	0	155	4376	20	75.3	75.09	0.29
4-bit MULTIPLIER	359	12/10	15	1111	741	100	93.12	86.80	3.37
CIRCUIT A	1983	38/34	86	3842	5060	200	86.44	83.61	1.24
CIRCUIT B	2723	64/64	98	3636	5856	200	71.26	72.57	3.65

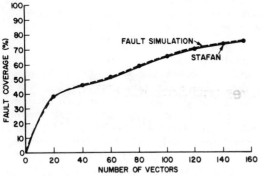

Fig. 7   Fault coverage for a 64-bit ALU.

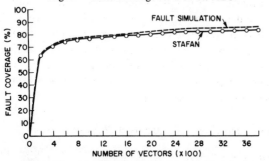

Fig. 8   Fault coverage for CIRCUIT A, a sequential circuit with 1983 gates.

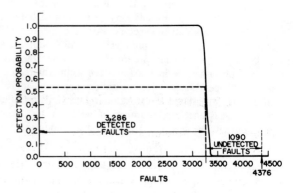

Fig. 9   Detection probabilities for faults in 64 bit ALU after simulating 155-vectors.

In order to determine how accurately the detection probabilities would predict the undetected faults, all the 4,376 faults of the 64-bit ALU were graded in terms of their cumulative detection probabilities as obtained after simulating through 155 vectors. The faults ordered in terms of decreasing detection probability are shown in Fig. 9. Since the estimated fault coverage was 75.09 percent, the first 3,286 faults in this graded list were assumed to be detected. This left the 1,090 faults, with detection probabilities less than 0.53, as undetected. Comparison with fault simulation results showed that of these 1,090 faults, 1,036 were indeed undetected. Also among the 3,286 faults that were assumed to be detected by STAFAN, only 46 faults were actually reported as undetected by fault simulator.

5.0 CONCLUSION

We started this paper with a promise to present an alternative to fault simulation which can handle large circuit complexity. We believe we have kept that promise. Preliminary trials have indicated good results not only for fault coverage but also for finding the faults that remain undetected by a vector set. There are, however, two parameters, α and β, used in the analysis which must be determined empirically. While β can be determined by a comparison with fault simulation results, α may, in general, depend upon the structure of circuit. Fortunately, in the experiments that we have conducted, the results seem to be only mildly affected by the value of α.

REFERENCES

[1] P. Goel, "Test generation costs analysis and projections," *Proceedings of 17th Design Automation Conference*, Minneapolis, Minnesota, June 23-25, 1980, pp. 77-84.

[2] V. D. Agrawal, "Sampling techniques for determining fault coverage in LSI circuits," *Journal of Digital Systems*, Vol. V, pp. 189-202, Fall 1981.

[3] J. Savir, G. Ditlow, and P. H. Bardell, "Random Pattern Testability," Thirteenth Annual Fault-Tolerant Computing Symposium, Milano, Italy, June 28-30, 1983, pp. 80-89.

[4] M. Abramovici, P. R. Menon, and D. T. Miller, "Critical Path Tracing — An Alternative to Fault Simulation," *Proceedings of 20th Design Automation Conference*, Miami Beach, Florida, June 27-29, 1983, pp. 214-220.

[5] J. E. Stephenson and J. Grason, "A testability measure for register transfer level digital circuits," Sixth International Fault Tolerant Computing Symposium, Pittsburgh, Pennsylvania, June 1976, *Digest of Papers*, pp. 101, 107.

[6] L. H. Goldstein, "Controllability/Observability Analysis of Digital Circuits," *IEEE Transactions on Circuits and Systems*, Vol. CAS-26, September 1979, pp. 685-693.

[7] S. G. Chappell, C. H. Elmendorf, and L. D. Schmidt, "LAMP: Logic-Circuit Simulator," *Bell System Technical Journal*, Vol. 53, October 1974, pp. 1451-1476.

[8] K. S. Trivedi, *Probability and Statistics with Reliability, Queueing, and Computer Science Applications*, Englewood Cliffs, N.J.: Prentice-Hall, 1982.

[9] A. K. Bose, et al., "A Fault Simulator for MOS LSI Circuits," *Proceedings of 19th Design Automation Conference*, Las Vegas, Nevada, June 14-16, 1982, pp. 400-409.

Chapter Four: Testing Regular Structures and Specific Circuits

There are four fundamental strategies for generating test patterns for logic circuits. The first is called the "structural" strategy, which is based on providing tests to detect the presence of specific hardware faults taken from a set of predefined fault effects. The second approach is called the "functional" strategy, which seeks to provide a sequence of input data with corresponding output responses that will verify the function of a circuit. The third scheme, known as the "pseudorandom" approach, uses a random-number generator (e.g., a feedback shift-register) to generate the test patterns. A counter can be employed to generate the test patterns, in which case the scheme is an "exhaustive" one.

4.1 Functional Testing

Functional tests are particularly useful when the internal details of a circuit are unknown and structural tests cannot be derived. Since gate-level fault models may not cover all physical failures, functional patterns are useful in checking out a circuit at operating speeds. Pseudorandom patterns have most of the advantages stated above; however, functional tests can be used to check out specific functions, which may appear with high probability in only a very long random pattern.

In the development of functional tests, two issues are extremely important. First, under failure, it could happen that the correct function is produced, but in addition, an incorrect event happens. For example, in a decoder, the correct line is activated, and in addition, another line is also activated. Second, under failure, the behavior of a combinational circuit could become sequential. This is particularly true in CMOS circuits when stuck-open failures happen because of floating gates [1].

4.2 Testing Iterative Arrays

The study of iterative logic arrays (ILAs) is becoming more important, since VLSI technology has made array-type realizations of general logical functions more attractive. This structural regularity in design is also known to facilitate the testing problem.

The problem of fault detection and location in ILAs of combinational cells was first studied by Kautz [2]. He assumed that all possible cell inputs must be applied to each cell to test it completely and that a fault in a cell may affect the cell outputs in any arbitrary manner. In general, the number of tests required to test a one-dimensional ILA of p cells is dependent on p, and is at most $pmn(n-1)$, where m is the number of columns and n is the number of rows of the cell flow table.

Methods to test an arbitrarily long but finite one-dimensional ILA of p cells with a fixed constant number of tests, independent of p, are discussed in [3] and [4]. Such arrays are referred to as C-testable.* Fault detection in bilateral arrays of combinational cells has been studied [5]. Testing two-dimensional iterative arrays (e.g., array multipliers), tree structures, and bit-sliced processors has also been reported in the literature.

4.3 Fault Detection in Programmable Logic Arrays

By virtue of its regular structure, which simplifies the design of irregular combinational logic, the programmable logic array (PLA) has found wide acceptance as an important logic component [6]. Most of the efforts in testing PLAs fall into three general categories [7]:

i) Pseudorandom or heuristic generation of test sets which attempts to cover all or most of a predefined set of faults [8].

ii) Augmentation of the PLA with additional hardware to simplify test-set generation and response evaluation for off-line testing [9], or built-in self-test [10].

iii) Introduction of new rules at the logic synthesis level to obtain logic structures that lend themselves to concurrent (or on-line) as well as off-line testing. Augmentation with additional hardware may also be required [11].

A subject closely related to on-line PLA testing is the design of totally self-checking checkers in PLAs [12]. For an extensive reference list on PLA testing the reader is referred to section 16 of the Bibliography.

4.4 Reprints

The purpose of functional testing is to validate the correct functional operation of digital systems according to their specification. Functional testing reduces the test generation complexity, and generates tests that do not depend on the implementation of the integrated circuit. For a brief overview see the tutorial paper by Abadir and

---

*For example, a ripple carry adder, in which a cell is assumed to be a full-adder, can be tested with eight patterns.

Rebhbati that is reprinted in Chapter One. Some of the functional testing techniques are reviewed in the paper by Su and Lin [13]

The iterative structure of an array is an attractive feature for logic design as well as for testing. Friedman, in the first reprinted paper considers one-dimensional unilateral combinational iterative arrays consisting of p cells. He considers the properties of such systems that enable them to be tested with a fixed constant number of tests independent of p, the number of cells in the system. Such systems are referred to as C-testable.

For analysis purposes, bit-sliced systems can be viewed as ILAs, which are one-dimensional cascades of identical cells. In the second reprinted paper, Sridhar and Hayes present a method for making an arbitrary ILA C-testable. To simplify response verification, they introduce I-testability, which ensures that identical test responses can be obtained from every cell in an ILA. The application of C- and I-testability to the design of easily testable and self-testing bit-sliced microcomputers are investigated.

Because of the regularity in its iterative structure, an array multiplier is well suited for implementation on a VLSI circuit chip. The paper by Shen and Ferguson [14] shows that conventional designs of array multipliers are not C-testable. They present the design of a 16×16 carry-save array multiplier which is C-testable.

The use of microprocessors in digital systems has been rapidly expanding. Thus the need has become acute for sound theoretical tools so that efficient, thorough, and cost-effective test programs can be developed to detect faults in microprocessors. In his paper, Przybylski [15] describes the validation and testing efforts of the MIPs project. MIPS is a high-performance single-chip microprocessor, designed and implemented at Stanford over a period of three years.

The paper by Brahme and Abraham [16] presents a functional-level graph model for the microprocessor. Their graph model is an improved version of the Thatte and Abraham model which is surveyed in the Abadir and Reghbati paper, reprinted in Chapter One. This more recent procedure to generate tests can be automated and is independent of the microprocessor implementation details.

Recent advances in VLSI technology have encouraged the use of multicomputer systems with a large number of processing elements (PEs) and memory modules (MMs). In such systems, various techniques are utilized to support restructurable data paths between the PEs and MMs. Thus, the intercommunication is becoming an increasingly complex but inevitable issue.

The reprinted paper by Agrawal surveys the testing and fault-tolerance issues of interconnection networks. An important class of networks is the (N,K) shuffle/exchange network. Such a network has N input terminals, N output terminals, and consists of a cascade of K identical $N \times N$ shuffle/exchange stages. In their paper, Lee and Shen [17] construct a test sequence of constant length which when applied to an (N,K) shuffle/exchange network will fully test the entire network.

The demands on test procedures required for testing today's semiconductor memories are that they detect faults of a wide variety and a complex nature and that they minimize the total testing time. Many functional testing techniques have been proposed in the literature and are now being used in industry. The paper by Abadir and Reghbati [18] surveys these techniques and compares their degree of fault-coverage and time-complexity.

Standard self-testing techniques (see Chapter Five) are not applicable to memories. The paper by You and Hayes [19] employs on-chip generation of test patterns whose aim is complete coverage of the most common failure modes. To reduce the overall testing time, several storage subarrays are tested in parallel. Built-in self-test of embedded memories is discussed in Sun and Wang [20].

The PLA is attractive in VLSI owing to its memory-like array structure. The reprinted paper by Fujiwara and Kinoshita describes how a PLA can be augmented so that its test patterns and responses do not depend on the personality of the PLA. For the augmented PLAs, "universal" test sets to detect faults are presented. Built-in testing techniques for PLAs are described in Daehn and Mucha [21], and the reprinted paper by Hua, Jou and Abraham.

Digital signal processing (DSP) has been used for scientific applications for more than two decades, but it is a relatively new tool for testing. The applications of DSP are becoming so numerous and so advantageous that nearly all production testing of analog and mixed digital-analog devices and circuits is expected to involve DSP. The reprinted paper by Mahoney shows how DSP can reduce the test time while also reducing the hardware required for analog testing.

Some IC chips like D/A and A/D converters are mixed digital-analog devices. Another example of such a device is the CODEC which provides bidirectional signal transformation between analog voice wave-forms and digital data. The reprinted paper by Shirachi describes five methods for testing CODECs. The basic CODEC test approach can be extended with some modifications to other types of circuits, such as digital filters or modems in which synchronized analog and digital wave-forms are required for both signal transmission and device control.

References

[1] R.L. Wadsack, "Fault Modeling and Logic Simulation of CMOS and MOS Integrated Circuits," *BSTJ*, Vol. 57, No. 3, 1978, pp. 1449–1474.

[2] W.H. Kautz, "Testing for Faults in Cellular Logic Arrays," in *Proc. of 8th Ann. Symp. Switching & Automata Theory*, 1967, pp. 161–174.

[3] A.D. Friedman, "Easily-Testable Iterative Systems," *IEEE Trans. Comput.*, Vol. C-22, No. 12, Dec. 1973, pp. 1061–1064.

[4] F.J.O. Dias, "Truth-Table Verification of an Iterative Logic Array," *IEEE Trans. Comput.*, Vol. C-25, 1976, pp. 605–613.

[5] F.G. Gray and R.A. Thompson, "Fault-Detection in Bilateral Arrays of Combinational Cells," *IEEE Trans. Comput.*, Vol. C-27, No. 12, Dec. 1978, pp. 1206–1213.

[6] H. Fleisher and L.I. Maissel, "An Introduction to Array Logic," *IBM J. Res. Dev.*, Vol. 19, March 1975, pp. 98–109.

[7] H.K. Reghbati, "Fault Detection in Programmable Logic Arrays," *SFU-CMPT Tech. Rept. 85-11*, 1985.

[8] E.B. Eichelberger and E. Lindbloom, "A Heuristic Test-Pattern Generator for PLAs," *IBM J. Res. Dev.*, Vol. 24, 1980, pp. 15–22.

[9] H. Fujiwara and K. Kinoshita, "A Design of PLAs with Universal Tests," *IEEE Trans. Comput.*, Vol. C-30, No. 11, Nov. 1981, pp. 823–828.

[10] R. Treuer, H. Fujiwara, and V.K. Agrawal, "Implementing a Built-In Self-Test PLA Design," *IEEE Design & Test*, Vol. 2, No. 2, April 1985, pp. 37–49.

[11] G.P. Mak et al., "The Design of PLAs with Concurrent Error Detection," *Proc. FTCS-12*, 1982, pp. 303–310.

[12] K. Son and D.K. Pradhan, "Completely Self-Checking Checkers in PLAs," *Proc. IEEE Test Conf.*, 1980, pp. 231–237.

[13] S.Y.H. Su and T. Lin, "Functional Testing Techniques for Digital LSI/VLSI Systems," *Proc. of ACM-IEEE Design Automation Conf.*, 1984, pp. 517–528.

[14] J.P. Shen and F.J. Ferguson, "The Design of Easily Testable VLSI Array Multipliers," *IEEE Trans. Comput.*, Vol. C-33, No. 6, June 1984, pp. 554–560.

[15] S. Przybylski, "The Design Verification and Testing of MIPS," *Proc. of MIT VLSI Conf.*, 1984, pp. 100–109.

[16] D. Brahme and J.A. Abraham, "Functional Testing of Microprocessors," *IEEE Trans. Comput.*, Vol. C-33, No. 6, June 1984, pp. 475–485.

[17] D.C.H. Lee and J.P. Shen, "Easily-Testable (N,K) Shuffle/Exchange Networks," *Proc. of Int. Conf. Par. Proc.*, 1983, pp. 65–70.

[18] M.S. Abadir and H.K. Reghbati, "Functional Testing of Semiconductor Random Access Memories," *ACM Computing Surveys*, Vol. 15, No. 3, Sept. 1983, pp. 175–198.

[19] Y. You and J.P. Hayes, "A Self-Testing Dynamic RAM Chip," *Proc. of MIT VLSI Conf.*, 1984, pp. 159–168.

[20] Z. Sun and L.T. Wang, "Self-Testing of Embedded RAMs," *Proc. IEEE Test Conf.*, 1984, pp. 148–156.

[21] W. Daehn and J. Mucha, "A Hardware Approach to Self-Testing of Large PLAs," *IEEE Trans. Comput.*, Vol. C-30, No. 11, Nov. 1981, pp. 829–832.

Easily Testable Iterative Systems

ARTHUR D. FRIEDMAN, MEMBER, IEEE

Abstract—It has been shown that the number of tests required to detect all faults in a one-dimensional unilateral combinational iterative array consisting of p cells will, in general, be proportional to p. In this paper we consider properties of such systems that enable them to be tested with a fixed constant number of tests independent of p, the number of cells in the system. Such systems are referred to as C-testable. Necessary and sufficient conditions on the basic cell state table are derived for an iterative system to be C-testable. It is shown that an arbitrary N-state cell table can be augmented by the addition of, at most, one row and less than $[\log_2 N]^2$ columns (for $N > 2$) so as to be C-testable.

Index Terms—Fault detection, iterative systems.

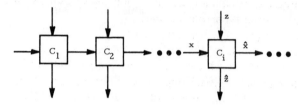

Fig. 1. A one-dimensional array.

THE study of the diagnosis of iterative systems[1] has potential usefulness as a first step in furthering our understanding of the diagnosis of more general systems of interconnected modules. In addition it is interesting in its own right since technological advances have made array-type realizations of general logical functions more attractive.

The testing of iterative arrays consists of detecting and locating faults in them by applying test inputs to their accessible input terminals and observing the outputs on accessible output terminals. The accessible inputs and outputs are usually associated with the boundaries of the array, although some input and output terminals of individual cells may also be accessible. The cells in the array are assumed to be identical and the array is assumed to contain p cells, where p may be arbitrarily large, but finite. We also assume that the array contains at most one faulty cell and that the fault is permanent (i.e., not transient).

An array is testable if it is possible to detect the presence of any faulty cell in it, independent of the size of the array. We shall be concerned with the problem of testing arrays of identical cells, independent of the size of the array. Although any given array (defined by a specific value of p) can be tested for all detectable faults by applying all combinations of inputs, this does not enable us to determine whether an array of arbitrary size is testable. Thus we must determine from the properties of an individual cell whether an arbitrary array of such cells can be tested.

The problem of testing iterative arrays was first studied by Kautz [1], who assumed that an individual cell can be tested for all its possible faults only by applying *all* possible cell inputs to that cell. It was also assumed that at most one cell was faulty and the fault in a cell may affect the cell outputs in any arbitrary way. These assumptions are referred to as the *general fault assumptions*.

The *restricted fault assumptions* are that all faults that can occur in the single faulty cell can be tested by applying a specified set of inputs (usually smaller than the set of all inputs) and that each fault will affect the cell outputs in a particular known manner.

We assume that the cells in an array contain combinational logic only, and the intercell connections in the array are unilateral, that is, signals travel in only one sense in each dimensional direction. A one-dimensional array is assumed to be as shown in Fig. 1.

Each cell receives an input x from its left-hand neighbor and an external input z. It generates an external output $\hat{z}$ and transmits an output $\hat{x}$ to its right-hand neighbor. The controllable inputs consist of the x input to the (leftmost) cell and the z inputs to all cells in the array. All $\hat{z}$ outputs and the $\hat{x}$ output of the rightmost cell are observable. We assume that the z input to cell i is independent of the z input to cell j for $i \neq j$. The behavior of a cell can be described by a table, which has a row for each x input (also referred to as state) and a column for each z input. The entries in the table consist of pairs of cell outputs $(\hat{x}, \hat{z})$ for each combination (x,z) of cell inputs. The entries in row x_i, column z_j will be represented functionally as $[\hat{x}(x_i,z_j), \hat{z}(x_i,z_j)]$. Note that the variables are not necessarily binary and may represent combinations of binary variables.

The necessary and sufficient conditions given by Kautz for testing a cellular array with a single faulty cell are as follows.

Condition 1: It should be possible to apply to the input terminals of a cell, a complete set of tests for detecting all faults in the cell, independent of the position of the cell in the array. (All inputs under the general fault assumptions.)

Condition 2: For each such test, it should be possible to propagate the effect of the fault to an observable output.

These results imply that a one-dimensional iterative system can be tested if and only if no two rows of the basic cell-state table are identical, and each state appears as an entry in the table at least once.

Manuscript received August 28, 1972; revised June 17, 1973. This work was sponsored by the Joint Services Electronics Program through the Air Force Office of Scientific Research/AFSC under Contract F44620-71-C-0067.

The author is with the Department of Electrical Engineering and Computer Science, University of Southern California, Los Angeles, Calif. 90007.

[1] An iterative system is the class of all iterative arrays with the same basic cell, but of arbitrary size. The term iterative array, with arbitrary size implied, is used in the same sense in this paper.

In general the number of tests required to test a one-dimensional array of p cells is a function of p, and can be shown to be at most $pMN(N-1)$ tests where M is the number of columns in the state table of a basic cell and N is the number of rows in the state table. Kautz also showed that an array can be tested with MN tests if and only if every column of the state table is a permutation column (i.e., each state appears exactly once as a next-state entry in each column).

In this paper we will attempt to determine necessary and sufficient conditions on the basic cell for iterative systems to be testable with a fixed constant number of tests independent of p, the number of cells in the array under the general fault assumption of Kautz. Such systems will be called C-testable. We will restrict ourselves to the case where the basic cell has no $\hat{z}$ output and hence to be detected, a fault must be propagated (through an arbitrary number of cells) to the right-hand border $\hat{x}$ output. Most of the results are easily extended to the case where the basic cell has a $\hat{z}$ output and faults need only be propagated to such an output. Consequently the analysis of this paper may be considered as a limiting worst case.

A fault that causes a change in the transition from state i with input z to the erroneous state k instead of the correct next-state j will be denoted by $i^z j/k$. The next state of the state table from initial state i and input sequence $z = z_1 z_2 \cdots z_n$ is denoted by $\hat{x}(i, z)$.

Theorem 1: In order to test all cells in an array of arbitrary length p for the fault $i^z j/k$ with a constant number of tests independent of p, assuming no $\hat{z}$ outputs, it is necessary and sufficient that: a) there exists a sequence R of length $q-1$ such that $\hat{x}(i,zR) = i$. (This condition is needed to insure that the input combination (i,z) can be applied to every qth cell in the array by a single test, and hence every cell in the array by q tests;) and b) the (arbitrarily long) input sequence $(Rz)^* = RzRzRz \cdots$ must be able to propagate the effects of the fault $i^z j/k$ to the rightmost cell (i.e., $\hat{x}(j,(Rz)^*) \neq \hat{x}(k,(Rz)^*)$). $(Rz)^*$ is called a *propagating sequence* for the pair j/k.

Proof: The sufficiency of these conditions is obvious. The proof of necessity is contained in the Appendix. □

A fault that can be tested in all cells by a constant number of tests will be said to be C-testable. A C-testable iterative system is one for which all faults are C-testable.

In order to determine whether a particular fault $a^z b/c$ for a given table is C-testable it is helpful to consider two distinct cases. (We denote the combined condition $\hat{x}(j,z) = r$ and $\hat{x}(k,z) = s$ by $\hat{x}(j/k,z) = r/s$.)

Case 1: There exists a sequence R such that $\hat{x}(b/c,Rz) = b/c$ and $\hat{x}(b/c,R) = a/d$. Such a sequence can be determined from an ordered pair graph with a node for each distinct ordered pair of states $b/c (b \neq c)$ and a branch labeled z from b/c to $\hat{x}(b/c,z)$ for all inputs z. A directed cycle from b/c to b/c with the last branch labeled z defines such a sequence. Since the ordered pair graph has $N(N-1)$ nodes (where N is the number of states) then $q < N(N-1)$ where q is the length of R.

Example 1: For the state table M_1 and the ordered pair graph of M_1 shown in Fig. 2 consider the fault $2^1 2/1$. A finite testing sequence $R1$ would correspond to a cycle on the

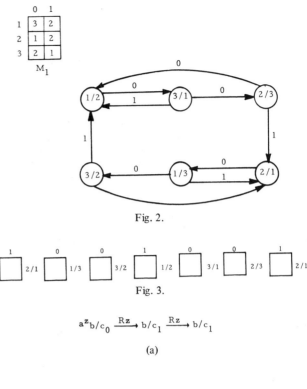

Fig. 2.

Fig. 3.

$$a^z b/c_0 \xrightarrow{Rz} b/c_1 \xrightarrow{Rz} b/c_1$$

(a)

$$a^z b/c_1 \xrightarrow{Rz} b/c_2 \xrightarrow{Rz} b/c_3 \xrightarrow{Rz} \cdots \xrightarrow{Rz} b/c_j \xrightarrow{Rz} b/c_t$$

(b)

Fig. 4.

pair graph from 2/1 to 2/1 with the last transition labeled 1 from a node 2/y. The only such transition to 2/1 is from 2/3. Hence R must be a path from 2/1 to 2/3. The shortest such path is defined by the input sequence 0 0 1 0 0. The cycle $1R$ of length 6 is shown in Fig. 3.

By applying the six tests defined by shifting this sequence to the left we can test all cells for the fault $2^1 2/1$. The set of tests also tests each cell for the faults $2^0 1/3$, $1^0 3/2$, $3^1 1/2$, $1^0 3/1$, and $3^0 2/3$. □

Case 2: There does not exist any sequence R such that $\hat{x}(b/c,Rz) = b/c$ and $\hat{x}(b/c,R) = a/d$. However, there does exist a sequence R such that $\hat{x}(b/c,R) = a/d$ and $\hat{x}(b/c,Rz) = b/c_1, c_1 \neq c$ and $(Rz)^*$ is a propagating sequence from b/c. An example of this situation is illustrated in Fig. 4(a). The most general case is illustrated in Fig. 4(b). Here $c_i \neq b$ for all c_i, and the c_i are all distinct for $i = 1, 2, 3, \cdots, j$ and $c_t = c_i$ for some $i = 1, \cdots, j$.

In the worst case such a sequence R could be determined from a graph of $n-1$ ordered pairs of the type $b/c_1 \cdot b/c_2 \cdots b/c_j$ where $c_i \neq b$ for all i and either all c_i are distinct or the last c_i is identical to some previous one and all others are distinct.

If $c_t = c_1$ the sequence Rz is defined by a cycle on this graph where for all nodes in the cycle all c_i are distinct. There are precisely $N!$ nodes that satisfy this constraint. If $c_t \neq c_1$ the sequence Rz is defined by a path from a node where all c_i are distinct to a node in which the last c, c_j is identical to some previous one but all the others are distinct.

Since the number of nodes in such a graph is $(N-1)N!$ then $(N-1)N!$ is an upper bound on the length of R.

Example 2: For the state table M_2 consider the fault $1^1 2/1$. Let us try to find a testing sequence $R1$ such that $\hat{x}(2/1,R1) = 2/1$ and $\hat{x}(2/1,R) = 1/y$ for $y = 2, 3$, or 4. From the ordered pair graph (Fig. 5) there is no 1 transition to $2/1$ since the only 1 transitions are to $2/4$. We must therefore try to find a sequence R such that $2/1 \xrightarrow{R1} 2/4 \xrightarrow{R1} 2/4$ which would be identified by a path on a graph consisting of nodes that represent two-state pairs $(2/1, 2/4) \xrightarrow{R1} (2/4, 2/4)$. Such a path is $(2/1, 2/4) \xrightarrow{0} (1/4, 1/3) \xrightarrow{1} (2/4, 2/4)$. All cells can therefore be tested for the fault $1^1 2/1$ by two tests defined by the sequence $R = (01)^*$ and shown in Fig. 6.

The preceding concepts can also be used to show that a fault is not C-testable.

Example 3: For the state table M_3 consider the fault $1^1 1/3$. From the pair graph (Fig. 7) there is no 1 transition to $1/3$. Hence a testing sequence must be a sequence $R1$ that propagates $1/3$ and is such that $\hat{x}(1/3,R) = 1/y$ where $1R$ also propagates $1/y$. In order for $1R$ to propagate $1/y$, $y = 3$. Since $\hat{x}(1/3,1) = 1/2$, R must be a propagating sequence for both $1/3$ and $1/2$. To propagate $1/2$ the first input of R must be 3. Thus there is no such R and this fault is not C-testable. □

Note that the ordered pair graph of M_3 is strongly connected, but the table is not C-testable. Conversely the table M_2 is C-testable even though its pair graph is not strongly connected. Thus a strongly connected pair graph is neither a necessary or sufficient condition for C-testability.

The following easily proven theorem gives a sufficient condition for C-testability.

Theorem 2: Consider a state table M with input columns I. Let I_1 and I_2 be subsets of I such that $I_1 \cup I_2 = I$. These subsets define a column decomposition of M into M_1 and M_2. If M_1 and M_2 are C-testable then M is C-testable. □

From Theorem 2 and a previous result of Kautz it follows that in order to determine whether an iterative system can be finitely tested, it is only necessary to examine faults in columns that are not permutation columns.

The following theorems give conditions for cells that are not C-testable and sufficient conditions for C-testability.

Theorem 3:

a) A table with a constant column S (i.e., $\hat{x}(i,S) = j$ for all states i) is not C-testable for the fault $i^S j/k$.

b) A table with three or more rows, a trap row $i(\hat{x}(i,z) = i$ for all z), and an entry to that trap row $\hat{x}(j,S) = i, j \neq i$ is not C-testable for the fault $j^S i/k$, $k \neq i, j$.

Proof:

a) Consider the fault $i^S j/k$. In order for this fault to be C-testable there must exist a propagating sequence $(RS)^*$ for j/k. But if $\hat{x}(i,S) = j$ for all i no sequence containing S is a propagating sequence.

b) Consider the fault $j^S i/k$ where $i \neq j$. In order for this fault to be C-testable there must exist a sequence R such that $\hat{x}(i/k,R) = j/m$ for some $m \neq j$. But since $\hat{x}(i,z) = i$ for all x there is no such sequence R.

Theorem 4: A table is C-testable if: a) the pair graph is

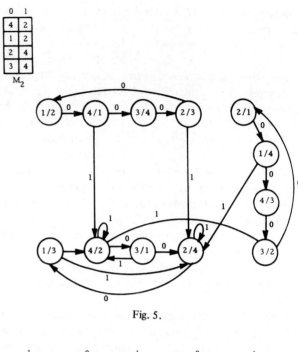

Fig. 5.

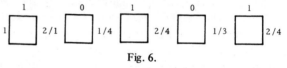

Fig. 6.

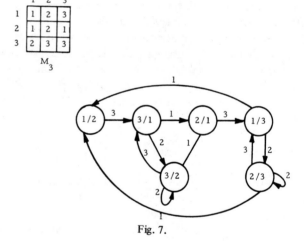

Fig. 7.

strongly connected; b) the table has no constant columns; and c) for every ordered state triplet j, k, q there exists a column S such that $\hat{x}(k,S) = \hat{x}(q,S) \neq \hat{x}(j,S)$.

Proof: Consider an arbitrary fault $i^z j/k$. Since the ordered pair graph is strongly connected there is a path from j/k to every other pair.

There must be some state m such that $\hat{x}(i/m,z) = j/k$ or $\hat{x}(i/m,z) = j/q$ for some state $q \neq j$ since z is not a constant column. But by the assumptions of the lemma there is a column S such that $\hat{x}(j/k,S) = \hat{x}(j/q,S)$. Let R be the sequence defined by a path from $\hat{x}(j/k,C)$ to i/m. Then the sequence SRz defines a test for $i^z j/k$ and this fault is C-testable. Thus any arbitrary fault is C-testable and the state table is also.

Example 4:

	1	2	3	4
1	4	1	3	4
2	1	1	1	1
3	2	2	3	1
4	3	2	1	4

M_4

From Theorem 4, state table M_4 is C-testable since the pair graph can be shown to be strongly connected, and all other conditions of Theorem 4 are satisfied. □

From Theorem 4 it follows that any table can be augmented so as to be C-testable.

Corollary: Any N-state table can be made C-testable by adding at most Q columns (where $Q \leq \log_2 N(\log_2 N + 1)/2$) and 1 row.

Proof: By augmenting the table it is possible to make it satisfy Theorem 4. Consider an $N \times m$ table. By adding a row $N + 1$ and columns $m + 1$, $m + 2$ as shown in Fig. 8 the resulting table has no constant columns and the pair graph is strongly connected. An additional set of columns can then be added to insure that there are no trap rows and that for every ordered set of three states i, j, k there exists an input S such that $\hat{x}(i,S) = \hat{x}(j,S) \neq \hat{x}(k,S)$. The number of additional columns required to insure this can be shown to be bounded by $S_0(S_0 + 1)$ where S_0 is the smallest integer $\geq \log_2 N$.[2] The additional row is not required if the original table has no constant columns. If the original ordered pair graph is strongly connected, column $m + 1$ is not required.

Example 5: Table M_3 can be augmented to be C-testable by adding one column (as shown in state table M_3') which is required to satisfy condition c) of Theorem 4 if $k = 2$, $q = 3$, $j = 1$.

	1	2	3	4
1	1	2	3	1
2	1	2	1	2
3	2	3	3	2

M_3'

A test for the fault $1^1 1/3$ is defined by the sequence $(1R)^*$ where $R = 4\ 3\ 4\ 3$ and is shown in Fig. 9.

The preceding results can easily be extended to tables with z outputs and to the case of restricted fault assumptions. The external output enables relaxation of propagating conditions since i/i must only be propagated to an external output. The restricted fault assumptions enable us to ignore faults that cannot occur. However, the problem of C-testability is much more difficult for two-dimensional arrays.

---

[2] The set of all columns must be what has been defined as a (2, 1) separating system [5], which has been shown to require at most $S_0(S_0 + 1)/2$ columns.

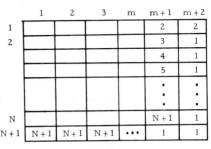

Fig. 8.

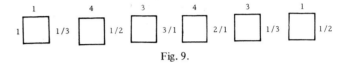

Fig. 9.

APPENDIX

PROOF OF THEOREM 1 NECESSITY

Suppose all cells can be tested for the fault $i^z j/k$ by a constant number of tests. In order to test an arbitrarily large number of cells p with a fixed number of tests, there must exist a single sequence that tests an arbitrarily large set of cells, $\{1, k_1, k_2, \cdots, k_r, \cdots\}$ for this fault. This single sequence must be defined by

$$(zR_1 zR_2 zR_3 \cdots zR_r \cdots)$$

and must have the property that $\hat{x}(j/k, R_q) = i/m_q$, $m_q \neq i$ in order to apply the input i to each of the cells to be tested for the fault and to propagate the effects of the fault. In addition the effects of this fault in different cells must be propagated simultaneously.

Consider the subsequence $zR_1 zR_2$. R_2 must be able to propagate j/k, and $j/q_1 = \hat{x}(j/k, R_1 z)$. If $q_1 = k$ then R_2 can equal R_1 and $(zR_1)^*$ satisfies the condition of the theorem. In a similar manner R_3 must propagate j/k, $j/q_1 = \hat{x}(j/k, R_1 zR_2 z), j/q_2 = \hat{x}(j/k, R_2 z)$. If the state table has N states $S = (S_1, S_2, \cdots, S_N)$ and R_w must propagate j/k, j/q_1, $j/q_2 \cdots j/q_{w-1}$ then the set of states $\{k, q_1, q_2 \cdots q_{w-1}\}$ is a subset of S and therefore the set of propagation conditions for R_t is identical to those for R_r for some $t < 2^N$ and some $r < t$. Then the sequence $(zR_r zR_{r+1} \cdots zR_t)^*$ satisfies the conditions of the theorem.

REFERENCES

[1] W. H. Kautz, "Testing for faults in cellular logic arrays," in *Proc. 8th Annu. Symp. Switching and Automata Theory*, 1967, pp. 161-174.
[2] P. R. Menon and A. D. Friedman, "Fault detection in iterative logic arrays," *IEEE Trans. Comput.*, vol. C-20, pp. 524-535, May 1971.
[3] A. D. Friedman and P. R. Menon, *Fault Detection in Digital Circuits*. Englewood Cliffs, N.J.: Prentice-Hall, 1971.
[4] O. C. Boeiens, "Fault diagnosis in iterative arrays," Ph.D. dissertation, Stevens Inst. Technol., Hoboken, N.J., 1970.
[5] A. D. Friedman, R. L. Graham, and J. D. Ullman, "Universal single transition time asynchronous state assignments," *IEEE Trans. Comput.*, vol. C-18, pp. 541-547, June 1969.

Arthur D. Friedman (S'61-M'65), for a photograph and biography see page 399 of the April 1973 issue of this TRANSACTIONS.

Design of Easily Testable Bit-Sliced Systems

THIRUMALAI SRIDHAR, MEMBER, IEEE, AND JOHN P. HAYES, SENIOR MEMBER, IEEE

Abstract—Bit-sliced systems are formed by interconnecting identical slices or cells to form a one-dimensional iterative logic array (ILA). This paper presents several design techniques for constructing easily testable bit-sliced systems. Properties of ILA's that simplify their testing are examined. C-testable ILA's, which require a constant number of test patterns independent of the array size, are characterized, and a method for making an arbitrary ILA C-testable is presented. A new testability concept for arrays called I-testability is introduced. I-testability ensures that identical test responses can be obtained from every cell in an ILA, and thus simplifies response verification. I-testable ILA's are characterized, as well as CI-testable arrays, which are simultaneously C- and I-testable. A method of making an arbitrary ILA CI-testable is presented. The application of C- and I-testing to the design of bit-sliced (micro-) computers is investigated. For this purpose a family of easily testable processor slices is described. The design of a self-testing CPU based on I-testing is discussed, and compared with a more conventional self-testing design.

Index Terms—Bit-sliced systems, design for testability, fault modeling, iterative logic arrays, self-testing, test generation.

I. INTRODUCTION

A DIGITAL system S^n is said to be *bit-sliced* if it can be realized as a one-dimensional array or cascade of N identical modules of type S^k, as shown in Fig. 1, where $n = Nk$. S^k is a module or *slice* that performs a set of operations I on k-bit operands or data words. The bit-sliced system S^n performs the same set of operations I on n-bit words. The best known example of S^n is a bit-sliced processor in which S^k is a k-bit (micro-) processor slice and k is typically 2, 4, or 8 [1]. Bit-slicing makes it easy to customize the data word size of a system. For example, using a typical 4-bit processor slice like the AMD 2901 [2] as the basic component, 8-, 16-, and 32-bit processors can easily be built. These processors, implemented using the same basic 4-bit slice, have a common set of control signals or (micro-) instructions I. Thus, bit-slicing helps in designing families of computers that have different word sizes, but use the same basic software. Because it employs identical components with uniform interconnections, bit-slicing is also useful in the design and layout of complex IC chips. For example, it is used extensively in the processor part of the Caltech OM-2 VLSI computer [3].

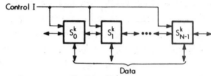

Fig. 1. General structure of a bit-sliced system composed of N identical k-bit slices S^k.

Conventional approaches to test pattern generation, both nonheuristic and heuristic, are generally inadequate when applied to VLSI devices such as microprocessors and microcomputers. Most nonheuristic testing methods have two serious limitations: 1) they treat low-level devices such as gates and flip-flops as primitive elements, and 2) they employ correspondingly low-level fault models such as the line stuck-at-0/1 model [4]. While heuristic testing methods are extensively used for testing complex systems, their effectiveness is usually very difficult to quantify [5]. Bit-sliced systems, on the other hand, lend themselves to exact nonheuristic testing approaches [6]. The short word size of the individual slices permits the use of high-level circuit models and powerful functional fault models. Using these models it is possible to construct test sets of near-minimal size that are guaranteed to detect all faults of interest. Moreover, the regular interconnection structure that characterizes bit slicing simplifies testing of the entire bit-sliced array. This paper investigates some new ways of exploiting these properties in the design of easily testable and self-testing digital systems.

Only a few attempts have been made in the past to use bit-slicing in the design of self-testing computers. In 1962 Forbes *et al.* at IBM designed the DX-1, an experimental self-diagnosable computer in which the CPU is partitioned into two identical slices each capable of testing the other [7]. An extension of this approach has been investigated by Ciompi and Simoncini [8]. Wakerly has designed a self-testing minicomputer that makes extensive use of error-detecting codes in a bit-sliced framework [9]. Recently, Könemann *et al.* have proposed the use of built-in pseudorandom test pattern generators and signature analyzers to make an individual bit slice self-testing [10]. None of the foregoing approaches guarantees complete detection (100 percent fault coverage) of all likely physical faults in the systems under consideration. Furthermore, where explicit nonrandom test patterns are used, considerable computational effort is required to construct these patterns and their responses.

For analysis purposes, bit-sliced systems can be viewed as *iterative logic arrays* (ILA's), which are one-dimensional cascades of identical cells. The properties of ILA's of simple cells have been studied in the context of switching theory [4],

Manuscript received December 12, 1980; revised June 2, 1981. This work was supported by the Joint Services Electronics Program under Contract F44620-76-C-0061, the National Science Foundation under Grant MCS78-26153, and the Naval Electronic Systems Command under Contract N00039-80-C-0641 (VHSIC Program).

T. Sridhar was with the Department of Electrical Engineering, University of Southern California, Los Angeles, CA 90007. He is now with Texas Instruments, Inc., Dallas, TX 75265.

J. P. Hayes is with the Department of Electrical Engineering, University of Southern California, Los Angeles, CA 90007.

[11]. In earlier works [6], [18] we have shown that bit-sliced processors can have a very desirable ILA property called *C-testability*. An ILA is *C-testable* if it can be tested with a constant number of test patterns independent of array size [12]. Some theoretical aspects of *C*-testability in unilateral arrays of combinational cells were examined in [6]. In Section II of this paper we study *C*-testability in more general ILA's of bilateral and sequential cells that are useful for modeling practical bit-sliced systems. As will be seen later, *C*-testable bit-sliced systems are very attractive in practice since they usually possess small test sets that are relatively easy to generate.

In Section III we introduce a new property in ILA's called *I-testability*. An array is *I-testable* if the test responses from every cell of the ILA can be made identical. Thus, the test response of an *I*-testable ILA can easily be verified using simple equality checkers. This approach is shown to be very useful in designing self-testing systems. *CI-testable* ILA's which have the properties of both *C*-testability and *I*-testability at the same time, are also examined in this section. *CI*-testability is characterized and a design modification scheme to make an ILA *CI*-testable is presented.

Some practical applications of the above ILA properties in designing easily testable bit-sliced systems are discussed in Section IV. A family of easily testable processor cells that closely resemble the AMD 2901 processor slice is described. Using high-level circuit and fault models, it is shown that an array of such processor cells is *CI*-testable. Furthermore, tests for an individual cell can readily be extended to tests for an array. The test requirements of another practical bit-sliced device, a microprogram sequencer are also considered. Finally, a self-testing (micro-) computer design based on *I*-testing is outlined and compared with conventional self-testing methods.

II. C-Testable Arrays

In this section we investigate *C*-testability in ILA's of the type shown in Fig. 2, which closely model the structure of bit-sliced systems. Such ILA's are called *bilateral*, since information flows between cells in both the left and right directions; in a *unilateral* ILA the cells communicate in one direction only. We first discuss arrays of combinational cells and then apply the results directly to a useful class of sequential arrays.

The problems of testing ILA's were first studied in detail by Kautz, who derived general testability conditions for unilateral combinational ILA's [13]. The concept of *C*-testability was introduced by Friedman [12], who analyzed a class of combinational arrays in which only the horizontal outputs of the rightmost cell are observable. Recently, various authors [14], [15] have studied *C*-testable ILA's having direct observable outputs from every cell. Our earlier work on *C*-testable ILA's [6] includes a characterization of *C*-testability, and a design modification scheme to make any ILA *C*-testable. The foregoing works deal with combinational unilateral ILA's only. This section extends the results of [6] to bilateral arrays.

The functional fault model used here is the same as that used

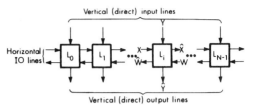

Fig. 2. General structure of a bilateral iterative logic array (ILA).

by Kautz, Friedman, and others. A *fault* can change the function realized by a combinational cell L to any other function provided that L remains combinational. Thus, to test an individual cell it is necessary and sufficient to apply all cell input patterns to L. It is assumed that at most one cell of an array can be faulty at any time. Although we construct tests for an array D based on this single-fault assumption, it can be shown that a single-fault test set T_D for D also verifies the truth table of D [16]. In other words, T_D also detects all detectable multiple faults in D. D is *C*-testable if $|T_D|$, i.e., the size of T_D, is independent of the number of cells in D.

Combinational Arrays

We first consider bilateral arrays of combinational cells with the structure shown in Fig. 2. Each cell has independent vertical input and output lines, as well as horizontal input and output lines to neighboring cells. The horizontal signal sets of each cell are denoted by X, W, $\hat{X}$, and $\hat{W}$, while the vertical input and output signal sets are Y and $\hat{Y}$, respectively; see Fig. 2. Arbitrary members of X, $\hat{X}$, W, $\hat{W}$, Y, and $\hat{Y}$ are denoted by x, $\hat{x}$, w, $\hat{w}$, y, and $\hat{y}$, respectively. Because of the interconnection structure, $X = \hat{X}$ and $W = \hat{W}$. Let $|X| = |\hat{X}| = n$, $|W| = |\hat{W}| = p$, and $|Y| = m$. It is customary [4] to represent the behavior of an ILA cell by a modified truth table called a *flow table*. Flow tables can be defined in several ways for bilateral ILA's. An $n \times mp$ table whose rows represent X and whose columns represent $Y \times W$ is called the *X flow table*. For simplicity, a member (a, b) of the Cartesian product $A \times B$ will sometimes be written as ab. Thus, a typical entry e in row x_i and column $y_j w_k$ of the X flow table is written as $(\hat{x}_a, \hat{y}_c \hat{w}_b)$, where $\hat{x}_a$ is the $\hat{X}$ output function $\hat{x}(x_i, y_j w_k)$, $\hat{w}_b$ is the $\hat{W}$ output function $\hat{w}(x_i, y_j w_k)$, and $\hat{y}_c$ is the direct output function $\hat{y}(x_i, y_j w_k)$. The members of the set $X \times \hat{W}$ represent the input/output signal pairs appearing at the left boundary of a cell in the array, and are called *states* of a cell. A table in which $X \times \hat{W}$ is the set of rows or states, and Y is the set of columns is called the $X\hat{W}$ *flow table* of the cell. An entry in this table is a possibly empty set of pairs, each consisting of next state and output function values. The $X\hat{W}$ flow table is easily derived from the X flow table. For example, consider the entry $e = (\hat{x}_a, \hat{y}_c \hat{w}_b)$ in row x_i and column $y_j w_k$ of the X flow table. The corresponding entry in the $X\hat{W}$ flow table has the next state $\hat{x}_a w_k$ and the $\hat{Y}$ output value $\hat{y}_c$ in row $x_i \hat{w}_b$ and column y_j. Each next state entry in the $X\hat{W}$ flow table also specifies a (state) *transition* of the form $\tau: x_i \hat{w}_b \xrightarrow{y_j} \hat{x}_a w_k$. Another table called the $\hat{X}W$ *flow tabl* in which the set $\hat{X} \times W$ is the set of rows (or states) and Y is the set of columns is defined similarly. Fig. 3 gives examples of the various flow tables. Here multiple entries are separated by semicolons, and the symbol ϕ indicates a null entry.

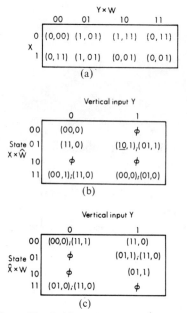

Fig. 3. (a) X flow table of a bilateral cell. (b) $X\hat{W}$ flow table. (c) $\hat{X}W$ flow table.

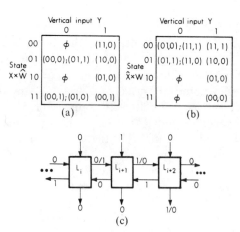

Fig. 4. (a) $X\hat{W}$ flow table of a cell L. (b) $\hat{X}W$ flow table of L. (c) Application of a distinguishing sequence to an array of L-type cells.

Let $a_0, b_0 b_1 \cdots b_{N-1}, c_{N-1}$ denote the input pattern applied to a bilateral array, where a_0 is the X input signal to the leftmost cell L_0, b_i is the Y input signal to L_i, for $i = 0, 1, \cdots, N - 1$, and c_{N-1} is the W input signal to the rightmost cell L_{N-1}. The $\hat{Y}$ output pattern S of the array can easily be determined from the $X\hat{W}$ flow table. S is the output sequence defined by the $X\hat{W}$ flow table for the input sequence $b_0 b_1 \cdots b_{N-1}$ with $a_0 c'$ and $a' c_{N-1}$ as the initial and final states, respectively. The corresponding sequence of next states gives the X and W signals appearing at the cell boundaries. Certain types of state transitions are useful for characterizing C-testability in ILA's. A transition $\tau : s \xrightarrow{y} \hat{s}$ of flow table F is called a *repeatable transition* of F if there exists an input sequence (sometimes called a *transfer sequence*) that drives F from $\hat{s}$ back to s. F is called a *repeatable transition* or *RT flow table* if every transition of F is a repeatable transition. For example, the $X\hat{W}$ flow table of Fig. 3(b) is not an RT flow table, since the underlined transition $01 \xrightarrow{1} 10$ is not repeatable. On the other hand, the $X\hat{W}$ flow table of Fig. 4(a) is an RT flow table. Since $X\hat{W}$ and $\hat{X}W$ flow tables are complementary, an $\hat{X}W$ flow table is an RT flow table if and only if the corresponding $X\hat{W}$ flow table is an RT flow table.

Consider an array of N bilateral cells L with an $X\hat{W}$ flow table F of size $np \times m$. According to the fault model described earlier, to test this array completely it is necessary and sufficient to verify all npm transitions of F in every cell. To verify some transition $\tau : x_i \hat{w}_b \xrightarrow{y_j} \hat{x}_a w_k$ in cell L_i of the ILA, it is necessary to apply the signals x_i, y_j, and w_k to the appropriate input lines of L_i. At the same time, the $\hat{X}$, $\hat{W}$, and $\hat{Y}$ output signals of L_i should be observed at the primary output lines of the ILA. Let q be the number of array input patterns required to verify τ in every cell of the array. C-testability requires q to be a constant independent of the array size N. Suppose that τ is not a repeatable transition. Then while verifying τ in cell L_0, say, it is not possible to verify τ in any other cell at the same time. This is because the initial state $x_i \hat{w}_b$ of τ is not reachable from the state $\hat{x}_a w_k$. Hence, to verify τ in every cell we need at least N array input patterns, one for each cell. In other words q depends on N. Thus, we have the following necessary condition for C-testability.

Theorem 1: A bilateral ILA is C-testable only if the $X\hat{W}$ cell flow table is an RT flow table.

In order to verify the $\hat{X}$ and $\hat{W}$ output signals of a cell, we make use of distinguishing sequences for the states of the $X\hat{W}$ and $\hat{X}W$ flow tables, respectively. These sequences are defined by extending the notion of a pairwise state distinguishing sequence for a unilateral cell [4], [6]. In the case of an $X\hat{W}$ flow table F, a *distinguishing sequence* $DS(x_a \hat{w}_c, x_b \hat{w}_c)$ for a pair of states $x_a \hat{w}_c$ and $x_b \hat{w}_c$ is a Y input sequence which, when applied to F, produces two different $\hat{Y}$ output sequences with $x_a \hat{w}_c$ and $x_b \hat{w}_c$ as the initial states. A flow table is said to be *reduced* with respect to the $\hat{Y}$ outputs if there exist distinguishing sequences for every $x_a, x_b \in X$ and for every $\hat{w}_c \in \hat{W}$.

Example 1: Consider the $X\hat{W}$ flow table shown in Fig. 4(a). The possible distinguishing sequences are $DS(00, 10) = 10$ and $DS(01, 11) = 0$. Fig. 4(c) demonstrates how the sequence $DS(00, 10)$ can be used to verify the $\hat{X}$ output signal of any cell in an array. The $\hat{X}$ output signal of the cell L_i, in which the transition $\tau : 01 \xrightarrow{0} 00$ is being verified, is propagated to the $\hat{Y}$ outputs by applying signals 1 and 0, corresponding to the sequence $DS(00, 10)$, to the Y input lines of the cells L_{i+1} and L_{i+2} as shown. Any change in the $\hat{X}$ output of L_i due to a fault f in L_i is detected at the $\hat{Y}$ output of L_{i+2}. A change from a to b, where $a, b \in \{0, 1\}$, in the signal value on a line induced by fault f is indicated as a/b in Fig. 4(c). Fig. 4(b) gives the $\hat{X}W$ flow table of the cell L. It is evident that this flow table also has distinguishing sequences to verify the $\hat{W}$ output signal of a cell, and hence is reduced. □

For testing C-testable arrays, we introduce a class of input test patterns called C-tests that are periodic with respect to the cell inputs. A detailed treatment of C-tests appears in the Appendix. Roughly speaking, a C-test for some transition, say $\tau : x_a \hat{w}_c \xrightarrow{y_j} \hat{x}_b w_d$ in an $X\hat{W}$ flow table, is an input pattern of the form $C_{ix}(\tau) = x_a \hat{w}_c, (y_j D_{ix} R_{ix})^*$, where D_{ix} is a distinguishing sequence and R_{ix} is a transfer sequence that restores the cell state to $x_a \hat{w}_c$. The asterisk superscript indicates that the pattern $y_j D_{ix} R_{ix}$ is applied repeatedly; thus $y_j D_{ix} R_{ix} y_j D_{ix} R_{ix} y_j D_{ix} \cdots$ to the Y inputs of the array. Hence, $C_{ix}(\tau)$ verifies the $\hat{X}$ output signal produced by the transition τ in cells $L_0, L_{k_i}, L_{2k_i}, L_{3k_i}, \cdots$ spaced at fixed intervals along the array. The period k_i is the number of cells spanned by the Y subpattern $y_j D_{ix} R_{ix}$. We can completely test τ in all cells by means of a set of C-tests $CT_x(\tau)$ obtained from $C_{ix}(\tau)$ by shifting $y_j D_{ix} R_{ix}$ up to k_i cell positions, and by allowing D_{ix} to range over all members of a set of distinguishing sequences $SDS(\hat{x}_b w_d)$. The periodic structure of $C_{ix}(\tau)$ implies that $CT_x(\tau)$ is of fixed size, independent of the length of the array being tested. Similarly, C-tests $C_{jw}(\tau)$ and $CT_w(\tau)$ for verifying the $\hat{W}$ output signals produced by τ in all cells can be constructed. This leads to the following results which are proven formally in the Appendix.

Theorem 2: There exist C-tests that verify every transition τ in the $X\hat{W}$ and $\hat{X}W$ flow tables of a cell, if both flow tables are reduced RT flow tables.

Theorem 2 implies that if a flow table is a reduced RT table, then the corresponding ILA is C-testable using a set of C-tests.

Theorem 3: A bilateral ILA with reduced $X\hat{W}$ and $\hat{X}W$ cell flow tables is C-testable if and only if the tables are RT flow tables.

Example 1 (continued): The $X\hat{W}$ and $\hat{X}W$ flow tables of the cell L in Fig. 4 are reduced RT flow tables. Hence, an array D of L-type cells is C-testable using C-tests. For example, consider the transition $\tau: 01 \xrightarrow{0} 00$ in the $X\hat{W}$ flow table. A C-test for τ in $CT_x(\tau)$ is $C_{1x}(\tau) = 01, (01010)^*$; see Fig. 4(c). If the array length is 12, for example, then $C_{1x}(\tau) = 01, 010100101001$, where the length of the periodic Y pattern is adjusted to match the number of cells present in D. C_{1x} applies the transition τ to the cells $L_0, L_5, L_{10}, \cdots$, and verifies the corresponding $\hat{X}$ output signals. Let $s^q[C_{1x}(\tau)]$ denote the C-test obtained by shifting the test pattern applied to D by $C_{1x}(\tau)$ q positions to the left. Thus, $s^1[C_{1x}(\tau)] = 00, (10100)^*$ and verifies the $\hat{X}$ output signal of cells $L_4, L_9, L_{14}, \cdots$. Similarly, the other three shifted versions of $C_{1x}(\tau)$, $s^2[C_{1x}(\tau)]$, $s^3[C_{1x}(\tau)]$, $s^4[C_{1x}(\tau)]$, test τ in the remaining cells of the array. Thus, the $\hat{X}$ output entry in τ is verified in every cell of D with only 5 array input patterns (C-tests) independent of the size of D. □

Theorem 3 suggests that any bilateral array can be made C-testable by converting its $X\hat{W}$ and $\hat{X}W$ flow tables to reduced RT tables. Fig. 5 indicates the proposed modification scheme. Two new columns y_a and y_b are added to the original $X\hat{W}$ flow table F. Next state entries in the new columns are chosen so that every state of F is reachable from every other state. Consequently, the modified $X\hat{W}$ flow table is an RT flow table. The output entries a_i, for $i = 0, 1, \cdots, n - 1$, and a'_j, for $j = 0, 1, \cdots, p - 1$, are also chosen to make both the $X\hat{W}$ and $\hat{X}W$ flow tables reduced with respect to the $\hat{Y}$ output. Let Q_a be an input sequence obtained by repeating y_a n_1 times, that is, $Q_a = y_a^{n_1}$. Similarly, let $Q_b = y_b^{n_2}$. The a_i and a'_j entries are then selected to make Q_a and Q_b minimum-length distin-

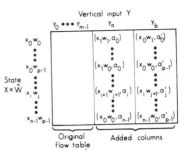

Fig. 5. Flow table modification to make any bilateral ILA C-testable.

guishing sequences for the $X\hat{W}$ and $\hat{X}W$ flow tables, respectively, with $n_1 = \lceil \log_2 n \rceil$ and $n_2 = \lceil \log_2 p \rceil$. The construction of the a_i's and a'_j's using shift register sequences is described in [16]. The next theorem is proven in the Appendix.

Theorem 4: A bilateral array of cells whose $X\hat{W}$ and $\hat{X}W$ flow tables are modified as shown in Fig. 5 is C-testable.

Sequential Arrays

Consider the class of sequential arrays with the basic cell L shown in Fig. 6. L is a synchronous sequential machine containing two modules: a combinational module M_1 whose output is the next state $\hat{S}$, and a memory module M_2. The internal state S is directly presented at the $\hat{Y}$ output, thus L is a Moore-type machine [4]. Many practical ILA circuits such as registers, shift registers, and counters belong to this class. We show next that the results obtained above for combinational arrays are directly applicable to this class of sequential arrays.

The fault model defined earlier for single-module combinational cells is extended here to the 2-module sequential cell L. A fault in a module M_i of L can change the function realized by M_i to any other function provided the number of internal states of M_i is not increased. As before, we assume that at most one module is faulty in an array. Thus, to test L it is necessary and sufficient to verify the truth table of M_1 and the state table of M_2. We also assume that the memory module M_2 of any cell L_j can be independently initialized to any state $s \in S$ via the Y input lines or other external lines.

Let P be an array of L-type cells, and let D be the combinational subarray formed from the combinational modules M_1 in P. P can be tested completely by testing its subarray D and the memory modules in every cell of P. The subarray D is tested like any other combinational ILA. For this purpose, the necessary S inputs to M_1 in each cell L_j are obtained by appropriately initializing the state of M_2 in L_j. To test the memory module M_2, we need to verify its state table, in which each next state entry should equal the corresponding present input $\hat{s} \in \hat{S}$. Most of these entries are automatically verified while testing D. This is because the output of M_2 is directly observable at $\hat{Y}$, and all possible $\hat{S}$ inputs to M_2 must be generated while testing M_1 exhaustively according to the fault model. Any remaining untested entries of the state table of M_2 can readily be verified by applying appropriate initializing input patterns to P. Thus, we have the following result.

Theorem 5: An array of type L sequential cells is C-testable if and only if its combinational subarray is C-testable.

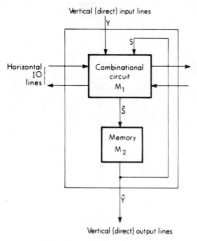

Fig. 6. A sequential cell L.

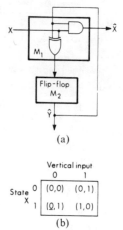

Fig. 7. (a) A 1-bit counter cell. (b) The X flow table of the incrementer module M_1.

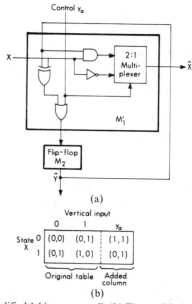

Fig. 8. (a) Modified 1-bit counter cell. (b) The modified X flow table for the incrementer logic.

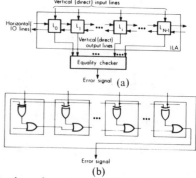

Fig. 9. (a) A scheme for response verification in an I-testable iterative logic array. (b) An ILA implementation of the equality checker.

It follows that all results obtained previously for combinational arrays are directly applicable to this special class of sequential arrays.

Example 2: Consider an N-bit binary counter array composed of N 1-bit counter cells L with the structure of Fig. 7(a). The combinational module M_1 of L is a 1-bit incrementer, whose X flow table F appears in Fig. 7(b). F is a reduced but not RT flow table, since it contains the nonrepeatable transition which is underlined. Hence, the counter array is not C-testable. It can be made C-testable by modifying F to make it an RT flow table. This can be done by adding a new column y_a, as shown in Fig. 8(b). Fig. 8(a) shows the modifications needed in the counter cell itself in order to realize the flow table of Fig. 8(b). If the new control variable y_a is 1, then the entries in the added column are selected; whereas if y_a is 0, the entries in the original table are effective. Any sequential cell with the structure of Fig. 6 can be similarly redesigned to make it C-testable. □

III. I-Testable Arrays

In the previous section we studied input test pattern generation for ILA's. In this section we examine output response verification. For this purpose we introduce a new property of ILA's called I-testability which greatly simplifies response checking. Let T_D be a test set that tests an ILA D completely under the usual single-cell fault assumptions. D is I-testable with respect to T_D if the expected responses to T_D appearing at the vertical outputs of every cell of D are identical. The input patterns forming T_D are called I-tests. The responses of D to the I-tests of T_D can easily be verified by comparing the signals appearing on the vertical output lines from the individual cells. This comparison can be done by means of an equality checking circuit, as shown in Fig. 9(a). Thus, I-testing in which I-tests and equality checkers are used, eliminates the need to store explicitly the expected responses to T_D. The checking circuit of Fig. 9(a) can itself be designed as an ILA, as shown in Fig. 9(b), and can easily be integrated into the original ILA D. Practical applications of I-testing in designing self-testing bit-sliced systems are discussed in Section IV.

We begin by analyzing I-testability in a basic class of ILA's, namely unilateral ILA's of identical combinational cells; see Fig. 10. Practical array circuits such as ripple-carry adders and incrementers used in the processors and microprogram sequencers discussed in the next section belong to this class. Later we also discuss the property of CI-testability in such unilateral arrays. A *CI-testable ILA* is both C-testable and I-testable with respect to some test set T_D, i.e., every test pattern in T_D

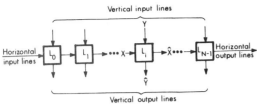

Fig. 10. A combinational unilateral ILA.

is a C-test as well as an I-test. The concepts and results presented here for combinational unilateral ILA's can readily be extended to more general bilateral and sequential ILA's.

I-Testability

Let L be an ILA cell with the input/output signal sets X, $\hat{X}$, Y, and $\hat{Y}$ shown in Fig. 10. Let F be the $n \times m$ X flow table of L, in which rows represent X and columns represent Y. It is useful to decompose F into $q = |\hat{Y}|$ disjoint $n \times m$ flow tables $F_0, F_1, \cdots, F_{q-1}$, called *partitioned flow tables*, such that F_k contains all the entries or transitions of F for which $\hat{y}(x_i, y_j) = \hat{y}_k$. The null symbol ϕ is entered in F_k to denote each entry of F for which $\hat{y}(x_i, y_j) \neq \hat{y}_k$.

Example 3: The flow table F of a 1-bit full adder cell and its two partitioned flow tables F_0 and F_1 are shown in Fig. 11. X is the carry input, Y is the 2-bit data input, $\hat{X}$ is the carry output, and $\hat{Y}$ is the sum output. □

To test the ILA of Fig. 10 completely with the basic cell L treated as a single module, it is necessary and sufficient to verify all mn entries of the flow table F for every cell of the ILA. In other words, all non-ϕ entries of each partitioned flow table F_k must be verified for every cell. While verifying an entry of F_k in some cell L_j using I-tests, every other cell in the ILA should be confined to non-ϕ entries appearing in F_k; otherwise the expected vertical output signals from every cell will not be identical.

We are interested in obtaining necessary and sufficient conditions for an ILA to be I-testable. Our approach is to modify the following well-known testability criteria for ILA's.

Theorem 6 [13]: An ILA is testable if and only if

1) every state appears as a next state entry in the cell flow table F at least once, and

2) no two rows of F are identical.

For the ILA to be I-testable the conditions of Theorem 6 must be modified to apply individually to every partitioned flow table F_k, for $k = 0, 1, 2, \cdots, q - 1$. The first condition of Theorem 6 ensures that every X input signal is applicable to every cell of the ILA. In a partitioned flow table F_k it is not necessary to have every state appear as a next state entry because not all X input signals may produce the $\hat{Y}$ output signal $\hat{y}_k$. If, say, $x_a \in X$ does not produce the output $\hat{y}_k$, then the row in F_k corresponding to x_a contains ϕ entries only, and hence can be deleted from F_k. Thus, condition 1) of Theorem 6 is modified for I-testability as follows. Every undeleted state should appear at least once as a next state entry in F_k for all k.

The second condition of Theorem 6 states that any faulty $\hat{X}$ output signal of a cell must eventually be propagated to some observable output line of the ILA. This propagation condition may be modified for I-testability by introducing a distingu-

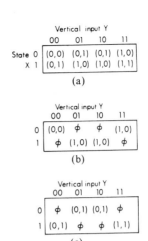

Fig. 11. (a) Flow table F for a 1-bit full adder cell. (b) Partitioned flow table F_0 for $\hat{y} = 0$. (c) Partitioned flow table F_1 for $\hat{y} = 1$.

ishability criterion called *I-distinguishability*. Roughly speaking, a state x_i of a partitioned flow table is said to be *I-distinguishable* from another state x_j, if there is a Y input signal that distinguishes x_i from x_j by producing different $\hat{Y}$ or $\hat{X}$ output signals. (A more precise definition is given in the Appendix.)

Example 3 (continued): In the full adder flow table F_0 of Fig. 11(b) state $x_0 = 0$ is I-distinguishable from $x_1 = 1$ due to columns $y_0 = 00$ and $y_3 = 11$. For instance, the Y input signal 00 produces different $\hat{Y}$ output signals for states x_0 and x_1. x_1 is I-distinguishable from x_0 because of columns $y_1 = 01$ and $y_2 = 10$. □

Note that cases exist where x_a is I-distinguishable from x_b, but x_b is not I-distinguishable from x_a. The next result is proven in the Appendix.

Theorem 7: An ILA is I-testable if and only if in each partitioned cell flow table F_k

1) every (undeleted) state appears at least once as a next state entry, and

2) every state that appears at least once as a next state entry is I-distinguishable from every other state.

The following bounds on the number of I-tests $|T_D|$ for an I-testable ILA are easily established [16]:

$$mn \leq |T_D| \leq mn(n-1)(N-1) + mn.$$

Example 3 (continued): Consider again an N-bit ripple-carry adder ILA. It is evident that the two partitioned flow tables F_0 and F_1 of Fig. 11 both satisfy the conditions of Theorem 7. Hence, the adder array is I-testable. I-tests for verifying the entries in F_0 and F_1 can easily be constructed. For example, consider the entry $(0, 0)$ in row $x_0 = 0$ and column $y_0 = 00$ of F_0. An I-test that checks this entry in all cells of the ILA is x_0, y_0^N, which applies x_0 to the horizontal input of the leftmost cell, and y_0 to the vertical inputs of all N cells. It can easily be shown that the minimum number of I-tests for this adder array is $6 + 2N$. Fig. 12 shows a 3-cell ripple carry adder and the corresponding minimal set of 12 I-tests. □

Example 4: The 1-bit incrementer cell M_1 of Fig. 7 has the flow table shown in Fig. 13. The partitioned flow table F_0 satisfies both conditions of Theorem 7, whereas F_1 satisfies only condition 2). Hence, an incrementer ILA is not I-testable. □

(a)

Cell transition $\tau: x_i \xrightarrow{Y_k} \hat{x}_i$	I-test(s) for τ $X_0, Y_0 Y_1 Y_2$	Cells tested
$0 \xrightarrow{00} 0$	0,000000	L_0, L_1, L_2
$0 \xrightarrow{01} 0$	0,010101	L_0, L_1, L_2
$0 \xrightarrow{10} 0$	0,101010	L_0, L_1, L_2
$0 \xrightarrow{11} 1$	0,110101 0,001101 0,000011	L_0 L_1 L_2
$1 \xrightarrow{00} 0$	1,000101 1,110001 1,111100	L_0 L_1 L_2
$1 \xrightarrow{01} 1$	1,010101	L_0, L_1, L_2
$1 \xrightarrow{10} 1$	1,101010	L_0, L_1, L_2
$1 \xrightarrow{11} 1$	1,111111	L_0, L_1, L_2

(b)

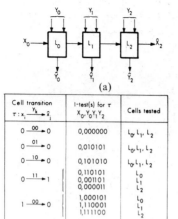

Fig. 12. (a) A 3-bit ripple-carry adder ILA. (b) A minimal set of 12 I-tests for this adder.

Vertical input Y

	0	1
State X 0	(0,0)	(0,1)
1	(0,1)	(1,0)

F

	0	1
0	(0,0)	ϕ
1	ϕ	(1,0)

F_0

	0	1
0	ϕ	(0,1)
1	(0,1)	ϕ

F_1

Fig. 13. Flow tables F, F_0, and F_1 of a 1-bit incrementer cell.

CI-Testability

Let T_D be a test set that completely tests an array D of the type shown in Fig. 10. D is CI-testable if every test in T_D is an I-test and $|T_D|$ is independent of the size N of D. A CI-testable ILA must, of course, satisfy the conditions for C-testability and I-testability obtained earlier. During I-testing, every cell in D is confined to entries appearing in only one partitioned flow table at a time. Also, for the ILA to be C-testable the flow table being used must be an RT flow table (Theorem 1). Hence, we have the following necessary condition for CI-testability.

Theorem 8: An ILA is CI-testable only if all the partitioned flow tables are RT flow tables.

Example 3 (continued): The partitioned flow tables F_0 and F_1 of a full adder cell which appear in Fig. 11 are not RT flow tables. Hence, a ripple-carry adder ILA is not CI-testable. Nevertheless, it is C-testable [14], [16] and, as we showed earlier, also I-testable. However, there is no test set T_D that serves as a complete set of C-tests and I-tests simultaneously. □

The definitions of distinguishing sequence, C-test, and reduced flow table appearing in Section II can all be modified for CI-testability by adding the constraint that the $\hat{Y}$ output signals from every cell of the ILA be identical; see the Appendix for details. An *I-distinguishing sequence* $IDS(x_a, x_b)$ for the states x_a and x_b of a partitioned flow table F_k produces a constant output sequence of the form $\hat{y}_k^* = \hat{y}_k\hat{y}_k\hat{y}_k \cdots$ when x_a is the initial state. It produces a different $\hat{Y}$ output sequence when x_b is the initial state. F_k is *I-reduced* with respect to $\hat{Y}$ if $IDS(x_a, x_b)$ exists for every state x_a appearing as a next state entry in F_k, where x_b is any other state. For example, the partitioned flow tables F_0 and F_1 of the full adder cell of Example 3 are both I-reduced. If C-tests are required to produce identical $\hat{Y}$ outputs for all cells, the result is a class of fixed-length I-tests called *CI-tests*. An ILA is CI-testable if all faults in it can be detected by a set of CI-tests. The next theorem is analogous to Theorem 3 for C- testable arrays.

Theorem 9: A unilateral ILA with I-reduced partitioned cell flow tables is CI-testable if and only if the partitioned flow tables are all RT flow tables.

Let F be the $n \times m$ flow table of cell L in an ILA that is not CI-testable. It is possible to change F to F' by adding new columns so that the modified partitioned flow tables $\{F_k'\}$ are I-reduced RT flow tables. Thus, according to Theorem 9, an array of the modified cells is CI-testable. In the proposed modification scheme q new columns, one for each partitioned flow table, are introduced, as shown in Fig. 14. The next state entries in the new columns are such that all n states in every partitioned flow table are connected together in a single loop. Thus, the modified flow tables $\{F_k'\}$ are all strongly connected and so are RT flow tables. The output entries in the new columns ensure that the $\{F_k'\}$ are all I-reduced. Hence, an array of the modified cells is CI-testable.

Example 3 (continued): It was shown already that the flow tables F_0 and F_1 of a 1-bit full adder cell are I-reduced non-RT flow tables. We now add a new column y_a to the original flow table F, as depicted in Fig. 15(a). It is clear that the modified partitioned flow tables F_0' and F_1' in Fig. 15(b) and (c) are RT flow tables. Hence, an array of the modified full adder cells is CI-testable. Fig. 16 shows a hardware implementation of the full adder modified for CI-testability according to the above procedure. □

IV. APPLICATIONS

The analysis of the preceding sections indicates that C- and I-testability can greatly simplify the testing of ILA's. C-testability ensures small easily generated test sets, while I-testability simplifies the test verification process. Hence, C- and I-testability appear to be particularly attractive for the design of practical array-like circuits such as bit-sliced computers, where ease of testing is a primary design goal. To demonstrate this we analyze a family of bit-sliced processors introduced in [18] that closely resemble commercial processor slices. We also discuss the use of I-testing to design a self-testing bit-sliced computer.

Processors

Fig. 17 shows a cell C which constitutes a 1-bit general-purpose processor slice [18]. It performs essentially the same functions as the widely used 2901 4-bit microprocessor slice, namely addition, subtraction, shifting, and various logical operations such as AND, OR, and EXCLUSIVE-OR [2]. A simpler version of C without the shifting function is analyzed

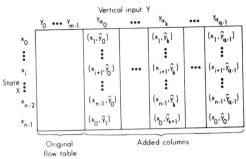

Fig. 14. Flow table modification to make a unilateral ILA CI-testable.

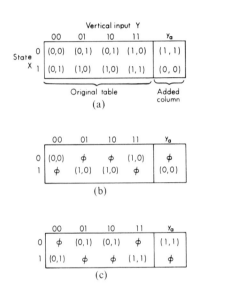

Fig. 15. The modified flow tables of a 1-bit full adder cell. (a) F', (b) F'_0, and (c) F'_1.

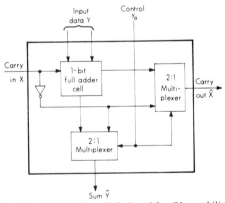

Fig. 16. A full adder cell designed for CI-testability.

in [6]. Although C is a 1-bit slice, it can easily be extended to larger slices [18]. Note that 1-bit processors are commercially available, e.g., the Motorola 14500 which, however, is not bit-sliced [17]. Like the Motorola 10800 and Texas Instruments 74S481 processors [1], C has only two working registers M_A or M_T. However, it is feasible to replace any single register M_A or M_T by a register file, if desired. Ripple-carry propagation only is used in C, since a carry-lookahead scheme destroys the ILA structure.

The circuit and fault models introduced in Sections II and III are used here in a more general form. A circuit U to be tested is treated as a well-defined network of register-level

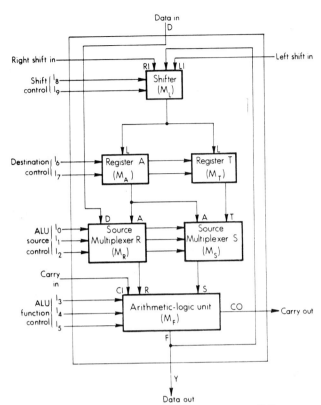

Fig. 17. An easily testable 1-bit processor cell C.

modules such as multiplexers, arithmetic-logic units (ALU's), and registers. A *module* is a black box whose input–output behavior is known, but whose internal structure or gate-level logic design is unspecified. A *fault* in a module M may change the function realized by M to any other function, provided the number of internal states of M is not increased. We also assume that at most one module of U can be faulty at a time.

To test U for functional faults of the foregoing type, the input–output behavior of every module must be completely verified. Because of the necessity of exhaustively testing the individual modules, we can only apply the fault model to relatively small modules. However, in the bit-sliced circuits of interest, the single-module fault assumption implies that a network of modules can be tested in an efficient nonexhaustive way. Exhaustive functional testing of the individual modules has the added advantage of detecting design errors in the modules. Thus, the test data obtained for U according to the above fault model can also be used during the design process to verify the correctness of the logic design of U.

The processor cell C is divided into six register-level modules: four combinational modules M_R, M_S, M_F, and M_L, and two (synchronous) sequential modules M_A and M_T; see Fig. 17. We are interested in constructing a test sequence T_C for all functional faults in C. If M_i is a combinational module in C with n input lines, then it is necessary and sufficient for its test set T_i to be the set of all 2^n input combinations to the module. If M_i is a sequential module, then the checking sequence approach [4] is used in constructing T_i. This approach can be used without difficulty to obtain short efficient test sequences for small sequential modules like M_A or M_T of C. Faults in a module M_i of C are detected by a test sequence T_i

which, when applied to the primary inputs of C, causes T_i to be applied to M_i and also causes the responses of M_i to be propagated to the observable outputs of C. Since at most one module is allowed to be faulty, a test sequence T_C for C can be obtained by concatenating the T_i''s of every module M_i in C; see [18] for details. The entire processor cell C can be tested by a test sequence T_C containing no more than 114 test patterns.

An N-bit processor array PA consisting of N identical copies of C in cascade has the structure shown in Fig. 1. The circuit layout of an NMOS chip that implements a 4-bit version of this processor array is shown in Fig. 18. This chip was designed by the authors using the Caltech/Xerox VLSI design software [3]. To test the array PA it is sufficient to apply the test sequence T_C to every cell in the array. In addition, the responses of the cells to T_C must be propagated to the observable outputs of PA. Every test pattern in T_C can easily be extended to a C-test for PA [18]. Hence, PA is C-testable with only 114 test patterns. We now show how the tests for an individual cell C can be modified to form I-tests or CI-tests for the array PA.

Consider, for instance, the register modules M_A and M_T of C. Any desired L input signal can be applied to these modules by transferring the external D signal to L via the multiplexer M_R, the ALU M_F, and the shifter M_L. Any suitable ALU logical operation, such as OR, may be used for this purpose. Also, since a logical operation involves no interaction between cells, the same L input signal can be applied to the inputs of M_A and M_T of every cell of PA simultaneously. Hence, a complete test sequence T_{AT} for M_A and M_T can be applied to every cell simultaneously. Now to observe the output responses of M_A and M_T to T_{AT}, we can propagate the output signals A and T to the observable output Y via the source multiplexers M_R and M_S, and the ALU. By setting the ALU to perform the EXCLUSIVE-OR operation $F = R \oplus S$, the signals A and T can both be observed at Y at the same time. Again due to the use of a logical operation for this propagation, the A and T signals of every cell of PA can be observed simultaneously at their respective Y outputs. The resulting output responses to T_{AT} at the Y outputs of all cells will be identical. Thus, the tests for the register modules M_A and M_T of C are easily extended into CI-tests for PA. A similar approach can be used to extend the tests for the modules M_R and M_S of C into CI-tests.

Now consider testing the ALU module M_F. Testing the ALU's logical operations is quite straightforward because they involve no interactions between adjacent cells. Any logical operation can be tested in all cells simultaneously by generating identical ALU input operands in every cell. Since the ALU output F is connected directly to the observable Y output, the test responses at every Y output must be identical. Thus, all logical operations can be verified with CI-tests. The remaining ALU operations are the two's complement arithmetic operations addition and subtraction. Basically, these operations use ripple-carry addition with a 1-bit full adder in each cell. (For subtraction, one input operand is complemented before addition.) An N-bit ripple-carry adder with a 1-bit full adder cell is C-testable with only 8 C-tests [14], [16] and is I-testable

Fig. 18. Chip layout of a 4-bit processor array of type C cells.

with a minimum of $6 + 2N$ I-tests; see Example 3 in Section III. This implies that the adder array has only six tests that are both C- and I-tests; the corresponding six CI-tests for PA are readily derived. It is also possible to convert the two C-tests that are not I-tests into CI-tests for PA by introducing additional test cycles as follows. The N-bit nonidentical Y output signals of PA produced by a non-I-test are not verified in the first clock cycle, but instead are propagated via the shifter M_L and stored internally, say, in register T. During the next clock cycle, this stored N-bit word is compared bit-by-bit with the expected N-bit word which is applied to the external D inputs. This comparison can be done in the ALU using the previously verified EXCLUSIVE-OR operation. Because two identical words are being compared, the Y output of every cell will be identical as desired in this second clock cycle. Thus, the ALU module of every cell can also be tested completely with CI-tests.

The remaining module of C, namely the shifter M_L, also involves interaction between adjacent cells. It can be tested with CI-tests only using one of the approaches discussed above for the ALU. Hence, the N-bit processor array PA is CI-testable. The number of CI-tests for PA can be shown to be no more than 168 [16].

The 1-bit processor slice C of Fig. 17 can be extended to a k-bit slice called C^k that more closely resembles commercial bit slices, without sacrificing its easily-testable features [18]. The word size can be expanded from 1 to k by using a straightforward replication technique. Every module M_i in C, except M_F, is replaced by k copies of M_i in C^k. The ALU module M_F is retained as a single module operating on k-bit operands so that any fast carry propagation scheme like carry lookahead can be implemented within the ALU. Also, the number of working registers in C^k may be increased. For example, the k-bit register A can be replaced by an $n \times k$ register file or scratchpad memory; the resulting cell is denoted by $C^{k,n}$. Test generation for C^k and $C^{k,n}$ is similar to that for C discussed earlier. It can be shown that a processor array of type C^k or $C^{k,n}$ cells is CI-testable in essentially the same way as the array PA.

Microprogram Sequencers

A 1-bit microprogram sequencer S based on the 2909 slice [2] appears in Fig. 19. Its function is to generate the microinstruction addresses required in a microprogrammed CPU. S consists of two combinational modules: a multiplexer M_1 and an OR-AND logic module M_2, and two sequential modules: a microprogram counter M_3 and a 4×1 stack M_4. The multiplexer M_1 selects the next microinstruction address from one of three sources: the vertical external input, the stack, or the microprogram counter. Thus, S is essentially a 1-bit version of the 4-bit 2909 slice, except that it does not include a latched data input line.

Test patterns for S are derived in much the same way as for C. A test sequence T_S for S can be obtained by concatenating the test sequences T'_1, T'_2, T'_3, and T'_4 for the individual modules. The total number of tests for S can be shown to be at most 451.

An N-bit microprogram sequencer MS is realized by connecting N identical copies of S in cascade. Consider the subarrays formed by the three modules M_1, M_2, and M_4. Since there is no communication between the cells via these modules, the tests for M_1, M_2, and M_4 can be applied to every cell of MS in parallel. The identical test responses of these modules are readily propagated to the vertical observable output Y, resulting in a set of I-tests for M_1, M_2, and M_4. The remaining module, the microprogram counter M_3, involves communication between adjacent cells of MS via the carry lines. The next state function of M_3 is implemented by a 1-bit incrementer of the kind discussed in Section III; see Example 4. Hence, the counter subarray CA formed by the counter modules of every cell is neither C-testable nor I-testable. Therefore, MS is also not C-testable. However, it can be made I-testable as follows. Let t be a non-I-test for subarray CA. The responses of CA to t at the B outputs of M_3 in every cell are not identical. This N-bit response can be compared bit-by-bit in the multiplexer module M_1 with the expected N-bit response applied externally at the D inputs. For this comparison, a special EXCLUSIVE-OR function $L = D \oplus B$ is introduced into M_1, and is selected by means of the fourth unused multiplexer control input combination. This internal comparison approach is similar to the one used in the processor array PA for testing the ALU's arithmetic operations and the shifter. Thus, the microprogram sequencer MS is I-testable and requires at most $449 + 2N$ I-tests.

Self-Testing Processor

We conclude by describing a bit-sliced CPU design that uses I-testability to achieve self-testing in an efficient manner. We modify a conventional bit-sliced design to use I-testing and equality checkers in all major subunits as illustrated in Fig. 9.

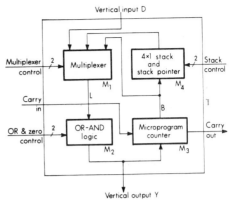

Fig. 19. A 1-bit microprogram sequencer cell S.

The overall structure of a bit-sliced CPU designed in this fashion is shown in Fig. 20. It consists of a bit-sliced execution unit (E-unit) and a microprogrammable instruction unit (I-unit). The E-unit is an I-testable processor array PA, and the I-unit comprises an I-testable bit-sliced microprogram sequencer MS, a control store CS which contains microprograms, and a read-only memory (ROM) MR for generating the starting address of a microprogram from the opcode of an external instruction. CPU's constructed from commercial bit-sliced families usually have the foregoing organization.

The I-testable arrays PA and MS discussed earlier in this section are used in the proposed design. Their output responses are verified by two equality checkers EC_1 and EC_2 integrated into PA and MS, respectively. Conventional duplication and comparison is used to make the remaining CPU components CS and MR self-testing; these components are both typically ROM's. The outputs of the duplicated ROM's are compared bit-by-bit using the two equality checkers EC_3 and EC_4 appearing in Fig. 20. Finally, the outputs of all four equality checkers $EC_1:EC_4$ can be combined by an OR gate to yield a single error signal for the CPU.

The test requirements for the various CPU components are as follows. As mentioned earlier, the processor array PA is CI-testable with 168 test patterns, while the microprogram sequencer MS is I-testable and requires $449 + 2N$ test patterns, where N is the microinstruction address length. These test patterns, being small in number, are stored in the control store CS itself. The equality checkers EC_1 and EC_2 verify the corresponding test responses. The duplicated units CS and MR are continuously tested by their respective equality checkers, and hence they do not need any specific test patterns. The four equality checkers $EC_1:EC_4$ can be implemented as ILA's, as shown in Fig. 9(b); they can also be tested by applying appropriate test patterns that are stored in CS. Tests for these equality checker ILA's are generated in the same way used for any other ILA. The number of tests required by an N-bit checker ILA is $4 + 4N$.

The test patterns for PA, MS, and $EC_1:EC_4$ stored in CS may be regarded as test programs at the microinstruction level. Two operating modes are assumed for the CPU: a normal mode in which the CPU executes its regular programs, and a test mode in which the CPU tests itself by executing the foregoing test programs. During the test mode, the presence of a fault is indicated by the equality checkers $EC_1:EC_4$. Since the

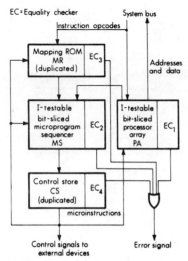

Fig. 20. A self-testing bit-sliced CPU.

number of test patterns for *PA*, *MS*, and EC_1:EC_4 are quite small, typically less than 1000, the time spent by the CPU in its test mode is also small. For example, with a CPU cycle time of 100 ns, the test mode duration will be less than 100 μs. Hence, it is possible to enter the test mode very frequently, say once every second. Idle periods during which the CPU is waiting for a memory or IO operation to be completed can be utilized for testing the CPU.

We now briefly compare the foregoing design scheme to that used by Wakerly [9] for a more conventional bit-sliced computer that is completely self-testing. Wakerly's design uses a modulo 15 residue code for checking a 16-bit CPU. This code includes a 4-bit check word which is processed by an extra 4-bit processor slice, to detect any error in a single 16-bit data word. A complex "fixup" circuit is used to overcome the inability of the arithmetic code to check logical operations. Thus, Wakerly's scheme uses more hardware than the design of Fig. 20. It also achieves less fault coverage since not every functional fault may appear as a single-word error. On the other hand, the use of a coding scheme ensures continuous error detection, and requires no special test mode. In [9] the nonbit-sliced microprogram sequencer is duplicated, while the control store ROM is not duplicated but is checked by a parity check code. The design proposed here, however, uses *I*-testing to test a nonduplicated bit-sliced microprogram sequencer, while the control store is duplicated. Also, the processor and microprogram sequencer arrays are tested periodically and not continuously as in Wakerly's design. In both designs duplication with comparison is used for unstructured or hard-to-handle units.

V. Conclusions

The basis of the design approach described here is to exploit some properties of bit-sliced networks that facilitate testing. The relatively low complexity of the individual slices allows a powerful functional fault model to be employed. Unlike the classical line stuck-at-zero/one fault model, the model used here does not require a detailed gate-level description of the circuit under test; such descriptions are often unavailable in practice. Circuit modules are tested exhaustively, but the complete circuit is tested in an efficient nonexhaustive manner. Test data generated according to our fault model has the added advantage of being useful for design verification during the initial logic design of the slice.

The regular array-like interconnections characteristic of bit-sliced systems allow the tests for a single slice to be easily extended to tests for an array of arbitrary length. Particularly useful in this regard are the *C*-tests, which are constant in number and independent of array length, and *I*-tests, which produce identical output responses from every cell. We have shown that complete sequences of *I*-tests or *CI*-tests can be derived for processor arrays and microprogram sequencers that closely resemble commercial products. *I*-tests and *CI*-tests seem particularly well suited to the design of self-testing systems, since they allow test response verification using simple equality checkers. Thus, the overhead required in testing time and built-in test equipment is relatively small. While *I*-testing can only be used directly in ILA-structured systems, it is a special type of replication-with-comparison testing, where the replication is inherent in the use of multiple slices. As a result, *I*-testing can readily be combined with conventional self-testing designs that are based on replication and comparison.

Appendix

In this Appendix the concepts of *C*-test and *I*-test introduced in Sections II and III are developed in more formal fashion.

C-Tests

First, we generalize the definition of a distinguishing sequence for a cell flow table [6] to the bilateral case.

Definition 1: A *distinguishing sequence* $DS(x_a\hat{w}_c, x_b\hat{w}_c)$ for a pair of states $x_a\hat{w}_c$ and $x_b\hat{w}_c$ of an $X\hat{W}$ flow table F is a Y input sequence which, when applied to F produces two different $\hat{Y}$ output sequences with $x_a\hat{w}_c$ and $x_b\hat{w}_c$ as the initial states.

Definition 2: A *set of distinguishing sequences* $SDS(x_a\hat{w}_c)$ for a state $x_a\hat{w}_c$ of an $X\hat{W}$ flow table F is a set of Y input sequences, which when applied to F with $x_a\hat{w}_c$ as the initial state produces a set of $\hat{Y}$ output sequences different from the sets of output sequences produced with any other state $x_b\hat{w}_c$ for all $x_b \in X$, as the initial state. Hence, we have

$$SDS(x_a\hat{w}_c) = \{DS(x_a\hat{w}_c, x_b\hat{w}_c): x_b \in X \text{ and } x_b \neq x_a\}.$$

Definition 3: An $X\hat{W}$ flow table is *reduced* with respect to the $\hat{Y}$ output if there exists a $DS(x_a\hat{w}_c, x_b\hat{w}_c)$ for every $x_a, x_b \in X$ and for every $\hat{w}_c \in \hat{W}$.

Definition 4: Let D be an ILA whose $X\hat{W}$ flow table is a reduced RT table. Let $\tau : x_a\hat{w}_c \xrightarrow{y_j} \hat{x}_b w_d$ be any transition in the $X\hat{W}$ flow table. A *C-test* $C_{ix}(\tau)$ associated with this transition is a periodic input test pattern of the form $x_a\hat{w}_c(y_jD_{ix}R_{ix})^*$, where D_{ix} is a distinguishing sequence in $SDS(\hat{x}_bw_d)$, and R_{ix} is a transfer sequence that takes the cell state back to $x_a\hat{w}_c$.

$C_{ix}(\tau)$ verifies the $\hat{X}$ part of τ with respect to D_{ix} in the leftmost cell of D, and in all cells spaced periodically at intervals k_i along D, where k_i is the periodicity of $C_{ix}(\tau)$, i.e., the number of cells spanned by $y_jD_{ix}R_{ix}$. Thus, $C_{ix}(\tau)$ tests cells $L_0, L_{k_i}, L_{2k_i}, L_{3k_i}, \cdots$. Let $s^q[C_{ix}(\tau)]$ denote the input sequence obtained by shifting the test pattern applied to D by $C_{ix}(\tau)$ $q < k_i$ cell positions to the left. $s^q[C_{ix}(\tau)]$ verifies the $\hat{X}$ part of τ in the cells $L_{k_i-q}, L_{2k_i-q}, L_{3k_i-q}, \cdots$ using the distinguishing sequence D_{ix}. Let $CT_x(\tau)$ denote the set of $\sum_{i=1}^{r} k_i$ C-tests of the form $s^q[x_a\hat{w}_c,(y_jD_{ix}R_{ix})*]$ obtained by allowing D_{ix} to range over all r members of $SDS(\hat{x}_bw_d)$, and by allowing q to range over $0, 1, \cdots, k_i - 1$. $CT_x(\tau)$ completely verifies the $\hat{X}$ output signal produced by every cell of the array during the transition τ. Note that the size of $CT_x(\tau)$ is independent of the length of D. In a similar manner we can define the C-tests $C_{jw}(\tau)$ and $CT_w(\tau)$, which verify the $\hat{W}$ output signals produced by τ. Hence, $CT(\tau) = CT_x(\tau) \cup CT_w(\tau)$ forms a complete set of C-tests for τ in every cell of D. The size of $CT(\tau)$ is independent of the length of D.

Suppose that the $X\hat{W}$ and $\hat{X}W$ tables of some ILA cell are reduced RT flow tables. Then by Definition 3 there exist SDS's for every state of the flow tables. Also, since they are RT flow tables, the transfer sequnces required in the C-tests C_{ix} and C_{jw} exist. Hence, there are C-test sets $CT_x(\tau)$ and $CT_w(\tau)$ for every transition in the $X\hat{W}$ and $\hat{X}W$ flow table, as asserted by Theorem 2.

Theorem 3 states that having only repeatable transitions in reduced $X\hat{W}$ and $\hat{X}W$ flow tables is a necessary and sufficient condition for C-testability in bilateral ILA's. The necessity part follows directly from Theorem 1. To prove the sufficiency part, suppose that the $X\hat{W}$ and $\hat{X}W$ flow tables are reduced RT flow tables. As explained above there exist sets of C-tests $CT_x(\tau)$ and $CT_w(\tau)$ for every transition τ in these tables. The size of these C-test sets is independent of the array size, hence the array is C-testable. As before, let $|X| = |\hat{X}| = n, |W| = |\hat{W}| = p$, and $|Y| = m$. It can be shown [16] that the size of a complete set of C-tests CT_D for a C-testable array D is bounded as follows:

$$mnp \leq |CT_D| < 2mn^2p^2(n + p).$$

Finally, let us prove the correctness of the flow table modification scheme described in Fig. 5 (Theorem 4). The modified $X\hat{W}$ and $\hat{X}W$ flow tables are reduced and have $Q_a = y_a^{n_1}$ and $Q_b = y_b^{n_2}$ as their respective distinguishing sequences. Now we show that by using the new vertical inputs y_a and y_b, it is always possible to construct a transfer sequence $R:x_iw_j \rightarrow x_ew_f$ that takes a cell from any state x_iw_j to any other state x_ew_f. An input sequence $y_a^{m_1}$ takes x_iw_j to a new state $x_ew_{j'}$, where m_1 is such that $i + m_1 \equiv e \pmod{n}$, and j' is given by $j + m_1 \equiv j' \pmod{p}$. A second input sequence $y_b^{m_2}$ can take the intermediate state $x_ew_{j'}$ to the final state x_ew_f, where m_2 is such that $j' + m_2 \equiv f \pmod{p}$. Thus, the desired transfer sequence R is $y_a^{m_1} y_b^{m_2}$. Since the modified $X\hat{W}$ and $\hat{X}W$ flow tables are both reduced RT tables, Theorem 3 implies that an array of the modified cells is C-testable.

I- and CI-Tests

The concept of a pairwise state distinguishing sequence is modified for I-testable unilateral ILA's as follows.

Definition 5: A (fault-free) state x_a is I-distinguishable from another (faulty) state x_b if and only if there exists an input $y \in Y$ for flow table F_k such that the entries $e_a = (\hat{x}(x_a,y), \hat{y}(x_a,y))$ and $e_b = (\hat{x}(x_b,y), \hat{y}(x_b,y))$ satisfy either of the following conditions:
1) $e_a \neq \phi$ and $e_b = \phi$
2) $e_a \neq \phi, e_b \neq \phi$ and $\hat{x}(x_a,y) \neq \hat{x}(x_b,y)$.

Let x_a be I-distinguishable from x_b due to input y in F_k. Suppose that x_a is the expected $\hat{X}$ output signal from a cell L_j in the ILA during the verification of some entry of F_k for L_j. If we apply y to the vertical input of the cell L_{j+1} to the right of L_j, then a fault in L_j that changes x_a to x_b will be propagated either to the $\hat{Y}$ output of L_{j+1} if $\hat{y}(x_b,y) \neq \hat{y}_k$, or to the $\hat{X}$ output of L_{j+1} if $\hat{x}(x_a,y) \neq \hat{x}(x_b,y)$. Thus, the above distinguishability property helps in propagating fault signals on the $\hat{X}$ output lines of a cell to observable output of the ILA. I-distinguishability with respect to a fixed $\hat{y}$ value is also necessary to ensure that the expected vertical outputs in the propagating cell L_{j+1} is identical to that in the test cell L_j. Hence, for I-testability every state x_a of F_k, for all k, that appears as a next state entry in F_k should be I-distinguishable from every other (faulty) state, from which Theorem 7 follows.

We now modify slightly Definitions 1, 2, and 3 for C-testable ILA's, so that the requirements of I-testability are also satisfied. The following three definitions refer to a partitioned flow table F_k, but not to the complete flow table F.

Definition 6: An I-distinguishing sequence $IDS(x_a,x_b)$ for states x_a and x_b of a partitioned flow table F_k is a Y input sequence which when applied to F_k produces a constant output sequence $\hat{y}_k^* = \hat{y}_k\hat{y}_k\hat{y}_k \cdots$, when x_a is the initial state, and some different output sequence when x_b is the initial state.

From Definitions 1 and 6 it follows that $IDS(x_a,x_b) \neq IDS(x_b,x_a)$, whereas the ordinary distinguishing sequences [4] $DS(x_a,x_b)$ and $DS(x_b,x_a)$ are always the same.

Definition 7: A *set of I-distinguishing sequences* $SIDS(x_a)$ for a state x_a of F_k is a set of Y input sequences which, when applied to F_k with x_a as the initial state, produces a set of constant output sequences $\{\hat{y}_k^*, \hat{y}_k^*, \cdots, \hat{y}_k^*\}$ different from the sets of output sequences produced with any other state as the initial state. In general

$$SIDS(x_a) = \{IDS(x_a,x_b):x_b \in X \text{ and } x_b \neq x_a\}.$$

Definition 8: A partitioned flow table F_k is I-*reduced* with respect to the $\hat{Y}$ output if there exists an I-distinguishing sequence $IDS(x_a,x_b)$ for every pair of states (x_a,x_b), where x_a is any state that appears at least once as a next state entry in F_k and x_b is any other state.

Example 3 (continued): Once more consider the full adder flow tables of Fig. 11. Let the two states x_a and x_b in Definition 6 be $x_0 = 0$ and $x_1 = 1$. Then either of the input patterns $y_0 = 00$ or $y_3 = 11$ can serve as $IDS(x_0,x_1)$ in F_0, while either $y_1 = 01$ or $y_2 = 10$ can be $IDS(x_1,x_0)$. Hence, $SIDS(x_0) = \{IDS(x_0,x_1)\} = \{y_0\}$ or $\{y_3\}$, and $SIDS(x_1) = \{IDS(x_1,x_0)\} = \{y_1\}$ or $\{y_2\}$. Similarly, in F_1, $SIDS(x_0) = \{IDS(x_0,x_1)\} = \{y_1\}$ or $\{y_2\}$, and $SIDS(x_1) = \{IDS(x_1,x_0)\} = \{y_0\}$ or $\{y_3\}$. Thus, F_0 and F_1 are both I-reduced partitioned flow tables. □

Using the foregoing I-distinguishing sequences one can construct a class of fixed-length I-tests, called CI-tests, which are analogous to the C-tests of Definition 4.

Definition 9: Consider a transition $\tau: x_a \xrightarrow{y_j} x_c$ in a partitioned RT flow table F_k. A CI-test $CI_i(\tau)$ associated with this transition is a periodic input sequence of the form $x_a,(y_j D_i R_i)^*$, where D_i is a member of $SIDS(x_c)$ and R_i is a transfer sequence that drives F_k back to the state x_a.

By applying $CI_i(\tau)$ and its shifted versions for each distinguishing sequence D_i, the transition τ of F_k can be tested completely in every cell of the array. At the same time, the test response at the direct output lines of every cell is identical. A CI-test exists for every transition τ in F_k, if F_k is an I-reduced RT flow table. Let CIT_D be a set of CI-tests for an ILA D whose partitioned cell flow tables are all I-reduced RT flow tables. The following bounds on the size of CIT_D are easily established:

$$mn \leq |CIT_D| \leq mn(n-1)(2n-1).$$

REFERENCES

[1] J. P. Hayes, "A survey of bit-sliced computer design," *J. Digital Syst.*, to be published.
[2] Advanced Micro Devices, Inc., *Am2900 Bipolar Microprocessor Family*, Sunnyvale, CA, 1976.
[3] C. Mead and L. Conway, *Introduction to VLSI Systems*. Reading, MA: Addison-Wesley, 1980.
[4] A. D. Friedman and P. R. Menon, *Fault Detection in Digital Circuits*. Englewood Cliffs, NJ: Prentice-Hall, 1971.
[5] J. P. Hayes and E. J. McCluskey, "Testability considerations in microprocessor-based design," *Computer*, vol. 13, pp. 17–26, Mar. 1980.
[6] T. Sridhar and J. P. Hayes, "Testing bit-sliced microprocessors," in *Dig. 9th Fault-Tolerant Comput. Symp.*, Madison, WI, 1979, pp. 211–218.
[7] R. E. Forbes *et al.*, "A self-diagnosable computer," in *Proc. FJCC*, 1965, pp. 1073–1086.
[8] P. Ciompi and L. Simoncini, "Design of self-diagnosable minicomputers using bit-sliced microprocessors," *J. DA and FTC*, vol. 1, pp. 363–375, Oct. 1977.
[9] J. Wakerly, *Error-Detecting Codes, Self-Checking Circuits and Applications*. New York: North-Holland, 1978.
[10] B. Könemann *et al.*, "Built-in logic block observation techniques," in *Dig. 1979 Test Conf.*, Cherry Hill, NJ, 1979, pp. 37–41.
[11] F. C. Hennie, *Iterative Arrays of Logical Circuits*. Cambridge, MA: MIT Press, 1961.
[12] A. D. Friedman, "Easily testable iterative systems," *IEEE Trans. Comput.*, vol. C-22, pp. 1061–1064, Dec. 1973.
[13] W. H. Kautz, "Tesing for faults in combinational cellular logic arrays," in *Proc. 8th Symp. Switching Automat. Theory*, 1967, pp. 161–174.
[14] F. J. O. Dias, "Truth-table verification of an iterative logic array," *IEEE Trans. Comput.*, vol. C-25, pp. 605–613, June 1976.
[15] R. Parthasarathy and S. M. Reddy, "On fault diagnosis of iterative logic arrays," in *Proc. 17th Annu. Allerton Conf.*, Oct. 1979.
[16] T. Sridhar, "Easily testable bit-sliced digital systems," Ph.D. dissertation, Dep. Elec. Eng., Univ. of Southern California, Los Angeles, Aug. 1981.
[17] Motorola, Inc., *MC14500B Industrial Control Unit Handbook*, Phoenix, AZ, 1977.
[18] T. Sridhar and J. P. Hayes, "A functional approach to testing bit-sliced microprocessors," *IEEE Trans. Comput.*, vol. C-30, pp. 563–572, Aug. 1981.

Thirumalai Sridhar (S'77–M'81) received the B.E. degree in electronics engineering from Bangalore University, Bangalore, India, in 1973, the M.E. degree (with distinction) in electrical communication engineering from the Indian Institute of Science, Bangalore in 1975, and the Ph.D. degree in electrical engineering at the University of Southern California, Los Angeles in 1981.

From 1975 to 1977 he was with the Switching Research and Development Division of Indian Telephone Industries Ltd., Bangalore, where he designed microprocessor-based controllers for electronic switching systems. From 1977 to 1981 he has held positions as a Teaching and Research Assistant at the University of Southern California. From 1979 to 1980 he was also the recipient of a graduate fellowship from IBM Corporation. In August 1981 he joined the Central Research Laboratories, Texas Instruments, Dallas, TX, where he is a member of the Technical Staff of the VLSI Laboratory. His current interests are in the areas of design for testability, VSLI systems, fault-tolerant computing, microprocessor-based systems, and computer architecture.

Dr. Sridhar is a member of Eta Kappa Nu.

John P. Hayes (S'67–M'70–SM'81) received the B.E. degree from the National University of Ireland, Dublin in 1965, and the M.S. and Ph.D. degrees from the University of Illinois, Urbana, in 1967 and 1970, respectively, all in electrical engineering.

From 1965 to 1967 he was with the Digital Computer Laboratory of the University of Illinois where he participated in the design of the ILLIAC 3 computer. From 1967 to 1970 he was engaged in research in the area of fault diagnosis of digital systems at the Coordinated Science Laboratory of the University of Illinois. In 1970 he joined the Operations Research Group at the Shell Benelux Computing Center of the Royal Dutch/Shell Company in The Hague, where he was involved in mathematical programming and software development. Since 1972 he has been with the Departments of Electrical Engineering and Computer Science of the University of Southern California, where he is currently an Associate Professor. His research interests include fault-tolerant computing, computer architecture, VLSI design, and microprocessor/microcomputer-based systems. He was Technical Program Chairman of the 1977 International Conference on Fault-Tolerant Computing. He is the author of the book *Computer Architecture and Organization* (New York: McGraw-Hill, 1978), and Editor of the Computer Architecture and Systems Department of *Communications of the ACM*.

Dr. Hayes is a member of the Association for Computing Machinery and Sigma Xi.

Test length is independent of network size in this simple, straightforward methodology for testing MINs. It requires only four test sequences for single-fault diagnosis.

Testing and Fault Tolerance of Multistage Interconnection Networks

Dharma P. Agrawal, Wayne State University

Advances in LSI and VLSI technology are encouraging greater use of multiple-processor systems with processing elements to provide computational parallelism and memory modules to store the data required by the PEs. A simple connection between the PE and MM is usually sufficient in a uniprocessor system, but a system with a large number of PEs and MMs requires a more complex data path.

Multistage interconnection networks, or MINs, are useful in providing programmable data paths between functional modules in multiprocessor systems. The MINs are usually segmented into several stages, and the linkages between various stages are assigned so that any input can access any one of the outputs, and vice versa. Each stage connects inputs to appropriate links of the next stage so that the cumulative effect of all stages satisfies input-output connection requirements.

These networks are usually implemented with simple modular switches, and several MINs employing two-input two-output switching elements have been described in the literature.[1-5] Permutation capability and other issues related to MINs have also been widely covered, but little attention has been paid to the reliability of these networks. Since correct functioning of a multiple-processor system is dependent on the proper functioning of its interconnections, an MIN fault could lead to a catastrophic situation. But this can be easily avoided by taking necessary measures whenever a fault is detected.

In this article, we consider fault-detection-and-location techniques and fault-tolerant design schemes for a class of MINs implemented with 2×2 switching elements connecting a set of N PEs to N MMs. For simplicity, it is usually assumed that N is a power of two ($N \equiv 2^n$ or $n = \log_2 N$). Figure 1 illustrates the basic form and the two allowed states of a 2×2 SE. The logical level at the control input line decides whether it is in its straight (T) connected mode or exchange (X) state.

Each stage of the MIN is implemented with $N/2 = 2^{n-1}$ SEs; a three-stage network for $N = 8$ is given in Figure 2. (For simplicity, the control lines connected to the

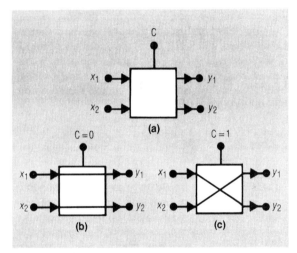

Figure 1. A switching element (a) with the control set at zero for a straight, or T, connection (b) and at one for an exchange, or X, connection (c).

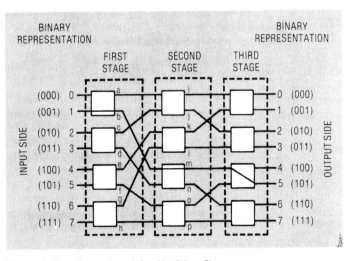

Figure 2. Baseline network for $N = 8$ ($n = 3$).

SEs have been omitted.) Unless otherwise specified, we will consider the MINs as consisting of $n = \log_2 N$ stages with each stage utilizing $N/2$ switches. The numbers assigned to the inputs and outputs are called the source and the destination tags, respectively. Because each SE has two outputs, SEs of the first stage connect each input to two outputs. In Figure 2, for example, input number 3 can be connected to outputs c and d of stage 1. In the second stage, input 3 can be connected to four outputs—i, j, m, and n. Finally, in the third stage, input 3 can be connected to any one of all eight outputs. In other words, the SEs are connected so that any input can be connected to any one of the eight outputs; there is a unique path from each input to each output,[1] and each path passes through n SEs with one SE in each stage.

Another advantage of these networks lies in the simplicity of control-signal computation. In fact, the binary representation of the destination tag provides the input-output connection of n SEs used in establishing the path. For example, in Figure 2 let us consider that input number 1 (001 in binary) is to be connected to output number 5 (101 in binary). Since the first and last bits of the destination tag are ones, the path uses the lower output line in the SEs of the first and the last stage. Since the second bit is zero, it indicates use of the upper link of the SE in the second stage. If we want to start establishing a connection from the destination side, we can use a similar routing scheme by generating the control signals on the basis of the input tag.[1]

Faults and the SE

Faults occur more frequently at an IC's terminals than within its logic circuitry,[6] and individual failures are more apt to be caused by a single fault than by multiple faults. Thus, our discussion assumes that the logic circuit within an SE is fault free and is limited to a single, permanent fault at one of its terminals.* This type of fault, usually known as a stuck-at-0 or stuck-at-1 fault (abbreviated as s-a-α with $\alpha\epsilon 0,1$), causes malfunctioning of the switch and precludes changing the logic level of the faulty terminal.

As shown in Figure 1, each SE has two inputs, two outputs, and one control line. This means that three different, single stuck-type terminal faults can occur in an SE:

(1) control line s-a-α ($\alpha\epsilon 0,1$),
(2) either input line s-a-α ($\alpha\epsilon 0,1$), or
(3) either output line s-a-α ($\alpha\epsilon 0,1$).

These are illustrated in Figures 3 through 5.

In Figure 3, when the control line is s-a-α, its state cannot be changed. In Figure 3a, with $\alpha = 0$, the SE will remain T-connected and send x_1 to y_1 and x_2 to y_2; the state of the SE cannot be altered to the X-mode even if we apply a logical 1 at the control line. In Figure 3b, with $\alpha = 1$, the situation is reversed. In other words, if we want to test an SE for s-a-T or s-a-X, we should try to change its state from T to X-mode or from X to T-mode, respectively. If the observed outputs differ from the fault-free values, the SE is faulty. This basic strategy for testing an SE can be extended to test an MIN, as discussed in the following sections.

If either of the two inputs of an SE is s-a-α ($\alpha\epsilon 0,1$), the input always stays at its α value regardless of the actual input, and the faulty input line can no longer be used to transfer data. As shown in Figure 4, sending either 0 or 1 at input line x_2 will transmit α to the output line. Thus, a string of 0's and 1's sent along x_2 will produce all α's ($\alpha\epsilon 0,1$) at the corresponding output line.

Figure 5 shows the situation when one of the two output lines of an SE is s-a-α ($\alpha\epsilon 0,1$). The logic level at output line y_2 will be observed as α and will be independent of the data sent to this line; hence, data cannot be transmitted through line y_2.

Fault diagnosis at an I/O link

The fault diagnosis problem consists of two steps: fault detection and fault location. The presence of a fault is detected first by applying known stimuli at the input termi-

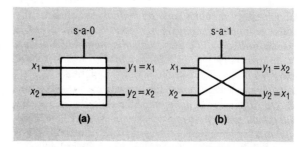

Figure 3. Switching element with control line stuck-at-α ($\alpha\epsilon 0, 1$).

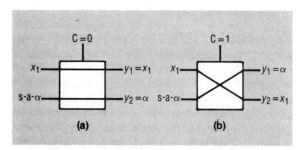

Figure 4. SE with input line x_2 s-a-α ($\alpha\epsilon 0, 1$).

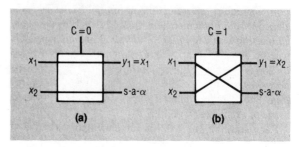

Figure 5. SE with output line y_2 s-a-α ($\alpha\epsilon 0, 1$).

---

*Logical faults within an IC can also be tested if they are equivalent to one of the three stuck-type faults considered.

nals and comparing actual output with expected values. Any mismatch indicates the presence of a fault. For an SE, since inputs x_1 and x_2 can be either 0 or 1, there are four possible values of inputs x_1x_2: 00, 01, 10, and 11. An N-input MIN has 2^N possible combinations of inputs and outputs. The simplest form of testing is to apply each combination and observe the corresponding output. For a large value of N, however, the total number of test inputs can be excessively large, making the procedure prohibitively time-consuming. Fortunately, it is well established[7] that all possible single faults can be detected by a subset of all possible input combinations. In selecting the appropriate test sequences, efforts are always directed toward minimizing the number of tests required to detect and locate all possible faults.

So and Narraway[8,9] have described an on-line fault diagnosis procedure for a switching network utilizing several stages of crossbar switches. One such simple switching module, illustrated in Figure 6, establishes the connection between an input and the desired output by closure of the crosspoints at various stages. So and Narraway define a permanently open switch as faulty and have developed a simple test procedure to detect this single fault in a network. The fault's existence is confirmed if testing a large but finite number of network paths fails to establish a connecting path between an input and an arbitrary output.

The Staran-like network shown in Figure 7 utilizes three stages of 2×2 crossbar switch modules. For example, the connection between input number 3 and output number 6 is achieved through one module of each stage marked by F. Any fault detected at output 6 may be due to the presence of an open circuit at any one of the three switches used by the path. To locate the faulty module, we have to test and ascertain proper functioning of any two of the three specific switches marked by F. Similarly, a faulty path in an n-stage network passes through n switches and thereby necessitates testing of $n-1$ modules. Since the procedure first requires locating the faulty path, the test sequence depends on pre-establishing the healthy paths, and the number of tests increases as the number of inputs increases.

It may be noted that So and Narraway are not concerned with one or more permanently closed switches. Since their test procedure detects only a permanently open switch, it is applicable to MINs implemented with SEs like those in

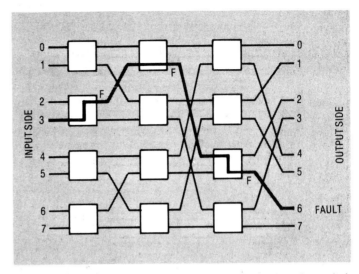

Figure 7. Testing a Staran-like network that uses 2 × 2 crossbar switch modules (see Figure 6).

Figure 1. But the fault translates to any one of the SE's inputs/outputs as an open circuit, thereby making it equivalent to an s-a-α fault. Our concern here is with the control-line fault and multiple faults, and the test procedure suggested by So and Narraway is not adequate for a general class of possible faults.

Feng and Wu recently introduced an excellent methodology for fault diagnosis of MINs that makes the sequence and number of tests independent of network size.[10] They used the simple strategy of applying 01 or 10 as inputs to each SE. Since one of the two SE inputs receives a logical 1 and the other a 0, the SE outputs should have different logical value; a discrepancy indicates the presence of a fault. The simplicity and effectiveness of their strategy of using 1-out-of-2 code[7] is demonstrated by the fact that they require only four tests for diagnosis of any single fault. A similar complementary input technique has also been utilized in designing an improved duplex system[11] and in fault diagnosis of microcomputer systems.[12,13]

In the fault model, Feng and Wu[10] consider only two of the 16 possible states of the 2×2 switch to be allowable states (see Figure 1), and the remaining 14 states are assumed to be faulty situations. Their test sequences consist of two phases: the first is useful for fault detection and the second helps in locating the fault. Both phases employ the same two tests. In the first phase, control lines of all the switches are kept at logical 0; that is, all the switches are in T-connected mode. The input test is selected so that every SE of the network gets a logical 1 at only one input terminal. In the second test, the control lines are maintained at the same level while complementary values are applied as test inputs. These two input test sequences are explicit in Figure 8a. Since each input of the second test is complementary to the first test, under normal conditions the outputs of the two tests are complementary. If any one of the links is s-a-α, the same logical level will be observed at one output line for both tests. Thus, to detect a single fault, the outputs of the first two tests are compared; noncomplementary values at any one of the output terminals indicates the presence of the fault. In Figure 8a, both tests

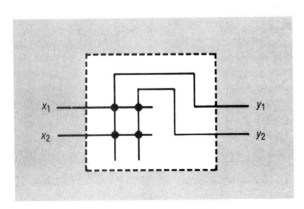

Figure 6. A 2 × 2 crossbar switch module.

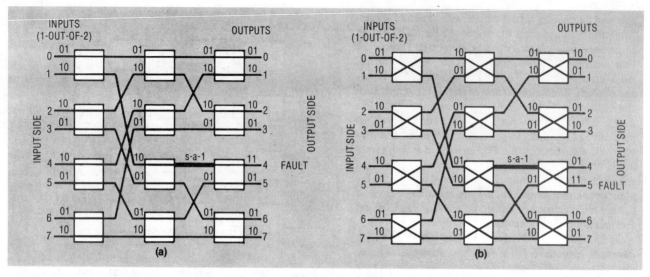

Figure 8. Input test sequence and corresponding output for detection of an s-a-α fault at one link: (a) all SEs are in T-mode; (b) all SEs are in X-mode. The faulty link is common to the faulty path in both phases of the test.

produce logic level 1 at output line 4, indicating a fault. The fault may actually have occurred at any link along the unique path connecting input 1 to the output 4, and additional tests are required to locate the fault.

To locate the faulty line, the control signals for all the SEs are changed to 1 (X-mode), and the same set of two test inputs is successively applied to the faulty network. In Figure 8b, a fault is again detected at output number 5. The fault can now be located by finding the common link used to establish the faulty paths of both versions of Figure 8. Thus, the link s-a-1 can be easily identified as the site of the fault.

These four tests can diagnose any single fault at any one link of the network, and the sequence is independent of the size of the MIN. This means that a sequence of four tests is adequate for detecting and locating any single link s-a-α of an n-stage MIN. The proper selection of the bit patterns makes the number of required tests independent of the size of the network or the value of N.

Control-line fault diagnosis in MINs

The control line of an SE, which could be s-a-0 or s-a-1, can be tested by verifying whether its state can be changed. The stimuli used for diagnosis of an s-a-α fault in Figure 8 can also be used to detect and locate a fault at one of the control lines.[14]

Testing is again conducted in two phases. The first phase consists of setting all the SEs in a T-connected mode by applying 0's at their control inputs. Two input tests (Figure 9a) are set up so that each SE gets a logical 1 at only one of its two inputs and outputs are compared with the expected values. Normally, we would expect one output of each SE to be at logical 1. An appropriate match

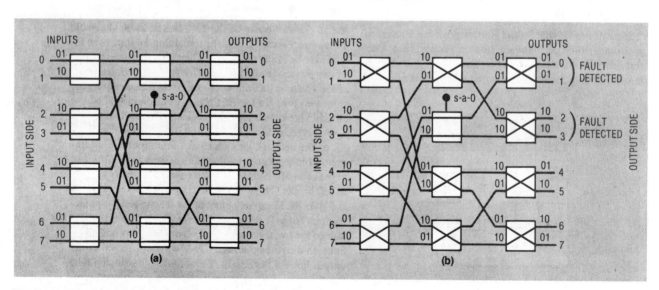

Figure 9. Testing for a control line s-a-α: (a) all SEs are in T-mode; (b) all SEs are in X-mode. Both phases are necessary to detect and locate a stuck-at-0 control line.

occurs if no fault is present or if any one of the control lines is s-a-0. The second situation is true because all the SEs are kept in T-mode; thus, if a control line is s-a-0, this becomes transparent and no fault is detected.

In the second phase, the same tests are applied with all SEs in exchange mode. No fault will be detected at the output if the network is completely fault free. But in the case of a single s-a-0 fault at one control line, output different from the expected value will be produced at any two of the SEs only. One such situation is shown in Figure 9b. The location of the fault or the faulty SE can also be determined by backtracking from the output links where faults have been noticed.

Note that any control line s-a-1 will be detected and located during the first phase of the test procedure and that faults due to any control line s-a-α are different from faults at any one of the connecting links.

On-line fault-diagnostic technique

The test sequences described above require setting specific control signals and keeping all SEs first in T-mode and then in X-mode. This means that to complete the test successfully the current control settings have to be altered. This can easily be avoided by using the current value of the control signals and the instantaneous input data bits to test the network.[14] In this approach the basic strategy is to employ two phases, with the first phase utilizing the current value of the control signals and thereby maintaining the status of all SEs unaltered. In other words, all SEs in T-mode are kept as they are, and the exchanged SEs are allowed to remain in the X state. In addition, the instantaneous value of the input data is used as the first test, and its complementary value is used as the second test. If there is no fault, each output will satisfy the 1-out-of-2 code in time domain. This implies that if the first input test produces a logical level 1 at an output terminal, the second test must send a 0, or vice versa. Any violation of this condition indicates the presence of a fault, as illustrated in Figure 10a.

The fault at output number 0 of Figure 10a will be observed when there is a fault at any link connecting input number 7 to output number 0. To locate the faulty link, the second phase utilizes the complementary states of all the SEs; the SEs in T-mode are changed to the X state and vice versa. The same two tests from phase one are applied in the second phase and the outputs are observed. The nonfaulty network produces 1-out-of-2 code in time domain, and a logical level of 1 is observed for one and only one of the two inputs. Any invalid output indicates the presence of a fault, as shown in Figure 10b. The faulty link can be easily identified by finding the common link used in the faulty paths of the two testing phases. The minimum number of required test sequences for a single fault diagnosis is still four.

Multiple stuck-type and other faults

If multiple faults are present, some faults may mask others so that it may not be possible to detect the existence of faults by using the test sequences for a single fault. One such example is shown in Figure 11 where the application of the two tests in both phases produces the correct 1-out-of-2 outputs. Thus in the case of multiple faults, these tests do not provide conclusive results and the two testing phases are not adequate. A simple and straightforward solution is to increase the number of testing phases and assign one phase for each stage. This procedure will test all $N/2$ SEs of a stage. Such a sequence of n tests has recently been used by Falavarjani and Pradhan for fault diagnosis of multiple stuck-type control-line faults.[15] The last stage is tested first by applying the input so that inputs to each SE of the last stage satisfy 1-out-of-2 code. The output is observed for correct functioning of the last stage. The procedure is repeated for successive stages to diagnose multiple control-line faults located in a single stage. This

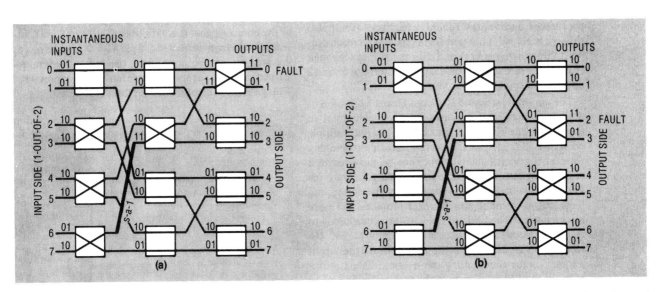

Figure 10. On-line testing for detection of an s-a-α fault at one of the links using (a) the control signal's instantaneous value and (b) complementary value. The faulty link is common to the faulty path in both phases.

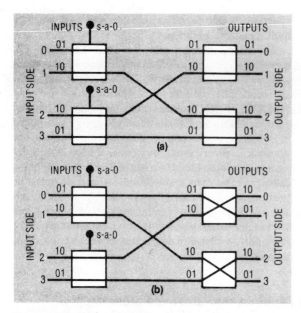

Figure 11. Fault-masking due to multiple faults: (a) all control lines = 0; (b) all control lines = 1.

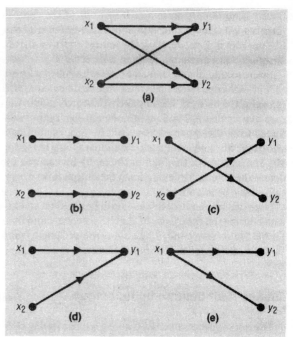

Figure 12. An SE graph model (a), with control s-a-0 (b), s-a-1 (c), y_2 s-a-α (d), and x_2 s-a-α (e).

technique is particularly useful in an LSI environment if a single LSI chip implementing a stage is found faulty and has to be replaced.

If multiple faults are located in different stages of the network and a mixture of control lines and link faults is present, the problem has to be treated in an altogether different way. Feng and Wu[10] have provided a systematic way to generate test sequences for detecting multiple faults at the control lines and connecting links. They have also given the upper bound for the number of tests needed for network fault diagnosis. Shen and Hayes[16] have used the fault model proposed by Opferman and Tsao-Wu[17] to study the effect of control-line faults on information transfer between various units. In their study, they show that the presence of a fault may prohibit a direct path between an input and an output. For example, in Figure 11, under assumed faults, input line 0 cannot be connected to outputs 2 and 3. If multiple passes through the network are allowed, and if outputs could be fed back to the inputs through a PE, we could start from input 0 and reach output 1 through the network and then go to input 1 through the PE. The second pass could connect input 1 to output 2 or 3, thereby providing the desired access.

The major objective of Shen and Hayes[16] has been to investigate possible communication from any input to any output line whenever faults are present and multiple passes are allowed. This idea is very useful in determining whether a graceful degradation[18] can be provided in a multiple-processor system. A generalized procedure for testing faults at either control lines or the connecting links, or their combination, has been proposed in a more recent work.[19] The basic strategy of the procedure is to assign one node to each input and output of the SE. Directed edges are drawn from the input nodes to the output nodes, thereby representing the possible data transfer paths in the SE (see Figure 12a). The connecting paths possible under different single s-a-α faults can be easily obtained. For example, if the control line is s-a-0, input 0

(1) cannot be connected to output 1 (0), and this leads to the reduced graph model shown in Figure 12b. Similar arguments can be used for other types of single faults, and the reduced graph models are shown in Figure 12c-e.

The other faults occurring in digital systems are bridge faults and shorted-diode faults.[7] One example of the AND-bridge fault, shown in Figure 13, introduces a fictitious AND gate and makes the two SE inputs equal to $x_1 x_2$. This forces the two outputs y_1 and y_2 to be equal to $x_1 x_2$ and independent of the control signal. Thus, $x_1 = x_2 = 0(1)$ makes $y_1 = y_2 = 0(1)$ as if no fault were present. In other words, the fault is transparent whenever x_1 and x_2 are both zeros or ones. As seen from Table 1, the bridge fault becomes detectable only when $x_1 = 0(1)$ and $x_2 = 1(0)$, forcing the outputs y_1 and y_2 to be 0 irrespective of the control signal (i.e., for both C = 0 and C = 1). Essentially, this can be treated as x_1 (x_2) s-a-0, and the rest of the analysis can be done using the test procedure described earlier. The shorted-diode faults are not present in TTL gates and therefore are not discussed here.

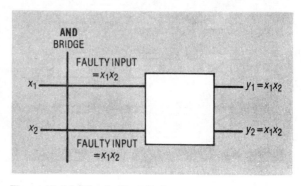

Figure 13. A bridge fault at the inputs.

Testing a Benes network

The Benes network[20] is designed with $2n - 1$ stages utilizing $N/2$ SEs each. In a Benes network, the first n stages are similar to the MINs discussed earlier. Although n stages allow any input to be connected to any one of the outputs, n stages are not enough to provide all possible input-output paths simultaneously. This versatile connectivity is achieved in a Benes network by $n - 1$ additional stages.

In a Benes network, any connecting link can be either s-a-0 or s-a-1 and can be tested by the technique presented earlier. The two test modes, consisting of two input test sequences each, can be applied to the Benes network (see Figures 14 and 15). The first mode will detect the fault and determine the faulty path. The second test mode helps in selecting, and thereby identifying, the faulty link in the faulty path.

Testing a control line s-a-α fault is slightly complicated because of several alternate paths between an input and any arbitrary output. This redundancy is due to the later

Table 1.
AND bridge fault and SE outputs y_1 and y_2 for all possible combinations of C and inputs x_1 and x_2
($y_1 = y_2 = x_1 x_2$ when the fault is present).

C x_1 x_2	OUTPUT WITH NO FAULT y_1 y_2	OUTPUT WITH AND BRIDGE FAULT y_1 y_2	FAULT DETECTED	AND BRIDGE FAULT EQUIVALENT TO
0 0 0	0 0	0 0	NO	
0 0 1	0 1	0 0	YES	x_1 s-a-0
0 1 0	1 0	0 0	YES	x_2 s-a-0
0 1 1	1 1	1 1	NO	
1 0 0	0 0	0 0	NO	
1 0 1	1 0	0 0	YES	x_1 s-a-0
1 1 0	0 1	0 0	YES	x_2 s-a-0
1 1 1	1 1	1 1	NO	

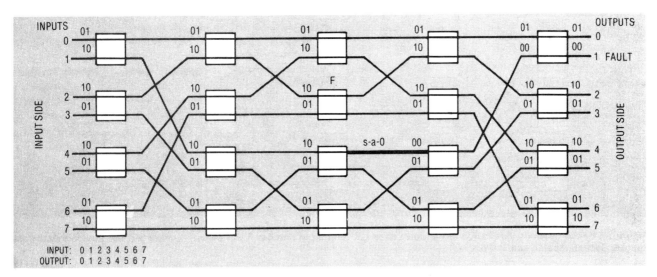

Figure 14. Benes network with all SEs parallel-connected.

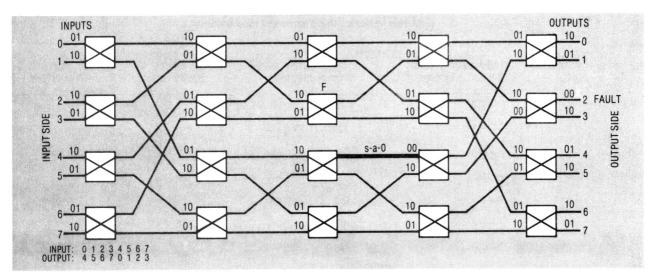

Figure 15. Benes network with all SEs cross-connected.

$(n-1)$ stages, and one such example illustrating multiple possible paths between input number 3 and output number 5 is shown in Figure 16. Opferman and Tsao-Wu[17] have provided a simple procedure for selecting some unique input-output permutation and for generating the control signal that is useful in testing any SE to determine if it is s-a-T or s-a-X. Their novel approach is based on looping dual input to output and then back to input. This process is continued until all inputs and outputs have been covered. Two inputs (or outputs) are called dual if their binary representations differ only in the least significant bit position. This means that if $b_{n-1} b_{n-2} \ldots b_2 b_1 b_0$ represents the first input (output) number in binary, its dual input (output) will be $b_{n-1} b_{n-2} \ldots b_2 b_1 \bar{b}_0$ where $\bar{b}_0$ represents the logical complement of b_0. Thus, 6 and 7 (110 and 111 respectively in binary) are dual of each other. The basic strategy is to select the input-output permutation so that only one loop for the complete network is obtained in the sense of the duals. One such input-output combination is given below[17]:

$$\begin{pmatrix} \text{Input side:} & 0 & 1 & 2 & 3 & 4 & 5 & 6 & 7 \\ \text{Output side:} & 0 & 2 & 7 & 4 & 3 & 6 & 5 & 1 \end{pmatrix}$$

As shown in Table 2, this input-output connection makes a single loop of the duals. It can be achieved by the control settings shown in Figure 17. All the SEs in the first stage are to be T- mode ($C=0$). In the second stage, the control setting for the four switches can be given as 0, 1, 0, and 0. The middle-stage control states can be given as 0, 1, 0, and 1. The control vector for the last two stages can be given as 0,0,1,1 and 0,1,1,1. $C=0$ indicates a T connection of the SE while $C=1$ provides an X connection.

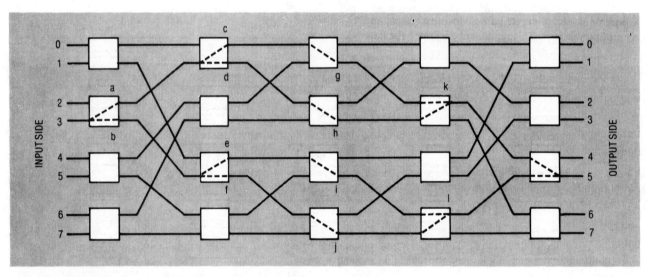

Figure 16. Benes network: illustration of multiple paths between input number 3 and output number 5. (Four possible paths are 3acgk5, 3adhk5, 3bell5, and 3bfjl5.)

Table 2.
Path traversal for a closed looping in the
Benes network of Figure 17.
(Arrow indicates the direction of loop traversal.)

PATH NO.	PATH CONNECTION FROM/TO INPUT NO. BINARY	(DECIMAL)	TO/FROM OUTPUT NO. BINARY	(DECIMAL)	NEXT DUAL IN THE LOOP INPUT BINARY (DEC.)	OUTPUT BINARY (DEC.)
1	→110	(6)	101	(5)		100 (4)
2	011	(3)	100	(4)	010 (2)	
3	010	(2)	111	(7)		110 (6)
4	101	(5)	110	(6)	100 (4)	
5	100	(4)	011	(3)		010 (2)
6	001	(1)	010	(2)	000 (0)	
7	000	(0)	000	(0)		001 (1)
8	111	(7)	001	(1)	110 (6)	

The control signals of Figure 17 test each SE for one of the two possible states, and a sequence of inputs is applied to test the network. If the desired permutation can be achieved, then each SE can be said to be working fault free in its selected mode of operation. It is worth mentioning that the fault becomes transparent if the control line for any one of the SEs is s-a-α and α happens to be the same as the selected value for that particular SE's control signal. The presence of a control-line fault can be detected only if α is different from the applied value. This will be indicated at the two outputs, and the faulty SE can be located by backtracking the network's control signals.

The second phase of testing is continued with all SEs connected in a complementary mode. The SEs receiving a 0 value at the control input now get a 1, and those with C = 1 are changed to C = 0. Basically, the operating mode of the SEs is changed from T- to X- mode, and vice versa. A Benes network with complementary control signals is given in Figure 18. The modifications in the control signals lead to another unique single loop in terms of duals and can be given as

Input side: $\begin{pmatrix} 0 & 1 & 2 & 3 & 4 & 5 & 6 & 7 \\ 3 & 1 & 5 & 6 & 7 & 2 & 0 & 4 \end{pmatrix}$
Output side:

If it is possible to achieve this permutation as well, then it can easily be concluded that there is not a single s-a-α fault at the control line of any one of the SEs. If such a fault is present, it can be detected at two of the outputs, and backtracking will locate the faulty SE. For example, if the control signal for switch F were s-a-1, the connection of Figure 17 could be achieved, but in Figure 18, F would be cross-connected. This would interchange the connections for output numbers 2 and 4, making the faulty SE easy to locate. This looping procedure can be extended further for diagnosis of multiple faults at the control lines.[17]

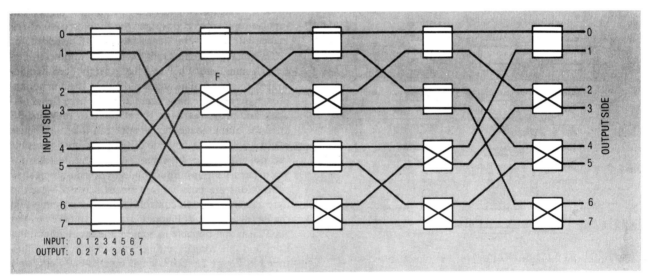

Figure 17. Benes network testing using a single-loop input-output connection.

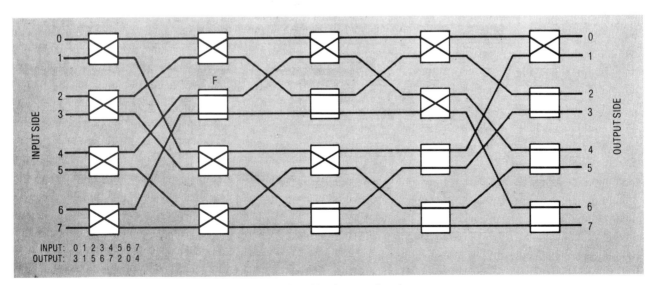

Figure 18. Benes network with complementary control signal (makes one loop).

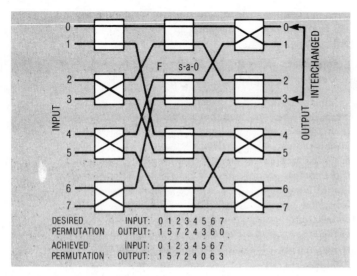

Figure 19. Control setting for the permutation in a baseline network. The desired output is impossible if F is s-a-0.

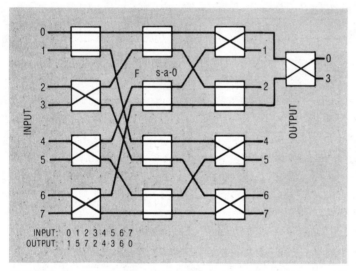

Figure 20. Required permutation of Figure 19 achieved with an additional SE.

Fault-tolerant design of MINs

A network is said to be fault tolerant if under certain faults it can establish a desired input-output permutation. This generally requires augmentation of the network redundancy. A faulty input or output link means that the line cannot be used for any data transmission, and additional stages will not help. But control-line faults can be tolerated, and this section is devoted to the design of networks that tolerate one s-a-α fault at the control line of any one of the SEs.

Figure 19 is an example of baseline,[1] wherein the desired input-output bijection becomes impossible if the SE control line marked by F is s-a-0. This fault interchanges the connections of outputs 0 and 3, but the required connection can still be established if one SE is added and connected as shown in Figure 20. This procedure for any MIN with n stages works only where the SE with a control-line fault is known a priori and where the bit patterns of all SE control signals are provided ahead of time. If the fault has to be tolerated in any arbitrary SE and for all possible input-output connections, then the addition of one SE is inadequate and several stages may have to be added.

The value of adding stages has been mentioned by Opferman and Tsao-Wu[17] and has recently been demonstrated by Sowrirajan and Reddy.[21] In their novel approach they use the basic characteristic of the Benes network and prove that with the addition of only one SE after the output stage the network can tolerate a single control-line fault at any one of the SEs. This is possible because redundancy provides several alternate paths for any arbitrary input-output connection. Thus, almost all connections are possible in a Benes network—even in the presence of a single control-line fault. For example, in the Benes network of Figure 17, if the SE marked by F is s-a-T, it is still possible to achieve the desired permutation, and one control setting for such bijection is illustrated in Figure 21. For a single control-line fault, only two permutations are not possible, and these are as follows:

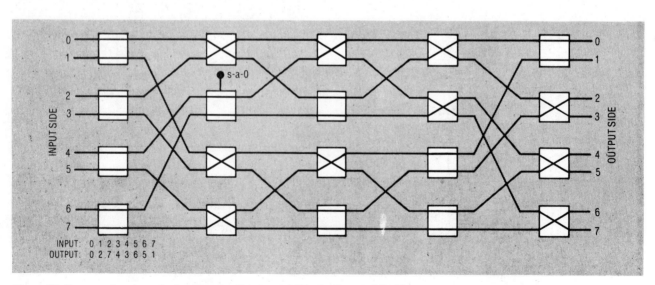

Figure 21. Benes network control setting for the permutation (under s-a-0 fault).

Case 1.
Input: $\begin{pmatrix} 0 & 1 & 2 & 3 & 4 & 5 & 6 & 7 \\ 0 & 1 & 2 & 3 & 4 & 5 & 6 & 7 \end{pmatrix}$

if any SE in the center stage is s-a-X.

Case 2.
Input: $\begin{pmatrix} 0 & 1 & 2 & 3 & 4 & 5 & 6 & 7 \\ 4 & 5 & 6 & 7 & 0 & 1 & 2 & 3 \end{pmatrix}$

if any SE in the center stage is s-a-T.

For these two cases, if the control signals are computed on the basis of a no-fault condition (as shown in Figures 14 and 15), all but two input-output permutations are possible. A single control-line fault interchanges only two output connections (outputs 2 and 6 in Figure 14 and 1 and 5 in Figure 15), and the desired bijection of all input-output pairs cannot be satisfied.

In the example of Figure 14, if the SE marked by F is s-a-1, the control setting for the desired connection interchanges outputs 2 and 6. The required bijection can be achieved if only one SE is added after the last stage.[21] This is called a fault-tolerant Benes network, and the desired permutation is obtained by first computing the control signals for a modified permutation with outputs 0 and 4 interchanged. The changes in the output alter the permutation to be provided by the faulty Benes network as follows:

Input: $\begin{pmatrix} 0 & 1 & 2 & 3 & 4 & 5 & 6 & 7 \\ 4 & 1 & 2 & 3 & 0 & 5 & 6 & 7 \end{pmatrix}$

This connection can now be established, and one possible control setting is shown in Figure 22. Finally, the outputs 0 and 4 are interchanged again by an additional SE, thereby satisfying the permutation requirements. A similar effect can be observed if the additional SE is placed before the input side of the Benes network. This requires the swapping of inputs 0 and 4 while computing the control signals. For example, if SE F of Figure 15 is s-a-0, the desired connection can be achieved by the modified fault-tolerant Benes network of Figure 23. Thus, the networks in Figures 22 and 23 are one-control-line fault tolerant. Design techniques for multiple-control-line fault-tolerant

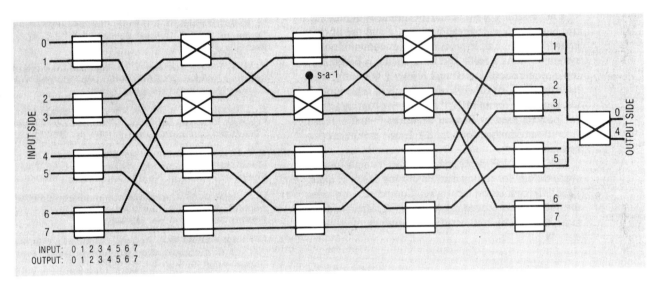

Figure 22. Modified fault-tolerant Benes network control setting for the permutation.

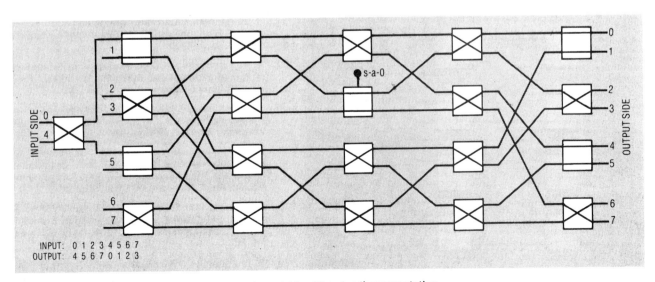

Figure 23. Modified fault-tolerant Benes network control setting for the permutation.

SEs in a Benes network require additional stages and have been described in the work of Sowrirajan and Reddy.

If the network consists of less than n stages, or if there is a link fault, and the network is to be used in a multiprocessor system, data transfer from one input to one of the outputs (or from one computer to another computer) might take more than one pass. In this case the outputs have to be fed back to the input through the computers connected at both ends of the network.[16] One such example is shown in Figure 24 where inputs 0,1,2, and 3 can be connected to any one of the four outputs 4,5,6, and 7 in one pass. Similarly, inputs 4,5,6, and 7 can be connected to any one of the four outputs 0,1,2, and 3. Thus, if the output can be fed back to the input side, any input can be connected to any one of the outputs in two passes at most. This type of interconnection assignment is preferable from the fault-tolerance viewpoint.

The versatility of MINs and the simplicity of their control have led to widespread acceptance and use of these networks in multiple-processor and communications systems.[22,23] As a result, network testing is becoming a topic of increased interest and research. Recently, there have been reports of work on an involved table-look-up test procedure for an MIN,[24] an adaptive routing scheme for packets used in Banyan networks,[25] and a reconfiguration technique for a fault-tolerant multimicroprocessor system.[26]

This article has described a simple, straightforward methodology for testing and diagnosing faults in multistage interconnection networks. Compared with other methods, its test procedures have one significant advantage: they make test length independent of network size. ■

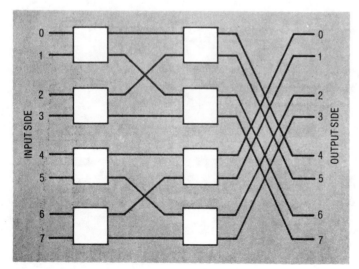

Figure 24. Multiple pass network.

References

1. C. L. Wu and T. Y. Feng, "On a Class of Multistage Interconnection Networks," *IEEE Trans. Computers*, Vol. C-29, No. 8, Aug. 1980, pp. 694-702.

2. D. K. Pradhan and K. L. Kodandapani, "A Uniform Representation of Single- and Multi-Stage Interconnection Networks used in SIMD Machines," *IEEE Trans. Computers*, Vol. C-29, No. 9, Sept. 1980, pp. 777-791.

3. M. A. Abidi and D. P. Agrawal, "On Conflict-Free Permutations in Multistage Interconnection Networks," *Journal Digital Systems*, Vol. 5, No. 2, Summer 1980, pp. 115-134.

4. M. A. Abidi and D. P. Agrawal, "Two Single-Pass Permutations in Multistage Interconnection Networks," *Proc. 1980 Conf. Information Science and Systems*, Mar. 26-28, pp. 516-522.

5. C. L. Wu and T. Y. Feng, "The Reverse-Exchange Interconnection Network," *IEEE Trans. Computers*, Vol. C-29, No. 9, Sept. 1980, pp. 801-811.

6. C. W. Weiss, "Bounds on the Length of Terminal Stuck-fault Tests," *IEEE Trans. Computers*, Vol. C-21, No. 3, Mar. 1972, pp. 305-309.

7. M. A. Breuer and A. D. Friedman, *Diagnosis and Reliable Design of Digital Systems*, Computer Science Press, Rockville, Md., 1976.

8. K. M. So and J. J. Narraway, "On-line Fault Diagnosis of Switching Networks," *IEEE Trans. Circuit Systems*, Vol. CAS-26, No. 7, July 1979, pp. 575-583.

9. J. J. Narraway and K. M. So, "Fault Diagnosis in Inter-Processor Switching Networks," *Proc. IEEE Int'l Conf. Circuits and Computers*, Oct. 1-3, 1980, pp. 750-753.

10. T. Y. Feng and C. L. Wu, "Fault-Diagnosis for a Class of Multistage Interconnection Networks," *IEEE Trans. Computers*, Vol. C-30, No. 10, Oct. 1981, pp. 743-758.

11. D. P. Agrawal, "A Duplex System with Improved Performance," *Proc. 1979 Conf. Information Science and Systems*, March 28-30, pp. 333-336.

12. D. P. Agrawal and V. K. Agarwal, "On-line Fault Detection and Correction in Microprocessor Systems," *Proc. Compcon Fall 79*, Sept. 4-7, pp. 92-99.

13. D. P. Agrawal and V. K. Agarwal, "On-Line Bus Fault Diagnosis in Microprocessor Systems," *Journal Digital Systems*, Vol. 4, No. 4, Winter 1980, pp. 377-391.

14. D. P. Agrawal, "Automated Testing of Computer Networks," *Proc. 1980 Int'l Conf. Circuits and Computers*, Oct. 1-3, pp. 717-720.

15. K. M. Falavarjani and D. K. Pradhan, "Fault-Diagnosis of Parallel Processor Interconnection Networks," *Proc. Fault-Tolerant Computing Symp.*, June 1981.

16. J. P. Shen and J. P. Hayes, "Fault Tolerance of a Class of Connecting Networks," *Proc. 7th Symp. Computer Architecture*, May 6-8, 1980, pp. 61-71.

17. D. C. Opferman and N. T. Tsao-Wu, "On a Class of Rearrangeable Switching Networks—Part II: Enumeration Studies and Fault Diagnosis," *Bell Syst. Tech. Journal*, Vol. 50, No. 5, May-June 1971, pp. 1601-1618.

18. B. R. Borgerson and R. F. Freitag, "A Reliability Model of Graceful Degradation and Standby-sparing Systems," *IEEE Trans. Computers*, Vol. C-24, May 1975, pp. 517-525.

19. D. P. Agrawal, "Dynamic Accessibility Testing and Optimum Availability Considerations of Multistage Networks," (communicated for publication).

20. V. E. Benes, *Mathematical Theory of Connecting Networks and Telephone Traffic,* Academic Press, New York, 1965.
21. S. Sowrirajan and S. M. Reddy, "A Design for Fault-Tolerant Full Connection Networks," *Proc. Conf. Int'l Science and Systems*, March 26-28, 1980, pp. 536-540.
22. D. P. Agrawal, T. Y. Feng, and C. L. Wu, "A Survey of Communication Processor Systems," *Proc. Compsac 78*, Nov. 13-16, pp. 663-673.
23. T. Y. Feng and D. P. Agrawal, "A Study of Communication Processor Systems," Tech. Rep. R.A.D.C., TR 79-310, Dec. 1979, 179 pp.
24. B. D. Rathi and M. Malek, "Fault Diagnosis of Interconnection Networks," *Proc. Symp. Distributed Data Acquisition, Computing and Control*, Dec. 3-5, 1980, pp. 110-119.
25. A. R. Tripathi and G. J. Lipovski, "Packet Switching in Banyan Networks," *Proc. 6th Ann. Symp. Computer Architecture*, April 1979, pp. 160-167.
26. V. P. Nelson, "Fault Tolerance in Reconfigurable Multiprocessor Systems," *Proc. Compsac 80,* Oct. 27-31, pp. 372-380.

Dharma P. Agrawal is an associate professor in the Department of Electrical and Computer Engineering at Wayne State University. His research interests include parallel/distributed processing, computer architecture, fault-tolerant computing, and information retrieval. He has been on the faculties at Southern Methodist University; Federal Institute of Technology, Lausanne, Switzerland; Roorkee University and M.N.R. Engineering College, Allahabad, India.

Agrawal has served as a referee for various journals and conferences. He was a member of the program committees for Compcon Fall 79 and the Sixth Symposium on Computer Arithmetic. Currently, he is a member and secretary of the IEEE Computer Society Publications Board. He is also serving as a guest editor for the *IEEE Transactions on Computers.* He is listed in *Who's Who in the Midwest* and *1981 Outstanding Young Men of America.* He is a senior member of the IEEE and a member of the ACM and Sigma Xi.

Agrawal received the BE degree in electrical engineering from the Ravishankar University, Raipur, India; the ME (Hons) degree in electronics and communication engineering from the University of Roorkee, Roorkee, India; and the DSc in electrical engineering from the Ecole Polytechnique Federale de Lausanne, Switzerland.

April 1982

A Design of Programmable Logic Arrays with Universal Tests

HIDEO FUJIWARA, MEMBER, IEEE, AND KOZO KINOSHITA, MEMBER, IEEE

Abstract—In this paper the problem of fault detection in easily testable programmable logic arrays (PLA's) is discussed. The easily testable PLA's will be designed by adding extra logic. These augmented PLA's have the following features: 1) for a PLA with n inputs and m columns (product terms), there exists a "universal" test set such that the test patterns and responses do not depend on the function of the PLA, but depend only on the size of the PLA (the values n and m); 2) the number of tests is of order $n + m$. For the augmented PLA's, universal test sets to detect faults in PLA's are presented. The types of faults considered here are single and multiple stuck faults and crosspoint faults in PLA's. Fault location and repair of PLA's are also considered.

Index Terms—Easily testable design, fault detection, fault location, logic circuits, programmable logic arrays (PLA's), universal test sets.

I. INTRODUCTION

WITH the increasing circuit density in a single large-scale integrated (LSI) circuit chip, the difficulty of testing the circuits is becoming apparent. In order to overcome this problem, methods have been suggested in which test points and additional logic are used for the purpose of easing the test generation problem [1]–[6]. Designing easily testable circuits, one should pay attention to the following features: 1) the cost of generating test patterns is low, that is, the computation time for test generation is short; and 2) the cost of testing the circuits is low, that is, the length of test sequence is short.

This paper is concerned with the problem of fault detection and location in the easily testable programmable logic arrays (PLA's) which have the above mentioned features. The PLA, which is conceptually a two-level AND-OR, is attractive in LSI due to its memory-like array structure. A method is presented to augment PLA's by adding extra logic so that the augmented PLA's have the following easily testable features: 1) for a PLA with n inputs and m columns (product terms), there exists a "universal" test set such that the test patterns and responses do not depend on the function of the PLA, but depend only on the size of the PLA (the values n and m); 2) the number of tests is of order $n + m$. Since the augmented PLA's have the universal test set, the test generation of PLA's is no more necessary, and the cost of test generation is considerably reduced to almost zero. The augmented PLA's introduced in this paper are similar to the PLA which was independently obtained by Hong and Ostapko [3]. However, since the augmented PLA's considered in this paper need less additional hardware than [3] and also since no test sequence for multiple faults appears in [3], we will present universal test sequence for single and multiple faults in the PLA's. First, single-stuck faults and single crosspoint faults are considered, and the universal test sets to detect these faults in PLA's are presented. Then the type of faults are extended to multiple faults. Fault location and repair of PLA's are also considered where the faults are assumed to be multiple crosspoint faults.

II. PROGRAMMABLE LOGIC ARRAYS

A programmable logic array (PLA) consists of three main parts. These are the decoders, the AND, and the OR arrays. The decoders are usually implemented by single-bit decoders or double-bit decoders, as shown in Figs. 1 and 2. Both the AND array and the OR array are used to implement multioutput combinational logic with sum-of-products forms. Fig. 3 shows an example of a 4-input, 2-output PLA with single-bit decoders, which realizes two functions in the following:

$$f_1 = x_1 \lor x_4 \lor \bar{x}_2\bar{x}_3 \lor \bar{x}_1 x_2 x_3$$
$$f_2 = \bar{x}_2 \lor x_4 \lor x_1\bar{x}_3 \lor \bar{x}_1 x_2 x_3.$$

Fig. 4 shows another realization using a PLA with double-bit decoders.

In the following Sections III and IV, we present a method to augment PLA's with single-bit decoders or double-bit decoders by adding extra logic so that the augmented PLA's are easily testable PLA's with short "universal" test sequences. The types of faults considered in Sections III and IV are single faults in the PLA which are the stuck faults and the crosspoint faults. A crosspoint fault in a PLA is a fault such that the presence (absence) of a contact between a row and column of the PLA becomes the absence (presence) of the contact.

III. AUGMENTED PLA'S WITH SINGLE-BIT DECODERS

In order to design an easily testable PLA with single-bit decoders, we augment a given PLA by adding extra logic, that is, a shift register, two cascades of EOR's (EXCLUSIVE-OR's), and one column and one row to AND and OR arrays, respectively, as shown in Fig. 5. The connections of the added column in the AND array is arranged so that each row of the AND array has an odd number of connections. Similarly, the connections of the added row in the OR array is arranged so that each column has an odd number of connections. In the augmented

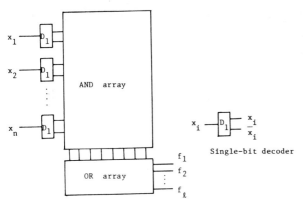

Fig. 1. PLA with single-bit decoders.

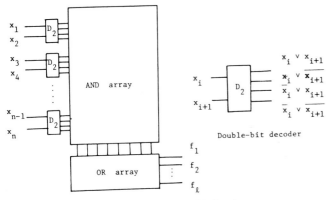

Fig. 2. PLA with double-bit decoders.

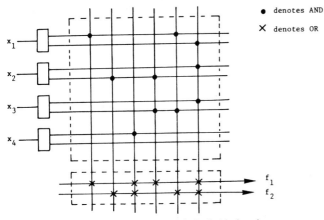

Fig. 3. Example of PLA with single-bit decoders.

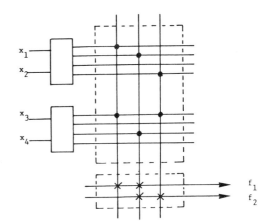

Fig. 4. Example of PLA with double-bit decoders.

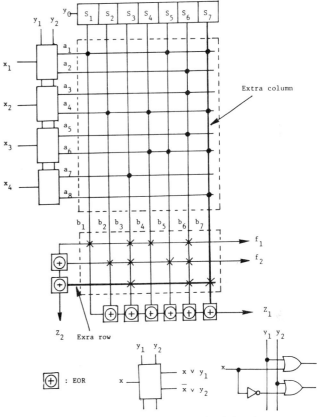

Fig. 5. Augmented PLA with single-bit decoders.

PLA each column (product term) b_i is ANDed by each variable S_i of the shift register as follows:

$$b_i = p_i \cdot S_i \qquad \text{for } i = 1, 2, \cdots, m$$

where p_i is a product term generated by the ith column of the AND array without the shift register and m is the number of columns. Fig. 5 shows a PLA augmented from the PLA shown in Fig. 3.

The augmented PLA has the following properties.

1) The shift register can be used to select an arbitrary column of the AND array by setting 1 to the selected column and 0 to all other columns. [See Fig. 6(a).]

2) The augmented decoders with control inputs y_1 and y_2 can be used to sensitize an arbitrary row of the AND array by setting 0 or 1 to the selected row and 1 to all other rows. [See Fig. 6(b).]

3) The cascade of EOR's below the OR array can be used as a parity checker to detect single errors in the sensitized row of the AND array. [See Fig. 6(c).]

4) The cascade of EOR's on the left of the OR array can be used as a parity checker to detect single errors in the sensitized column of the OR array. [See Fig. 6(d).]

Utilizing the above properties of the augmented PLA we can present a universal test set to detect single faults in the following:

1) stuck faults on the input or output lines of gates within the decoders, the AND array, and the OR array,

2) crosspoint faults in the AND and OR arrays.

Table I shows the test set $A_{n,m}$ to detect the above types of faults, where n is the number of inputs, m is the number of columns in the AND array, and

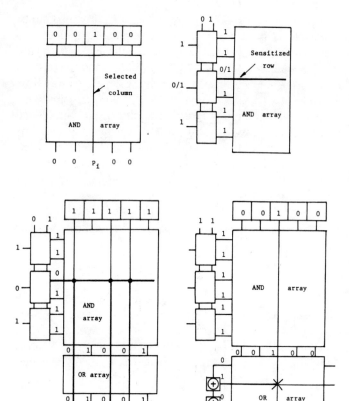

Fig. 6. Features of the augmented PLA.

TABLE I
UNIVERSAL TEST SET $A_{n,m}$

	x_1 ... x_i ... x_n	y_1 y_2	S_1 ... S_j ... S_m	Z_1 Z_2
I_1	$-$ .... $-$	$-$ $-$	0 ..... 0 0	0 0

For $j = 1, 2, \cdots, m$

	x_1 ... x_n	y_1 y_2	S_1 ... S_j ... S_m	Z_1 Z_2
I_{2j}^0	0 0	0 1	0 0 ... 0 1 0 0	1 1
I_{2j}^1	1 1	1 0	1 0 ... 0 1 0 . ?	1 1

For $i = 1, 2, \cdots, n$

	x_1 ... x_i ... x_n	y_1 y_2	S_1 ... S_m	Z_1 Z_2
I_{3i}^0	1 ..... 1 0 1 1	0 1	1 ..... 1 1	ϵ_m $-$
I_{3i}^1	0 ..... 0 1 0 0	1 0	1 ..... 1 1	ϵ_m $-$

$$\epsilon_m = \begin{cases} 0 & \text{if } m \text{ is odd} \\ 1 & \text{if } m \text{ is even} \end{cases}$$

and " $-$ " represents DON'T CARE.

For this test set $A_{n,m}$ we have the following theorem.

Theorem 1: Let $M_{n,m}$ be an augmented PLA with single-bit decoders which has n inputs and m columns in the AND array. For any $M_{n,m}$ the test set $A_{n,m}$ can detect all single stuck and crosspoint faults in the decoders, and AND array and the OR array.

Proof: When we apply test inputs I_{2j}^0 and I_{2j}^1, the jth column is set to 1 and other columns are all set to 0. Therefore, both I_{2j}^0 and I_{2j}^1 can detect any crosspoint fault on the jth column of the OR array by observing the output Z_2, and a stuck-at-0 fault on the jth column, stuck-at-1 faults on the other columns, and stuck-at-0 faults on the rows of the AND array by observing the output Z_1. Any stuck fault on the row of the OR array can be detected by some I_{2j}^0 and I_{2j}^1 ($1 \leq j \leq m$).

By applying test I_{3i}^0 (I_{3i}^1), the $(2i - 1)$th ($2i$th) row of the AND array is set to 0 and other rows are all set to 1. Therefore, test I_{3i}^0 (I_{3i}^1) can detect all crosspoint faults and a stuck-at-1 fault on the $2i$-1th ($2i$th) row in the AND array by observing the output Z_1.

Tests I_{2j}^0, I_{2j}^1 ($j = 1, 2, \cdots, m$) and I_{3i}^0, I_{3i}^1 ($i = 1, 2, \cdots, n$) can also detect all stuck faults in the decoders. The stuck-at-0 faults on the input lines of OR gates can be detected by I_{2j}^0 and I_{2j}^1 ($j = 1, 2, \cdots, m$). The stuck-at-1 faults on the input lines of OR gates can be detected by I_{3i}^0 and I_{3i}^1 ($i = 1, 2, \cdots, n$). The stuck-at-0 faults on the input lines x_i ($i = 1, 2, \cdots, n$), y_1, and y_2 can be detected by I_{2j}^0 and I_{2j}^1 ($j = 1, 2, \cdots, m$). The stuck-at-1 faults on the input lines x_i ($i = 1, 2, \cdots, n$), y_1, and y_2 can be detected by I_{3i}^0 and I_{3i}^1 ($i = 1, 2, \cdots, n$). Q.E.D.

Next, we will show that the test set $A_{n,m}$ can also detect any multiple stuck fault in the EXCLUSIVE-OR cascades under the fault assumption that permits only stuck-type faults on the external input and output lines of EOR gates, that is, no fault within EOR gates is considered.

We have the following lemma. Similar results appear in [12].

Lemma 1: For an EXCLUSIVE-OR cascade realization of a k-input linear function, all multiple stuck faults on the external lines of EOR gates can be detected by the following $k + 1$ tests:

$$t_0 = (0, 0, \cdots, 0)$$
$$t_1 = (1, 0, \cdots, 0)$$
$$t_2 = (0, 1, 0, \cdots, 0)$$

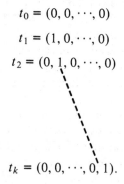

$$t_k = (0, 0, \cdots, 0, 1).$$

Proof: Any linear function of k or fewer variables can be expressed in the form

$$l_A(x_1, x_2, \cdots, x_k) = a_0 \oplus a_1 x_1 \oplus a_2 x_2 \oplus \cdots \oplus a_k x_k$$

where $a_i = 0$ or 1 for $i = 1, 2, \cdots, k$ and $A = (a_0, a_1, a_2, \cdots, a_k)$.

It can easily be shown that even if one or more stuck-at-0 or stuck-at-1 faults are introduced into the EXCLUSIVE-OR cascade, the resulting circuit still realizes a linear function. So, any faulty function of the EXCLUSIVE-OR cascade can be expressed in the above form. When $A = (0, 1, \cdots, 1)$, the linear function l_A represents the fault-free linear function l_k.

Let $l_B = b_0 \oplus b_1 x_1 \oplus \cdots \oplus b_k x_k$ be a linear function realized by a circuit under test where $b_i = 0$ or 1 for $i = 1, 2, \cdots, k$. Applying the vectors $t_0, t_1, \cdots, t_k$ to the equation

$$l_A(x_1, x_2, \cdots, x_k) = l_B(x_1, x_2, \cdots, x_k)$$

we obtain

$$a_0 = b_0$$
$$a_0 \oplus a_1 = b_0 \oplus b_1$$
$$\vdots$$
$$a_0 \oplus a_k = b_0 \oplus b_k$$

which implies

$$a_i = b_i \quad \text{for } i = 0, 1, 2, \cdots, k.$$

Therefore, we can uniquely determine the values a_i's, and thus distinguish whether l_B equals to the fault-free linear function or not. Q.E.D.

In Lemma 1 $k + 1$ tests $t_0, t_1, \cdots, t_k$ can easily be extended to k linearly independent vectors plus zero vector. Then we have the following lemma.

Lemma 2: If k input vectors are linearly independent, then these k vectors plus zero vector are sufficient to detect any multiple stuck fault on the external lines of EOR gates in a k-input EXCLUSIVE-OR cascade.

Let $M_{n,m}$ be an augmented PLA, and let C_1 and C_2 be the cascades of EOR's having the output Z_1 and Z_2, respectively, in the augmented PLA shown in Fig. 5. Let $M_{OR} = [a_{ij}]$ be a matrix of l rows and m columns where

$$a_{ij} = \begin{cases} 1 & \text{if there exists a link at the } (i,j)\text{th position} \\ & \text{of the OR array, and} \\ 0 & \text{otherwise.} \end{cases}$$

By Lemmas 1 and 2 we have the following theorem for the multiple faults in two cascades of EOR's, C_1 and C_2.

Theorem 2: The tests I_1, I_{2j}^0 $(j = 1, 2, \cdots, m)$ in $A_{n,m}$ are sufficient to detect all multiple stuck faults in C_1. If the column rank of matrix M_{OR} is equal to the number of inputs of the EXCLUSIVE-OR cascade C_2, then all the multiple stuck faults in C_2 can be detected by tests I_1, and I_{2j}^0 $(j = 1, 2, \cdots, m)$ in $A_{n,m}$.

Note that if the column rank of matrix M_{OR} is not equal to the number of inputs of the cascade C_2 although it hardly occurs, then it is not guaranteed that all multiple stuck faults in the cascade C_2 can be detected by the tests mentioned above. To overcome this problem, it might be necessary to add extra OR array so that the rank of M_{OR} is equal to the number of inputs of the EXCLUSIVE-OR cascade C_2. Note also that we permit only stuck-type faults on the external input and output lines of EOR gates. However, by adding extra array it is possible to generate all test patterns for the cascade C_1 and C_2. This technique was reported by Hong and Ostapko [3].

Using the test set $A_{n,m}$, we can construct a test sequence for the augmented PLA's as follows:

$$\alpha_{n,m} = I_1 I_{21}^0 I_{22}^0 \cdots I_{2m}^0 I_{21}^1 I_{22}^1 \cdots I_{2m}^1$$
$$U_1 U_2 \cdots U_{m-1} I_{31}^0 I_{32}^0 \cdots I_{3n}^0 I_{31}^1 I_{32}^1 \cdots I_{3n}^1$$

where the test pattern U_i $(i = 1, 2, \cdots, m - 1)$ is defined as

$$x_1 \cdots x_n y_1 y_2 S_1 S_2 \cdots S_i S_{i+1} \cdots S_m Z_1 Z_2$$
$$U_i = - \cdots - 1 1 1 1 \cdots 1 0 \cdots 0 \; \bar{\epsilon}_i.$$

The test sequence $\alpha_{n,m}$ can also detect the shift function of the shift register, as well as all single stuck and crosspoint faults in the PLA. The length of the test sequence is $2n + 3m$.

Now we have presented the test set and the test sequence for the augmented PLA's. Both the test set $A_{n,m}$ and the test sequence $\alpha_{n,m}$ have the following advantages of easy testability. The test set $A_{n,m}$ does not depend on the connection pattern of the PLA, but depends only on the values n and m, that is, the test patterns and responses are uniquely determined only by the size of the PLA. Therefore, the test set $A_{n,m}$ is "universal." We can also see that the test sequence $\alpha_{n,m}$ is universal. Hence, the test generation of the augmented PLA's is not more necessary. Moreover, the universal test sequence is a very short test sequence whose length is $2n + 3m$. In this way we can see that the augmented PLA's are very easily testable PLA's having a universal test sequence.

IV. AUGMENTED PLA'S WITH DOUBLE-BIT DECODERS

The PLA's with double-bit decoders can be similarly augmented by adding extra logic, that is, a shift register, two cascades of EOR's, and one column and one row to the AND and OR arrays, respectively, as shown in Fig. 7. Fig. 7 shows a PLA augmented from the PLA shown in Fig. 4. The augmented double-bit decoders have four control inputs y_1, y_2, y_3, and y_4, as shown in Fig. 7.

For the augmented PLA with double-bit decoders, we can similarly construct a universal test set $B_{n,m}$, as shown in Table II, and the following theorem holds.

Theorem 3: Let $N_{n,m}$ be an augmented PLA with double-bit decoders which has n inputs and m columns in the AND array. For any $N_{n,m}$, the test set $B_{n,m}$ can detect all single stuck and crosspoint faults in the decoders, the AND array, and the OR array.

Theorem 4: $B_{n,m}$ can detect all multiple stuck faults in the cascade C_1. If the column rank of M_{OR} is equal to the number of rows in the OR array, then all multiple stuck faults in the cascade C_2 can also be detected by $B_{n,m}$.

Using the test set $B_{n,m}$, we can construct the universal test sequence $\beta_{n,m}$ for the augmented PLA's as follows:

$$\beta_{n,m} = I_1 I_{21}^0 I_{22}^0 \cdots I_{2m}^0 I_{21}^1 I_{22}^1 \cdots I_{2m}^1 I_{21}^2 I_{22}^2 \cdots I_{2m}^2$$
$$I_{21}^3 I_{22}^3 \cdots I_{2m}^3 U_1 U_2 \cdots U_{m-1} I_{31}^0 I_{32}^0 \cdots I_{3\frac{n}{2}}^0$$
$$I_{31}^1 I_{32}^1 \cdots I_{3\frac{n}{2}}^1 I_{31}^2 I_{32}^2 \cdots I_{3\frac{n}{2}}^2 I_{31}^3 I_{32}^3 \cdots I_{3\frac{n}{2}}^3$$

where the test pattern U_i $(i = 1, 2, \cdots, m - 1)$ is defined as

$$x_1 \cdots x_n y_1 y_2 y_3 y_4 S_1 \cdots S_i S_{i+1} \cdots S_m \quad Z_1 Z_2$$
$$U_i = - \cdots - 1 1 1 1 1 \cdots 1 0 \cdots 0 \quad \bar{\epsilon}_i \; -.$$

The length of the test sequence is $2n + 5m$.

Note that we also state more general results for PLA with k-bit decoders. The universal test sequence might be constructed as follows:

$$I_1 I_{21}^0 \cdots I_{2m}^0 I_{21}^1 \cdots I_{2m}^1 \cdots I_{21}^{2^k-1} \cdots I_{2m}^{2^k-1}$$
$$U_1 U_2 \cdots U_{m-1} I_{31}^0 \cdots I_{3n/k}^0 I_{31}^1 \cdots I_{3nk}^1 \cdots I_{31}^{2^k-1}$$
$$\cdots I_{3n/k}^{2^k-1}$$

where I_{2j}^i's and I_{3j}^i's are the test patterns for k-bit decoders extended from 2-bit decoders. The total test length is $(2^k + 1)m + (n/k)2^k$.

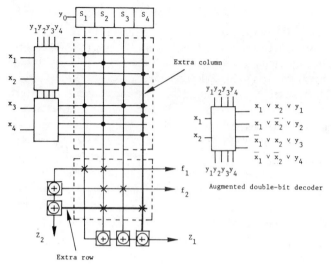

Fig. 7. Augmented PLA with double-bit decoders.

TABLE II
UNIVERSAL TEST SET $B_{n,m}$

	$x_1 x_2 \cdots x_{2i-1} x_{2i} \cdots x_{n-1} x_n$	$y_1 y_2 y_3 y_4$	$s_1 \cdots s_j \cdots s_m$	$z_1\ z_2$
I_1	$-\ -\ \cdots\ -\ \ \ -\ \cdots\ -\ -$	$-\ -\ -\ -$	$0 \cdots 0 \cdots 0$	$0\ \ 0$

For $j = 1, 2, \cdots, m$

	$x_1 x_2 \cdots x_{2i-1} x_{2i} \cdots x_{n-1} x_n$	$y_1 y_2 y_3 y_4$	$s_1 \cdots s_j \cdots s_m$	$z_1\ z_2$
I_{2j}^0	$0\ 0 \cdots 0\ 0\ \cdots\ 0\ 0$	$1\ 0\ 0\ 0$	$0 \cdots 1 \cdots 0$	$1\ \ 1$
I_{2j}^1	$0\ 1 \cdots 0\ 1\ \cdots\ 0\ 1$	$0\ 1\ 0\ 0$	$0 \cdots 1 \cdots 0$	$1\ \ 1$
I_{2j}^2	$1\ 0 \cdots 1\ 0\ \cdots\ 1\ 0$	$0\ 0\ 1\ 0$	$0 \cdots 1 \cdots 0$	$1\ \ 1$
I_{2j}^3	$1\ 1 \cdots 1\ 1\ \cdots\ 1\ 1$	$0\ 0\ 0\ 1$	$0 \cdots 1 \cdots 0$	$1\ \ 1$

For $i = 1, 2, \cdots, \frac{n}{2}$

	$x_1 x_2 \cdots x_{2i-1} x_{2i} \cdots x_{n-1} x_n$	$y_1 y_2 y_3 y_4$	$s_1 \cdots s_j \cdots s_m$	$z_1\ z_2$
I_{3i}^0	$1\ 1 \cdots 0\ 0\ \cdots\ 1\ 1$	$0\ 1\ 1\ 1$	$1 \cdots 1 \cdots 1$	$\epsilon_m\ \ -$
I_{3i}^1	$1\ 0 \cdots 0\ 1\ \cdots\ 1\ 0$	$1\ 0\ 1\ 1$	$1 \cdots 1 \cdots 1$	$\epsilon_m\ \ -$
I_{3i}^2	$0\ 1 \cdots 1\ 0\ \cdots\ 0\ 1$	$1\ 1\ 0\ 1$	$1 \cdots 1 \cdots 1$	$\epsilon_m\ \ -$
I_{3i}^3	$0\ 0 \cdots 1\ 1\ \cdots\ 0\ 0$	$1\ 1\ 1\ 0$	$1 \cdots 1 \cdots 1$	$\epsilon_m\ \ -$

V. MULTIPLE FAULT DETECTION

So far we have discussed the single fault detection problem for the augmented PLA's. In this section we extend the single fault model to the types of multiple faults in the following, and present the universal test set for them. Note that no more than one of the following multiple faults occurs simultaneously.

1) Multiple stuck faults on the primary outputs x_i ($i = 1, 2, \cdots, n$).
2) Multiple stuck faults on the control inputs y_i ($i = 1, 2, 3, 4$).
3) Multiple stuck faults on the rows in the AND array.
4) Multiple stuck faults on the columns in the AND and OR arrays.
5) Multiple stuck faults on the rows in the OR array.
6) Multiple stuck faults on the input and output lines of the EOR cascade C_1.
7) Multiple stuck faults on the input and output lines of the EOR cascade C_2 provided that the column rank of M_{OR} is equal to the number of rows in OR array.
8) Odd number of crosspoint faults on the columns of the OR array.
9) Odd number of crosspoint faults on the rows of the AND array.

For the class of the above mentioned multiple faults, we can show that the test sets $A_{n,m}$ and $B_{n,m}$ are also universal test sets for the augmented PLA's with single-bit decoders and double-bit decoders, respectively.

Now, the cascade of EOR's in the PLA is used as a parity checker to detect odd number of errors on the rows and columns of the AND and OR arrays, respectively. In the same way this approach can be extended to other multiple fault model by applying error detecting codes such as linear codes [10], where the augmented PLA's will be designed to have more than two cascades of EOR's.

VI. FAULT LOCATION AND REPAIR

In this section we consider the fault detection and repair of the augmented PLA's. For the augmented PLA's there exists a universal test sequence of fault location. The types of faults considered here are the multiple crosspoint faults in the AND and/or OR arrays.

Fault location test can be performed by identifying the configuration of the AND and OR arrays, that is, whether there exists a link between each row i and column j of the arrays. By applying the test pattern shown in Fig. 8, we can identify whether there exists a link or contact between row i and column j and the AND array. If the value of output Z_1 is 0 (1), then there exists a link (correspondingly, no link) at the (i,j)th position of the AND array.

For the OR array, the test pattern shown in Fig. 9 can detect the presence or absence of a link between row i and column j of the OR array. If the value of output f_i is 1 (0), then there exists a link (no link) at the (i,j)th position of the OR array. Note that the last row of the OR array does not have a direct output, however it is observable from the additional output Z_2.

Using these test patterns, we can completely determine the configuration of the AND and OR arrays. Therefore, all the multiple crosspoint faults can be found by observing the responses. The number of tests for the AND array is $2nm$, and the number of tests for the OR array is m. Hence, the total length of the fault location test sequence for the augmented PLA's is $2nm + m$.

Now after the fault location test the faulty PLA can be repaired as follows. For a field programmable logic array (FPLA), it is known that an arbitrary product term can be both logically and physically deleted and a new product term can be generated by using spare columns of the PLA, and thus the faulty PLA can be repaired using spare rows and/or columns [11]. If the PLA is a mask programmable logic array (MPLA), it is difficult to repair the PLA for itself. In this case we can repair the faulty PLA by using the well-known memory

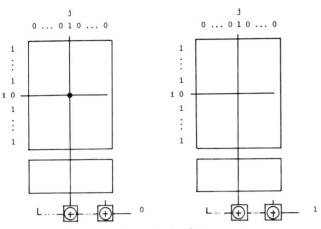

Fig. 8. Fault location test for AND array.

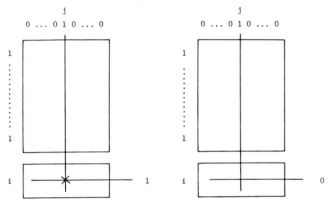

Fig. 9. Fault location test for OR array.

patch technique. This is a method to recover the function of the faulty MPLA by switching it to a spare FPLA when some errors occur.

VII. CONCLUSION

In this paper we have introduced a design of easily testable PLA's by adding extra logic and presented universal test sequences for them. These augmented PLA's have the very short universal test sequences such that the test patterns and responses are uniquely determined only by the size of the PLA's independently of the function of them. The length of the universal test sequences to detect single and multiple faults is of order $n + m$, and the length of the universal test sequences to locate multiple faults is of order nm, where n is the number of inputs and m is the number of columns of the AND array.

The augmented PLA in this paper, however, has a weak point such that there exist some stuck faults inside of EXCLUSIVE-OR gates which cannot be detected by the universal tests. This problem can be resolved by adding extra OR array such as Hong and Ostapko [3] to generate the test patterns for such faults.

Although we have not considered PLA's with flip-flops in this paper, we can augment them to have the universal test sets. This can be done by applying a scan-in and scan-out technique such as LSSD [5], [6].

ACKNOWLEDGMENT

The authors would like to thank Prof. H. Ozaki of Osaka University for his encouragement and support. The authors would also like to thank Dr. T. Sasao of Osaka University and Y. Fukui of Sharp Corporation for their useful discussions.

REFERENCES

[1] R. G. Bennetts and R. V. Scott, "Recent developments in the theory and practice of testable logic design," *Computer*, pp. 47–63, June 1976.
[2] S. M. Reddy, "Easily testable realizations for logic functions," *IEEE Trans. Comput.*, vol. C-21, pp. 1183–1188, Nov. 1974.
[3] S. J. Hong and D. L. Ostapko, "FITPLA: A programmable logic array for function independent testing," in *Dig. 10th Int. Symp. Fault-Tolerant Comput.*, Kyoto, Japan, June 1980, pp. 131–136.
[4] H. Fujiwara, K. Kinoshita, and H. Ozaki, "Universal test sets for programmable logic arrays," in *Dig. 10th Int. Symp. Fault-Tolerant Comput.*, Kyoto, Japan, June 1980, pp. 137–142.
[5] M. J. Williams and J. B. Angell, "Enhancing testability of large-scale integrated circuits via test points and additional logic," *IEEE Trans. Comput.*, vol. C-22, pp. 46–60, Jan. 1973.
[6] E. B. Eichelberger and T. W. Williams, "A logic design structure for LSI testability," in *Proc. 14th Des. Automat. Conf.*, June 1977, pp. 462–468.
[7] H. Fleisher and L. I. Maissel, "An introduction to array logic," *IBM J. Res. Develop.*, vol. 19, pp. 98–109, Mar. 1975.
[8] D. L. Ostapko and S. J. Hong, "Fault analysis and test generation for programmable logic arrays (PLA's)," in *Dig. 8th Int. Symp. Fault-Tolerant Comput.*, Toulouse, France, June 1978, pp. 83–89.
[9] V. K. Agarwal, "Multiple fault detection in programmable logic arrays," in *Dig. 9th Int. Symp. Fault-Tolerant Comput.*, Madison, WI, June 1979, pp. 227–234.
[10] W. W. Peterson and E. J. Weldon, Jr., *Error-Correcting Codes*. Cambridge, MA: MIT Press, 1972.
[11] *Signetics Field Programmable Logic Arrays*, Signetics, CA, Oct. 1977.
[12] K. K. Saluja and S. M. Reddy, "Fault detecting test sets for Reed-Muller canonic networks," *IEEE Trans. Comput.*, vol. C-24, pp. 995–998, Oct. 1975.

Hideo Fujiwara (S'70–M'74) was born in Nara, Japan, on February 9, 1946. He received the B.E., M.E., and Ph.D. degrees in electronic engineering from Osaka University, Osaka, Japan, in 1969, 1971, and 1974, respectively.

Since 1974 he has been with the Department of Electronic Engineering, Osaka University. His research interests are switching theory and logic design, and he specializes in the development of testing, testable logic design, and system diagnosis.

Dr. Fujiwara is a member of the Institute of Electronics and Communication Engineers of Japan and the Information Processing Society of Japan.

Kozo Kinoshita (S'58–M'64) was born in Osaka, Japan, on June 21, 1936. He received the B.E., M.E., and Ph.D. degrees in communication engineering from Osaka University, Osaka, Japan in 1959, 1961, and 1964, respectively.

From 1964 to 1966 he was an Assistant Professor and from 1967 to 1977 he was an Associate Professor of Electronic Engineering at Osaka University. Since 1978 he has been a Professor in the Department of Information and Behavioral Sciences, Hiroshima University. His fields of interest are logic design, fault diagnosis, and fault-tolerant design of digital systems.

Dr. Kinoshita is a member of the Institute of Electrical Engineers of Japan, the Institute of Electronics and Communication Engineers of Japan, and the Information Processing Society of Japan.

BUILT-IN TESTS FOR VLSI FINITE-STATE MACHINES

Kien A. Hua, Jing-Yang Jou and Jacob A. Abraham

Computer Systems Group
Coordinated Science Laboratory
University of Illinois
Urbana, Illinois 61801

ABSTRACT

The testing problem is becoming increasingly difficult to solve for VLSI technology, especially for sequential circuits. In this paper we present a solution to this problem which implements finite-state machines using the sequential structure of Programmable Logic Arrays (PLAs). Built-in tests for the structure are also designed to quickly test the finite-state machine. This is done for a very small additional hardware cost. The difficulty in interconnecting the test logic to the very compact naked PLA is also considered in this paper for the first time. A detailed layout example is given as well as overhead estimates for a wide range of PLA parameters.

1. INTRODUCTION

The introduction of LSI/VLSI technology has increased the difficulty of both designing and testing the complex systems which can be implemented on a chip. The testing problem especially, is felt to be the primary bottleneck to the widespread use of the next generation of VLSI chips. Design for testability has been proposed as a solution to this problem [1]. Specifically, since testing of combinational circuits is much simpler than testing sequential circuits, a widely used technique for testability is Level Sensitive Scan Design [2] [3] and variants of it [4] [5]. Here, all latches in the circuit are chained together as shift registers during test, thus providing serial access to them with a very small overhead in the number of pins. Thus the latches can be tested with a set of patterns and then can be used to test the combinational logic by scanning in test patterns and scanning out results. This approach clearly does not test a circuit at speed, and the test could take a long time due to the serial nature of the scan paths.

The controllers embedded within VLSI chips are finite-state machines (FSMs). These could be broken up for testability as mentioned above. However, we would like to take a different approach to designing testable FSMs. We will implement the FSM with a very regular structure which is easy to test, and we will embed the test pattern generator on-chip for rapid testing of the FSM. The regular
_____
ACKNOWLEDGEMENTS: This research was supported by the Semiconductor Research Corporation under contract SRC RSCH 83-01-014.

structure which will be used is a Programmable Logic Array (PLA) with feedback, and we will extend existing results in testing combinational PLAs [6] [7] [8] to those with feedback. The test pattern generator will be designed to take up only a relatively small amount of area in nMOS technology, our vehicle for implementing the example designs. This is an interesting problem since a straightforward implementation of proposed gate level designs in nMOS was found to take up excessive area.

The programmable nature of the PLA makes the design task much easier. An experiment in designing a small complex system, the IBM 7441 Buffered Terminal Control Unit using PLAs is described in [9]. This paper showed that the PLA approach exploits many of the benefits of LSI/VLSI without the high engineering design costs. In addition to making circuit design easier, due to the array-oriented structure, the PLA approach for LSI/VLSI also simplifies the testing problem.

Much effort has been directed to the problem of fault detection in the PLAs in recent years. Pseudo random number sequences [10] are attractive in many testing problems; unfortunately is not an effective approach for PLA testing due to the high fan-in in PLAs [1]. A PLA testing scheme using multiple parallel signature analyzers was proposed in [11]. This approach is suited for PLAs with a very large number of product lines compared to the number of input lines; however, typical PLAs have a ratio of number of product lines to number of input lines about four, and this approach will therefore require a high overhead in area.

A method of generating a minimal single fault detection test set from the product term specification of the PLA was presented in [12]. This testing scheme however is not appropriate for built-in tests. A universal test set for PLA was proposed in [6]. This scheme minimizes the test generation overhead. The test procedure, however, requires part of the test patterns generated externally and fed to the PLA inputs. This is impractical in a system environment where typical inputs to the FSM are buried within a chip and not easily controllable.

Another built-in test approach using nonlinear feedback shift registers for test pattern generation, and multiple input signature analyzers for test response evaluation, was proposed in [7]. This approach still takes up quite a large amount of area for the test logic and for the interconnections between the test logic and the naked

PLAs. The latter is because the PLA design is very compact. The size of a typical nMOS PLA cell is 16 lambda by 16 lambda, where lambda is half the minimum feature size [13]. The test logic must be very simple, or else a large area will have to be wasted for interconnections (Figure 1(a)). This problem is particularly crucial when the number of interconnections is large.

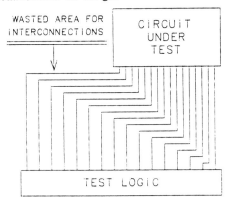

Figure 1(a)   The outputs from the test logic do not line up with the inputs to the circuit under test

In this paper, we will present a complete built-in test design scheme for VLSI finite-state machines using PLAs. Unlike other built-in test schemes for PLA, the difficulty in interconnecting the test logic and the naked PLA will be considered in this paper. We have done the layout for the test logic cells and example PLAs in nMOS technology to demonstrate that all the test logic cells line up perfectly with the PLA cells (Figure 1(b)). This eliminates the interconnection problem

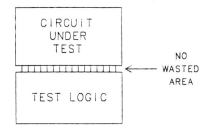

Figure 1(b)   The outputs from the test logic line up with the inputs to the circuit under test

discussed previously. Also, the test logic presented in this paper can be cascaded to implement testable PLAs of any size. Since our approach conserves the regularity of the PLA structure, it is very suitable for design automation.

2. FINITE STATE MACHINE WITH BUILT-IN TEST

2.1. Overall structure of the FSM with built-in test

The structure of our proposed Finite State Machine with Built-in Test (FSMBT) is shown in Figure 2. An extra product line and an extra output line are added to the AND and OR arrays, respectively (darker lines). The contacts on the extra product line are arranged so as to make the columns of the AND plane all have an odd number of non-contacts. Let n be the number of output lines of the PLA. If n is odd (even), we arrange the contacts on the extra output or parity line such that each row in the OR plane has an even (odd) number of contacts.

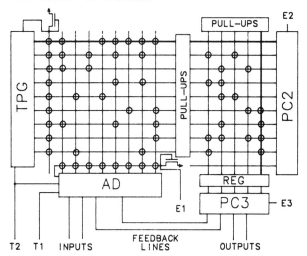

Figure 2   Structure of the FSMBT

Two control signals T1 and T2 are used for the test. During normal operation, T1=T2=0.

The Test Pattern Generator (TPG) will generate the test sequence for the OR plane when T2 is set to logical 1. When T2=0, the TPG is isolated from the naked PLA, and is reset to its initial state. The parity checkers PC2 and PC3 (Figure 2) are used for test response evaluation when the OR plane is under test.

Normally, T1=T2=0 and the Augmented Decoder (AD) functions as a normal single bit decoder network. During the test of the OR plane, T2 resets the AD to its initial state. When T1 is set to 1 and T2 is set to 0, the AD generates the test sequence for the testing of the AND plane. The NOR gate output E1 together with the checker PC2 (Figure 2) are used for test response evaluation when the AND plane is under test.

We summarize in Table I the different operation modes of the FSMBT.

Table I   Modes of operation

T2	T1	Mode of operation
0	0	Normal machine operation
0	1	Test the AND plane
1	0	Test the OR plane
1	1	Not used

2.2. Fault Coverage

The FSMBT covers any single fault, in the naked PLA or the test logic, that is of the following types:

(1) Short fault between two adjacent lines in the PLA.

(2) Stuck-at faults which cause a line to be stuck-at 1 or stuck-at 0.

(3) Contact fault due to a missing device at the cross point, or an extra device at the cross point where there should not be one.

As pointed out in [14], a great majority of physical failures are covered by shorts and stuck-at faults. The three types of faults considered in this paper are therefore sufficient to cover most of the physical failures in PLAs.

Although we assume a single fault model for simplicity in this paper, multiple stuck-at faults and short faults of the PLA lines (i.e., columns and rows) will also be detected using this scheme.

2.3. Test Pattern Generator

We employ the universal test set proposed in [6] as shown below for both the AND plane and the OR plane:

$$
\begin{aligned}
t_0 &= 0\;0\;0\;...\;0\;0 \\
t_1 &= 1\;0\;0\;...\;0\;0 \\
t_2 &= 0\;1\;0\;...\;0\;0 \\
t_3 &= 0\;0\;1\;...\;0\;0 \\
&\;\;\vdots \\
t_{n-1} &= 0\;0\;0\;...\;1\;0 \\
t_n &= 0\;0\;0\;...\;0\;1
\end{aligned}
$$

The Test Pattern Generator is shown in Figure 3. This, however, uses a shift register that is only half the size of the shift registers used in previous schemes employing this approach. Each shift register output is shared by two adjacent product lines by multiplexing. Our design not only saves on the number of transistors, but also eliminates the wasted area for interconnections of the TPG to the AND plane. If an n-bit shift register has to be used for an n-product term PLA, in order

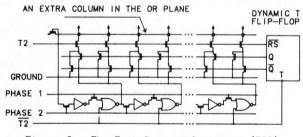

Figure 3 The Test Pattern Generator (TPG)

to have the outputs of the shift register line up with the product lines, we will have to fit at least four inverters in a width of 16 lambda. Using the design rules in [13], a pull-up generally takes up 6 lambda in width. It is therefore impossible to interconnect the four inverters within 16 lambda to make the two bits of the shift register. Our proposed TPG uses transmission gates for multiplexing to reduce the number of inverters to half of that required by previous schemes.

The checkers PC2 and PC3 can be used to detect any stuck-at fault or short fault in the TPG. We will describe these next.

If any output line i of the TPG is stuck-at-0, the test t_i can detect this. If it is stuck-at-1 the test t_0 can detect this. If any output line i of the TPG is shorted to the adjacent line i+1, the test t_i or t_{i+1} can detect this fault.

If the output of the feedback NOR gate (just an extra column in the OR plane) is stuck-at-1 or stuck-at-0, then PC2 will detect this since a constant logic value is shifted into the TPG.

The stuck fault at the input of the T flip-flop is covered by the stuck fault at the right-most output of the shift register (Figure 3). If the flip-flop output Q is stuck-at-1 (or 0), this can be detected when the input to the flip-flop is 0 (or 1) because two 1's will appear on the product lines and E2 will become 0 when the tests t_i, i > 0 are applied. Testing the stuck fault at $\overline{Q}$ is similar. If Q shorts to $\overline{Q}$, the TPG is isolated from the PC2 and E2 will be carrying a constant value all the time.

During the testing of the OR plane, T1=0 and T2=1 causes all the outputs of the AD to be preset to 0's and E1 should be 1. If T2 is stuck-at-0, the AD functions as a decoder network. Since half of the decoder outputs will have logic 1's, E1 will become 0. This detects T2 stuck-at-0. If T2 is stuck-at-1, all the columns in the AND plane will be floating (Figure 4). This makes E1=1 throughout the testing of the AND plane. This fault therefore can be detected during the testing of the AND plane by observing E1 (Figure 2).

2.4. Augmented Decoder (AD)

The Augmented Decoder (AD) is shown in detail in Figure 4. Again we use the multiplexing scheme described previously. Each output of the AD is connected to its normal decoder input under normal PLA operations. When T1 becomes logical 1, the output is either connected to ground or the shift register output depending on the state of the T flip-flop (Figure 4). The testing of the AD is similar to that of the TPG with some differences we will describe in the following.

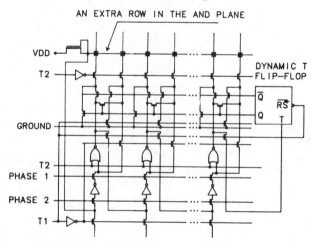

Figure 4 The Augmented Decoder (AD)

We have already described the detection of the stuck-at faults at T2 in section 2.3. During the testing of the AND plane, T1=1; if T1 is stuck-at-0, the machine is doing its normal operations, and is not in testing mode. E1 will therefore be 0 throughout the test. Since we expect E1=1 when the test t_0 is applied to the AND plane, the test t_0 will detect T1 stuck-at-0. During the testing of the OR plane, E1 should be 1 throughout the test under the fault free condition. If T1 is stuck-at-1, the AD functions as a test pattern generator. Any test patterns except t_0 from the AD will make E1=0. E1 thus will detect T1 stuck-at-1.

Since the decoders are embedded in the shift register and the AD, testing of the shift register will also detect any fault exists in the decoders.

2.5. Parity Checkers

Figure 5 shows the Exclusive-OR (XOR) gate and the parity checker implemented using the gate. Notice that we need two levels for the parity checkers because of the compactness of the PLA cells. This will also speed up the propagation of signals along the parity checkers significantly. If the number of product lines is very large, it might be necessary to employ more levels to obtain the desired speed.

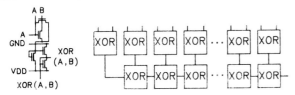

Figure 5 Parity checker implemented using the Exclusive-OR gate

The XOR gate implemented as shown in Figure 5 needs only three vectors, (00, 01, 10), to be tested [16]. The test set generated by the TPG therefore can also detect any single fault in the PC2 circuit.

As we have already described in Section 2.1, the contacts on the extra output or parity line are arranged such that each row in the OR plane has an even (odd) number of contacts if the total number of output lines of the PLA is odd (even). With such an arrangement, E3 can be tested for stuck-at faults. For instance, if n is odd, during the testing of the OR plane, E3 should be 0 under the fault free condition. This test for E3 stuck-at-1. During the testing of the AND plane, when the test pattern t_0 is applied, E3 should have logical 1 under the fault free condition since there are odd number of output lines. This tests for E3 stuck-at-0.

We assume in this paper that no output line is redundant. An output line is redundant if it is identical to another output line, or if it outputs a constant logical value regardless of the input combinations. We can always avoid such redundant outputs in practice. Many CAD programs are available for doing this. Although a regular PLA output line may be identical to the parity line (i.e., the extra output line), we can solve this problem by not making them adjacent. With this assumption, any stuck-at fault or short fault at the inputs of the PC3 can be detected during the testing of the OR plane. An output line i can be detected for stuck-at-1 (or stuck-at-0) when it is supposed to have value 0 (or 1). If an output line i shorts to the adjacent line i+1, this fault will be detected when the two lines carry opposite values.

If the detection of faults at the external lines of all the parity cells inside PC3 is desired, a totally self-checking (TSC) parity checker [15] can be used. This causes no extra area compared to PC3, but requires one extra pin.

2.6. Testing of the Output Register and the NOR planes

Stuck-at faults and short faults at the product lines, the output lines and the bit lines (i.e., output lines from the decoders) of the PLA are covered by the stuck-at faults and short faults of the inputs to PC2, the inputs to PC3 and the outputs of the AD respectively. When we test modules PC2, PC3 and AD, all the lines in the PLA planes are tested as well.

If a missing contact or an extra contact fault occurs on the column i of the AND plane, PC2 will detect this fault during the testing of the AND plane when the test pattern t_i is applied. Similarly, a contact fault on the rows of the OR plane can be detected by PC3 during the testing of the OR plane.

2.7. Test Responses

Let m/2 be the number of inputs, p be the number of product lines and n be the number of output lines. We summarize in Table II the test responses under the fault free condition. The total number of test patterns is m+p+2.

3. OVERHEAD AND COMPARISON WITH OTHER SCHEMES

We have done layouts for the test logic cells in order to study the actual overhead. A layout example with 32 inputs, 18 feedback paths, 190 product terms and 49 outputs as shown in Figure 6 was studied. This PLA is identical in size to one of the eight PLAs used in the BellMac-32A microprocessor described in [17]. The overhead for this example is 20%.

A built-in test example for PLA with 10 inputs, 30 outputs and 150 product terms was presented in [11]. The overhead was estimated in [11] to be 23%. This example has a ratio of 15 for number of product terms to number of inputs. The overhead will become worse when this ratio is lower as it is usually the case. The eight PLAs in the BellMac microprocessor [17], for instance, have this ratio ranging from 1.8 to 6.7. This scheme [11] also becomes impractical when the number of inputs is quite large, because 2^n test patterns would be required where n is the number of outputs from the decoders. In the example above (of a PLA similar to that used in the BellMac-32A), the number of test patterns required by [11] would be 10^{15}.

The testing scheme presented in [7] uses the

Table II Test responses under fault free conditions

Test mode		Product lines						Bit lines						Correct response		
T2	T1	P_1	P_2	P_3	...	P_{p-1}	P_p	B_1	B_2	B_3	...	B_{m-1}	B_m	E1	E2	E3
0	1	1	1	1	...	1	1	0	0	0	...	0	0	1	1	-
0	1	1	1	1	...	1	1	1	0	0	...	0	0	0	1	-
0	1	1	1	1	...	1	1	0	1	0	...	0	0	0	1	-
.	.	.	.	.		.	.	.	.	.		.	.	.	.	
0	1	1	1	1	...	1	1	0	0	0	...	1	0	0	1	-
0	1	1	1	1	...	1	1	0	0	0	...	0	1	0	1	-
1	0	0	0	0	...	0	0	0	0	0	...	0	0	1	0	0
1	0	1	0	0	...	0	0	0	0	0	...	0	0	1	1	e
1	0	0	1	0	...	0	0	0	0	0	...	0	0	1	1	e
.	.	.	.	.		.	.	.	.	.		.	.	.	.	.
1	0	0	0	0	...	1	0	0	0	0	...	0	0	1	1	e
1	0	0	0	0	...	0	1	0	0	0	...	0	0	1	1	e

e = 1 if n is even
e = 0 if n is odd
- is don't care

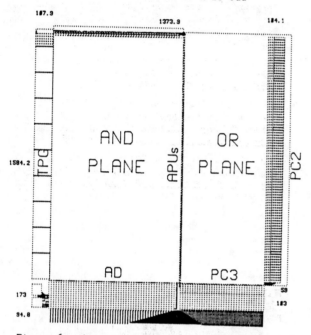

Figure 6 A layout example the overhead is 20%

built-in logic block observer (BILBO) scheme [5]. Since BILBO takes up even more area than the linear feedback shift register used in [11], the overhead for the scheme presented in [7] is generally worse than that for the scheme presented in [11]. For the example we use, we estimate the overhead to be 50% if the scheme presented in [7] is used.

The scheme we proposed in this paper is thus good for VLSI finite state machines when large digital systems are implemented using PLAs. Figure 7 shows the overhead estimates for a wide range of PLA parameters, where p is the number of product terms and N is the sum of the number of inputs (m), the number of outputs (n), and the number of feedback lines (f). We assumed that m=n=f in obtaining these graphs. It shows that the overhead becomes smaller as the size of the PLA increases.

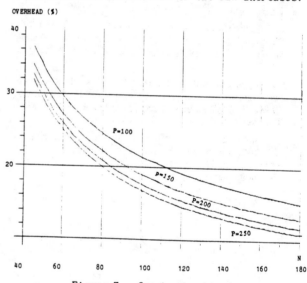

Figure 7 Overhead estimates

4. CONCLUSIONS

Using the universal test set allows a Finite State Machine implemented with a PLA to be tested quickly. By modifying the single-bit decoders, we get a test pattern generator at a low cost. The use of the feedback NOR gates of the Augmented Decoder also for test response evaluation saves a lot of hardware since another parity checker would otherwise have to be used. By using the multiplexing scheme, we also cut down the size of the shift registers in half compared with other PLA testing schemes.

We have thus achieved a FSMBT with very low overhead. Typical overhead ratios will be about 20%, which is reasonably small for VLSI chips. Our test logic cells can be cascaded to accommodate PLAs of any desired size since the proposed scheme matches the regularity of the PLA structure. Existing PLA synthesis programs can be modified for designing such FSMBTs.

Papers such as [9] and [17] have demonstrated the advantages of using PLAs in LSI/VLSI designs. Its is our intention in this paper to reinforce this point by showing how PLAs can solve the testing problem for large finite-state machines.

REFERENCES

[1] T. W. Williams and K. P. Parker, "Design for Testability - A Survey," IEEE Trans. Comput., Vol. c-31, No. 1, Jan. 1982, pp. 2-15.

[2] M. J. Williams and J. B. Angell, "Enhancing Testability of Large-Scale Integrated Circuits via Test Points and Additional Logic," IEEE Trans. Comput., Vol. c-22, Jan. 1973, pp. 46-60.

[3] E. B. Eichelberger and T. W. Williams, "A Logic Design Structure for LSI Testability," Proc. 14th Design Automation Conference, June 1977, pp. 462-468.

[4] J. H. Stewart, "Future Testing of Large LSI Circuit Cards," Dig. 1977 Semiconductor Test Symp., Oct. 1977, pp. 6-17.

[5] B. Koenemann, J. Mucha and G. Zwiehoff, "Built-in Logic Block Observation Techniques," Dig. 1979 Test Conf., Oct. 1979, pp. 37-41

[6] H. Fujiwara and K. Kinoshita, "A Design of Programmable Logic Arrays with Universal Tests," IEEE Trans. Comput., Vol. c-30, No. 11, Nov. 1981, pp. 823-828.

[7] W. Daehn and J. Mucha, "A Hardware Approach to Self-testing of Large Programmable Logic Arrays," IEEE Trans. Comput., Vol. c-30, No. 11, Nov. 1981, pp. 829-833.

[8] S. J. Hong and D. L. Ostapko, "FITPLA: A Programmable Logic Array for Function Independent Testing," Dig. 10th Int'l Symp. on Fault-Tolerant Computing, Oct. 1980, pp. 131-136.

[9] J. C. Logue, N. F. Brickman, F. Howley, J. W. Jones and W. W. Wu, "Hardware Implementation of A Small System in Programmable Logic Arrays," IBM J. Res. Develop., Mar. 1975, pp. 110-119.

[10] D. K. Bhavsar and R. W. Heckelman, "Self-testing by Polynomial Division," Proc. IEEE Int'l Test Conf., 1981, pp. 208-216.

[11] S. Z. Hassan and E. J. McCluskey, "Testing PLAs Using Multiple Parallel Signature Analyzers," Proc. 13th Int'l Symp. on Fault-Tolerant Computing, June 1983, pp. 422-425.

[12] P. Bose and J. A. Abraham, "Test Generation for Programmable Logic Arrays," IEEE 19th Design Automation Conf., 1982, pp. 574-580.

[13] C. Mead and L Conway, Introduction to VLSI Systems, Addison-Wesley, 1980.

[14] J. Galiay, Y. Crouzet and M. Vergniault, "Physical Versus Logical Fault Models MOS LSI Circuits: Impact on Their Testability," IEEE Trans. Comput., Vol. c-29, No. 6, June 1980, pp. 527-531.

[15] D. K. Pradhan and J. J. Stiffer, "Error-Correcting Codes and Self-Checking Circuits," IEEE Computer, Mar. 1980, pp. 27-37.

[16] Y. Crouzet, "The P. A. D.: A Self-Checking LSI Circuit for Fault-Detection in Microcomputers," Proc. 12th Int'l Symp. on Fault-Tolerant Computing, June 1982, pp. 55-62.

[17] H-F. S. Law and M. S. Shoji, "PLA Design for the BellMac-32A Microprocessor," Proc. 1982 Int'l Conf. on Circ. and Comput., Sept. 1982, pp. 161-164.

NEW TECHNIQUES FOR HIGH SPEED ANALOG TESTING

Matthew V. Mahoney

LTX Corporation
LTX Park at University Avenue
Westwood, MA 02090

Abstract

Analog transmission parameters like frequency response, intermodulation distortion, and envelope (group) delay distortion are vital to audio and communications circuits. They are seldom tested thoroughly, however, because of the test time required by conventional techniques. Moreover, comprehensive testing requires such a wide variety of specialized instrumentation that, in practice, the more difficult tests may be omitted entirely.

In this paper we consider a radically different approach to analog testing, one which greatly reduces the test time while reducing the variety of hardware required. By employing special Digital Signal Processing in a totally coherent, phase-locked test fixture, it is in fact possible to measure gain, harmonic distortion, IM distortion, frequency response, envelope delay distortion, and noise, all in a single brief test.

Introduction

Even though modern analog circuits may require literally hundreds of measurements, most of these tests are conducted one at a time, switching in or out different sources, filters, and meters. At the component and wafer level, thorough testing is thus impractical, both because of the disproportionate test time, and because of the variety of automated hardware required. The practical result is that components generally receive far less comprehensive testing than either the manufacturer or device user would like to see.

In principle, however, one ought to be able to gather all the information required in a time interval several orders of magnitude smaller than even the fastest conventional automated systems require. Likewise, there is no theoretical reason why such a wide variety of test circuitry is required to test circuit functions that, after all, apply mostly to analog transmission paths. If a test system could be designed which approached the theoretical limits, it would become feasible even at the wafer level to conduct fairly comprehensive tests on each circuit.

In this paper I would like to show three related techniques that make it possible for a computer-based test system to approach the minimum test time:

1. Coherence
2. Hardware Emulation
3. Multitone Testing

The first principle, coherence, requires that every signal source, every filter, every voltmeter, every circuit element, and the device under test, be precisely coordinated in timing, in a manner completely specified by the test engineer. Such a system is said to be _coherent_. It not only enables very high speed measurement of single parameters, but equally important, enables a number of seemingly _different_ tests to be conducted simultaneously--something not possible before.

Coherence is impractical with conventional analog instruments, however. To make it a reality requires another change: replace all conventional sources, filters and measurement circuits with _mathematical models_, and connect these models physically to the test device via special interface circuits that allow the device to "think" it is still connected to traditional hardware. This is what is meant by _emulated_ hardware. To the device, it appears that real instruments are in use, except that they are much faster and more precise. In reality, the instruments reside only in software, in the numerical domain, where there is no problem in coordinating all time and frequency relations as precisely as we choose.

Coherent, emulated hardware makes possible a third technique--<u>multitone testing</u>, whereby a number of different stimuli may be applied simultaneously, and yet with assurance that all response components are <u>orthogonal</u>, or statistically independent. One can measure IM distortion, for example, while simultaneously measuring frequency response and propogation delay. It is this, as much as anything, that accounts for the real increase in test throughput.

Let us now examine each of the principles a bit more thoroughly. A good place to start is with the generalized form of a coherent system in which the instruments are replaced by models based on Digital Signal Processing, or DSP (Figure 1).

Note carefully: in this structure, there are no conventional generators, filters, or meters. To the fullest extent possible, all instrumentation is implemented in the <u>numerical</u> domain, by mathematical models within the processor. The "processor" may be a special computer, or it may represent the standard test computer, assisted by DSP hardware for high speed computation. For maximum speed and simplicity, each model is built into the processor's operating system, and called by a simple mnemonic command, like "RMS," or "AVG."

The two interface elements act strictly as translators, bridging the gap between the numerical domain of the processor and the electrical domain of the device's input and output ports. These ports may be either analog or digital, so each interface actually has two parts (Figure 2).

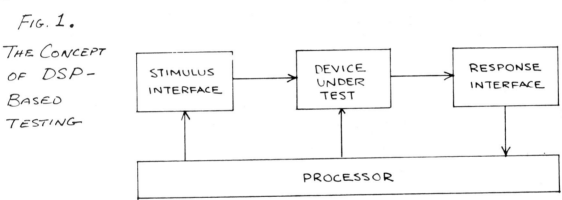

FIG. 1. THE CONCEPT OF DSP-BASED TESTING

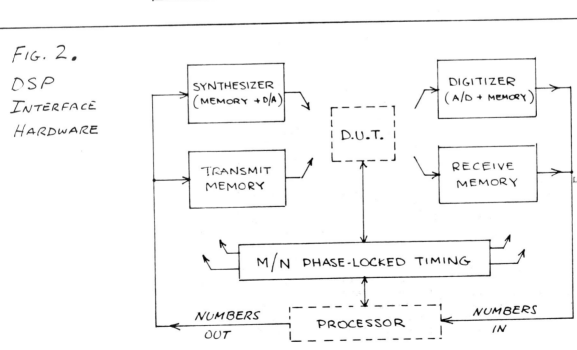

FIG. 2. DSP INTERFACE HARDWARE

Each unit contains its own pattern memory, commonly of 4096 words capacity. If the device input is analog, the appropriate interface is a combination of memory (RAM), D/A converter, and timing, referred to as a <u>synthesizer</u>. A more detailed diagram is shown in Figure 3.

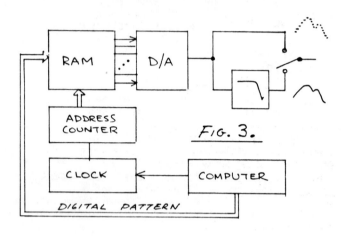

If the device input is digital, the test pattern is placed in another memory, the transmit or send RAM. In this case, the only processing required is to "format" the 1's and 0's into the proper electrical drive levels and digital time relations.

The output port may also be analog or digital. If digital, the bit pattern is collected in the <u>Receive</u> RAM, then sent as a block to the processor. If analog, the output waveform is sampled and <u>digitized</u> by an A/D converter into numerical form and, again, transferred as a data block. Figure 4 shows the diagram of a Digitizer.

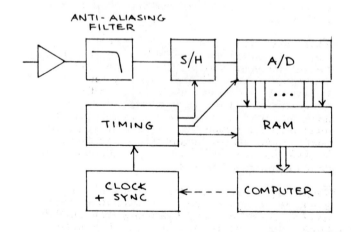

For the Synthesizer and Digitizer, each waveform sample is a word of 16 bits. Each analog pattern RAM thus has a capacity of 4K x 16 bits. For the Receive and Transmit RAMs, the word length may be larger or smaller, according to the number of digital pins.

This generalized structure can handle digital-to-digital, D/A, A/D, and all-analog circuits. For convenience, however, let us think of the numerical patterns in and out as representing <u>signals</u>--waveforms of any arbitrary shape and duration. Each signal, whether stimulus or response, is carried in the processor as a set of numbers, usually as a 1 x N word matrix. Test signals are synthesized by matrix math which simulate various signal sources; device response is conversely analyzed by matrix math which is modelled after the behavior of idealized test hardware.

Computation is always done at greater precision than that of the word itself. Each of these N words, or signal samples, might come from the digitizer at 16 bits, but the processing is done at 24 or 32 bits, to avoid accumulation of roundoff errors. A typical digitizer output signal or synthesizer input signal (Figure 5) contains 400 to 2000 samples of 16 bits each.

Fig. 5.

"MAT A" = SAMPLED WAVEFORM

+342.62 mV
−28.57 mV

1 2 3 4 5 6 7 8 9 10 11 12 etc. →
INDEX →

$$\text{MAT A} = [a_1, a_2, a_3, a_4, \cdots, a_N]$$

The dimension of MAT A = $1 \times N$

To simplify matters for the programmer, each signal matrix is denoted by a letter. The simulated "hardware" that will operate on that signal is denoted by the matrix operation. To illustrate, let A represent digitized waveform just entered into the processor memory from the digitizer. To find its RMS voltage, we can write

$$\text{MAT B} = \text{RMS}(A)$$

The "MAT" notation just warns the compiler that the argument, A, is not a single variable, but a 1 x N matrix. The answer, RMS voltage, is (in this example) just a single number.

In multitone testing, the Fast Fourier Transform, or FFT, is commonly used:

MAT B = FFT(A)

Here, both B and A are 1 x N matrices. B is a list of all the frequency components, expressed in rectangular complex form.

If phase is not important to the problem, the programmer may use the magnitude form:

MAT B = MAG FFT(A)

This time, B is just a list of the magnitudes of the frequency components, from DC up to the Nyquist frequency. If you were to plot this data list from left to right, you would have the same picture that you would see on a spectrum analyzer, except with better resolution. The MAG FFT, in effect, simulates a bank of bandpass filters.

To <u>synthesize</u> a complex waveform, we work in reverse. First, we create a list of all the frequency components we want to exist in the final, time-varying waveform. If this list is labelled as matrix B, then we obtain the time function by writing

MAT A = INVERSE FFT(B)

The time-varying waveform, A, may be then sent to the synthesizer to be converted to an analog waveform. If the synthesizer memory is looped, the waveform will be continuous and periodic.

We will use the FFT shortly in multitone testing. First, however, let us say a bit more about coherence.

Coherent Measurement

Consider a trivial measurement--voltage output at 1000 Hz (Figure 6).

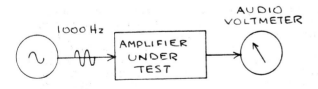

FIG. 6.

Once the signal source and Device Under Test (DUT) have settled to a steady state, the audio voltmeter may begin its measurement. In principle, only one output cycle is required to get a correct measurement, either RMS or absolute-average responding. (A half-cycle is possible if the DUT has no non-linearity.)

$$V_{RMS} = \sqrt{\frac{1}{t_2 - t_1} \int_{t_1}^{t_2} V_{in}^2 \, dt}$$

$$V_{ABS \atop AVG} = \frac{1}{t_2 - t_1} \int_{t_1}^{t_2} |V_{in}| \, dt$$

Please note: there is almost no audio voltmeter on the market which actually works the way the equations indicate. If one did, it would have a detector (squaring or absolute) followed by a <u>timed integrator</u>. The difference, $t_2 - t_1$, would equal one cycle (to include possible non-linear distortion) and the <u>measurement</u> time would be 1 ms at 1000 Hz.

But this would require a <u>coherent</u> system, in which both the signal source and AVM are accurately timed. Most AVMs are designed for <u>non</u>-coherent testing. They use long time-constant averaging circuits instead of integrators. The measurement time is more likely to be 25 to 100 ms. This is but one example of why coherence enables fast testing: it permits the test engineer to utilize the exact equations, i.e., to simulate an ideal measurement circuit.

The Unit Test Period

Coherence is important for a second reason. Not only does it permit a very short measurement interval, but it also enables a number of different measurements to be made in parallel during this same interval.

To measure the gain of the circuit in Figure 6, one could place a 1 KHz bandpass filter between the device and the audio voltmeter. At the same time, a second filter and AVM pair could be used to measure the second harmonic, and so on. As long as each meter was exactly timed over a whole number of input cycles, each measurement would be correct, and the entire group would take no longer than the gain measurement alone.

The time required to make this entire group of measurements is dictated by the slowest element--in this case, the 1000 Hz path. For simplicity, we may thus consider all voltmeters to integrate over one common interval, the so-called <u>Test Period</u>, with no sacrifice of speed. Correct measurements can be made over any integral multiple of some basic period (1 ms in this example), but our interest obviously lies with the shortest interval. Because this is the "unit" multiple, this period is commonly clled the <u>Unit Test Period</u>, or UTP.

Unfortunately, coherent voltmeters would not buy us any time savings if real analog filters were used; the 1 KHz bandpass probably takes 25 ms to settle, far more if it is a very sharp filter. Here is a second reason why emulated hardware is important: besides facilitating coherence, it eliminates the switching and settling time of conventional hardware. Ideally, the UTP should be the dominant component of test time. To approach this condition, one must digitize directly at the device output (Figure 7).

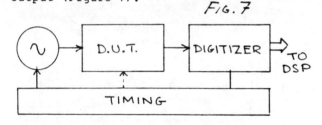

FIG. 7

In this case, we collect precisely N samples, spanning precisely 1 ms. The FFT acts both as the bank of filters and the bank of voltmeters. Moreover, it can proceed "off line," while the device is undergoing a different group of tests. With fast processing, there may be no penalty of computation time, whatsoever.

To conduct this test, the synthesizer begins to apply the 1000 Hz tone, and the DUT is allowed to settle. (This is very quick for just an amplifer.) The digitizer is then gated on for N samples, timed to span exactly one Unit Test Period, or 1 ms. If we knew with assurance that nothing beyond the fifth harmonic appeared in the DUT output, N could be as small as 11 samples--i.e., the sampling rate could be 11 ks/s. Typically, an FFT algorithm takes power of two samples to be correct, so N would be set to 16, and Fs = 16 ks/s.

Now let us expand the "unit" test to include more items. For example, we might apply 1000 and 3500 Hz. There is no physical change in the connections of Figure 7. The synthesizer is simply fed a new pattern (for two tones instead of one), and the FFT is asked to look for a few additional components--say, the gain at <u>two</u> frequencies, maybe the relative phase delay, and some IM products (e.g., 2500, 4500, 6000, 8000, etc.).

Be careful, however. The UTP is no longer 1 ms, or 1 cycle. It must be a time interval over which <u>both</u> tones produce a whole number of cycles, and over which <u>all IM products produce a whole number of cycles</u>. Furthermore, over this same UTP, the number of samples, N, must also be a value which meets all sampling and FFT requirements.

Let M represent the whole number of cycles for any specified tone. Each tone bears the relation

$$\frac{M}{N} = \frac{F_T}{F_S}$$

where F_T is the tone frequency and F_S is the sampling frequency. M and N are the <u>smallest integers that satisfy this ratio</u>.

In the single tone example, the ratio is 1000/16000, or 1/16. M = 1 cycle and N = 16 samples in the UTP.

If F_T = 3500 Hz, however, the story is different. For simplicity, assume that the device is band-limited to 8 KHz, so F_S may remain at 16 ks/s, as before. The ratio is

$$\frac{F_T}{F_S} = \frac{3500}{16000} \rightarrow \frac{7}{32} = \frac{M}{N}$$

M is now 7 cycles (at 3500 Hz) and N has doubled to 32 samples. During the UTP, the 1000 Hz component now completes two cycles.

The UTP is given by

$$UTP = N/F_S$$
$$= 32/16 \text{ ks/s}$$
$$= 2 \text{ ms}$$

But wait! Is this enough? Previously we indicated that non-linearity up to the fifth order might be tested. This means that the IM product,

$$(4 \times 1000) - (3500) = 500 \text{ Hz}$$

should also be considered. For this, M/N = 500/16,000, or 1/32. Luckily, 32 samples will still do the job.

It turns out that for any number of tones, if all the M_i have been made integers as required, the IM products will also have integer M. You need not bother to check all the combinations.

With good practice, <u>all</u> components are periodic over the UTP. If the device is an A/D converter, quantization distortion is intrinsically periodic over the UTP. The only component not so periodic is random noise.

The Primitive Frequency.

Ignoring random noise, the device response is thus a complex wave with a period equal to the UTP. Being periodic, it has a discrete frequency spectrum based on multiples of a fundamental frequency which is the reciprocal of the UTP. In most cases, this is not equal to any of the input test frequencies. To distinguish it from the other "fundamentals," it is often termed the primitive frequency, denoted by Δf, or simply Δ.

$$\Delta = 1/UTP$$
$$\Delta = F_s/N$$

In the above 2-tone test, $\Delta = 16000/32$, or 500 Hz. The IM products, as well as the test tones themselves are all multiples of Δ. In fact, each tone is given by

$$F_T = M * \Delta$$

For example, the previous M's were 2 for the low tone, and 7 for the high tone. The frequencies are thus 2Δ and 7Δ, or 1000 Hz and 3500 Hz.

M-Over-N Phase Lock

In coherent testing, it is not the frequencies that must be precise so much as the ratios. For example, a typical M/N ratio for digital audio testing is 23/1024. For a standard sampling rate of 44100 s/s, this gives a test tone frequency of

990.5273438 Hz

It is improbable that any conventional ATE audio source could be set to this frequency, let alone hold a consistent phase relationship. But with emulated hardware (i.e., using a digitizer and/or synthesizer) we would have only to construct a digital representation of 23 cycles of a pure sinusoid, over 1024 samples. With the synthesizer, digitizer and device all clocked together, everything stays in step and the test tone-- relative to F --will be exact.

It is often convenient to run one of these elements at a different clock rate than the others. To permit this, yet retain coherence, the different clocks will be locked by a phase-locked loop, or PLL (Figure 8).

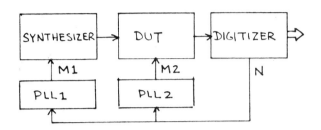

FIG. 8

These loops contain two or more registers to permit integer/ratio synchronization. Since these integers are related to M and N, this type of synchronization is often termed "M-over-N" synchronization.

There is one other period to consider: the period of the matrix operation, which serves to filter, measure or otherwise analyze the digitizer device response. For simple functions like RMS or average, this may be controlled software:

MAT B = AVG(A(45 to 244))

This example finds the average of that part of signal A from the 45th sample to the 244th--200 samples in all.

The classical Discrete Fourier Transform (DFT) measures phase and magnitude of only one spectral line at a time but has the advantage of adapting to any N. The FFT, in contrast, provides a look at the complete spectrum, but has the disadvantage of requiring N to be a power of 2. One must therefore be careful in using the FFT to ensure that the response components are jointly periodic with the FFT.

This condition is not satisfied simply because you set the digitizer to collect (say) 1024 samples--or because you fill out a shorter sequence with zeroes to make 1024 samples. You must ensure that the Unit Test Period has a value of N that either equals N(FFT) or is a submultiple.

If N(UTP) matches N(FFT), one may speak just of N samples without confusion. The FFT then creates bins, or little spectral boxes into which energy may fall--if any is present. These bins are spaced by Δ Hz. The lowest is 0 Hz, or DC, and the highest is the Nyquist frequency:

$$\text{MAXIMUM } M = N/2$$
$$\text{MAXIMUM } F_T = \Delta(N/2)$$
$$= F_s/2$$

Unfortunately, each bin can receive energy from nearly <u>any</u> place in the spectrum. That is, each one of the Fourier "filters" that supplies a bin, has a broad frequency response (Figure 9).

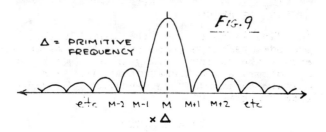

The main lobe has a width of 2Δ, while the other are Δ wide. To ensure that the only energy recorded in a bin is valid--i.e., actually occured at the 100% point--one must ensure that the UTP matches the FFT in sample length. This restricts the signal spectrum to sharp spectral lines spaced exactly by Δ. Then, <u>each Fourier filter can pass one line correctly, while blocking all others</u>.

So far we have seen how to create composite test signls whose components--and those of the device response--all produce whole numbers of cycles over a controlled interval--the UTP. We have also seen how to look at any particular output component, while completely blocking the others. Let us now put that to work in <u>Multitone Testing</u>.

Multitone Technique

By combining the previous techniques in a coherent test system, it is practical to measure virtually all the common transmission parameters at once. The technique is to apply all the frequencies of interest at once, as symbolically shown in Figure 10.

You recognize that coordination of real analog hardware would represent an impossible task. Emulated hardware presents no problem, however. Each of the K "tones" is locked into step with its neighbors because they are part of the digital pattern in the synthesizer RAM. The filters, of course, are actual Fourier filters, each with its own phase-indicating voltmeter.

Insofar as the synthesizer is concerned, K, the number of input tones, can be anything from 1 (a pure sinusoid) up to N/2. Remember, however, that the inputs must be chosen to avoid ambiguity at the output. We would not want to apply any frequency that would fall on the harmonic of another, for example. Here are a few guidelines.

1. The primitive frequency, Δ, is never part of the input set.

2. No M is a close multiple of any other M (a ninth harmonic might be okay, but not a second or third).

3. No M should be the sum or difference of any other two M's.

4. The peak value of the multitone complex must not overload the test device. Good practice is to set the input amplitude so that the peaks are the same as the peak value of a conventional (sine wave) test.

As a matter of interest, choosing <u>odd</u> Ms guarantees condition 3. <u>Prime</u> Ms will ensure both #2 and #3, explaining why prime multiples occur so often in tests of this kind.

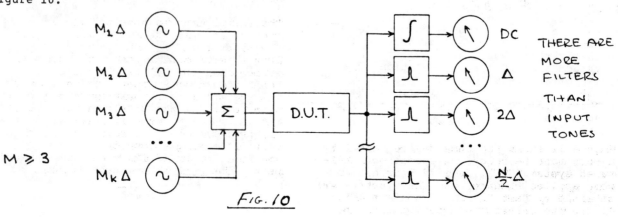

Figures 11 and 12 compare the results of testing an analog filter first by conventional hardware, then by coherent multitone techniques. Figure 11 shows the gain versus frequency response of a commercial telephone-grade filter (Intel 2912). The data for this plot was obtained by a conventional sweep method which, in ATE, actually means moving the input frequency from one discrete setting to the next, each time pausing long enough for the filter and the audio voltmeter to achieve steady-state AC response. In this test the average pause was 30 msec; since 79 test frequencies were used, the data for the curve took 2.4 seconds to gather.

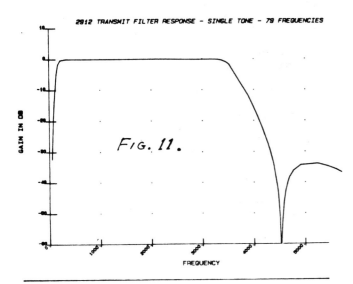

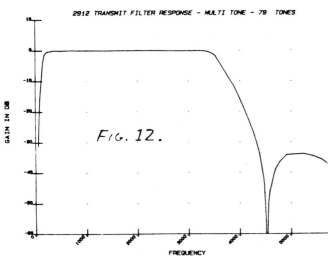

Figure 12 shows the same device tested by a multitone technique on a coherent, DSP-based system (LTX). All 79 frequencies were applied at once, and the spectrum was obtained by Fast Fourier Transform (FFT). If the two curves are superimposed, the differences are found to be smaller than the device-to-device variations, and the procedure is quite usable for production or inspection. The total time to exercise the filter, to execute the FFT, and to prepare the data for plotting was approximately 175 ms, versus 2.4 seconds for the "swept" response.

In addition to saving time, the FFT also provides relative phase delay (not shown) and, since no input falls on a harmonic of any other, a measurement of non-linear distortion at many frequencies may be made from the same FFT.

Random Noise

A useful trick for measuring random noise is to choose a UTP which is a submultiple of the FFT period. For example, the <u>Unit</u> N might be 256 sample, but we might gather 1024 samples, or 4 UTP, for the FFT. This produces (in this example) 4 times as many bins as there are periodic spectral components. For every bin in which an expected component might fall, there are 3 bins which are empty--unless non-periodic components like noise are present. To estimate the device noise, sum up the total power in these special lines, and multiply (in this example) by 4/3.

Limits of Accuracy

Assuming we measure only at frequencies that avoid interference, how accurately can we measure the individual tones after passing through the DUT? To answer this, assume that the synthesizer distortion is less than that of the DUT, and that the gain and phase errors of the fixture are removed by calibrating the synthesizer and digitizer directly together. In this case, the limit to digitizer accuracy is primarily the A/D non-linearity and quantization error as a function of K, the number of tones in the multi-tone input.

The synthesizer is presumed always to operate at full scale, and attenuation is placed at the DUT input to provide the largest multitone signal that does not quite drive the DUT to its clipping level. For K = 4 or more tones, the peak-to-RMS voltage ratio of a phase-optimized multitone signal is almost always between 10 and 12. With K tones of equal amplitude, all periodic over the same UTP, the RMS value of a single tone is, at the smallest, about $1/12K$ times the composite peak. The <u>peak</u> of one tone is thus

$$1/\sqrt{6K}$$

times the peak of the complex.

For 4 or more tones the signal-to-noise ratio of any individual frequency measurement is <u>reduced</u> by the amount

$$\approx 7.8 + 10 \log K \text{ (decibels)}$$

with respect to the signal-to-noise ratio obtained with the same synthesizer and digitizer used to measure one frequency at a time.

Signal-To-Noise

Let n represent the useful number of bits in the digitizer (usually 14 for a 16-bit A/D, or 13 for a 14-bit A/D). Assume the digitizer input gain is binary-ranged, and that the minimum signal thus operates at half-scale. The quantization noise falling into any frequency slot is generally smaller than the (individual) signal tone amplitude by at least

$$6n - 10 \log K \text{ (decibels)}$$

Amplitude Estimation Error

Under the same conditions, the A/D errors will cause an uncertainty in the measurement via FFT of any one of the dominant or strong lines generally less than the limit

$$\pm 11 \sqrt{K}/2^n \text{ (decibels)}$$

The uncertainty of weaker lines will increase roughly in inverse proportion to the amplitude.

Phase Estimation Error

Using the complex FFT to analyze lines, both the real and imaginary parts of a frequency component are subject to amplitude uncertainty. From the above equation one may approximate the phase uncertainty as

$$\pm 73 \sqrt{K}/2^n \text{ (degrees)}$$

Note that these equations pertain to tones which have not been attenuated in passing through the DUT. Roughly speaking, the uncertainty of an answer grows in inverse proportion to the amount of attenuation. A tone attenuated 20 dB (10 to 1, in voltage) would have a measurement uncertainty about ten times the amount given by the last two equations.

Calibration

These equations apply only to a well-calibrated system. In the analog-to-analog case this is easily accomplished by running the same multitone test twice: once at the <u>input</u> to the test device, and then at the <u>output</u>. It is only the difference in amplitude flatness and phase flatness which is recorder.

Summary

Coherent, multitone testing is presently used in production testing of newer and more complex circuits, particularly those that have built-in filters (combo codecs, DTMF receivers, audio reconstruction filters, etc.). Using this technique, the test speed approaches the limit for the type of DUT, yet allows simultaneous measurement of frequency response, envelope delay, IM distortion and harmonic distortion. The accuracy diminishes as the number of tones used, but with good A/D conversion of 12 bits or more, tests with 20 to 100 tones still exhibit enough accuracy to satisy most telecommunications standards. Because most of the analysis takes place mathematically, the resulting fixture is extremely versatile as well as very fast. The fixture may be used directly to evaluate D/A and A/D circuits via spectral means as well.

References

[1] M. Mahoney, "Multitone Testing—A Technique For Fast Automatic Measurement," Proc. Electronics Test and Measurement Conference, October 1982 (Morgan-Grampian, Publisher).

[2] B. Blesser, "Digitization of Audio," Journal of the A.E.S., October 1978, Volume 26, Number 10.

[3] R. Talimbiras, "Some Considerations in the Design of Wide-Dynamic-Range Audio Digitizing Systems," Proc. A.E.S. 57th convention, May 1977.

[4] B. M. Gordon, "Digital Sampling and Recovery of Analog Signals," EEE Magazine, May 1970.

[5] S. Stearns, <u>Digital Signal Analysis</u>, Hayden Book Co., New Jersey, 1975.

[6] Recommendations 0.81 and 0.131, <u>CCITT Yellow Book</u>, International Telecommunication Union, Geneva, 1981.

[7] "Transmission Parameters...," Bell System Publication 41009, AT&T, May 1975.

[8] F. Brglez, "Digital Signal Processing Considerations in Filter-Codec Testing," Proc. IEEE Test Conference, October, 1981.

CODEC TESTING USING SYNCHRONIZED ANALOG AND DIGITAL SIGNALS

Dr. Douglas K. Shirachi

Zehntel Automation Systems
2625 Shadelands Drive
Walnut Creek, California 94598

ABSTRACT

This paper describes the test requirements for operational verification of a CODEC circuit used for telecommunications printed circuit boards and illustrates how the test engineer may utilize five different in-circuit test methods to optimize maximum fault coverage with minimum test time.

INTRODUCTION

The purpose of a CODEC (coder/decoder) used for telephone communications systems is to provide bidirectional signal transformation between analog voice waveforms at the subscriber telephone handset and serial, digital data for the time-division-multiplexed (TDM) highway of a central office switch as shown in Figure 1. Conceptually, the CODEC acts as both an analog-to-digital and digital-to-analog convertor with filters to provide for noise rejection as well as optimum signal bandwidth utilization. The analog/digital conversion circuitry also combines nonlinear, amplification circuitry to compensate for line losses and to maximize the signal-to-noise ratio while minimizing the digital word size.

Looking at the internal functions of a CODEC device as shown in Figure 2, an analog voice signal with a 4 KHz bandwidth is first filtered by a bandpass filter with a passband between 200 Hz and 3.4 KHz. This filter also removes 50/60 Hz power line noise contamination. The resultant band-limited signal is then transformed by a nonlinear digital coding scheme into a digital PCM word which is in turn transmitted from the CODEC. By applying the Bell u-Law transfer characteristic as shown in Figure 3, a 78 dB amplitude range can be encoded by a 8-bit digital word but would require a 13-bit word if a linear conversion process were utilized.

Test requirements impose special signal coordination capabilities from the tester such as the ability to generate as well as measure both digital and analog waveform characteristics and precise synchronization of digital clock and data strobe signals. The waveform characteristics which are presented to the CODEC are analog voice signals with a 4 KHz bandwidth while the output is an eight-bit digital word sampled at an 8 KHz digitizing rate, giving an effective digital transmission rate of 64 K bits/second.

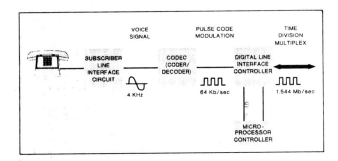

Figure 1. Digital Telephone Interface

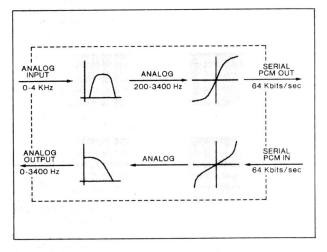

Figure 2. CODEC Functional Diagram

Reprinted from *The Proceedings of the 1984 Test Conference*, 1984, pages 447-454. Copyright © by The Institute of Electrical and Electronics Engineers, Inc.

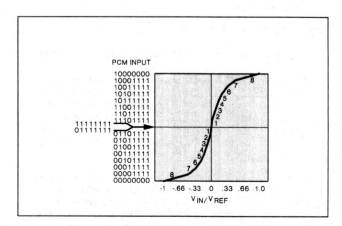

Figure 3. u-Law Transfer Characteristics

When a single CODEC is configured to be multiplexed with other CODECs (24 CODECs for the Bell D2 configuration and 32 CODECs for the CCITT CEPT configuration), then each bit of a PCM digital word is clocked out at a Bell 1.544 MHz or CCITT 2.048 MHz shift rate which appears as a data burst for each individual CODEC channel. The digitizing rate for any single channel remains at the 8 KHz sampling rate; however, the data strobe duty cycle for any particular channel changes from 50 percent to less than 2 percent.

A practical test plan should identify and examine those functions which are most important to the operation of the device-under-test in the given application, which may be restricted to a few operational verification tests, and determine the combination of test conditions which return the highest probabilities of functional integrity per unit of test time. Furthermore, the test plan may also require verification of conformance to Bell or CCITT standards as part of a product performance test. A balance in the test strategy between maximum test plan coverage for correct function and minimum total testing time.

Two test strategies come to mind when considering the utilization of both subsystem functional and in-circuit component tests. The first test strategy is to conduct an in-circuit board test to detect assembly errors prior to a functional test which is used for performance verification. The advantages of in-circuit testing are used to diagnose faults by testing individual components while the functional test technique can detect timing anomalies as well as incompatible circuit interaction at the board level. The second test strategy uses a functional test as a prescreening process for verifying proper board function, perhaps at the same time applying environmental thermal stress to detect marginal performance, and then using in-circuit test to perform component fault diagnosis only for boards which fail the functional test.

This paper discusses and demonstrates test solutions using synchronized analog and digital signals as well as IEEE-488 instruments for detailed inspection of the CODECs. The test techniques presented support the two test strategies described above by allowing the test engineer to apply a number of different test plans in accordance with the test strategy and equipment configuration. These methods also support configurational differences between manufacturers of CODEC circuitry who have their own unique packaging arrangements as well as timing constraints.

PRACTICE OF CODEC TESTING

In-circuit testing of CODECs should not duplicate the detailed component test function, but should determine if the device has maintained its operational functionality after assembly onto a printed-circuit board. Therefore, testing CODEC functional parameters such as absolute gain, gain tracking, signal-to-noise ratio, harmonic distortion, idle-channel noise, voltage-PCM transfer function, channel crosstalk, phase shift, and power supply rejection ratio are not necessary during in-circuit testing. Confirming basic operations such as the correct transformation between analog voltages and digital PCM codes and vice versa, or the measurement of analog waveform passbands through the filter elements may be all that is necessary to provide cost-effective, efficient solutions to testing.

Test Techniques

Test methodologies will be presented for testing the input/output functions of both the Coder and Decoder sections along with their signal conditioning filters. These methodologies are described in Table I and in Figure 4.

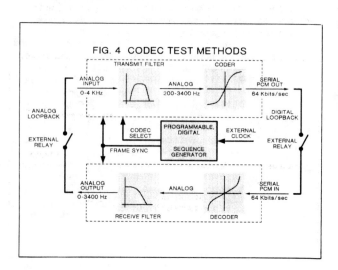

Figure 4. CODEC Test Methods

Table I. Codec Test Methods

Method	Input	Output
Coder	analog	digital
Decoder	digital	analog
Analog Loopback	digital	digital
Digital Loopback	analog	analog
Functional Instruments	IEEE-488	IEEE-488

The first two methods investigate individual packaging functions and are appropriate test techniques for a traditional in-circuit test approach. The next two techniques can be considered "quasi" functional tests since the test signal path between stimulus and response may traverse a number of different IC's, but not dynamically test the complete board operation. However, an acceptable test response may require that all of the IC devices within a defined board-level subsystem show operational compatibility amongst themselves. The last method allows for test configurations where detailed performance parameters such as harmonic distortion which are specific to the device being tested must be verified.

Coder Testing

Verification of the Coder and Transmit Filter sections of the CODEC is accomplished with an analog input/digital output method. If the Transmit Filter is packaged separately from the Coder, the Coder and Transmit Filter functions are tested separately with the Transmit Filter tested as an analog input/analog output device as shown in Figure 5.

An example of a filter test for an Intel 2912A PCM Transmit/Receive Filter as tested on a Zehntel in-circuit test system, is shown below. A low voltage, analog AC stimulus of 1.59 KHz (AC2) using a 10 ohm (R1) series resistance is presented to the filter input. The peak output voltage is measured and accepted if it is within a 0.4 volt tolerance band.

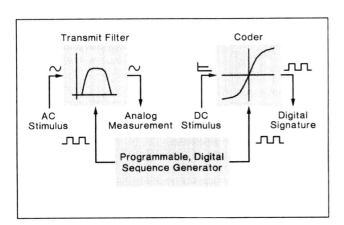

Figure 5. Transmitter Test Configuration

"INTEL 2912A FILTER"

Transmit Test

TEST V STIM V;

1.8-2.2V PK .5V AC2 R1 (INP)-(16)/(15),(11),(13);

Measure 1.8-2.2 volt peak output voltage at VFx (pin 16) and input 0 .5 volt, 1.59 KHz AC stimulus through a 10 ohm series resistor into filter input network (INP). Stimulus and measurements referenced to analog ground (pin 15) and digital ground (pin 11). PDN (pin 13) is set low (powered up state).

The Coder section can be given a DC voltage as its input and the appropriate PCM code, measured as a digital signature such as a Cyclic Redundancy Check (CRC), can be verified. Changing the amplitude of the DC input will change the output digital signature so that the non-linear transfer characteristics of the Coder can also be checked. For the test program shown below, a DC analog stimulus is presented to the Transmit Filter, control protocols which provide timing signals as shown in Figure 6 are defined and activated, and digital signatures are measured and compared to the acceptance criteria. Four different values for DC input voltages were chosen for transfer function verification. These voltages were required to be sufficiently large to be near the extreme ends of the semi-logrithmic transfer function curve; otherwise, the small-signal amplification characteristics of the curve places a premium on low-voltage stimulus accuracy. The CODEC clock line is synchronized to an external clock oscillator which is either a Bell 1.544 MHz or CCITT 2.048 MHz master clock reference. In this example, an Intel 2911A PCM CODEC - A Law is being tested, and a CCITT 32 channel system is simulated.

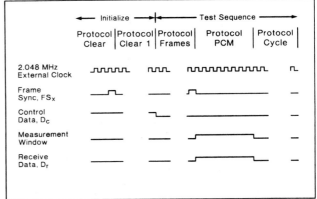

Figure 6. CODEC Timing Diagram

"INTEL 2911A CODEC"

TEST DIGITAL STIM V;

    Stimulate with analog and digital signals.

EXT CLK X;

    Synchronize to 2.048 MHz external clock.

CONTROL PROTOCOL CLEAR (8)

    Pattern length = 8 clock cycles long. Initialize CODEC with Frame Synchronization pulse to initiate TDM multiplexing. Clear internal registers.

CONTROL PROTOCOL CLEAR1 (8)

    Initialize CODEC to clear internal registers.

CONTROL PROTOCOL FRAMES (8)

    Generate Frame Sync pulse and select CODEC to be tested.

CONTROL PROTOCOL CYCLE(8)

    Timing for unused PCM time slots.

CONTROL PROTOCOL PCM (8)

    Measurement window is opened for 8 clock cycles to measure 8-bit PCM word transmitted after Frame Sync pulse is initiated.

SUBR DCHI

    Initialize CODEC.

CALL DCHI;

    CODEC initialization procedure.

STARTLOOP;

    STARTLOOP/ENDLOOP sequence repeats test until system timeout. Allows for about 500 stimulus measurement cycles.

INVOKE FRAMES;

    Generate Frame Sync pulse.

INVOKE PCM;

    Measure digital PCM.

REPEAT 30 TIMES;

    Fill remaining time slots with extra timing cycles.

INVOKE CYCLE;

END REPEAT;

ENDLOOP;

2.52V R1 (3)/(12),(5);

    Input 2.52 volt DC stimulus through a 10 ohm series resistor into Coder input (pin 3), referenced to digital ground (pin 12) and analog ground (pin5).

CRC HAC74; M(13) TO F16;

    Measure digital signature (hexidecimal AC74) at Dx output (pin 13) until F16 measurement window edge (16 msec.).

Verification of the Power Down (PDN) and Time Slot Transmit (TSx) outputs can be performed by specifying measurement parameters for these lines in a similar fashion to the CODEC PCM signatures.

Because of manufacturing tolerances between CODEC ICs and the necessity for remote sensing of signals through a test fixture, slight ground shifts or DC voltage offsets may occur, necessitating the use of alternative CRC signatures for a particular test. An example of a test is shown below.

2.85V R1 (3)/(12),(5);

    Input a 2.85 volt DC stimulus through a 10 ohm series resistor into Coder input (pin 3), referenced to digital ground (pin 12) and analog ground (pin 5).

CRC H0481 OR H34E2 OR H14A0; M(13) TO F16;

    Three different measurement signatures are valid at the Dx output (pin 13), and the OR statements accommodate these valid signatures.

<u>Decoder Testing</u>

The operation of the Decoder circuitry is verified by stimulating the input of the Decoder with a serial, digital sequence representing the PCM code for a desired DC output voltage. An AC signal can be simulated by a sequence of PCM codes representing time-sampled values of the desired waveform. Using a programmable, digital sequence generator, creating and presenting PCM codes in their proper order with appropriate timing becomes a reasonably easy task. Measurement of the resultant analog output can be performed using a digital voltmeter synchronized within a preprogrammed measurement time window. In the example shown below, the PCM code representing a 3 volt amplitude is presented to the decoder at the proper moment during the simulated TDM timeslot. The resultant conversion to an analog voltage is measured with a 0.13 volt tolerance by the test system's digital voltmeter.

"INTEL 2911A CODEC"

TEST V STIM DIGITAL;

    Stimulate with analog and digital signals.

EXT CLK X;

    Synchronize to 2.048 MHz external clock.

CONTROL PROTOCOL FRAMESR (8);

   Generate Frame Sync pulse with 0 volt PCM code as input into Dr (pin 7).

CONTROL PROTOCOL CYCLER (8);

   Timing cycles for unused PCM time slots.

CONTROL PROTOCOL PCMA (8);

   Presents digital PCM code for 3.0 volts (10101011).

STARTLOOP;

INVOKE FRAMESR;

   Generate Frame Sync pulse.

INVOKE PCMA;

   Generate 3.0 volt PCM code.

REPEAT 30 TIMES;

   Fill unused PCM time slots.

INVOKE CYCLER;

END REPEAT;

ENDLOOP;

2.93-3.06V (9)/(12),(5);

   Measure 2.93-3.06 volts DC at the Decoder analog output VFr (pin 9), referenced to digital ground (pin 12) and analog ground (pin 5)

Other voltages can be measured by substituting the appropriate digital PCM codes into CONTROL PROTOCOL PCMA or by defining additional CONTROL PROTOCOLS with the desired PCM codes and invoking them in sequence.

A test for the Receive Filter is shown below. It is similar to that for the Transmit Filter and differs only by the pin assignments and test voltages.

"INTEL 2912A FILTER"

   Receive test.

TEST V STIM V;

.8-1.2V PK 1V AC2 R1 (10)-(6)/(15),(11),13;

   Measure 0.8-1.2 volt peak output voltage at Receive Filter output (pin 6), referenced to analog ground (pin 15) and digital ground (pin 11). Input 1 volt, 1.59 KHz AC stimulus through a 10 ohm series resistor into Receive Filter input (pin 10). Power Down line (pin 13) is placed in low state (CODEC maintains powered up state).

Analog Loopback Method

The digital input/digital output method is called the Analog Loopback method because the analog output of the Decoder is fed back to the analog input of the Coder through an external relay. By connecting the analog output of the Coder to the analog input of the Decoder, the complete CODEC can be tested as a digital device as shown in Figure 7. A programmable, digital sequence generator is used to program the digital PCM input to the Decoder in the same manner as in the previous method, but in the present case, a digital CRC signature is measured at the digital output of the Coder rather than an analog parameter at the Decoder. A measurement node can still be placed at the analog output of the Decoder to monitor its output signal and to verify that the Decoder is operating properly.

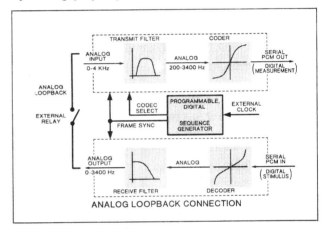

Figure 7. Analog Loopback Connection

Recently, a new generation of CODECs has been created which integrates the Transmit Filter, Receive Filter, Coder and Decoder into a single device; such an example is the AMI S3507 Single Chip Combo Codec With Filters which is the subject of the next example test plan. To test the AC operation of the filter circuitry, a sequence of digital PCM codes which simulate a 400 Hz squarewave input stimulus sampled at an 8 KHz rate can be presented to the CODEC Decoder section. After waiting the conversion time specified by the device manufacturer, the identical digital PCM bit sequence should appear at the output of the Coder section. A 400 Hz squarewave input stimulus was chosen to insure that the fundamental frequency and the important harmonics of this waveform lie within the passband of the Transmit Filter and Receive Filter.

The CODEC also requires digital strobe and clock signals to synchronize the signal conversions, switched capacitor filter timing, and PCM pulse widths. The rising edge of the Receive Strobe clocks the PCM data into the Decoder's input register, and the rising edge of the Transmit Strobe clocks the PCM data out the the Coder's output register. The receive and transmit clocks (a single, external 1.544 MHz clock) control the timing interval of each PCM data bit. This CODEC test simulates a 24 channel Bell configuration.

In the following program listing, a sequence of control protocols define a square wave by providing positive, zero and negative signal amplitudes. The square wave is composed of 42 periodic waveform samples created by 20 positive amplitude samples, 20 negative amplitude samples, and 2 zero amplitude samples.

"AMI S3507 CODEC"

    CODEC Test: Analog Loopback (Digital In - Digital Out)

SET RELAY 5; WAIT 300MS;

    Connect analog loopback through user relay.

DIGITAL;

EXT CLK X*;

    Sync to falling edge of external 1.544 MHz clock.

CONTROL PROTOCOL CLEAR (193);

    Clear CODEC internal registers.

CONTROL PROTOCOL ZERO (193);

    Input zero amplitude (PCM = 11111111).

CONTROL PROTOCOL POS (193);

    Input positive amplitude (PCM = 10001100)

CONTROL PROTOCOL NEG (193);

    Input negative amplitude (PCM = 00001100)

INVOKE CLEAR;

    Clear registers and commence 400 Hz squarewave.

INVOKE ZERO;

    1 zero sample.

REPEAT 20 TIMES;

INVOKE POS;

    20 positive samples.

END REPEAT;

INVOKE ZERO;

    1 zero sample.

REPEAT 20 TIMES;

INVOKE NEG;

    20 negative samples.

END REPEAT;

INVOKE ZERO;

    1 zero sample.

CRC H923E;M(6) TO F16;

    Measure digital signature (hexidecimal 923E) at PCM OUT (pin 6) until F16 measurement window edge (16 msec.).

The device can also be tested for other squarewave amplitudes by changing the positive and negative PCM codes along with the expected digital CRC signatures.

Digital Loopback Method

The converse of the above method is called Digital Loopback method which consists of wiring the digital PCM output of the Coder to the digital input of the Decoder through an external relay. Analog AC or DC signals (depending upon filter integration into the device packaging) can be programmed to stimulate the Coder, and the resultant analog signal measured to verify proper CODEC operation. Again, an intermediate node can be used to measure the CRC signature at the digital loopback connection if it is desirable to differentiate between Coder or Decoder malfunctions.

The example shown below test the Intel 2912A PCM Filter and Intel 2911A PCM CODEC as a single functional unit using an external 2.048 MHz clock oscillator. The 0.5 volt, 1.59 KHz sinusoidal analog stimulus is presented to the input resistor network of the Transmit Filter and the peak analog voltage is measured at the Receive Filter output.

"CODEC ANALOG LOOPBACK"

    Intel 2912/2911 subsystem test.

TEST V STIM DIGITAL

SET RELAY 5; WAIT 300 MS;

    Connect Dx to Dr via external user relay.

EXT CLK X;

    Synchronize to 2.048 MHz external clock.

CONTROL PROTOCOL CLEARF (8);

    Provide Frame Sync pulse to initiate TDM multiplexing process. Clear internal registers.

CONTROL PROTOCOL CLEARF1 (8);

    Initialize CODEC to clear internal registers.

CONTROL PROTOCOL FRAMESF (8);

    Provide Frame Sync pulse, select CODEC device to be tested.

CONTROL PROTOCOL CYCLEF (8);

    Timing for unused PCM time slots.

```
SUBR DCHIF;

INVOKE CLEARF;

    Initialize CODEC.

REPEAT 31 TIMES;

    Clear internal registers for 1 frame.

INVOKE CLEARF1;

END REPEAT;

CALL DCHIF;

    Initialize CODEC internal registers.

STARTLOOP;

INVOKE FRAMESF;

    Generate Frame Sync.

REPEAT 31 TIMES;

    Fill remaining time slots with extra timing
    cycles.

INVOKE CYCLEF;

END REPEAT;

ENDLOOP;

1.6-2.0V PK .5V AC2 R2
(INP)-(6)/(11),(15),(12),(5);
```

Measure 1.6-2.0 volt peak output voltage at Receive driver output (pin 6). Input 0 .5 volt, 1.59 KHz AC stimulus through a 10 ohm series resistor filter input network (INP), referenced to analog ground (2912 pin 15, 2911 pin 5) and digital ground (2912 pin 11, 2911 pin 12).

Functional Instruments Method

Most in-circuit testers support functional IEEE-488 instruments as an option to supplement standard tester capabilities. The IEEE-488 instrument bus permits flexibility for configuring additional test instruments to perform detailed analyses of certain device parameters which are difficult to measure by the standard features of the tester such as PCM time-slot scheduling or waveform harmonic distortion. Any instrument which supports the IEEE-488 standard is a candidate for inclusion into the standard in-circuit board test system.

The test program described here is an example of using an IEEE-controlled function generator and counter/timer to test the Coder section of a CODEC by applying a sinusoidal stimulus and measuring the resultant PCM frequency.

```
"IEEE-488 CODEC TEST"

    CODEC test using IEEE-488 (GPIB) instruments

SET RELAY 11; WAIT 300MS;

    Enable external 1.544 MHz oscillator.

TEST GPIB STIM GPIB;

CLEAR GPIB ALL;

    Clear GPIB interface.

TALK 6,("G1B0C0D0F1E3A200E1H");

    Set function generator with 4 volt p-p, 1 KHz
    sinusoid.

TRIGGER 6;

    Trigger function generator.

TALK 5,("C");

    Clear counter/timer.

TALK 5,("F0R3A+");

    Measure CODER output frequency.
```

SUMMARY

Five methods for testing CODECs have been described which permit the test engineer flexibility in designing test plans which integrate the following features: (1) adaptation to test strategies which integrate functional and in-circuit test methods, (2) alternatives for optimizing between maximum fault coverage and minimum test time, and (3) allowance for configurational differences between IC manufacturers.

One may extend the basic CODEC test approach with slight modifications to other types of circuits such as digital filters or modems where synchronized analog and digital waveforms are required for both signal transmission and device control. The key element required for comprehensive testing is the ability to precisely synchronize and cue the presentations of analog and digital signals in such a manner that the device being tested believes that it is operating in its intended circuit environment rather than the simulated configuration of a tester.

ACKNOWLEDGEMENTS

The author would like to thank Jeff Phillips of Zehntel, Inc., Paul Smith of Zehntel, Ltd., and Klaus Schwerbrock of Zehntel, GmbH. for their cooperation and contributions to this paper.

REFERENCES

[1] AMI Telecommunications Design Manual, American Microsystems, Inc.

[2] Phillips, Jeff, "Testing CODEC Networks", Test!, March, 1984.

[3] Schwerbrock, Klaus, "In-Circuit Test von CODECs", Zehntel Applikations-Notiz, Number 123, January, 1984.

[4] Shirachi, Douglas and Brian Crosby, "CODECs - A Challenge to In-Circuit Testing", Electronics Test, June, 1984.

[5] Smith, Jim, "Digital Telephony and the Integrated Circuit CODEC", Hybrid Products Databook, National Semiconductor, Inc.

[6] "Telecommunications", Signetics Analog Applications Manual.

Chapter Five: Testable Design and Built-In Self-Test

The definitions of testability are numerous; some are formal, others informal. An informal definition of testability follows: A logic circuit is "testable" if a set of test patterns that guarantees detection, and location if required, of a predefined set of fault conditions can be generated, evaluated, and applied cost effectively. A more formal definition of testability is based on the concepts of the "controllability" and "observability" features of a circuit design. Note that an internal node is controlled from the primary inputs and observed at the primary outputs and the process of testing relies on the ability both to control and to observe each node in the circuit.

The inclusion of on-chip circuitry for testing is called built-in self-test (BIST). Such a circuitry is responsible for test generation, response evaluation, and test application. All BIST methods have some associated cost. Since the self-test circuitry uses some chip area, the chip's yield and reliability will both be affected. These costs are compensated for by a reduced testing and maintenance cost. The added cost due to the BIST circuitry should be less than the saving in life-cycle costs so that the use of the BIST is justified.

5.1 Design for Testability

Historically, design engineers have used logic and design verification simulators whereas test engineers have used fault simulators and test-pattern generators. The historical separation of design and testing has resulted in logic designs that made little or no concession to testability requirements. With the advent of VLSI, the separation of design and test engineering should be reexamined.

Structured design techniques for testability, such as level-sensitive scan design (LSSD), have been extremely successful in producing designs that are testable. An interesting question is whether these techniques can exploit new technologies, such as CMOS, and work is needed in this particular area. The trade-offs of hardware overhead for testability with test-generation time also need to be explored because the time for test generation is a one-time overhead, whereas additional hardware goes onto every copy of the system. Also promising is research into the development of regular structures, such as arrays, which are easy to test.

5.2 Built-in Self-Test

Self-testing schemes can be categorized as concurrent or nonconcurrent. A concurrent (on-line) scheme aims at detecting failures during the normal circuit operation (see section 1.7), while in a nonconcurrent (off-line) strategy, the normal operation is suspended before the test can be applied.

A means for economically handling the problems of test pattern generation and response evaluation are of extreme importance to the successful implementation of an (off-line) BIST scheme. One effective method for managing these problems is the built-in logic block observer (BILBO) technique, in which the test pattern source is a linear feedback shift register (LFSR), and the response evaluator is a multiple input signature register (MISR).

The BILBO technique is quite suitable for chip-level self-test. An extension of such a scheme to the case of multichip modules is the STUMPS strategy. The source of test patterns is, again, an LFSR. The latch outputs of the source are connected to the scan-in ports of the individual chips. The scan-out ports of the chips are connected through AND gates to the MISR. The AND gates are controlled by enable/disable lines. These control lines allow any subset of the chips to be selected for testing. When all AND gates are enabled, the outcome of the test indicates whether or not the multichip module operates correctly. If the multichip module fails the test, it is possible to rerun the test with only a subset of the chips selected, by properly enabling the desired AND gates and disabling the rest. Thus, by a simple binary sort, the faulty chip can be isolated.

A drawback of the previous scheme is that testing the interconnections, either those between the chip pins and bonding pads or those between chips, is difficult. This problem is resolved if an external or boundary scan path is included on the chip.

5.3 Reprints

A VLSI circuit is "testable" if a set of test patterns can be generated, evaluated, and applied so that predefined levels of performance, defined in terms of fault-detection, fault-location, and test application criteria, are satisfied within a predefined cost budget and time scale. The reprinted paper by Williams and Parker surveys some of the design for testability (DFT) approaches that have been put forward. Most DFT practices are built upon the concept that if the value in all the circuit latches can be controlled and observed in a straightforward manner, then the testing problem of a general sequential circuit can be reduced to that of a combinational logic network. For a more detailed discussion of such techniques the reader is referred to the paper by McCluskey [1].

Potential problems in the design of tests to detect stuck-open faults in CMOS circuits are identified in the paper by Reddy et al. [2], who give methods to modify certain given circuits and to design new circuits that are testable for stuck-open faults.

To test for a fault, one "controls" and "observes." Controllability is a measure of the degree to which one can control the logic states of the internal nodes from the primary inputs (chip input pins). Observability is a measure of the degree to which one can discern the logic states of the internal nodes at the primary outputs. The paper by Goldstein [3] presents a method for analyzing digital circuits in terms of six functions that characterize combinational and sequential controllability and observability. Savir, in his paper reprinted here, shows that in general no conclusion can be drawn regarding the testability of the circuit, based strictly on the examination of the controllability and observability figures. More discussion on testability measures is given in the reprinted paper by Agrawal and Mercer.

Savir, in his second reprinted paper, introduces the notion of syndrome-testing, which is based on counting the number of ones realized by a Boolean function and comparing it to the fault-free count. Since there may be circuits and faults for which these fault-free and faulty syndromes are the same, modifications to produce a testable design are generally required. Markowsky shows in his reprinted paper that a combinational circuit can always be modified to produce a single-fault, syndrome-testable circuit.

In general, syndrome-testable combinational circuits require some pin penalty and maybe some logic for producing the testable design. Savir shows in his third reprinted paper how this penalty associated with the testable design can be traded off with the extra running time of the syndrome-test procedure.

The method of syndrome testing is applicable to VLSI testing since it does not require test generation and fault simulation. Barzilai et al. in their reprinted paper show how a VLSI chip can be electronically partitioned into multi-input, multi-output macros in test mode. The macros are syndrome tested in sequence. These researchers investigated ways to reduce the number of references needed for testing, by using weighted syndrome sums. A self-test architecture based on their approach is also presented.

The approach discussed in the paper by Miller and Muzio [4] is closely related to Savir's notion of syndrome testability. They present a new characterization of syndrome testability and an alternative method for dealing with syndrome-untestable faults.

The increasing complexity of VLSI systems has made the amount of data to be handled during their testing very large. One way of alleviating the problem is to employ efficient data-compression techniques. LFSRs can be used for compressing the test response data. This compressed response is known as the "signature." For a tutorial introduction to signature analysis, the reader is referred to the paper by Abadir and Reghbati in Chapter One. The paper by Smith [5] characterizes classes of dependent errors that are likely to occur and determines the likelihood of their detection. To improve the fault detection probability, Hassan and McCluskey propose to use more than one signature [6].

Hassan [7] proposes a method for using signature analysis to test sequential machines. In order to make a given sequential machine signature testable, it must be augmented by introducing an extra input and some logic.

Parity testing, verification testing, combined LFSR and shift register (LFSR/SR) testing, and condensed LFSR testing are among the other techniques that have been recently proposed for self-testing. A discussion of parity testing is given in the paper by Carter [8]. It is based on calculating the odd parity realized by a Boolean equation. Both parity and syndrome testing are referred to as syndrome driver counter (SDC) testing, since SDCs are used for test generation. SDC testing schemes can be considered as self-testing techniques based on functional partitioning, since they are concerned with circuit input-output functional (not structural) dependence by Boolean equations.

Verification testing is based on structural partitioning to generate exhaustive test patterns by constant weight counters (CWCs). It is referred to as CWC testing. There are two advantages in using the structural rather than the actual functional dependence: better fault coverage and less computation.

Combined LFSR/SR testing is based on linear codes to generate test patterns for pseudo-exhaustive testing [9,10]. Recently Akers has advocated a different approach to this problem. In his paper [11] he examines the possibility of introducing linear sums of the test signals for achieving locally exhaustive testing.

The condensed LFSR testing technique [12] increases VLSI testing efficiency by condensing the length of test patterns based on linear codes and structural partitioning. It uses LFSRs to generate test patterns and is applicable to any given combinational network. Since it does not rewire the original network inputs during self-testing, the possibility of undetected faults on some inputs is eliminated.

The BILBO is a universal element for use in a scan path and/or a self-test environment. The element is capable of performing a number of different functions according to the values placed on two mode-control lines. The survey paper by Williams and Parker and the paper by

Koenemann, Zwiehoff, and Bosch, both reprinted in this chapter, describe the BILBO. For a detailed discussion of the BIST techniques the reader is referred to the two reprinted papers by McCluskey.

Eichelberger and Lindbloom [13] face the questions of test coverage and diagnosis for random-pattern LSSD logic self-test. First, they discuss a general structure for self-test; then circuit modifications that render logic designs random-pattern testable are described. Next, modifications for some particular random-pattern resistant networks are discussed. Also given is a procedure for practicing net-level diagnosis with LSSD random-pattern self-test.

In Chapter Four, testing of iterative logic arrays (ILAs) was discussed. We saw that C-testable iterative arrays have very simple test structures, independent of the length of the arrays. Aboulhamid and Cerny show in their reprinted paper that all C-testable arrays are also pI-testable (partitioned I-testable). An ILA is pI-testable if it can be partitioned into blocks of equal length such that all blocks have the same test response, and such that the number of tests is independent of the ILA's size. If an ILA is pI-testable, in many cases it yields rather simple built-in testing structures, both for the test generator and for the response verifier.

5.3.1 BIST and Industry

The initial practical development of scan-design and BIST techniques was carried out largely by electronics systems companies, such as IBM and HP, with captive integrated circuit (IC) manufacturing capabilities, or by universities [14,15]—not by the IC makers themselves. This had two effects. The first was that individual companies produced their own scan-design variants (as you can see in the paper by Williams and Parker reprinted here) and the second was that implementation of scan techniques at the printed-circuit board (PCB) level, using standard cataloged devices, was considered to be costly.

The situation is now changing [16]. For example, the microbit (for the microprocessor built-in test) concept was developed recently at Siemens [17]. Microbit is a general concept for testing printed-circuit boards that contain a microprocessor or any microprocessor-based system. In this self-testing technique, two BILBOs plus a small amount of test circuitry must be added to a microcomputer board. Also, a special test program is stored in the processor's programmable read-only memory. The program sequentially tests all portions of the microcomputer.

As another example, Storage Technology Corporation has developed an LSI self-testing method using LSSD circuits already designed onto a gate-array chip [18]. Pseudorandom test patterns are generated by combining some of the LSSD shift-register latches into a shift register of maximal length. Then the pattern responses of the on-chip logic are analyzed by a BILBO-like built-in signature register.

IC manufacturers are beginning to produce standard devices with in-built scan facilities, the first of which was announced by Advanced Micro Devices (AMD) and Monolithic Memories [19]. This is the Advanced Micro Devices Am29818 serial shadow register (SSR) device. The Monolithic Memories device is the SN54/74S818 and, although internally different from the Am29818, both devices are interchangable, pin for pin. The SSR device is the first off-the-shelf general purpose device with in-built scan facilities. AMD anticipates the production of further devices, such as memories, PLAs, and gate arrays, that utilize the SSR concept [20]. An example of the use of such devices for testing is given in the paper by Agrawal [21].

Kuban and Bruce of Motorola [22] describe one of the first applications of the BIST approach to a commercial microcomputer, in which they obtained very high fault coverage for a very low cost in silicon area. Another success story is the LOCST (LSSD on-chip self-test) technique at IBM, which utilizes on-chip pseudorandom-pattern generation, on-chip signature analysis, a boundary scan feature, and an on-chip monitor test controller. This method is described in the reprinted paper by LeBlanc.

References

[1] E.J. McCluskey, "A Survey of Design for Testability Scan Techniques," *VLSI Design,* Vol. 5, No. 12, Dec. 1984, pp. 38–61.

[2] S.M. Reddy et al., "On Testable Design of CMOS Logic Circuits," *Proc. IEEE Test Conf.,* 1983, pp. 435–445.

[3] L.H. Goldstein, "Controllability/Observability Analysis for Digital Circuits," *IEEE Trans. on Circuits and Systems,* Vol. CAS-26, No. 9, Sept. 1979, pp. 685–693.

[4] D.M. Miller and J.C. Muzio, "Spectral Fault Signatures for Single Stuck-at Faults in Combinational Networks," *IEEE Trans. Comput.,* Vol. C-33, No. 8, Aug. 1984, pp. 765–769.

[5] J.E. Smith, "Measures of the Effectiveness of Fault Signature Analysis," *IEEE Trans. on Comput.,* Vol. C-29, No. 6, June 1980, pp. 510–514.

[6] S.Z. Hassan and E.J. McCluskey, "Increased Fault Coverage Through Multiple Signatures," *Proc. of FTCS-14,* IEEE, 1984, pp. 354–359.

[7] S.Z. Hassan, "Signature Testing of Sequential Machines," *IEEE Trans. on Comput.,* Vol. C-33, No. 8, Aug. 1984, pp. 762–764.

[8] W.C. Carter, "The Ubiquitous Parity Bit," *Proc. of FTCS-12,* 1982, pp. 289–296.

[9] D.T. Tang and C.L. Chen, "Logic Test Pattern Generation Using Linear Codes," *Proc. of FTCS-13,* IEEE, 1983, pp. 222–226.

[10] Z. Barzilai, D. Coppersmith, and A. Rosenberg, "Exhaustive Generation of Bit Patterns with Applications to VLSI Self-Testing," *IEEE Trans. Comput.*, Vol. C-32, No. 2, Feb. 1983, pp. 190-194.

[11] S. Akers, "On the Use of Linear Sums in Exhaustive Testing," *Proc. FTCS-15*, 1985, pp. 148-153.

[12] L-T. Wang and E.J. McCluskey, "A New Condensed Linear Feedback Shift Register Design for VLSI/System Testing," *Proc. of FTCS-14*, IEEE, 1984, pp. 360-365.

[13] E.B. Eichelberger and E. Lindbloom, "Random-Pattern Coverage Enhancement and Diagnosis for LSSD Logic Self-Test," *IBM J.R.D.*, Vol. 27, No. 3, 1983, pp. 265-272.

[14] J-L. Rainard and Y-J. Verney, "A 16-Bit Self-Testing Multiplier," *IEEE J. Solid-State Circuits*, Vol. SC-16, No. 3, June 1981, pp. 174-179.

[15] M.M. Tsao et al., "C.Fast: A Fault Tolerant and Self Testing Microprocessor," *Proc. of the CMU Conference on VLSI Systems and Computations*, Computer Science Press, Rockville, MD, 1981, pp. 357-366.

[16] D.R. Resnick, "Testability and Maintainability with a New 6K Gate Array," *VLSI Design*, March/April 1983, pp. 442-451.

[17] P.P. Fasang, "Microbit Brings Self-Testing on Board Complex Microcomputers," *Electronics Magazine*, McGraw-Hill, New York, NY, March 10, 1983, pp. 116-119.

[18] D. Komonytsky, "Synthesis of Techniques Creates Complete System Self-Test," *Electronics Magazine*, McGraw-Hill, New York, NY, March 10, 1983, pp. 110-115.

[19] F. Lee, V. Coli, and W. Miller, "On Chip Circuitry Reveals System's Logic States," *Electronic Design*, Vol. 31, No. 8, April 14, 1983, pp. 119-124.

[20] AMD, "On-Chip Diagnostics Handbook," *Advanced Micro Devices*, Sunnyvale, Calif., March 1985.

[21] O. Agrawal, "Registered PROMs Help Identify System Faults," *Computer Design*, April 1985, pp. 157-162.

[22] J.R. Kuban and W.C. Bruce, "Self-Testing the Motorola MC6804P2," *IEEE Design & Test*, Vol. 1, No. 2, May 1984, pp. 33-41.

Design for Testability—A Survey

THOMAS W. WILLIAMS, MEMBER, IEEE, AND KENNETH P. PARKER, MEMBER, IEEE

Invited Paper

Abstract—This paper discusses the basics of design for testability. A short review of testing is given along with some reasons why one should test. The different techniques of design for testability are discussed in detail. These include techniques which can be applied to today's technologies and techniques which have been recently introduced and will soon appear in new designs.

I. INTRODUCTION

INTEGRATED Circuit Technology is now moving from Large-Scale Integration (LSI) to Very-Large-Scale Integration (VLSI). This increase in gate count, which now can be as much as factors of three to five times, has also brought a decrease in gate costs, along with improvements in performance. All these attributes of VLSI are welcomed by the industry. However, a problem never adequately solved by LSI is still with us and is getting much worse: the problem of determining, in a cost-effective way, whether a component, module, or board has been manufactured correctly [1]–[3], [52]–[68].

The testing problem has two major facets:

1) test generation [74]–[99]
2) test verification [100]–[114].

Test generation is the process of enumerating stimuli for a circuit which will demonstrate its correct operation. Test verification is the process of proving that a set of tests are effective towards this end. To date, formal proof has been impossible in practice. Fault simulation has been our best alternative, yielding a quantitative measure of test effectiveness. With the vast increase in circuit density, the ability to generate test patterns automatically and conduct fault simulation with these patterns has drastically waned. As a result, some manufacturers are foregoing these more rigorous approaches and are accepting the risks of shipping a defective product. One general approach to addressing this problem is embodied in a collection of techniques known as "Design for Testability" [12]–[35].

Design for Testability initially attracted interest in connection with LSI designs. Today, in the context of VLSI, the phrase is gaining even more currency. The collection of techniques that comprise Design for Testability are, in some cases, general guidelines; in other cases, they are hard and fast design rules. Together, they can be regarded essentially as a menu of techniques, each with its associated cost of implementation and return on investment. The purpose of this paper is to present the basic concepts in testing, beginning with the fault models and carrying through to the different techniques associated with Design for Testability which are known today in the public sector. The design for testability techniques are divided into two categories [10]. The first category is that of the ad hoc technique for solving the testing problem. These techniques solve a problem for a given design and are not generally applicable to all designs. This is contrasted with the second category of structured approaches. These techniques are generally applicable and usually involve a set of design rules by which designs are implemented. The objective of a structured approach is to reduce the sequential complexity of a network to aid test generation and test verification.

The first ad hoc approach is partitioning [13], [17], [23], [26]. Partitioning is the ability to disconnect one portion of a network from another portion of a network in order to make testing easier. The next approach which is used at the board level is that of adding extra test points [23], [24]. The third ad hoc approach is that of Bus Architecture Systems [12], [27]. This is similar to the partitioning approach and allows one to divide and conquer—that is, to be able to reduce the network to smaller subnetworks which are much more manageable. These subnetworks are not necessarily designed with any design for testability in mind. The forth technique which bridges both the structured approach and the ad hoc approach is that of Signature Analysis [12], [27], [33], [55]. Signature Analysis requires some design rules at the board level, but is not directed at the same objective as the structure approaches are—that is, the ability to observe and control the state variables of a sequential machine.

For structured approaches, there are essentially four categories which will be discussed—the first of which is a multiplexer technique [14], [21], Random Access Scan, that has been recently published and has been used, to some extent, by others before. The next techniques are those of the Level-Sensitive Scan Design (LSSD) [16], [18]–[20], [34], [35] approach and the Scan Path approach which will be discussed in detail. These techniques allow the test generation problem to be completely reduced to one of generating tests for combinational logic. Another approach which will be discussed is that of the Scan/Set Logic [31]. This is similar to the LSSD approach and the Scan Path approach since shift registers are used to load and unload data. However, these shift registers are not part of the system data path and all system latches are not necessarily controllable and observable via the shift register. The fourth approach which will be discussed is that of Built-In Logic Block Observation (BILBO) [25] which has just recently been proposed. This technique has the attributes of both the LSSD network and Scan Path network, the ability to separate the network into combinational and sequential parts, and has the attribute of Signature Analysis—that is, employing linear feedback shift registers.

For each of the techniques described under the structured approach, the constraints, as well as various ways in which

---

Manuscript received June 14, 1982; revised September 15, 1982.
T. W. Williams is with IBM, General Technology Division, Boulder, CO 80302.
K. P. Parker is with Hewlett-Packard, Loveland Instrument Division, Loveland, CO 80537.

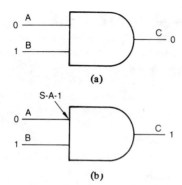

Fig. 1. Test for input stuck at fault. (a) Fault-free AND gate (good machine). (b) Faulty AND gate (faulty machine).

they can be exploited in design, manufacturing, testing, and field servicing will be described. The basic storage devices and the general logic structure resulting from the design constraints will be described in detail. The important question of how much it costs in logic gates and operating speed will be discussed qualitatively. All the structured approaches essentially allow the controllability and observability of the state variables in the sequential machine. In essence, then, test generation and fault simulation can be directed more at a combinational network, rather than at a sequential network.

A. Definitions and Assumptions

A model of faults which is used throughout the industry that does not take into account all possible defects, but is a more global type of model, is the Stuck-At model. The Stuck-At model [1]–[3], [9], [11] assumes that a logic gate input or output is fixed to either a logic 0 or a logic 1. Fig. 1(a) shows an AND gate which is fault-free. Fig. 1(b) shows an AND gate with input "A," Stuck-At-1 (S-A-1).

The faulty AND gate perceives the "A" input as 1, irrespective of the logic value placed on the input. The pattern applied to the fault-free AND gates in Fig. 1 has an output value of 0 since the input is 0 on the "A" input and 1 on the "B" input, and the AND'ing of those two leads to a 0 on the output. The pattern in Fig. 1(b) shows an output of 1, since the "A" input is perceived as a 1 even though a 0 is applied to that input. The 1 on the "B" input is perceived as a 1, and the results are AND'ed together to give a 1 output. Therefore, the pattern shown in Fig. 1(a) and (b) is a test for the "A" input, S-A-1, since there is a difference between the faulty gate (faulty machine) and the good gate (good machine). This pattern 01 on the "A" and "B" inputs, respectively, is considered a test because the good machine responds differently from the faulty machine. If they had the same response then that pattern would not have constituted a test for that fault.

If a network contained N nets, any net may be good, Stuck-At 1 or Stuck-At 0; thus all possible network state combinations would be 3^N. A network with 100 nets, then, would contain 5×10^{47} different combinations of faults. This would be far too many faults to assume. The run time of any program trying to generate tests or fault simulate tests for this kind of design would be impractical.

Therefore, the industry, for many years, has clung to the single Stuck-At fault assumption. That is, a good machine will have no faults. The faulty machines that are assumed will have one, and only one, of the stuck faults. In other words, all faults taken two at a time are not assumed, nor are all faults taken three at a time, etc. History has proven that the single Stuck-At fault assumption, in prior technologies, has been adequate. However, there could be some problems in LSI—particularly with CMOS using the single Stuck-At fault assumption.

The problem with CMOS is that there are a number of faults which could change a combinational network into a sequential network. Therefore, the combinational patterns are no longer effective in testing the network in all cases. It still remains to be seen whether, in fact, the single Stuck-At fault assumption will survive the CMOS problems.

Also, the single Stuck-At fault assumption does not, in general, cover the bridging faults [43] that may occur. Historically again, bridging faults have been detected by having a high level—that is, in the high 90 percent—single Stuck-At fault coverage, where the single Stuck-At fault coverage is defined to be the number of faults that are tested divided by the number of faults that are assumed.

B. The VLSI Testing Problem

The VLSI testing problem is the sum of a number of problems. All the problems, in the final analysis, relate to the cost of doing business (dealt with in the following section). There are two basic problem areas:

1) test generation
2) test verification via fault simulation.

With respect to test generation, the problem is that as logic networks get larger, the ability to generate tests automatically is becoming more and more difficult.

The second facet of the VLSI testing problem is the difficulty in fault simulating the test patterns. Fault simulation is that process by which the fault coverage is determined for a specific set of input test patterns. In particular, at the conclusion of the fault simulation, every fault that is detected by the given pattern set is listed. For a given logic network with 1000 two-input logic gates, the maximum number of single Stuck-At faults which can be assumed is 6000. Some reduction in the number of single Stuck-At faults can be achieved by fault equivalencing [36], [38], [41], [42], [47]. However, the number of single Stuck-At faults needed to be assumed is about 3000. Fault simulation, then, is the process of applying every given test pattern to a fault-free machine and to each of the 3000 copies of the good machine containing one, and only one, of the single Stuck-At faults. Thus fault simulation, with respect to run time, is similar to doing 3001 good machine simulations.

Techniques are available to reduce the complexity of fault simulation, however, it still is a very time-consuming, and hence, expensive task [96], [104], [105], [107], [110], [112]–[114].

It has been observed that the computer run time to do test [80] generation and fault simulation is approximately proportional to the number of logic gates to the power of 3;[1] hence, small increases in gate count will yield quickly increasing run times. Equation (1)

---

[1]The value of the exponent given here (3) is perhaps pessimistic in some cases. Other analyses have used the value 2 instead. A quick rationale goes as follows: with a linear increase k in circuit size comes an attendant linear increase in the number of failure mechanisms (now yielding k squared increase in work). Also, as circuits become larger, they tend to become more strongly connected such that a given block is effected by more blocks and even itself. This causes more work to be done in a range we feel to be k cubed. This fairly nebulous concept of connectivity seems to be the cause for debate on whether the exponent should be 3 or some other value.

$$T = KN^3 \qquad (1)$$

shows this relationship, where T is computer run time, N is the number of gates, and K is the proportionality constant. The relationship does not take into account the falloff in automatic test generation capability due to sequential complexity of the network. It has been observed that computer run time just for fault simulation is proportional to N^2 without even considering the test generation phase.

When one talks about testing, the topic of functional testing always comes up as a feasible way to test a network. Theoretically, to do a complete functional test ("exhaustive" testing) seems to imply that all entries in a Karnaugh map (or excitation table) must be tested for a 1 or a 0. This means that if a network has N inputs and is purely combinational, then 2^N patterns are required to do a complete functional test. Furthermore, if a network has N inputs with M latches, at a minimum it takes 2^{N+M} patterns to do a complete functional test. Rarely is that minimum ever obtainable; and in fact, the number of tests required to do a complete functional test is very much higher than that. With LSI, this may be a network with $N = 25$ and $M = 50$, or 2^{75} patterns, which is approximately 3.8×10^{22}. Assuming one had the patterns and applied them at an application rate of 1 μs per pattern, the test time would be over a billion years (10^9).

C. Cost of Testing

One might ask why so much attention is now being given to the level of testability at chip and board levels. The bottom line is the cost of doing business. A standard among people familiar with the testing process is: If it costs $0.30 to detect a fault at the chip level, then it would cost $3 to detect that same fault when it was embedded at the board level; $30 when it is embedded at the system level; and $300 when it is embedded at the system level but has to be found in the field. Thus if a fault can be detected at a chip or board level, then significantly larger costs per fault can be avoided at subsequent levels of packaging.

With VLSI and the inadequacy of automatic test generation and fault simulation, there is considerable difficulty in obtaining a level of testability required to achieve acceptable defect levels. If the defect level of boards is too high, the cost of field repairs is also too high. These costs, and in some cases, the inability to obtain a sufficient test, have led to the need to have "Design for Testability."

II. DESIGN FOR TESTABILITY

There are two key concepts in Design for Testability: controllability and observability. Control and observation of a network are central to implementing its test procedure. For example, consider the case of the simple AND block in Fig. 1. In order to be able to test the "A" input Stuck-At 1, it was necessary to control the "A" input to 0 and the "B" input to 1 and be able to observe the "C" output to determine whether a 0 was observed or a 1 was observed. The 0 is the result of the good machine, and the 1 would be the result, if you had a faulty machine. If this AND block is embedded into a much larger sequential network, the requirement of being able to control the "A" and "B" inputs to 0 and 1, respectively, and being able to observe the output "C," be it through some other logic blocks, still remains. Therein lies part of the problem of being able to generate tests for a network.

Because of the need to determine if a network has the attributes of controllability and observability that are desired, a number of programs have been written which essentially give analytic measures of controllability and observability for different nets in a given sequential network [69]-[73].

After observing the results of one of these programs in a given network, the logic designer can then determine whether some of the techniques, which will be described later, can be applied to this network to ease the testing problem. For example, test points may be added at critical points which are not observable or which are not controllable, or some of the techniques of Scan Path or LSSD can be used to initialize certain latches in the machine to avoid the difficulties of controllability associated with sequential machines. The popularity of such tools is continuing to grow, and a number of companies are now embarking upon their own controllability/observability measures.

III. AD HOC DESIGN FOR TESTABILITY [10]

Testing has moved from the afterthought position that it used to occupy to part of the design environment in LSI and VLSI. When testing was part of the afterthought, it was a very expensive process. Products were discarded because there was no adequate way to test them in production quantities.

There are two basic approaches which are prevalent today in the industry to help solve the testing problem. The first approach categorized here is Ad Hoc, and the second approach is categorized as a Structured Approach. The Ad Hoc techniques are those techniques which can be applied to a given product, but are not directed at solving the general sequential problem. They usually do offer relief, and their cost is probably lower than the cost of the Structured Approaches. The Structured Approaches, on the other hand, are trying to solve the general problem with a design methodology, such that when the designer has completed his design from one of these particular approaches, the results will be test generation and fault simulation at acceptable costs. Structured Approaches lend themselves more easily to design automation. Again, the main difference between the two approaches is probably the cost of implementation and hence, the return on investment for this extra cost. In the Ad Hoc approaches, the job of doing test generation and fault simulation are usually not as simple or as straightforward as they would be with the Structured Approaches, as we shall see shortly.

A number of techniques have evolved from MSI to LSI and now into VLSI that fall under the category of the ad hoc approaches of "Design for Testability." These techniques are usually solved at the board level and do not necessarily require changes in the logic design in order to accomplish them.

A. Partitioning

Because the task of test pattern generation and fault simulation is proportional to the number of logic gates to the third power, a significant amount of effort has been directed at approaches called "Divide and Conquer."

There are a number of ways in which the partitioning approach to Design for Testability can be implemented. The first is to mechanical partition by dividing a network in half. In essence, this would reduce the test generation and fault simulation tasks by 8 for two boards. Unfortunately, having two boards rather than one board can be a significant cost disadvantage and defeats the purpose of integration.

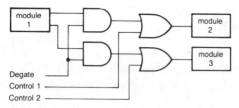

Fig. 2. Use of degating logic for logical partioning.

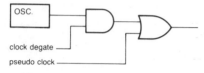

Fig. 3. Degating lines for oscillator.

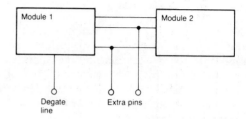

Fig. 4. Test points used as both inputs and outputs.

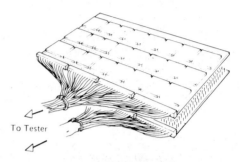

Fig. 5. "Bed of Nails" test.

Another approach that helps the partitioning problem, as well as helping one to "Divide and Conquer" is to use jumper wires. These wires would go off the board and then back on the board, so that the tester and the test generator can control and observe these nets directly. However, this could mean a significant number of I/O contacts at the board level which could also get very costly.

Degating is another technique for separating modules on a board. For example, in Fig. 2, a degating line goes to two AND blocks that are driven from Module 1. The results of those two AND blocks go to two independent OR blocks—one controlled by Control Line 1, the other with Control Line 2. The output of the OR block from Control Line 1 goes into Module 2, and the output of Control Line 2 goes into Module 3. When the degate line is at the 0 value, the two Control Lines, 1 and 2, can be used to drive directly into Modules 2 and 3. Therefore, complete controllability of the inputs to Modules 2 and 3 can be obtained by using these control lines. If those two nets happen to be very difficult nets to control, as pointed out, say, by a testability measure program, then this would be a very cost-effective way of controlling those two nets and hence, being able to derive the tests at a very reasonable cost.

A classical example of degating logic is that associated with an oscillator, as shown in Fig. 3. In general, if an oscillator is free-running on a board, driving logic, it is very difficult, and sometimes impossible, to synchronize the tester with the activity of the logic board. As a result, degating logic can be used here to block the oscillator and have a pseudo-clock line which can be controlled by the tester, so that the dc testing of all the logic on that board can be synchronized. All of these techniques require a number of extra primary inputs and primary outputs and possibly extra modules to perform the degating.

B. Test Points

Another approach to help the controllability and observability of a sequential network is to use test points [23], [24]. If a test point is used as a primary input to the network, then that can function to enhance controllability. If a test point is used as a primary output, then that is used to enhance the observability of a network. In some cases, a single pin can be used as both an input and an output.

For example, in Fig. 4, Module 1 has a degate function, so that the output of those two pins on the module could go to noncontrolling values. Thus the external pins which are dotted into those nets could control those nets and drive Module 2.

On the other hand, if the degate function is at the opposite value, then the output of Module 1 can be observed on these external pins. Thus the enhancement of controllability and observability can be accommodated by adding pins which can act as both inputs and outputs under certain degating conditions.

Another technique which can be used for controllability is to have a pin which, in one mode, implies system operation, and in another mode takes N inputs and gates them to a decoder. The 2^N outputs of the decoder are used to control certain nets to values which otherwise would be difficult to obtain. By so doing, the controllability of the network is enhanced.

As mentioned before, predictability is an issue which is as important as controllability and observability. Again, test points can be used here. For example, a CLEAR or PRESET function for all memory elements can be used. Thus the sequential machine can be put into a known state with very few patterns.

Another technique which falls into the category of test points and is very widely used is that of the "Bed of Nails" [31] tester, Fig. 5. The Bed of Nails tester probes the underside of a board to give a larger number of points for observability and controllability. This is in addition to the normal tester contact to the board under test. The drawback of this technique is that the tester must have enough test points to be able to control and observe each one of these nails on the Bed of Nails tester. Also, there are extra loads which are placed on the nets and this can cause some drive and receive problems. Furthermore, the mechanical fixture which will hold the Bed of Nails has to be constructed, so that the normal forces on the probes are sufficient to guarantee reliable contacts. Another application for the Bed of Nails testing is to do "drive/sense nails" [31] or "*in situ*" or "in-circuit" testing, which, effectively, is the technique of testing each chip on the board independently of the other chips on the board. For each chip, the appropriate nails and/or primary inputs are driven so as to prevent one chip from being driven by the other chips on the board. Once this state has been established, the isolated chip on the board can now be tested. In this case, the resolution to the failing

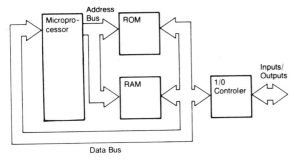

Fig. 6. Bus structured microcomputer.

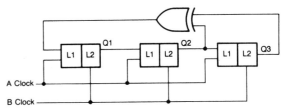

Fig. 7. Counting capabilities of a linear feedback shift register.

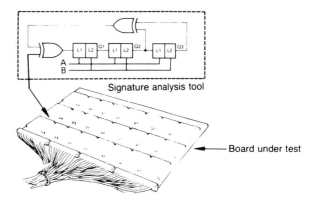

Fig. 8. Use of signature analysis tool.

chip is much better than edge connector tests, however, there is some exposure to incomplete testing of interconnections and care must be taken not to damage the circuit when overdriving it. Design for testability in a Bed of Nails environment must take the issues of contact reliability, multiplicity, and electrical loading into account.

C. Bus Architecture

An approach that has been used very successfully to attack the partitioning problem by the microcomputer designers is to use a bus structured architecture. This architecture allows access to critical buses which go to many different modules on the computer board. For example, in Fig. 6, you can see that the data bus is involved with both the microprocessor module, the ROM module, the RAM module, and the I/O Controller module. If there is external access to the data bus and three of the four modules can be turned off the data bus—that is, their outputs can be put into a high-impedance state (three-state driver)—then the data bus could be used to drive the fourth module, as if it were a primary input (or primary output) to that particular module. Similarly, with the address bus, access again must be controlled externally to the board, and thus the address bus can be very useful to controlling test patterns to the microcomputer board. These buses, in essence, partition the board in a unique way, so that testing of subunits can be accomplished. A drawback of bus-structured designs comes with faults on the bus itself. If a bus wire is stuck, any module or the bus trace itself may be the culprit. Normal testing is done by deducing the location of a fault from voltage information. Isolating a bus failure may require current measurements, which are much more difficult to do.

D. Signature Analysis

This technique for testing, introduced in 1977 [27], [33], [55] is heavily reliant on planning done in the design stage. That is why this technique falls between the Ad Hoc and the Structured Approaches for Design for Testability, since some care must be taken at the board level in order to ensure proper operation of this Signature Analysis of the board [12]. Signature Analysis is well-suited to bus structure architectures, as previously mentioned and in particular, those associated with microcomputers. This will become more apparent shortly.

The integral part of the Signature Analysis approach is that of a linear feedback shift register [8]. Fig. 7 shows an example of a 3-bit linear feedback shift register. This linear feedback shift register is made up of three shift register latches. Each one is represented by a combination of an $L1$ latch and an $L2$ latch. These can be thought of as the master latch being the $L1$ latch and the slave latch being the $L2$ latch. An "A" clock clocks all the $L1$ latches, and a "B" clock clocks all the $L2$ latches, so that turning the "A" and "B" clocks on and off independently will shift the shift register 1-bit position to the right. Furthermore, this linear shift register has an EXCLUSIVE-OR gate which takes the output, $Q2$, the second bit in the shift register, and EXCLUSIVE-OR's it with the third bit in the shift register, $Q3$. The result of that EXCLUSIVE-OR is the input to the first shift register. A single clock could be used for this shift register, which is generally the case, however, this concept will be used shortly when some of the structured design approaches are discussed which use two nonoverlapping clocks. Fig. 7 shows how this linear feedback shift register will count for different initial values.

For longer shift registers, the maximal length linear feedback configurations can be obtained by consulting tables [8] to determine where to tap off the linear feedback shift register to perform the EXCLUSIVE-OR function. Of course, only EXCLUSIVE-OR blocks can be used, otherwise, the linearity would not be preserved.

The key to Signature Analysis is to design a network which can stimulate itself. A good example of such a network would be microprocessor-based boards, since they can stimulate themselves using the intelligence of the processor driven by the memory on the board.

The Signature Analysis procedure is one which has the shift register in the Signature Analysis tool, which is external to the board and not part of the board in any way, synchronized with the clocking that occurs on the board, see Fig. 8. A probe is used to probe a particular net on the board. The result of that probe is EXCLUSIVE-OR'ed into the linear feedback shift register. Of course, it is important that the linear feedback shift register be initialized to the same starting place every time, and that the clocking sequence be a fixed number, so that the tests can be repeated. The board must also have some initialization, so that its response will be repeated as well.

After a fixed number of clock periods—let's assume 50—a particular value will be stored in $Q1$, $Q2$, and $Q3$. It is not necessarily the value that would have occurred if the linear feedback shift register was just counted 50 times—Modulo 7.

The value will be changed, because the values coming from the board via the probe will not necessarily be a continuous string of 1's; there will be 1's intermixed with 0's.

The place where the shift register stops on the Signature Analysis Tool—that is, the values for $Q1$, $Q2$, and $Q3$ is the Signature for that particular node for the good machine. The question is: If there were errors present at one or more points in the string of 50 observations of that particular net of the board, would the value stored in the shift register for $Q1$, $Q2$, and $Q3$ be different than the one for the good machine? It has been shown that with a 16-bit linear feedback shift register, the probability of detecting one or more errors is extremely high [55]. In essence, the signature, or "residue," is the remainder of the data stream after division by an irreduceable polynomial. There is considerable data compression—that is, after the results of a number of shifting operations, the test data are reduced to 16 bits, or, in the case of Fig. 8, 3 bits. Thus the result of the Signature Analysis tool is basically a Go/No-Go for the output for that particular module.

If the bad output for that module were allowed to cycle around through a number of other modules on the board and then feed back into this particular module, it would not be clear after examining all the nodes in the loop which module was defective—whether it was the module whose output was being observed, or whether it was another module upstream in the path. This gives rise to two requirements for Signature Analysis. First of all, closed-loop paths must be broken at the board level. Second, the best place to start probing with Signature Analysis is with a "kernel" of logic. In other words, on a microprocessor-based board, one would start with the outputs of the microprocessor itself and then build up from that particular point, once it has been determined that the microprocessor is good.

This breaking of closed loops is a tenant of Design for Testability and for Signature Analysis. There is a little overhead for implementing Signature Analysis. Some ROM space would be required (to stimulate the self-test), as well as extra jumpers, in order to break closed loops on the board. Once this is done, however, the test can be obtained for very little cost. The only question that remains is about the quality of the tests—that is, how good are the tests that are being generated, do they cover all the faults, etc.

Unfortunately, the logic models—for example, microprocessors—are not readily available to the board user. Even if a microprocessor logic model were available, they would not be able to do a complete fault simulation of the patterns because it would be too large. Hence, Signature Analysis may be the best that could be done for this particular board with the given inputs which the designer has. Presently, large numbers of users are currently using the Signature Analysis technique to test boards containing LSI and VLSI components.

IV. STRUCTURED DESIGN FOR TESTABILITY

Today, with the utilization of LSI and VLSI technology, it has become apparent that even more care will have to be taken in the design stage in order to ensure testability and produceability of digital networks. This has led to rigorous and highly structured design practices. These efforts are being spearheaded not by the makers of LSI/VLSI devices but by electronics firms which possess captive IC facilities and the manufacturers of large main-frame computers.

Most structured design practices [14]-[16], [18]-[21], [25], [31], [32], [34], [35] are built upon the concept that if the

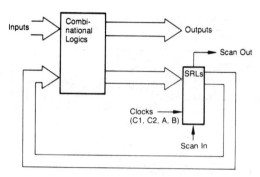

Fig. 9. Classical model of a sequential network utilizing a shift register for storage.

values in all the latches can be controlled to any specific value, and if they can be observed with a very straightforward operation then the test generation, and possibly, the fault task, can be reduced to that of doing test generation and fault simulation for a combinational logic network. A control signal can switch the memory elements from their normal mode of operation to a mode that makes them controllable and observable.

It appears from the literature that several companies, such as IBM, Fujitsu Ltd., Sperry-Univac, and Nippon Electric Co., Ltd. [14]-[16], [18]-[21], [31], [32], [35] have been dedicating formidable amounts of resources toward Structured Design for Testability. One notes simply by scanning the literature on testing, that many of the practical concepts and tools for testing were developed by main-frame manufacturers who do not lack for processor power. It is significant, then, that these companies, with their resources, have recognized that unstructured designs lead to unacceptable testing problems. Presently, IBM has extensively documented its efforts in Structured Design for Testability, and these are reviewed first.

A. Level-Sensitive Scan Design (LSSD)

With the concept that the memory elements in an IC can be threaded together into a shift register, the memory elements values can be both controlled and observed. Fig. 9 shows the familiar generalized sequential circuit model modified to use a shift register. This technique enhances both controllability and observability, allowing us to augment testing by controlling inputs and internal states, and easily examining internal state behavior. An apparent disadvantage is the serialization of the test, potentially costing more time for actually running a test.

LSSD is IBM's discipline for structural design for testability. "Scan" refers to the ability to shift into or out of any state of the network. "Level-sensitive" refers to constraints on circuit excitation, logic depth, and the handling of clocked circuitry. A key element in the design is the "shift register latch" (SRL) such as can be implemented in Fig. 10. Such a circuit is immune to most anomalies in the ac characteristics of the clock, requiring only that it remain high (sample) at least long enough to stabilize the feedback loop, before being returned to the low (hold) state [18], [19]. The lines D and C form the normal mode memory function while lines I, A, B, and $L2$ comprise additional circuitry for the shift register function.

The shift registers are threaded by connecting I to $L2$ and operated by clocking lines A and B in two-phase fashion. Fig. 11 shows four modules threaded for shift register action. Now note in Fig. 11 that each module could be an SRL or, one level up, a board containing threaded IC's, etc. Each level of pack-

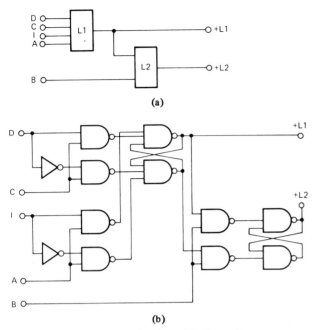

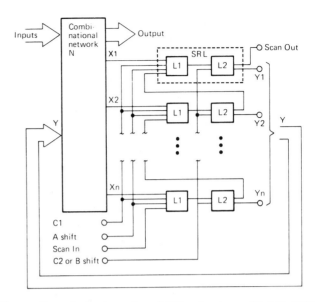

Fig. 10. Shift register latch (SRL). (a) Symbolic representation. (b) Implementation in AND-INVERT gates.

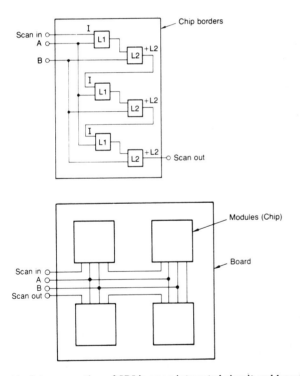

Fig. 11. Interconnection of SRL's on an integrated circuit and board.

aging requires the same four additional lines to implement the shift register scan feature. Fig. 12 depicts a general structure for an LSSD subsystem with a two-phase system clock. Additional rules concerning the gating of clocks, etc., are given by Williams and Eichelberger [18], [19]. Also, it is not practical to implement RAM with SRL memory, so additional procedures are required to handle embedded RAM circuitry [20].

Given that an LSSD structure is achieved, what are the rewards? It turns out that the network can now be thought of as purely combinational, where tests are applied via primary

Fig. 12. General structure of an LSSD subsystem with two system clocks.

inputs and shift register outputs. The testing of combinational circuits is a well understood and (barely) tractable problem. Now techniques such as the D-Algorithm [93] compiled code Boolean simulation [2], [74], [106], [107], and adaptive random test generation [87], [95], [98] are again viable approaches to the testing problem. Further, as small subsystems are tested, their aggregates into larger systems are also testable by cataloging the position of each testable subsystem in the shift register chain. System tests become (ideally) simple concatenations of subsystem tests. Though ideals are rarely achieved, the potential for solving otherwise hopeless testing problems is very encouraging.

In considering the cost performance impacts, there are a number of negative impacts associated with the LSSD design philosophy. First of all, the shift register latches in the shift register are, logically, two or three times as complex as simple latches. Up to four additional primary inputs/outputs are required at each package level for control of the shift registers. External asynchronous input signals must not change more than once every clock cycle. Finally, all timing within the subsystem is controlled by externally generated clock signals.

In terms of additional complexity of the shift register hold latches, the overhead from experience has been in the range of 4 to 20 percent. The difference is due to the extent to which the system designer made use of the $L2$ latches for system function. It has been reported in the IBM System 38 literature that 85 percent of the $L2$ latches were used for system function. This drastically reduces the overhead associated with this design technique.

With respect to the primary inputs/outputs that are required to operate the shift register, this can be reduced significantly by making functional use of some of the pins. For example, the scan-out pin could be a functional output of an SRL for that particular chip. Also, overall performance of the subsystem may be degraded by the clocking requirement, but the effect should be small.

The LSSD structured design approach for Design for Testability eliminates or alleviates some of the problems in designing, manufacturing and maintaining LSI systems at a reasonable cost.

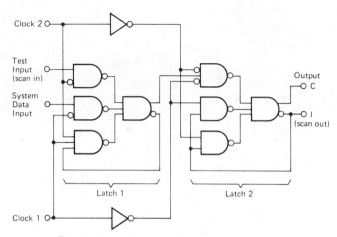

Fig. 13. Raceless D-type flip-flop with Scan Path.

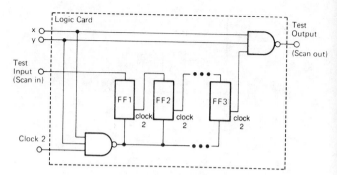

Fig. 14. Configuration of Scan Path on Card.

B. Scan Path

In 1975, a survey paper of test generation systems in Japan was presented by members of Nippon Electric Co., Ltd. [21]. In that survey paper, a technique they described as Scan Path was presented. The Scan Path technique has the same objectives as the LSSD approach which has just been described. The Scan Path technique similarities and differences to the LSSD approach will be presented.

The memory elements that are used in the Scan Path approach are shown in Fig. 13. This memory element is called a raceless D-type flip-flop with Scan Path.

In system operation, Clock 2 is at a logic value of 1 for the entire period. This, in essence, blocks the test or scan input from affecting the values in the first latch. This D-type flip-flop really contains two latches. Also, by having Clock 2 at a logic value of 1, the values in Latch 2 are not disturbed.

Clock 1 is the sole clock in system operation for this D-type flip-flop. When Clock 1 is at a value of 0, the System Data Input can be loaded into Latch 1. As long as Clock 1 is 0 for sufficient time to latch up the data, it can then turn off. As it turns off, it then will make Latch 2 sensitive to the data output of Latch 1. As long as Clock 1 is equal to a 1 so that data can be latched up into Latch 2, reliable operation will occur. This assumes that as long as the output of Latch 2 does not come around and feed the system data input to Latch 1 and change it during the time that the inputs to both Latch 1 and Latch 2 are active. The period of time that this can occur is related to the delay of the inverter block for Clock 1. A similar phenomenon will occur with Clock 2 and its associated inverter block. This race condition is the exposure to the use of only one system clock.

This points out a significant difference between the Scan Path approach and the LSSD approach. One of the basic principles of the LSSD approach is level-sensitive operation—the ability to operate the clocks in such a fashion that no races will exist. In the LSSD approach, a separate clock is required for Latch 1 from the clock that operates Latch 2.

In terms of the scanning function, the D-type flip-flop with Scan Path has its own scan input called test input. This is clocked into the $L1$ latch by Clock 2 when Clock 2 is a 0, and the results of the $L1$ latch are clocked into Latch 2 when Clock 2 is a 1. Again, this applies to master/slave operation of Latch 1 and Latch 2 with its associated race with proper attention to delays this race will not be a problem.

Another feature of the Scan Path approach is the configuration used at the logic card level. Modules on the logic card are all connected up into a serial scan path, such that for each card, there is one scan path. In addition, there are gates for selecting a particular card in a subsystem. In Fig. 14, when X and Y are both equal to 1—that is the selection mechanism—Clock 2 will then be allowed to shift data through the scan path. Any other time, Clock 2 will be blocked, and its output will be blocked. The reason for blocking the output is that a number of card outputs can then be put together; thus the blocking function will put their output to noncontrolling values, so that a particular card can have unique control of the unique test output for that system.

It has been reported by the Nippon Electric Company that they have used the Scan Path approach, plus partitioning which will be described next, for systems with 100 000 blocks or more. This was for the FLT-700 System, which is a large processor system.

The partitioning technique is one which automatically separates the combinational network into smaller subnetworks, so that the test generator can do test generation for the small subnetworks, rather than the larger networks. A partition is automatically generated by backtracing from the D-type flip-flops, through the combinational logic, until it encounters a D-type flip-flop in the backtrace (or primary input). Some care must be taken so that the partitions do not get too large.

To that end, the Nippon Electric Company approach has used a controlled D-type flip-flop to block the backtracing of certain partitions when they become too high. This is another facet of Design for Testability—that is, the introduction of extra flip-flops totally independent of function, in order to control the partitioning algorithm.

Other than the lack of the level sensitive attribute to the Scan Path approach, the technique is very similar to the LSSD approach. The introduction of the Scan Path approach was the first practical implementation of shift registers for testing which was incorporated in a total system.

C. Scan/Set Logic

A technique similar to Scan Path and LSSD, but not exactly the same, is the Scan/Set technique put forth by Sperry-Univac [31]. The basic concept of this technique is to have shift registers, as in Scan Path or in LSSD, but these shift registers are not in the data path. That is, they are not in the system data path; they are independent of all the system latches. Fig. 15 shows an example of the Scan/Set Logic, referred to as bit serial logic.

The basic concept is that the sequential network can be

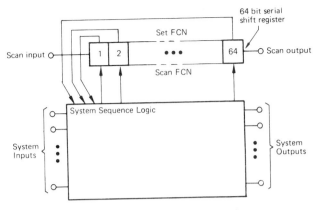

Fig. 15. Scan/Set Logic (bit-serial).

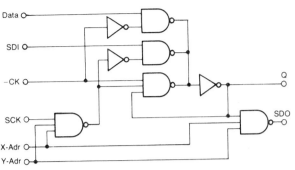

Fig. 16. Polarity-hold-type addressable latch.

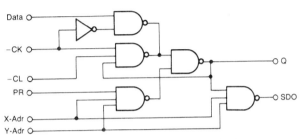

Fig. 17. Set/Reset type addressable latch.

sampled at up to 64 points. These points can be loaded into the 64-bit shift register with a single clock. Once the 64 bits are loaded, a shifting process will occur, and the data will be scanned out through the scan-out pin. In the case of the set function, the 64 bits can be funneled into the system logic, and then the appropriate clocking structure required to load data into the system latches is required in this system logic. Furthermore, the set function could also be used to control different paths to ease the testing function.

In general, this serial Scan/Set Logic would be integrated onto the same chip that contrains sequential system logic. However, some applications have been put forth where the bit serial Scan/Set Logic was off-chip, and the bit-serial Scan/Set Logic only sampled outputs or drove inputs to facilitate in-circuit testing.

Recently, Motorola has come forth with a chip which is T^2L and which has I^2L logic integrated on that same chip. This has the Scan/Set Logic bit serial shift registers built in I^2L. The T^2L portion of the chip is a gate array, and the I^2L is on the chip, whether the customer wants it or not. It is up to the customer to use the bit-serial logic if he chooses.

At this point, it should be explained that if all the latches within the system sequential network are not both scanned and set, then the test generation function is not necessarily reduced to a total combinational test generation function and fault simulation function. However, this technique will greatly reduce the task of test generation and fault simulation.

Again, the Scan/Set technique has the same objectives as Scan Path and LSSD—that is, controllability and observability. However, in terms of its implementation, it is not required that the set function set all system latches, or that the scan function scan all system latches. This design flexibility would have a reflection in the software support required to implement such a technique.

Another advantage of this technique is that the scan function can occur during system operation—that is, the sampling pulse to the 64-bit serial shift register can occur while system clocks are being applied to the system sequential logic, so that a snapshot of the sequential machine can be obtained and off-loaded without any degradation in system performance.

D. Random-Access Scan

Another technique similar to the Scan Path technique and LSSD is the Random-Access Scan technique put forth by Fujitsu [14]. This technique has the same objective as Scan Path and LSSD—that is, to have complete controllability and observability of all internal latches. Thus the test generation function can be reduced to that of combinational test generation and combinational fault simulation as well.

Random-Access Scan differs from the other two techniques in that shift registers are not employed. What is employed is an addressing scheme which allows each latch to be uniquely selected, so that it can be either controlled or observed. The mechanism for addressing is very similar to that of a Random-Access Memory, and hence, its name.

Figs. 16 and 17 show the two basic latch configurations that are required for the Random-Access Scan approach. Fig. 16 is a single latch which has added to it an extra data port which is a Scan Data In port (SDI). These data are clocked into the latch by the SCK clock. The SCK clock can only affect this latch, if both the X and Y addresses are one. Furthermore, when the X address and Y address are one, then the Scan Data Out (SDO) point can be observed. System data labeled Data in Figs. 16 and 17 are loaded into this latch by the system clock labeled CK.

The set/reset-type addressable latch in Fig. 17 does not have a scan clock to load data into the system latch. This latch is first cleared by the CL line, and the CL line is connected to other latches that are also set/reset-type addressable latches. This, then, places the output value Q to a 0 value. A preset is directed at those latches that are required to be set to a 1 for that particular test. This preset is directed by addressing each one of those latches and applying the preset pulse labeled PR. The output of the latch Q will then go to a 1. The observability mechanism for Scan Data Out is exactly the same as for the latch shown in Fig. 16.

Fig. 18 gives an overall view of the system configuration of the Random-Access Scan approach. Notice that, basically, there is a Y address, an X address, a decoder, the addressable storage elements, which are the memory elements or latches, and the sequential machine, system clocks, and CLEAR function. There is also an SDI which is the input for a given latch, an SDO which is the output data for that given latch, and a scan clock. There is also one logic gate necessary to create the preset function.

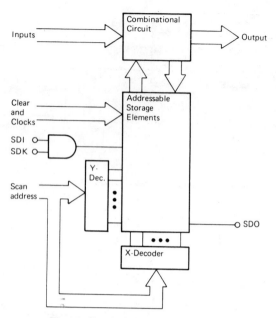

Fig. 18. Random-Access Scan network.

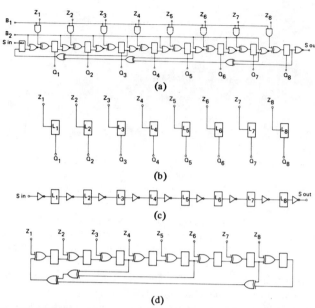

Fig. 19. BILBO and its different modes. (a) General form of BILBO register. (b) $B_1 B_2 = 11$ system orientation mode. (c) $B_1 B_2 = 00$ linear shift register mode. (d) $B_1 B_2 = 10$ signature analysis register with m multiple inputs ($Z_1, Z_2, \cdots, Z_8$).

The Random-Access Scan technique allows the observability and controllability of all system latches. In addition, any point in the combinational network can be observed with the addition of one gate per observation point, as well as one address in the address gate, per observation point.

While the Scan Path approach and the LSSD approach require two latches for every point which needs to be observed, the overhead for Random-Access Scan is about three to four gates per storage element. In terms of primary inputs/outputs, the overhead is between 10 and 20. This pin overhead can be diminished by using the serial scan approach for the X and Y address counter, which would lead to 6 primary inputs/outputs.

V. Self-Testing and Built-In Tests

As a natural outgrowth of the Structured Design approach for "Design for Testability," Self-Tests and Built-In Tests have been getting considerably more attention. Four techniques will be discussed, which fall into this category, BILBO, Syndrome Testing, Testing by Verifying Walsh Testing Coefficients, and Autonomous Testing. Each of these techniques will be described.

A. Built-In Logic Block Observation, BILBO

A technique recently presented takes the Scan Path and LSSD concept and integrates it with the Signature Analysis concept. The end result is a technique for Built-In Logic Block Observation, BILBO [25].

Fig. 19 gives the form of an 8-bit BILBO register. The block labeled L_i ($i = 1, 2, \cdots, 8$) are the system latches. B_1 and B_2 are control values for controlling the different functions that the BILBO register can perform. S_{IN} is the scan-in input to the 8-bit register, and S_{OUT} is the scan-out for the 8-bit register. Q_i ($i = 1, 2, \cdots, 8$) are the output values for the eight system latches. Z_i ($i = 1, 2, \cdots, 8$) are the inputs from the combinational logic. The structure that this network will be embedded into will be discussed shortly.

There are three primary modes of operation for this register, as well as one secondary mode of operation for this register. The first is shown in Fig. 19(b)—that is, with B_1 and B_2 equal to 11. This is a Basic System Operation mode, in which the Z_i values are loaded into the L_i, and the outputs are available on Q_i for system operation. This would be your normal register function.

When $B_1 B_2$ equals 00, the BILBO register takes on the form of a linear shift register, as shown in Fig. 19(c). Scan-in input to the left, through some inverters, and basically lining up the eight registers into a single scan path, until the scan-out is reached. This is similar to Scan Path and LSSD.

The third mode is when $B_1 B_2$ equals 10. In this mode, the BILBO register takes on the attributes of a linear feedback shift register of maximal length with multiple linear inputs. This is very similar to a Signature Analysis register, except that there is more than one input. In this situation, there are eight unique inputs. Thus after a certain number of shift clocks, say, 100, there would be a unique signature left in the BILBO register for the good machine. This good machine signature could be off-loaded from the register by changing from Mode $B_1 B_2 = 10$ to Mode $B_1 B_2 = 00$, in which case a shift register operation would exist, and the signature then could be observed from the scan-out primary output.

The fourth function that the BILBO register can perform is $B_1 B_2$ equal to 01, which would force a reset on the register. (This is not depicted in Fig. 19.)

The BILBO registers are used in the system operation, as shown in Fig. 20. Basically, a BILBO register with combinational logic and another BILBO register with combinational logic, as well as the output of the second combinational logic network can feed back into the input of the first BILBO regis-

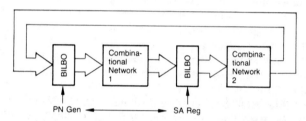

Fig. 20. Use of BILBO registers to test combinational Network 1.

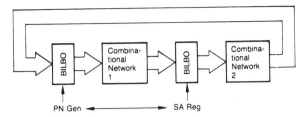

Fig. 21. Use of BILBO registers to test combinational Network 2.

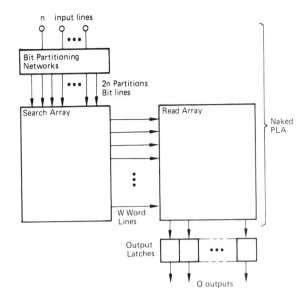

Fig. 22. PLA model.

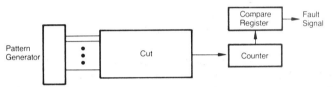

Fig. 23. Syndrome test structure.

ter. The BILBO approach takes one other fact into account, and that is that, in general, combinational logic is highly susceptible to random patterns. Thus if the inputs to the BILBO register, $Z_1, Z_2, \cdots, Z_8$, can be controlled to fixed values, such that the BILBO register is in the maximal length linear feedback shift register mode (Signature Analysis) it will output a sequence of patterns which are very close to random patterns. Thus random patterns can be generated quite readily from this register. These sequences are called Pseudo Random Patterns (PN).

If, in the first operation, this BILBO register on the left in Fig. 20 is used as the PN generator—that is, its data inputs are held to fixed values—then the output of that BILBO register will be random patterns. This will then do a reasonable test, if sufficient numbers of patterns are applied, of the Combinational Logic Network 1. The results of this test can be stored in a Signature Analysis register approach with multiple inputs to the BILBO register on the right. After a fixed number of patterns have been applied, the signature is scanned out of the BILBO register on the right for good machine compliance. If that is successfully completed, then the roles are reversed, and the BILBO register on the right will be used as a PN sequence generator; the BILBO register on the left will then be used as a Signature Analysis register with multiple inputs from Combinational Logic Network 2, see Fig. 21. In this mode, the Combinational Logic Network 2 will have random patterns applied to its inputs and its outputs stored in the BILBO register on the far left. Thus the testing of the combinational logic networks 1 and 2 can be completed at very high speeds by only applying the shift clocks, while the two BILBO registers are in the Signature Analysis mode. At the conclusion of the tests, off-loading of patterns can occur, and determination of good machine operation can be made.

This technique solves the problem of test generation and fault simulation if the combinational networks are susceptible to random patterns. There are some known networks which are not susceptible to random patterns. They are Programmable Logic Arrays (PLA's), see Fig. 22. The reason for this is that the fan-in in PLA's is too large. If an AND gate in the search array had 20 inputs, then each random pattern would have $1/2^{20}$ probability of coming up with the correct input pattern. On the other hand, random combinational logic networks with maximum fan-in of 4 can do quite well with random patterns.

The BILBO technique solves another problem and that is of test data volume. In LSSD, Scan Path, Scan/Set, or Random-Access Scan, a considerable amount of test data volume is involved with the shifting in and out. With BIBLO, if 100 patterns are run between scan-outs, the test data volume may be reduced by a factor of 100. The overhead for this technique is higher than for LSSD since about two EXCLUSIVE-OR's must be used per latch position. Also, there is more delay in the system data path (one or two gate delays). If VLSI has the huge number of logic gates available than this may be a very efficient way to use them.

B. Syndrome Testing

Recently, a technique was shown which could be used to test a network with fairly minor changes to the network. The technique is Syndrome Testing. The technique requires that all 2^n patterns be applied to the input of the network and then the number of 1's on the output be counted [115], [116].

Testing is done by comparing the number of 1's for the good machine to the number of 1's for the faulty machine. If there is a difference, the fault(s) in the faulty machine are detected (or Syndrome testable). To be more formal the Syndrome is:

Definition 1: The *Syndrome S* of a Boolean function is defined as

$$S = \frac{K}{2^n}$$

where K is the number of minterns realized by the function, and n is the number of binary input lines to the Boolean function.

Not all Boolean functions are totally Syndrome testable for all the single stuck-at-faults. Procedures are given in [115] with a minimal or near minimal number of primary inputs to make the networks Syndrome testable. In a number of "real networks" (i.e., SN74181, etc.) the numbers of extra primary inputs needed was at most one (<5 percent) and not more than two gates (<4 percent) were needed. An extension [116] to this work was published which showed a way of making a network Syndrome testable by adding extra inputs. This resulted in a somewhat longer test sequence. This is accomplished by holding some input constant while applying all 2^k inputs ($k < n$) then holding others constant and applying 2^l input patterns to l inputs. Whether the network is modified or not, the test data volume for a Syndrome testable design is extremely low. The general test setup is shown in Fig. 23.

The structure requires a pattern generator which applies all possible patterns once, a counter to count the 1's, and a com-

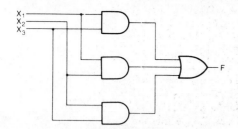

Fig. 24. Function to be tested with Walsh coefficients.

TABLE I
EXAMPLES OF WALSH FUNCTIONS AND WALSH COEFFICIENTS

$X_1 X_2 X_3$	W_2	$W_{1,3}$	F	$W_2 F$	$W_{1,3} F$	W_{ALL}	$W_{ALL} F$
0 0 0	−1	+1	0	+1	−1	+1	+1
0 0 1	−1	−1	0	+1	+1	−1	−1
0 1 0	+1	+1	0	−1	−1	−1	−1
0 1 1	+1	−1	1	+1	−1	+1	−1
1 0 0	−1	−1	0	+1	+1	−1	−1
1 0 1	−1	+1	1	−1	+1	+1	−1
1 1 0	+1	−1	1	−1	−1	+1	−1
1 1 1	+1	+1	1	+1	+1	−1	+1

$C_{ALL} = 4$

pare network. The overhead quoted is necessary to make the CUT Syndrome testable and does not include the pattern generator, counter, or compare register.

C. Testing by Verifying Walsh Coefficients

A technique which is similar to Syndrome Testing, in that it requires all possible input patterns be applied to the combinational network, is testing by verifying Walsh coefficients [117]. This technique only checks two of the Walsh coefficients and then makes conclusions about the network with respect to stuck-at-faults.

In order to calculate the Walsh coefficients, the logical value 0 (1) is associated with the arithmetic value −1(+1). There are 2^n Walsh functions. W_0 is defined to be 1, W_i is derived from all possible (arithmetic) products of the subject of independent input variables selected for that Walsh function. Table I shows the Walsh function for W_2, $W_{1,3}$, then $W_2 F$, $W_{1,3} F$, finally W_{all} and $W_{all} F$. These values are calculated for the network in Fig. 24. If the values are summed for $W_{all} F$, the Walsh coefficient C_{all} is calculated. The Walsh coefficient C_0 is just $W_0 F$ summed. This is equivalent to the Syndrome in magnitude times 2^n. If $C_{all} \neq 0$ then all stuck-at-faults on primary inputs will be detected by measuring C_{all}. If the fault is present $C_{all} = 0$. If the network has $C_{all} = 0$ it can be easily modified such that $C_{all} \neq 0$. If the network has reconvergent fan-out then further checks need to be made (the number of inverters in each path has a certain property); see [117]. If these are successful, then by checking C_{all} and C_0, all the single stuck-at-faults can be detected. Some design constraints maybe needed to make sure that the network is testable by measuring C_{all} and C_0. Fig. 25 shows the network needed to determine C_{all} and C_0. The value p is the parity of the driving counter and the response counter is an up/down counter. Note, two passes must be made of the driving counter, one for C_{all} and one for C_0.

D. Autonomous Testing

The fourth technique which will be discussed in the area of self-test/built-in-test is Autonomous Testing [118]. Autonomous Testing like Syndrome Testing and testing Walsh coefficients requires all possible patterns be applied to the network inputs. However, with Autonomous Testing the outputs of

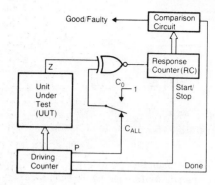

Fig. 25. Tester for veryfying C_0 and C_{all} Walsh coefficients.

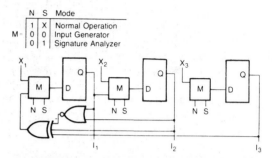

Fig. 26. Reconfigurable 3-bit LFSR module.

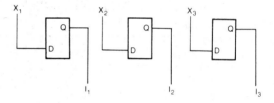

N = 1: Normal Operation

Fig. 27. Reconfigurable 3-bit LFSR module.

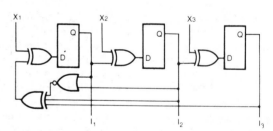

N = 0, S = 1: Signature Analyzer

Fig. 28. Reconfigurable 3-bit LFSR module.

the network must be checked for each pattern against the value for the good machine. The results is that irrespective of the fault model Autonomous Testing will detect the faults (assuming the faulty machine does not turn into a sequential machine from a combinational machine). In order to help the network apply its own patterns and accumulate the results of the tests rather than observing every pattern for 2^n input patterns, a structure similar to BILBO register is used. This register has some unique attributes and is shown in Figs. 26–29. If a combinational network has 100 inputs, the network must be modified such that the subnetwork can be verified and, thus, the whole network will be tested.

Two approaches to partitioning are presented in the paper "Design for Autonomous Test" [118]. The first is to use

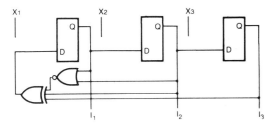

Fig. 29. Reconfigurable 3-bit LFSR module.

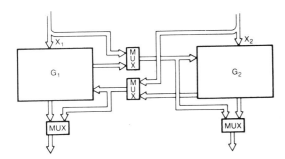

Fig. 30. Autonomous Testing—general network.

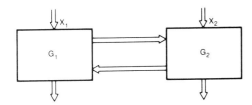

Fig. 31. Autonomous Testing—functional mode.

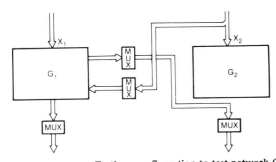

Fig. 32. Autonomous Testing—configuration to test network G_1.

multiplexers to separate the network and the second is a Sensitized Partitioning to separate the network. Fig. 30 shows the general network with multiplexers, Fig. 31 shows the network in functional mode, and Fig. 32 shows the network in a mode to test subnetwork G_1. This approach could involve a significant gate overhead to implement in some networks. Thus the Sensitized Partitioning approach is put forth. For example, the 74181 ALU/Function Generator is partitioned using the Sensitized Partitioning. By inspecting the network, two types of subnetworks can be partitioned out, four subnetworks N_1, one subnetwork N_2 (Figs. 33 and 34). By further inspection, all the L_i outputs of network N_1 can be tested by holding $S_2 = S_3 =$ low. Further, all the H_i outputs of network N_1 can be tested by holding $S_0 = S_1 =$ high, since sensitized paths exist through the subnetwork N_2. Thus far fewer than 2^n input patterns can be applied to the network to test it.

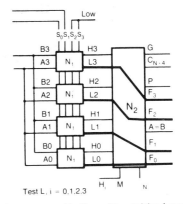

Fig. 33. Autonomous Testing with sensitized partitioning.

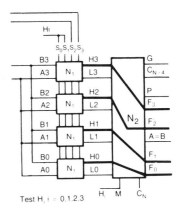

Fig. 34. Autonomous Testing with sensitized partitioning.

VI. CONCLUSION

The area of Design for Testability is becoming a popular topic by necessity. Those users of LSI/VLSI which do not have their own captive IC facilities are at the mercy of the vendors for information. And, until the vendor information is drastically changed, the Ad Hoc approaches to design for testability will be the only answer.

In that segment of the industry which can afford to implement the Structured Design for Testability approach, there is considerable hope of getting quality test patterns at a very modest cost. Furthermore, many innovative techniques are appearing in the Structured Approach and probably will continue as we meander through VLSI and into more dense technologies.

There is a new opportunity arriving in the form of gate arrays that allow low volume users access to VLSI technology. If they choose, structured design disciplines can be utilized. Perhaps "Silicon Foundries" of the future will offer a combined package of structured, testable modules and support software to automatically provide the user with finished parts AND tests.

ACKNOWLEDGMENT

The authors wish to thank D. J. Brown for his helpful comments and suggestions. The assistance of Ms. B. Fletcher, Ms. C. Mendoza, Ms. L. Clark, Ms. J. Allen, and J. Smith in preparing this manuscript for publication was invaluable.

REFERENCES

General References and Surveys

[1] M. A. Breuer, Ed., *Diagnosis and Reliable Design of Digital Systems.* Rockville, MD: Computer Science Press, 1976.
[2] H. Y. Chang, E. G. Manning, and G. Metze, *Fault Diagnosis of*

[3] A. D. Friedman and P. R. Menon, *Fault Detection in Digital Circuits.* Englewood Cliffs, NJ: Prentice-Hall, 1971.
Digital Systems. New York: Wiley-Interscience, 1970.
[4] F. C. Hennie, *Finite State Models for Logical Machines.* New York: Wiley, 1968.
[5] P. G. Kovijanic, in "A new look at test generation and verification," in *Proc. 14th Design Automation Conf.*, IEEE Pub. 77CH1216-1C, pp. 58-63, June 1977.
[6] E. I. Muehldorf, "Designing LSI logic for testability," in *Dig. Papers, 1976 Ann. Semiconductor Test Symp.*, IEEE Pub. 76CH1179-1C, pp. 45-49, Oct. 1976.
[7] E. I. Muehldorf and A. D. Savkar, "LSI logic testing—An overview," *IEEE Trans. Comput.*, vol. C-30, no. 1, pp. 1-17, Jan. 1981.
[8] W. W. Peterson and E. J. Weldon, *Error Correcting Codes.* Cambridge, MA: MIT Press, 1972.
[9] A. K. Susskind, "Diagnostics for logic networks," *IEEE Spectrum*, vol. 10, pp. 40-47, Oct. 1973.
[10] T. W. Williams and K. P. Parker, "Testing logic networks and design for testability," *Computer*, pp. 9-21, Oct. 1979.
[11] IEEE, Inc., *IEEE Standard Dictionary of Electrical and Electronics Terms.* New York: Wiley-Interscience, 1972.

Designing for Testability

[12] "A designer's guide to signature analysis," Hewlett-Packard Application Note 222, Hewlett Packard, 5301 Stevens Creek Blvd., Santa Clara, CA 95050.
[13] S. B. Akers, "Partitioning for testability," *J. Des. Automat. Fault-Tolerant Comput.*, vol. 1, no. 2, Feb. 1977.
[14] H. Ando, "Testing VLSI with random access scan," in *Dig. Papers Compcon 80*, IEEE Pub. 80CH1491-OC, pp. 50-52, Feb. 1980.
[15] P. Bottorff and E. I. Muehldorf, "Impact of LSI on complex digital circuit board testing," *Electro 77*, New York, NY, Apr. 1977.
[16] S. DasGupta, E. B. Eichelberger, and T. W. Williams, "LSI chip design for testability," in *Dig. Tech. Papers, 1978 Int. Solid-State Circuits Conf.* (San Francisco, CA, Feb. 1978), pp. 216-217.
[17] "Designing digital circuits for testability," Hewlett-Packard Application Note 210-4, Hewlett Packard, Loveland, CO 80537.
[18] E. B. Eichelberger and T. W. Williams, "A logic design structure for LSI testability," *J. Des. Automat. Fault-Tolerant Comput.*, vol. 2, no. 2, pp. 165-178, May 1978.
[19] —, "A logic design structure for LSI testing," in *Proc. 14th Design Automation Conf.*, IEEE Pub. 77CH1216-1C, pp. 462-468, June 1977.
[20] E. B. Eichelberger, E. J. Muehldorf, R. G. Walter, and T. W. Williams, "A logic design structure for testing internal arrays," in *Proc. 3rd USA-Japan Computer Conf.* (San Francisco, CA, Oct. 1978), pp. 266-272.
[21] S. Funatsu, N. Wakatsuki, and T. Arima, "Test generation systems in Japan," in *Proc. 12th Design Automation Symp.*, pp. 114-122, June 1975.
[22] H. C. Godoy, G. B. Franklin, and P. S. Bottoroff, "Automatic checking of logic design structure for compliance with testability groundrules," in *Proc. 14th Design Automation Conf.*, IEEE Pub. 77CH1216-1C, pp. 469-478, June 1977.
[23] J. P. Hayes, "On modifying logic networks to improve their diagnosability," *IEEE Trans. Comput.*, vol. C-23, pp. 56-62, Jan. 1974.
[24] J. P. Hayes and A. D. Friedman, "Test point placement to simplify fault detection," in *FTC-3, Dig. Papers, 1973 Symp. on Fault-Tolerant Computing*, pp. 73-78, June 1973.
[25] B. Koenemann, J. Mucha, and G. Zwiehoff, "Built-in logic block observation techniques," in *Dig. Papers, 1979 Test Conf.*, IEEE Pub. 79CH1509-9C, pp. 37-41, Oct. 1979.
[26] M. D. Lippman and E. S. Donn, "Design forethought promotes easier testing of microcomputer boards," *Electronics*, pp. 113-119, Jan. 18, 1979.
[27] H. J. Nadig, "Signature analysis-concepts, examples, and guidelines," *Hewlett-Packard J.*, pp. 15-21, May 1977.
[28] M. Neil and R. Goodner, "Designing a serviceman's needs into microprocessor based systems," *Electronics*, pp. 122-128, Mar. 1, 1979.
[29] S. M. Reddy, "Easily testable realization for logic functions," *IEEE Trans. Comput.*, vol. C-21, pp. 1183-1188, Nov. 1972.
[30] K. K. Saliya and S. M. Reddy, "On minimally testable logic networks," *IEEE Trans. Comput.*, vol. C-23, pp. 1204-1207, Nov. 1974.
[31] J. H. Stewart, "Future testing of large LSI circuit cards," in *Dig. Papers 1977 Semiconductor Test Symp.*, IEEE Pub. 77CH1261-7C, pp. 6-17, Oct. 1977.
[32] A. Toth and C. Holt, "Automated data base-driven digital testing," *Computer*, pp. 13-19, Jan. 1974.
[33] E. White, "Signature analysis, enhancing the serviceability of microprocessor-based industrial products," in *Proc. 4th IECI Annual Conf.*, IEEE Pub. 78CH1312-8, pp. 68-76, Mar. 1978.
[34] M.J.Y. Williams and J. B. Angell, "Enhancing testability of large scale integrated circuits via test points and additional logic," *IEEE Trans. Comput.*, vol. C-22, pp. 46-60, Jan. 1973.
[35] T. W. Williams, "Utilization of a structured design for reliability and serviceability," in *Dig., Government Microcircuits Applications Conf.* (Monterey, CA, Nov. 1978), pp. 441-444.

Faults and Fault Modeling

[36] R. Boute and E. J. McCluskey, "Fault equivalence in sequential machines," in *Proc. Symp. on Computers and Automata* (Polytech. Inst. Brooklyn, Apr. 13-15, 1971), pp. 483-507.
[37] R. T. Boute, "Optimal and near-optimal checking experiments for output faults in sequential machines," *IEEE Trans. Comput.*, vol. C-23, no. 11, pp. 1207-1213, Nov. 1974.
[38] —, "Equivalence and dominance relations between output faults in sequential machines," Tech. Rep. 38, SU-SEL-72-052, Stanford Univ., Stanford, CA, Nov. 1972.
[39] F.J.O. Dias, "Fault masking in combinational logic circuits," *IEEE Trans. Comput.*, vol. C-24, pp. 476-482, May 1975.
[40] J. P. Hayes, "A NAND model for fault diagnosis in combinational logic networks," *IEEE Trans. Comput.*, vol. C-20, pp. 1496-1506, Dec. 1971.
[41] E. J. McCluskey and F. W. Clegg, "Fault equivalence in combinational logic networks," *IEEE Trans. Comput.*, vol. C-20, pp. 1286-1293, Nov. 1971.
[42] K.C.Y. Mei, "Fault dominance in combinational circuits," Tech. Note 2, Digital Systems Lab., Stanford Univ., Aug. 1970.
[43] —, "Bridging and stuck-at faults," *IEEE Trans. Comput.*, vol. C-23, no. 7, pp. 720-727, July 1974.
[44] R. C. Ogus, "The probability of a correct output from a combinational circuit," *IEEE Trans. Comput.*, vol. C-24, no. 5, pp. 534-544, May 1975.
[45] K. P. Parker and E. J. McCluskey, "Analysis of logic circuits with faults using input signal probabilities," *IEEE Trans. Comput.*, vol. C-24, no. 5, pp. 573-578, May 1975.
[46] K. K. Saliya and S. M. Reddy, "Fault detecting test sets for Reed-Muller canonic networks," *IEEE Trans. Comput.*, pp. 995-998, Oct. 1975.
[47] D. R. Schertz and G. Metze, "A new representation for faults in combinational digital circuits," *IEEE Trans. Comput.*, vol. C-21, no. 8, pp. 858-866, Aug. 1972.
[48] J. J. Shedletsky and E. J. McCluskey, "The error latency of a fault in a sequential digital circuit," *IEEE Trans. Comput.*, vol. C-25, no. 6, pp. 655-659, June 1976.
[49] —, "The error latency of a fault in a combinational digital circuit," in *FTCS-5, Dig. Papers, 5th Int. Symp. on Fault Tolerant Computing* (Paris, France, June 1975), pp. 210-214.
[50] K. To, "Fault folding for irredundant and redundant combinational circuits," *IEEE Trans. Comput.*, vol. C-22, no. 11, pp. 1008-1015, Nov. 1973.
[51] D. T. Wang, "Properties of faults and criticalities of values under tests for combinational networks," *IEEE Trans. Comput.*, vol. C-24, no. 7, pp. 746-750, July 1975.

Testing and Fault Location

[52] R. P. Batni and C. R. Kime, "A module level testing approach for combinational networks," *IEEE Trans. Comput.*, vol. C-25, no. 6, pp. 594-600, June 1976.
[53] S. Bisset, "Exhaustive testing of microprocessors and related devices: A practical solution," in *Dig. Papers, 1977 Semiconductor Test Symp.*, pp. 38-41, Oct. 1977.
[54] R. J. Czepiel, S. H. Foreman, and R. J. Prilik, "System for logic, parametric and analog testing," in *Dig. Papers, 1976 Semiconductor Test Symp.*, pp. 54-69, Oct. 1976.
[55] R. A. Frohwerk, "Signature analysis: A new digital field service method," *Hewlett-Packard J.*, pp. 2-8, May 1977.
[56] B. A. Grimmer, "Test techniques for circuit boards containing large memories and microprocessors," in *Dig. Papers, 1976 Semiconductor Test Symp.*, pp. 16-21, Oct. 1976.
[57] W. A. Groves, "Rapid digital fault isolation with FASTRACE," *Hewlett-Packard J.*, pp. 8-13, Mar. 1979.
[58] J. P. Hayes, "Rapid count testing for combinational logic circuits," *IEEE Trans. Comput.*, vol. C-25, no. 6, pp. 613-620, June 1976.
[59] —, "Detection of pattern sensitive faults in random access memories," *IEEE Trans. Comput.*, vol. C-24, no. 2, Feb. 1975, pp. 150-160.
[60] —, "Testing logic circuits by transition counting," in *FTC-5, Dig. Papers, 5th Int. Symp. on Fault Tolerant Computing* (Paris, France, June 1975), pp. 215-219.
[61] J. T. Healy, "Economic realities of testing microprocessors," in *Dig. Papers, 1977 Semiconductor Test Symp.*, pp. 47-52, Oct. 1977.
[62] E. C. Lee, "A simple concept in microprocessor testing," in *Dig.*

Papers, *1976 Semiconductor Test Symp.*, IEEE Pub. 76CH1179-1C, pp. 13-15, Oct. 1976.

[63] J. Losq, "Referenceless random testing," in *FTCS-6, Dig. Papers, 6th Int. Symp. on Fault-Tolerant Computing* (Pittsburgh, PA, June 21-23, 1976), pp. 81-86.

[64] S. Palmquist and D. Chapman, "Expanding the boundaries of LSI testing with an advanced pattern controller," in *Dig. Papers, 1976 Semicondctor Test Symp.*, pp. 70-75, Oct. 1976.

[65] K. P. Parker, "Compact testing: Testing with compressed data," in *FTCS-6, Dig. Papers, 6th Int. Symp. on Fault-Tolerant Computing* (Pittsburgh, PA, June 21-23, 1976).

[66] J. J. Shedletsky, "A rationale for the random testing of combinational digital circuits," in *Dig. Papers, Compcon 75 Fall Meet.* (Washington, DC, Sept. 9-11, 1975), pp. 5-9.

[67] V. P. Strini, "Fault location in a semiconductor random access memory unit," *IEEE Trans. Comput.*, vol. C-27, no. 4, pp. 379-385, Apr. 1978.

[68] C. W. Weller, in "An engineering approach to IC test system maintenance," in *Dig. Papers, 1977 Semiconductor Test Symp.*, pp. 144-145, Oct. 1977.

Testability Measures

[69] W. J. Dejka, "Measure of testability in device and system design," in *Proc. 20th Midwest Symp. Circuits Syst.*, pp. 39-52, Aug. 1977.

[70] L. H. Goldstein, "Controllability/observability analysis of digital circuits," *IEEE Trans. Circuits Syst.*, vol. CAS-26, no. 9, pp. 685-693, Sept. 1979.

[71] W. L. Keiner and R. P. West, "Testability measures," presented at AUTOTESTCON '77, Nov. 1977.

[72] P. G. Kovijanic, "testability analysis," in *Dig. Papers, 1979 Test Conf.*, IEEE Pub. 79CH1509-9C, pp. 310-316, Oct. 1979.

[73] J. E. Stephenson and J. Grason, "A testability measure for register transfer level digital circuits," in *Proc. 6th Fault Tolerant Computing Symp.*, pp. 101-107, June 1976.

Test Generation

[74] V. Agrawal and P. Agrawal, "An automatic test generation system for ILLIAC IV logic boards," *IEEE Trans. Comput.*, vol. C-C-21, no. 9, pp. 1015-1017, Sept. 1972.

[75] D. B. Armstrong, "On finding a nearly minimal set of fault detection tests for combinational logic nets," *IEEE Trans. Electron. Comput.*, vol. EC-15, no. 1, pp. 66-73, Feb. 1966.

[76] R. Betancourt, "Derivation of minimum test sets for unate logical circuits," *IEEE Trans. Comput.*, vol. C-20, no. 11, pp. 1264-1269, Nov. 1973.

[77] D. C. Bossen and S. J. Hong, "Cause and effect analysis for multiple fault detection in combinational networks," *IEEE Trans. Comput.*, vol. C-20, no. 11, pp. 1252-1257, Nov. 1971.

[78] P. S. Bottorff et al., "Test generation for large networks," in *Proc. 14th Design Automation Conf.*, IEEE Pub. 77CH1216-1C, pp. 479-485, June 1977.

[79] R. D. Edlred, "Test routines based on symbolic logic statements," *J. Assoc. Comput. Mach.*, vol. 6, no. 1, pp. 33-36, 1959.

[80] P. Goel, "Test generation costs analysis and projections," presented at the 17th Design Automation Conf., Minneapolis, MN, 1980.

[81] E. P. Hsieh et al., "Delay test generation," in *Proc. 14th Design Automation Conf.*, IEEE Pub. 77CH1216-1C, pp. 486-491, June 1977.

[82] C. T. Ku and G. M. Masson, "The Boolean difference and multiple fault analysis," *IEEE Trans. Comput.*, vol. C-24, no. 7, pp. 691-695, July 1975.

[83] E. I. Muehldorf, "Test pattern generation as a part of the total design process," in *LSI and Boards: Dig. Papers, 1978 Ann. Semiconductor Test Symp.*, pp. 4-7, Oct. 1978.

[84] E. I. Muehldorf and T. W. Williams, "Optimized stuck fault test patterns for PLA macros," in *Dig. Papers, 1977 Semiconductor Test Symp.*, IEEE Pub. 77CH1216-7C, pp. 89-101, Oct. 1977.

[85] M. R. Page, "Generation of diagnostic tests using prime implicants," Coordinated Science Lab. Rep. R-414, University of Illinois, Urbana, May 1969.

[86] S. G. Papaioannou, "Optimal test generation in combinational networks by pseudo Boolean programming," *IEEE Trans. Comput.*, vol. C-26, no. 6, pp. 553-560, June 1977.

[87] K. P. Parker, "Adaptive random test generation," *J. Des. Automat. Fault Tolerant Comput.*, vol. 1, no. 1, pp. 62-83, Oct. 1976.

[88] —, "Probabilistic test generation," Tech. Note 18, Digital Systems Laboratory, Stanford University, Stanford, CA, Jan. 1973.

[89] J. F. Poage and E. J. McCluskey, "Derivation of optimum tests for sequential machines," in *Proc. 5th Ann. Symp. on Switching Circuit Theory and Logic Design*, pp. 95-110, 1964.

[90] —, "Derivation of optimum tests to detect faults in combinational circuits," in *Mathematical Theory of Automation.* New York: Polytechnic Press, 1963.

[91] G. R. Putzolu and J. P. Roth, "A heuristic algorithm for testing of asynchronous circuits," *IEEE Trans. Comput.*, vol. C-20, no. 6, pp. 639-647, June 1971.

[92] J. P. Roth, W. G. Bouricius, and P. R. Schneider, "Programmed algorithms to compute tests to detect and distinguish between failures in logic circuits," *IEEE Trans. Electron. Comput.*, vol. EC-16, pp. 567-580, Oct. 1967.

[93] J. P. Roth, "Diagnosis of automata failures: A calculus and a method," *IBM J. Res. Devel.*, no. 10, pp. 278-281, Oct. 1966.

[94] P. R. Schneider, "On the necessity to examine D-chairs in diagnostic test generation—An example," *IBM J. Res. Develop.*, no. 11, p. 114, Nov. 1967.

[95] H. D. Schnurmann, E. Lindbloom, R. G. Carpenter, "The weighted random test pattern generation," *IEEE Trans. Comput.*, vol. C-24, no. 7, pp. 695-700, July 1975.

[96] E. F. Sellers, M. Y. Hsiao, and L. W. Bearnson, "Analyzing errors with the Boolean difference," *IEEE Trans. Comput.*, vol. C-17, no. 7, pp. 676-683, July 1968.

[97] D. T. Wang, "An algorithm for the detection of tests sets for combinational logic networks," *IEEE Trans. Comput.*, vol. C-25, no. 7, pp. 742-746, July 1975.

[98] T. W. Williams and E. E. Eichelberger, "Random patterns within a structured sequential logic design," in *Dig. Papers, 1977 Semiconductor Test Symp.*, IEEE Pub. 77CH1261-7C, pp. 19-27, Oct. 1977.

[99] S. S. Yau and S. C. Yang, "Multiple fault detection for combinational logic circuits," *IEEE Trans. Comput.*, vol. C-24, no. 5, pp. 233-242, May 1975.

Simulation

[100] D. B. Armstrong, "A deductive method for simulating faults in logic circuits," *IEEE Trans. Comput.*, vol. C-22, no. 5, pp. 464-471, May 1972.

[101] M. A. Breuer, "Functional partitioning and simulation of digital circuits," *IEEE Trans. Comput.*, vol. C-19, no. 11, pp. 1038-1046, Nov. 1970.

[102] H.Y.P. Chiang et al., "Comparison of parallel and deductive fault simulation," *IEEE Trans. Comput.*, vol. C-23, no. 11, pp. 1132-1138, Nov. 1974.

[103] E. B. Eichelberger, "Hazard detection in combinational and sequential switching circuits," *IBM J. Res. Devel.*, Mar. 1965.

[104] E. Manning and H. Y. Chang, "Functional technique for efficient digital fault simulation," in *IEEE Int. Conv. Dig.*, p. 194, 1968.

[105] K. P. Parker, "Software simulator speeds digital board test generation," *Hewlett-Packard J.*, pp. 13-19, Mar. 1979.

[106] S. Seshu, "On an improved diagnosis program," *IEEE Trans. Electron. Comput.*, vol. EC-12, no. 1, pp. 76-79, Feb. 1965.

[107] S. Seshu and D. N. Freeman, "The diagnosis of asynchronous sequential switching systems," *IRE Trans, Electron. Compat.*, vol. EC-11, no. 8, pp. 459-465, Aug. 1962.

[108] T. M. Storey and J. W. Barry, "Delay test simulation," in *Proc. 14th Design Automation Conf.*, IEEE Pub. 77CH1216-1C, pp. 491-494, June 1977.

[109] S. A. Szygenda and E. W. Thompson, "Modeling and digital simulation for design verification diagnosis," *IEEE Trans. Comput.*, vol. C-25, no. 12, pp. 1242-1253, Dec. 1976.

[110] S. A. Szygenda, "TEGAS2—Anatomy of a general purpose test generation and simulation system for digital logic," in *Proc. 9th Design Automation Workshop*, pp. 116-127, 1972.

[111] S. A. Szygenda, D. M. Rouse, and E. W. Thompson, "A model for implementation of a universal time delay simulation for large digital networks," in *AFIPS Conf. Proc.*, vol. 36, pp. 207-216, 1970.

[112] E. G. Ulrich and T. Baker, "Concurrent simulation of nearly identical digital networks," *Computer*, vol. 7, no. 4, pp. 39-44, Apr. 1974.

[113] —, "The concurrent simulation of nearly identical digital networks," in *Proc. 10th Design Automation Workshop*, pp. 145-150, June 1973.

[114] E. G. Ulrich, T. Baker, and L. R. Williams, "Fault test analysis techniques based on simulation," in *Proc. 9th Design Automation Workshop*, pp. 111-115, 1972.

[115] J. Savir, "Syndrome–Testable design of combinational circuits," *IEEE Trans. Comput.*, vol. C-29, pp. 442-451, June 1980 (corrections: Nov. 1980).

[116] —, "Syndrome–Testing of 'syndrome-untestable' combinational circuits," *IEEE Trans. Comput.*, vol. C-30, pp. 606-608, Aug. 1981.

[117] A. K. Susskind, "Testing by verifying Walsh coefficients," in *Proc. 11th Ann. Symp. on Fault-Tolerant Computing* (Portland, MA), pp. 206-208, June 1981.

[118] E. J. McCluskey and S. Bozorgui-Nesbat, "Design for autonomous test," *IEEE Trans. Comput.*, vol. C-30, pp. 866-875, Nov. 1981.

Good Controllability and Observability Do Not Guarantee Good Testability

JACOB SAVIR

Abstract—In this paper we show that good controllability and observability do not guarantee good testability. In fact, one can easily find examples of faults that are difficult or impossible to detect, although both the controllability and observability figures are good.

Index Terms—Controllability, deterministic testing, observability, random testing, testability.

Introduction

The problem of analyzing the testability of a digital circuit has long been recognized to be an important one. With the levels of integration existing today, the cost of testing and diagnosis have become so large, that they are a significant part of the cost of the product. In order to reduce this cost it is crucial to have highly testable circuits. Since test generation and fault simulation consume a lot of computer time in present day densities, it is worthwhile to be able to predict whether or not the testing task is going to be easy.

A few testability measures and programs that implement them have been reported to date [1]–[5]. The limitations of these measures are:

1) The controllability, observability, and testability measures are not an accurate measure to the "ease of testing."

2) They fail to report testability problems in the presence of reconverging fanout.

3) The testability measures are defined such that "good controllability and observability figures usually imply good testability," which is not true in many cases.

4) Because the measures are not a true reflection of the ease of testing, they may guide the test designer to introduce hardware real estate (to enhance testability) in the wrong place.

In this paper we elaborate on these issues. The paper should not be regarded as a "new testability measure," but rather as an attempt to point out the limitations of the existing methods, and the kind of emphasis necessary from the future ones to come.

The discussion is restricted to combinational circuits and stuck-at-faults.

I. Definitions and Properties

Let C be a combinational circuit with n inputs, $x_1, x_2, \cdots, x_n$, and m outputs, $F_1, F_2, \cdots, F_m$. Let $\vec{x} = (x_1, x_2, \cdots, x_n)$. Let $g(\vec{x})$ be a line in the circuit. We denote by $g/0$ ($g/1$) the fault g stuck-at-zero (g stuck-at-one).

The controllability of a line in the circuit is a measure of how easy it is to set the line to a given value. Similarly, the observability of a line is a measure of how easy it is to observe its value. We define the controllability and observability of a fault in the following way.

Manuscript received June 5, 1982; revised November 10, 1982.
The author is with the IBM Thomas J. Watson Research Center, Yorktown Heights, NY 10598.

Definition 1: The *controllability of a fault* g/i, $i \in \{0, 1\}$ is the fraction of input vectors that will set the value of that line to $\bar{i}$. In other words, the controllability of a fault g/i is the probability that an input picked at random will set the value of line g to $\bar{i}$.

Definition 2: The *observability of a fault* g/i, $i \in \{0, 1\}$, is the fraction of input vectors that will propagate the effect of this fault to a primary output. In other words, the observability of a fault is the probability that an input picked at random will propagate the effect of this fault to a primary output.

The input vectors that detect the fault $g/0$ can be obtained by solving the Boolean equation

$$g(\vec{x}) \sum_{j=1}^{m} \frac{\partial F_j}{\partial g} = 1, \tag{1}$$

and the input vectors that detect the fault $g/1$ can be obtained by solving the equation

$$\bar{g}(\vec{x}) \sum_{j=1}^{m} \frac{\partial F_j}{\partial g} = 1 \tag{2}$$

where the summation symbol means the Boolean sum (OR operation). Thus, according to definitions 1 and 2, the controllability of the fault $g/0$, $c(g/0)$, is the fraction of input combinations that yield $g(\vec{x}) = 1$, and the controllability of the fault $g/1$, $c(g/1)$, is the fraction of input vectors that yield $\bar{g}(\vec{x}) = 1$. Similarly, the observability o of either fault ($g/0$, or $g/1$) is the fraction of the input vectors that yield

$$\sum_{j=1}^{m} \frac{\partial F_j}{\partial g} = 1. \tag{3}$$

It is worth while to note that the notion of the *syndrome* [6] of a function F, denoted by $S(F)$, is exactly the fraction of the input vectors that yield $F = 1$. Thus, we can take advantage of syndrome relations to compute controllability and observability figures. In particular, the following relations hold:

$$c(g/0) = S(g(\vec{x})), \tag{4}$$

$$c(g/1) = S(\bar{g}(\vec{x})), \tag{5}$$

and

$$o(g/0) = o(g/1) = S\left(\sum_{j=1}^{m} \frac{\partial F_j}{\partial g}\right). \tag{6}$$

Definition 3: The testability of a fault g/i, $i \in \{0, 1\}$ is the fraction of the input vectors that detect the fault.

In other words, the testability of a fault is the probability that an input picked at random will detect the fault.

According to Definition 3 and the notion of the syndrome, we can relate the testabilities $t(g/0)$ and $t(g/1)$ of the faults $g/0$ and $g/1$ to the following syndrome relations:

$$t(g/0) = S\left(g(\vec{x}) \sum_{j=1}^{m} \frac{\partial F_j}{\partial g}\right), \tag{7}$$

$$t(g/1) = S\left(\bar{g}(\vec{x}) \sum_{j=1}^{m} \frac{\partial F_j}{\partial g}\right). \tag{8}$$

The definitions listed above, and the properties of the syndrome, further imply the following relations between controllability, observability, and testability of a fault:

Property 1:

$$0 \leq c(g/i), o(g/i) \leq 1, \text{ for } i \in \{0, 1\} \tag{9}$$

Property 2:
$$c(g/0) = 1 - c(g/1) \quad (10)$$

Property 3:
$$o(g/0) = o(g/1) \quad (11)$$

Property 4:
$$t(g/i) \leq \text{Min}[c(g/i), o(g/i)], i \in \{0, 1\} \quad (12)$$

Property 5: If $g(\tilde{x})$ and $\sum_{j=1}^{m} \frac{\partial F_j}{\partial g}$ are independent (namely, both depend on different inputs), then

$$t(g/i) = c(g/i)o(g/i), i \in \{0, 1\}. \quad (13)$$

Note, that if a circuit is a tree, then Property 5 holds. However, we can even come up with a stronger statement. Let cone(g) be the collection of lines feeding (directly, or indirectly) the line g. Then Property 5 holds for line g if either

i) Every path originating in line $l \in$ cone(g) passes through line g, or

ii) Every path originating in line $l \in$ cone(g), and not passing through g, reaches the outputs $F_j, j = i_1, i_2, \cdots, i_k$, where $g \notin$ cone (F_j) for all $j = i_1, i_2, \cdots, i_k$.

The important point to observe here is that the tests that detect a given fault lie in the *intersection* of the set of input vectors that control the fault, and the collection of vectors that observe the fault. Thus, it is possible to have large controllability and observability sets, and small testability sets. So, no conclusion, based solely on controllability and observability, can be drawn, in general, regarding the testability of the circuit. Moreover, some testability measures proposed in the past define the testability as either the geometric mean of the controllability and the observability [1], [2], the square root of sum of squares of the controllability and the observability figures, or as a product of the two [5]. These definitions may lead to false conclusions. In fact, using an arbitrary function of the product of observability and controllability as a measure of testability may lead to a result which is either vastly greater or smaller than the actual testability.

II. EXAMPLES

Example 1: Consider the circuit of Fig. 1, with the fault $g/1$. We have

$$g = x_2$$

and

$$F = x_1 g + \bar{x}_2.$$

Thus, the controllability, observability, and testability are given by

$$c(g/1) = S(\bar{g}) = S(\bar{x}_2) = 1/2,$$
$$o(g/1) = S\left(\frac{\partial F}{\partial g}\right) = S(x_1 x_2) = 1/4,$$

and

$$t(g/1) = S\left(\bar{g}\frac{\partial F}{\partial g}\right) = S(\bar{x}_2 x_1 x_2) = 0.$$

In this example, for the specified fault, we have 50 percent controllability, 25 percent observability figure, and 0 testability figure. Notice that the product of the controllability and observability is much larger than the testability figure. The reconverging fanout, in this example, yields disjoint controllability and observability sets which explains the 0 testability figure. Obviously, the circuit of Fig. 1 is redundant.

Example 2: Consider the circuit of Fig. 2, with fault $g/0$. We have

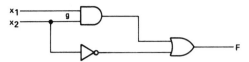

Fig. 1. The circuit of Example 1.

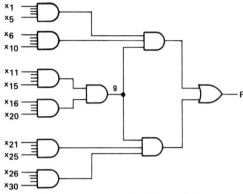

Fig. 2. The circuit of Example 2.

$$g = \prod_{j=11}^{20} x_j,$$
$$F = g\left(\prod_{j=1}^{10} x_j + \prod_{j=21}^{30} x_j\right)$$

where Πx_j means Boolean multiplication (AND operation).

$$\frac{\partial F}{\partial g} = \prod_{j=1}^{10} x_j + \prod_{j=21}^{30} x_j.$$

Thus,

$$c(g/0) = S(g) = 2^{-10},$$

and

$$o(g/0) = S\left(\frac{\partial F}{\partial g}\right) = 2^{-10} + 2^{-10} - 2^{-20} \cong 2^{-9}.$$

Since the line g meets the condition of Property 5, we have

$$t(g/0) = c(g/0)o(g/0) \cong 2^{-19}.$$

One could argue that neither the controllability nor the observability figures are too bad (application of, say, 10 000 random inputs are either very likely to control the fault, or observe the fault), but the testability figure is very low (it requires an application of about one million random inputs to have a good chance of detecting the fault).

Example 3: Consider the circuit of Fig. 3, with fault $g/1$. We have

$$g = \overline{x_2 x_3},$$
$$c(g/1) = S(x_2 x_3) = 1/4,$$
$$\frac{\partial F_1}{\partial g} = 0$$
$$F_2 = \bar{g} + x_3 x_4,$$
$$\frac{\partial F_2}{\partial g} = \overline{x_3 x_4},$$
$$F_3 = \bar{g} + x_4 x_5,$$
$$\frac{\partial F_3}{\partial g} = \overline{x_4 x_5},$$
$$o(g/1) = S(\overline{x_3 x_4} + \overline{x_4 x_5}) = S(\bar{x}_3 + \bar{x}_4 + \bar{x}_5) = 7/8,$$
$$t(g/1) = S[x_2 x_3(\bar{x}_3 + \bar{x}_4 + \bar{x}_5)] = S[x_2 x_3(\bar{x}_4 + \bar{x}_5)] = 3/16.$$

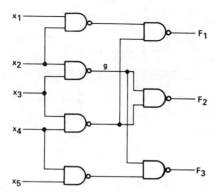

Fig. 3. The circuit of Example 3.

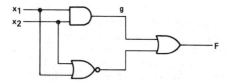

Fig. 4. The circuit of Example 4.

Note, that in this case

$$t(g/1) < c(g/1)o(g/1),$$

namely, the testability figure is worse than the product of the controllability and observability figures.

Example 4: Consider the circuit of Fig. 4, with fault $g/0$. We have

$$g = x_1 x_2,$$
$$F = g + \bar{x}_1 \bar{x}_2,$$
$$c(g/0) = S(g) = 1/4,$$
$$o(g/0) = S\left(\frac{\partial F}{\partial g}\right) = S(x_1 + x_2) = 3/4,$$
$$t(g/0) = S\left(g \frac{\partial F}{\partial g}\right) S[x_1 x_2 (x_1 + x_2)] = 1/4.$$

This is an example for which the testability is larger than the product of the controllability and observability.

III. Summary and Conclusions

In this paper we have shown that, in general, no conclusion can be drawn regarding the testability of the circuit, based strictly on the examination of the controllability and observability figures. The reason for that is that the testability is governed by the overlap between the input vectors that control the fault, and those that observe the fault. Thus, it is possible to have large controllability and observability sets, and, still, poor testability sets. We have shown that the testability of a fault is at best the smaller between the controllability and observability figures. We have shown examples where the controllability and observability figures were reasonably good, while the testability figure was poor, or even 0.

This paper should by no means be regarded as a tool for computing testability figures. The problem of computing the testability is in general NP-complete. The thrust of this paper is to show that the widely used notions of controllability and observability may be insufficient in analyzing testability issues. In particular this is critical when very high fault coverages are required. In this case, it is important to detect every detectable fault. It is necessary, therefore to have a testability estimation engine which besides being simple and efficient, has to be able to identify correctly testing bottle necks.

It is important to note that the definitions of controllability, observability, and testability apply both to random testing and deterministic testing. Because of its probabilistic interpretation, it is evident that it applies to random testing (and therefore to any method based on random application of patterns, like, for example, self-test with random patterns). It also applies to deterministic testing, since it addresses the size of the class of input patterns that either control, observe, or detect the fault. The larger the class of test vectors, the easier it will be for any deterministic test generator to locate one.

A subsequent paper will report on efforts to come up with testability prediction tools which trade off complexity with accuracy, and have the potential of meeting the challenges listed above.

References

[1] J. E. Stephenson and J. Grason, "A testability measure for register transfer level digital circuits," in *Proc. 1976 Int. Symp. Fault Tolerant Comput.*, Pittsburgh, PA, June 1976, pp. 101–107.

[2] J. Grason, "TMEAS, a testability measurement program," in *Proc. 16th Design Automat. Conf.*, San Diego, CA, June 1979, pp. 156–161.

[3] L. H. Goldstein, "Controllability/observability analysis of digital circuits," *IEEE Trans. Circuits Syst.*, vol. CAS-26, pp. 685–693, Sept. 1979.

[4] P. G. Kovijanic, "Computer-aided testability analysis," in *Proc. 1979 Autotescon*, pp. 292–294.

[5] R. G. Bennetts *et al.*, "CAMALOT: A computer aided measure for logic testability," *Proc. Inst. Elec. Eng.*, vol. 128-E pp. 177–189, Sept. 1981.

[6] J. Savir, "Syndrome-testable design of combinational circuits," *IEEE Trans. Comput.*, vol. C-29, pp. 442–451, June 1980; see also *IEEE Trans. Comput.*, vol. C-29, pp. 1012–1013, Nov. 1980.

TESTABILITY MEASURES -- WHAT DO THEY TELL US?

Vishwani D. Agrawal

Melvin Ray Mercer

Bell Laboratories
Murray Hill, New Jersey 07974

ABSTRACT

Methods are developed for interpreting the information contained in testability measures. Two types of inferences are sought. First, the relationship between the testability measure for a fault and its detectability is investigated. Second, the question, how testability values can be used to predict the test length required to achieve some given fault coverage, is attempted. Real LSI circuits are used as illustrations.

INTRODUCTION

As logic circuits become more dense, and more complex, the problem of test generation, in general, exceeds the capacity of existing computing resources [1]. One well accepted approach for reducing testing difficulties is to consider circuit testability as early as possible in the design cycle. Several algorithms for measuring testability in digital circuits have been proposed [2]-[6]. Each of these algorithms assumes that testability is an inherent property of the circuit based solely upon the circuit structure (or topology). This assumption allows estimation of the circuit's testability <u>before</u> test generation is begun. One basic requirement of a testability analysis algorithm is that it should be computationally simpler than the test generation. If the testability calculation is fast and if it accurately predicts the difficulty of test generation, then problem circuits can be identified early in their design cycle, and remedial actions can be taken as part of the design. If the testability results are accurate for individual internal circuit nodes, then areas of poor testability can be identified. This information can be used to direct circuit modifications which improve testability.

Clearly good, computationally fast and accurate testability measures would be useful tools for digital circuit designers. Some of the currently used testability measure programs have been described in [7]-[9]. This paper does not suggest a new testability algorithm. It does not suggest modifications or enhancements to existing testability measures. Instead, we ask what useful information comes from a testability measure and we propose exact statistical metrics to evaluate the quality of information produced by a testability measure.

The examples use results from SCOAP, but the statistical criteria can be applied to any testability measure. Only recently, the workers in this area have started looking into interpretations of testability measures. Previous work [10]-[13] on evaluating testability measures has been aimed at comparing the composite testability of several circuits with actual test generation experience. In contrast, we compare detailed testability predictions for the faults within one circuit.

The key point here is that the <u>utility</u> of a designer's response to a testability measure will be determined by the <u>quality</u> of information in the testability measure's results. An automobile driver may be able to fairly accurately judge his speed by the level of engine noise and external visual perception if his speedometer does not work. In contrast, the rate of change of the amount of gasoline in the fuel tank is a poor indicator of the car's speed. The utility of a measure itself may also depend upon how easy it is to compute and to analyze.

The statistical calculations proposed below evaluate the utility of testability measures for the associated designer task.

LIMITATIONS OF TESTABILITY MEASURES

In this discussion we will use one of the testability measures, namely, SCOAP [5], [8]. Similar arguments can be applied to other measures. SCOAP produces six values for each node in a logic circuit: combinational zero controllability CC0, combinational one controllability CC1, combinational observability CO, sequential zero controllability SC0, sequential one controllability SC1, and sequential observability SO. The values of each of these variables grow larger as the required testing effort increases. These values "provide a quantitative measure of the difficulty of controlling and observing the logic value of internal nodes from considerations of circuit topology alone [5]". The values are *estimates* because of the simplifying assumptions made by the algorithm. For example, Figure 1 shows two realizations for a very simple digital circuit -- an inverter. The SCOAP algorithm produces different (CC0) values for the two realizations because it assumes all inputs to all logic elements are independent. In this case, the assumption is not valid for the NAND gate realization.

All testability measures have similar simplifying assumptions so that all results are estimates. The amount of information degradation resulting from such

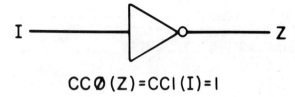

$CC0(Z) = CCI(I) = I$

a. Inverter realization

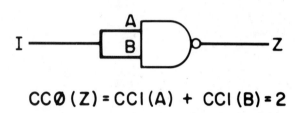

$CC0(Z) = CCI(A) + CCI(B) = 2$

b. NAND realization

Fig. 1 Two realizations of an inverter.

assumptions will vary with the circuit designs themselves and with the way the values are interpreted.

A second limitation of testability measures involves the restricted information source on which they rely. As shown in Figure 2, testability measures produce their results on the basis of circuit structure only. The actual testing process also involves a test pattern generator which is not considered by the testability measure. Clearly, the actual expense and difficulty of testing will depend heavily on the ability and sophistication of the test pattern generator as well as the circuit structure.

Other limitations exist, but the point is by now clear. Since the raw information from testability measures is not exact, some evaluation of the utility of testability values is necessary. This utility may vary considerably based upon the way in which the values are used.

WHAT DESIGNERS WANT TO KNOW

Usually, the test generation process goes in the following way. A set of single stuck-at faults is postulated for nodes in the circuit. Sufficient tests are generated to detect most of these faults. The results are: (1) a set of test patterns, (2) the fraction of detected faults (which serves as a figure of merit for the test set), and (3) the set of detected (or undetected) faults.

Two questions seem to be the most common from logic designers who are potential users of testability measures. The first question is: "Will this device require an inordinate amount of time, level of effort, and/or test length in order to provide acceptable testing?" If the answer to this question is "no", then the second question is never asked. For the case of potential problem designs, the next question is: "What are the problem areas in the design where a modification can ease the testing problem?"

The first question seems to be a simpler one than the second because the first requires some general

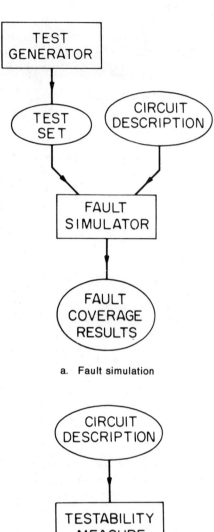

a. Fault simulation

b. Testability estimation

Fig. 2 Testability measures and fault simulation

information about the entire design. In contrast, the second question requires detailed information about individual components of the design.

Testability measures usually produce values for each node in the design. This information might be useful for answering the second question. These details must be combined in some way in order to provide information at the higher, overall design level implied in the first question.

Unfortunately, both questions are too vague to allow a meaningful statistical evaluation of the utility of testability measures. However, similar questions can be formulated which allow exact statistical evaluation.

FAULT-SPECIFIC TESTABILITY DATA

Every design eventually produces the results described in the last section--a set of test patterns and detected/undetected fault sets. The statistical measures presented below evaluate the accuracy with which these results can be predicted based upon testability analysis. All these measures compare the testability predictions with actual test pattern generation results for particular design cases.

SCOAP does not directly produce fault-specific data. At a point p in the circuit, there will be three combinational values:

1) CC0(p),
2) CC1(p),
and 3) CO (p).

Note that if p is stuck-at-zero, then to detect the fault it will be necessary to drive p to a logic one-- CC1(p) -- and also to observe the resulting logic value at p -- CO(p). Assuming that these tasks are independent, the total effort is:

testability(*p stuck at zero*) = $CC1(p) + CO(p)$.

Similarly:

testability(*p stuck at one*) = $CC0(p) + CO(p)$.

In this way, simple addition can be used to collect fault-specific measures from SCOAP results. If desired, the sequential SCOAP values (SC0, SC1, and SO) can also be included, but from our experience it seems that the combinational numbers are always much larger than the sequential numbers so that including the sequential parts in a simple summation makes little or no difference in the final result. Other ways of including sequential values in the above formulation were not investigated.

With the simple mapping described above, a numerical value can be assigned to every circuit fault based on SCOAP's testability values. The actual test generation process produces information about which faults are actually detected and which faults remain undetected. These two pieces of information can be compared statistically to measure the extent of agreement between testability predictions and actual fault detection results.

PROBABILITY OF DETECTION

Whether or not a fault is detectable is a deterministic phenomenon. In particular, all the nonredundant faults *are* potentially detectable. In practice, however, a test generation procedure has limits with respect to human effort, CPU time, number of test vectors, etc., and certain (non-redundant) faults will remain undetected. Under given circumstances, therefore, a fault will have only a certain probability of being detected. Since testability measures are often regarded as measures of the effort of test generation, we assume that the probability of

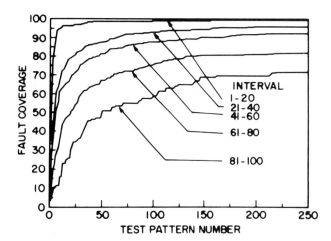

Fig. 3 Fault detection given testability for a large combinational circuit

detection (for a given test generation procedure) should be related to the testability measure.

CONDITIONAL PROBABILITY OF DETECTION GIVEN TESTABILITY VALUE

We ask: Is the probability of detection of a fault in fact dependent on its testability value? In order to answer this question, a set of testability intervals is selected. Each fault is assigned to exactly one testability interval based upon its testability value. The fraction of detected faults in each testability interval corresponds to the probability of detection of faults in the given testability interval. This statistical measure can be evaluated many times as the test patterns are applied during fault simulation. The result is a family of fault coverage plots where each member of the family corresponds to one testability interval.

Figure 3 shows an example of such results for a large combinational circuit with over 3000 faults. Since the testability values in SCOAP increase with required test generation effort the faults with low testability values are quickly detected, and their probabilities of detection quickly approach one. In contrast, faults with the greatest testability values are detected more slowly, and their final probability of detection is only about 0.7 . For this example, the testability information seems to be good in that the fault detection performance consistently improves as testability improves. However, note that even the least testable faults have a .7 probability of detection. Thus, if a designer tries to improve circuit testability by reducing the testability values for all faults in the highest interval, then 70% of the effort is wasted.

Figure 4 shows a second example of fault detection performance given testability intervals. The two circuit sizes are almost identical, but this circuit is sequential in nature. For this case, the difference in fault detection performance is not consistently indicated by its testability value -- the plots cross one another, and faults in the testability interval 401-600 have a <u>higher</u> final probability of detection than those in the interval 201-400. Larger testability intervals would probably produce results more like those of Figure 3.

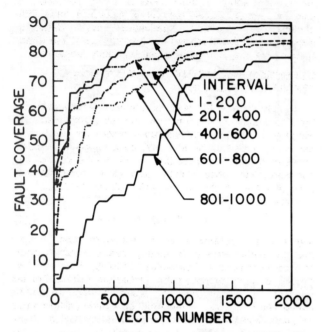

Fig. 4 Fault detection given testability for a large sequential circuit

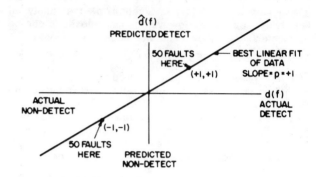

Fig. 5 An example of perfect correlation

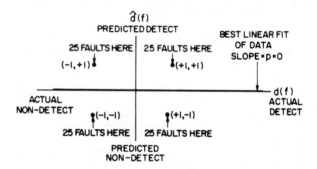

Fig. 6 An example of zero correlation

Generally, the conditional probability of detection given testability value indicates how well the testability measure predicts the detectability of fault sets with similar testability values. This statistical measure also suggests what fraction of the faults in each testability interval will be detected.

Different testability measures or different versions of the same testability measure can be compared via this analysis. The measure which most effectively differentiates faults with high probabilities of detection from those with low probabilities of detection is the most desirable testability measure.

This statistical tool also can provide a figure of merit for a testability measure by indicating what fraction of faults in the least testable interval are in fact undetected after the application of all test patterns. An ideal testability measure would result in 100% of all of the most testable faults and 0% of the least testable faults being detected (for a suitable selection of testability intervals).

COEFFICIENT OF CORRELATION BETWEEN TESTABILITY AND DETECTION

A second statistical calculation provides one simple numerical value which is an estimator of the testability measure's ability to predict which individual faults will be detected and which individual faults will not be detected.

One random variable $-d(f_i)-$ is associated with each fault, f_i, in the circuit based on *detection results*. If a fault is detected during simulation, the value $+1$ is assigned to $d(f_i)$, if the fault is not detected, the value -1 is assigned.

A second random variable $-\hat{d}(f_i)-$ is associated with each fault in the circuit based on *testability value*.

If a fault's testability is below a threshold to be discussed later, then a predicted detection is inferred, and $\hat{d}(f_i)$ is assigned the value $+1$. If a fault's testability is above the threshold, then a predicted non-detect is inferred, and $\hat{d}(f_i)$ is assigned the value -1.

The coefficient of correlation (ρ) between $d(f)$ and $\hat{d}(f)$ is a statistical estimate of the testability measure's performance in predicting whether or not faults will be detected. The value of ρ will vary with the selected threshold. In our analysis, all possible threshold values are considered, and the threshold which maximizes ρ is used.

Figure 5 shows one possible case where the two random variables are perfectly correlated and ρ has the value $+1$. Because (for this example) d and $\hat{d}$ have zero mean and unit variance, ρ may be interpreted as the slope of the best least-mean-square linear fit to the data. In this case, the testability measure is a perfect predictor of fault detection.

Figure 6 shows a second case where the two random variables are uncorrelated and ρ has the value 0. Again d and $\hat{d}$ have zero mean and unit variance, and the best linear fit is a line with zero slope (ρ). In this case, the testability measure gives absolutely no indication of fault detection.

In actual circuits, the value of ρ will be expected to have a value between $+1$ and 0. The closer ρ is to $+1$, the better the testability measure's ability to predict individual fault detections.

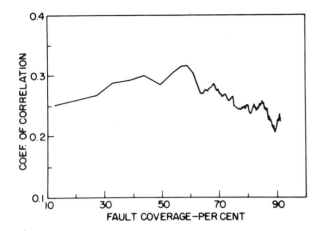

Fig. 7 Coefficient of correlation versus fault coverage for a large combinational circuit

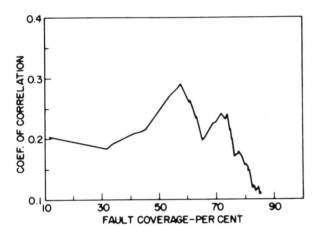

Fig. 8 Coefficient of correlation versus fault coverage for a large sequential circuit

The statistical measurement described above can be made at selected points in the test pattern sequence during fault simulation. The results can be plotted as ρ versus the circuit's fault coverage.

Figure 7 shows such a plot for the same combinational circuit used to produce Figure 3. Figure 8 shows a similar plot for the same sequential circuit used to produce Figure 4. Note that in both cases, the value of ρ is always relatively small (less than 0.4).

Figures 7 and 8 indicate that for these two circuits and their test sequences, the testability measure does a relatively poor job of predicting which <u>individual</u> faults will be detected and which <u>individual</u> faults will remain undetected. On the other hand, Figures 3 and 4 indicate that for the same circuits and test sequences, the same testability measure does a better job of predicting which <u>fault sets</u> have higher probabilities of detection and which <u>fault sets</u> have lower probabilities of detection.

Such data suggest that there is a level of resolution at which testability measures can provide useful information. Attempts to extract more detailed information about a fault have a small likelihood of success. In particular, testability measures have little chance of predicting individual fault detection performance. Even so, general information at the entire circuit level <u>may</u> be available via testability measures.

DETECTION PROBABILITY AS A FUNCTION OF TESTABILITY MEASURE

In this section we give an analysis which predicts fault coverage information from the testability measure. Suppose a fault in a circuit has a testability measure t. Let its detection probability $p(t)$ be a function bounded between 0 and 1. For SCOAP, t will be the sum of the combinational observability and the combinational controllability of the faulty line. Since $1 \leq t \leq \infty$, we assume

$$p(t) = e^{-\alpha t}, \qquad (1)$$

where α is a parameter to be discussed later. Notice that an infinite value of testability measure corresponds to a zero probability of detection. In SCOAP the effort of testing is represented by the computed values. Let us consider a circuit with N faults having testability measures $t_1, t_2, ..., t_N$. If the fault detection performance of each test vector is statistically independent, then, after applying v vectors, the probability of detecting the ith fault is

$$1 - \left[1 - p(t_i)\right]^v$$

The fault coverage is then given by

$$f(v) = \frac{1}{N} \sum_{i=1}^{N} \left\{ 1 - \left[1 - p(t_i)\right]^v \right\}$$

$$= 1 - \frac{1}{N} \sum_{i=1}^{N} \left[1 - p(t_i)\right]^v \qquad (2)$$

Substituting (1) into (2), we get

$$f(v) = 1 - \frac{1}{N} \sum_{i=1}^{N} \left[1 - e^{-\alpha t_i}\right]^v \qquad (3)$$

The value of α should depend upon the test generation procedure. In other words, α should relate the test generator to the testability measure which is a function of circuit topology alone.

We will determine α by an experimental procedure. Suppose we have an ordered set of v vectors for a circuit such that the ith fault, which has a testability measure t_i, is detected on the v_ith vector. Then the probability of this event is

$$\left[1 - p(t_i)\right]^{v_i - 1} p(t_i).$$

We define the likelihood function or the joint probability of detecting the faults in this particular order as

$$L = \prod_{i=1}^{N} \left[1 - p(t_i)\right]^{v_i - 1} p(t_i)$$

The value of α that maximizes this probability is called its maximum likelihood estimate. Substituting from (1), and taking logarithm, we get

$$\ln L = -\alpha \sum_{i=1}^{N} t_i + \sum_{i=1}^{N} (v_i - 1) \ln \left[1 - e^{-\alpha t_i}\right]$$

After differentiating with respect to α and setting the derivative to zero, we obtain the maximum likelihood estimate of α as follows:

$$\frac{1}{L}\frac{dL}{d\alpha} = -\sum_{i=1}^{N} t_i + \sum_{i=1}^{N} \frac{(v_i-1) e^{-\alpha t_i} t_i}{1-e^{-\alpha t_i}} = 0$$

$$\text{or} \quad \sum_{i=1}^{N} t_i \left[1 - \frac{(v_i-1) e^{-\alpha t_i}}{1-e^{-\alpha t_i}}\right] = 0 \quad (4)$$

For $\alpha = 0$, the left hand side (LHS) in (4) is $-\infty$. Also, for $\alpha = \infty$, LHS $= \sum_{i=1}^{N} t_i > 0$. Thus (4) has a solution in the range $0 < \alpha < \infty$.

For the 3,000 gate circuit considered in the previous discussion, a set of 2,000 vectors was generated. Fault simulation determined the vector number (v_i) on which the ith fault was detected. Of the 3949 nonredundant faults 3371 faults were detected by the vector set. The SCOAP program was used to calculate t_i for each fault. Solving (4) numerically, α was obtained as 0.012. This value when substituted in (3) gave estimated fault coverage of 87 percent at 2,000 vectors, 92 percent at 4,000 vectors, and 95 percent at 10,000 vectors.

CONCLUSIONS

It seems clear that good testability information is of significant use to logic designers especially if it is available early in the design cycle. Existing testability measures do provide some information about the circuit as it is implemented. However, care and good engineering judgement should be used in the interpretation of testability results.

For the design cases which have been presented, some conclusions can be drawn. First, the testability data provide a relatively poor indication of whether or not an individual fault will be detected by a given test. Second, the testability data contain some indication of the probability of detection of a fault. The quality (or resolution) of these data seems to vary from one circuit to another. Finally, statistical tools have been presented to estimate the required testing effort for given testability values. The useful applications of this estimate require further investigation.

The statistical tools described here should be useful for any design effort which employs testability measures as part of the design process. They can be used to indicate the level of confidence which should be given to testability results. These same tools can also be used to compare various testability measures in terms of the quality of information which they produce.

REFERENCES

[1] O. H. Ibarra and S. K. Sahni, "Polynomially Complete Fault Detection Problems," *IEEE Trans on Computers*, Vol. C-24, pp. 242-249, March 1975.

[2] H. Y. Chang and G. W. Heimbigner, "Controllability, Observability, and Maintenance Engineering Technique (COMET)," *Bell Syst. Tech. Journal*, Vol. 53, pp. 1505-1534, October 1974.

[3] J. E. Stephenson and J. Grason, "A Testability Measure for Register Transfer Level Digital Circuits," Sixth International Fault Tolerant Computing Symposium, 1976, *Digest of Papers*, pp. 101-107.

[4] J. A. Dussault, "A Testability Measure," IEEE Semiconductor Test Conference, Cherry Hill, New Jersey, Oct.-Nov., 1978, *Digest of Papers*, pp. 113-116.

[5] L. H. Goldstein, "Controllability/Observability Analysis of Digital Circuits," *IEEE Trans. on Circuits and Systems*, Vol. CAS-26, pp. 685-693, September 1979.

[6] P. G. Kovijanic, "Testability Analysis," IEEE Semiconductor Test Conference, Cherry Hill, New Jersey, October 23-25, 1979, *Digest of Papers*, pp. 310-316.

[7] J. Grason, "TMEAS, A Testability Measurement Program," *Proceedings of 16th Design Automation Conference*, San Diego, CA, June 25-27, 1979, pp. 156-161.

[8] L. H. Goldstein and E. L. Thigpen, "SCOAP: Sandia Controllability/Observability Analysis Program," *Proceedings of 17th Design Automation Conference*, Minneapolis, MN, June 1980, pp. 190-196.

[9] R. G. Bennetts, C. M. Maunder, and G. D. Robinson, "CAMELOT: a computer-aided measure for logic testability," *Proceedings of International Conference on and Circuits and Computers*, Port Chester, N.Y, Oct. 1-3, 1980, pp. 1162-1165, also *IEE Proc.*, Vol. 128, Pt.E, p. 177-189, September 1981.

[10] P. G. Kovijanic, "Single Testability Figure of Merit," IEEE Semiconductor Test Conference, Philadelphia, PA., Oct. 27-29, 1981, *Digest of Papers*, pp.521-529.

[11] B. Dunning and P. Kovijanic, "Demonstration of a figure of merit for inherent testability," IEEE Autotescon, Orlando, Florida, Oct. 19-21, 1981, pp. 515-520.

[12] S. Takasaki, "Testability Measure Analysis at the Functional Level," Fifth Annual IEEE Workshop on Design for Testability, Vail Colorado, April 21,22, 1982.

[13] J. Hickman and Theo Powell, "Pre Analysis of LSI Gate-Array Designs," 1982 IEEE Computer Elements Workshop, New York, N. Y., May 20-21, 1982.

Syndrome-Testable Design of Combinational Circuits

JACOB SAVIR, MEMBER, IEEE

Abstract—Classical testing of combinational circuits requires a list of the fault-free response of the circuit to the test set. For most practical circuits implemented today the large storage requirement for such a list makes such a test procedure very expensive. Moreover, the computational cost to generate the test set increases exponentially with the circuit size.

In this paper we describe a method of designing combinational circuits in such a way that their test procedure will require the knowledge of only one characteristic of the fault-free circuit, called the syndrome. This solves the storage problem associated with the test procedure. The syndrome-test procedure does not require test vector generation, and thus the expensive stage of test generation and fault simulation is eliminated.

Index Terms—Combinational circuit, fan-out-free circuit, minterm, prime implicant, single fault, stuck-at fault.

I. INTRODUCTION

IN THIS paper we restrict ourselves to the class of permanent stuck-at faults, i.e., a faulty line appears logically as if it is either stuck at a constant 0, or at a constant 1. The classical approach [6] to the problem of testing combinational circuits was to design a test set which detects all faults from a prescribed set, and store both the test set and the expected output to the individual test vectors. Whenever a circuit was tested, the actual response and the expected output were compared to determine whether or not the circuit under test (CUT) was fault-free. For most practical circuits manufactured today, this approach requires test generation, possibly coupled with fault simulation, and high data volumes for implementation. The cost and time required to implement the classical approach grows quickly with the levels of integration. In this paper we show that by using a new method it is possible to reduce the storage requirement considerably at a very low cost, and avoid the expensive stage of test generation.

Recently, a few other approaches to the testing problem have been presented in the literature. In [3] the method of transition counts is described. This method has the advantage of drastically reducing the storage requirement, but it is difficult to determine a proper test input sequence. In [5] a referenceless testing method is described. Its disadvantage lies in requiring very long test patterns to achieve a reasonable detection probability. In [8], a practical test procedure which employs testing for some functional attributes of combinational circuits is described. It is shown in [8] that by testing for specific functional properties it is possible to determine whether or not some classes of combinational circuits suffer any stuck-at faults.

In this paper we show a method of designing combinational circuits so that the storage requirement for implementing the test procedure will be restricted to only one number, called the syndrome of the circuit, which is based on the number of minterms realized by the switching function. Since a fault-free and faulty circuit do not necessarily have different syndromes, special design of the combinational circuit is required in order to make it syndrome-testable. We say that a circuit is syndrome-testable if the syndrome of any faulty version of a circuit, induced by a single stuck-at fault, does not equal the syndrome of the fault-free circuit. Thus, the test procedure consists of applying all input combinations to the CUT and recording its syndrome (usually implemented by a counter). If the actual syndrome equals the expected syndrome, the circuit is fault-free; otherwise a fault is detected and the procedure stops. This approach has both the advantage of requiring only one reference and the advantage of avoiding the test generation process. In order to reduce the test length for circuits with large number of inputs (for instance more than 20), the combinational circuit is partitioned into subcircuits and designed so that each subcircuit is syndrome-testable. In this way, each subcircuit can be tested separately in no more than 1 s.

The penalty paid for producing a syndrome-testable design is a slight increase in the number of pins.

In Section II various syndrome properties of combinational circuits are described. Section III describes the proposed testable design. The paper concludes with a brief summary.

II. SYNDROME—PROPERTIES OF COMBINATIONAL CIRCUITS

Definition 1: The *syndrome* of a Boolean function is defined as $S = K/2^n$, where K is the number of minterms realized by the function and n the number of binary input lines.

The syndrome is a functional property. Thus, various realizations of the same function will have the same syndrome. Clearly, $0 \leq S \leq 1$, where the boundaries are attained by the constant functions.

The syndrome of various n-input gates is shown in Fig. 1.

It is useful to find the input-output syndrome relations between various interconnected sections of a logic circuit. These input-output relations are especially simple when the associated sections have unshared (disjoint) inputs.

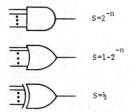

Fig. 1. Syndrome of various n-input gates.

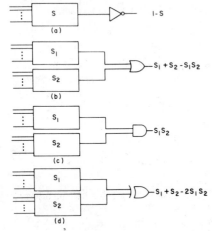

Fig. 2. Input-output syndrome relations.

Lemma 1: When the inputs are unshared, the input-output syndrome relations of the networks terminating in an INVERTER, OR gate, AND gate, or EXCLUSIVE-OR gate are given in Fig. 2.

Proof: We will prove the input-output syndrome relation of the network terminating in an OR gate. The other proofs are similar.

Denote by (n, K, S) the three-tuple describing the number of inputs, the number of minterms, and the syndrome realized by a given logic block. Let (n_1, K_1, S_1) and (n_2, K_2, S_2) be the three-tuples describing the two interconnected blocks in Fig. 2(b).

Since the network is terminated in an OR gate, its output is 1 whenever any of the inputs to the OR gate is 1. Thus, the number of minterms K realized by the network is

$$K = K_1 2^{n_2} + K_2 2^{n_1} - K_1 K_2.$$

Hence,

$$S = \frac{K}{2^{n_1+n_2}} = \frac{K_1}{2^{n_1}} + \frac{K_2}{2^{n_2}} - \frac{K_1}{2^{n_1}} \cdot \frac{K_2}{2^{n_2}}.$$

$$S = S_1 + S_2 - S_1 S_2. \quad \text{Q.E.D.}$$

Notice that the input-output syndrome relations shown in Fig. 2 are similar to the input-output signal probability relations reported in [9].

Lemma 1 is useful when determining the syndrome of a fan-out-free combinational circuit. It also provides an algebraic tool to find the number of minterms realized by a fan-out-free network. Traditionally, one had to use mapping tools or equivalent methods to find the number of minterms realized by a given function, which were quite cumbersome.

Example 1: Find the syndrome and number of minterms realized by the fan-out-free circuit of Fig. 3 without using a

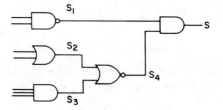

Fig. 3. Example 1.

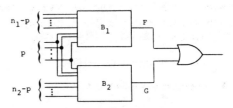

Fig. 4. Two blocks having shared inputs.

Karnaugh map or equivalent-normal form. We have

$$S_1 = 1 - 2^{-2} = \frac{3}{4}, S_2 = 1 - 2^{-2} = \frac{3}{4}, S_3 = 2^{-3} = \frac{1}{8}.$$

Hence,

$$S_4 = 1 - (S_2 + S_3 - S_2 S_3) = \frac{7}{32}, S = S_1 S_4 = \frac{21}{128},$$

$$K = S \cdot 2^n = 21.$$

Let $K(F)$ and $S(F)$ denote the number of minterms and the syndrome of a switching function F, respectively. In the case of interconnected blocks with shared (conjoint) inputs, the following lemma is very useful.

Lemma 2: Let two blocks be interconnected as shown in Fig. 4. Let the Boolean functions realized by blocks B_1 and B_2 be F and G, respectively. Then

$$S(F + G) = S(F) + S(G) - S(FG). \quad (1)$$

Proof: Consider the network of Fig. 4. Let the number of inputs to the blocks realizing the functions F and G be n_1 and n_2, respectively. Also, let p be the number of inputs which are common to both blocks. Thus,

$$K(F + G) = K(F) \cdot 2^{n_2-p} + K(G) \cdot 2^{n_1-p} - K(FG),$$

$$S(F + G) = \frac{K(F + G)}{2^{n_1+n_2-p}} = \frac{K(F)}{2^{n_1}} + \frac{K(G)}{2^{n_2}} - \frac{K(FG)}{2^{n_1+n_2-p}},$$

$$S(F + G) = S(F) + S(G) - S(FG). \quad \text{Q.E.D.}$$

Note, that Lemma 2 reduces to Lemma 1 in the disjoint case.

From Lemma 2 the following relations are implied:

$$S(FG) = 1 - S(\overline{FG}) = S(F) + S(G) + S(\bar{F}\bar{G}) - 1 \quad (2)$$

$$S(F \oplus G) = S(F\bar{G}) + S(\bar{F}G) \quad (3)$$

the syndrome of any combinational circuit can now be calculated. Let x_i be an input to a combinational circuit and let the sum of products of the output function be expressed in the form

$$F = Ax_i + B\overline{x}_i + C$$

where A, B, and C do not depend on the input x_i. Using relations (1)–(3) and Lemma 1 we can easily show that

$$S(F) = \frac{S(A\overline{C}) + S(B\overline{C})}{2} + S(C). \quad (4)$$

Thus, by repeatedly applying (4), and each time eliminating another input, the output syndrome can be calculated.

Example 2: Find the syndrome of the function

$$F = \overline{xy}z + wxz + xy + \overline{vw}y\overline{z}.$$

Solution: We can describe F in the form

$$F = A_1 x + B_1 \overline{x} + C_1$$

where

$$A_1 = wz + y$$
$$B_1 = \overline{y}z$$
$$C_1 = \overline{vw}y\overline{z}.$$

Thus,

$$A_1\overline{C}_1 = wz + wy + yv + yz$$
$$B_1\overline{C}_1 = \overline{y}z.$$

Note that $S(x_1^* x_2^* \cdots x_n^*) = 2^{-n}$ and $S(x_1^* + x_2^* + \cdots + x_n^*) = 1 - 2^{-n}$ where $x_i^* \in \{x_i, \overline{x}_i\}$, $i = 1, 2, \cdots, n$.
Thus $S(C_1) = 1/16$, $S(B_1\overline{C}_1) = 1/4$

$$S(F) = \frac{S(A_1\overline{C}_1) + 1/4}{2} + \frac{1}{16}. \quad (5)$$

Now, we express $A_4\overline{C}_1$ in the form

$$A_1\overline{C}_1 = A_2 y + B_2 \overline{y} + C_2$$

where

$$A_2 = w + v + z$$
$$B_2 = 0$$
$$C_2 = wz.$$

Thus

$$A_2\overline{C}_2 = w\overline{z} + \overline{w}v + \overline{w}z$$
$$B_2\overline{C}_2 = 0$$
$$S(A_1\overline{C}_1) = \frac{S(A_2\overline{C}_2)}{2} + \frac{1}{4}. \quad (6)$$

Now we express $A_2\overline{C}_2$ in the form

$$A_2\overline{C}_2 + A_3 z + B_3 \overline{z} + C_3$$

where

$$A_3 = \overline{w}$$
$$B_2 = w$$
$$C_3 = \overline{w}v.$$

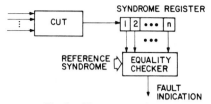

Fig. 5. The test procedure.

Thus,

$$A_3\overline{C}_3 = \overline{wv}, \ S(A_3\overline{C}_3) = \frac{1}{4}$$

$$B_3\overline{C}_3 = w, \ S(B_3\overline{C}_3) = \frac{1}{2}$$

$$S(A_2\overline{C}_2) = \frac{\frac{1}{4} + \frac{1}{2}}{2} + \frac{1}{4} + \frac{5}{8}. \quad (7)$$

By using (7), (6), and (5) we derive the desired syndrome to be

$$S(F) = \frac{15}{32}.$$

The syndrome calculation is described in the following recursive procedure.

Procedure 1

Step 1: Express the output function in a sum of products form.

Step 2: Use (4) to eliminate one input variable from the syndrome calculation.

Step 3: If all the input variables have been eliminated accumulate the partial syndrome calculations and calculate the desired syndrome, otherwise go to Step 2.

III. THE TESTABLE DESIGN

We are given a switching function to be realized with logic gates. Our purpose is to reach a syndrome-testable realization, namely, to have a design such that no fault can cause the circuit to have the same syndrome as the fault-free circuit. We would like to achieve this goal while inserting the minimum number of extra inputs.

The class of faults we consider in this paper is the single stuck-at type. However, many multiple faults will be covered by the proposed design. In Section III-A we consider two-level circuits which are the most simple to design. Here, whenever we refer to a two-level circuit we implicitly imply an AND–OR circuit. The results can be easily extended to OR–AND circuits. As a consequence, our design will be applicable to the important class of programmable logic arrays (PLA's). In Section III-B we treat the class of general combinational circuits.

The test procedure for the syndrome-testable circuits is shown in Fig. 5. Each possible input combination is applied to the CUT exactly once. The syndrome-register is a counter which counts the number of ones appearing on the output of the CUT. The equality checker checks the register's contents with the expected syndrome. If the syndromes are equal, the CUT is reported to be fault-free; otherwise a fault is detected and the CUT is declared faulty. Note that the only difference between the syndrome and the ones-count is the implicit location of the binary point in the syndrome register. If the binary point is considered to be on the left-hand side of the number stored in the syndrome register, that number represents the syndrome, while if the binary point is viewed as being on the right-hand side of the number, this number represents the ones-count.

A. Syndrome—Testable Design of Two-Level Circuits

Definition 2: The function $F(x_1, x_2, \cdots, x_i, \cdots, x_n)$ is said to be *positive* (*negative*) in the variable x_i, if there exists a disjunctive or conjunctive expression for it in which x_i appears only in uncomplemented (complemented) form.

Definition 3: The function $F(x_1, x_2, \cdots, x_i, \cdots, x_n)$ is said to be *unate* in x_i, if it is either positive or negative in x_i.

Lemma 3: A two-level irredundant circuit which realizes a unate function in all its variables is syndrome-testable.

Proof: It is sufficient to consider the set of stuck-at faults at the inputs to the AND gates and the fan-out branches. Any stuck-at-0 fault at the input to an AND gate causes the prime implicant (PI) realized by the AND gate to vanish. Any stuck-at-1 fault at the input of an AND gate corresponds to a growth term which covers the original PI and its adjacent neighbor. Since we assume that the circuit is irredundant both faults change the number of minterms realized by the function, and thus change its syndrome.

In order to complete the proof we have to consider the influence of stuck-at faults at the fan-out branches on the syndrome. Assume without loss of generality that the function is positive in x_i and that x_i is a fan-out branch. Thus, the function F can be expressed in the form

$$F = Ax_i + B$$

where both A and B do not depend on x_i. The expression A has two or more terms. A stuck-at-0 at the input x_i causes all the PI's associated with x_i to vanish. A stuck-at-1 at input x_i causes all the PI's associated with x_i to cover a greater number of minterms. By similar argument to that presented earlier we conclude that a fault at input x_i changes the syndrome of the function. Q.E.D.

Definition 4: Two paths that emanate from a common line and reconverge at some forward point are said to be *reconverging paths* [1].

Definition 5: The *inversion parity* [4] of a reconverging path is equal to the number of inversions modulo 2 in the path, i.e., the number of inversions modulo 2 between the point of divergence and the point of reconvergence.

In the case of two-level circuits, the points of divergence can only be at the input branches, and the point of reconvergence at the output line. The inverters can appear only at the inputs to the AND gates.

Lemma 4: There exist two-level irredundant circuits which are not syndrome-testable.

Proof: By Lemma 3, a circuit which is not syndrome-testable can not be unate in all its variables. Hence, there exists at least one input, say x_i, from which at least two reconverging paths emanate with nonequal inversion parities. Thus, the function F can be expressed in the form

$$F = Ax_i + B\bar{x}_i + C$$

where A, B, and C do not depend on variable x_i, and both A and B include at least one term. Thus, a stuck-at fault (stuck-at-0 or stuck-at-1) at the fan-out branch x_i will cause the faulty syndrome to be identical to the fault-free syndrome if and only if

$$S(A\bar{C}) = S(B\bar{C}), \tag{8}$$

i.e., if and only if the number of minterms added by the growth terms equal the number of minterms eliminated by the vanishing terms. Only stuck-at faults at input fan-out lines which have property (8) can make the circuit syndrome-untestable.
Q.E.D.

Example 3: Consider the function

$$F = xz + y\bar{z}.$$

the fault-free syndrome is $1/2$. The faulty syndrome induced by the fault $z/0$ (we denote by $b/0$ line b stuck-at-0 and by $b/1$ line b stuck-at-1) is also $1/2$. Thus, F is not syndrome-testable.

It is already obvious at this point that two-level combinational circuits can be made syndrome-testable by controlling the "size" of the PI's with extra input insertion. For example, the function F of Example 3 may be made syndrome-testable by inserting one extra input w to realize the new function $F = wxz + y\bar{z}$. During normal operation the input w is fed with a constant logic 1, while for testing purposes it is used as a valid input. The function F is now syndrome-testable since $S(wx) \neq S(y)$.

Lemma 5: Every two-level irredundant combinational circuit can be made syndrome-testable by attaching control (extra) inputs to the AND gates.

Proof: See Appendix A.

From Lemmas 3 and 4 it is evident that the candidates for syndrome-untestability are the nonunate input lines. In order to keep track of the actual lines in which the function F is syndrome-untestable we define the set

$$T = \{x_i | F \text{ is syndrome-untestable in } x_i\}.$$

Clearly,

$$T \subseteq \{x_i | F \text{ is nonunate in } x_i\}.$$

Our goal is to reach a syndrome-testable realization of a given function while adding the minimum number of extra inputs. We choose to add the control inputs in an uncomplemented form (complemented inputs are acceptable as well). By Lemma 3 the modified function will always be syndrome-testable in the control inputs.

Procedure 2 describes an algorithm for designing syndrome-testable two-level combinational circuits. The syndrome-testable design is achieved by modifying the original redundant sum of products. The modification requires an introduction of a nearly minimal number of control inputs. We use the following notation in describing the algorithm:

- C set of control lines.
- PI_i prime implicant number i.
- **PI**   set of prime implicants.
- j a number specifying how many times the function has been modified so far.
- $F^{(j)}$ the jth modified function.
- $T^{(j)}$ the set T which corresponds to function $F^{(j)}$.
- FLAG   an identifier describing whether the function should be modified with an existing control variable or with a new one.

Procedure 2

Step 1—Initialization:

$j = 0$
$C = \phi$
FLAG $= 0$
PI $= \{PI_1, PI_2, \cdots, PI_k\}$
Derive $T^{(0)}$

Step 2:

$$F^{(j)} = \sum_{i=1}^{k} PI_i \quad (\sum \text{ denotes Boolean sum}).$$

If $T^{(j)} = \phi$ — stop, $F^{(j)}$ is syndrome-testable.
If $C = \phi$ go to Step 4.

Step 3: For each $c_n \in C$ and $PI_j \in$ **PI** derive $T^{(nj)}$ for function

$$\sum_{i \neq j} PI_i + c_n \cdot PI_j.$$

Let $|T^{(\alpha\beta)}| = \min_{n,j} \{|T^{(nj)}|\}$ (when the minimum is attained by several members choose one arbitrarily).
If $|T^{(\alpha\beta)}| = |T^{(j)}|$ go to Step 4.
Otherwise FLAG $= 0$, go to Step 5.

Step 4: Add new input: $C \leftarrow C \cup \{c_\delta\}$.
For each $PI_j \in$ **PI** derive $T^{(\delta j)}$ for the function

$$\sum_{i \neq j} PI_i + c_\delta \cdot PI_j.$$

Let $|T^{(\delta\mu)}| = \min_{j} \{|T^{(\delta j)}|\}$.
If $|T^{(\delta\mu)}| = |T^{(j)}|$ then FLAG $= 0$, otherwise FLAG $= 1$.

Step 5:

$j \leftarrow j + 1$
If FLAG $= 0$ then
$T^{(j)} = T^{(\alpha\beta)}; PI_\beta \leftarrow c_\alpha \cdot PI_\beta$
If FLAG $= 1$ then
$T^{(j)} = T^{(\delta\mu)}; PI_\mu \leftarrow c_\delta \cdot PI_\mu$
Go to Step 2.

By Lemma 5, Procedure 1 is guaranteed to halt. Although,

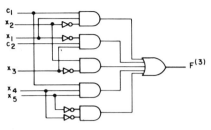

Fig. 6. The syndrome-testable design for Example 4.

at each iteration of the algorithm the optimal input insertion (or one of the optimal insertions in the case of multiple choices) is obtained, an overall optimal solution is not guaranteed. The reason for that is that local optimization does not necessarily lead to global optimization.

Example 4: Find the syndrome-testable design with nearly minimum number of control inputs for the function

$$F = x_1\bar{x}_2 + \bar{x}_1 x_3 + x_2\bar{x}_3 + x_4 x_5 + \bar{x}_4\bar{x}_5.$$

Solution: Let $PI_1 = x_1\bar{x}_2$, $PI_2 = \bar{x}_1 x_3$, $PI_3 = x_2\bar{x}_3$, $PI_4 = x_4 x_5$, $PI_5 = \bar{x}_4\bar{x}_5$.

First pass of Procedure 1 yields

$$T^{(0)} = \{x_1, x_2, x_3, x_4, x_5\}$$

$$F^{(0)} = F$$

$$T^{(1)} = \{x_3, x_4, x_5\}$$

$$F^{(1)} = c_1 x_1 \bar{x}_2 + \bar{x}_1 x_3 + x_2 \bar{x}_3 + x_4 x_5 + \bar{x}_4 \bar{x}_5.$$

Second pass of the algorithm gives

$$T^{(2)} = \{x_3\}$$

$$F^{(2)} = c_1 x_1 \bar{x}_2 + \bar{x}_1 x_3 + x_2 \bar{x}_3 + c_1 x_4 x_5 + \bar{x}_4 \bar{x}_5.$$

Third pass of Procedure 1 yields

$$T^{(3)} = \phi$$

$$F^{(3)} = c_1 x_1 \bar{x}_2 + c_2 \bar{x}_1 x_3 + x_2 \bar{x}_3 + c_1 x_4 x_5 + \bar{x}_4 \bar{x}_5.$$

The testable design is shown in Fig. 6.

The only difficulty involving the design arises when n is large. Since the test procedure requires the application of all 2^n input combinations, the design will be worthwhile for $n \leq 20$ (which is equivalent to about 1 s of testing with a 1 MHz machine). However, for $n > 20$, the problem can be overcome by designing the circuit with subcircuits, as shown in Fig. 7, such that every subcircuit has no more than 20 inputs. This improvement will cost in $m = \lceil q/20 \rceil$ extra outputs, where q is the number of inputs to the AND gates (the extra OR gates do not constitute a problem in present day technology). Each subcircuit must be designed to be syndrome-testable, and can be tested by no more than 2^{20} inputs. Thus, the total test length will be approximately $m2^{20}$, which is equivalent to about m seconds of testing. Note that faults at the inputs of the OR gate generating the objective functions are not covered when the individual subcircuits are syndrome-tested. However, if we let the OR gates of the individual subcircuits have a tristate capability we can syndrome-test the OR gate which generates the objective function by directly exercising its inputs and thus test for the uncovered faults mentioned above.

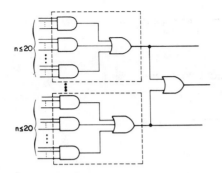

Fig. 7. The proposed partition for large two-level circuits. The input lines may be shared between subcircuits, although not shown explicitly.

B. Syndrome—Testable Design of General Combinational Circuits

Lemma 6: Every fan-out-free inrredundant combinational circuit, composed of AND, OR, NAND, NOR, and NOT gates is syndrome-testable.

Proof: In a fan-out-free circuit there is a unique path from any line to the circuit output. Thus, if we label a given line by g, the output F will be either positive or negative in g, depending on the inversion parity of the path originating at g and terminating at the output line. Assume without loss of generality that F is positive in g. We can, therefore, represent F as

$$F = Ag + b$$

where

$$g = g(x_{i_1}, x_{i_2}, \cdots, x_{i_n}),$$

and both A and B do not depend on $x_{i_1}, x_{i_2}, \cdots, x_{i_n}$.

A stuck-at-0 fault on line g eliminates all the minterms associated with Ag, and does not create any new ones. Therefore, the circuit is syndrome-testable with respect to $g/0$. In order to prove that $g/1$ is syndrome-testable, we use a contrapositive approach. Assume that $g/1$ is syndrome-untestable. Thus, all the minterms covered by $A\bar{g}$ are already covered by F. Therefore F can be represented by

$$F = Ag + A\bar{g} + B = A + B$$

which is independent of g, contradicting the previous assumption that F is positive in g. Q.E.D.

Lemma 7: Let $g = g(x_{i_1}, x_{i_1}, \cdots, x_{i_n})$ be a line in a general combinational circuit. Let the equivalent sum of products of the function F with respect to line g be

$$F = Ag + B\bar{g} + C.$$

Then the fault $g/0$ is syndrome-untestable if and only if

$$S(A\bar{C}g) = S(B\bar{C}g) \quad (9)$$

and, the fault $g/1$ is syndrome-untestable if and only if

$$S(A\bar{C}\bar{g}) = S(B\bar{C}\bar{g}). \quad (10)$$

Proof: We prove relation (9). The proof of (10) is similar. The fault $g/0$ is syndrome-untestable if and only if

$$S(F) = S(B + C).$$

Using Lemmas 1 and 2, we have

$$S(F) = S(A\bar{C}g) + S(B\bar{C}g) + S(C)$$

$$S(B + C) = S(B\bar{C}) + S(C).$$

Thus, the fault $g/0$ is syndrome-untestable if and only if

$$S(A\bar{C}g) + S(B\bar{C}g) = S(B\bar{C}),$$

or

$$S(A\bar{C}g) = S(B\bar{C}\bar{g}). \quad \text{Q.E.D.}$$

Corollary 1: Let g be any line in an irredundant combinational circuit composed of AND, OR, NAND, NOR, and NOT gates. If there exists only one path from g to the circuit output, the function F is syndrome-testable with respect to faults in g.

Proof: Directly from Lemmas 6 and 7.

EXCLUSIVE-OR and EQUIVALENCE gates are source of problems to syndrome testing because of their inherent nonunateness. The following lemma displays the conditions necessary so that the faults on the inputs of an EXCLUSIVE-OR gate be syndrome-untestable. The corresponding conditions for an EQUIVALENCE gate can be obtained from Lemmas 1 and 8.

Lemma 8: Let $F = g \oplus h$, where $g = g(x_1, x_2, \cdots, x_n)$ and $h = h(x_1, x_2, \cdots, x_n)$. The fault $h/0$ is syndrome-untestable if and only if

$$S(gh) = S(\bar{g}h) \quad (11)$$

and the fault $h/1$ is syndrome-untestable if and only if

$$S(g\bar{h}) = S(\bar{g}\bar{h}). \quad (12)$$

Proof: We prove (11). The proof of relation (12) is similar. The fault $h/0$ is syndrome-untestable if and only if

$$S(g \oplus h) = S(g).$$

Thus,

$$S(g\bar{h}) + S(\bar{g}h) = S(g)$$

$$S(\bar{g}h) = S(gh). \quad \text{Q.E.D.}$$

Corollary 2: Let $F = g \oplus h$, where $g = g(x_{i_1}, x_{i_2}, \cdots, x_{i_u})$, $h = h(y_{i_1}, y_{i_2}, \cdots, y_{i_v})$ and $\{x_{i_1}, \cdots, x_{i_u}\} \cap \{y_{i_1}, \cdots, y_{i_v}\} = \phi$. Then, faults on line $h(h/0$ or $h/1)$ are syndrome-untestable if and only if

$$S(g) = 1/2. \quad (13)$$

Proof: Directly from Lemmas 1 and 8.

It is evident from Corollary 2 that attaching an extra input to an EXCLUSIVE-OR gate will not fix the syndrome-untestable condition since relation (13) will hold for the new input. There are two basic approaches to fix the syndrome-untestable condition with respect to inputs of an EXCLUSIVE-OR gate. One way of handling the problem is by breaking the inherent symmetry of the gate. This can be done when the actual implementation of the gate is known. For example, in the case where the EXCLUSIVE-OR gate is implemented by NAND gates, the output can be made syndrome-testable with respect

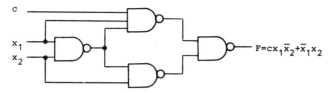

Fig. 8. A syndrome-testable modification of an EXCLUSIVE-OR gate.

Fig. 9. Using an extra AND gate and an extra input to fix the syndrome-untestable condition in h ($h/0$ or $h/1$).

to faults at the inputs by adding a control input c as shown in Fig. 8.

The second approach is to add a control input via an extra AND gate to the input lines of the EXCLUSIVE-OR gate as shown in Fig. 9.

The problem is how to modify the design of a general combinational circuit so that it will be syndrome-testable. We try to do it by means of control input insertion and/or a small amount of extra logic. The control inputs connected to AND and NAND gates should be applied with a constant 1 under normal operation, while the control inputs attached to OR and NOR gates should be applied with a constant 0 under normal operation. Any control input connected to AND or NAND gates cannot also be connected to OR or NOR gates.

From the previous lemmas and corollaries it is clear that the candidate lines for syndrome-untestability are the fan-out lines, which have unequal inversion parities along its reconverging branches, the lines that feed them (directly or indirectly) and inputs and outputs of EXCLUSIVE-OR gates. The following procedure describes the modification algorithm.

Procedure 3

Step 1: Find the set of lines in which the function is syndrome-untestable by finding the equivalent sum of products with respect to the candidate lines for syndrome-untestability.

Step 2: Use Procedure 2 with the following modifications:

  PI → Set of AND, NAND, OR, NOR gates.

$\Sigma_{i \neq j} PI_i + c \cdot PI_j$ → The function obtained by attaching the control input c to gate number j. For the faults at the inputs of EXCLUSIVE-OR gates, use either the method of Fig. 8 or Fig. 9.

As of today the author still lacks a proof that every multilevel combinational circuit can be modified to be syndrome-testable by extra input insertions. However, our experience with small size circuits show that Procedure 3 does generate a syndrome-testable design.

In order to make the syndrome-test procedure feasible it is necessary to partition the general combinational circuit to subcircuits, in such a way that the total testing time will be of reasonable size. In order to do that we use the previous concept that the time required for exhaustive testing of a subcircuit which has no more than 20 inputs is acceptable. Fig. 10 shows the partitioning method.

Fig. 10 shows an imbedded subcircuit M_k and its interconnection with previous subcircuits. Subcircuit M_k might in general receive inputs from two sources: primary inputs to the circuit and output from some other subcircuits (like M_i and M_j in Fig. 10). Since the inputs of the latter kind, namely those which are outputs of previous subcircuits, may in general depend on more than 20 inputs, an exhaustive exercise of subcircuit M_k will require more testing time than what we consider to be acceptable. Therefore, for those cases where the subcircuit M_k depends directly or indirectly on more than 20 inputs, the outputs of subcircuits like M_i and M_j in Fig. 10, may be designed to have a tristate capability. Thus, if we provide an extra I/O pin at these tristate outputs, we may have the capability of both observing the syndrome-test results when subcircuits like M_i and M_j are tested, and of directly exercising the input lines of subcircuit M_k when the outputs of M_i and M_j are held at the high impedance state. In this way, every subcircuit can be tested in no more than 2^{20} inputs, which is equivalent to about 1 s of testing time. Therefore, a circuit which is partitioned to m subcircuits may be syndrome-tested in no more than m seconds. The pin-penalty paid for this partitioning is one pin for each tristate output. We have calculated the partitioning pin-penalty for uniform trees [2] of up to 1024 inputs and found it to be no more than 7 percent.

Since the conditions of syndrome-untestability are very strict [i.e., relations (9) and (10)] it is believed that, in general, very few lines will meet them. Thus, only a few extra inputs will be required to achieve a syndrome-testable design. Table I displays a few MSI combinational logic, their number of pins and the number of extra pins needed to make them syndrome-testable. As seen from Table I, the number of extra pins does not exceed 1 (or 5 percent) for these functions. This is not surprising because one extra input can invalidate the syndrome-untestable condition in various portions of the circuit. Because of the strictness of the syndrome-untestable condition, it is feasible to correct untestable conditions of several faults by means of only one additional input. Fig. 11 shows a syndrome-testable modification of SN74181. The modification requires one extra input and two extra AND gates. Note that the extra AND gates could have been avoided if we assumed a certain realization of the EXCLUSIVE-OR gates (see Fig. 8).

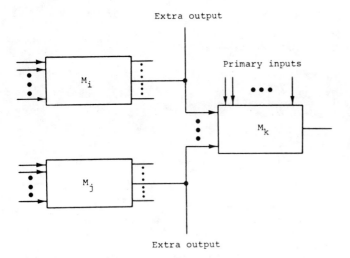

Fig. 10. The proposed partitioning for general combinational circuits. The outputs of M_i and M_j are designed to have a tristate capability.

TABLE I
THE NUMBER OF CONTROL INPUTS NECESSARY IN ORDER TO MAKE THE CIRCUITS SYNDROME-TESTABLE AS A FUNCTION OF NUMBER OF PINS FOR VARIOUS MSI COMBINATIONAL LOGIC

Function	# Pins	# Control inputs
Dual 4-line-to-1-line Data selectors/Mux SN74153	16	1
Data selectors/Mux SN74150	24	1
Mux SN74151	16	1
Dual carry save adders SN74183	14	1
Look-ahead carry generators SN74182	16	0
Arithmetic logic units SN74181	24	1
Total	110	5

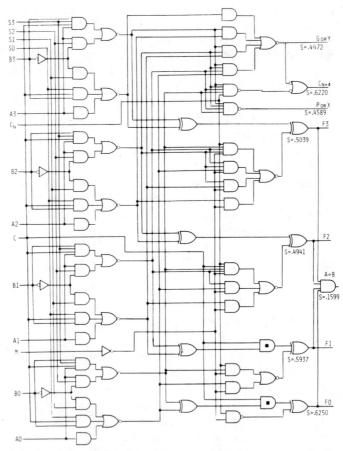

Fig. 11. Syndrome-testable design of arithmetic logic units, SN74181, requiring one extra input C and two extra gates (marked). The output syndromes are rounded to four decimal digits.

IV. SUMMARY AND CONCLUSIONS

A new approach to the design of testable combinational circuits was presented in this paper. The design method is to modify given realizations by inserting extra I/O so that the final circuit will be syndrome-testable. A procedure that produces a nearly minimal number of extra input insertions was described. It was also shown how to partition very large combinational circuits to subcircuits such that the total testing time will be of acceptable length. Although we restricted ourselves, in this paper, to single-output networks, the ideas easily generalize to multiple-output networks as well.

It is well known that one of the most severe restrictions on IC manufacturers is the number of pins per chip. Although the testable design requires an increase in the number of pins, it is believed that the pin overhead is very low because of the strictness of the syndrome-untestable condition.

In this paper we have mainly used extra input lines to

achieve a syndrome-testable design. It may be possible, though, to use also extra logic in producing the testable design. This, in fact, may be very attractive since, in general, chip area is more available than I/O pins. We are currently investigating the possibility of trading off some of the extra I/O pins with extra logic. It is important to mention that the syndrome-testable design can be superimposed on LSSD (level-sensitive scan design) [10]. If this is done, the extra I/O pins required by the syndrome-testable design are actually implemented by extra SRL's (shift register latches) connected to the scan path. Thus, the I/O pin-penalty paid by the syndrome-testable design is traded off with extra logic.

APPENDIX A

Before proving Lemma 5, we first prove the following lemma.

Lemma 9: Let T be the set of syndrome-untestable input lines. Let $T^{(j)}$ denote the set T of the jth modified function, $j = 0, 1, 2, \cdots$, as defined previously in the paper. Let $x_i \in T^{(0)}$ and $x_j \notin T^{(0)}$, be two input lines to the two-level circuit. If, by attaching a control input c_1 to a (set of) PI's, the condition of syndrome-untestability in x_i is corrected and in x_j is spoiled, namely $x_i \notin T^{(1)}$ and $x_j \in T^{(1)}$, then by attaching another control input c_2 to the same (set of) PI's the spoiled condition in x_j is corrected again, namely, $x_i, x_j \notin T^{(2)}$.

Proof: The function F can be expressed in the form

$$F^{(0)} = \alpha_1 x_i + \alpha_2 \bar{x}_i + \alpha_3 x_j + \alpha_4 \bar{x}_j + \alpha_5 x_i x_j + \alpha_6 x_i \bar{x}_j + \alpha_7 \bar{x}_i x_j + \alpha_8 \bar{x}_i \bar{x}_j + \alpha_9$$

where α_k do not depend on x_i and x_j for all $k = 1, 2, \cdots, 9$. Since $x_i \in T^{(0)}$, then as a consequence of Lemma 4 we have

$$S[(\alpha_1 + \alpha_5)\bar{\alpha}_3\bar{\alpha}_9] + S[(\alpha_1 + \alpha_6)\bar{\alpha}_4\bar{\alpha}_9]$$
$$= S[(\alpha_2 + \alpha_7)\bar{\alpha}_3\bar{\alpha}_9] + S[(\alpha_2 + \alpha_8)\bar{\alpha}_4\bar{\alpha}_9]. \quad (14)$$

Since $x_j \notin T^{(0)}$, then as a consequence of Lemma 4 we have

$$S[(\alpha_3 + \alpha_5)\bar{\alpha}_1\bar{\alpha}_9] + S[(\alpha_3 + \alpha_7)\bar{\alpha}_2\bar{\alpha}_9]$$
$$\neq S[(\alpha_4 + \alpha_6)\bar{\alpha}_1\bar{\alpha}_9] + S[(\alpha_4 + \alpha_8)\bar{\alpha}_2\bar{\alpha}_9]. \quad (15)$$

In order to correct the syndrome-untestable condition in x_i we modify the function in the form (other ways are possible also)

$$F^{(1)} = c_1 \alpha_1 x_i + \alpha_2 \bar{x}_i + \alpha_3 x_j + \alpha_4 \bar{x}_j + \alpha_5 x_i x_j + \alpha_6 x_i \bar{x}_j + \alpha_7 \bar{x}_i x_j + \alpha_8 \bar{x}_i \bar{x}_j + \alpha_9.$$

According to the assumption stated in the Lemma, $F^{(1)}$ is syndrome-testable in x_i and untestable in x_j. This is translated to the following two conditions (by applying relation (8) to $F^{(1)}$): Since $x_i \notin T^{(1)}$ we have

$$S[(\alpha_1 + \alpha_5)\bar{\alpha}_3\bar{\alpha}_9] + S[(\alpha_1 + \alpha_6)\bar{\alpha}_4\bar{\alpha}_9]$$
$$- \frac{1}{2}[S(\alpha_1\bar{\alpha}_3\bar{\alpha}_5\bar{\alpha}_9) + S(\alpha_1\bar{\alpha}_4\bar{\alpha}_6\bar{\alpha}_9)]$$
$$\neq S[(\alpha_2 + \alpha_7)\bar{\alpha}_3\bar{\alpha}_9] + S[(\alpha_2 + \alpha_8)\bar{\alpha}_4\bar{\alpha}_9]. \quad (16)$$

Since $x_j \in T^{(1)}$ we have

$$S[(\alpha_3 + \alpha_5)\bar{\alpha}_1\bar{\alpha}_9] + S[(\alpha_3 + \alpha_7)\bar{\alpha}_2\bar{\alpha}_9]$$
$$+ \frac{1}{2} S[(\alpha_3 + \alpha_5)\alpha_1\bar{\alpha}_9]$$
$$= S[(\alpha_4 + \alpha_6)\bar{\alpha}_1\bar{\alpha}_9] + S[(\alpha_4 + \alpha_8)\bar{\alpha}_2\bar{\alpha}_9]$$
$$+ \frac{1}{2} S[(\alpha_4 + \alpha_6)\alpha_1\bar{\alpha}_9] \quad (17)$$

by combining relations (14) and (16) we get that

$$\alpha_1\bar{\alpha}_3\bar{\alpha}_5\bar{\alpha}_9 \neq 0 \text{ or } \alpha_1\bar{\alpha}_4\bar{\alpha}_6\bar{\alpha}_9 \neq 0 \quad (18)$$

by combining relations (15) and (17) we get that

$$S[(\alpha_3 + \alpha_5)\alpha_1\bar{\alpha}_9] \neq S[(\alpha_4 + \alpha_6)\alpha_1\bar{\alpha}_9]. \quad (19)$$

We have to show now that if the function is modified again as

$$F^{(2)} = c_1 c_2 \alpha_1 x_i + \alpha_2 \bar{x}_i + \alpha_3 x_j + \alpha_4 \bar{x}_j + \alpha_5 x_i x_j + \alpha_6 x_i \bar{x}_j + \alpha_7 \bar{x}_i x_j + \alpha_8 \bar{x}_i \bar{x}_j + \alpha_9$$

it will be syndrome-testable in both x_i and x_j. In order to do this we show that the condition for syndrome-untestability in x_i and x_j does not hold for $F^{(2)}$. The condition for syndrome-untestability in x_i for the function $F^{(2)}$ becomes

$$S[(\alpha_1 + \alpha_5)\bar{\alpha}_3\bar{\alpha}_9] + S[(\alpha_1 + \alpha_6)\bar{\alpha}_4\bar{\alpha}_9]$$
$$- \frac{3}{4}[S(\alpha_1\bar{\alpha}_3\bar{\alpha}_5\bar{\alpha}_9) + S(\alpha_1\bar{\alpha}_4\bar{\alpha}_6\bar{\alpha}_9)]$$
$$= S[(\alpha_2 + \alpha_7)\bar{\alpha}_3\bar{\alpha}_9] + S[(\alpha_2 + \alpha_8)\bar{\alpha}_4\bar{\alpha}_9]. \quad (20)$$

From relations (14) and (18) we easily see that (20) is false. The condition for syndrome-untestability in x_j for the function $F^{(2)}$ becomes

$$S[(\alpha_3 + \alpha_5)\bar{\alpha}_1\bar{\alpha}_9] + S[(\alpha_3 + \alpha_7)\bar{\alpha}_2\bar{\alpha}_9]$$
$$+ \frac{3}{4} S[(\alpha_3 + \alpha_5)\alpha_1\bar{\alpha}_9]$$
$$= S[(\alpha_4 + \alpha_6)\bar{\alpha}_1\bar{\alpha}_9] + S[(\alpha_4 + \alpha_8)\bar{\alpha}_2\bar{\alpha}_9]$$
$$+ \frac{3}{4} S[(\alpha_4 + \alpha_6)\alpha_1\bar{\alpha}_9]. \quad (21)$$

From relations (17) and (19) we can easily see that (21) is also false. Thus, $F^{(2)}$ is syndrome-testable in both x_i and x_j.
Q.E.D.

Proof of Lemma 5: Let T be the set of syndrome-untestable input lines. From Lemma 9 we know that by adding at most 2 extra inputs we can reduce the cardinality of T by at least one. Thus, by adding at most $2|T|$ control inputs the function can be modified to be syndrome-testable. Q.E.D.

Acknowledgment

The author wishes to thank Prof. E. J. McCluskey, and the members of the Center for Reliable Computing at Stanford University for their helpful criticism. Thanks are also due to the members of the VLSI DA Group at the T.J. Watson Research Center, Yorktown Heights, NY. The referees comments are, also, greatly appreciated.

References

[1] D. E. Armstrong, "On finding a nearly minimal set of fault detection tests for combinational logic nets," *IEEE Trans. Electron. Comput.*, vol. EC-15, pp. 66–73, Feb. 1966.

[2] J. P. Hayes, "On realization of Boolean functions requiring a minimal or near-minimal number of tests," *IEEE Trans. Comput.*, vol. C-20, pp. 1506–1513, Dec. 1971.

[3] ——, "Transition count testing of combinational circuits," *IEEE Trans. Comput.*, vol. C-25, pp. 613–620, June 1976.

[4] I. Kohavi and Z. Kohavi, "Detection of multiple faults in combinational logic networks," *IEEE Trans. Comput.*, vol. C-21, pp. 556–568, June 1972.

[5] J. Losq, "Referenceless random testing," in *Proc. 6th Ann. Symp. on Fault-Tolerant Comput.*, Pittsburgh, PA, pp. 108–113, June 21–23, 1976.

[6] J. P. Roth, W. G. Bouricius, and P. R. Schneider, "Programmed algorithms to compute tests to detect and distinguish between failures in logic circuits," *IEEE Trans. Electron. Comput.*, vol. EC-16, pp. 567–580, Oct. 1967.

[7] J. Savir, "Syndrome-testable design of combinational circuit," in *Proc. 9th Ann. Int. Symp. on Fault-Tolerant Computing*, June 1979, pp. 137–140.

[8] A. Tzidon, I. Berger, and M. Yoeli, "A practical approach to fault detection in combinational networks," *IEEE Trans. Comput.*, vol. C-27, pp. 968–971, Oct. 1978.

[9] K. P. Parker and E. J. McCluskey, "Probabilistic treatment of general combinational networks," *IEEE Trans. Comput.*, vol. C-24, pp. 668–670, June 1975.

[10] E. B. Eichelberger and T. W. Williams, "A logic design structure for LSI testability," in *Proc. 14th Ann. Design Automation Conf.*, June 1977, pp. 462–468.

Jacob Savir (M'78) was born in Bucharest, Romania, on December 13, 1946. He received the B.Sc. (cum laude) and M.Sc. degrees in electrical engineering from the Technion—Israel Institute of Technology, Haifa, Israel, in 1968 and 1973, respectively, the M.S. degree in statistics and Ph.D. degree in electrical engineering from Stanford University, Stanford, CA, in 1977 and 1978, respectively.

In the summer of 1967 he was a student trainee in ASEA, Sweden, working on high-power control systems. From 1968 to 1972 he served as an electronic engineer in the Israel Defence Forces. From 1972 to 1974 he worked as a Research Assistant at the Department of Electrical Engineering, Technion, Israel. From 1974 to 1975 he was a project engineer in the Israel Ministry of Defence. From 1975 to 1978 he worked, first as a Ph.D. student and later as an IBM Post-Doctoral Fellow, with Prof. E. J. McCluskey on intermittent fault test strategies and testable design of digital systems. He joined IBM in 1978 at the T. J. Watson Research Center, Yorktown Heights, NY, where he is currently a research staff member, investigating DA approaches to VLSI. His research interests include, testing and diagnosis of failures, testable design, fault simulation and reliability of digital systems. He authored several papers in the field of testing for logic faults and testable design of digital circuits.

Dr. Savir is a member of Sigma Xi.

Correction to "Syndrome-Testable Design of Combinational Circuits"

> During the production process of the paper "Syndrome-Testable Design of Combinational Circuits" by Dr. Jacob Savir, which appeared in the June 1980 issue of this TRANSACTIONS (vol. C-29, pp. 442–451), several corrections supplied by the author were inadvertently omitted from the final version of the paper. We apologize to Dr. Savir for this omission. The following material is the correct version of his Example 2 and Lemma 7, which appear on pages 444 and 447, respectively, of the June issue. Several other errata, which were supplied by the author but were also not corrected in the paper, are included as well.

On p. 444, second column, 21st line, the phrase "such that no fault" should read: "such that no single fault."

On p. 445, second column, the 4th line should read: "paths emanate with unequal inversion parities."

On p. 446, second column, the 11th line should read: "First pass of Procedure 2 yields."

On p. 446, second column, the 19th line should read: "Third pass of Procedure 2 yields."

On p. 447, first column, the 13th line should read: "$F = Ag + B$."

Example 2: Find the syndrome of the function

$$F = \bar{x}\bar{y}z + wxz + xy + \bar{v}\bar{w}y\bar{z}.$$

Solution: We can describe F in the form

$$F = A_1 x + B_1 \bar{x} + C_1$$

where

$$A_1 = wz + y$$
$$B_1 = \bar{y}z$$
$$C_1 = \bar{v}\bar{w}y\bar{z}.$$

Thus,

$$A_1 \bar{C}_1 = wz + wy + yv + yz$$
$$B_1 \bar{C}_1 = \bar{y}z.$$

Note that $S(x_1^* x_2^* \cdots x_n^*) = 2^{-n}$ and $S(x_1^* + x_2^* + \cdots + x_n^*) = 1 - 2^{-n}$ where $x_i^* \in \{x_i, \bar{x}_i\}$, $i = 1, 2, \cdots, n$.

Thus, $S(C_1) = 1/16$, $S(B_1 \bar{C}_1) = 1/4$

$$S(F) = \frac{S(A_1\bar{C}_1) + 1/4}{2} + \frac{1}{16}. \qquad (5)$$

Now, we express $A_1 \bar{C}_1$ in the form

$$A_1 \bar{C}_1 = A_2 y + B_2 \bar{y} + C_2$$

where

$$A_2 = w + v + z$$
$$B_2 = 0$$
$$C_2 = wz.$$

Thus

$$A_2 \bar{C}_2 = w\bar{z} + \bar{w}v + \bar{w}z$$

$$B_2 \bar{C}_2 = 0$$

$$S(A_1 \bar{C}_1) = \frac{S(A_2 \bar{C}_2)}{2} + \frac{1}{4}. \qquad (6)$$

Now we express $A_2 \bar{C}_2$ in the form

$$A_2 \bar{C}_2 = A_3 z + B_3 \bar{z} + C_3$$

where

$$A_3 = \bar{w}$$
$$B_3 = w$$
$$C_3 = \bar{w}v.$$

Thus,

$$A_3 \bar{C}_3 = \bar{w}\bar{v}, \; S(A_3 \bar{C}_3) = \frac{1}{4}$$

$$B_3 \bar{C}_3 = w, \; S(B_3 \bar{C}_3) = \frac{1}{2}$$

$$S(A_2 \bar{C}_2) = \frac{\frac{1}{4} + \frac{1}{2}}{2} + \frac{1}{4} = \frac{5}{8}. \qquad (7)$$

By using (7), (6), and (5) we derive the desired syndrome to be

$$S(F) = \frac{15}{32}.$$

Lemma 7: Let $g = g(x_{i1}, x_{i2}, \cdots, x_{in})$ be a line in a general combinational circuit. Let the equivalent sum of products of the function F with respect to line g be

$$F = Ag + B\bar{g} + C.$$

Then the fault $g/0$ is syndrome-untestable if and only if

$$S(A\bar{C}g) = S(B\bar{C}g) \qquad (9)$$

and, the fault $g/1$ is syndrome-untestable if and only if

$$S(A\bar{C}\bar{g}) = S(B\bar{C}\bar{g}). \qquad (10)$$

Proof: We prove relation (9). The proof of (10) is similar. The fault $g/0$ is syndrome-untestable if and only if

$$S(F) = S(B + C).$$

Using Lemmas 1 and 2, we have

$$S(F) = S(A\bar{C}g) + S(B\bar{C}\bar{g}) + S(C)$$
$$S(B + C) = S(B\bar{C}) + S(C).$$

Thus, the fault $g/0$ is syndrome-untestable if and only if

$$S(A\bar{C}g) + S(B\bar{C}\bar{g}) = S(B\bar{C}),$$

or

$$S(A\bar{C}g) = S(B\bar{C}g).$$

Q.E.D.

Syndrome-Testability Can be Achieved by Circuit Modification

GEORGE MARKOWSKY

Abstract—In [1] and [2] Savir developed many facets of syndrome-testing (checking the number of minterms realized by a circuit against the number realized by a fault-free version of that circuit) and presented evidence showing that syndrome-testing can be used in many practical circuits to detect all single faults. In some cases, where syndrome-testing did not detect all single stuck-at-faults, Savir showed that by the addition of a small number of additional "control" inputs and gates one would get a function which is syndrome-testable for all single stuck-at faults, and yet which realizes the original function when the "control" inputs are fed appropriate values. However, he left open the question of whether one could always modify a circuit to achieve syndrome-testability. In this correspondence we show that a combinatorial circuit can always be modified to produce a single-fault, syndrome-testable circuit.

Index Terms—Circuit modification, stuck-at-faults, syndrome testability.

Syndrome testing offers the possibility of simple and thorough testing of circuits and even the possibility of self-testing. Unfortunately, a circuit need not be syndrome-testable. Savir [1], [2] showed that many circuits of "practical" interest are either syndrome-testable or can be made syndrome-testable by the addition of a small number of additional control lines. This correspondence shows that a circuit can always be made syndrome-testable by adding "enough" control lines. Since there does not exist a handy and useful characterization of circuits of "practical" interest, the approach used here is based on no particular property of the circuit being analyzed, and consequently it is not surprising that the technique we use to prove the main results is not of practical interest. The chief point of this correspondence is to demonstrate that syndrome-testability can always be achieved. It is my hope that knowledge of this fact will spur people to analyze further the problem of modifying "practical" circuits to achieve syndrome-testability. The reader is urged to consult Savir's papers for basic information on syndrome.

Definition: Let $x_1, \cdots, x_n$ be Boolean variables and F a Boolean function in some subset of these variables. Let K be the number of assignments of values to the variables which give F the value of 1. Then the *syndrome* of F, $S(F)$ is defined to be the quantity $K/2^n$. Finally, the *cosyndrome* of F, $C(F)$ is defined to be the quantity $1 - S(F)$. □

Fig. 1 is a schematic representation of an OR gate in a typical combinatorial circuit. Fig. 2 shows how it can be modified by adding additional AND gates and various numbers of additional control variables. For a NOR gate, it is clear that the solution would be the same as for the OR gate since the syndrome of NOR circuit is simply the cosyndrome of the corresponding OR circuit. For AND and NAND circuits we use additional OR gates, rather than AND gates.

Theorem 1: The procedure described above can always be tailored so that the circuit (of Fig. 2) ending at OUT is single-fault syndrome-testable assuming that each of the circuits ending at the l_i are single-fault syndrome-testable.

Proof: The remarks made above show that the result for the OR case would imply the result for the NOR case. Furthermore, by substituting cosyndrome for syndrome in the argument below one gets the result in the AND case and hence the NAND case. Thus, we only give the proof in the OR case, i.e., the case of Figs. 1 and 2.

Let $B_i(*B_i)$ denote the function realized by the circuits feeding line $l_i(*l_i)$ and let $B_i^f(*B_i^f)$ denote the same circuit having a single fault f somewhere in it. It is also reasonable to assume $B_i \not\equiv 0, 1$ to avoid discussing trivialities. Let $B(*B)$ denote the function realized at OUT in Fig. 1 (2) and $B^f(*B^f)$ the function which results from a single fault f somewhere in it.

Let $\delta = \max \{\delta_i\}$, where δ_i is the number of variables which eventually feed into line l_i. Finally, let $k_1 < k_2 < \cdots < k_m$ be chosen so that $2^{-(k_i+\delta)} > 10m2^{-k_{i+1}}$. For consistency let $k_0 = 0$.

The proof is inductive in nature and begins by noticing that simple gates are syndrome-testable. Next, we observe that any single stuck-at fault in OUT or any $*l_i$ is syndrome-testable. By induction we assume that $S(B_i^f) \neq S(B_i)$ for all single faults occurring in the circuit feeding l_i. It immediately follows that $S(*B_i^f) \neq S(*B_i)$ for all single faults occurring in the circuit feeding $*l_i$, since either the fault occurs in B_i, whence $S(*B_i^f) = S(B_i^f)/2^{k_i}$ and $S(*B_i) = S(B_i)/2^{k_i}$, or the fault occurs in some c_{ij} or $*l_i$ and either $S(*B_i^f) = 0, 1$ or $S(B_i)/2^{k_i-1}$, while $S(*B_i) = S(B_i)/2^{k_i} \neq 0, 1$.

In order to prove that $S(*B) \neq S(*B^f)$ for all f we proceed as follows. Let i_0 be such that $*B_i^f \equiv *B_i$ for $1 \leq i < i_0$, but $*B_{i_0}^f \not\equiv *B_{i_0}$.

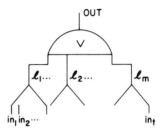

Fig. 1. A schematic representation of part of a combinatorial circuit.

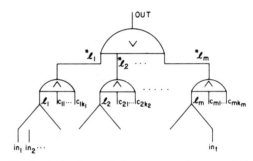

Fig. 2. A circuit modified to ensure syndrome testability.

Since at most $k_{i_0} + \delta$ lines feed l_{i_0}, we must have $|S(*B_{i_0}^f) - S(*B_{i_0})| \geq 2^{-(k_{i_0}+\delta)}$ because the syndromes of these two subcircuits have denominators $\leq 2^{-(k_{i_0}+\delta)}$ and numerators which differ by at least 1. We consider two cases as follows.

1) $S(*B_{i_0}^f) \geq S(*B_{i_0}) + 2^{-(k_{i_0}+\delta)}$: In this case since the syndrome of a disjunction of circuits is not greater than the sum of the syndromes of the subcircuits

$$S(*B) \leq S(\bigvee_{i \leq i_0} *B_i) + S(*B_{i_0})$$
$$+ S(\bigvee_{i > i_0} *B_i) \leq S(A) + S(*B_{i_0}) + m2^{-k_{i_0}+1}$$

where $A = \bigvee_{i \leq i_0} *B_i$, since for $i > i_0$ $S(*B_i) \leq 2^{-k_i} \leq 2^{-k_{i_0}+1}$.

However, $m2^{-k_{i_0}+1} \leq (1/10)2^{-(k_{i_0}+\delta)}$ because of the way we selected the k_i.

Conversely, $S(*B^f) \geq S(A) + S(*B_{i_0}^f) - S(A \wedge *B_{i_0}^f)$, since the quantity on the right-hand side is exactly the syndrome of $S(A \vee *B_{i_0}^f)$, and is based on the fact that the number of elements in the union of two sets is the sum of the number of elements in each set minus the number in their intersection.

Now we use the fact that the syndrome of a disjunction of circuits is not greater than the sum of the syndromes of the circuits to obtain

$$S(A \wedge *B_{i_0}^f) \leq \sum_{i=1}^{i_0} S(*B_i \wedge *B_{i_0}^f) \leq m2^{-k_1}2^{-(k_{i_0}-1)}$$
$$\leq (1/5)2^{-(k_{i_0}+\delta)}.$$

The fact that each $S(*B_i \wedge *B_{i_0}^f) \leq 2^{-k_i}2^{-(k_{i_0}-1)} \leq 2^{-k_1}2^{-(k_{i_0}-1)}$ follows since f is a single fault, we have at least $k_i + k_{i_0}$ *independent* control lines and k_1 is the smallest of the k_i's. Note that we used the fact that f was a single-fault to observe that at most one of the control lines $c_{i_01}, \cdots, c_{i_0k_{i_0}}$ is stuck-at 1 (a control line stuck-at 0 is equivalent to a $*l_i$ stuck-at 0). Thus, we have $S(*B) \leq S(A) + S(*B_{i_0}) + (1/10)2^{-(k_{i_0}+\delta)}$, but $S(*B^f) \geq S(A) + S(*B_{i_0}) + (4/5)2^{-(k_{i_0}+\delta)}$. Thus, $S(*B^f) \neq S(*B)$.

2) $S(*B_{i_0}) \geq S(*B_{i_0}^f) + 2^{-(k_{i_0}+\delta)}$: In this case arguing as above we get that

$$S(*B) \geq S(A) + S(*B_{i_0}^f) + (9/10)2^{-(k_{i_0}+\delta)}$$

and

$$S(*B^f) \leq S(A) + S(*B_{i_0}^f) + m2^{-(k_{i_0+1}-1)}$$
$$\leq S(A) + S(*B_{i_0}^f) + (1/5)2^{-(k_{i_0}+\delta)}$$

where we use the fact that f is a single-fault to conclude that $S(*B_i^f) \leq 2^{-(k_i-1)}$. Thus, again $S(*B^f) \neq S(*B)$. □

The above proof actually proves a lot more than Theorem 1. This stronger result is given in Theorem 2.

Theorem 2: The procedure of Theorem 1 produces a circuit with the following properties:

1) if f and g are single faults such that for some i, $S(*B_i^f) \neq S(*B_i^g)$, then $S(*B^f) \neq S(*B^g)$;

2) if f is any multiple fault involving only the l_i and any lines feeding the l_i such that for some i, $S(B_i^f) \neq S(B_i)$, then $S(*B^f) \neq S(*B)$.

Proof: Both 1) and 2) above are proved essentially the same way that Theorem 1 is proved. For 1) we pick i_0 minimal with the property that $S(*B_{i_0}^f) \neq S(*B_{i_0}^g)$. As before, these two values must differ by at least $2^{-(k_{i_0}+\delta)}$ and the rest of the argument in Theorem 1 goes through essentially unchanged (1/10 and 9/10 become 1/5 and 4/5, respectively). For 2) the argument is similar since all the bounds in Theorem 1 are based on the number of control lines and we only use the fact that $S(B_i^f)$ is between 1 and 0, which is true regardless of whether f is a single or multiple fault. Indeed, the only time we use the assumption that f is a single fault is in limiting its effect on the control lines. □

Theorem 2 shows that we have a limited ability to diagnose single faults and detect multiple faults. Diagnosing all single faults is not something one would expect from syndrome testing since one cannot do it for simple $\vee$ and $\wedge$ gates. In particular, the modifications described above will not produce circuits in which every single fault is syndrome diagnosable. Diagnosing multiple faults is also not feasible for two additional reasons: 1) the large amount of storage necessary to store the syndrome values for each multiple fault defeats the whole purpose of syndrome testing; 2) in modifying a circuit by adding control lines to achieve a syndrome-testable design there is always a multiple fault in the control lines which yields the original non-syndrome-testable circuit.

The question of multiple fault detection is not fully resolved. The above argument will handle multiple faults that are not "too bad," i.e., where the number of faults allowed is bounded by a suitable function of the number of lines. To detect all multiple faults would seem to require an even more complicated analysis and more control lines. For this reason we have chosen not to pursue this problem further.

ACKNOWLEDGMENT

The author would like to thank J. Savir for introducing him to this problem and supplying him with some background information and helpful comments.

REFERENCES

[1] J. Savir, "Syndrome-testable design of combinational circuits," *IEEE Trans. Comput.*, vol. C-29, pp. 442–451, June 1980.

[2] —, "Syndrome-testing of syndrome-untestable combinational circuits," IBM Res. Rep. RC8112, Feb. 7, 1980.

Syndrome-Testing of "Syndrome-Untestable" Combinational Circuits

JACOB SAVIR

Abstract—In [1] and [2] a method of designing syndrome-testable combinational circuits was described. It was shown that, in general, syndrome-testable combinational circuits require some pin-penalty and maybe some logic for producing the testable design.

In this correspondence we show a method of syndrome-testing circuits which are not "syndrome-testable." The idea is to perform multiple constrained syndrome-tests on various portions of the circuits in such a way that an overall full syndrome-test coverage will be achieved. Thus, with this method the extra pin-penalty associated with the testable design is traded off with the extra running time of the syndrome-test procedure.

Index Terms—Inversion parity, reconvergent fan-out, unate function.

I. INTRODUCTION

The syndrome of a Boolean function is a count of the number of minterms it covers, normalized with respect to the number of all possible input combinations. Syndrome-testing [1], [2] is a method which observes the syndrome of the combinational circuit and decides whether or not it is fault-free, as opposed to traditional methods of testing which try to verify point-by-point correct response to the test-set. Syndrome-testing has a very low storage requirement for implementation and avoids the expensive stage of test generation.

However, not every combinational circuit is syndrome-testable. We say that a circuit is syndrome-testable if no single stuck-at fault can cause the faulty circuit to realize the same syndrome as the fault-free circuit. It was observed earlier [1]–[3] that combinational circuits which do not have reconvergent fan-out with unequal inversions parities are syndrome-testable. Also, formulas which enable identification of all the syndrome-untestable lines were developed in [1] and [2]. Furthermore, a procedure was developed [1], [2] which enables modifying syndrome-untestable circuits, by inserting extra I/O pins, and making them syndrome-testable. Thus, with a small penalty of extra I/O pins, and maybe some extra logic, every combinational circuit can be modified to be syndrome-testable [5].

There are cases where the penalty of extra I/O pins is too high a price to pay. In these cases syndrome-testing is still a feasible approach, except that multiple syndrome-test runs will be required to guarantee 100 percent syndrome-test coverage. The way a complete syndrome-test coverage is achieved is by performing several constrained syndrome-test runs to cover all possible "syndrome-untestable" faults. The constrained syndrome-test is a syndrome-test performed on a subset of the input space, while the other inputs are held at a constant value. The constraints on the input set are chosen in such a way that the output of the constrained function becomes unate with respect to the tested line, and therefore syndrome-testable.

In Section II we review basic results concerning syndrome-testing. Section III presents the algorithm for syndrome-testing circuits which are "syndrome-untestable," and Section IV concludes with a brief summary.

Manuscript received July 7, 1980; revised February 5, 1981.
The author is with the IBM T. J. Watson Research Center, Yorktown Heights, NY 10598.

The circuits are assumed to be irredundant and only single-stuck-at faults are considered in this correspondence.

II. REVIEW OF BASIC RESULTS CONCERNING SYNDROME-TESTING

A more detailed discussion of the results presented in this section can be found in [1], [2], and [5].

Definition 1: The *syndrome*, S, of a Boolean function is defined as

$$S = \frac{K}{2^n}$$

where K is the number of minterms realized by the function, and n the number of binary input lines.

Let $F = Ag + B\bar{g} + C$ be the sum of product representation of the function F with respect to the line $g = g(x_1, x_2, \cdots, x_n)$, where A, B, and C do not depend on g. Also, let $S(F)$ denote the syndrome of the function F. We have the following results.

Result 1: The fault g stuck-at-zero ($g/0$) is syndrome-untestable if and only if

$$S(A\bar{C}g) = S(B\bar{C}g).$$

Result 2: The fault g stuck-at-one ($g/1$) is syndrome-untestable if and only if

$$S(A\bar{C}\bar{g}) = S(B\bar{C}\bar{g}).$$

Result 3: If the function F is unate in line g, the faults on g are syndrome-testable.

Result 4: Every combinational circuit may be made syndrome-testable by attaching to it extra I/O pins and logic [5].

For circuits which were designed to be syndrome-testable, the test procedure will go as follows. Apply all possible input combinations to the circuit under test (each input combination is applied exactly once) and count the number of ones appearing at its output. Notice that the only difference between the syndrome and the ones-count is the location of the binary point. Thus, there is no essential difference between the syndrome and the ones-count, and a binary counter can serve the purpose of measuring the syndrome. If the syndrome stored in the counter by the time the test has been completed is equal to the fault-free syndrome the circuit is declared fault-free; otherwise the circuit is faulty. It should be noted that in order to make the syndrome-test procedure of acceptable length, large circuits with many inputs must be partitioned to subcircuits such that each subcircuit will have no more than 20–25 inputs. Each partition then is designed to be syndrome-testable. The reason for choosing a partition which does not have more than 25 inputs is to restrict the syndrome-testing time to a few seconds.

III. TESTING "SYNDROME-UNTESTABLE" CIRCUITS

As mentioned earlier, there are occasions where it is practically impossible to pay the penalty of extra I/O pins to produce a syndrome-testable design. In many cases the chip's (or module's) pins are all used up and there is no more room left for insertion of test inputs and outputs. The technique to be described next is applicable for syndrome-testing circuits which fall within this category.

Theorem 1: Let $F = Ag + B\bar{g} + C$ be an irredundant sum of products representation[1] of the output F with respect to line g. Then there exists an assignment to a subset of the input space for which ($A \ne 0, B = 0$) or ($A \ne 0, B = 1$) or ($B \ne 0, A = 0$) or ($B \ne 0, A = 1$).

Proof: We prove that there exists an assignment to a subset of the input space for which $A \ne 0, B = 0$. The proofs of all the other cases are similar. We first prove that there exists an assignment which yields $B = 0$. We prove this by a contrapositive approach: assume that such an assignment does not exist. Then B must be identically equal to 1 ($B \equiv 1$). Under these conditions we have

$$F = Ag + 1\bar{g} + C = A + C + \bar{g}$$

which is negative in g. This contradicts the theorem's assumption that the representation of F is irredundant.

Now we show that among the assignments which yield $B = 0$, there exists at least one that will also impose $A \ne 0$. In order to prove this last statement, assume, again, that such an assignment does not exist. Then the Boolean expression A must be covered by B. In this case the function F reduces to (by letting $A = AB$)

$$F = ABg + B\bar{g} + C = B\bar{g} + A + C$$

which is negative in g. This implies, again, that the representation of F is redundant which contradicts the theorem's assumption.

Q.E.D.

The idea behind Theorem 1 is that there exist assignments to a subset of the input space which will make the residual function, namely the Boolean function obtained after these assignments are applied to the input space, unate in line g. Therefore, if line g was syndrome-untestable in the original circuit, it is definitely syndrome-testable in the residual circuit.

Example 1: Consider the combinational circuit described in Fig. 1(a). All the boldface lines can be shown to be syndrome-untestable. Fig. 1(b) shows an assignment to two of the input lines which makes the residual function unate in all the boldface lines and, therefore, syndrome-testable. So, by performing a constrained syndrome-test procedure, namely by syndrome-testing the residual circuit, one can decide whether or not any one of the boldface lines in Fig. 1(a) is faulty.

Definition 2: A *constrained syndrome-test procedure* is a syndrome-test on a restricted number of input lines, while the rest of the inputs are held at a constant value.

The procedure for syndrome-testing circuits which are not "syndrome-testable" are given below.

Procedure 1

Step 1: Perform a syndrome-test procedure on the original circuit. If the recorded syndrome is different from the fault-free syndrome-stop-the circuit is faulty. Otherwise, go to Step 2.

Step 2: For each fan-out line which is syndrome-untestable find a set of input values which will make the residual function unate with respect to this line. Perform a constrained syndrome test-procedure on the circuit. If the recorded residual syndrome is different from the fault-free one, the circuit is faulty; otherwise, the circuit is fault-free.

Step 2 suggests to perform a separate constrained syndrome-test procedure for every syndrome-untestable fan-out line.[2] Usually, the number of syndrome-untestable fan-out lines is small [1], [2], and the amount of additional testing time required is not significant. However, it is possible to select constraints that will be capable of

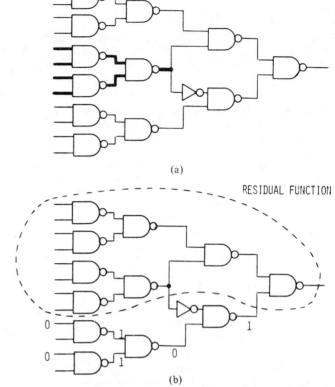

Fig. 1. (a) All the boldface lines are syndrome-untestable. (b) All the syndrome-untestable lines shown in (a) are now syndrome-testable in the residual function.

yielding a syndrome-test which covers more than one untestable line at a time. In fact, it is possible to minimize the number of constrained syndrome-test runs required by going through a covering procedure [4]. In order to minimize the number of constrained syndrome-test runs the following table is generated. You list the syndrome-untestable fan-out lines as column headings and input constraints which make the function unate with respect to the syndrome-untestable lines as row headings. An x is entered at the ijth entry if the input constraint listed in row i makes the function unate in the line listed in column j. Then a minimal number of rows must be selected to cover all possible columns. Although the covering procedure is known to be prohibitive in general, this is not the case here, since the number of syndrome-untestable fan-out lines is usually small.

However, it is useful to have a more practical method of nearly-minimizing the number of constrained syndrome-test runs. This approximate method will generate only a nearly-minimal number of constrained syndrome-test runs, but will require less computation.

Definition 3: A *maximally-unate constraint* (with respect to a set of syndrome-untestable lines) is a constraint on a subset of the input lines which makes the residual function either unate or independent of each syndrome-untestable line, in such a way that removal of any input constraint will leave the residual function nonunate in at least one of the syndrome-untestable lines.

Example 2: Consider the function

$$F = x_1\bar{x}_2 + x_2\bar{x}_3 + x_3\bar{x}_1 + x_4\bar{x}_5 + \bar{x}_4 x_5$$

which is syndrome-untestable in all input lines. The constraint $x_1 = 0$ is a simple constraint with a residual function $F_{r_1} = x_2 + x_3 + x_4\bar{x}_5 + \bar{x}_4 x_5$. The constraint $x_1 = x_4 = 0$ is a maximally unate constraint with a residual function $F_{r_2} = x_2 + x_3 + x_5$. Notice that removal of either $x_1 = 0$ or $x_4 = 0$ will leave the residual function F_{r_2} nonunate in at least one of the syndrome-untestable lines. The constraint $x_1 = x_4 = x_5 = 0$ is not maximally unate with $F_{r_3} = x_2 + x_3$ because the removal of $x_5 = 0$ will yield a unate residual function.

A syndrome-test procedure performed under a maximally-unate constraint will cover all the faults in the syndrome-untestable lines

---

[1] The sum of products representation is obtained by cutting the line under consideration and applying to it a "pseudoinput" g. The primary output f is then expressed, as a sum of products, in terms of the primary inputs $x_1, \cdots, x_n$ and this pseudoinput g. For examples, consult [6]. An irredundant sum of products is one that does not contain redundant literals. For example, $x_1 g + x_2 \bar{g}$ is irredundant, while $x_1 g + \bar{g} + x_2$ is redundant since it can be rewritten as $\bar{g} + x_1 + x_2$.

[2] In this way all the syndrome-untestable lines that feed this fan-out line directly or indirectly will also be covered.

for which the residual function is unate. Therefore, a set of maximally-unate constraints for which the residual functions cover each syndrome-untestable line at least once, defines the collection of constrained syndrome-test runs. The following procedure outlines an approximate but practical way to nearly-minimize the number of constrained syndrome test runs.

Procedure 2

Step 1: Find the set of syndrome-untestable fan-out lines.

Step 2: Find a maximally-unate constraint which covers as many unmarked syndrome-untestable fan-out lines as possible. Mark all the syndrome-untestable lines which are now covered by the residual function defined by this maximally-unate constraint. Record this maximally-unate constraint. If all the syndrome-untestable lines are marked-stop, the collection of the maximally-unate constraints recorded so far defines the constrained syndrome-test runs. Otherwise, repeat Step 2.

Several clarifications are now called for. First, note that the syndrome-untestable fan-out lines are practically a small fraction of the fan-out lines which have unequal inversion parities along its reconvergent paths. These fan-out lines are those whose Boolean functions satisfy result 1 or 2 (quoted earlier in the paper). Thus, in Step 1 we start off with a relatively small number of lines. Next, there are several ways of finding the maximally unate constraints. By definition, one could have computed the expression A or B of the sum of product representation of f with respect to a syndrome-untestable fan-out line g, and then solve equations of the kind $A = 0$ or $A = 1$. A solution of that would have produced a unate constraint with regard to g. Usually, such a solution has a number of DON'T CARE values assigned to the input lines. By repeating this procedure against the next syndrome-untestable fan-out line, one could extend the existing constraint by assigning values to the yet unassigned inputs to become either unate or independent in this new line. This process is continued until the function is either unate or independent in any of the syndrome-untestable fan-out lines. Another way of achieving this same goal is by assigning desensitized values to those fan-out branches emanating from the syndrome-untestable fan-out line and backtracking to the input lines to find a consistent set of input values which will justify this configuration. This operation is similar to the consistency operation in the D-algorithm except that it works on desensitized values as opposed to sensitized ones. By repeating this process against other syndrome-untestable lines a maximally unate constraint is achieved. Thus, to conclude, the complexity of computing a maximally unate constraint is similar to the consistency operation in the D-algorithm, with the exception that it is performed on a very restricted number of lines.

Example 3: Consider the two-level circuit whose output function is

$$F = x_1 x_2 x_3 + \bar{x}_1 \bar{x}_2 \bar{x}_3 + x_4 \bar{x}_5 + \bar{x}_4 x_5 + x_6 \bar{x}_7 + x_7 \bar{x}_8 + x_8 \bar{x}_6 + x_9 x_{10}.$$

Procedure 2 yields the following step-wise partial results.

Step 1: The set of syndrome-untestable lines is (by using Results 1 and 2)

$$T = \{x_1, x_2, x_3, x_4, x_5, x_6, x_7, x_8\}.$$

Step 2: $x_1 = 0, x_4 = 0, x_6 = 0$ (maximally-unate constraint)

$F_{r_1} = \bar{x}_2 \bar{x}_3 + x_5 + x_7 + x_8 + x_9 x_{10}$ (residual function)

$T = \{x_1 \check{x}_2, \check{x}_3, x_4, \check{x}_5, x_6, \check{x}_7, \check{x}_8\}$ (marked syndrome-untestable lines covered by F_{r_1}).

Step 2 (Repeated): $x_2 = 0, x_5 = 0, x_7 = 0$ (maximally-unate constraint)

$F_{r_2} = \bar{x}_1 \bar{x}_3 + x_4 + x_6 + x_8 + x_9 x_{10}$ (residual function)

$T = \{\check{x}_1, \check{x}_2, \check{x}_3, \check{x}_4, \check{x}_5, \check{x}_6, \check{x}_7, \check{x}_8\}$ (marked "syndrome-untestable" lines covered by F_{r_1} and F_{r_2}).

To conclude, this circuit can be tested with a complete syndrome-test coverage in two constrained syndrome-test runs

$$x_1 = x_4 = x_6 = 0 \tag{1}$$

$$x_2 = x_5 = x_7 = 0. \tag{2}$$

IV. Summary and Conclusions

In this correspondence we showed a method of overcoming the problem of syndrome-testing circuits which are not syndrome-testable by the original definition. The penalty of extra I/O pins in producing a syndrome-testable design can be avoided if an increase in the syndrome-testing time is acceptable. The increase in the testing time is due to separate syndrome-test runs performed on residual portions of the circuit obtained after some of the input lines are attached to constant binary values. Since all the residual portions can be selected to be unate in the "syndrome-untestable" lines, the faults on these lines are detected when they are syndrome-tested.

The minimization of the number of syndrome-test runs necessary to cover all syndrome-untestable lines can be achieved by going through a prime-implicant-like covering procedure. A method of nearly-minimizing the number of syndrome-test runs, which is computationally feasible was described. In many cases the nearly-minimal situation is actually the minimum one.

References

[1] J. Savir, "Syndrome-testable design of combinational circuits," in *Dig. 9th Int. Symp. Fault-Tolerant Comput.*, vol. 9, June 1979, pp. 137–141.

[2] ——, "Syndrome-testable design of combinational circuits," *IEEE Trans. Comput.*, vol. C-29, pp. 442–451, June 1980; also *IEEE Trans. Comput.*, vol. C-29, pp. 1012–1013, Nov. 1980, for corrections.

[3] A. Tzidon, I. Berger, and M. Yoeli, "A practical approach to fault detection in combinational networks," *IEEE Trans. Comput.*, vol. C-27, pp. 968–971, Oct. 1978.

[4] E. J. McCluskey, *Introduction to the Theory of Switching Circuits*. New York: McGraw-Hill, 1965.

[5] G. Markowsky, "Syndrome-testability can be achieved by circuit modification," IBM Res. Rep. RC8299, June 1980.

[6] S. S. Yau et al., "An efficient algorithm for generating complete test sets for combinational logic nets," *IEEE Trans. Comput.*, vol. C-20, pp. 1245–1251, Nov. 1971.

The Weighted Syndrome Sums Approach to VLSI Testing

ZEEV BARZILAI, JACOB SAVIR, GEORGE MARKOWSKY, AND MERLIN G. SMITH

Abstract—With the advent of VLSI, testing has become one of the most costly, complicated, and time consuming problems. The method of syndrome-testing is applicable toward VLSI testing since it does not require test generation and fault simulation. It can also be considered as a vehicle for self-testing. In order to employ syndrome-testing in VLSI, we electronically partition the chip into macros in test mode. The macros are then syndrome tested in sequence.

In this paper we show the means to syndrome-test macros. We examine the size of the syndrome driver counter and establish a method of determining its minimal length. The problem of minimizing the number of syndrome references needed for testing is also investigated. It is shown that it is always possible to use one weighted syndrome sum as reference for each and every macro. The question of weighted sum syndrome-testability is addressed and methods to achieve it are discussed. A self-test architecture based on these concepts is described.

Index Terms—Partitioning, self-testing, syndrome-testable design, syndrome-testing.

I. Introduction

In recent years considerable attention has been given to LSI/VLSI testing and testable design. Digital circuit manufacturers are well aware today of the need to design testability fixtures early in the design stage, or otherwise they will have to pay a higher testing bill later in the process. Using traditional testing schemes for VLSI requires high test generation and fault simulation times. New techniques are required.

Syndrome-testing [2], [4]–[6] is a step to achieve this goal. The notion of syndrome-testing is based on counting the number of ones realized by a Boolean function and comparing it to the fault-free count. Since there may be circuits and faults for which these fault-free and faulty syndromes are the same, modifications to produce a testable design are generally required. The syndrome-testable design, thus ensures that all the faulty syndromes will differ from the fault-free one by adding a small amount of I/O and logic. In a VLSI environment the chip is electronically partitioned into macros, in test mode, in order to reduce the overall test time.

In this paper we assume that the chip has been designed according to the Level Sensitive Scan Design (LSSD) [1] rules. From a testing standpoint it means that, basically, the testing problem has been reduced to testing the combinational circuitry between Shift Register Latches (SRL). Thus this task of testing the combinational logic can be accomplished by means of syndrome-testing.

In this paper we investigate a cost effective scheme for VLSI testing. Our proposal is to use syndrome-testing to test each and every macro of the chip. In Section II we consider the problem of parallel syndrome-testing of all functions involved in a macro. In Section III we determine the minimum counter size necessary to drive a multi-input–multioutput macro, where all macro outputs are tested in parallel. In Section IV we investigate ways to further reduce the number of references needed for testing by using weighted syndrome sums. The weighted syndrome sums approach has a natural appeal to self-testing because of the enormous test data savings that it may offer. The question of untestability with regard to weighted sums is then addressed, and some ways to overcome it are presented. Section V presents the self-test architecture. The paper concludes with a brief summary.

Manuscript received November 21, 1980; revised June 4, 1981.
The authors are with IBM Thomas J. Watson Research Center, Yorktown Heights, NY 10598.

II. Syndrome-Testing of Multiple Output Circuits

Since the notion of syndrome-testing requires the application of all possible input combinations to the Circuit Under Test (CUT), a VLSI is partitioned to limit the test time to some acceptable level. Thus, assume that the VLSI chip has been partitioned into R macros (syndrome partitions), where each macro is a multiinput–multioutput digital circuit. Usually, short test times will mean smaller macros and, therefore, more macros, while longer test times will usually result in larger and fewer macros. It is important to note that there are tradeoffs between macro sizes, test times, and partitioning penalties, i.e., extra hardware and I/O pins. Fig. 1(a) describes the partitioning of the chip into macros, and Fig. 1(b) illustrates schematically a typical macro.

When we go to syndrome-test the chip we must make sure that all macros are tested properly. Those macros which have disjoint sets of inputs can always be tested in parallel. In the other cases where the macros are interconnected a sequential syndrome test procedure may be used.

The question arises as to how one should syndrome-test a given macro. If we separately syndrome-test each output function involved in the macro, this will require the repetition of the syndrome-test procedure as many times as there are output functions in the macro. This can be a very time-consuming process because of the multiplicity of the syndrome-test runs. Another way to achieve this same purpose is to try to syndrome-test all the output functions in parallel, namely, to use only one syndrome-test procedure exercising the totality of the macro inputs and obtaining all output syndromes in one pass. The following theorem shows that by using the method of parallel syndrome-testing, the true syndromes are recorded at the outputs of the macro.

Theorem 1: Let $y_i = f(x_{i1}, x_{i2}, \cdots, x_{ik_i})$, $i = 1, 2, \cdots, m$ be the output functions realized by the macro. Let p be the length of the counter which drives the input combinations to the macro, $p \geq k_i$ $\forall i$. Then by running a syndrome-test procedure which exercises all 2^p possible combinations of the counter, the true syndrome appears at each output y_i.

Proof: Let M_i be the number of minterms in the function y_i, $i = 1, 2, \cdots, m$. Then, by definition the syndrome realized by output y_i is given by

$$S_i = \frac{M_i}{2^{k_i}}.$$

Let the inputs $x_{i1}, x_{i2}, \cdots, x_{ik_i}$ be connected to the bits $b_{i1}, b_{i2}, \cdots, b_{ik_i}$ of the counter. When the counter steps through all its possible input combinations each vector $\bar{x} = (x_{i1}, x_{i2}, \cdots, x_{ik_i})$ is applied 2^{p-k_i} times. Thus, the normalized ones' count appearing at the output y_i is given by

$$\frac{2^{p-k_i} M_i}{2^p} = \frac{M_i}{2^{k_i}} = S_i, \qquad i = 1, 2, \ldots, m$$

which is the correct syndrome. Q.E.D.

III. The Syndrome Driver Counter Size

One of the key questions regarding parallel syndrome-testing of multiinput–multioutput macros is the minimum size of the input counter which drives the exhaustive test vectors, called the Syndrome Driver Counter (SDC). The length of the SDC falls within the range described by the following theorem.

Theorem 2: The length of the SDC L is bounded by

$$\max_i \{k_i\} \leq L \leq n, \qquad i = 1, 2, \cdots, m$$

where k_i is the number of inputs feeding the output function y_i, and n is the total number of inputs to the macro.

Proof: Since there are n input lines to the macro, the length of

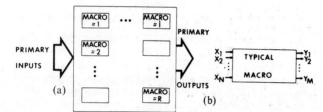

Fig. 1. (a) A VLSI chip partitioned into macros. (b) A typical macro with n inputs and m outputs.

the SDC is bounded from above by n. Also, since the SDC must be capable of driving each and every output function, it is clear that its length is bounded from below by max $\{k_i\}$. Q.E.D.

Note that both the lower bound and the upper bound are attainable. The lower bound, for example, is attainable in the case where all the input sets driving the output functions are disjoint, while the upper bound is attainable when one of the output functions is dependent upon all input variables.

It is of utmost importance to be able to determine the minimal size of the SDC because it has a direct effect on the test time required to syndrome-test the macro. Let the minimal size SDC be q. Let, also, the bits of the SDC be denoted by $b_1, b_2, \cdots, b_q$.

The problem of determining the minimal size SDC is, therefore, the problem of constructing a function $G: \{x_1, x_2, \cdots, x_n\} \rightarrow \{b_1, b_2, \cdots, b_q\}$ where $G(x_i) = b_j$ if and only if there is a direct connection between input x_i and bit counter b_j, and certain relationships (discussed below) between the x_i's are taken into account. Note that G is a surjection or an onto mapping.

Definition 1: An input pair (x_i, x_j) is said to be *adjacent* if there exists at least one output function y_i, $i = 1, 2, \cdots, m$ which depends on both x_i and x_j.

In order to minimize the size of the SDC it is necessary to connect as many input lines to the same counter bit. A pair of input lines may be connected to the same counter bit if and only if they are not adjacent. We next show a constructive method of finding the function G.

Definition 2: A pair of inputs (x_i, x_j) is said to be *nonadjacent* if they are not adjacent.

We use the nonadjacency (NA) graph to create the function G. The NA graph is defined to have n vertices, designated by $x_1, x_2, \cdots, x_n$ to correspond to the n input lines. There is an arc between node x_i and node x_j if and only if the pair (x_i, x_j) is nonadjacent. By analyzing the NA graph it is possible to identify a maximal class of inputs that can be connected to the same counter bit. From the definition of the NA graph it is evident that the input lines $x_{i1}, x_{i2}, \cdots, x_{ik}$ can be connected to the same counter bit if they form a clique, i.e., the subgraph induced by $x_{i1}, x_{i2}, \cdots, x_{ik}$ is a complete graph.

Each clique in the NA graph defines a collection of inputs that can be connected to a single counter bit. Our objective is, therefore, to find all possible maximal cliques and then go through a prime-implicant-like covering procedure [3] to determine the minimum set of such cliques that cover all possible inputs. Each maximal clique obtained in the minimal cover will correspond to one bit counter. Thus the number of maximal cliques appearing in the minimal cover determine the minimal size of the SDC. Unfortunately, this problem is well known to be NP-complete and there is no known algorithm for solving it that is efficient in all cases. In many practical cases, though, it is possible to get good answers efficiently.

The function G is established in the following way. Let the set of maximal cliques appearing in the minimal cover be

$$\{MC_i | MC_i = \{x_{i1}, x_{i2}, \cdots, x_{ik_i}\}, \quad i = 1, 2, \cdots, q\}.$$

The bit counters are then defined by referring each such MC to a different bit, namely,

$$MC_i \Leftrightarrow b_i, \quad i = 1, 2, \cdots, q.$$

Note that in general the cliques are not disjoint. In particular, suppose input x_i is included in several cliques, i.e.,

$$x_i \in \{MC_{j1}, MC_{j2}, \cdots, MC_{jl}\}$$

then there are l choices for defining $G(x_i)$, namely,

$$G_r(x_i) = b_{jr}, \quad r = 1, 2, \cdots, l.$$

Note, therefore, that there are multiple G functions possible, all of them leading to the same minimum counter size.

Example 1: Consider an eight input, five output macro, whose output functions are given by

$$y_1 = f_1(x_1, x_2, x_3, x_4)$$
$$y_2 = f_2(x_6, x_7, x_8)$$
$$y_3 = f_3(x_3, x_4, x_5, x_6)$$
$$y_4 = f_4(x_3, x_5, x_6, x_7)$$
$$y_5 = f_5(x_1, x_4, x_7, x_8).$$

According to Theorem 2, we can already identify the range of the SDC length

$$4 \leq L \leq 8.$$

To determine the minimum size SDC q we next create the NA graph (Fig. 2). The set of MC's for the graph of Fig. 2 is

$$\{\{x_2, x_5, x_8\}, \{x_1, x_5\}, \{x_1, x_6\}, \{x_2, x_6\}, \{x_2, x_7\}, \{x_3, x_8\}, \{x_4\}\}.$$

Operating a minimal covering procedure on these sets yields three possible covers

$$A_1 = \{\{x_2, x_5, x_8\}, \{x_1, x_6\}, \{x_2, x_7\}, \{x_3, x_8\}, \{x_4\}\}$$
$$A_2 = \{\{x_1, x_5\}, \{x_2, x_6\}, \{x_2, x_7\}, \{x_3, x_8\}, \{x_4\}\}$$
$$A_3 = \{\{x_1, x_5\}, \{x_1, x_6\}, \{x_2, x_7\}, \{x_3, x_8\}, \{x_4\}\}.$$

Note that the minimal size SDC for this example is 5. Note also that the set A_1 defines 4 possible G functions, as opposed to sets A_2 and A_3 which each define 2 functions.

Using the set A_2, the two possible G functions are

$G_1(x_1) = G_1(x_5) = b_1$		$G_2(x_1) = G_2(x_5) = b_1$
$G_1(x_2) = G_1(x_6) = b_2$		$G_2(x_6) = b_2$
$G_1(x_7) = b_3$	or	$G_2(x_2) = G_2(x_7) = b_3$
$G_1(x_3) = G_1(x_8) = b_4$		$G_2(x_3) = G_2(x_8) = b_4$
$G_1(x_4) = b_5$		$G_2(x_4) = b_5$

where

$$\{x_1, x_5\} \Leftrightarrow b_1$$
$$\{x_2, x_6\} \Leftrightarrow b_2$$
$$\{x_2, x_7\} \Leftrightarrow b_3$$
$$\{x_3, x_8\} \Leftrightarrow b_4$$
$$\{x_4\} \Leftrightarrow b_5.$$

The connection between the minimal size SDC and the macro inputs, as defined by the function G_1, is shown in Fig. 3.

IV. THE WEIGHTED SYNDROME SUMS

In Section II we have shown that it is possible to syndrome-test a multiinput–multioutput macro in one test procedure. A brute force implementation of this parallel syndrome-testing is to use m references one for each output syndrome. In a VLSI environment this may mean tens or hundreds of syndrome reference words. Since this trend in increasing density is expected to grow in the future, it may be attractive to try to reduce the number of references. Furthermore, one of the major alternatives to VLSI testing is to use a built-in test. This

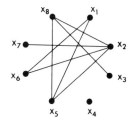

Fig. 2. The NA graph for Example 1.

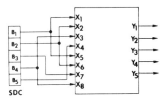

Fig. 3. The connection diagram between the SDC and the macro inputs as implied by the function G_1.

reduction in the number of output references makes the notion of syndrome-testing a very attractive approach to self-testing because it reduces the area overhead needed for implementation.

In this section we show a method of combining output syndromes in such a way as to reduce dramatically the number of references needed for syndrome-testing.

Let $\vec{S} = (S_1, S_2, \cdots, S_m)$ be a vector representing the output syndromes of the macro. Let also the function

$$g = \sum_{i=1}^{m} w_i Z_i$$

be a linear combination of the variables $Z_1, \cdots, Z_m$ with coefficients $w_1, w_2, \cdots, w_m$, where w_i is an integer, $i = 1, 2, \cdots, m$.

Definition 3: A weighted syndrome sum, (WSS), for an m-output macro with coefficients w_i, $i = 1, 2, \cdots, m$ is defined as

$$\text{WSS} = g(\vec{S}).$$

Our intention is to use as references one or more WSS's instead of the complete collection of output syndromes. From here on we assume that each output function has been designed to be syndrome-testable [4], [5]. It is important to note that when using WSS's as references, as opposed to a complete collection of output syndromes, there may be undetectable faults, in the sense of the following definition.

Definition 4: Let the faulty syndromes induced by a fault f be denoted by S_i^f, $i = 1, 2, \cdots, m$. Let also $\Delta \vec{S} = \vec{S} - \vec{S}^f$. Then the fault f is said to be *weighted sum syndrome untestable* (WSSU) if

$$g(\Delta \vec{S}) = \sum_{i=1}^{m} w_i \Delta S_i = 0.$$

The following example demonstrates the concept of WSSU.

Example 2: Consider the circuit of Fig. 4. Suppose we use only one weighted sum with coefficients 2 and 3, namely let

$$g = 2Z_1 + 3Z_2.$$

Let input line x_3 be stuck at zero. Then the fault-free and faulty syndromes of the outputs are

$$S_1 = \frac{5}{16} \quad S_1^f = \frac{1}{8}$$

$$S_2 = \frac{3}{8} \quad S_2^f = \frac{1}{2}$$

Thus

$$\Delta S_1 = \frac{3}{16} \quad \Delta S_2 = -\frac{1}{8}$$

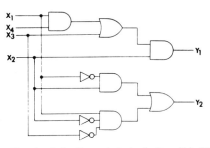

Fig. 4. The circuit for Example 2; the fault $x_3/0$ is WSSU.

and, therefore,

$$g(\Delta \vec{S}) = 2\Delta S_1 + 3\Delta S_3 = 0.$$

According to Definition 4 the fault $x_3/0$ is WSSU.

We would like to emphasize that as long as all coefficients are nonzero a fault which affects only one output function can never yield a WSSU condition, since every such fault will change the corresponding syndrome and therefore change the WSS. The choice of coefficients for a WSS is very important. In fact, the following theorem shows that we can always find coefficients for a WSS such that every single fault is WSS-testable.

Theorem 3: Suppose we are given a macro with output lines $y_1, \cdots, y_m$ and that line y_i is fed by $k_i \geq 1$ input lines. Then the WSS, with coefficients

$$w_i = 2^{\left(\sum_{j=1}^{i} k_j\right) + i}$$

detects all single faults.

Proof: We need only show that if some $\Delta \vec{S}_i \neq \vec{0}$, then $\sum_{i=1}^{m} w_i \Delta S_i \neq 0$. Let $i_o = \max \{i \mid \Delta S_i \neq 0\}$. Note that $|\Delta S_{i_0}| \geq 2^{-k_{i_0}}$

$$|w_i \Delta S_i| \geq 2^{\left(\sum_{j=1}^{i} k_j\right) + i}$$

for all $i < i_o$ and $\Delta S_i = 0$ for all $i > i_o$. Thus

$$\left| \sum_{i=1}^{m} w_i \Delta S_i \right| \geq |w_{i_0} \Delta S_{i_0}| - \sum_{i < i_c} |w_i \Delta S_i| \geq 2^{\left(\sum_{j=1}^{i_0-1} k_j\right) + i_0}$$

$$- \sum_{i < i_0} 2^{\left(\sum_{j=1}^{i} k_j\right) + i} > 0.$$

Q.E.D.

Note that Theorem 3 is not all that useful since the coefficients are so large, which forces us to store larger numbers. Indeed, the storage requirements are roughly the same as if we were to store the syndrome values for each output line individually. The point of Theorem 3 is to show that we can always get by with one WSS if necessary. Thus the WSS approach is no worse than simply storing the individual syndrome values. The benefit of the WSS approach is that often we can find a *small* number (ideally one) of WSS's with *small* coefficients which work for a given circuit. Below we discuss two approaches which can be used to find economical WSS schemes in many practical cases. Both cases depend on knowing something about the macro's structure. As noted earlier, if there is no overlap of various input lines any WSS having all its coefficients nonzero will get the job done. Theorem 4 generalizes this approach.

Theorem 4: If we pick r WSS's, defined by the weighted sums

$$g_i = \sum_{j=1}^{n} w_{ij} Z_j, \quad i = 1, \cdots, r$$

such that any $r \times r$ submatrix selected from the matrix $W = [w_{ij}]_{i=1,\cdots,r; j=1,\cdots,m}$ is nonsingular, then a fault may be WSSU only if it affects at least $r + 1$ outputs.

Proof: A fault f will be WSSU if

$$\sum_{j=1}^{m} w_{ij}\Delta S_j = 0, \quad i = 1, 2, \cdots, r.$$

We have to prove that at least $r + 1$ ΔS_j's are nonzero in the solution space. Or, if we express it differently, we have to prove that at most m-r-1 ΔS_j's are zeros in the solution space (excluding the trivial solution which corresponds to the fault-free condition). The argument goes as follows. Assume that there are m-$r + k$-1, $k \geq 1$, ΔS_j's which are zero. By substituting zeros for those ΔS_j's we are left with a homogeneous system of r equations in $r + 1$-k unknowns. Since all the columns of the matrix W are linearly independent because of the way we have chosen the coefficients w_{ij}, the rank of the matrix associated with the above homogeneous system is $r + 1$-k. Therefore, the solution for this homogeneous system is the trivial solution. This means that the only way we might have more than m-r-1 ΔS_j's which are zeros is by having a trivial solution which we have excluded. Q.E.D.

Corollary 1: A fault f may be WSSU only if it affects a line in an $r + 1$th degree overlap of the circuitry feeding the macro outputs.

Proof: Directly from Theorem 4.

It is very important to mention that both Theorem 4 and Corollary 1 refer to a necessary but not sufficient condition for a fault to be WSSU. Theorem 4 describes mathematical conditions for which an WSSU may exist. However, it is not guaranteed that a pattern of faulty syndromes occurring in the mathematical solution will in fact physically exist in the circuit. Thus it is very likely that most of the faults that may occur in the $r + 1$th degree overlap will end up being weighted sum syndrome-testable. Another approach is based on the number of different values which can occur as syndrome values at each output of the macro. The details are spelled out in Theorem 5.

Theorem 5: Suppose we are given a macro with output lines $y_1, \cdots, y_m$ such that d_i different syndrome values can appear at y_i as the result of a single fault or normal operation. There exists a set of r WSS's having n bit coefficients which detect all single faults if

$$\prod_{i=1}^{m} d_i < 2^{nr} + 1.$$

More generally, such a set exists if the number of different $\Delta\vec{S}$'s is $<2^{nr}$.

Proof: If we allow our coefficients to have n bits, we have a choice of 2^{nm} different WSS's. Given r of these WSS's we wish them to have the property that for each $\Delta\vec{S}$, some $j = 1, \cdots, r$ has the property that $g_j(\Delta\vec{S}) \neq 0$. Altogether we have 2^{nmr} different systems of r WSS's having n-bit coefficients. Of these at most $2^{n(m-1)r}$ have the property that $g_i(\Delta\vec{S}) = 0$ for all $i = 1, \cdots, m$. To see this, note that since some $\Delta S_{i_0} \neq 0$, once we specify the coefficients w_{jt}, $t \neq i_o$, there is at most one choice for w_{ji_0} (there may not be any since we are dealing with integers). Observe that there can be at most

$$\prod_{i=1}^{m} d_i - 1$$

different $\Delta\vec{S}$'s. Thus the number of inadequate systems is at most

$$\left(\prod_{i=1}^{m} d_i - 1\right) 2^{n(m-1)r}.$$

As long as this number is $<2^{nmr}$, i.e.,

$$\prod_{i=1}^{m} d_i < 2^{nr} + 1$$

there must be at least one adequate system. The more general statement follows in the same way. Q.E.D.

Note that Theorem 5 does not furnish an efficient way to find an adequate system. Also, the upper bound is not the tightest possible, but it does give some idea of what can be done. In general, there may not be

$$\prod_{i=1}^{m} d_i - 1$$

different $\Delta\vec{S}$'s since the values are not necessarily independent. There is a rough estimate which might be used to estimate d_i based on the following observation. Given n inputs, there are 2^{2^n} different functions but only 2^n different syndrome values. Thus if a circuit has f different single faults, one could estimate the number of different syndrome values by $\log_2(f)$.

Example 3: Fig. 5 illustrates a binary to two's complement conversion circuit with eight inputs and six outputs. Consider the weighted sum defined by

$$g = \sum_{i=0}^{5} 2^i Z_i.$$

Table I displays the output syndromes and the corresponding values of the weighted syndrome sums for all single faults that may affect two or more outputs. We denote by $b/0$ line b stuck-at 0, and by $b/1$ line b stuck-at 1. There are 45 rows in Table I corresponding to the fault-free case and all relevant faulty cases. The first six columns of the table correspond to all output lines, while the last column refers to the value of the weighted sum. Each entry of the table in the first six columns represents the corresponding syndrome value, while the entries in the last column display the value of weighted syndrome sum. As seen from Table I none of the faulty weighted syndrome sums are identical to the fault-free one, and thus the circuit is not WSSU.

Example 4 demonstrates a WSSU case and suggests possible modifications to produce a weighted sum syndrome testable circuit.

Example 4: Fig. 6 describes a 4-input–4-output circuit. Suppose we choose two weighted sums defined by

$$g_1 = \sum_{i=0}^{3} 2^i z_i$$

$$g_2 = \sum_{i=0}^{3} 2^{3-i} z_i.$$

From the previous discussion we know that the potential WSSU faults must lie in a third degree overlap between the output functions. Thus the only candidates for consideration are lines x_1 and x_3.

The candidate WSSU faults yield the following weighted syndrome sums:

Fault-free case	$WSS_1 = 2.5$	$WSS_2 = 4.25$
$x_1/0$	$WSS_1 = 2.5$	$WSS_2 = 4.25$
$x_1/1$	$WSS_1 = 2.5$	$WSS_2 = 4.25$
$x_3/0$	$WSS_1 = 3.0$	$WSS_2 = 2.25$
$x_3/1$	$WSS_1 = 2.0$	$WSS_2 = 6.25$

Thus the faults $x_1/0$ and $x_1/1$ are WSSU.

In order to correct for this untestable condition we modify one of the functions in which x_1 is involved. In this example we decide to add an extra input to the AND gate realizing Y_o. This will correct the WSSU condition as evidenced by the following results:

Fault-free case	$WSS_1 = 2.375$	$WSS_2 = 3.25$
$x_1/0$	$WSS_1 = 2.5$	$WSS_2 = 4.25$
$x_1/1$	$WSS_1 = 2.25$	$WSS_2 = 2.25$

If we check the weighted syndrome sums induced by $x_3/0$ and $x_3/1$ we may find that these faults are still testable under the above modification.

Note that if we are using a WSS for a given macro, adding extra inputs and AND gates are essentially equivalent to dividing the appropriate coefficient of the WSS by a power of 2.

V. The Syndrome Self-Test Architecture

Fig. 7 illustrates the fundamental architecture of a chip in self-test mode. The chip has been partitioned to R syndrome partitions. We assume that each syndrome partition has LSSD chains on both inputs and outputs. The Syndrome Driver Counters (SDC) are a simple modification of input SRL chains. These SDC's work as regular SRL's in functional mode and as pure counters (or as linear feedback

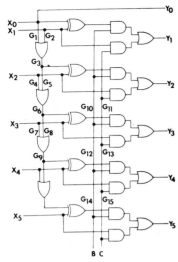

Fig. 5. A binary to two's complement conversion circuit.

TABLE I
THE SYNDROMES AND WEIGHTED SYNDROME SUMS OF VARIOUS FAULTS IN THE CIRCUIT OF FIG. 5

	Y0	Y1	Y2	Y3	Y4	Y5	WEIGHTED SUM
FAULT FREE	.5	.4375	.46875	.484375	.4921875	.49609375	30.875
X0/0	.0	.375	.4375	.46875	.484375	.4921875	29.75
X0/1	1.	.5	.5	.5	.5	.5	32.0
X1/0	.5	.25	.4375	.46875	.484375	.4921875	30.0
X1/1	.5	.625	.5	.5	.5	.5	31.75
X2/0	.5	.4375	.375	.46875	.484375	.4921875	30.125
X2/1	.5	.4375	.5625	.5	.5	.5	31.625
X3/0	.5	.4375	.46875	.4375	.484375	.4921875	30.25
X3/1	.5	.4375	.46875	.53125	.5	.5	31.5
X4/0	.5	.4375	.46875	.484375	.46875	.4921875	30.375
X4/1	.5	.4375	.46875	.484375	.515625	.5	31.375
B /0	.5	.25	.25	.25	.25	.25	16.0
B /1	.5	.625	.6875	.71875	.734375	.7421875	45.75
C /0	.5	.25	.25	.25	.25	.25	16.0
C /1	.5	.625	.6875	.71875	.734375	.7421875	45.75
G1/0	.5	.4375	.4375	.46875	.484375	.4921875	30.375
G1/1	.5	.4375	.5	.5	.5	.5	31.375
G2/0	.5	.4375	.4375	.46875	.484375	.4921875	30.375
G2/1	.5	.4375	.5	.5	.5	.5	31.375
G3/0	.5	.4375	.375	.4375	.46875	.484375	29.375
G3/1	.5	.4375	.5	.5	.5	.5	31.375
G4/0	.5	.4375	.46875	.46875	.484375	.4921875	30.5
G4/1	.5	.4375	.46875	.5	.5	.5	31.25
G5/0	.5	.4375	.46875	.4375	.46875	.484375	29.75
G5/1	.5	.4375	.46875	.5	.5	.5	31.25
G6/0	.5	.4375	.46875	.375	.4375	.46875	28.25
G6/1	.5	.4375	.46875	.5	.5	.5	31.25
G7/0	.5	.4375	.46875	.484375	.484375	.4921875	30.625
G7/1	.5	.4375	.46875	.484375	.5	.5	31.125
G8/0	.5	.4375	.46875	.484375	.4375	.46875	29.125
G8/1	.5	.4375	.46875	.484375	.5	.5	31.125
G9/0	.5	.4375	.46875	.484375	.375	.4375	27.125
G9/1	.5	.4375	.46875	.484375	.5	.5	31.125
G10/0	.5	.25	.25	.484375	.4921875	.49609375	29.625
G10/1	.5	.625	.6875	.484375	.4921875	.49609375	32.125
G11/0	.5	.25	.25	.484375	.4921875	.49609375	29.625
G11/1	.5	.625	.6875	.484375	.4921875	.49609375	32.125
G12/0	.5	.25	.25	.25	.4921875	.49609375	27.75
G12/1	.5	.625	.6875	.71875	.4921875	.49609375	34.0
G13/0	.5	.25	.25	.25	.4921875	.49609375	27.75
G13/1	.5	.625	.6875	.71875	.4921875	.49609375	34.0
G14/0	.5	.25	.25	.25	.25	.49609375	23.875
G14/1	.5	.625	.6875	.71875	.4921875	.49609375	34.0
G15/0	.5	.25	.25	.25	.25	.49609375	23.875
G15/1	.5	.625	.6875	.71875	.4921875	.49609375	34.0

shift registers) in test mode. The SDC's are responsible for exercising the syndrome partitions by going through all possible combinations.

The multiplexer (MUX) selects one out of R syndrome partitions by connecting all its outputs to the syndrome measurement circuitry. Thus, the syndrome partitions are tested in sequence. If τ_i is the time needed for testing the ith syndrome partition

$$\sum_{i=1}^{R} \tau_i$$

is the total test time. The syndrome measurement circuitry depends on which testing approach is being used. The inputs to the measurement circuitry are the outputs of the syndrome partition currently under test. It provides as output the actual syndromes to be compared with the references stored in the Read Only Storage (ROS). The comparator compares the output of the syndrome measurement circuitry to the reference stored in the ROS. A fault indication bit is turned on if a discrepancy is observed. For the weighted syndrome

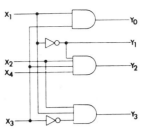

Fig. 6. An example which is WSSU for the functions defined in Example 4.

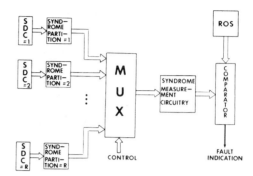

Fig. 7. The fundamental architecture for self-test.

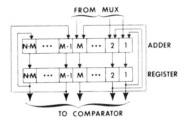

Fig. 8. Syndrome measurement circuitry for the weighted syndrome sums scheme

sums approach, with coefficients which are powers of 2, namely $w_i = 2^i$, the syndrome measurement circuitry is composed of an adder and a register (Fig. 8). If an upper bound to the number of inputs (outputs) to the syndrome partitions is $n(m)$, then the lengths of both the adder and the register is $n + m$. The weighting coefficients are implemented by connecting the MUX outputs to the proper adder inputs.

VI. SUMMARY AND CONCLUSIONS

The use of weighted syndrome sums in VLSI testing has been investigated in this paper. The design for weighted sum syndrome testability require partitioning into macros and selecting either one or several weighted syndrome sums for test reference. Thus the number of references needed for test implementation is roughly proportional to the number of macros. This low storage requirement has a natural application in self-test systems based on syndrome measurements.

REFERENCES

[1] E. B. Eichelberger and T. W. Williams, "A logic design structure for LSI testability," in *Proc. 14th Annu. Design Automation Conf.*, pp. 462–468, June 1977.

[2] G. Markowsky, "Syndrome testability can be achieved by circuit modification," *IEEE Trans. Comput.*, vol. C-30, pp. 604–606, Aug. 1981.

[3] E. J. McCluskey, *Introduction to the Theory of Switching Circuits*, New York: McGraw-Hill, 1965.

[4] J. Savir, "Syndrome-testable design of combinational circuits," in *Proc. 9th Int. Symp. on Fault-Tolerant Computing*, pp. 137–140, June 1979.

[5] ——, "Syndrome-testable design of combinational circuits," *IEEE Trans. Comput.*, vol. C-29, pp. 442–451, June 1980; also see *IEEE Trans. Comput.*, vol. C-29, pp. 1012–1013, Nov. 1980.

[6] ——, "Syndrome-testing of 'syndrome-untestable' combinational circuits," *IEEE Trans. Comput.*, vol. C-30, pp. 606–608, Aug. 1981.

BUILT-IN TESTING: STATE-OF-THE-ART

Bernd Koenemann
Honeywell Inc.
Corporate Solid State Laboratory
12001 State Highway 55
Plymouth, MN 55441

Gunther Zwiehoff, Robert Bosch - GmbH, Reutlingen, West Germany

Abstract

The ever increasing complexity of integrated circuits and the pervasiveness of their application have stimulated a growing interest in the issues of microelectronics testability, maintainability, reliability and safety. This paper will give an overview of techniques that have been developed for the built-in self-test of digital electronic systems. The focus will be on compact off-line self-test.

Introduction

The progress being made in the microelectronics industry is setting new standards for industrial innovation. The growth curves show that the functional complexity of monolithic integrated circuits is continuing to increase at an enormous pace. Mastering the complexity has become a field of international competition in academia and industry. In the course of this development, the area of design for test has experienced a change from a state of neglection to rather substantial research and development action. Testability is no longer perceived as a disputable burden but is beginning to be accepted as a desirable design quality.

Beyond being a design quality, testability has a crucial impact on important product qualities. The pervasiveness of microelectronics is beginning to confront us in practically every facet of our lives. The society and the individual's well being become more and more dependent on the proper functioning of integrated circuits and systems built thereof. Microelectronics is replacing familiar and proven equipment in traditionally non-electronic domains. From this point of view it should be obvious that

- maintainability
- reliability
- safety

are key qualites to be expected from microelectronic products. Testability is an instrument for achieving the desired qualities.

Built-in self-test has emerged as a powerful approach to embedding extensive test functions into system components at a reasonable cost. This paper will give an overview of some built-in self-test techniques.

Background

Design for test as a structured design discipline has been applied to large mainframe computer systems for quite some time. The key element of the structured methods is the strict separation of storage elements and combinational circuits for test purposes. All storage elements except memory arrays are accessible to the tester, i.e. can be loaded and read directly from the tester. The term "scanning" has become the familiar term for describing the loading and reading of storage elements for testing purposes[1]. The class of design styles providing the scanning capabilities can best be called scan-designs.

In order to keep the extra system I/O connections for the scanning capabilities low, it is necessary to map the access paths to the storage elements on either existing data/address bus I/Os or on very few special test I/Os. In other words, the access to the storage elements must be serialized. Many methods of serialization are conceivable and have been proposed and used. The most popular approach is the serial scan-path design style[2,3]: the scanning function is implemented by connecting all storage elements into a serial shift register. Embedded large storage arrays (RAM, ROM) are not efficiently testable by serial scan approaches. Hence, it is recommended to provide a separate test bus for accessing the embedded arrays. Figure 1 illustrates the concept of serial scan designs in diagrammatical form.

An important side-effect of the serial scan design approach is that each subsystem or component has logically the same standardized test interface. The scan-path can be extended across component boundaries by simply plugging the scan interfaces together. At the system level, a test and maintainence processor can be attached to the test interfaces as shown in figure 2 for a multi-chip system.

Reprinted from *Curriculum for Testing Technology*, 1983, pages 83-89. Copyright © 1983 by The Institute of Electrical and Electronics Engineers, Inc.

SERIAL SCAN DESIGN

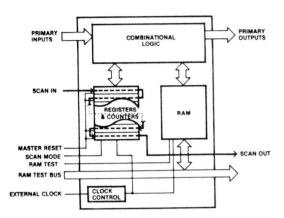

Figure 1: Concept of serial scan design

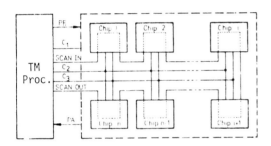

Figure 2: Multi-chip system with test and maintenance processor

By inclusion of the test and maintenance processor the system is made self-testing. It is an off-line self-test because the system has to be in a special test mode for testing. For mainframe systems the test and maintenance processor can be of the size of a minicomputer in order to handle the voluminous and complex test stimuli and compare patterns.

It is obvious that for smaller, ultimately chip level, systems a much more compact approach to test stimuli generation and response comparison is required.

Techniques for Compact Built-in Self-Test

Self-Test Strategies

Before entering the discussion of compact self-test techniques in particular, it is useful to review some basic strategies in general. Self-test strategies vary in degree of centralization vs. decentralization of the active test functions, and to what extent existing system functions are exploited for the test functions. In the example of the mainframe computer system with a built-in test and maintenance (TM) processor, the active test functions are centralized in the TM processor. The scan-path and chip addressing provide passive aids for simplifying and partitioning the test task for the TM processor. Figure 3 shows the strategy of centralized self-testing: the TM processor sequentially addresses modules and performs the testing.

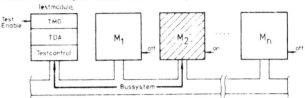

Figure 3: Centralized Self-Test Strategy

Typical examples for system level centralized self-test are described in[4,5]

A diametrically opposing strategy would distribute the test logic completely, such that each module contains its own self-test logic, implementing a completely decentralized strategy. The decentralized strategy is illustrated in figure 4:

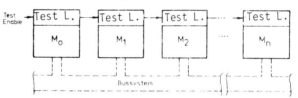

Figure 4: Decentralized Self-Test Strategy

The test enable signal disconnects the modules from the inter-module connection network and initiates the self-test of the modules. The modules test themselves in parallel. The inter-module connection network is tested separately. Because of the recurring overhead for the distributed test logic, compact self-test techniques are used. The partially decentralized AAFIS approach[6] constitutes a pivotal breakthrough in practical self-test hardware development. The BILBO concept[7] is an extension of the AAFIS ideas.

The third basic self-test strategy is called bootstrapping. This strategy is based on the utilization of the intelligence of microprocessor systems for self-test purposes. The concept of bootstrapping is illustrated in figure 5:

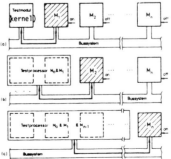

Figure 5: Bootstrapping Self-Test Strategy

The test begins with the self-test of a kernel. The kernel then starts testing a first module, M_1. If the test is passed, M_1 is added to the kernel to form a more powerful test-processor. By incorporating each newly tested module into the test-processor, the power of the test-processor is continuously increasing. A good description of bootstrapping is given in [8].

No matter which strategy is used, a self-testing circuit requires that the functions of Test Pattern Generation (TPG) and Test Data Evaluation (TDE) are incorporated into the circuit. Compact implementations are needed for both functions.

Compact Test Data Evaluation

During Test Data Evaluation (TDE), the responses of the circuit under test are compared with the known good responses. Any mismatch indicates a problem in the circuit under test. The comparison can occur either bit by bit or on a block basis.

The bit by bit approach requires that the known good responses, i.e. the compare data, are available bit by bit. This means that the compare data must be either entirely stored in the TDE (stored response approach) or are generated on-line by an algorithmic response pattern generator. The first method requires an excessive amount of storage space, while the second method implies either functional duplication of the circuit under test or is restricted to very simple response pattern types. In other words, the bit by bit approach is not practical for most cases.

The block-wise approach works with data compression techniques: a large data block is compressed into a codeword. The comparison task is thus reduced to comparing codewords. Ideally, one codeword is used for the whole response pattern data sequence. The problem is to use coding techniques that minimize the probability of aliasing. Aliasing occurs if the response given by the circuit under test contains errors (i.e. it differs from the known good response), but the codeword is identical to that of the known good response. In the case of aliasing, faults in the circuit under test remain undetected. In order to reduce the probability of aliasing, the codes used for data compression must have good error detecting properties.

Many different coding techniques have been proposed and investigated[9-12]. The most widely used technique is called signature analysis[13]. It is based on cyclic coding schemes well known from error detection methods in block data transfer applications[14]. The linear cyclic codes are very easily implemented in hardware. The basic circuit configuration is a linear feedback shift register. Figure 6 shows a parallel signature analyzer.

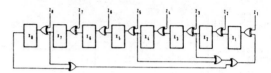

Figure 6: Signature Register with Parallel Data Inputs

The aliasing probability of signature register has been the subject of several papers[15-18]. If one assumes that all possible errors in the evaluated data block are equally likely, it can be shown by algebraic methods that the probability for aliasing to occur in an m bit signature is always smaller than 2^{-m}.

$$P_A \leq 2^{-m} \qquad (1)$$

The assumption of equally likely errors is, however, not adequate[15,16]. It can, for instance, be shown that under this assumption the last m bits in the evaluated data block are as good an indicator of errors as the signature is.

Aliasing in a signature analyzer is triggered if errors occurring at different clock cycles have certain algebraic properties. A thorough theoretical analysis of signature analysis would require to correlate the algebraic conditions for aliasing with algebraic properties of the test stimulus sequence, the circuit structure under test, and a fault probability distribution. To date, only experimental results are available to verify signature analysis. Both intuition and a simple quantitative example indicate that signature analyzers work best, if the feedback taps are chosen according to the rules for maximum length pseudo-random number generation. Extensive fault-simulation runs done by the authors and others[19,20] verify the validity of equation (1) empirically. It is important to note that the relationship holds regardless of whether pseudo random or deterministic test stimuli are used.

Thus, despite a lack of conclusive theoretical proof, signature analysis can be considered as a working and practical test data evaluation method. The probability of aliasing is bounded as formulated in equation (1), and can be made arbitrarily small by increasing the number of bits in the signature. For all practical purposes, the test data evaluation problems seem to be solved by using signature anlaysis. The real challenge of compact built-in test therefore is in the area of compact test pattern generation.

Compact Test Pattern Generation

Compact testing is typically associated with random or pseudo-random test pattern generation[21-23]. Some test data evaluation techniques indeed are based on statistical properties of the test pattern sequence and its correlation with the test response sequence[24]. The signature analysis technique is, however, not dependent on random or pseudo-random test patterns, but can also be used in conjunction with other types of test patterns. In general, three types of test pattern generation schemes can be utilized, namely stored pattern TPG, compact hardware TPG, and microcoded TPG. These schemes are best applicable to different types of test objects.

Stored pattern TPG is the most flexible approach, yet requires an excessive amount of storage space if used exclusively. Stored pattern TPG should, however, not be completely ignored for compact testing approaches. Stored patterns can be used as seeds for pseudo-random pattern generation and/or for changing the linear dependencies in pseudo-random pattern generators. The pseudo-random and/or other TPG hardware is in such an approach used to expand the stored pattern data resources.

Compact hardware TPG has received a very wide attention recently, with most of the attention given to pseudo-random TPG. Pseudo-random TPG, like signature analysis, is easily implemented with linear feedback shift registers. The major problem to be solved is to identify the relationship between the length of pseudo-random test patterns and the fault coverage for any given circuit under test. One approach to alleviating this problem is to use circuit partitioning and exhaustive testing of the partitions[25]. Pseudo-random TPG is used as a compact means of generating exhaustive test-sequences. The method is applicable to sequential circuits as well as combinational circuits.

Most approaches based on pseudo-random TPG, however, do not require exhaustive patterns. The fault-coverage for pseudo-random patterns applied to combinational networks has been investigated thoroughly. The theoretical prediction of path sensitization probabilities[26] has yielded an insight into how circuit properties (logic depth and average fan-in) influence the pseudo-random fault coverage. Empirical data[27] validate the theoretical predictions. The data indicate that pseudo-random testing of combinational logic is possible with very reasonable test pattern sequence lengths.

One of the major breakthroughs that has lead to the current renewed interest in built-in test, is the recognition that scan designs remove the sequentiality from the circuits under test, and, hence, provide an ideal set-up for pseudo-random testing[7,28]. The overhead of adding a pseudo-random TPG and a signature analyser for TDE to a circuit with scan-design is minimal (typically 1% or less). The resulting structured approach to built-in test will be discussed in more detail in section 4 of this paper.

The combination of pseudo-random TPG with scan-design is not necessary. The experience with pseudo-random testing of sequential circuits is, however, limited. Successes have been reported for some cases[6,29]. The approach reported in[30] leaves it to the user of a gate array with built-in test, whether or not to apply scan design to the internal logic on the array.

Another type of useful compact hardware TPG are counters. Counters are very often already available in a design, and need not be separately provided like pseudo-random TPGs. The counters can be used for stepping through the address space of storage arrays (RAM, ROM). In a typical bootstrapping self-test approach, the program counter and a signature analyzer (as check sum generator) form the kernel of the self-test: the program counter is initialized and then steps through the address space of the microcode ROM. The ROM contents are captured in the signature analyzer[8,3]. The microcode ROM itself contains special microcode for testing. A counter can also be used as a hardware aid in microcoded RAM tests.

Microcode, obviously, is a valuable source for deterministic test-patterns in microprocessor based systems. The access to all computational resources at the microcode level can be exploited to partition the test task. Special partitioning features can be implemented in microcode to further simplify testing. Diagnostic microcode has been used for a long time in system test applications. Very compact test code can be written by combining diagnostic microcode with hardware based TPG (pseudo-random, counters)[32]

Towards a Structured Methodology for Built-in Self-Test

Methodology Objectives

Design for test has long been, and in many cases still is, more black magic than an engineering discipline. Without a structured approach, it is left to the ingenuity of the individual designers to make design for test decisions ad hoc. Only regorous approaches, like the scan design techniques, alleviate the designers from the burden of having to invent new concepts for every design. At the same time, massive CAD support becomes available with the rigorous approach.

Built-in self test is on the way of becoming available in a strucutred methodology. Not all questions are answered yet, but dramatic progress has been made in the past few years. The objectives of the structured methodology are to provide the designs with transparent, standardized test interface rules, simple design for testability guidelines, and the type of CAD support we are used to from scan-designs. Ideally the CAD tools will be able to more or less automatically add the built-in test features at the push of a button.

Standard Test and Maintenance Interface

The standard test and maintenance interface is a direct extension of the scan path design interface. The system is partitioned into test and maintenance units (TMUs). Each TMU is in turn broken into modules. The TMUs are equipped with local compact built-in self test hardware. Figure 7 depicts the overall structure of a system designed with a standard test and maintenance interface.

Each module in a TMU is connected to a local Test and Maintenance bus, TM bus, via a conventional scan-design interface.

resolution at the TMU level. The BILBO concept[7] implements a design approach for a local TM processor.

The global TM bus in turn furnishes the global TM processor with a conventional scan design interface to the TMUs. The global TM processor extends the scan-design interface to the outside world for remote diagnostics.

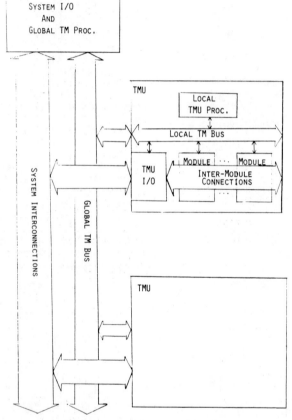

Figure 7: Standard Test and Maintenance Interface Structure

The important extension at the TMU level is the inclusion of the TMU I/O into the scan design: all TMU I/O pins must be directly scannable. The extension of the scan-path to I/O pins is called boundary scan in[33]. Boundary scan allows for complete isolation of the internal circuitry within the TMU from the system level interconnection. The system interconnection network is easily tested by modifying the boundary scan facilities to allow for data sampling at the TMU inputs and for test pattern stimulation at the TMU outputs[30,34].

The local TMU process contains simple test mode control logic and hardware for pseudo-random TPG and TDE by signature analysis. By virtue of the isolation capabilities provided by boundary scan, the TMUs can test themselves in parallel. Error propagation across the TMU boundaries is prevented by the boundary scan design. Hence, the local self-test immediately provides diagnostic

Design for Pseudo-Random Testability

The standard test and maintenance interface design prepares the circuits under test for pseudo-random testing. The scan design separates combinational circuit elements from storage elements and interfaces to the local TM processors. In order to make that scheme efficient, the combinational networks under test must be susceptible to pseudo-random testing. The theoretical and practical experience with pseudo-random testing[26,27] indicate that large PLAs are not easily testable. While PLAs can be made self-testable[35] by modified BILBO's, such an approach is not generally compatible with the standard test and maintenance interface rules. A different approach has been proposed in[33]. By adding segment select circuitry, PLA's can be made pseudo-random testable.

Reference[33] also describes techniques for enhancing the pseudo-random testability of other combinational networks.

CAD Support for Built-in Testing

The acceptance of the scan-design techniques has been largely due to the CAD support possible with this design style. It is desirable to eventually have available a built-in self test approach with similarly powerful CAD support. The standard test and maintenance interface provides a good, stable background for high leverage CAD tool development. The high degree of standardization makes it possible to have most testability features automatically added to a design by the CAD system.

A more difficult problem is to quantitatively analyze the pseudo-random testability of combinational circuits. Work in this area is going on as evident from reference^{33}. Such an analysis tool will not only help to determine difficult to test circuit nodes, but eventually guide the automatic insertion of test points for increased testability. Also, the required length of pseudo-random test sequences can be calculated.

Another desirable CAD tool is a simulator that is powerful enough to simulate a large system over many clock cycles to determine the good signatures. Hardware implementations of fast cycle simulation algorithms certainly are capable of handling the required huge simulation tasks^{36,37}

Summary

Built-in testing is coming of age. A considerable wealth of experience has been gathered, and has lead to the conception of powerful approaches for off-line built-in self test. The universities can contribute to the further progress of the state-of-the-art by researching the conditions for pseudo-random testability, alternative compact test pattern generation schemes, and required CAD support tools.

References

[1] W.C. Carter et al., "Design of Serviceability Features for the IBM System/360", IBM Journal, vol. 8, April 1964, pp. 115-126.

[2] H. Huelters, "Verfahren und Anordnung zur Fehlendiagnose bei taktgesteuerten Geraeten", German Patent 2111493, March 1971.

[3] A.C. Hirtle et al., "Data Processing System Having Auxiliary Register Storage", U.S. Patent 3,582,902, June 1971.

[4] N.C. Berglund, "Processor Development in the LSI Environment", IBM S/38 Tech. Dev., 1978, pp. 7-10.

[5] J. Reilly et al., "Processor Controller for the IBM 3081", IBM Journal Res. Dev., Jan. 1982, pp.22-29.

[6] N. Benowitz et al., "An Advanced Fault Isolation System for Digital Logic", IEEE Trans. Comp., vol. C-24, May 1975, pp. 489-497.

[7] B. Könemann et. al, "Built'In Logic Block Observation Techniques", 1979 Test Conf., Digest of Papers, Oct. 1979, pp. 37-41.

[8] J. Boney and E. Rupp, "Let Your Next Microcomputer Check Itself and Cut Down Your Testing Overhead", Electronic Design, vol. 27(18), Sept. 1979, pp. 100-105.

[9] J.P. Hayes, "Check Sum Test Methods", Proc. FT.CS-6, June 1976, pp. 114-119.

[10] J.P. Hayes, "Transition Count Testing of Combinational Logic Circuits", IEEE Trans. Comp., vol. C-25, June 1976, pp. 613-620.

[11] J. Savin, "Syndrome-Testable Design of Combinational Circuits", IEEE Trans. Comp., June 1980, pp. 442-451.

[12] A.K. Sussleind, "Testing by Verifying Walsh Coefficients, Proc. FTCS-11, June 1981, pp. 206-208.

[13] R.A. Frohwerk, "Signature Analysis: A New Digital Field Service Methods", Hewlett-Packard Journal, May 1977, pp. 2-8.

[14] W.W. Peterson, "Encoding and Error-Correction Procedures for the Base-Chauduri Codes", IRE Trans. Inform. Theory, vol. IT-6, 1960, pp. 459-470.

[15] J.E. Smith, "Measures of the Effectiveness of Fault Signature Analysis", IEEE Trans. Comp., vol. C-29, June 1980, pp. 510-514.

[16] W.C. Carter, "The Ubiquitous Parity Bit", Proc. FTCS-12, June 1982, pp. 289-296.

[17] H.J. Nadig, "Testing a Microprocessor Product Using a Signature Analysis," 1978 Test Conf., Digest of Papers, Oct. 1978, pp. 159-169.

[18] R. David, "Feedback Shift Register Testing", Proc. FTCS-8, June 1978, pp. 103-107.

[19] G. Zwiehoff et al., "Experiemente mit eimem Simulationsmodell für selbsttestende ICs", NTG-Fachberichte Bd. 68, VDE-Verlag Gmbh, Berlin, 1979, pp. 105-108.

[20] T. Sridhar et al., "Analysis and Simulation of Parallel Signature Analyzers", 1982 Test Conf., Digest of Papers, Nov. 1982, pp. 656-661.

[21] K.P. Parker, "Compact Testing: Testing with Compact Data", Proc. FTCS-6, June 1976, pp. 93-98.

[22] J. Losq, "Referenceless Random Testing", Proc. FTCS-6, June 1976, pp. 108-113.

[23] R. David and G. Blanchet, "About Random Fault Detection of Combinational Networks", IEEE Trans. Comp., vol. C-25, June 1976, pp. 659-664.

[24] S. Ohteru et al., "Digital Test System Using Statistical Method", Proc. FTCS-10, Oct. 1980, pp. 179-181.

[25] E.J. McCluskey and S. Bozorgui-Nesbat, "Design for Autonomous Test", IEEE Trans Comp., vol. C-30, Nov. 1981, pp. 866-876.

[26] P. Agrawal and V.D. Agrawal, "Probabilistic Analysis of Random Test Generation Method for Irredundant Combinatorial Logic Networks", IEEE Trans. Comp. C-24, 1975, 691-695.

[27] T.W. Williams and E.B. Eichelberger, "Random Patterns Within a Structured Sequential Logic Design", 1977 Test Symp., Digest of Papers, Oct. 1977, pp. 19-27.

[28] H. Eiki et al., "Autonomous Testing and its Application to Testable Design of Logic Circuits", Proc. FTCS-10, Oct. 1980, pp. 173-178.

[29] B. Könemann et al., "Built-In Test for Complex Digital Integrated Circuits", IEEE Journal Soldi State Circ., vol. SC-15, June 1980, pp. 315-319.

[30] D.R. Resnick, "Testability and Maintainability with a New 6K Gate Array", VLSI Design, March/April 1983, pp. 34-38.

[31] U. Theus and H. Leutiger, "A Self-Testing ROM Device", Proc. ISSCC, Feb. 1981, pp. 176-177.

[32] G. Zwiehoff, "Ein Selbsttestkonzept fur hochintegrierte Digital Schaltungen", Ph.D. thesis, RWTH Aachen, West Germany, 1982

[33] E.B. Eichelberger and E. Lindbloom, "Random-Pattern Coverage Enhancement and Diagnosis for LSSD Logic Self-Test", IBM Journal Res. Dev., vol. 27, May 1983, pp. 265-272.

[34] D. Komonytsky, "LSI Self-Test Using Level-Sensitive Scan Design and Signature Analysis", 1982 Test Conf., Digest of Papers, Nov. 1982, pp. 414-424.

[35] W. Daehn and J. Mucha, "A Hardware Approach to Self-Testing of Large Programmable Logic Arrays", IEEE Trans. Corp., vol. C-30, Nov. 1981, pp. 829-833.

[36] G.F. Pfister, "The Yorktown Simulation Engine: Introduction", Proc. 19th Design Automation Conference, June 1982, pp. 51-54.

[37] T. Sasaki et al., "HAL: A Block Level Hardware Logic Simulator", Proc. 20th Design Automation Conference, June 1983, pp. 150-156.

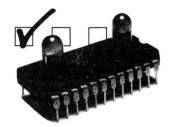

Built-In Self-Test Techniques

BIST allows thorough testing at reasonable cost. The chip overhead due to BIST can be minimized by a wise choice of implementation techniques.

Edward J. McCluskey, Stanford University

Testing a circuit requires the application of a test stimulus and the comparison of the actual circuit response with the correct response. General-purpose testers, though commonly used for this purpose, are very expensive; and tester cost is not the only difficulty encountered in using an external tester. There are also problems with:

(1) *Time.* The turnaround time to generate test patterns, the time taken to apply the test patterns, and the computation time are growing too large.

(2) *Volume.* The number of test patterns is becoming too large to be handled efficiently by the tester hardware.

Several techniques have been proposed for reducing the complexity of external testing by moving some or all of the tester functions onto the chip itself or onto the board on which the chips are mounted. This article presents techniques for generating test patterns and evaluating output responses in built-in self test designs. These techniques are intended to solve the problems listed above as well as to reduce the tester cost. A companion article, "Built-In Self-Test Structures," discusses the structures used to integrate the test and functional circuitry (see pp. 29-36 this issue).

General BIST attributes. The inclusion of on-chip circuitry for testing is called "built-in self test" (BIST), "built-in test" (BIT), "self test," "autonomous test," "in-situ test," or "self-verification." There is some ambiguity in the use of these terms. In particular, BIT and self-test are sometimes used to mean implicit testing (concurrent checking or monitoring) or system-level, periodic testing.[1] The discussion here is restricted to explicit test (wafer sort, production test, board test, maintenance test, repair test) techniques.

Any test method must consist of: (1) a strategy for generating the inputs to be applied, (2) a strategy for evaluating the output responses, and (3) the implementation mechanisms. This article presents a survey of the different techniques available for each of these items.

Table 1 lists the most important attributes for evaluating BIST structures. All techniques must be able to detect single stuck faults in the functional circuitry. Most methods can detect some other faults as well. Ex-

---

Summary

A system that includes self-test features must have facilities for generating test patterns and analyzing the resultant circuit response. This article surveys the structures that are used to implement these self-test functions. The various techniques used to convert the system bistables into test scan paths are discussed. The addition of bistables associated with the I/O bonding pads so that the pads can be accessed via a scan path (external or boundary scan path) is described. Most designs use linear-feedback shift registers for both test pattern generation and response analysis. The various linear-feedback shift register designs for pseudorandom or pseudoexhaustive input test pattern generation and for output response signature analysis are presented.

**Table 1.
BIST attributes.**

Fault characteristics	Fault classes tested
	    Interchip wiring and chip I/O connections
	    Single stuck faults in functional circuitry
	    Combinational faults in functional circuitry
	    Sequential faults in functional circuitry
	    Delay faults
	Fault coverage—percentage of faults known to be detected

Cost characteristics	Area overhead—additional active area, interconnect area Pin overhead—additional pins required for testing Performance penalty—added path delays Yield decrease—due to increased area Reliability reduction—due to increased area
Other characteristics	Generality—degree to which BIST structure is function-dependent Time required to execute test Diagnostic resolution Engineering changes—effect on BIST structure Functional circuitry—scan path, random pattern design changes

haustive or pseudoexhaustive techniques can detect all *combinational faults,* faults that do not change a combinational circuit into a sequential circuit. Of increasing concern are *sequential faults,* faults (such as MOS stuck-open faults) that do cause combinational circuits to exhibit sequential behavior. An important distinction between techniques is the difficulty of accurately determining the set of faults detected. A related issue is that of calculating the fault coverage—the percentage of faults detected for each class of fault.

All BIST methods have some associated cost. Since the self-test circuitry uses chip area, there is a decrease both in the yield and in the reliability. These costs are compensated, however, by the reduced testing and maintenance cost. The added cost due to the BIST circuitry should be less than the savings in life-cycle cost in order to justify the use of BIST.

The ideal BIST structure would be completely general in the sense that it would be applicable for any kind of functional circuitry. This ideal is never reached. For example, a technique that works well for random combinational logic may not be suitable for a chip with RAM. Due to the pattern sensitivity of RAMs, specially generated patterns must be used to test them. One of the most general techniques for testing chips without RAMs uses pseudorandom patterns as test inputs. The pseudorandom patterns are in-

> **The ideal BIST structure would be completely general and applicable to any functional circuit. This ideal is never reached.**

dependent of the functional circuitry. However, some logic structures are difficult to test with random patterns and, to avoid them, redesign of the functional circuit may be required. An important aspect of a BIST structure is the amount of "custom design" that is required. For very high volume parts, it may be best to optimize the BIST structure with custom circuitry. For lower volume parts, a standard technique may be best.

Input test stimulus generation

Test vectors can be generated manually or by a test pattern generation program and stored (*off-line* test pattern generation) or they can be calculated while they are being applied (*concurrent* test pattern generation). In theory it would be possible to generate test vectors off line and store them in an on-chip ROM. This has not, however, been an attractive scheme: it does nothing to reduce the cost of test pattern generation and it requires a very large ROM. All of the BIST methods described here rely on concurrent test pattern generation.

A number of techniques have been proposed for concurrent test pattern generation. If the chip includes a processor and memory, it is possible to use a test program to generate appropriate signals to stimulate the circuitry.[2] This test program may be stored in ROM; but this should not be confused with the technique of storing the actual test patterns in ROM. Gordon and Nadig[3] have suggested that it is sufficient that this program be written to force state changes or to "wiggle" the circuit nodes. For an evaluation of this technique see Hughes et al.[4] More systematic approaches to writing test programs are described in Hayes and McCluskey.[5] The test programs usually rely on a *functional* test approach: they typically are written to exercise the functionality of the various system components. They can be based on the system diagnostic programs. Some reconfiguration of the circuit during test mode (to permit initialization and perhaps to break feedback loops) may be necessary to ensure good fault coverage.[3]

Two concurrent test pattern generation approaches that do not depend on the availability of an instruction processor have been proposed. Since these methods can be used with or without an instruction processor, they are more general than the test program approach. One of these methods—*pseudorandom* testing—uses a set of pseudorandomly generated patterns as test patterns. The other—*exhaustive* testing—uses all possible input combinations as test patterns. The random technique has the advantage of being applicable to sequential as well as to combinational circuits; however, there are difficulties in determining the required test sequence length and fault coverage. Some circuits may require modification to obtain adequate coverage with reasonable test lengths.[6] Exhaustive testing eliminates the need for a fault model and fault simulation, although for large numbers of inputs this technique may require too much time. Often the circuit is naturally segmented into subcircuits, each of which has few enough inputs so that exhaustive testing is practical. Some additional work may be required to identify these segments and to specify the tests for them. For the few circuits that do not have satisfactory natural segments, some form of circuit artificial segmentation via the test patterns is required. This can require additional computation, but not as much as full automatic test pattern generation for the unsegmented circuit.[7]

An important problem connected with concurrent random test pattern generation is determining the length of the random sequence that is required to obtain a satisfactory fault coverage. One straightforward technique uses fault simulation to determine fault coverage. The difficulty with simulation is its high cost. An analytical method of estimating fault coverage would be preferable. A number of approaches have been suggested,[8,9] but none of the techniques yet developed is capable of getting accurate fault coverage estimates without a great deal of computation.

Another approach is aimed at discovering those faults that are hard to detect with random patterns. Three schemes for coping with these "random-pattern resistant" faults have been proposed. One would generate test patterns for these faults (using a deterministic technique such as the D algorithm) and store these patterns in a ROM. Thus, some few patterns would be obtained from the ROM and the remaining patterns would be randomly generated.[10] The second scheme modifies the network being tested so that none of its faults are random-pattern resistant.[6,10] A third approach modifies the pseudorandom pattern generator to produce either more zeros than ones or more ones than zeros. This technique depends on the fact that faults that resist detection with patterns having equal distributions of ones and zeros will not be resistant to patterns with other distributions of the ones and zeros.[11]

Random testing. If an instruction processor is available, the random test patterns can be generated by a program. A thorough discussion of random number programs is given in Knuth.[12] A simple program for random test pattern generation is presented in Brack.[13] For random test pattern generation it is not so important that the numbers used be strictly random, since they are not being used for statistical purposes.

Random test patterns can also be generated by means of a simple circuit called an autonomous linear-feedback shift register or ALFSR. An ALFSR is a series connection of delay elements (D flip-flops) with no external inputs and with all feedback provided by means of exclusive-or gates (XORs). A four-stage ALFSR is shown in Figure 1a and the general standard form of ALFSR is shown in Figure 1b. The symbol h_i in Figure 1b indicates the possible presence of a feedback connection from the output of each stage. If $h_i = 1$, there is feedback from stage i; and if $h_i = 0$, the stage i output is not connected to the XOR feedback network. The ALFSR can be specified by just listing the values of the h_i or by specifying the generating function as shown in Figure 1.

Another possible realization of an ALFSR, called a "modular realization," is shown in Figure 2. There are as many XOR gates in the modular realization as there are feedback taps in the standard circuit. The gates are placed in the "reverse" positions from the locations of the feedback taps. If in the standard LFSR there are m

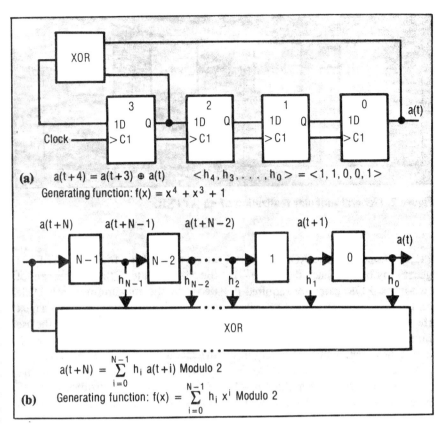

Figure 1. Standard form of autonomous linear-feedback shift register, or ALFSR: (a) four-stage circuit; (b) N-stage circuit.

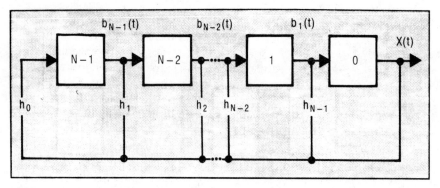

Figure 2. General modular realization of an ALFSR.

"taps" (inputs to the XOR network generating the feedback signal), $m-1$ two-input XOR gates are required if an iterative structure is used to realize the XOR network. This is the minimum-gate realization. It is slower than a tree network, which also requires $m-1$ gates, but has a delay of $\log m$ gate propagations rather than $m-1$ gate delays. The modular circuit also requires $m-1$ XOR gates. It has a delay of only one gate propagation. For circuits with more than two feedback signals, faster operation always results with the modular rather than the standard LFSR.

The sequence of states for the ALFSR of Figure 1 is shown in Table 2. Note that the sequence repeats after 15 ($2^n - 1$) clocks. This is the maximum period for a four-stage ALFSR; the all-zero state of the register cannot occur in the maximum-length cycle since an all-zero state always has a next state that is also all zeros due to the use of XORs to form the feedback signal. In general, the maximum period for an n-stage ALFSR is $2^n - 1$. There are maximum-length realizations for all values of n. The generating function corresponding to a maximum-length ALFSR is called a *primitive polynomial*. Tables of primitive polynomials can be found in Golumb,[14] Peterson,[15] and many other publications.

Of course, the signals produced by an ALFSR are not really random since they are produced by a fixed circuit. As previously discussed, truly random signals are not required for test pattern generation. What is necessary are signals that produce the same types of test patterns as random signals. The output of an ALFSR can be shown to possess many of the properties of random signals. The sequences produced by maximum-length ALFSRs are called *pseudorandom* sequences or *pseudonoise* sequences to distinguish them from truly random sequences. For test pattern generation, pseudorandom sequences are better than random sequences since the pseudorandom sequences can be be reproduced for simulation. One period of the output sequence produced by the ALFSR of Figure 1a is:

(0 0 0 1 1 1 1 0 1 0 1 1 0 0 1)

The five-stage ALFSR with feedback connections given by $H = <100101>$ has the following output sequence:

(1 1 1 1 1 0 0 0 1 1 0 1 1 1 0 1 0 1 0 0 0 0 1 0 0 1 0 1 1 0 0)

The randomness characteristics of pseudorandom sequences are discussed and proved in Golumb.[14]

Exhaustive and pseudoexhaustive testing. The application of all 2^n input combinations to the (combinational) circuit being tested is called *exhaustive testing*. Any binary counter can be used to develop these signals. Since the order of generation of the combinations is not important, it may be more efficient to use an ALFSR modified so that it cycles through all states. To do this it is necessary to modify the ALFSR so that the all-zero state is included in the state sequence.[16,17]

Exhaustive testing provides a thorough test, but can require a prohibitively long test time for networks with many (20 or more) inputs. With the techniques described below, however, it is possible to reduce the test time to a practical value while retaining many of the advantages of exhaustive testing. These *pseudoexhaustive* techniques apply all possible inputs to portions of the circuit under test rather than to the entire circuit. The first such technique, called *verification testing*,[7] is applicable to multiple-output circuits in which none of the outputs depends on all of the inputs.

Most combinational networks have more than one output. In many cases each of the outputs depends on only a subset of the inputs. For example, the parity generator network of the TI SN54/74LS630 has 23 inputs and six output functions, but each output depends on only 10 of the inputs. It may not be practical to exhaustively test the outputs by applying all combinations of the network inputs (2^{23} for this example). However, it may be possible to exhaustively test each output by applying all combinations of only those inputs on which the output depends. For the SN74LS630, each output can be exhaustively tested with $2^{10} = 1024$ input patterns, and all six outputs can be tested one after another with $(6)(1024) = 6144$ patterns. In fact, for this circuit it is possible, by an appropriate choice of input patterns, to apply all possible input combinations to each output concurrently rather than serial-

**Table 2.
State sequence for Figure 1a.**

State	Q_1	Q_2	Q_3	Q_4
0	1	0	0	0
1	1	1	0	0
2	1	1	1	0
3	1	1	1	1
4	0	1	1	1
5	1	0	1	1
6	0	1	0	1
7	1	0	1	0
8	1	1	0	1
9	0	1	1	0
10	0	0	1	1
11	1	0	0	1
12	0	1	0	0
13	0	0	1	0
14	0	0	0	1
15 = 0	1	0	0	0

ly. Thus, with only 1024 rather than (6)(1024) test patterns, each output can be tested exhaustively by using the verification testing techniques described in McCluskey[7] and Barzilai et al.[18]

It has been shown in McCluskey[19] that any network for which no output depends on all inputs can be tested pseudoexhaustively with fewer than 2^n test patterns. This paper also derives a specification for test sets that consist of constant-weight vectors. These test sets are suitable for concurrent test pattern generation since they are simply generated by constant-weight counters. The constant-weight test set is shown to be a minimum-length test set for many networks. In fact, it may be more economical to use an ALFSR to generate the verification test inputs even though the test lengths are longer than constant-weight tests. Techniques for designing ALFSR circuits for this purpose are described in Wang and McCluskey.[20]

It is still possible that a verification test set, even though smaller than 2^n, is too long. Also, there are many circuits with an output that depends on all inputs. Such circuits require 2^n inputs for exhaustive test, and this may be too large a set. In other words, there are circuits for which the verification test approach does not result in a satisfactory test procedure. A pseudoexhaustive test is still possible for such circuits, but it is necessary to resort to a partitioning or segmentation technique.

Such a procedure is described in McCluskey and Bozorgui-Nesbat.[17] The technique presented in that paper relies on exhaustive testing, but divides the circuit into segments or partitions to avoid excessively long input test sequences. It differs from previous attempts along these lines in that the partitions may divide the signal path through the circuit rather than just separate the signal paths from one another. While it is possible to use multiplexers to enforce the segmentation, they are not necessary. A partitioning method that does not alter the functional circuitry is also described in McCluskey and Bozorgui-Nesbat.[17]

Output response analysis

On-chip storage of a fault dictionary (all test inputs with the correct output responses)[21] requires too much memory to be a practical technique. The simplest practical method for analyzing the output response is to match the outputs of two identical circuits. Identical circuits may be available either because the function being designed naturally leads to replicated subfunctions[22] or because the functional circuitry is duplicated redundantly for concurrent checking.[23] If

The choice of a compaction technique is influenced by the amount of circuitry to implement it and by the loss of "effective fault coverage."

identical outputs are not available, it is necessary to resort to some technique for compacting the response pattern. Techniques for reducing the volume of output data were originally developed in connection with portable testers. Their use is called *compact testing*.[24] This technique is sometimes also called *response compression*. The term *compression* is used in communication theory to mean that there is no loss of information. Since an important issue in connection with reducing the amount of output response data is the fact that some information is always lost, it is preferable to use the term *compaction* to emphasize this aspect of the process. In compact testing, the output response pattern is passed through a circuit, called a compacter, that has fewer output bits than input bits. The output of the compacter is called the *signature* of the test response. The aim is to reduce the number of bits that must be examined to determine whether the circuit under test is faulty.

Compaction techniques. Many portable testers use *transition counting* as a compaction technique. This involves counting the number of transitions (0 following a 1 or 1 following a 0) in the response sequence.[25] Transition counting has not received serious considera-

tion for BIST since recent research has developed better methods.

The choice of a compaction technique is influenced mainly by two factors: (1) the amount of circuitry required to implement the technique and (2) the loss of "effective fault coverage." In general, a fault goes undetected if none of the input test patterns produces an incorrect circuit output in the presence of the fault. With output response compaction it is also possible for a fault to fail to be detected even though the output response differs from the fault-free response. This happens whenever the output response from a faulty circuit produces a signature that is identical to the signature of a fault-free circuit. This phenomenon is called *aliasing*. A faulty circuit test output response signature that is identical to the fault-free signature is called an *alias*.

Many compaction schemes have been studied. These techniques can be grouped into three classes:

- Parity techniques.
- Counting techniques.
- Linear feedback shift register, or LFSR, techniques.

A comparison of parity techniques, LFSR techniques, and combined parity and LFSR techniques is given in Benowitz et al.[26] for pseudorandom test patterns. No advantages were discovered for the use of parity techniques. Carter discusses the use of parity techniques in connection with exhaustive testing, and demonstrates high values for stuck-fault detection.[27,28]

The only counting technique that has been seriously considered for BIST is called *syndrome analysis*.[29] This technique is applicable only to exhaustive testing and requires counting the number of ones in the output response stream. It has been shown that it is possible to detect any single stuck fault in the circuit using this method, although some circuit modification may be required.[30] A generalization of syndrome testing that uses Walsh coefficients has been studied,[31] but its practicality has yet to be demonstrated.

Since all of the actual implementations of BIST using compaction rely

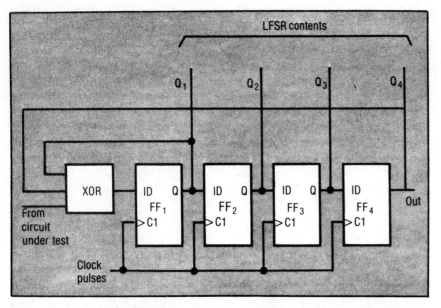

Figure 3. A four-stage linear-feedback shift register used as a serial signature analyzer.

**Table 3.
Aliasing waveforms for LFSR of Figure 3.**

Time:	0	1	2	3	4	5	6	7
Fault-free output sequence:	0	1	1	0	1	0	0	1
Faulty output sequence:	0	0	0	0	0	0	0	1
Error sequence:	0	1	1	0	1	0	0	0
$<H> = <1\ 1\ 0\ 1>$								
Other alias error sequences:	1	1	0	1	0	0	0	0
	1	1	0	1	1	1	0	1
	1	0	1	1	1	0	0	0
	1	0	0	0	0	0	0	1

on LFSR techniques, these are described in more detail below.

The compaction techniques all require that the fault-free signature for the circuit be known. This can be found by (fault-free) simulation of the design or by measurement on an actual circuit that has been verified to be fault-free by some other method.

Signature analysis. The most popular BIST compaction circuit is an LFSR with its input equal to the output response of the circuit under test. This circuit, shown in Figure 3, was called a "cyclic code checker" when it was first proposed in Benowitz et al.[26] This method of output response compaction is most often called *signature analysis,* a term coined by Hewlett-Packard to describe its use in their product, the 5004A Signature Analyzer.[32] The term "signature" describes the LFSR contents after the response pattern of the circuit being tested is shifted into the LFSR.

The usefulness of signature analysis depends on the fact that the final state of the LFSR flip-flops, the signature, depends on the bit pattern that is applied at the input. If a fault causes the output bit sequence to change, this usually results in a different signature in the LFSR. However, aliasing can occur. It is possible for a fault to cause an output bit sequence that produces the same final LFSR contents as the fault-free circuit. In this case the fault is not detected. The output sequences that have this property depend on the structure of the LFSR used. They are characterized in Benowitz et al.[26] in terms of division of the Galois field polynomial representation of the LFSR and the output response sequences.

Table 3 illustrates the alias phenomenon for the LFSR of Figure 3. Both the faulty and the fault-free sequences shown in the table leave the same signature—100—in the LFSR. The error sequence is determined as the (bitwise) sum modulo 2 of the faulty and fault-free sequences. It can be shown that aliasing occurs whenever the error sequence is equal to the LFSR feedback vector $<H>$ or to the sum (bitwise modulo 2) of shifted versions of $<H>$.

Any compaction technique can cause some loss of effective fault coverage due to aliasing. Experience with the HP product and simulation studies of BIST designs[33] have not discovered any signature analysis applications for which aliasing is a problem. However, it has not been possible to derive any general properties of the effective fault coverage obtained by signature analysis. As discussed in connection with Table 3, it is possible to characterize the alias error sequences; but no simple relationship between the circuit faults and the resulting errors has yet been found. The problem of determining effective coverage loss for LFSR compaction is discussed in Smith.[34]

Two methods are suggested in Benowitz et al.[26] for signature analysis for a multiple-output circuit-under-test. One such method, *serial signature analysis,* uses a multiplexer to direct each of the outputs to the LFSR in turn. A circuit for this is shown in Figure 4a. With this scheme the input test patterns must be applied to the network m times for an m output network.

The other technique, the *parallel signature analyzer,* compacts K network outputs in parallel using a K-bit parallel code checker, as shown in Figure 4b. The parallel technique requires each test pattern to be applied only m/K times. Network outputs are connected to the LFSR through XOR gates added to the shift lines between stages. XOR gates also connect the network output to the first LFSR stage. The design of a four-stage parallel signature analyzer is detailed in Figure 5.

In general, the parallel signature analyzer is faster but requires more added circuitry than the serial signature analyzer.[26] A detailed comparison of these two techniques in a particular system is presented in Benowitz et al.[35] Fault coverage data, derived by hardware fault insertion, are also reported. Another study of fault coverage for a parallel signature analyzer is reported in Konemann et al.[36] Physical fault insertion was carried out on an experimental microprocessor with an 8-bit parallel signature analyzer. In this paper, the term "multiple-input signature register" is used. The serial signature analyzer is not discussed.

Besides requiring more hardware than the serial analyzer, the parallel signature analyzer has an additional alias source: an error in output Z_j at time t_i followed by an error in output Z_{j+1} at time t_{i+1} has no effect on the signature. More generally, an error in output Z_j at time t_i followed by an error in output Z_{j+h} at time t_{i+h} has no effect on the signature.[37-39]

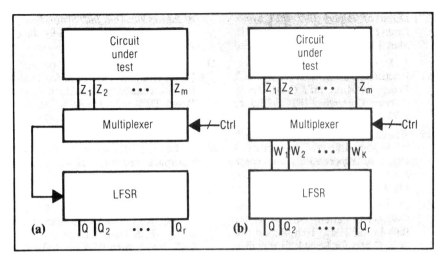

Figure 4. Connection of multi-outputs to LFSR signature analyzer: (a) serial signature analysis using a multiplexer; (b) parallel signature analysis.

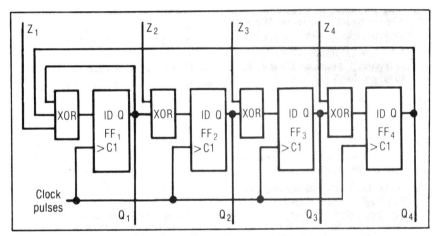

Figure 5. A four-stage LFSR configured as a parallel signature analyzer.

By far the most common methods for built-in test pattern generation and output response analysis in current designs use pseudorandom input patterns and either serial or parallel signature analysis. Major issues for future designs include the questions of whether detailed fault simulation is economically justified and whether the single-stuck fault model is adequate for VLSI chips. Signature analysis will probably continue to be the chosen technique for output response analysis. In situations for which single stuck faults and detailed fault simulation are used, pseudorandom input pattern generation will probably continue as the dominant technique. For other situations, some alternative input generation method such as the pseudoexhaustive may prove most efficient. □

Acknowledgment

This work was supported in part by the National Science Foundation under Grant No. MCS-8200129. The author wishes to thank Joseph McCluskey for his help in preparing this paper.

References

1. J. B. Clary and R. A. Sacane, "Self-Testing Computers," *Computer*, Vol. 12, No. 10, Oct. 1979, pp. 49-59.
2. J. R. Kuban and W. C. Bruce, "Self-Testing the Motorola MC6804P2," *IEEE Design & Test*, Vol. 1, No. 2, May 1984, pp. 33-41.
3. G. Gordon and H. Nadig, "Hexadecimal Signatures Identify Trouble-spots in Microprocessor Systems," *Electronics*, Mar. 1977, pp. 89-96.
4. J. L. A. Hughes, S. Mourad, and E. J. McCluskey, "An Experimental Study Comparing 74LS181 Test Sets," *Digest of Papers Compcon Spring 85*, Feb. 1985, pp. 384-387.
5. J. P. Hayes and E. J. McCluskey, "Testability Considerations in Microprocessor-Based Design," *Computer*, Vol. 13, No. 3, Mar. 1980, pp. 17-26.
6. E. B. Eichelberger and E. Lindbloom, "Random-Pattern Coverage Enhancement and Diagnosis for LSSD Logic Self-Test," *IBM J. Research and Development*, Vol. 27, No. 3, May 1983, pp. 265-272.
7. E. J. McCluskey, "Verification Testing—A Pseudoexhaustive Test Technique," *IEEE Trans. Computers*, Vol. C-33, No. 6, June 1984, pp. 541-546.
8. J. -C. Rault, "A Graph Theoretical and Probabilistic Approach to the Fault Detection of Digital Circuits,"

Digest of Papers 1971 Int'l Symp. Fault-Tolerant Computing (FTC-1), Mar. 1971, pp. 26-29.

9. J. J. Shedletsky, "Random Testing: Practicality vs. Verified Effectiveness," *Proc. 7th Ann. Int'l Conf. Fault-Tolerant Computing* (FTCS-7), June 1977, pp. 175-179.

10. J. Savir, G. Ditlow, and P. H. Bardell, "Random Pattern Testability," *Digest of Papers 13th Ann. Symp. Fault-Tolerant Computing*, (FTCS-13), June 1983, pp. 80-89.

11. C. Chin and E. J. McCluskey, "Weighted Pattern Generation for Built-in Self Test," Tech. Report No. 84-7, Center for Reliable Computing, Computer Systems Laboratory, Stanford University, Stanford, Calif., Aug. 1984.

12. D. E. Knuth, *The Art of Computer Programming*, Vol. 2 (Seminumerical Algorithms), Addison-Wesley, Reading, Mass., 1969.

13. J. W. Brack, "Random Numbers with Software," *Machine Design*, Feb. 1979, pp. 76-77.

14. S. W. Golumb, *Shift Register Sequences*, Aegean Park Press, Laguna Hills, Calif., 1982.

15. W. W. Peterson and E. J. Weldon, *Error-Correcting Codes*, 2nd ed., The Colonial Press, 1972.

16. G. H. De Visme, *Binary Sequences*, The English Universities Press, Ltd., London, 1971.

17. E. J. McCluskey and S. Bozorgui-Nesbat, "Design for Autonomous Test," *IEEE Trans. Computers*, Vol. C-30, No. 11, Nov. 1981, pp. 866-875.

18. Z. Barzilai et al., "The Weighted Syndrome Sums Approach to VLSI Testing," *IEEE Trans. Computers*, Vol. C-30, No. 12, Dec. 1981, pp. 996-1000.

19. E. J. McCluskey, "Built-in Verification Test," *Digest of Papers Int'l Test Conf. IEEE*, Nov. 1982, pp. 183-190.

20. L.-T. Wang and E. J. McCluskey, "A New Condensed Linear Feedback Shift Register Design for VLSI/System Testing," *Digest of Papers 14th Int'l Conf. Fault-Tolerant Computing* (FTCS-14), June 1984, pp. 360-365.

21. M. A. Breuer and A. Friedman, "Design to Simplify Testing," *Diagnosis and Reliability of Digital Systems*, Computer Science Press, Woodland Hills, Calif., 1976, pp. 291-303.

22. T. Sridhar and J. P. Hayes, "Testing Bit-Sliced Microprocessors," *Digest of Papers 9th Ann. Int'l Symp. Fault-Tolerant Computing* (FTCS-9), June 1979, pp. 211-218.

23. R. M. Sedmak, "Design for Self-Verification: An Approach for Dealing with Testability Problems in VLSI-Based Designs," *Digest of Papers 1979 Test Conf. IEEE*, Oct. 1979, pp. 112-124.

24. J. Losq, "Efficiency of Random Compact Testing," *IEEE Trans. Computers*, Vol. C-27, No. 6, June 1978, pp. 516-525.

25. J. P. Hayes, "Transition Count Testing of Combinational Logic Circuits," *IEEE Trans. Computers*, Vol. C-27, No. 6, June 1976, pp. 613-620.

26. N. Benowitz et al., "An Advanced Fault Isolation System For Digital Logic," *IEEE Trans. Computers*, Vol. C-24, No. 5, May 1975, pp. 489-497.

27. W. C. Carter, "The Ubiquitous Parity Bit," *Digest of Papers 12th Ann. Int'l Symp. Fault-Tolerant Computing* (FTCS-12), June 1982, pp. 289-296.

28. W. C. Carter, "Signature Testing with Guaranteed Bounds for Fault Coverage," *Digest of Papers Int'l Test Conf. IEEE*, Nov. 1982, pp. 75-82.

29. J. Savir, "Syndrome-Testable Design of Combinational Circuits," *IEEE Trans. Computers*, Vol. C-29, No. 6, June 1980, pp. 442-451.

30. J. Savir, "Syndrome-Testing of 'Syndrome-Untestable' Combinational Circuits," *IEEE Trans. Computers*, Vol. C-30, No. 8, Aug. 1981, pp. 606-608.

31. A. Susskind, "Testability and Reliability of LSI," RADC Report RADC-TR-80-384, Rome Air Development Center, Griffiss Air Force Base, New York, Jan. 1981.

32. A. Y. Chan, "Easy-to-Use Signature Analyzer Accurately Troubleshoots Complex Logic Circuits," *Hewlett-Packard J.*, May 1977, pp. 9-14.

33. C. C. Perkins et al., "Design for In-Situ Chip Testing with a Compact Tester," *Digest of Papers 1980 Test Conf. IEEE*, Nov. 1980, pp. 29-41.

34. J. E. Smith, "Measure of the Effectiveness of Fault Signature Analysis," *IEEE Trans. Computers*, Vol. C-29, No. 6, June 1980, pp. 510-514.

35. N. Benowitz, D. F. Calhoun, and G. W. K. Lee, "Fault Detection/Isolation Results from AAFIS Hardware Built-in Test," *NAECON '76 Record*, 1976, pp. 215-222.

36. B. Konemann, J. Mucha, and G. Zwiehoff, "Built-in Test for Complex Digital Integrated Circuits," *IEEE J. Solid-State Circuits*, Vol. SC-15, No. 3, June 1980, pp. 315-318.

37. S. Z. Hassan, "Algebraic Analysis of Parallel Signature Analyzers," Tech. Report No. 82-5, Center for Reliable Computing, Computer Systems Laboratory, Stanford University, Stanford, Calif., June 1982.

38. T. Sridhar et al., "Analysis and Simulation of Parallel Signature Analyzers," *Digest of Papers Int'l Test Conf. IEEE*, Nov. 1982, pp. 656-661.

39. S. Z. Hassan, D. J. Lu, and E. J. McCluskey, "Parallel Signature Analyzers—Detection Capability and Extensions," *Digest of Papers Compcon Spring 83*, Feb. 1983, pp. 440-445.

Edward J. McCluskey is at Stanford University where he is a professor of electrical engineering and computer science as well as director of the Center for Reliable Computing. He received his ScD degree from MIT. McCluskey served as the first president of the IEEE Computer Society. He received the IEEE Centennial Medal and the first IEEE Computer Society Technical Achievement Award in Testing. His book, *Logic Design Principles*, will be published by Prentice-Hall in 1986. Current consulting for Lockheed, Siemens, NCR, and IBM on design for testability contributed to the preparation of this article. He is also on the Board of Directors of Synthesized Computer Systems, Inc.

McCluskey can be contacted at the Stanford Center for Reliable Computing, ERL-460, Stanford, CA 94305.

Built-In Self-Test Structures

Current designs for random logic chips use a variety of BIST architectures—from highly structured approaches for semicustom parts to mixed techniques for custom devices.

Edward J. McCluskey, Stanford University

The preceding article, "Built-In Self-Test Technqiues," describes techniques for generating test inputs and evaluating output responses. This article surveys the methods used for combining the test and functional circuitry. Since most BIST methods use some scan-path features, scan techniques are presented before discussing the BIST structures. See McCluskey[1] for a more complete discussion of scan path structures, which were introduced in connection with external testers but are now used extensively for BIST.

Scan path techniques

The most common forms of scan techniques are based on converting the system bistable elements into a shift register called a *scan path*. With the circuit in test mode, it is possible to shift an arbitrary test pattern into the bistable elements. By returning the circuit to normal mode for one clock period, the bistable element contents and primary input signals act as inputs to the combinational circuitry, and new values are stored in the bistable elements. If the circuit is then placed into test mode, it is possible to shift out the contents of the bistable elements and compare these contents with the correct response.

It is assumed that the circuit is constructed of D flip-flops interconnected by combinational circuits and that all of the flip-flops are controlled by a single system clock. These assumptions mean that the circuit can be considered to have the general structure shown in Figure 1. Drawing the circuit in this form is done to simplify the following discussion, but is not meant to imply any restrictions on the designer other than those just stated. The same general results hold true if other types of flip-flops (S-R, J-K, T,...) are used. Since the extension to other flip-flops is straightforward, the details will not be presented here; they can be found in Liu.[2]

In the following discussion of various scan-path techniques, it is important to distinguish between latches and flip-flops. A latch has the transparency property: if the data input changes while the clock is active, the latch output will follow the data input change. A flip-flop changes only when the clock input makes a specific transition—the active transition. The flip-flop output takes on the value present at the data input when this active transition occurs. Subsequent changes of

Summary

A system that includes self-test features must have facilities for generating test patterns and analyzing the resultant circuit response. This article surveys the structures that are used to implement these self-test functions. The various techniques used to convert the system bistables into test scan paths are discussed. The addition of bistables associated with the I/O bonding pads so that the pads can be accessed via a scan path (external or boundary scan path) is described. Most designs use linear-feedback shift registers for both test pattern generation and response analysis. The various linear-feedback shift register designs for pseudorandom or pseudoexhaustive input test pattern generation and for output response signature analysis are presented.

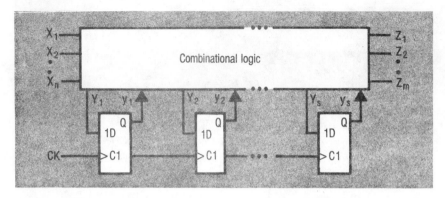

Figure 1. General structure of circuit for scan path discussion.

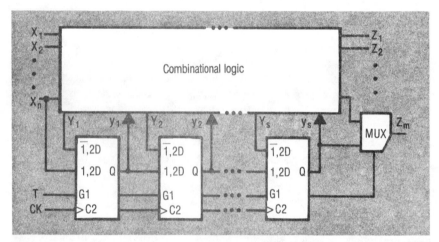

Figure 2. Stanford scan-path architecture—MD flip-flops used to provide scan path.

the data input have no effect until the next active transition of the clock. An important characteristic of flip-flops is that a shift register can be constructed by connecting the output of one flip-flop directly to the data input of the next flip-flop; the conversion of a latch register to a shift register requires an extra latch between each register stage. The symbols used for bistable logic elements follow IEEE Standard 91-1984.

Stanford scan-path design. The first published description of a scan-path design-for-testability structure was a paper based on the Stanford PhD research of Michael Williams.[3] In this technique each of the circuit flip-flops is replaced by a multiplexed data or MD flip-flop as shown in Figure 2. A multiplexer is placed at the data input to permit a selection of two different data inputs—$\overline{1},2D$ (normal system operation) and 1, 2D (test mode). The choice of data input is based on the value of the control input, T. When T=0, data is gated from the $\overline{1},2D$ input upon an active clock transition. Data is taken from 1,2D if T is equal to 1.

The modification of the basic circuit structure of Figure 1 to obtain a scan path architecture using MD flip-flops is shown in Figure 2. One additional input, the T input, has been added. For normal operation, T is equal to 0, and the circuit is connected as in Figure 1. The upper data inputs ($Y_1...Y_s$) act as the flip-flop D inputs. In order to test the circuit, T is set equal to 1. The lower data inputs become the flip-flop D inputs. Thus $D_i = Q_{i-1}$ for i from 2 to s, and a shift register is formed. The primary input X_n is connected to D_1 becoming the shift register input and Q_s, the shift register output, appears at the primary output Z_m.

Testing of the combinational logic is accomplished by (1) setting T = 1 (scan mode); (2) shifting the test pattern y_j values into the flip-flops; (3) setting the corresponding test values on the X_i inputs; (4) setting T = 0 and, after a sufficient time for the combinational logic to settle, checking the output Z_k values; (5) applying a clock signal to CK; and (6) setting T = 1 and shifting out the flip-flop contents via Z_m. The next y_j test pattern can be shifted in at the same time. The y_j values shifted out are compared with the good response values for y_j. The flip-flops must also be tested. This is accomplished by shifting a 0 through a string of 1's and then a 1 through a string of 0's through the shift register to verify the possibility of shifting both a 1 and a 0 into each flip-flop.

Two-port flip-flop designs. A basic requirement of the scan path technique is that it be possible to gate data into the system flip-flops from two different sources. One method of doing this is to add multiplexers to the system flip-flops as shown in Figure 2. Another possibility is to replace each system flip-flop by a *two-port flip-flop*, a flip-flop having two control inputs with the data source determined by which of the controls is pulsed. When a pulse is applied to C1, data is entered from D1; and when a pulse occurs at C2, data is entered from D2.

Figure 3 shows the structure of a network with two-port flip-flops used to provide the scan path. The testing procedure is basically the same as that described in connection with Figure 2. In the circuit of Figure 3, changing between test mode and normal mode is accomplished by changing the clocking rather than by changing the mode signal.

Latch-based structures. Some systems are designed using latches rather than flip-flops as the bistable elements. For latch-based systems it is not possible to directly reconfigure the system bistable elements into a shift register for test purposes. The most popular technique for introducing scan path testability into latch-based systems is

IBM's level-sensitive scan design. LSSD requires the use of extra latches to allow the system latches to be connected into a shift register. The Univac scan-set technique, described by Stewart,[4,5] avoids the necessity of configuring the system latches into a shift register by using a separate test-data shift register. This register can load test data in parallel to or from the system latches. Univac has also proposed the use of multiplexers to scan out latch contents. Fujitsu and Amdahl avoid the use of test shift registers entirely. They use a combination of demultiplexing and multiplexing to set and scan out the system latches.

Level-sensitive scan design. A scan-path design method, called level-sensitive scan design, for latch-based systems was presented in Eichelberger.[6] LSSD is the standard design technique in current use at IBM. In this method each system latch is replaced by a two-port latch (L1 latch), and a second (single-port) latch (L2 latch) is added to permit reconfiguration of the system latches into a shift register for test purposes. A two-port latch is directly analagous to a two-port flip-flop. It is a latch with two data inputs, each of which is controlled by a separate clock. The reason that this method is called level-sensitive has to do with the design of the latches to be hazard free and thus not dependent on the clock rise and fall times for correct operation.

A general structure for a system designed using the LSSD technique is shown in Figure 4. Examples of some specific designs using this structure are presented in Das Gupta.[7] During normal operation the system is clocked with two interleaved nonoverlapping pulse trains applied to the CK1 and CK2 inputs. Other possible structures are discussed in Eichelberger.[6] The way a system designed using this technique is tested is very similar to testing a system using two-port flip-flops. The differences when using LSSD are that the test vector v_j values are scanned in via SDI by applying pulses alternately to the test clock input TCK (called the A clock in some

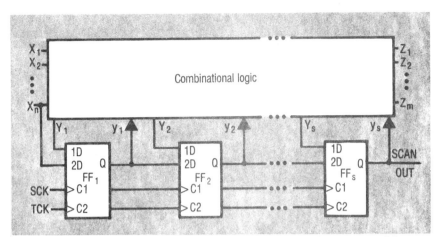

Figure 3. General structure of circuit using two-port flip-flops to provide scan path.

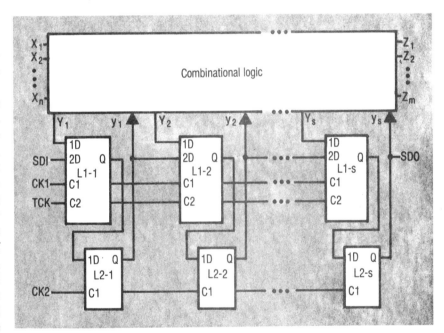

Figure 4. General structure of circuit using two-port latches to provide scan path—LSSD double latch design. SDI is scanned-in test data, SDO scanned-out test data. TCK is test clock.

LSSD papers) and the system clock input CK2 (also called the B clock). The new values of y_j are entered into the corresponding L1 latches by applying one clock pulse to the system clock CK1. The new y_j values are scanned out by applying clock pulses alternately to CK2 and TCK.

The external scan path

The scan path techniques just described make the test pattern generation problem much easier: combinational circuit tests are simpler to derive than sequential circuit tests. The test application difficulties still remain. The amount of test data that must be stored is reduced, but the time to apply a single pattern is increased due to the need to scan data in and out of the chip. A considerable reduction in tester complexity can be obtained by adding an external scan path to the design, as shown in Zasio.[8] The *external* or *boundary scan path* is formed by associating a flip-flop or latch pair with each of the I/O bonding pads. The resulting structure is shown in

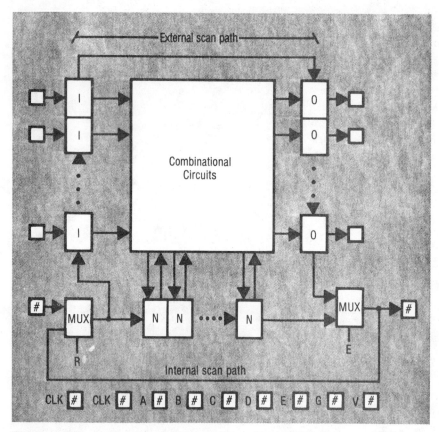

Figure 5. Structure with both internal and external scan paths.

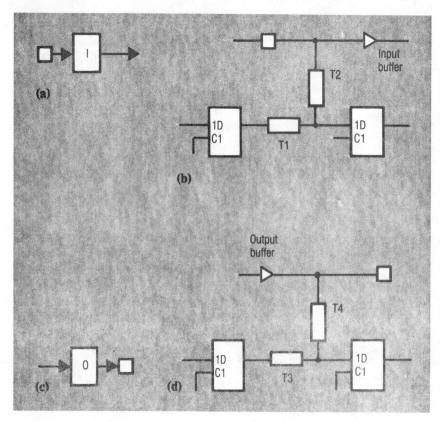

Figure 6. Details of I/O bonding-pad scan-path elements: (a) input pad symbol; (b) input pad circuit; (c) output pad symbol; (d) output pad circuit.

Figure 5, and the details of the circuitry associated with the bonding pads are shown in Figure 6.

The advantages of the external scan ring structure are as follows:

(1) It is not necessary to probe all of the I/O pins during wafer sort.
(2) The board interconnections can be easily tested.
(3) A simple test of the operating speed is possible.

The Zasio paper describes the testing of a 256-pin chip by probing only 13 pins during wafer sort. Seven of the probed pins are used only for test: two test clocks, three mode controls, a scan-in pin, and a scan-out pin. (Two test clocks are required since the design is latch based and uses a two-phase clock.) The other probed pins are power, ground, and up to four system clocks.

The testing of the functional circuitry is done by shifting the test patterns into the internal and external scan paths and using the system clock to transfer the new state into the internal path. The new state and output response are transferred off the chip by means of the external and internal scan paths.

The board interconnections can be tested by shifting data onto the output pads via the external scan path, capturing the signals present at the input pads in the corresponding external scan path flip-flops, and then shifting out the captured data to verify that the correct signals were received.

The chip speed is checked by connecting the internal path output back to its input by means of the two multiplexers shown in Figure 5. An inversion is provided so that a ring oscillator is formed. By measuring the speed of oscillation, a measure of the chip operating speed is obtained. A similar technique is also used on the external scan path. Other discussions of the use of an I/O scan path are presented in Komonytsky[9] and Resnick.[10]

Built-in self-test architectures

Several schemes for incorporating BIST techniques into a design have been proposed. They differ in the requirements placed on the functional circuitry and on the mechanisms used to generate test inputs and analyze test output response. The different possibilities are listed in Table 1. Not all combinations of mechanisms are viable; for example, if the functional circuit doesn't have scan paths, it isn't possible to reconfigure the scan path into a test pattern generator. A discussion of the details of several different BIST architectures follows.

The only commercial semicustom part currently announced uses an external scan path that is reconfigured for test stimulus and response analysis. No scan path is required in the functional circuitry. This part is the SCX6260 CMOS gate array of National Semiconductor. The design of this array was done at CDC and is described in Resnick[10]; it is produced also by Motorola.

External test chip and no scan paths. The earliest detailed study of a BIST structure for random logic chips is described in Benowitz.[11] Figure 7 shows the architecture considered in this study. Each module has a response analysis circuit and an input multiplexer added to it. The multiplexer chooses between the normal system inputs and an external LFSR that generates test patterns. A number of possibilities were studied for the response analyzers. The results of simulation studies that compare cost, test time, and fault coverage for several implementations are presented in the paper. Parity check circuits were found to be the fastest but also the most expensive and poorest in fault coverage. This paper was the first to suggest the use of an LFSR for output response analysis. Both serial and parallel implementations were simulated. In the simulations, the fault coverage was determined by simulating stuck faults at circuit I/Os. The only assumptions made about the circuit-under-test are that it not contain RAM. Sequential as well as combinational logic circuits can be present in the circuit-under-test.

A similar structure is described in Perkins.[12] This structure makes use of a tester circuit that is external to the chip, but could be located on the same board. Pseudorandom patterns are applied in parallel to the chip inputs, and a parallel signature analyzer receives both the chip outputs as well as the test inputs. A special feature of this design is its ability to be configured by means of a tester circuit register to apply input signals to the appropriate chip input pins. Testing of RAM and ROM chips is discussed. The most novel feature of the structure is the inclusion of a source resistance detector at each logic input and an output current detector at each logic output. The main purpose of these detectors is to watch over the integrity of the interconnection system.

Internal scan path and external test chip. The inclusion of an internal scan path in the functional circuitry permits more careful control of the fault coverage. A design for BIST structure for chips with internal scan paths mounted on a multichip module is presented in Bardell.[13] This design, shown in Figure 8, uses a special test chip added to the module. The test chip contains a shift register pattern generator and a parallel signature analyzer. The scan paths on each chip are loaded in parallel from the pattern generator. The system clocks are then

Table 1. Choices for BIST architectures.

Functional circuitry
- No scan path
- Internal scan path only
- Both internal and external scan paths

Input test pattern generation
- Done in separate chip on the same module
- Internal scan path reconfigured
- External scan path reconfigured

Output response analysis
- Done in separate chip on the same module
- Internal scan path reconfigured
- External scan path reconfigured
- Done by checking circuits

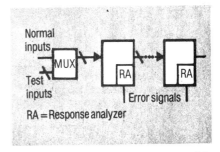

Figure 7. Architecture for BIST with no scan path in functional circuits and an external LFSR for test pattern generation.

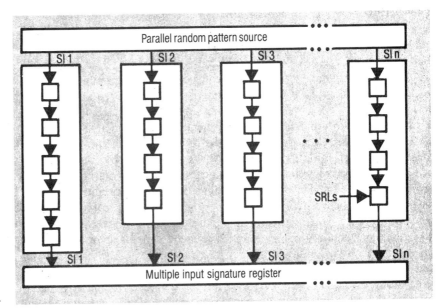

Figure 8. Multichip module BIST structure using chip scan paths.

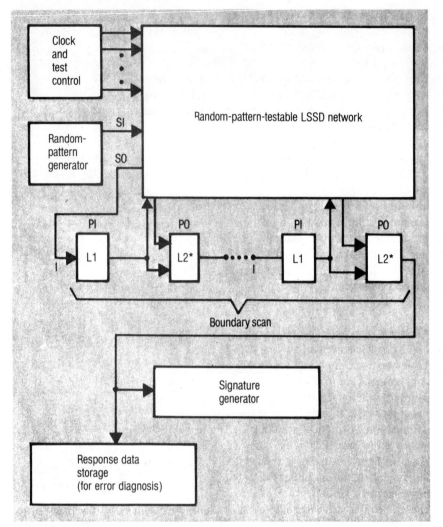

Figure 9. BIST structure using internal and external scan paths in conjunction with an external test chip.

pulsed, and the test results are scanned out to the parallel signature analyzer. New test patterns can be scanned in at the same time that the test results are being scanned out.

Internal and external scan paths with an external test chip. A drawback of the previous scheme is the difficulty of testing the interconnections, either those between the chip pins and bonding pads or those between chips. This is made much easier by the inclusion of an external or boundary scan path on the chip. An external scan path consists of a scan path flip-flop connected to each of the bonding pads. A structure that makes use of an external scan path in connection with an external test chip was proposed in Eichelberger[14] and is shown in Figure 9. The scan input of the external scan path is connected to the scan-out point of the internal scan path. Pseudorandom test patterns are generated in a pattern generator circuit and are scanned into the combined scan path. The system clocks are pulsed, and the scan path latches are scanned out into the serial signature analyzer circuit. The resulting signature is then compared in the analyzer with a precalculated fault-free circuit signature in order to generate a failure signal. The scan-out point is also connected to a pin so that, in case of a failure, intermediate signatures can be examined externally for diagnostic purposes.

BIST structure using register reconfiguration. A concern with BIST designs is the amount of extra circuitry required. One technique for reducing the extra circuitry is to make use of the flip-flops or latches already in the design for test generation and analysis. The system registers are redesigned so that they can function as pattern generators or signature analyzers for test purposes.

The structure described in Konemann[15,16] applies to circuits that can be partitioned into independent modules. Each module is assumed to have its own input and output registers, or such registers are added to the circuit where necessary. No precise definition of a module is given, nor is the problem of identifying modules discussed. It is assumed that the circuit modularity is evident. The registers are redesigned so that for testing purposes they can act as either shift registers or parallel signature analyzers. The redesigned register is called a BILBO (built-in logic block observer).

The technique just described is most suitable for circuits that can be partitioned so that the input and output registers of the resulting modules can be reconfigured separately. They can be used for scan path designs, but the existence of scan paths is not a requirement. There are other reconfiguration techniques that are suitable for scan path designs.

BIST structure using external scan-path reconfiguration. A technique in which the external scan path is reconfigured into a linear-feedback shift register for test pattern generation and into another LFSR for signature analysis is described in Komonytsky[9,17] and Resnick.[10] The basic design of these circuits incorporates both an internal scan path and an external scan path for the I/O pads. The BIST modification provides for the reconfiguration of the input scan-path latches into an LSFR for use as a pseudorandom pattern generator and the reconfiguration of the output scan-path latches into a signature analyzer. A test pattern is generated in the LFSR, scanned into the internal latches, the system clock is pulsed, and the resulting latch contents are scanned out to the signature analyzer. The details of the design are presented in the references cited.

Other BIST designs using LFSRs for test vector generation and signature analysis are described in Eiki,[18] Heckelman,[19] and Fasang.[20]

BIST structure using concurrent checking circuits. For systems that include concurrent checking circuits, it is possible to use this circuitry to verify the response during explicit (off-line) testing. Thus, the necessity of implementing a separate response analysis circuit such as a signature analyzer is avoided. This approach to BIST is described in Sedmak.[21,22] Since the checking circuitry recommended in this paper involves duplication of the functional circuitry and comparison of the outputs of the two implementations, this technique avoids the alias problem and consequent loss of effective fault coverage of a signature analyzer. Figure 10 shows a suggested BIST implementation using this scheme. (The duplicate circuitry is actually realized in complementary form to reduce design and common mode faults.) The test patterns are shifted serially into the input registers from the ISG (input stimulus generator) circuits, with are added for test pattern generation. The ISG cycles through all input patterns, thus generating exhaustive test patterns. The contents of the two input registers are checked by the compare 1 circuit. The output response of the combinational circuitry is checked by the compare 2 circuit. Another design that uses the concurrent check circuits for BIST is described in Lu.[23]

There is a large variety of BIST architectures in current designs. The choice of technique is very dependent on the production volume expected for the part. Semicustom parts rely on very structured approaches that reduce designer time at the expense of higher overhead. Custom de-

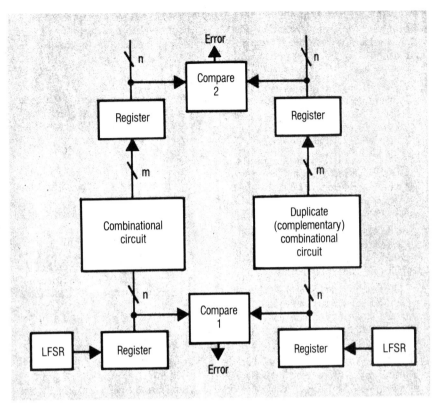

Figure 10. BIST using concurrent checking compare circuits for output response analysis.

signs use methods that require less area overhead but more engineering time since they may be a mixture of various techniques. In the future this situation will probably continue. The main trends will be the introduction of more facilities for BIST into semicustom chip architectures and increasing use of external scan paths.

The methods discussed here are directed at random designs. BIST techniques are also being introduced for more structured circuits such as RAMs and PLAs. While these are important, space does not permit a discussion of them here. BIST for RAMs is discussed in Sun.[24] Self-testing PLAs are considered in Fujiwara.[25] □

Acknowledgment

This work was supported in part by the National Science Foundation under Grant No. MCS-8200129. The author wishes to thank Joseph McCluskey for his help in preparing this paper.

References

1. E. J. McCluskey, "A Survey of Design for Testability Scan Techniques," *VLSI Design,* Vol. V, No. 12, Dec. 1984, pp. 38-61.

2. Richard Liu, "Two Port J-K Flip-Flops for Scan Path Design Center for Reliable Computing Technical Report, Stanford University, Stanford, Calif.

3. M. J. Y. Williams and J. B. Angel, "Enhancing Testability of Large Scale Integrated Circuits via Test Points and Additional Logic," *IEEE Trans. Computers,* C-22, No. 1, Jan. 1973, pp. 46-60.

4. J. H. Stewart, "Future Testing of Large LSI Circuit Cards," *Digest 1977 Semiconductor Test Symp.,* Cherry Hill, N.J., Oct. 1977, pp. 6-15.

5. J. H. Stewart, "Application of Scan/Set for Error Detection and Diagnostics," *Digest 1978 Semiconductor Test Conf.,* Cherry Hill, N.J., Oct. 1978, pp. 152-158.

6. E. B. Eichelberger and T. W. Williams, "A Logic Design Structure for LSI Testability," *Proc. 14th Design Automation Conf.,* New Orleans, June 1977, pp. 462-468.

7. S. Das Gupta, E. B. Eichelberger, and T. W. Williams, "LSI Chip Design for Testability," *Proc. IEEE Solid-State Circuits Conf.,* San Francisco, Calif., Feb. 1978, pp. 216-217.

8. J. J. Zasio, "Shifting Away From Probes for Wafer Test," *Digest Compcon Spring 83,* San Francisco, Calif., Feb. 1983, pp. 395-398.

9. D. Komonytsky, "LSI Self-Test Using Level-Sensitive Scan Design and Signature Analysis," *Digest 1982 Int'l Test Conf.,* Philadelphia, Pa., Nov. 1982, pp. 414-424.

10. D. R. Resnick, "Testability and Maintainability with a New 6K Gate Array," *VLSI Design,* Mar./Apr. 1983, pp. 34-38.

11. N. Benowitz et al., "An Advanced Fault Isolation System for Digital Logic," *IEEE Trans. Computers,* C-24, No. 5, May 1975, pp. 489-497.

12. C. C. Perkins et al., "Design for In-Situ Chip Testing with a Compact Tester," *Digest 1980 Test Conf.,* Philadelphia, Pa., Nov. 1980, pp. 29-41.

13. P. H. Bardell and W. H. McAnney, "Self-Testing of Multichip Logic Modules," *Digest 1982 Int'l Test Conf.,* Philadelphia, Pa., Nov. 1982, pp. 200-204.

14. E. B. Eichelberger and E. Lindbloom, "Random-Pattern Coverage Enhancement and Diagnosis for LSSD Logic Self-Test," *IBM J. Res. Develop.,* Vol. 27, No. 3, May 1983, pp. 265-272.

15. B. Konemann, J. Mucha, and G. Zwiehoff, "Built-in Logic Block Observation Technique," *Digest 1979 Test Conf.,* Cherry Hill, N.J., Oct. 1979, pp. 37-41.

16. B. Konemann, J. Mucha, and G. Zwiehoff, "Built-in Test for Complex Digital Integrated Circuits," *IEEE J. Solid-State Circuits,* Vol. SC-15, No. 3, June 1980, pp. 315-318.

17. D. Komonytsky, "Synthesis of Techniques Creates Complete System Self-Test," *Electronics,* Mar. 1983, pp. 110-115.

18. H. Eiki, K. Inagaki, and S. Yajima, "Autonomous Testing and its Application to Testable Design of Logic Circuits," *Proc. 10th Int'l Symp. Fault-Tolerant Computing* (FTCS-10), Kyoto, Japan, Oct. 1980, pp. 173-178.

19. R. W. Heckelman and D. K. Bhavsar, "Self-Testing VLSI," *Proc. IEEE Solid State Circuits Conf.,* Feb. 1981, pp. 174-175.

20. P. P. Fasang, "BIDCO, Built-in Digital Circuit Observer," *Digest 1980 Test Conf.,* Philadelphia, Pa., Nov. 1980, pp. 261-266.

21. R. M. Sedmak, "Design for Self-Verification: An Approach for Dealing with Testability Problems in VLSI-based Designs," *Digest 1979 Test Conf.,* Cherry Hill, N.J., Oct. 1979, pp. 112-124.

22. R. M. Sedmak, "Implementation Techniques for Self-Verification," *Digest 1980 Test Conf.,* Philadelphia, Pa., Nov. 1980, pp. 267-278.

23. D. J. Lu and E. J. McCluskey, "Recurrent Test Patterns," *Proc. 1983 Int'l Test Conf.,* Philadelphia, Pa., Oct. 1983, pp. 76-82.

24. Z. Sun and L.-T. Wang, "Self-Testing of Embedded RAMS," *Proc. 1984 Int'l Test Conf.,* Philadelphia, Pa., Oct. 1984, pp. 148-156.

25. H. Fujiwara, "A New PLA Design for Universal Testability," *IEEE Trans. Computers,* Aug. 1984, pp. 745-750.

Edward J. McCluskey is the author of another article in this issue. His photo and biographical sketch appear on p. 28.

Built-In Testing of One-Dimensional Unilateral Iterative Arrays

E. M. ABOULHAMID AND E. CERNY

Abstract — It has been shown in the literature that C-testable iterative arrays have very simple test structures, independent of the length of the arrays. We show in this work that all C-testable arrays are also pI-testable, which is a property yielding, in many cases, rather simple built-in-testing structures, both for the test generator and for the response verifier.

Index Terms — Built-in testing, C-testability, iterative logic arrays, testability, test generation.

I. INTRODUCTION

The great complexity of LSI circuits has profoundly affected the efficiency of testing such devices, by increasing the difficulty of deriving tests having sufficient fault coverage, and by increasing the length of the test sequences required. Techniques such as LSSD [5] improve both of these factors. Nevertheless, with LSSD, all individual tests and their responses must be shifted serially in and out of the circuit under test (CUT), which makes testing time relatively long. This situation can be improved by adding logic that is dedicated to performing the majority of test functions on the chip. Hence, we obtain a design with built-in-testing (BIT) features.

BIT usually means that testing is performed "off-line," in a special test mode, as in LSSD. The additional logic used in such structures consists primarily of tests generators and response verifiers. The signatures produced by the response verifiers may be observed at the end of the test by means such as LSSD. The difference from normal LSSD lies in the fact that only the final signature must be collected, and only a few additional tests have to be performed by the LSSD technique.

A classification of test techniques and a review of BIT techniques can be found in [2]; many particular cases are discussed in [1], [8], [9], [13]. In this work we are interested in *off-line BIT of unilateral one-dimensional iterative arrays (ILA)*, where each cell is a combinational circuit. We present test generator and response verifier structures to be added to an ILA to provide for BIT. We assume that at most one cell can be faulty at any time; however, depending on the cell's structure and the form of the tests, the complete truth table of the array may be verified, provided that the tests conform to conditions stipulated in [4]. The only constraint on the type of faults within a cell is that the cell must remain memoryless. We consider exhaustive testing of each cell in an ILA. The number of test vectors is assumed to be independent of the number of cells forming the array.

Our approach is based on the following notions [4], [6], [10]–[12]. An ILA is C-testable if it can be tested with a constant number of tests, independent of the ILA's size. An ILA is I-testable if the test responses from every cell of the ILA can be made identical. An ILA is CI-testable if it is I-testable with a constant number of tests, independent of the ILA's size.

An ILA can be C-testable and I-testable but this does not imply that it is CI-testable without structural modification. Even if the cell's structure is changed to accommodate CI-testability, there is still some difficulty in constructing a test generator useable within the context of BIT. Namely, the test derived for CI-testable arrays may have a different loop test length for each of the test vectors [4], [10]. This irregularity has an adverse influence on the structure of the test generator, and on the distribution of the tests to cell inputs. Hence, the objective of this work is to find an approach that yields test vectors having an equal loop length. The price paid for this may be an increase in the number of test vectors. Such an increase is not very critical since the tests are applied rapidly through BIT, rather than one at a time from the outside by means such as LSSD.

Suppose now that an ILA is partitioned into cell blocks of equal length, each block thus forming a "super-cell." If the basic cell is C-testable but not CI-testable, is it possible to find a super-cell (i.e., the smallest feasible block size) which is CI-testable? If we limit ourselves to testing exhaustively the cell (rather than the super-cell), the answer is yes, and one possible solution is what we call pI-*testability* [3]. An ILA is pI-testable (partitioned I-testable) if it can be partitioned into blocks of equal length such that all blocks have the same test response (but cells within a block need not), and such that the number of tests is independent of the ILA's size. The loop test length is then equal to the block length.

II. THEORETICAL BACKGROUND

First, we review some necessary concepts on ILA testing; for more details, see [4], [7], [10]. An ILA consists of N interconnected identical cells. A cell is characterized by its flow table of n rows and m columns. The rows correspond to the different states of the cell, and the columns identify the input symbols.

Let $x = (x_1, \cdots, x_p)$, $s = (s_1, \cdots, s_q)$, and $z = (z_1, \cdots, z_k)$ represent, respectively, the input, state, and output variables of a cell such that

$$x_j, s_j, z_j \in \{0, 1\} \quad \text{and} \quad 2^{p-1} < m \leq 2^p, \quad 2^{q-1} < n \leq 2^q.$$

A set of identifying sequences (SIS) of a cell is a set of input sequences $SIS = \{X_1, X_2, \cdots, X_r\}$, such that for any initial state s_i the application of all sequences $s_i X_j, X_j \in SIS$ to the cell results in a set of output responses different from that obtained by starting in any other initial state. It can be shown that every reduced flow table has an SIS. If $\|SIS\| = r = 1$ then the unique sequence is called a distinguishing sequence (DS). The SIS of a cell can thus be used to detect those faults of the cell that affect its state variables.

In order to test exhaustively a particular transition $t: s_i \xrightarrow{x/z} s_j$, the following set of tests can be applied. To each $X_j \in SIS$ concatenate a sequence T_j which ensures transferring the cell to its initial state s_i, such that the sequence xX_jT_j has the same length for all $j = 1, \cdots, r$. This set of tests $\{xX_jT_j\}$ is called "a loop test associated with transition t," i.e., $Lp(t)$. Furthermore, it can be shown that an $Lp(t)$ exists for every transition in a reduced flow table if the table has strongly connected components. To test exhaustively a single transition t in the first, $|Lp(t)| + 1, 2|Lp(t)| + 1$, etc., cells in an ILA, the input symbols forming each sequence are applied to the first $|Lp(t)|$ cells, and then repeated as many times as the size of the ILA requires. Procedure 1 tests exhaustively every cell in the entire ILA.

Procedure 1:
 For $i := 1$ **to** mn **do**
 begin

Apply $Lp(t_i)$ to ILA;
Apply $|Lp(t_i)| - 1$ rotations of each sequence of $Lp(t_i)$ to ILA;
end.

Here, mn is the number of possible transition and $|Lp(t_i)|$ is the length of each sequence in the $Lp(t_i)$.

If r is the cardinality of the corresponding *SIS*, then the number of tests needed to test the ILA is $r * \sum_{i=1}^{mn} |Lp(t_i)|$. If the number of tests is independent of the size of the ILA, then the array is *C*-testable.

Lemma: If an ILA is *C*-testable, then there exist loop tests $Lp(t)$ whose length $cL = |Lp(t)|$ is the same for all transitions of the cell.

Proof: Let $Lp(t_1), Lp(t_2), \cdots, Lp(t_{mn})$ be the loop tests obtained as in [4] or as shown above. Then new loop tests $Lp'(t_i)$, $i = 1, \cdots, mn$, can be constructed by repeating the same test $Lp(t_i)$ such that their lengths are the same and equal to the least common multiple of the original lengths $|Lp(t_i)|$, $i = 1, \cdots, mn$.

If these new loop tests $Lp'(t_i)$ are used for testing the ILA as indicated in Procedure 1 before, then all cells separated by distance cL have the same responses. This is stated formally in the following theorem.

Theorem: If an ILA is *C*-testable, then it is *pI*-testable with $r \cdot mn \cdot cL$ tests, where $r = \|SIS\|$, m = number of states, n = number of input symbols, and cL is the common loop test length.

Note that in many cases there exist solutions in which cL is smaller than the least common multiple mentioned in the proof of the theorem.

Example 1: Consider the cell in Fig. 1 [4], with $SIS = \{1, 01\}$. Without taking into consideration *pI*-testability, an ILA composed of such cells would require 56 tests [10]. By enforcing *pI*-testability through the preceding lemma, there are 96 tests obtained by rotation (5 times each) from the following elementary tests. The tested transition is followed by a particular sequence from *SIS* (bold face) and by a transfer sequence.

```
A⁰ B¹ C¹ C¹ C¹ C⁰ A     A¹ D¹ D¹ D¹ D¹ D⁰ A
B⁰ B¹ C⁰ A⁰ B⁰ B⁰ B     B¹ C¹ C⁰ A⁰ B⁰ B⁰ B
C⁰ A¹ D⁰ A⁰ B¹ C¹ C     C¹ C¹ C¹ C¹ C¹ C¹ C
D⁰ A¹ D¹ D¹ D¹ D¹ D     D¹ D¹ D¹ D¹ D¹ D¹ D

A⁰ B⁰ B¹ C¹ C¹ C⁰ A     A¹ D⁰ A¹ D¹ D¹ D⁰ A
B⁰ B⁰ B¹ C¹ C⁰ A⁰ B     B¹ C⁰ A¹ D⁰ B⁰ B⁰ B
C⁰ A⁰ B¹ C⁰ A⁰ B¹ C     C¹ C⁰ A¹ D⁰ A⁰ B¹ C
D⁰ A⁰ B¹ C⁰ A¹ D¹ D     D¹ D⁰ A¹ D¹ D¹ D¹ D
```

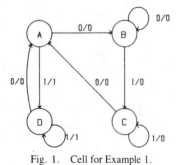

Fig. 1. Cell for Example 1.

III. BUILT-IN TESTING

In what follows we assume that $m = 2^p$, $n = 2^q$. We use a classical BIT scheme where normal ILA inputs are multiplexed with those coming from the test generator. The response verifier produces a "signature" which is multiplexed with the normal ILA outputs and brought into an output register for eventual external verification (by LSSD, for instance). The number of bits in the register is equal to the number of ILA outputs, increased by 1 if comparison of the state of the last cell with that of the first cell is desired. The output register must be constructed in such a way that when entering the test mode it is reset to zero, and then whenever an output of the response verifier takes the value 1, the corresponding bit in the register is set and remains in that state for the rest of the test period. Faultless behavior of the ILA is signaled by an all-zeros signature at the end of the test.

The test generator contains a phase counter which controls stepping through the set of identifying sequences (not necessary if the *SIS* degenerates into a *DS*), a generator of all transitions of the first ILA cell (i.e., its present state and inputs), and two combinational blocks. The first one generates an identifying sequence which corresponds to the state of the phase counter; the second one then produces the necessary transfer sequences. The first cL sections of the input register (a section comprises the inputs to one cell) must, in general, be connected to form a circular shift register, and be enlarged by the bits required to store the states of all the cells within one cL cycle. The required $cL - 1$ circular shifts are generated by a shift counter.

The phase, transition, and shift counters are linked by their carry signals to form a synchronized unit of $\lfloor \log_2 r \rfloor + p + q + \lfloor \log_2(cL) \rfloor$ bits. In practice, not all the components will necessarily appear in the generator.

All blocks of the ILA produce the same outputs under test; hence, the response verifier can be implemented as a series of two-input EXCLUSIVE-OR gates (XOR) placed so that each output bit of cell i in block j is compared to the corresponding output bit of cell i in block $j + 1$. Note that there are as many outputs of the verifier as there are normal CUT outputs. They are brought to the output register described previously.

There are several modes in which the BIT system operates:

i) Normal: The test generator and the response verifier are disconnected, and the CUT operates on its normal inputs.

ii) CUT test: The output register is reset, tests are generated, and the verifier and the output register accumulate test responses.

iii) LSSD mode: The final test signature (and the final state of the test generator) is recovered and further tests are administered from the outside to complete the verification of the registers, XOR gates, and multiplexers, if necessary.

In the case that the cycle length cL is too long for practical purposes, the generator can be quite considerably simplified if the verifier is permitted to observe the ILA state variables. Any input symbol is then a *DS*. The verifier must now compare the states between two neighboring blocks, rather than just their outputs.

The comparator connections must be modified also if we do not want to increase the number of bits in the output register. Therefore, the outputs of the first block are compared to those of the last block, the outputs of the second one with those of the one before last, etc. This leaves half of the bits in the output register free to latch the outputs of the state comparators connected in a similar way as the output comparators. The fault coverage is reduced by those faults which affect in the same manner a pair of blocks whose outputs are compared. Nevertheless, all single-cell faults are detected provided that the next state of the last cell is compared to the present state of the first cell (as supplied by the generator).

We now assume that state observation is provided when necessary. If the cell's flow table corresponds to a Eulerian graph, then there exists a transition cycle that passes exactly once through all transitions, i.e., a Eulerian cycle. This cycle can be decomposed into a number of simple cycles such that each transition belongs to one and only one of them. By selecting as the transfer sequences either the Eulerian cycle or the subcycles, and if necessary observing the cell states, exactly mn tests are required to test the ILA.

Under these conditions, three approaches can be defined for designing the test generator.

Approach 1: The entire Eulerian cycle is stored in a shift register of $mn(\log_2 m + \log_2 n)$ bits.

Approach 2: Decomposition of an *E*-cycle into subcycles.

If it is desired to have cL smaller than the length of the *E*-cycle (resulting in a simpler test generator), the subcycles are used instead of the *E*-cycle. The $Lp(t)$ can be chosen to be equal to the subcycle containing the transition t, and completed so that all $Lp(t)$'s have the same length cL. The circular shift phase of the test may not be necessary here, for if we test exhaustively the first cell in the cL cycle, then all circular shifts of the tests are automatically performed. That is, all cells in the cycle are tested.

Example 2: Let us first consider the case of an adder and then extend the approach to an entire ALU. An adder has a Eulerian flow table and all inputs are a *DS* (Fig. 2). This allows us to construct a BIT adder as follows. The subcycles of an *E*-cycle are

$$0 \xrightarrow{00} 0;\ 0 \xrightarrow{01} 0;\ 0 \xrightarrow{10} 0;\ 1 \xrightarrow{11} 1$$
$$1 \xrightarrow{01} 1;\ 1 \xrightarrow{10} 1;\ 0 \xrightarrow{11} 1 \xrightarrow{00} 0.$$

Therefore, $cL = 2$. If c designates the state, a, b the inputs of the first cell, and α, β the inputs of the second cell in the $Lp(t)$ of length 2, then we generate a, b, c exhaustively and α, β are generated from a, b, c as $\alpha = maj(a, \bar{b}, c)$, $\beta = maj(\bar{a}, b, c)$. There is no need for a circular shift register.

Consider now an ALU ILA constructed of one-bit-wide cells [11]. Table I shows the coding of the control variables (s_0, s_1, s_2).

The arithmetic operations $a + b$, $a - b$, and $b - a$ require that the intercell states be taken into consideration using a distinguishing sequence *DS*. The logic operations induce no intercell dependence, and their exhaustive test set consists of four vectors. These are covered, however, by the test sequence for $a + b$. Therefore, the adder test subcircuit remains as before; and the subtraction operations $b - a$ ($a - b$) require that the inputs a (b) and α (β) be complemented for the adder test sequence to remain applicable.

The resulting test generator is shown in Fig. 3, an 8-bit ALU with the response verifier is in Fig. 4. The implementation of the ALU cell must be such that the carry-out is equal to the carry-in signal for all logic operations; otherwise, the state comparator on the right of Fig. 4 would produce an incorrect output during testing of the logic operations. Alternatively, the output of the state comparator could be disabled during these subtests.

Approach 3: In certain cases, the generator can become simpler if we generate an $Lp(t)$ test for a representative transition of each Eulerian subcycle, all of such $Lp(t)$'s having the same length cL. Then we rotate each such test $cL - 1$ times for the block of cL cells. This approach may require more tests than the minimum obtainable through approaches 1 or 2 because the tests belonging to the cycles shorter than cL will be generated by the circular shift register more than once.

Example 1 (continued): Without state observation there was a need for 96 tests, and the generator was rather complex. If we use state observation, then only 8 tests are needed as follows. The subcycles of the *E*-cycle are

$$A \xrightarrow{1} D \xrightarrow{1} D \xrightarrow{0} A;\quad A \xrightarrow{0} B \xrightarrow{1} C \xrightarrow{0} A;$$
$$B \xrightarrow{0} B;\quad C \xrightarrow{1} C.$$

TABLE I
ALU OPERATION CODING

s_0	s_1	s_2	Operation
0	0	0	$a + b$
0	0	1	$a - b$
0	1	0	$b - a$
0	1	1	a OR b
1	0	0	a AND b
1	0	1	$\bar{a}$ AND b
1	1	0	$a \oplus b$
1	1	1	$\overline{a \oplus b}$

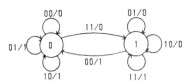

Fig. 2. State diagram of a full adder cell.

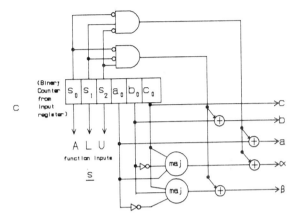

Fig. 3. ALU test generator.

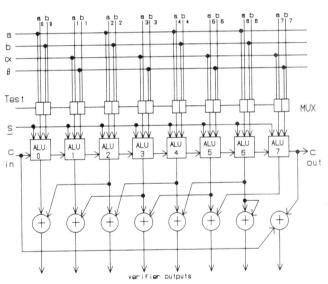

Fig. 4. The ALU with its response verifier (normal outputs not shown).

The length-equalized $Lp(t)$'s with $cL = 3$ are

$$A\ ^1\ D\ ^1\ D\ ^0\ A \qquad A\ ^0\ B\ ^1\ C\ ^0\ A$$
$$B\ ^0\ B\ ^0\ B\ ^0\ B \qquad C\ ^1\ C\ ^1\ C\ ^1\ C.$$

For instance, by selecting representative transitions $D1$, $A0$, $B0$, $C1$, one from each $Lp(t)$ cycle, the rest of each cycle can be obtained combinatorially from those transitions, as in Example 2. The remaining tests are obtained from these four basic ones by circular shifts (2shifts/basic test), producing the following 12 tests:

(1) $A\ ^1\ D\ ^1\ D\ ^0\ A$ (2) $A\ ^0\ B\ ^1\ C\ ^0\ A$

(3) $B\ ^1\ C\ ^0\ A\ ^0\ B$ (4) $C\ ^0\ A\ ^0\ B\ ^1\ C$

(5) $D\ ^0\ A\ ^1\ D\ ^1\ D$ (6) $D\ ^1\ D\ ^0\ A\ ^1\ D$

(7) $B\ ^0\ B\ ^0\ B\ ^0\ B$ (8) $B\ ^0\ B\ ^0\ B\ ^0\ B$

(9) $B\ ^0\ B\ ^0\ B\ ^0\ B$ (10) $C\ ^1\ C\ ^1\ C\ ^1\ C$

(11) $C\ ^1\ C\ ^1\ C\ ^1\ C$ (12) $C\ ^1\ C\ ^1\ C\ ^1\ C.$

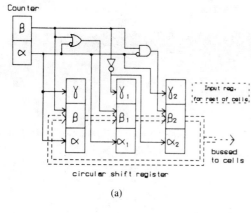

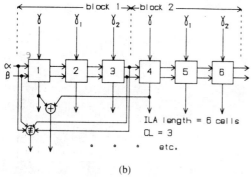

Fig. 5. BIT arrangement for Example 1; (a) test generator; (b) response verifier.

Consider the following binary coding of the states A, B, C, and D: $A = 00$, $B = 01$, $C = 10$, $D = 11$. The representative transitions of each subcycle will be coded as (α, β, γ) thus

$D1$: 1 1 1 $A0$: 0 0 0 $B0$: 0 1 0 $C1$: 1 0 1.

Let $\alpha_1 \beta_1 \gamma_1$ and $\alpha_2 \beta_2 \gamma_2$ be the second and third transitions of the $Lp(t)$ cycle. These can be now generated combinatorially from α, β, γ. The generator consists of a 2-bit modulo-4 counter generating α and β, and a combinational network [Fig. 5(a)] realizing the functions

$\gamma = a$, $a_1 = a$, $\beta_1 = \bar{a} + \beta$, $\gamma_1 = \bar{\beta}$, $a_2 = \bar{\beta}$, $\beta_2 = \bar{a}\beta$, $\gamma_2 = a$.

The corresponding verifier is depicted in Fig. 5(b).

IV. Conclusion

In this presentation we have introduced the concept of *pI*-testability of iterative logic arrays. It has been shown that all *C*-testable arrays are *pI*-testable, without any change in the structure of the cell. A minor modification to the test generation procedure of [4] was introduced, in order to make the tests satisfy the conditions of *pI*-testability. The cost is some increase in the number of tests generated as compared to the original method; however, the gain is a considerable simplification in the structure of the test generator and the response verifier. No change in the cell structure is required. The output verifier compresses the responses of the ILA under test through a set of comparators and XOR gates, producing a final one-word signature.

Finally, let us note that the *pI*-testability approach is relevant to current VLSI trends, where iterative arrays are becoming more interesting due to their regular structure. It would thus be useful to extend the results obtained here to two-dimensional arrays and to systolic arrays.

References

[1] V. K. Agrawal and E. Cerny, "Store and generate built-in-testing approach," in *Proc. FTCS 11*, June 1981, pp. 35–40.
[2] M. G. Buehler and M. W. Sievers, "Off line, built-in test techniques for VLSI circuits," *Computer*, pp. 69–82, June 1982.
[3] E. Cerny and E. M. Aboulhamid, "Built-in-testing of *pI*-testable iterative arrays," in *Proc. FTCS 13*, June 1983, pp. 33–36.
[4] F. J. O. Dias, "Truth-table verification of an iterative logic array," *IEEE Trans. Comput.*, vol. C-25, pp. 605–613, June 1976.
[5] E. B. Eichelberger and T. W. Williams, "A logic design for LSI testability," in *Proc. 14th Design Automat. Conf.*, June 1977, pp. 462–468.
[6] A. D. Friedman, "Easily testable iterative systems," *IEEE Trans. Comput.*, vol. C-22, pp. 1061–1064, Dec. 1973.
[7] F. C. Hennie, *Finite-State Models For Logical Machines.* New York: Wiley, 1968.
[8] E. J. McCluskey and S. Bozorgui-Nesbat, "Design for autonomous test," in *Proc. IEEE Test Conf.*, 1980, pp. 15–21.
[9] J. Savir, "Syndrome testable design of combinational circuits," *IEEE Trans. Comput.*, vol. C-29, pp. 442–550, June 1980.
[10] T. Sridhar and J. P. Hayes, "Testing bit-sliced systems microprocessors," in *Proc. FTCS 9*, 1979, pp. 211–218.
[11] ——, "A functional approach to testing bit-sliced microprocessors," *IEEE Trans. Comput.*, vol. C-30, pp. 563–571, Aug. 1981.
[12] ——, "Design of easily testable bit-sliced systems," *IEEE Trans. Comput.*, vol. C-30, pp. 842–854, Nov. 1981.
[13] A. K. Susskind, "Testing by verifying Walsh coefficients," in *Proc. FTCS 11*, June 1981, pp. 206–208.

LOCST: A Built-In Self-Test Technique

With its low hardware cost, simple implementation and excellent coverage, this technique promises to meet the needs of a variety of VLSI environments.

Johnny J. LeBlanc, IBM Federal Systems Division

The advent of very large scale integration technologies has increased interest in built-in self-test as a technique for achieving effective and economical testing of VLSI components. As used in this article, the term "built-in self-test" refers to the capability of a device to generate its own test pattern set and to compress the test results into a compact pass-fail indication. Many built-in self-test techniques have been proposed over the past 10 years, ranging from self-oscillation to functional pattern testing of microprogrammed devices to random-pattern testing (for examples, see papers by Mucha et al.,[1] Sedmak,[2] and McCluskey et al.[3]). These various techniques provide different capabilities for defect detection and self-test execution time. They also impose different requirements for implementation and control.

Benefits to be gained from self-test, however, are common to all implementation techniques and include

- reduced test pattern storage requirements,
- reduced test time, and
- defect isolation to the chip level.

Since test patterns are generated automatically, only self-test initialization, control, and pass-fail comparison patterns need be stored, significantly reducing pattern storage requirements. Test time is reduced because one can use simple hardware devices (e.g., counters or linear-feedback shift registers) to control test execution, rather than retrieving test patterns from storage devices (e.g., disks) and applying them to the component under test. When components with built-in self-test are mounted on higher-level packages, the self-test pass-fail indication provides defect isolation to the chip level (e.g., during card repair testing).

At the IBM Federal Systems Division we have implemented a VLSI built-in self-test technique, which can be incorporated at very low hardware cost into any chip conforming to level-sensitive scan design (LSSD) rules, on three VLSI signal-processing chips. Our method (designated LSSD on-chip self-test, or LOCST) uses on-chip pseudorandom-pattern generation

---

Summary

A built-in self-test technique utilizing on-chip pseudorandom-pattern generation, on-chip signature analysis, a "boundary scan" feature, and an on-chip monitor test controller has been implemented on three VLSI chips by the IBM Federal Systems Division. This method (designated LSSD on-chip self-test, or LOCST) uses existing level-sensitive scan design strings to serially scan random test patterns to the chip's combinational logic and to collect test results. On-chip pseudorandom-pattern generation and signature analysis compression are provided via existing latches, which are configured into linear-feedback shift registers during the self-test operation. The LOCST technique is controlled through the on-chip monitor, IBM FSD's standard VLSI test interface/controller. Boundary scan latches are provided on all primary inputs and primary outputs to maximize self-test effectiveness and to facilitate chip I/O testing.

Stuck-fault simulation using statistical fault analysis was used to evaluate test coverage effectiveness. Total test coverage values of 81.5, 85.3, and 88.6 percent were achieved for the three chips with less than 5000 random-pattern sequences. Outstanding test coverage (>97%) was achieved for the interior logic of the chips. The advantages of this technique, namely very low hardware overhead cost (<2%), design-independent implementation, and effective static testing, make LOCST an attractive and powerful technique.

and on-chip signature analysis result compression. This is not a new self-test method; LOCST utilizes the serial-scan, random-pattern test technique pioneered by Eichelberger et al.[4,5] and Bardell et al.[6] of IBM. This article (1) details the adaptation of this technique to our existing chip testability architecture, (2) details the implementation of LOCST on three VLSI chips designed and fabricated by IBM FSD, and (3) discloses the results of the test coverage evaluations performed on these three chips. (For a thorough understanding of the principles of serial-scan, random-pattern testing, I strongly recommend a review of references 4, 5, and 6 and also a very comprehensive paper by Komonytsky.[7])

Standard FSD VLSI testability features

For a better understanding of the self-test architecture chosen for LOCST, a discussion of design features typical to IBM FSD's products is warranted. Figure 1 illustrates the three standard testability features incorporated in our VLSI products. They include

- level-sensitive scan design,
- "boundary scan" latches, and
- a standard maintenance interface, the on-chip monitor, or OCM.

All chips are designed following IBM's LSSD rules (see Eichelberger and Williams[8]) to ensure high test coverage and high diagnostic resolution during chip manufacture testing. "Boundary scan" is a requirement that all primary inputs (PIs) feed directly into shift register latches (SRLs, or LSSD latches) and all primary outputs (POs) are fed directly from SRLs. Boundary scan greatly simplifies chip-to chip interconnect testing and also provides an ideal buffer between LSSD VLSI products and non-LSSD vendor components, thereby reducing the complexity of testing "mixed-technology" cards.

The OCM is a standard maintenance interface for our VLSI chips (Figure 2). It consists of seven lines: two for data transfer, four for control, and one for error reporting. The OCM maintenance bus can be configured as either a ring, a star, or a multidrop network, depending on system maintenance requirements. The four major functions of the OCM are

- scan string control,
- error monitoring and reporting,
- chip configuration control, and
- clock event control: run/stop, single cycle, and stop on error.

During LSSD testing (chip manufacture testing), scan strings are accessed via either dedicated or shared PIs and POs. (Note: The OCM is not used as a test aid during LSSD testing; it is simply logic to be tested by LSSD test patterns.) During card and system test, however, chip scan strings are accessed via the OCM interface.

The error detection hardware depicted in Figure 1 consists of on-chip error checkers used for on-line system error detection and/or fault isolation (described by Bossen and Hsiao[9]). When these checkers are triggered by

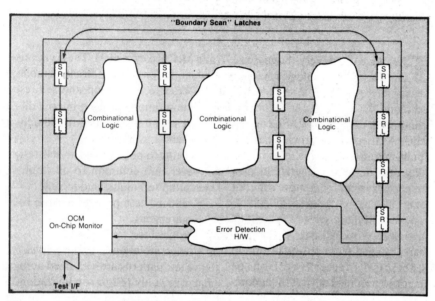

Figure 1. Standard VLSI features.

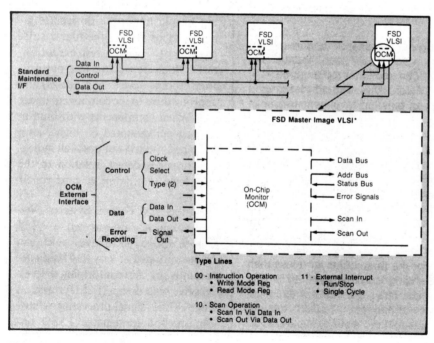

Figure 2. On-chip monitor.

an on-chip error, an attention signal is sent to the system maintenance processor through the OCM interface. The system maintenance processor reads internal chip error registers (or writes internal chip mode control registers) via OCM "instructions."

LOCST architecture

The basic self-test methodology used in LOCST is to (1) place pseudorandom data into all chip LSSD latches via serial scan, (2) activate system clocks for a single cycle to capture the results of the random-pattern stimuli through the chip's combinational logic, and (3) compress the captured test results into a pass-fail signature. With the existing testability features (LSSD, boundary scan, OCM) on each chip, it was a simple matter to incorporate a self-test capability.

To perform the pseudorandom-pattern-generation and signature-compression operations while in LOCST self-test mode, functional SRLs are reconfigured into linear-feedback shift registers, or LFSRs. The pseudorandom-pattern generator, or PRPG, is 20 bits in length, and the signature analyzer (SA) is 16 bits in length (see Figure 3). It should be noted that the devices shown in Figure 3 operate as normal serial-scan latches *and* as linear-feedback shift registers. The transformation from normal serial-scan mode to LFSR mode is controlled by multiplexing the scan inputs with a self-test enable signal (controlled via the OCM interface). The parallel data ports of these latches are not modified in any way. During self-test the data port clocks (system clocks) are disabled to prevent outside data from disrupting the deterministic sequences of the LFSRs.

The feedback polynomial for the PRPG was chosen because it is the least expensive "maximal-length" 20-bit LFSR implementation in terms of XOR gates required. For the LOCST implementation, the characteristic polynomial of the PRPG and the SA is fixed. Differing test pattern sequences can be obtained by altering the initial value (or "seed") of the PRPG. The feedback polynomial for the SA was chosen because of its proven performance (see Frohwerk[10] and Smith,[11] for example). The result of using a 20-bit PRPG and a 16-bit SA is a self-test capability with $2^{20} - 1$ possible random-pattern sequences and a very low probability of signature analysis fault masking (approximately $1/2^{16}$ or 0.0015 percent).

A high-level block diagram of the LOCST implementation structure is shown in Figure 4. In self-test mode the initial 20 SRLs of the chip's scan strings are configured into a PRPG LFSR, and the last 16 SRLs are configured into an SA LFSR. For normal LSSD chip manufacture testing, a chip usually contains several scan strings —each accessible from chip input and output pins. During LOCST testing, however, all scan strings except the one containing the OCM latches are configured into a single scan string. (Note: Random test patterns are scanned into the single scan string under OCM control. SRLs that are part of the OCM and any chip clock generation circuitry *cannot* be included in the LOCST scan string since self-test control and clock control cannot be disrupted by random data.)

The following is a description of the LOCST sequence:

(1) Initialize all internal latches: scan known data into all SRLs; this includes scanning "seeds" into PRPG and SA registers.

(2) Activate self-test mode: enable PRPG and SA registers; disables system clocks on input boundary SRLs and LFSRs.

(3) Perform self-test operation:

 (a) Apply scan clocks until entire scan string (up to the SA LFSR) if filled with pseudorandom patterns. This step also scans test data into the SA LFSR for test result compression.

 (b) Activate system clocks for single-cycle operation.

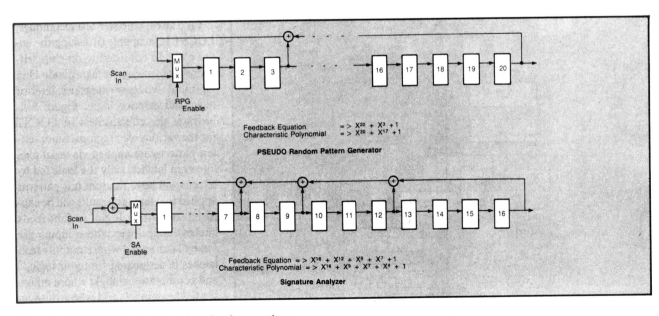

Figure 3. Linear-feedback shift register implementations.

(c) Repeat (a) and (b) until finished.

(4) Read out test result signature and compare with known "good" value.

The "good" value from step 4 can be obtained in two ways; (1) simulation of the entire self-test sequence, or (2) the "golden chip" approach (that is, determine what the "good" value is by performing the LOCST self-test operation on chips which have passed all other forms of manufacture and functional testing). Due to the high cost of the first method, the second is currently being used. If the correct "good" signature value were known (via simulation) during the chip design phase, a hardware comparator could be placed on the chip to provide an immediate pass-fail indication. Our implementations of LOCST require that the 16-bit signature be read by an external processor for comparison against the stored 16-bit "good" value.

The entire LOCST self-test operation is controlled by an external processor via the OCM interface. The external processor may be a chip or card tester or a system maintenance processor, depending on the testing environment. The OCM provides the following self-test control functions:

- PRPG and SA enable control,
- scan access to internal SRLs for random-pattern insertion and test result compression,
- chip clock control for single-cycle operation (if on-chip clock generation is used), and
- access to self-test results via direct register read or via scan.

If a chip does not have an OCM, control of these functions must be provided by some other means.

The data port clocks of input SRLs (i.e., boundary scan LSSD latches fed directly by primary inputs) are inhibited during self-test mode to prevent unknown data from corrupting the self-test sequence. If the input latch clocks are not disabled, known values *must* be ensured on chip PIs during self-test execution.

LOCST limitations

Like all on-chip self-test techniques, LOCST is incapable of testing the entire chip. In considering on-chip self-test effectiveness, we can divide chip logic into two basic categories: interior logic and exterior logic. Figure 5 illustrates the effectiveness of LOCST for the various chip regions. Since self-test patterns are applied via serial scan into chip latches, only the logic fed by latches will have random test patterns applied to it and test results will be captured only for logic which feeds latches. Chip logic whose inputs are fed by latches and whose outputs feed latches is designated "interior logic," and combinational logic whose inputs are fed by chip PIs and whose outputs feed chip POs is designated "exterior logic."

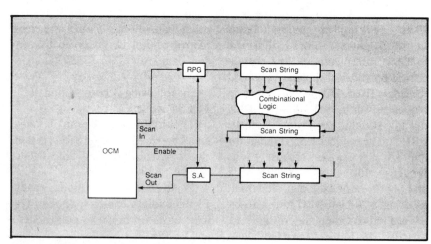

Figure 4. LOCST architectures.

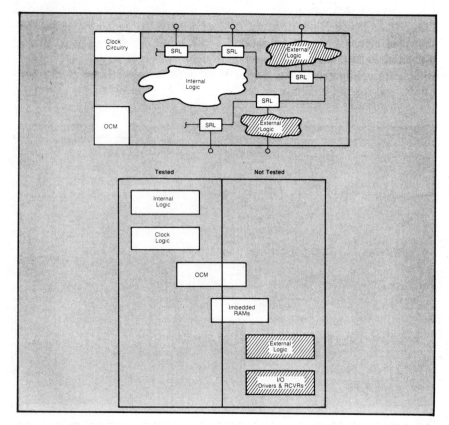

Figure 5. LOCST effectiveness.

Obviously, external logic is completely untestable by the LOCST technique. The importance of boundary scan to on-chip self-test also should be obvious. The larger the percentage of exterior logic on a chip, the less effective on-chip self-test becomes. In the ideal case with 100 percent boundary scan, the only exterior logic would be I/O drivers and receivers (and with 100 percent boundary scan, I/O drivers and receivers would be very easy to test!)

Types of chip logic that do not clearly fall into the categories of interior or exterior are the OCM logic and embedded RAMs. Since the OCM controls the self-test operation, internal OCM logic is not tested by random patterns during self-test. Rather, the OCM is tested to the extent that all OCM functions needed to perform the self-test operation will have been exercised (i.e., scan control, clock control, loading self-test registers, etc.). Remaining OCM functions are tested by exercise of the OCM's remaining instruction set. RAMs embedded in a chip will not be completely tested by the LOCST self-test technique. Special RAM self-test circuitry would be needed to provide effective testing with random patterns. This topic is not addressed here.

The locations of the PPG and SA LFSRs are not illustrated because this would require a detailed scan string diagram. As mentioned previously, the PRPG and SA LFSRs utilize existing functional latches. The two other chips, B and C, when configured with a vendor multiply chip, perform digital filtering functions. Like Chip A, Chips B and C are primarily arithmetic data pipelines. All three chips are now incorporated in signal-processing systems.

To determine the testing effectiveness of the LOCST technique on these three chips, we performed fault simulation of the self-test procedure. Fault simulation provides a test coverage value upon which self-test effectiveness is based. The fault simulation was based on the classical stuck-fault model. Full fault simulation of the LOCST operation would have been too costly, so we followed this methodology:

- We used a statistical random sample of the full stuck-fault list. Test coverage results therefore have a 95 percent confidence level.
- Since no significant ($\ll 1\%$) error masking occurs due to the LFSR compression of the test results,[10-12] simulation of the serial compression activity of the SA LFSR was not performed. If the detection of a fault is observed at an SRL, it is assumed that this fault will be detected after LFSR compression.

We generated pseudorandom patterns placed in the latches during fault simulation via a PL/I program, using the same characteristic polynomial as the PRPG LFSR implemented on the chips (see Figure 3). A plot of test coverage vs. the number of self-test sequences for Chip A is presented in Figure 7. A total chip test coverage of 88.6 percent was achieved (with 95 percent confidence) with 3000 self-test sequences. Figure 8a displays the coverage evaluation results for Chip A in a different manner. Here Chip A's logic is divided into three categories (interior logic, exterior logic, and OCM logic) to highlight the LOCST testing effectiveness for each. LOCST test effectiveness for all three chips is summarized in Figure 8.

Implementions and coverage evaluation

The LOCST technique has been implemented on three VLSI chips used for signal-processing applications. The three chips—hereafter called Chip A,

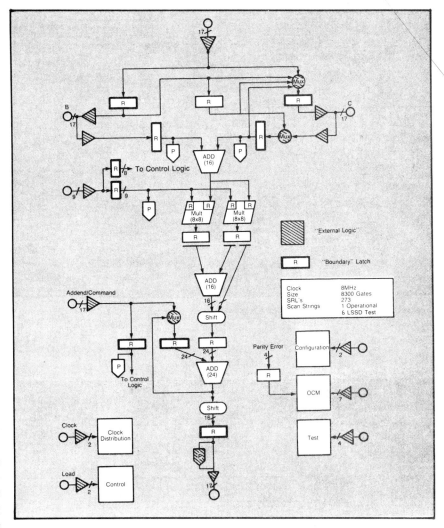

Figure 6. Signal-processing Chip A.

Chip B, and Chip C—were designed and fabricated in 1982. The addition of the LOCST capability (i.e., LFSRs for PRPG and SA functions and OCM self-test control logic) represents a hardware overhead of less than two percent. (Note: This figure does not include LSSD overhead or OCM overhead, as these features are included whether or not LOCST is implemented. Total testability overhead is in the 10-15 percent range.)

One of the three chips, Chip A, performs front-end signal-processing functions requiring high-rate, multiply-intensive algorithms such as finite-impulse response filtering, linear beam-forming, and complex band-shifting operations. Chip A performs these functions by utilizing a simple add-multiply-add pipelined data structure. A high-level diagram of Chip A is shown in Figure 6.

Overall test coverage values of 88.6, 81.5, and 85.3 percent (mean values of a 95 percent confidence interval) were obtained for the three chips respectively. Very good coverage (>97%) was obtained for the interior logic of all three chips with relatively few random-pattern loads (<5000). Test coverage obtained by deterministic LSSD test pattern generation was greater than 99 percent for all three chips. Whether or not test coverage comparable to that of LSSD testing could be obtained if more random-pattern loads were simulated (e.g., 10K, 100K, or 1M) was not evaluated because of the limited budget of this evaluation task.

LOCST execution time

In addition to providing high test coverage, a self-test technique should execute in a relatively short period of time. Table 1 presents the equation for calculating LOCST execution times and the predicted test times for the three FSD chips. For the assumed scan rate (based on existing FSD scan controllers) and the number of self-test sequences (based on the presented test coverage evaluation), subsecond execution times are achieved for all three chips.

If a large number of random-pattern loads is required to achieve adequate test coverage results, if the scan rate is slow (e.g., 1 MHz or less), or if a chip contains a large number of SRLs, LOCST self-test times may become quite large (minutes). An alternative to the basic LOCST implementation is to use many parallel scan strings feeding a multiple-input signature register, or MISR. This modification, illustrated in Figure 9, reduces the number of serial shifts required to fill all chip SRLs with random test data, thereby reducing the overall LOCST test time.

Self-test environments

One of the greatest potentials of self-test is the possibility of eliminating the need to produce a unique test pattern set for each test environment. The major test environments are

- chip manufacture test,
- card test,
- operational system test, and
- field return test (repair test).

The lack of defect diagnostic information is the key reason that self-test is not considered a viable technique for chip manufacture testing. But ongoing research is investigating the use of self-test techniques for LSI devices in the chip manufacture environment. A very promising technique using random-pattern testing for diagnosing failures has been developed by F. Mo-

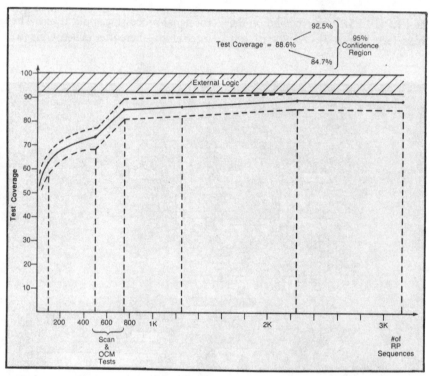

Figure 7. Test coverage results for Chip A.

Table 1.
LOCST test time.

	NO. OF RPs	NO. OF SRLs	TEST TIME
Chip A	2K	213	0.43s
Chip B	500	230	0.12s
Chip C	3K	223	0.67s

Test time = (No. of RPs/scan rate) × No. of SRLs
Scan rate = 1 MHz
RP = random-pattern sequence

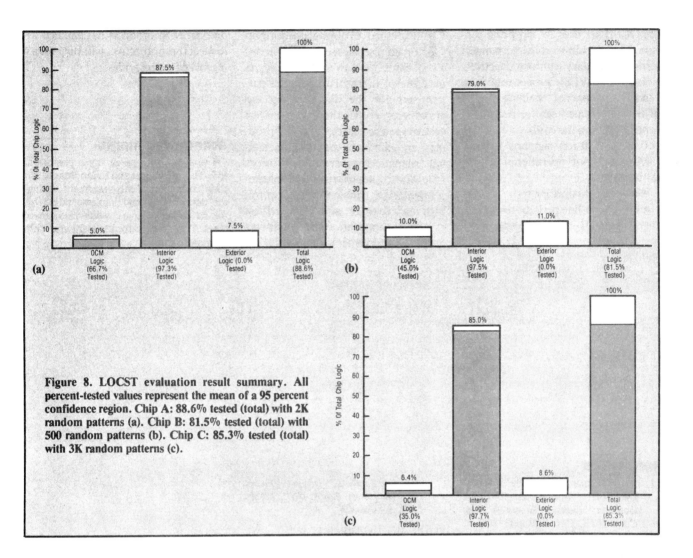

Figure 8. LOCST evaluation result summary. All percent-tested values represent the mean of a 95 percent confidence region. Chip A: 88.6% tested (total) with 2K random patterns (a). Chip B: 81.5% tested (total) with 500 random patterns (b). Chip C: 85.3% tested (total) with 3K random patterns (c).

tika et al.[13] of IBM Kingston. Presently, LOCST does not replace LSSD testing in the FSD chip manufacture test environment but is used as a supplemental chip-testing technique. As a minimum, since it provides a rapid pass-fail indication, self-testing would be useful in a production test environment to provide efficient preliminary screening of product.

The inclusion of several 10,000-gate VLSI components onto cards that have historically contained 5000 to 8000 gates of logic posed a serious problem to traditional card test methodologies. On-chip self-test offers a very effective solution. LOCST is used to verify that the FSD VLSI components on a card are defect-free. All FSD VLSI components are accessed via their OCM interface, requiring only seven card connector pins. Chip boundary scan latches (accessed via

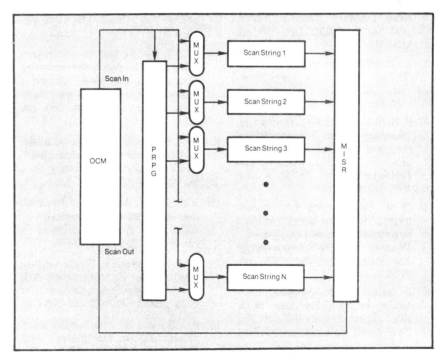

Figure 9. LOCST modification for faster execution.

November 1984 463

the OCM) are used to apply and capture data for chip-to-chip interconnect testing. Boundary scan also effectively isolates FSD VLSI components from vendor components, enabling the use of traditional methods for testing the vendor logic on the card.

On-chip self-test supports the following types of operational system testing:
- system initialization test,
- system on-line periodic test, and
- system off-line fault localization test.

The objectives of implementing an on-chip self-test capability in our VLSI chips were to substantially reduce the plethora of unique test pattern sets for the differing test environments, reduce the volume of test vectors required to test our VLSI products, and eliminate the need for manual test pattern generation. Priorities of low hardware overhead, simple implementation, simple self-test control, high test coverage, and short self-test execution time were of prime importance. The implementation of LOCST on our VLSI products has enabled us to meet these objectives without violating any of the priorities. □

Acknowledgments

I would like to express my appreciation to E. B. Eichelberger and E. Lindbloom for their counsel and encouragement during the development and implementation of this self-test technique. I would also like to thank Tina Nguyen for her assistance with the fault simulation activity.

References

1. B. Koenemann, J. Mucha, and G. Zwiehoff, "Built-In Logic Block Observer," *Digest of Papers 1979 Test Conf. IEEE,* Oct. 1979, pp. 37-41.
2. R. M. Sedmak, "Design for Self-Verification: An Approach for Dealing with Testability Problems in VLSI-Based Designs," *Digest of Papers 1979 Test Conf. IEEE,* Oct. 1979, pp. 112-120.
3. E. J. McCluskey and S. Bozorgui-Nesbat, "Design for Autonomous Test," *IEEE Trans. Computers,* Vol. C-30, No. 11, Nov. 1981, pp. 866-875.
4. E. B. Eichelberger and E. Lindbloom, "Random-Pattern Coverage Enhancement and Diagnosis for LSSD Logic Self-Test," *IBM J. Research and Development,* Vol. 27, No. 3, May 1983, pp. 265-272.
5. T. W. Williams and E. B. Eichelberger, "Random Patterns Within a Structured Sequential Logic Design," *Digest of Papers 1977 Semiconductor Test Symp. IEEE,* Oct. 1977, pp. 19-26.
6. P. Bardell and W. McAnney, "Self-Testing of Multichip Logic Modules," *Digest of Papers 1982 Int'l Test Conf. IEEE,* Nov. 1982, pp. 200-204.
7. D. Komonytsky, "LSI Self-Test Using Level Sensitive Scan Design and Signature Analysis," *Digest of Papers 1982 Int'l Test Conf. IEEE,* Nov. 1982, pp. 414-424.
8. E. B. Eichelberger and T. W. Williams, "A Logic Design Structure for LSI Testability," *J. Design Automation and Fault Tolerant Computing,* Vol. 2, No. 2, May 1978, pp. 165-178.
9. D. C. Bossen and M. Y. Hsiao, "ED/FI: A Technique for Improving Computer System RAS," *Digest of Papers 11th Ann. Int'l Symp. Fault-Tolerant Computing,* June 1981, pp. 2-7.
10. R. A. Frohwerk, "Signature Analysis: A New Digital Field Service Method," Hewlett-Packard Application Note 222-2, pp. 9-15.
11. J. E. Smith, "Measures of the Effectiveness of Fault Signature Analysis," *IEEE Trans. Computers,* Vol. C-29, No. 6, June 1980, pp. 510-514.
12. D. K. Bhavsar and R. W. Heckelman, "Self-Testing by Polynomial Division," *Digest of Papers 1981 Int'l Test Conf. IEEE,* Oct. 1981, pp. 208-216.
13. F. Motika, et al., "An LSSD Pseudo Random Pattern Test System," *Int'l Test Conf. 1983 Proc. IEEE.,* Oct. 1983, pp. 283-288.

Johnny J. LeBlanc is a staff engineer working in the System/Subsystem Testability department of the IBM Federal Systems Division in Manassas, Virginia, where he is responsible for the development of VLSI design for testability methodologies, including VLSI test generation, built-in test, and on-chip self-test.

LeBlanc received his BS in electrical engineering from the University of Southwestern Louisiana in 1976 and an MS in electrical engineering from North Carolina State University in 1978.

Questions concerning this article may be addressed to LeBlanc at Bldg. 400/044, IBM Federal Systems Division, 9500 Godwin Drive, Manassas, Va. 22110.

Chapter Six: VLSI Test Systems and the Future

During the design phase, the primary concern is to validate the logical design and its implementation. Emphasis is on design validation rather than on proving or disproving the existence of a fault. This emphasis changes once the design reaches the manufacturing or field-use phase. The emphasis in the latter phases is on testing for faults. Following a discussion of VLSI test systems, we will look at the emerging expert-systems approach to the computer-aided design and test of integrated circuits.

6.1 VLSI Test Systems

Each type of tester is known generically as an automatic test equipment (ATE). The complete test system (including software support, management system, etc.) is known as an automatic test system (ATS). Traditional ATE systems cannot accommodate newly designed, technologically advanced devices with large pin counts and high data rates. Advanced 16-bit and 32-bit microprocessors will grow in pin count size to more than 100 pins, will operate at data rates up to 30 MHz, and will have gate delays as short as 1.5 nanoseconds. Even faster bipolar and CMOS gate arrays and programmable array logic (PAL) devices will have up to 256 pins and data rates higher than 50 MHz. For test rates above 40 MHz, the test system must utilize 100K ECL circuits. At the present time there are no general purpose test systems operating above 100 MHz. This limit is set by the access time of the memories used for storing the binary test patterns.

VLSI test systems that can simultaneously test on two or three test heads are definitely more cost effective. Within each test head, it is important to have multiple, high speed, DC voltage and current measuring circuits, since DC measurements can restrict the testing throughput of large pin count IC's. Costs can also be reduced by software that makes a VLSI test system easy to program. For a detailed discussion of the economics of test systems the reader is referred to the paper by Bowers and Pratt [1].

The test equipment industry is faced with three major challenges. One formidable task is to produce testers operating up to 500 MHz which will be required for future silicon and GaAs VHSIC's. Another challenge is the test problem presented by wafer-scale silicon IC's. The third fundamental problem is due to the inductance, capacitance, and transmission line reflections that exist between the IC under test and the pin driver and detection circuits housed in the test heads.

In order to meet these challenges, IC's designed for laser repair, easier testability, or self-test, will become mandatory. Contact-free laser techniques [2] may become the dominant measurement method for characterizing GaAs IC's operating at rates above 1 GHz.

6.2 Expert Systems for VLSI CADT

Artificial intelligence (AI) techniques offer one possible avenue toward the development of new CADT tools to handle the complexities of future VLSI systems. One technique that has arisen from research in AI is the knowledge-based approach to designing a system for a particular function. The essence of this approach is to ask what knowledge—what facts and reasoning abilities—a human expert uses in a task and to develop data structures and code that represent this knowledge explicitly. This contrasts with systems that possess such knowledge only implicitly [3,4,5,6].

The knowledge-based approach to VLSI CADT has the following interrelated advantages over more traditional techniques for organizing software:

- It is easier to incrementally improve the system's capabilities.
- It is easier for the system to explain what it is doing and why.
- It is easier for a human expert to determine what is incorrect or incomplete about the system's knowledge and explain how to fix it.
- It is easier to interactively use a human expert's abilities.

Several knowledge-based VLSI CADT systems have been implemented. For example, Weaver [7] is a channel/switchbox routing program, that simultaneously considers all the important routing metrics. It consists of a set of knowledge-based expert modules organized around a communicating medium called a "black board." Each expert decides, based on its own knowledge and metric criteria, what should be done next. A "focus of attention expert" decides which expert should be allowed to give advice at a given time.

As a second example, Talib [8] synthesizes NMOS layouts for cells of as many as 100 transistors, given a netlist and cell boundary constraints. Talib employs hierarchical planning and successive refinement to minimize the search space. It recognizes design situations and applies common topological structures stored in its knowledge base. Two projects are under way at Carnegie-Mellon University to combine Talib, Weaver, and Mason (a heuristic floor planner) into a complete system for physical design. These projects are knowledge-based expert systems, one is interactive while the other is automatic.

An example of an interactive knowledge-based consultant for VLSI design, is the VLSI Expert Editor (Vexed) which is under development at Rutgers University [9]. It provides interactive aid to the user in implementing a circuit given its functional specification. Initially, the circuit is represented as a single "black box" module, whose functional specifications are typed by the user. Together, the user and Vexed refine this top level circuit module into submodules, representing the major functional blocks of the circuit. Each submodule is in turn refined into sub-sub-modules, and so on, until the design is finished.

A second current thrust of knowledge-based systems is the development of an intelligent aid in debugging VLSI circuits. When the circuit does not perform correctly, the task is to determine whether the failure is due to a design or manufacturing error and to localize the cause of the failure. The system should generate and rank hypotheses on possible sources of the circuit failure by reasoning back from output failure symptoms to plausible internal faults.

For further reading on "AI Techniques in Design and Test" the reader is referred to the recent volume of Artificial Intelligence (North Holland, Volume 24, 1984), and the forthcoming special issue of Design & Test (IEEE, Volume 2, No. 4, August 1985).

6.3 Reprints

The basic purpose of an ATE system is to drive the inputs and to monitor the outputs of a device under test (DUT). ATE speed and timing accuracy must cause DUT input transitions and output strobes to occur precisely at preprogrammed times [10]. A successful ATS design with advanced performance levels requires significant contributions from all aspects of hardware and software technologies. The first three reprinted articles in this chapter describe the current state-of-the-art in automatic test systems design.

The potential subnanosecond timing measurement accuracy of modern VLSI test systems are presenting serious problems. The reprinted paper by Barber shows that there is an urgent need for test system manufacturers to develop 256-pin test heads that can place high impedance comparator circuits within 10 cm of the MOS devices under test.

The future of VLSI design is superbly described in the reprinted paper by Séquin. It is evident that complexity management schemes, expert systems, and new integrated circuit technologies will affect the VLSI design scene in fundamental ways. The applicability of knowledge-based systems in the areas of test generation and design-for-testability is shown in the paper by Bending [11] and the reprinted paper by Horstmann, respectively. Diagnostic reasoning for troubleshooting of digital systems is described in the reprinted paper by Davis.

References

[1] G.H. Bowers, Jr., and B.G. Pratt, "Low-Cost Testers: Are They Really Low Cost?," *IEEE Design & Test*, Vol. 2, No. 3, June 1985, pp. 20–28.

[2] R.K. Jain, "Picosecond Optical Techniques Offer a New Dimension for Microelectronic Test," *Test and Measurement World*, June 1984, pp. 40–53.

[3] W.S. Adolph and H.K. Reghbati, "Application of AI Techniques to VLSI Design," *SFU-CMPT Tech. Rep. 85-12*, 1985.

[4] H. Brown, et al., "Palladio: An Exploratory Environment for Circuit Design," *IEEE Computer*, Vol. 16, No. 12, Dec. 1983, pp. 41–56.

[5] M. Stefik and L. Conway, "Towards the Principled Engineering of Knowledge," *AI Magazine*, Vol. 16, No. 3, 1983, pp. 4–16.

[6] H.K. Reghbati and N. Cercone, "Knowledge Representation for Computer-Aided Design and Test of VLSI Systems," in *Knowledge Representation*, Springer Verlag, 1986 (to appear).

[7] R. Joobbani and D. Siewiorek, "Weaver: A Knowledge-Based Routing Expert," *Proc. of IEEE Design Automation Conference*, 1985, pp. 266–272.

[8] J. Kim and J. McDermott, "Talib: An IC Layout Design Assistant," *Proc. National Conference on Artificial Intelligence*, 1983, pp. 197–201.

[9] T.M. Mitchell, et al., "A Knowledge-Based Approach to Design," *Tech. Rep. LCSR-TR-65*, Rutgers University, 1985.

[10] R. Garcia, "The Fairchild Sentry 50 Tester: Establishing New ATE Performance Limits," *IEEE Design & Test*, Vol. 1, No. 2, May 1984, pp. 101–109.

[11] M.J. Bending, "Hitest: A Knowledge-Based Test Generation System," *IEEE Design & Test*, Vol. 1, No. 2, May 1984, pp. 83–92.

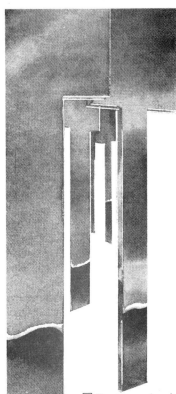

SPECIAL REPORT

VLSI test gear keeps pace with chip advances

More powerful test equipment copes with faster speeds, higher pin counts of very large-scale integrated circuits

By Howard Bierman, *Senior Editor*

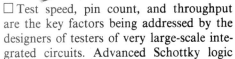

☐ Test speed, pin count, and throughput are the key factors being addressed by the designers of testers of very large-scale integrated circuits. Advanced Schottky logic and emitter-coupled-logic arrays demand 50-megahertz and higher speeds; increased gate counts demand increased pin counts; and, since throughput is determined by the number of functional tests required to exercise logic, denser chips demand longer test times. Such requirements call for significant advances in the tester state of the art.

Today's VLSI testers can handle up to 300-pin channels, provide complex levels of functional testing, and make provision for interfacing with computer-aided-design and -test equipment. Since a primary function of the VLSI tester is to time-test patterns, signal-timing relationships can be defined precisely with better than 1-nanosecond accuracy. And signal timing is program-selectable and can be changed in real time from one test cycle to the next, up to a maximum cycle rate of 100 MHz.

Fast-moving field

VLSI testers are experiencing a rapid and wide-ranging pace of development, just as the chips they are testing are evolving rapidly. This review of the major players in the field is accompanied by two other articles. Beginning on page 129, Eugene Hnatek, of Viking Laboratories Inc., Mountain View, Calif., delineates the technological trends shaping VLSI testers. Then, starting on page 135, Dean Johnson, of Fairchild Camera & Instruments Corp.'s San Jose, Calif., Analog Test Systems Group, presents a solution to the problem of testing high-performance chips that mix analog and digital signals.

As these articles make clear, the new generation of digital test systems has focused on the pressing issue of design-for-testability. One object of design-for-testability is to reduce programming and test times; another objective is to limit the skyrocketing costs and excessive time needed for VSLI testing. Currently available digital VLSI testers follow the traditional shared-resources architecture in equipment supplied by Tektronix Inc., Cybernetics Technology Inc., GenRad Semiconductor Test Inc., Teradyne Inc., Fairchild's Digital Test division, Takeda-Riken America Inc., and Ando Corp. However, a new test-per-pin architecture was introduced last fall by Megatest Corp. at the International Test Conference. A comparison of key performance specifications for eight major VLSI testers is shown in the table.

A typical shared-resource architecture (Fig. 1a) includes a master clock generator, a number of timing generators (generally fewer than 20) followed by a complex switching matrix to distribute timing signals to waveform formatters, and pin-electronics drivers and comparators. Since a large number of different paths are possible through the switching matrix for ICs with high pin counts, signal delay (or skew) varies for different input/output pin combinations, making calibration or deskewing difficult. Another serious obstacle with shared resources is the lengthy programming time that is required to schedule the routing of timing signals through the switching matrix.

A straightforward arrangement is to provide every pin of the chip to be tested with its own testing resources, or test-per-pin architecture (Fig. 1b). Thus each pin is supplied with a programmable high-speed timing generator, waveform formatter, dc parametric unit, pin driver, pin comparator, and programmable current load. Since there is no longer a need to switch signal routing, greater accuracy is possible. Also, software is simpler and faster to develop, and the tester-per-pin modular structure permits higher pin-count sections to be added conveniently. Shared-resource architecture is less costly, obviously, since less hardware is demanded, but the tester-per-pin approach seems to become more appealing as the complexity of the chips being tested increases.

On-chip testing has surfaced as a viable way of testing VLSI ICs more effectively and will have to be taken into account by new generations of test gear. Called by various names, including scan-path, serial-scan, and level-

Reprinted from Electronics, April 19, 1984. Copyright © 1985, McGraw-Hill Inc. All rights reserved.

SPECIAL REPORT

sensitive-scan-design (LSSD), the technique structures the logic so that its response is independent both of the order in which inputs change and circuit delays between logic elements. Thus the IC is converted from a time-dependent sequential circuit to a combinational circuit. The key advantage is that combinational circuits are responsive to truth-table analysis, while sequential circuits are not, and truth tables can be generated quickly and efficiently by large mainframe computers. Although roughly 20% more silicon is required on the chip for the added test circuits, more than 98% of a chip design can be checked using the serial-scan technique.

Now available

The key to the 100-MHz test operand speed in Cybernetics Technology's Viking 200 VLSI tester is the high-speed computer architecture and manufacturing know-how acquired as a sponsored spinoff of Control Data Corp. The Viking 200, now scheduled for delivery from the Eden Prairie, Minn., company in late fall, will handle up to 256 I/O pins at rates of up to 100 MHz, with multiplex capabilities up to 200 MHz for emitter-coupled-logic characterization. For the purpose of production testing, a 50-MHz Viking 100 is being readied with 20-picosecond timing accuracy and 1-ps precision, capable of handling as many as 128 pins.

As shown in Fig. 2, the testers in the Viking series are composed of three basic sections: a system-control computer, a local network serving as the interconnecting test bus, and one or more test stations. The system controller, which is either a Digital Equipment Corp. VAX-11/730 or VAX-11/780, handles data management, communications, programming capability, and storage of test programs sent to the test stations over the local net. In addition, the system controller performs data analysis on test results.

A second level of computer control is contained at each test station, thus allowing it to operate independently of the system controller once it has been loaded with appropriate test programs. Since different circuit technologies dictate different signal and impedance levels, the pin-electronics section at each test station provides high voltage for TTL and MOS unipolar chips and low voltage for ECL bipolar parts. Instead of mechanical relays, electronic switching is used to provide faster and more reliable testing.

Test stations may be programmed in several languages, allowing users to select a familiar language or one most suitable for a specific test application. A common interface allows a routine written in one language to be called from a program written in another. Languages supported include the Abbreviated Test Language for All Systems (Atlas), Pascal, Fortran, and Comprehensive Tester Application Software (CTAS). The latter language enables inexperienced programmers to create additional test sequences in any of the supported languages, to rearrange the order of sequences, and to modify test specifications where necessary.

The S-3295, from Tektronix of Beaverton, Ore., is a 20- or 40-MHz VLSI test system supporting up to 256 pins and configured as 128 input and 128 output pins, which can be combined in pairs as I/O pins. To handle the diversity of logic types, pin-electronics cards contain provisions for testing MOS, TTL, and ECL ICs, as well as hybrid chips with mixed-logic families.

A high-speed pattern processor in the S-3295 is capable of compressing and recreating functional patterns during a test run, which reduces the system's storage requirements; in addition, other patterns can be switched in or out on a cycle-by-cycle basis. The pattern processor feeds test vectors to each IC pin by the pin-electronics cards. Each pin card provides two input and two output channels and handles I/O switching, output loading, and error recording. Each pin-card data channel is provided with 64-K of local memory, or 256-K per card.

Since fast ICs tend to have fewer pins than slower units, the S-3295 takes advantage of a multiplexing technique to combine the data streams of two adjacent drivers on a pin-electronics card. Although the number of driver circuits is cut in half, two driver pulses can be inserted within one cycle, thus exercising the device under test at double the standard test-system rate, or up to 40 MHz.

The DUT outputs are furnished with programmable active load circuits to allow ICs to be tested under actual operating voltages and currents. For ECL chips, a 50-ohm terminating resistor is connected and the programmable load is removed.

Skewing, or a shift in timing between two signals that should be coincident, is the result of propagation and cable delays, as well as stray capacitance along the signal paths. To overcome this deficiency, the S-3295 uses look-up tables to correct skew differences on every pin card;

SUPPLIERS OF TEST EQUIPMENT FOR VERY LARGE SCALE INTEGRATED CIRCUITS

Manufacturer	Model	Maximum speed (MHz)	Maximum number of pins	Timing accuracy (Ps)	Timing resolution (Ps)	Estimated selling price ($ millions)
Cybernetics	Viking 100	50	128	20	1	0.5–0.9
	Viking 200	100	256	20	1	0.9–1.9
GenRad	GR-18	40	288	900	125	0.9–2.2
Takeda Riken	3340	40	256	800	125	0.8–2.2
Tektronix	S-3295	40	128	500	100	0.8–1.3
Teradyne	J941	40	96	1,000	100	0.6–1.2
Megatest	Mega One	40	256	700	100	1.2–2.7
Ando	DIC 8035B	40	256	500	100	1–1.6
Fairchild	Sentry 50	50	256	600	40	0.85–2

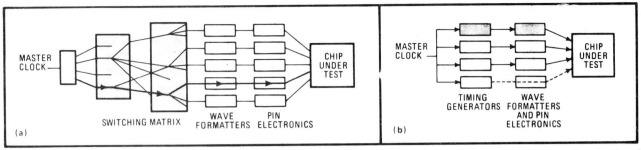

1. Different strokes. Traditional VLSI testers (a) feed a limited number of timing generators through a complex switching network to the pins of the chip being tested. In the alternative, tester-per-pin architecture (b), each pin is supplied with its own resources.

this produces pin-to-pin driver and comparator skew figures on the order of 500 ps. For testing serial-scan chips, the S-3295 includes an optional pattern generator backed by a 3-megabit serial memory (1 Mb each for force, compare, and mask).

Separate resources

A bold approach to VLSI testing, taken by Megatest Corp., San Jose, Calif., provides independent resources behind each pin rather than sharing resources, as commonly done in traditional tester designs. Critics of the Megatest concept argue that equipment costs are quite prohibitive and claim that the tester-per-pin architecture is a case of overkill. But proponents of the concept point to significant savings in software development costs and, more important, to the substantially faster test-program development times, which allow state-of-the-art chips to be brought to market sooner.

To test a logic chip, the Mega One [*Electronics*, Nov. 8, 1983, p. 101] feeds it signals simulating the voltage, current, and timing waveforms and states that it would encounter in its intended application. It then compares the output responses with the values anticipated by the designers. Shared-resource architecture makes use of time generators routed though a switching matrix to the specified pins of the DUT, where waveform conditioning creates appropriate test signals for each pin. For small-pin-count devices, shared resources are effective; for VLSI ICs with high pin counts, the test programmer becomes heavily burdened with such hardware details as relay closures, settling times, and path routing. In addition, as the switching matrix becomes more complex, skewing correction becomes more difficult and costly.

Megatest's Mega One provides each pin of the DUT with its own programmable timing generator, waveform formatter, dc parametric tester, pin driver, pin comparator, and programmable current load. The only high-speed signal routed though the test system is the system clock, distributed to all sections in parallel, without the need for a switching matrix.

Thus signal paths are short and dedicated, allowing precalibration of each pin individually. Since each pin is supplied with its own independent resources, test systems handling up to 512 pins are feasible. Timing and waveform shapes can be changed on the fly at a 40-MHz rate. The tester can operate at 40 MHz on any of its 256 pins and at 80 MHz on designated pins. Software is based on AT&T Bell Laboratories' Unix 4.2 operating system, with test programs written in standard Pascal using Megatest-supplied functions and procedures.

New entry

Early this year, Fairchild's long-awaited Sentry 50 tester was introduced, capable of a 50-MHz test speed (100 MHz multiplexed) with up to 256 I/O pins per test station. The system's distributed architecture is managed by a VAX/11-730 and a VMS operating system: up to 5 megabytes of main memory and 121 megabytes of disk storage, which can be doubled by adding a second disk, are available.

A key design feature of the San Jose, Calif., Digital Test division's Sentry 50 is the local memory section, which has a 64-K-word interleaved main memory and a 1-K-word high-speed subroutine memory. Multiple functional parameters can be passed from main memory to

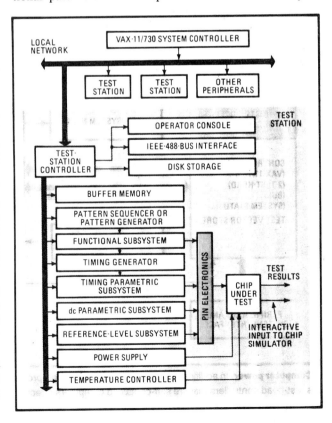

2. 100-MHz test station. A DEC VAX-11/730 takes care of data management, communications, and storage of test programs to be routed to the test stations of the Viking 200 tester. Over 100 taps along a one-mile run can be handled by the 10-MHz local network.

subroutine memory by means of a parameter-enable memory. On a cycle-to-cycle basis, this memory determines whether a particular pin is fed functional data from a subroutine memory as constant data or from the main memory as a parameter. Subroutines can be written to compress considerably the vector memory space required for bus-oriented parts, like microprocessors.

Pascal/50, a superset of standard Pascal, is the test-programming language and supports user-defined lists of tester pins. Three Pascal programming techniques are provided to optimize computational loading and cut communications tasks between the test-system and test-head controllers: data packets, elemental procedures, and pin lists. These techniques are said to raise throughputs by as much as three orders of magnitude. To test ICs housing test circuitry, such as the LSSD scheme, a serial-data memory is optional. This reconfigurable memory can generate serial-bit streams at a 50-MHz rate and can be programmed to provide one, two, four, or six channels, which can be routed to any tester pin.

Competitors abound

GenRad's latest entry in VLSI testers is the GR-18, built around a DEC PD-11/44 with an unmodified RSZ-11M operating system. The 40-MHz system from the Milpitas, Calif., subsidiary of GenRad Inc. can test and characterize a variety of technologies, including C-MOS, TTL, and ECL, as well as multichip modules with mixed technologies (such as MOS and TTL) and hybrids with mixed analog and digital functions. Up to 288 pins can be tested using two test heads programmed to operate in a combined mode; each test pin is provided with drive, compare, and load located in the test head. Either 16- or 12-system timing phases, with 125-ps resolution, are available on the fly.

Included in the GR-18 is a serial-data generator for efficient storage of long patterns used with scan-design and bus-oriented chips. A user-interface system provides test operators with easily understood menus to control the various systems modes. Network interfacing is by GenRad's GRnet, consisting of a pair of coaxial lines operating at 655-K-bytes per second.

Two Japan-based test-gear manufacturers supplying 40-MHz 256-pin units for VLSI chips are Takeda-Riken America, of Englewood Cliffs, N. J., with its Advantest T3340, and Ando of San Jose, Calif., with its DIC-8035B. Takeda-Riken's T3340 includes two types of pin electronics for either high-speed ECL chips or a high-voltage driver, as well as a high-impedance comparator for MOS and TTL circuits. A four-level automatic-calibration subsystem guarantees test-system accuracy to within 800 ps for the high-speed test station and 1.6 ns for the high-voltage station.

Each of the test pins can operate in any one of 24 drive modes with full waveform-format control, real-time sense control within a test cycle for microprocessor testing, and multiplexed I/O for memory testing. The data-clock rate can be selected in 1-ns increments from 1 kilohertz to 40 MHz, and the timing clock can be set within 125-ps resolution. The clock rate and each of the 32 phase clocks are selectable on the fly for 16 levels of timing values.

For LSSD or scan-path testing, the T3340 can generate primary patterns and scan vectors. Testing is done using an optional superbuffer memory to rewrite portions of the stimulus-and-expected-vector buffer memory. A hold mode is used to store the primary patterns while the serial pattern is being applied and exercised at the chip.

Ando's DIC-8035B 40-MHz system improves throughput with processors for parallel testing of two chips at the same time, the ability to perform up to eight simultaneous dc measurements using analog switching, and a 2.75-Mb/s program transfer speed. The resolution of the timing generator operating from a 500-MHz basic clock is 1 ns; the test rate varies from 1 to 25 ns.

Improved software is the area that Teradyne has pursued since the introduction of its J941 40-MHz tester several years ago. The latest extension to the Woodland Hills, Calif., Semiconductor Test division's Test Analysis Program (TAP) simplifies debugging by its ability to display pattern-generation memories and to allow the user to alter the test program through expanded symbolic editing capabilities. The user can review the contents of pattern memories in binary, octal, or user format and modify programmed formats and timing values. An automatic edge-lock program monitors edge-timing accuracy and automatically compensates for different channel propagation delays. □

3. Computer power. In addition to the VAX-11/730 distributed-processing unit, the Sentry 50 has test-head controllers that are self-contained computers directed by 68000 microprocessors. Up to two test heads can be supported for 256-pin capability.

A merger of CAD and CAT is breaking the VLSI test bottleneck

New computer-aided-design and -testing techniques ensure that chips are testable and cut the time needed to generate and execute test programs

by Eugene Hnatek, *Viking Laboratories Inc., Mountain View, Calif.*

☐ Testing has emerged as the biggest recurrent cost in the production of very large-scale integrated chips and as the bottleneck in the forward march of chip complexity and new-product development. For the complexity created by VLSI transistor counts and circuit functions has dramatically raised the amount of time needed to design and generate test programs.

Meanwhile, over the past 10 years there have been great advances in chip packaging and front-end wafer processing, which can at last be fully automated. Testing cannot—neither program generation nor design-for-test methods. In an effort to overcome this bottleneck, test engineers are now developing new techniques to ensure that chips are testable, to cope with their increasing complexity, and to reduce the amount of time needed to generate and execute test programs.

VLSI chips will require a new thought process. No longer isolated circuits, they are now systems-on-a-chip and geared to specific applications. Semicustom and custom circuits will coexist with a few standard integrated circuits, such as array processors, fast multipliers, signal-conditioning circuits, and combinations of memories and microprocessors. But most VLSI circuits will be custom in nature and characterized by many different designs (circuit, processing, and layout), low quantities of parts, sequential on-chip circuits, rapid change, a need for flexible and comprehensive test programs, high density, high pin (input/output) counts, and high speed.

The roles of design, systems, and test engineers are therefore changing. Design engineers must have access to technology menus listing complete systems-on-a-chip. They and design and test engineers will have to design test menus for these systems in partnership, concurrently with the circuit design, if these systems-on-a-chip are to be available in standard libraries.

To help users realize their designs quickly and efficiently, most major IC vendors have established remote design centers throughout the world. Many systems houses are also creating an internal capability for designing ICs and developing software. They will use IC vendors merely as silicon foundries to implement their designs in silicon, and this exponentially complicates the problem of generating tests. Each design will require its own test program, which must be compatible with the vendor's automatic test equipment, as well as the user's.

VLSI chips are being designed by groups of engineers, each working on a part of the chip. The exchange of data from a common data base is essential. To achieve the desired system design, VLSI circuits will be implemented

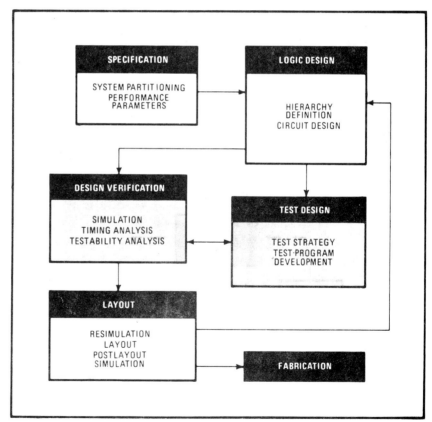

1. CAD/CAT merge. Interaction between the VLSI design and test groups, mapped out in this flow chart of an interleaved CAD-and-test-program system, must occur in the early specification stage in order to optimize test strategies and ensure device testability.

Electronics / April 19, 1984

Reprinted from Electronics, April 19, 1984. Copyright © 1985, McGraw-Hill Inc. All rights reserved.

with the best design and processing technology. They will contain mixed functions, both digital and analog, fabricated with one or more technologies: n-channel MOS, complementary-MOS, vertical-gate MOS, and bipolar processes. These realities also complicate test-program generation and call for a general-purpose mixed-technology tester.

Beyond redundancy

To increase yields and system reliability, extra elements (redundant circuits) will be common in chip designs. Externally, users of redundant chips cannot detect whether or not they have been repaired with redundant elements, so effective testing is impossible. To help users test these chips, vendors will have to use a silicon-signature technique to identify the redundant circuits used to repair a given circuit. Besides redundancy, VLSI chips will have either a design-for-testability or a built-in-test capability. External parts will be available for this and for on-chip error correction.

Such problems as escalating chip complexity and diversity will surely have to be solved through the clever coordination of automated work stations. Computer-aided design, manufacture, and test will all be required. Through mainframe computers, CAD systems used for chip design will have to be related intimately to the system generating automatic test programs. Ideally, the mainframe should have a resident translation program that would allow the automatic-test-generation program to create test vectors for any commercially available ATE the vendor may own. Figure 1 shows a flow chart of an interleaved CAD-and-test-program system.

Testing any logic chip involves simulating the voltage, current, and timing environment that it would find in a real system, sequencing it through a series of states, and then checking actual against expected responses. Testability is the main issue. In order to come up with a design characteristic that lets a chip's status be determined with confidence and with speed, all of the inputs must be controlled simultaneously, and many of the outputs must then be observed simultaneously, too.

Structural method

To make it easier for users to develop test programs, they will take a more application-oriented, as opposed to component-level, approach. In this structural (or menu-driven variation on a theme) method, the test engineer generates a skeleton, or master, program and then lets the CAD system develop specifics for multiple patterns. This technique reduces test-program development time for such generic product types as gate arrays that have hundreds and thousands of custom gate interconnection patterns for products of a given level of density. The test engineer first generates a test program for a family of chips and then goes on to adapt specific ones to fit the general concept.

It would be desirable if vendors provided source codes or test vectors, which dramatically cut the time needed to develop test programs but still take up engineering time—to make the vendor's software compatible with the user's ATE. For example, Digital Equipment Corp., in Maynard, Mass., receives 30% to 40% of its source codes from IC vendors. Because programs for different product lines have different formats, the company's engineers must still spend an average of 100 man hours per generic chip type to debug these source codes and adapt them to the DEC environment. Table 1 indicates typical development times for manually generated test programs.

The challenge of testing VLSI chips must first be approached with the attitude that the chip is basically untestable. Solid engineering decisions have to be made about what to test, what not to test, and what depth to test within the constraints of material, capital, and personnel resources. It is no longer possible to review a chip-specification or data sheet, define all possible test conditions, and implement that test.

Test generation involves a search for a sequence of input vectors that cause relevant faults to be detected

TABLE 1: DEVELOPMENT TIME FOR MANUALLY GENERATED TEST PROGRAMS

Chip type	Number of on-chip transistors	Test program development time (h)
Small- and medium-scale integrated circuits	200	120
Peripheral chips	2,000	320
Microprocessors	20,000	1,000
Very large-scale-integrated microprocessors	200,000	3,300

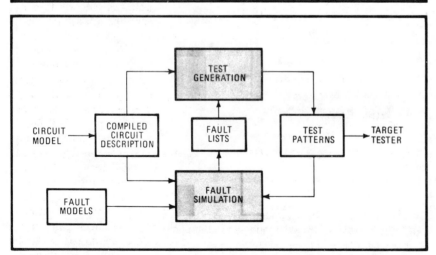

2. Design test. A typical computer-aided-test (CAT) system includes a test generator and fault simulator applied to a model of a VLSI circuit developed at the gate level. This figure represents the test-design block in Fig. 1.

on the primary chip outputs. A complete set of functional vectors does not imply an adequate test, however. With VLSI chips, test generation is complicated by buried flip-flops, asynchronous circuits, complex clock conditions, indeterminate states, the needs of circuit initialization, and nonfunctional inputs.

Test generation has a twofold goal: verifying the design of a chip and detecting faults through fault simulation, and the analysis of a given circuit's operation under fault conditions. Chip designers must analyze, classify, model, and test the physical failures of VLSI chips, taking into account the processing technology used to make them. The principles both of physical (chip layout) and logical design-for-testability must be applied, so the resulting circuit will be testable and cost-effective.

The effectiveness of fault simulation, the most critical step in generating VLSI test programs, is measured by the ensurance that the test patterns generated are accurate and that all faults have been detected. The result is usually a full fault-coverage vector set causing long test times. Automating this process with a common design data base, as well as pattern and test-program compression, can cut test times.

A computer-aided-test system is a set of tools to generate and evaluate test sequences for a component and tester by using logic and fault-logic modeling techniques. Figure 2 shows a typical CAT system architecture, an expansion of Fig. 1's test-design block. Logic and fault models have been the only practical ways of generating test programs with exhaustive test coverage for LSI chips, while keeping the number of test vectors to manageable levels.

By tradition, both pattern generation and fault simulation have been based upon quite simple fault models of "stuck-at" nodes or gate inputs and short circuits between nodes. A test pattern is applied to a computer simulation of the logic network. One output from a single gate is simulated as though it had gotten stuck at 1 or at 0. In this case, if the output does not perform as it had done back when no faults were simulated, the stuck-fault test has surely detected a fault.

Many CAT tools are available. The nomenclature has not been standardized, so the tools are sometimes confused and sometimes assumed to be interchangeable. All test tools help create test vectors (the digital inputs and correct outputs of a circuit used to verify correct operation) or test programs (the actual software for the test computer that performs the automatic test). Any logic simulator can verify test vectors created by the logic designer, but fault coverage of the test is not known. Some logic simulators include fault modeling, which gives designers a count of the faults detected by test vectors. The designers can then devise additional vectors to catch undetected faults.

Other CAT tools are used to help create test

TABLE 2: TYPICAL COMPUTER-AIDED-TESTING TOOLS

Tool	Logic simulation	Fault-simulation/ grading	Automatic-vector generation	Program generation	Comments
Tegas	Yes	Yes	Minimal	—	Models ambiguous and high-impedance states
Logicap	Yes	Yes	—	—	
Lasar	Yes	Yes	Minimal	Yes	Compatible with Teradyne testers; mainly used for printed-circuit boards
Newsim 2	Yes	Yes	—	—	Addresses oscillating faults, models ambiguous and high-impedance states
CATS				Yes	
Hilo-2	Yes	Yes			Features hierarchical-design approach with functional modeling
Hi test		Yes	Yes	Yes	Compatible with GenRad, LTX, Fairchild, and Teradyne automatic test equipment
Scoap		Yes			Determines circuit's level of testability. Will be incorporated in Tegas 6

Note: All fault and logic simulators model logic - 1 and logic - 0 states.

Creating test programs manually

Engineers develop test programs for integrated circuits in a sequence of steps described below:

1. From vendors and other sources, learn IC organization and electrical specifications, including:
   - Functional description.
   - Logic functions.
   - Controlling signals to each pin of the IC.
   - Block diagram of the internal structure.
   - Pinout.
   - Timing diagram for each critical sequence.
2. Develop the test strategy.
3. Generate test code from documentation:
   - Observe signals and compare them with functional truth and logic diagrams.
   - Start with simple and basic functional and dc tests to confirm that ICs are behaving properly.
   - Include additional tests to exercise remaining IC pins.
   - Create more complex pattern and timing tests.
   - Debug and refine software.
4. Perform characterization testing to determine which patterns are effective in weeding out faults and which should be used in 100% testing.
5. Generate 100% inspection test programs.
6. Release test program.

programs. Typically, these tools translate from a higher-order test language into the detailed language of the particular computer the ATE uses. At the moment, however, there is no completely automated method of taking the logic diagram, which performs logic and fault simulation, as well as testability analysis, and then generating the ATE test vectors.

Table 2 summarizes the available CAT tools. Such logic simulators as Tegas and Logcap, the ones most widely used by IC designers, provide fault simulation, which can be used to grade manually generated test vectors but can be very expensive. (Simulating a 1,000-node network for 1,000 patterns costs $10,000 to $30,000.) The quality of fault coverage of the patterns depends entirely on the designer's skill.

The most noticeable feature of test programs produced by current CAT systems is the absence of pattern structure or intuitive meaning. To handle test problems with the complexity of current VLSI chips, CAT architectures

Design-for-test: making sure the VLSI chip can be tested

Design-for-test is the process of making a deliberate design effort to ensure that very large-scale integrated circuits can be tested thoroughly and successfully with minimum effort and cost. It is the most effective way of cutting test-development and -production costs.

Design-for-test for VLSI chips was much in evidence during the IC discussion sessions at 1984's International Solid State Circuits Conference, 1983's International Test Conference, and 1983's Fault-Tolerant-Computing Conference. At ISSCC, Toshiba introduced a 256-K complementary-MOS electrically erasable programmable read-only memory with on-chip test circuits; Texas Instruments described a VLSI communications processor designed for testability; and Siemens unveiled a 256-K dynamic random-access memory with redundancy test capability.

The 1983 Fault-Tolerant-Computing Conference featured three papers on built-in test, and 1983's International Test Conference included 15 papers on that subject. The University of Michigan has used on-chip generation of test sequences to produce a 64-K dynamic RAM that tests both the cell partitions and individual cells within an area of the cell array. The U. S. Government's Very High-Speed Integrated Circuit (VHSIC) program is pushing testability, evidenced by an allocation of more than $40 million for built-in

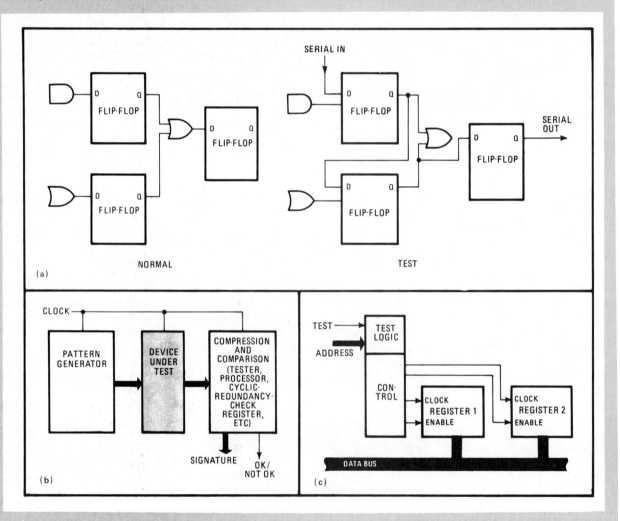

must develop techniques similar to those used by human experts (see "Creating test programs manually," p. 131). At present, no CAT system generates a full vector set without manual intervention.

Test engineers can generate tests that are compact in source form and make use of all the repetition and pattern-manipulation facilities of the ATE. Normally, these tests use more patterns than current CAT systems can generate. Test engineers can also generate tests for circuits much larger than those that current CAT systems can handle. Future CAT systems must contain reference libraries of testing techniques and solutions that can produce and implement test strategies. Such systems will be interactive, so that users can choose alternative algorithms for all stages of test production.

Level-sensitive-scan-detection and other scan techniques (see "Design-for-test: making sure the VLSI chip can be tested," below) promote economical design with low test-generation costs for LSI circuits. With current scan designs, the upper limit of CAT appears to be a few

---

and fault-tolerant test development. Recently introduced commercial ICs with design-for-test or built-in-test circuitry include Motorola's 6802, Intel's 2920, National Semiconductor's SCX series, Advanced Micro Devices' AM 29818, and Monolithic Memories' 54/74S818.

All design-for-test methods ensure that a design has enough observability and controllability to provide for complete and efficient testing. An observable node can easily be read from an IC output; that is, the user can easily determine that node's state. Controllability is the ease of controlling (or forcing) a node to a particular value. Complete testing of a logic network involves forcing every node to each logic state and verifying that the node "took" it.

Current design-for-test techniques include serial scan, level-sensitive scan detection (LSSD), signature analysis, unstructured design-for-test, and addressable registers. Scan-design approaches change the difficult problem of test generation for sequential circuits into the much easier task of generating tests for combinational circuits. They also reduce the need to run tests at system speed. With scan-set testing, more than 98% of a given circuit can be tested without resorting to more complex methods.

Serial-scan is a test mode that reconfigures all a logic network's flip-flops into serial shift registers, shown in (a). This design-for-test technique makes all flip-flop inputs quite accessible and all flip-flop outputs observable. Since the rest of the logic is only combinational, test vectors can be generated easily, though tediously. Since algorithms have been designed to create vectors to test combinational logic, the test-vector generation task can be computerized straightforwardly.

When serial scan is added to a logic network using a synchronous single-clock design, a multiplexer is needed on each flip-flop's input. In the worst-case—random-logic design with many clock sources—a second multiplexer and control line are added as well. In IC design, the added wiring associated with the control, clock, and serial string signals has an impact on chip area: an additional 5% for current-mode-logic designs and as much as an additional 50% for random-logic C-MOS.

LSSD is the serial-scan technique that is used by both Sperry and IBM. Instead of the flip-flop employed in serial scan, this technique makes use of a level-sensitive latch as the basic memory element. The serial chain requires the addition of a second latch and clock phase. Since both phases of the clock are independently controlled, users can test for on-chip delays by changing clock phasing and speed with LSSD. The major difference between it and serial scan is that LSSD is more applicable to MOS circuits, since the latch is implemented with only one transistor, and latch-type designs are more common.

The characteristics of the serial scan and the LSSD design-for-test styles are similar. The design-for-test rules are easy to apply, and if they are followed throughout the design, test vectors can be generated automatically and at a low cost compared with manual generation.

Signature-analysis techniques, shown in (b), involve the generation and application of many parallel test vectors, generated manually or randomly and applied to the device. As a result of the large number of patterns, all output vectors are not compared directly with good results, but rather compressed. The compressed result is then compared with the desired one. The technique can be implemented on chip for an effective self-test.

An addressable-register design-for-test, shown in (c), has the same objective as the serial-scan technique—easy access to all storage elements—but achieves it with a parallel bus rather than a serial-shift register. Each storage element (flip-flop, register, and so on) gets an address, so the tester can view all chip storage as elements in a memory with a specific location. The tester then has access to any storage element in one or two clock cycles. Logic must be added either in the control section or at each storage element to map the storage elements onto addresses. If the architecture has a central control block, this addition of logic can be provided for easily and with little area impact; if the logic must be added at each storage element, the area impact could be huge.

Addressable register characteristics are similar to serial scan. Vectors must be generated manually, although the job will be easier than it is for other design-for-test methods, since accessibility and controllability are good. Pin overhead is low and consistent with serial-scan techniques.

Unstructured (or ad hoc) design-for-test—conventional logic design with the later addition of design-for-test—is the dominant technique for today's IC designers. A few rules are applied, like requiring the logic circuit to reset to a known state, breaking long counters, and adding test nodes and pins where needed. In essence, designers must only satisfy their own requirements to create an initial vector set.

Test time is moderate to high, with typical vector counts of several thousand to 100,000. Pin overhead and area penalty are low, since only the features needed for testability are added.

SPECIAL REPORT

thousand gates. Scan paths can limit a circuit's performance, and they do not eliminate the need for real-time functional testing. "Requirements for CAT systems" lists the needs of CAT systems; "Problems with CAT" summarizes the shortcomings of the current CAT systems.

Software verification

Once a complete VLSI test vector set has come into existence, the test engineer must then verify that it does everything that it is supposed to do: exercise and test the chip in a way that is both accurate and complete. With test software that has been generated manually, every step (test vector) in the sequence is comprehensible and checked by the test engineer for errors that have been generated manually.

By contrast, the complexity of VLSI test programs and CAD/CAT-generated test vectors makes it unrealistic to check each step in the test sequence manually. Fault-simulation and logic-modeling errors could show up in the automatically generated truth table. The test program must therefore be checked or verified.

One way of verifying the integrity of software that has been generated by fault simulation and logic modeling is to run such programs against a discrete implementation of the VLSI chip. First, the test vectors are fed as inputs to the discrete implementation, and then the outputs are monitored for functionality.

Another method is to use software simulation to exercise the input, output, and associated routines under conditions approximating those of actual test operations, without using the ATE itself. The problem here is that errors contained in the CAD/CAT process for logic modeling and fault simulation are further simulated—and thus they are also compounded.

If software-simulation verification is to be used in an effective manner, the test program must be broken down into chunks that are comprehensible as well as bite-sized. Simulation is used only at those points where the possibilities for uncovering errors are greatest. It is at this stage that experienced circuit designers and test engineers

---

Problems with CAT

- Lack of complete documentation on available computer-aided-test systems.
- Test programs lack pattern and structure when algorithms are generated without knowledge of the circuit's overall structure and function.
- Simulation techniques must be developed to permit evaluation of complex patterns without the need to simulate all signal changes.
- Current systems can handle only several thousand equivalent gates.
- Lack of functional-test generation and related design-for-testability considerations.
- Lack of fault-simulation systems with the ability to address multilevel models and functional faults.

Requirements for CAT systems

- Ability to generate program from computer-aided-design logic simulation, fault simulation, and fault verification.
- High flexibility for different circuit functions (both analog and digital) and CAD-to-tester interfaces.
- Easy translation from CAD systems to automatic-test-equipment systems, like those from Fairchild, GenRad, Teradyne, Takeda-Riken, and Accutest.
- Cost effectiveness.
- Rapid program generation based on menu philosophy for gate arrays and custom-cell circuits.
- User-friendly interface.
- A reference library of previous test techniques.
- Interface manipulation of alternative algorithms by test engineer.
- Pattern compression through host computer, performed in conjunction with a CAD system.
- Standard interface and standardized software.
- Design-for-testability implemented in very large-scale integrated circuits. .

---

will succeed in generating meaningful tests that are compact, cost-effective, and also capable of uncovering critical design errors.

When small- and medium-scale integrated circuits were developed, circuit-design engineers were isolated from test engineers. That isolation created anxiety and friction among both groups and reduced efficiency as well. A circuit designer was not required to know very much about testing or testability. However, a test engineer was expected to perform as a circuit and system designer, a components engineer, a reliability/quality-assurance engineer, a software expert and programmer, an ATE expert, a failure analyst, a physicist, a chemist, a metallurgist, and a statistician.

Of necessity, the prevailing philosophy of design and test has shifted. From the first, VLSI circuit designs must be conceived with testability in mind and must also provide internal circuit-node points for testing. During the period of initial specification and design, a closer working relationship between design and test engineers is vitally necessary to give testability and design-for-testability equal priority with logical functionality.

The user-vendor interface must change as well. IC vendors covered their LSI chips with a shroud of secrecy, supposedly for competitive reasons. Not enough logic details were given to synthesize the data base, and test programs and vectors were withheld.

Users had little help developing the tests they needed to ensure the electrical integrity of the chips they bought, so they resorted to a brute-force approach based on a functional description of the chip. An open relationship between vendors and users—a relationship based on mutual trust, understanding, and respect—is critical to overcome the enormous burden of VLSI testing. □

VLSI testers ramp up capabilities for mixed-signal chips and hybrids

Full digital timing checks coupled with fast analog measurement check parts that mix analog and digital signals, like data converters and codecs

by Dean Johnson, *Fairchild Camera & Instrument Corp., Analog Test Systems Group, San Jose, Calif.*

☐ High-performance mixed-signal integrated circuits, such as data converters and codecs, require a high degree of flexibility in automatic test equipment in order to reduce the need for user-customized hardware add-ons—and that requirement applies to software as much as to hardware. Moreover, the ATE should be capable of performing fast and highly accurate analog, ac and dc parametric, and functional testing. Digital signal-processing techniques combined with automatic dc calibration and full timing provisions provide the necessary capabilities.

In addition to testing data converters and codecs, a mixed-signal tester should be equally effective in testing traditional analog ICs, such as operational amplifiers, filters, comparators, voltage regulators, and audio circuits. The system architecture should easily interface with external auxiliary equipment, such as probers, handlers, environmental chambers, and bus-controlled instrumentation. Finally, the ATE should exhibit high throughput and expandability and be capable of accommodating new processing techniques.

Mixed-signal testing

The digital portion of a tester must be able to handle four types of devices: mixed-signal ICs, mixed-signal hybrids, and small- and medium-scale MOS and bipolar ICs (table). Many of the key test parameters for these chips are associated with timing and measurement accuracy. Mixed-signal ICs and hybrids, in particular, demand such sophisticated test capabilities as precise timing and accurate ac parametric measurements. The primary testing requirements for SSI and MSI chips are fast dc and ac parametric measurement.

An ATE system drives the inputs and monitors the outputs of the device under test. The system exercises the DUT by transferring test-vector data from system memory to the test head, which interfaces directly with the chip. These test vectors, which define the input data, expected output data, and all associated timing, are transferred to the DUT at the system functional-test rate. Since the system measures the time differential between input transitions and output strobes to determine DUT propagation time, speed and timing accuracy are essential, making these events occur precisely at preprogrammed times.

The test-system timing environment generally is the responsibility of a subsystem composed of a crystal-controlled master clock and several timing generators. An example, shown in Fig. 1, is the General-Purpose Digital Option (GPDO) of the Fairchild series 80 tester (see "Analog and digital ATE in a single socket," p. 137). Here, the real-time clock and address-control unit determines the rate at which a program sequencer executes microcoded instructions stored in an instruction memory.

Operations executed by these instructions include the selection of pin-data formats, pin-data sources, and test periods, as well as the triggering of test-vector burst generation. The master clock also provides the time base for system- and pin-timing generators that create timing edges for controlling the events at the pins of the DUT.

Timing generators

The system-timing generator may be connected under program control to one or more device pins, either directly or as a delayed timing source connected to inputs from the pin-timing generator. The GPDO may have up to four system-timing generators. The system-timing parameters can be changed on the fly under software control. However, the pin-timing generator operates independently of other timing generators and cannot be changed on the fly.

In the GPDO, the pin-timing generators provide seven timing edges for each DUT pin during each test period. The series 80's central processing unit controls these edges. Each timing edge has several timing ranges, with the fastest ranges providing the greatest resolution. Maximum resolution is governed by the master clock's period and a vernier delay circuit with 8-bit resolution.

Two of the seven timing edges are used by the pin-

KEY DEVICE CHARACTERISTICS IN DIGITAL TESTING				
Device type	Pin count	Speed	Timing requirements	
			Measurement (ns)	Accuracy (ns)
Mixed signal hybrids	64–512	100 kHz–5 MHz	5–15	1
Mixed signal ICs	16–48	500 kHz–10 MHz	3–8	1
Small and medium scale MOS ICs	16–48	5–25 MHz	1–5	0.5
Small and medium scale bipolar ICs	16–48	10–70 MHz	0.5–2	0.1

Reprinted from Electronics, April 19, 1984. Copyright © 1985, McGraw-Hill Inc. All rights reserved.

data-formatting logic to control the pin driver's leading and trailing edges; two edges control pin-driver transitions to and from the high-impedance output state used for DUT input/output pin testing, and two control the comparator strobe. The last edge is used to make the comparison data coincide with the expected outputs of the DUT, whose pin-response edges are generated by similar timing generators.

When combined, as in Fig. 2, system- and pin-timing generators bring unique advantages to a mixed-signal ATE system. With an analog chip (containing digital registers) that operates at a few kilohertz, one test requirement might be to check the registers' setup and hold time, which is on the order of 2 to 10 nanoseconds. If the test equipment has only a system-timing generator, it can be set up to provide the required timing accuracy at the low frequency, but the same accuracy at digital frequencies is unacceptable because resolution is lower.

Another way

An alternative is to establish the accuracy of the system-timing generator for the shortest range and to slave a number of pin-timing generators to it. Using the system-timing generator as a coarse control, timing accuracy now remains the same for both millisecond- and nanosecond-edge placement. The actual measurement is performed by a time-measurement unit (TMU), which can measure the delays, widths, and rise times of test-device pin-data pulses.

If the system provides independent timing for each device pin, then it can test devices with a different timing requirement for each pin. An example of this would be testing an IC combining MOS, TTL, and analog circuits.

In a mixed-signal system, the accuracy of signal-amplitude and offset measurements is influenced chiefly by the calibration methods used and the hardware architecture. In the GPDO, the system CPU supports automated calibration of all device-pin voltages, timing delays, and pin-driver ramps. Autocalibration is performed at system initialization, when the test program is loaded, and periodically during testing to align dc voltages, timing deskews, and ramp circuits.

Dc calibration uses internal standards traceable to National Bureau of Standards references. For the timing standard, the system uses a length of high-quality coaxial cable with an expanded Teflon dielectric. In addition, a measurement is performed on the cable to determine the gain of the time-measurement circuit.

Traditional test systems use a number of common resources multiplexed to selected pins as required to measure signal amplitude or offset. Pin voltages routed from reference voltage sources on the individual pin-electronics cards may be inaccurate because of slight circuit differences between cards. An alternative to this approach is to use one digital-to-analog converter per pin for all test functions. This architecture allows a more flexible and more accurate stimulus capability.

Another desirable feature for mixed-signal testing is programmable rise and fall times. Without this capability, the pin-driver circuits would either be too fast for C-MOS circuits or too slow for testing fast TTL circuits.

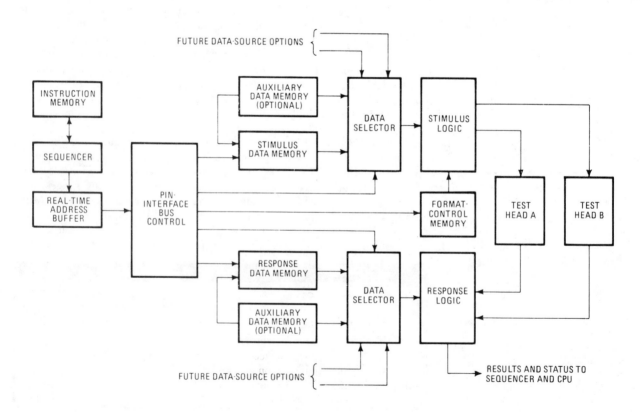

1. Identical architecture. Split-pin capability is feasible with the functional flow architecture in both the stimulus and response sections of the General-Purpose Digital Option of the series 80 tester. The function blocks (tinted) are replicated on the test-head interface cards.

Analog and digital ATE in a single socket

The General-Purpose Digital Option is a Fairchild series 80 subsystem that fully supports testing of digital and mixed-signal (combined analog and digital) integrated circuits. The GPDO hardware has three major functional parts: control and common resources, the test-head interface, and the test head.

The control-and-common-resources (CCR) module contains the control logic, system timing generators, and control memory that govern the overall functioning of the GPDO. Each test-head-interface (THIF) module contains the test-data memories, timing generators, and waveform formatters for each pin of the device under test, up to a maximum of 32 DUT pins. These two modules are housed in series 80 mainframe cabinets.

The number of GPDO cabinets required is a function of the number of DUT pins supported. The first cabinet houses the CCR module and one THIF module, supporting 32 pins. Additional cabinets house added THIF modules.

The GPDO manipulator-mount test head contains the pin-electronics drivers and comparators, supporting power supplies, and optional pulse generators. These units are housed in a half-bay cabinet, which also forms the base for a manipulator that supports the test-head module for use with handlers and probers.

The present design of the GPDO test head supports up to 64 DUT pins. Future designs will support ICs with up to 256 pins. The test head also supports the required analog test functions for mixed-signal chips.

With programmable rise and fall times, the test program can assign these parameters as required for each device pin. Until recently, this feature was available only in digital ATE equipment.

Split I/O pins are another advanced digital test-system technique that is highly valuable in mixed-signal ATE. This capability allows each test-head pin to be defined either as a device input, output, or both. To be exploited fully, split I/O must be supported by a system stimulus-and-response architecture, which, under program control, can be applied to each pin as required.

Split I/O advantages

A split I/O system can increase effective pin counts by as much as 50%. For example, no IC is partitioned to have an equal number of inputs and outputs. Instead, a typical low-pin-count chip has a 30 : 70 ratio of output to input pins—or vice versa for chips with large pin counts. Thus a 60-pin system can in all probability test an IC with from 96 to 100 pins using the split I/O capability. Without timing-per-pin and multiple-bit-per-pin capabilities, however, split I/O cannot fully support the stimulus and response data needs.

In the architecture of the stimulus and response circuitry shown in Fig. 1, the data memory, data selector, and logic functions for the stimulus side of the diagram are almost a mirror image of the response side, the format control memory being the only exception. Under program control, either the stimulus or the response side can be applied as required to any device pin to make it an input or an output. If the DUT has a pin that during the course of its operation functions sometimes as an input and sometimes as an output, then both the stimulus and response sides can be connected to that pin, with each side activated as required. Because the two sides are identical, no functionality at all is lost in the split-pin mode of operation.

The same is true of the timing architecture shown in Fig. 2. Again, the pin-timing generator, multiplexer, and format logic for the stimulus side of the diagram are the same as that for the response side. Consequently, either input or output timing signals can be applied as required to any pin, and no performance penalty is incurred.

Finally, a true split I/O feature should also be supported by a multiple-bit-per-pin capability that offers each test-head pin access to several bits, each serving a different function. In Fig. 2, for example, the stimulus and data memories each store test vectors on a per-pin basis. During testing, each memory supplies either one of two data bits to its associated device pin channel for each and every period. In the stimulus data memory, one bit specifies either a high or low state for the pin driver, and the second bit controls the pin-driver high/low impedance states. Similarly, the response data memory contains two bits used to compare the expected device output state and to mask don't-care DUT outputs.

The system's error memory continuously collects device fail or error data from the DUT output pins during real-time testing. When the test is completed, this data is transferred to the system CPU for analysis. The memory can also serve as a recorder for a learning mode in which the response of a known-good device is stored and compared with other chips of the same type.

System software

Except during real-time testing, the GPDO operates under the direct control of the software executed by the series 80 CPU. The programming language used for the CPU is a natural extension of the series 80 Analog Factor language, with all existing syntax preserved. New commands have been added to control real-time test execution and to specify test configurations, signal-path and logic setups, and stimulus and response test vectors.

Testing throughput is largely a function of system software and its associated architecture. The auxiliary memories shown in Fig. 1 provide up to 256-K of storage depth to the GPDO test-vector memories. The auxiliary memory can be downloaded directly into the pattern memory when switching between test heads holding different devices requiring different test programs. Alternately, upon command from the instruction memory, the auxiliary memory can issue a burst comprising from 2 to 256,000 vectors. Using the auxiliary memory in the burst mode effectively increases test-cycle frequency from 12.5 MHz to about 40 MHz. Both capabilities have the advantage of higher speeds (a 10-million-vector-per-second rate) over systems that reload local memory through direct memory access at a 2-megabyte-per-second rate. Since the GPDO test-vector memories are reconfigurable, they can be used to perform what is called deep serial testing. For example, many logic devices with small pin counts typically include memory circuits that are either very deep and narrow (such as 4-K by 1 bit) or shallow and very wide (such as 1-K by 4 bits). In such cases, the test-vector pattern can be configured by the ATE user to accommodate the unique testing requirements of the device.

A set of instructions stored in the instruction memory controls the operations of the GPDO. This microcode is loaded into the memory by the CPU through direct memory access and is executed by the control sequencer. Each 128-bit microcode word is partitioned into several fields controlling the operations to be performed during a given cycle and the section of the next instruction field.

Two methods for configuring instruction memory are in current use. One approach is to use an instruction for each test vector. However, to avoid prohibitive costs, the number of bits per pin must be limited, at the penalty of handling less powerful instructions. The GPDO approach uses one instruction to set up pin-test conditions and another instruction to trigger a burst of test vectors out of local memory. This method conserves memory space, while retaining a very powerful instruction set.

ATE system designers usually partition hardware to reach a balance between ease of manufacture, optimum performance, and ease of maintenance. Some ATE vendors arrive at this balance by partitioning tester functions. This approach, for example, might group all timing-associated hardware on a circuit card or family of cards. The GPDO approach is to partition hardware by pin to provide maximum serviceability, while also optimizing performance.

All stimulus and response hardware associated with a pin or a pair of pins is found in a card-cage assembly located in the GPDO mainframe. The card cage houses three types of pin-oriented circuit cards: test-head interface, pin-timing control, and pin-interface control. The test-head interface cards each consist of local memory, pin timing generators, first-level TMU, multiplexer, formatter, and comparison logic. Each of these cards services two system pins.

The pin-timing control card, one for each pin, has a second-level multiplexer and control circuits that multiplex and buffer timing signals to the test-head interfaces. The pin-interface control cards, also one for each pin, buffer address and control lines to the test-head interface.

Similar partitioning exists in the test head, which is a separate enclosure. The test head (Fig. 3) houses the pin-electronics assembly and the DUT interface circuits. These circuits are packaged as a single auxiliary test-head interface board plus pin-electronics cards, one for each two pins of the test head. The auxiliary test-head interface board contains the analog references, system probes, a peripheral I/O interface, and an ac scanner. Each pin-electronics card carries the driver, comparator, a programmable load, and I/O matrixes for two pins.

Therefore, any pin-related problem can quickly be traced to one of the cards in the card-cage assembly. If there is a question as to whether a problem exists on the mainframe card or the test-head card, signals at the mainframe I/O and the DUT connection points can be

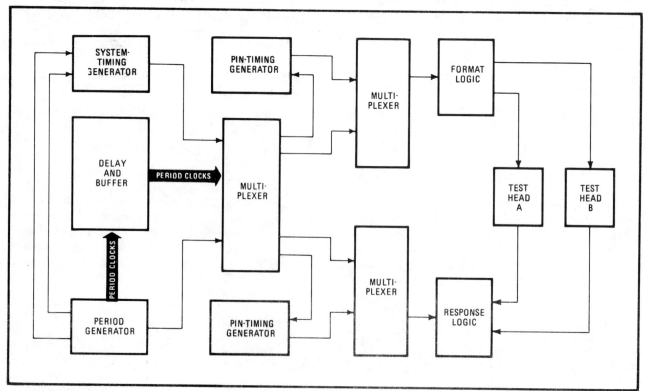

2. Combining generators. A system-timing generator provides a wide timing range but poor accuracy; pin-timing generators offer high accuracy but poor range. Combining the two in the GPDO results in high accuracy with a wide timing range.

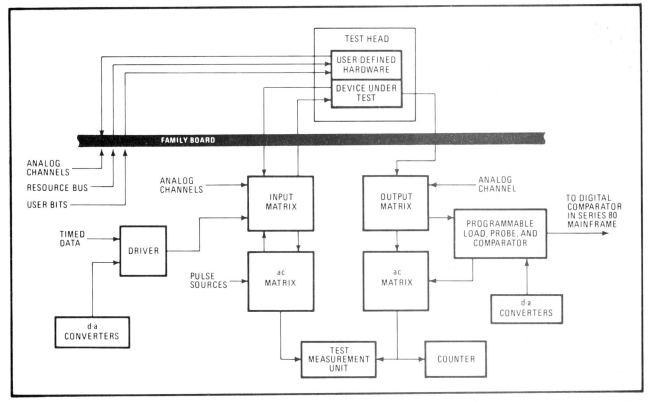

3. Test head. The test head provides all the necessary connection points for attaching any user-defined hardware to the analog portion of the integrated circuit being tested. An audio card is available to test telecommunications circuits.

multiplexed back through the time-measurement system to check the inputs and outputs of any board in order to isolate the problem source.

The test head interfaces directly with the user and thus receives the most abuse. Consequently, about 90% of any system malfunction will be pin-related and quickly isolated to the appropriate board.

Test-head design

One of the most critical aspects of advanced ATE design is the test head, of course. This is particularly true for a mixed-signal system, which must merge the two seemingly incompatible worlds of analog and digital signals. Without capabilities for digital and ac parametric testing, an analog test system generally does not need a test head. Instead, the great diversity of analog testing requirements calls for the support of an equally diverse assortment of user-customized load boards—not only for each different device but for each different application of a single IC as well.

To test purely analog chips, such as operational amplifiers, the user needed only to design a 5-by-5-inch printed-circuit board containing a few discrete components and several other op amps. Other devices, such as analog switches and regulators, required even less circuitry. General-purpose signals were fed to the board and DUT through long cables from the tester.

Today's analog chips are much more complicated and require more sophisticated tester support hardware. For example, a d-a converter requires about a square foot of pc board occupied by complex circuitry. Moreover, the nature of the analog signals involved requires the tester resources to be closer to the DUT. To measure leakage current, for instance, the distance between a picoammeter and the DUT must be as short as possible. Thus the increasingly stringent requirements of evolving analog devices have forced analog ATE designers to move their system resources closer and closer to the DUT—where digital resources have been all along.

This trend has culminated in the GPDO test head, a true mixed-signal design that minimizes the need for user-customized hardware and for the first time combines advanced test capabilities for single-insertion, single-shot ac and dc parametric analog and digital testing. The test head has two 50-ohm ports per pin; to these ports, a variety of external devices can be connected, including pulse generators, function generators, waveform analyzers, counters, and rf generators.

For testing the analog portion of an IC, a test head provides all the necessary connection points for attaching any user-defined hardware. Depending on the degree of accuracy and performance desired, interconnections can be made using anything from plain wire to semirigid technology, such as a microstrip board.

Both high-quality digital and analog support circuitry are available to the DUT socket through the expansion bus. An example of the combined power of sophisticated digital and analog test circuits is an audio card that allows the GPDO to test telecommunications circuits. For this application, the GPDO provides digital accuracies at analog frequencies. This capability is equally valuable for testing voltage-to-frequency and frequency-to-voltage converters that handle analog signal frequencies down to 10 hertz. □

Fundamental Timing Problems in Testing MOS VLSI on Modern ATE

Modern ATE systems can be adjusted to fractions of nanoseconds but cannot achieve this accuracy on MOS devices. Interim measures will help until test technology improves.

Mark R. Barber, AT&T Bell Laboratories

Reprinted from *The IEEE Design & Test*, Volume 1, Number 3, August 1984, pages 90-97. Copyright © 1984 by The Institute of Electrical and Electronics Engineers, Inc.

High-speed NMOS and CMOS VLSI circuits with large numbers of contacts present serious test problems when accurate timing measurements are required on output waveforms. Without appropriate corrective measures, these timing measurements can be in error by as much as 10 ns, even though the automatic test equipment can be adjusted to subnanosecond accuracy.

This was not a problem during the 1970's when integrated circuits generally had fewer that 60 pins. Then, IC test systems were designed with high-impedance measurement circuits only a few centimeters from the device output pins. In the 1980's, however, the number of pins on devices under test has grown to 128 and beyond, and transistor sizes below two microns have led to higher speed digital signals. Test system manufacturers have not been able to match these developments. If they had, we would have miniature, high-frequency, high-impedance comparator circuits that could be placed within five to 10 cm of the devices under test. Instead, most new VLSI test systems connect pulse drivers and high-impedance comparator circuits to DUT pins through 50- or 90- ohm transmission lines that may be as long as 50 cm when 256 comparators are required. Because MOS output circuits are generally mismatched to the transmission lines, test engineers and quality control personnel are forced to use transmission line theory and a well-known MOSFET model to analyze the waveform reflection and attenuation phenomena that occur during timing measurements. (Figure 1 illustrates the size of a modern 256-pin test head.)

Test specifications sometimes call for timing measurements to be performed with specified lumped capacitive loads on the order of 25 to 100 pF. While this was possible with the small, low-capacitance, 60-pin test heads manufactured during the 1970's, it is impossible when a 50-cm transmission line leading to a comparator presents a total capacitance exceeding 50 to 60 pF. Instead, the expected time delays on the capacitive loads must be calculated using data from the imperfect test system.

This article shows how timing corrections can be incorporated into cur-

---

Summary

Most manufacturers of VLSI test equipment are designing 256-pin test heads with transmission lines leading from the device under test to driver-comparator circuits that are sometimes 50 cm away. While this is acceptable for testing ECL circuits whose outputs are designed to drive 50-ohm loads, serious problems arise with MOS devices designed to drive capacitive loads. Unless timing measurements are made at the 50-percent level of the initial voltage step at the comparators and theoretical corrections are applied, MOS measurements can be in error by as much as 10 ns even though modern automatic test equipment can be adjusted to subnanosecond accuracy. The author points out that there is an urgent need for manufacturers to develop more compact test heads if silicon and gallium arsenide very high speed ICs are to be properly tested.

rent automatic test programs. But, to meet future needs, VLSI test system manufacturers must develop 256-pin test heads that present less than 20 pF of capacitance per DUT pin. This will be essential to properly test future Si and GaAs very-high-speed integrated circuits.

Test system pin electronics

Modern VLSI test systems contain anywhere from 128- to 256-pin electronic circuits mounted roughly 50 cm from the DUT, with each circuit connected to the DUT pins through various kinds of transmission lines to preserve waveform fidelity. The transmission line impedance is seldom exactly uniform as a signal incident on the DUT traverses the pin electronics circuit, passes along a coaxial cable, crosses a load board, and finally enters a DUT socket or a tungsten probe. A typical pin electronics circuit and transmission line leading to a DUT is illustrated in Figure 2a. The basic equations governing the capacitance per unit length and velocity in each of the four sections of the transmission line, together with an equation for the overall average characteristic impedance are as follows:[1]

$$C = 33.3(\epsilon_r)^{1/2}/Z_o \text{ pF/cm},$$
$$\epsilon_r = \text{dielectric constant}$$

$$\text{Velocity} = 1000/(CZ_o) \text{ cm/ns},$$
$$\text{if } C = \text{pF/cm}$$

$$\text{Avg } Z_o = 1000\tau_{ns}/(C_{TOTAL} - C_{PE}),$$
$$\text{if } C_{TOTAL} = \text{pF}$$

For example, in a 50-ohm system, it is difficult to maintain a constant impedance in the load and socket board sections of the transmission line; hence, the average value typically might be 60 ohms.

An ideal measurement system (Figure 2b) incorporates a high-impedance, unity gain buffer amplifier with an output impedance Z_o placed within a few centimeters of the DUT. The ideal measurement system might also include an optional specified load capacitor on the order of 25 to 100 pF. Note that we have to institute delays in the ideal measurement system so that the timing of the comparator strobe in

Figure 1. A typical 256-pin test head shown in the raised position above a wafer prober.

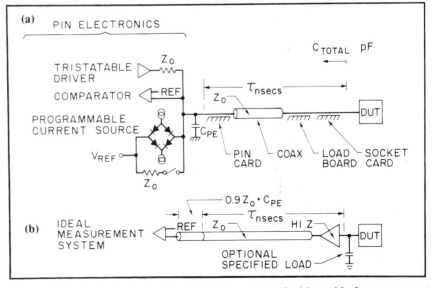

Figure 2. Typical pin electronics circuit (a) compared with an ideal measurement system (b) in which time delays ensure identical comparator strobe timing.

Figure 2b matches that of Figure 2a. The delay 0.9 ($Z_o \cdot C_{PE}$) represents the effect of the pin electronics capacitance at the end of the transmission line (see Barber[2]).

When a voltage driver in the pin electronics circuit is activated, a pulse with a voltage equal to one-half of the programmed value will be launched down the transmission line. When the pulse reaches the high-impedance MOS input circuits on the DUT, it will double in size to the original programmed value. A reflected pulse will return to the source, where it will be largely absorbed in the matching resistor. Although the lumped pin electronics capacitance (C_{PE}), shown in Figure 2, will cause a small echo from the source, its main effect will be to delay the time at which the programmed source voltage arrives at the DUT.[2] This delay, together with the transmission line delay (τ), will automatically be taken into account by the test system when time delay measurements are made. Therefore, stimulating the DUT with an accurately timed, high fidelity pulse from the test system is generally no problem.

For the most part, however, the high-impedance MOS output driver circuits in the DUT are not matched to the transmission lines, and a series of pulses will repeatedly echo between the pin electronics and the DUT. Therefore, DUT output resistances must be measured to calculate the attenuation of the pulses reaching the comparators and to analyze the waveform phenomena involved.

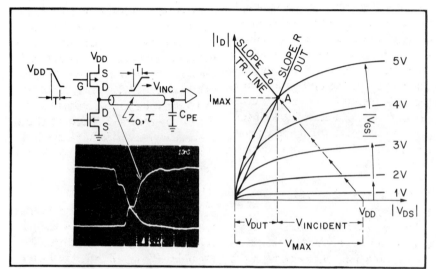

Figure 3. Determining the correct CMOS DUT output resistance (R) to be used in calculating the voltage incident of the pin electronics: an approximately linear voltage rise for 3ns followed by a 2τ delay before the first reflection from the comparator is observed; where the DUT operating point rises at a uniform rate to point A along a load line with a slope corresponding to the impedance of the transmission line.

DUT output resistance measurements

MOS output circuits are generally of the "pull-up" and "pull-down" variety as illustrated in Figures 3 and

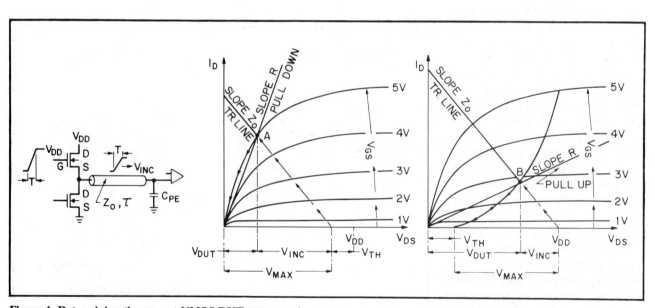

Figure 4. Determining the correct NMOS DUT output resistances to be used in calculating the voltage incident on the pin electronics: (a) the motion of the operating point for pull down devices; (b) a demonstration of the increase in output impedance for pull-up devices.

IEEE DESIGN & TEST

4. When the output switches to a high or low state, a voltage pulse will travel down a transmission line into the pin electronics circuit. Therefore, the DUT will be driving a purely resistive load for a period of 2τ seconds, after which time the first reflection arrives back at the DUT. To determine the proper reference level at which the comparator should measure pulse delays, it is necessary to determine the magnitude of the incident voltage on the transmission line; this requires acknowledge of DUT output resistance.

CMOS output circuits. The photograph in Figure 3 shows an approximately linear voltage rise for 3 ns followed by a 2τ delay (measured from the beginning of the ramp) before the first reflection from the comparator is observed. This implies that in this typical CMOS output circuit the current rises linearly with time. Referencing the theoretical MOSFET characteristic on the right in Figure 3, we see that the DUT operating point rises at a uniform rate to point A along a load line with a slope corresponding to the transmission line impedance. After subsequent reflections, the operating point for CMOS devices will move down the I-V curve toward the point of origin. The figure also shows how to calculate the voltage incident on the transmission line ($V_{INC} = I_{MAX} \cdot Z_o$) and the effective DUT output resistance (R), which we define as the slope of the line connecting the origin to the point A where $I = I_{MAX}$. This resistance is higher than that measured from the slope of the I-V characteristics near the origin. Therefore, the correct data for calculating R is obtained by applying a load Z_o and measuring the voltage across it. Preferably, the test system should provide such a resistive load connected to a programmable reference voltage in the pin electronics, as illustrated in Figure 5. The voltage V_L shown in Figure 5 could be measured using the dc measurement system, or it could be a dyamic pulse height measurement using the comparator.

NMOS output circuits. The technique for measuring the output impedance of NMOS pull-up circuits is the same as for CMOS. The output impedances of NMOS pull-up devices are generally high because the V_{GS} excursion is reduced as shown in Figure 4b. From the operating point B onward, the following equation is satisfied:

$$V_{GS} = V_{DD} - V_{INC}$$

For NMOS pull-down circuits, the output voltage falls from a value following a pull-up, which will generally be V_{DD} minus one MOS threshold voltage drop, where V_{DD} is the voltage excursion on the NMOS pull-up gate. The motion of the operating point is shown in Figure 4a. When measuring the output resistance of the NMOS pull-down circuit, the reference voltage shown in Figure 5 must therefore be set to $V_{DD} - V_{TH}$.

The comparator voltage waveform

Having determined the DUT resistance—and therefore the voltage on the transmission line incident on the comparator—the voltage as a function of time at the comparator can be calculated using the transmission line theory described in Ramo and Whinnery[1] and Barber.[2] The results for CMOS devices have been calculated for several transmission line impedances and are plotted in Figure 6. The calculations assume that the incident voltage (V_{INC}) rises linearly to its maximum value in $T = 3$ ns; the transmission line delay (τ) is 3 ns, and the pin electronics capacitance (C_{PE}) introduces a delay of 1 ns.

The initial voltage step at the comparator shown in Figure 6 is roughly twice the incident voltage[1]; however, after several reflections, the comparator voltage eventually rises to V_{MAX} equal to V_{DD} in the case of CMOS or falls from ($V_{DD} - V_{TH}$) for NMOS pull-down devices. The voltage scale can easily be changed to other appropriate values of V_{MAX}, and, in the case of falling voltages, the voltage scales should be reversed. Both of these changes are illustrated in Figure 6.

In general, the initial voltage step at the comparator is given by

$$2V_{INC} = 2V_{MAX} \cdot Z_o / (R_{DUT} + Z_o)$$

Therefore, new comparator reference levels must be computed when performing timing measurements. For example, if $R_{DUT}/Z_o = 2.5$ and a timing measurement is required at a level half-way up the initial wavefront at the comparator when $V_{MAX} = 5$ volts, then the comparator reference determined using the above equation

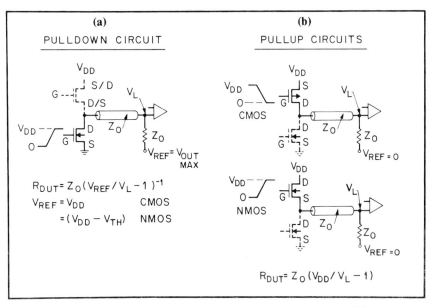

Figure 5. Dynamic or static technique for measuring resistance of pull-down (a) and pull-up (b) of CMOS and NMOS circuits.

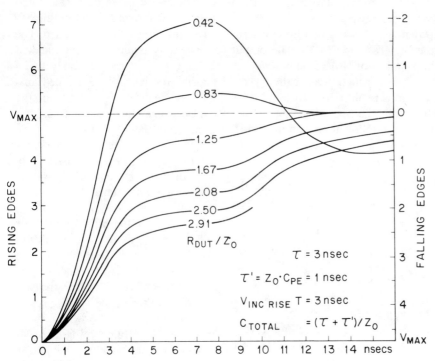

Figure 6. Comparator voltage calculations for rising and falling voltages from CMOS devices. The left and right vertical axes illustrate two different voltage scales.

should be set to $V_{INC} = 1.43$ volts. In the case of a voltage falling to zero from $V_{MAX} = 4.75$ volts, the corresponding 50-percent reference level would be 3.39 volts. To simplify the task for test engineers, it is convenient to tabulate the corrected comparator reference levels for different maximum output voltage excursions. This is illustrated in Table 1.

It should be obvious from Figure 6 that timing errors as large as 10 ns can occur if the incorrect comparator reference levels are used when measuring output delays from MOS devices. These errors will occur in spite of the fact that the test system drivers and comparators can be deskewed and adjusted to subnanosecond accuracy.

The comparator voltages shown in Figure 6 begin at time zero, which corresponds to the arrival of the first linear voltage ramp from the DUT after a delay τ. Normally, the timing of the comparator strobe will be ad-

Table 1.
Timing corrections to be added to comparator measurements to determine CMOS DUT output delays with 25 pF load capacitors. Z_0 was assumed to be 60 ohms.

R_{DUT}	V_{LOAD}	$V_{MAX} = 4.75$				$V_{MAX} = 5.00$			
		V_{REF} RISING	ADD ns	V_{REF} FALLING	ADD ns	V_{REF} RISING	ADD ns	V_{REF} FALLING	ADD ns
50	3.5		2.2		0.6		2.0		0.8
	2.5	2.59	1.5	2.16	1.3	2.72	1.4	2.28	1.4
	1.5		0.9		1.9		0.8		2.0
75	3.5		2.7		0.8		2.5		1.0
	2.5	2.11	1.7	2.64	1.6	2.22	1.6	2.78	1.6
	1.5		1.0		2.4		1.0		2.5
100	3.5		3.3		1.0		3.0		1.2
	2.5	1.78	2.1	2.97	1.9	1.87	2.0	3.13	2.0
	1.5		1.3		2.9		1.2		3.0
125	3.5		3.9		1.2		3.6		1.4
	2.5	1.54	2.5	3.21	2.2	1.62	2.4	3.38	2.4
	1.5		1.5		3.5		1.4		3.6
150	3.5		4.4		1.3		4.1		1.6
	2.5	1.36	2.9	3.39	2.6	1.43	2.7	3.57	2.7
	1.5		1.6		3.9		1.6		4.1
175	3.5		4.8		1.5		4.5		1.8
	2.5	1.21	3.2	3.54	2.8	1.28	3.0	3.72	3.0
	1.5		1.8		4.3		1.8		4.5
500	3.5		12.2		3.8		11.3		4.3
	2.5	0.51	7.7	4.24	6.9	0.54	7.3	4.46	7.3
	1.5		4.6		10.9		4.3		11.3

justed in the VLSI test system to compensate for the delays introduced by pin electronics capacitance (C_{PE}) at the driver, the transmission lines connected to the DUT input and output terminals, and the delay due to C_{PE} when the pulse arrives at the comparator.

Time delays with a capacitive load

MOS device specifications often call for testing with a specified load capacitance, normally 25 to 100 pF, at the DUT output terminals. With larger VLSI test systems, where transmission lines may add 30 to 50 pF in addition to C_{PE}, the effect of a 25-pF load must be deduced theoretically. If a lumped capacitor is added on the load board to increase the total capacitance to 100 pF, the waveform at the comparator will still exhibit a typically distorted shape due to transmission line reflections. Therefore, a good testing strategy is to use the test system to measure the delays through the internal DUT circuits into purely resistive loads (Z_o). Theoretical calculations can then be used to provide the additional delays caused by the specified loading capacitors on the DUT output circuits when observed by ideal measurement systems such as the one described in Figure 2b.

Calculation of capacitive load delay. DUT output voltages as functions of time on 25-pF or 100-pF load capacitors connected to CMOS devices are shown in Figure 7. (A description of the calculation for this figure can be found in Barber.[2]) Note that a new set of capacitor voltage curves must be generated if DUT output resistances are defined for transmission line impedance values other than 60 ohms.

The time $0.9(Z_o \cdot C_{PE})$ ns in Figure 7 corresponds to the start of the voltage rise when the capacitive load voltage is observed by the ideal measurement system shown in Figure 2b. Time zero therefore corresponds to the delay τ, which marks the beginning of the voltage rise at the comparator in the nonideal system shown in Figure 2a.

Calculation of capacitive load response from test system measurements. Figures 6 and 7 can be used to calculate the overall circuit time delays to specified voltage levels when CMOS output circuits are loaded by specified 25- or 100-pF capacitors. First, the DUT output impedances (R_{DUT}) must be measured as shown in Figure 5. Second, the DUT delay times are measured into purely resistive loads Z_o with comparator reference levels set to 50 percent of the initial wavefronts. Third, the overall delays to specific load capacitor voltages are calculated by noting the differences between the delay times in Figures 6 and 7 and adding these differences to the delay times measured in step two.

The time delay to the 50-percent level of the initial voltage step on the comparator shown in Figure 6 is very closely equal to

$$t = 0.9(Z_o \cdot C_{PE}) + T/2$$

Therefore, the time delay corrections are simply the delays from the zero voltage point to a specified voltage level in Figure 7 minus $T/2$. The quantity $Z_o \cdot C_{PE}$ does not enter into the calculations.

The 50-percent levels on the initial wavefronts arriving at the comparators were chosen as the basis for the computed time delay corrections because they will shift by an amount $\Delta T/2$ when the rise times (T) of the voltages on the transmission lines differ from the assumed value of 3 ns by ΔT. Because the capacitor waveforms will also shift by $\Delta T/2$ (see Barber[2]), the method described above, which uses the time differences between the curves in Figures 6 and 7, will apply for DUT output rise times varying several nanoseconds on either side of the assumed value of 3 ns.

Table 1 shows how correction factors can be prepared for the benefit of test engineers. It gives capacitive delay factors applicable to the 1.5-, 2.5-, and 3.5-volt levels for CMOS devices having output voltage pulse heights of 4.75 and 5.00 volts. Figure 8 summarizes the delay measurement process described above and shows how to calculate a load delay using an ideal measurement system from a comparator measurement. Note that the total delay in this particular example is measured from the 50-percent point on the DUT input waveform since this is usually the point programmed by the test system voltage drivers.

High-speed pass fail testing. Table 1 can also be used for setting comparator strobes in high-speed production testing. In this case, the strobe

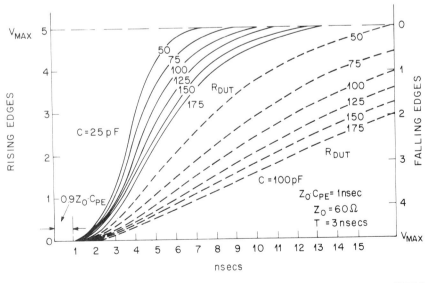

Figure 7. Load capacitor calculations for rising and falling voltages from CMOS devices. An ideal measurement system is assumed with specified loads of 25 pF and 100 pF. The left and right vertical axes illustrate two different voltage scales.

August 1984

must be positioned earlier in time than when using a load capacitor at the input to an ideal, high-impedance measurement system. For example, consider a 100-ohm DUT output that is defined to pass a test when the voltage rises above 3.5 V on a 25-pF load capacitor after a delay t. Using a typical 60-ohm transmission line measurement system the strobe must be placed at $(t-3.3)$ ns, with the reference level set to 1.78 V.

A problem with low DUT output impedances. It may be impossible to specify reference levels equal to 50 percent of the initial wavefronts at the comparators when the ratio R_{DUT}/Z_o falls below one to three. If the waveform for this ratio were to be shown in Figure 6, it would appear as a damped sinusoidal oscillation about V_{MAX}, with the first negative excursion falling to the 50-percent level of the initial wavefront. In this case, errors greater than 10 ns can be obtained when performing time delay measurements or high-speed pass/fail testing.

Timing corrections for various probes, sockets, and transmission lines. Transmission line delays in the test system, including delays in probes and sockets leading to and from the DUT, should be automatically subtracted by the system when comparator measurements are made. This can be achieved by introducing a fixed time delay of a few nanoseconds in the comparator strobe circuits. A simple way to determine these delays is to replace the DUT with a very short section of transmission line and to use a comparator to measure the delay at the 50-percent V_{MAX} level of the driver pulse, if that is to be the DUT input time reference. If the transmission line impedance does not exactly match the driver impedance, then the comparator level must be adjusted slightly.[2]

If we assume that the test system drives the DUT with linear ramp voltages and if DUT outputs are also linear ramps, but with different rise times, then the overall transmission line delays will consist of four components. A signal leaving a test system driver will be delayed[2] by approximately $Z_o \cdot C_{PE}/2$ plus the delay (τ) of the transmission line leading to the DUT input. After leaving a DUT output terminal, the signal travels along a second transmission line with delay τ and is then delayed by approximately $Z_o \cdot C_{PE}$ when it reaches the comparator. The delays introduced by the pin electronics capacitance at the driver and comparator are explained in Barber.[2] Note that the simple calibration procedure described above will determine the sum of the four delay components.

Time-to-tristate measurements. A measurement commonly performed on CMOS circuits is the time-to-tristate. This measurement is made us-

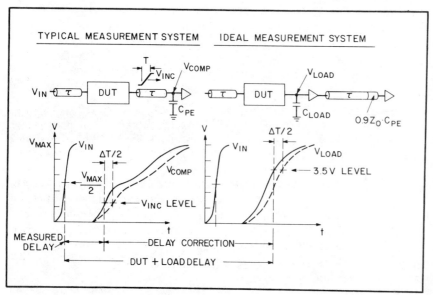

Figure 8. Determination of DUT output delay on a load capacitor at the 3.5-V level from a comparator measurement. Dashed curves show the effect of an increase ΔT in DUT output risetime. The two graphs should be superimposed, but have been separated for clarity.

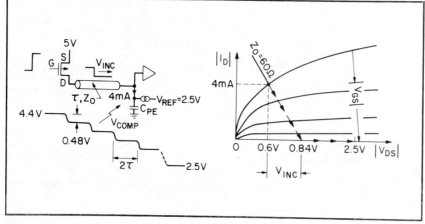

Figure 9. Typical transmission line phenomena occurring during a time-to-tristate measurement.

ing the programmable current source shown in Figure 2. For example, consider a PMOS transistor pulling up the output while the current source is programmed to sink 4 mA with a reference voltage of 2.5 volts. When the PMOS transistor is turned-off, a step voltage will be sent down the transmission line toward the comparator. Once again, the situation can be illustrated using the load line technique shown in Figure 9. When the PMOS transistor is turned-off, a step waveform of 0.24 volts will move toward the comparator, where it will double in amplitude and a series of reflections will follow. The voltage at the comparator will therefore fall linearly at a rate $Z_o \cdot i/\tau$ (neglecting C_{PE}) until the 2.5-volt reference is reached. The time-to-tristate measurement should be performed when the first step voltage arrives at the comparator.

A hardware solution to the timing measurement problem. A compact, low-capacitance, pin electronics circuit is shown in Figure 10—a circuit that should eliminate most of the timing errors due to transmission line effects. A high-impedance buffer amplifier is placed next to the DUT with a transmission line matched to the buffer output leading to a comparator that can be several meters away. Unity gain buffers, recommended previously,[3,4] are now available with slew rates of 2V/ns and input capacities of 8 pF. Drive pulses are transmitted along a second transmission line from a matched driver circuit that could also be several meters away. A high-speed, 1.5-pF FET switch is mounted near the DUT to isolate the capacitance of the driver transmission line when output measurements are being made on DUT I/O pins.

Instead of placing a programmable, diode bridge current source near the DUT for time-to-tristate measurements, as shown in Figure 2, it may only be necessary to use a 10k-ohm resistor connected to a ±20V, high-speed programmable voltage source. A 2-mA current should be sufficient for quickly charging or discharging the DUT pin when the total load capacitance is less than 20 pF. A 10k-ohm resistor is selected because this value is 10 to 100 times larger than the impedance of typical MOS output circuits.

Using these techniques only one buffer amplifier, one FET switch, one resistor, and one SPDT relay need to be placed near each DUT pin, preferably within 10 cm. In this way it should be possible to obtain DUT load capacitances under 20 pF and avoid the transmission line reflection problems that exist in present-day 256-pin test heads. ∎

Acknowledgments

I am indebted to C. T. Garrenton, S. M. Kang, D. Murphy, J. M. Niklos, and R. G. Pajak of the Bell Laboratories staff for providing simulated and measured data on CMOS output driver characteristics. I am also grateful to A. K. Stevens and other members of the staff of Takeda Riken America, Inc., who provided data for this project. Useful discussions were also held with the staff of GenRad Semiconductor Test, Inc.

References

1. S. Ramo and J. R. Whinnery, "Fields and Waves in Modern Radio," Ch. 1 and 7, John Wiley and Sons, New York, 1953.
2. M. R. Barber, "Subnanosecond Timing Measurements on MOS Devices Using Modern VLSI Test Systems," *IEEE Int'l Test Conf. Proc.*, Oct. 1983, pp. 170-180.
3. "J941 VLSI Test System Description," Teradyne Inc., Woodland Hills, Calif., Jan. 1984, pp. 14-15.
4. Y. Kuramitsu and Y. Gamo, "A Suitable Test System for Gate Arrays," *IEEE Int'l Test Conf. Proc.*, Oct. 1983, pp. 21-24.

Mark R. Barber supervises the Microprocessor Testing Group at AT&T Bell Laboratories. He previously worked on underwater acoustic signal processing, solid state microwave devices, semiconductor memory design, and the development of IC test equipment. Barber, who is an IEEE fellow, received the BE(hons) degree from the University of New Zealand in 1955, and a PhD from Cambridge University, England, in 1959.

His address is AT&T Bell Laboratories, 600 Mountain Ave., Murray Hill, NJ 07974.

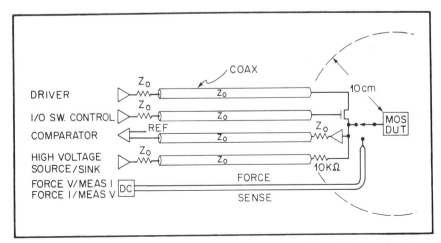

Figure 10. Proposed compact, low-capacitance, low-power, pin electronics circuit with driver, comparator, dc measurement unit, etc., located remotely.

Managing VLSI Complexity: An Outlook

CARLO H. SÉQUIN, FELLOW, IEEE

Reprinted from *The Proceedings of The IEEE*, Volume 71, Number 1, January 1983, pages 149-166. Copyright © 1983 by The Institute of Electrical and Electronics Engineers, Inc.

Abstract—The nature of complexity in the context of VLSI circuits is examined, and similarities with the complexity problem in large software systems are discussed. Lessons learned in software engineering are reviewed, and the applicability to VLSI systems design is investigated. Additional difficulties arising in integrated circuits such as those resulting from their two-dimensionality and from the required interconnections are discussed. The positive aspects of VLSI complexity as a way to increase performance and reduce chip size are reviewed.

With this discussion as a basis, the evolution of VLSI system design environments is outlined for the near-term, medium-term, and long-term future. The changing role of the designer is discussed. Recommendations are made for enhancements to our engineering curriculums which would provide the next generation of designers with skills relevant to managing VLSI complexity.

I. Introduction

VERY-LARGE-SCALE-INTEGRATION (VLSI) will soon make it economically viable to place 1 000 000 devices on a single chip, and the technological evolution will continue to double this number every 1–2 years for at least another decade [1]. According to G. Moore [2], the major hurdle faced in the construction of ever larger integrated systems is a *complexity barrier*. In order to exploit fully the technological potential of VLSI, new ways of managing the information associated with the design of a VLSI chip must be developed.

Why is it that complexity and its management have suddenly become such popular concerns? Mankind is routinely building systems with more than a million components: skyscrapers, air planes, telephone systems.... System complexity and the associated engineering issues do not seem to differ markedly for alternative implementations, e.g., whether a circuit is contained within a cabinet, on a printed-circuit board, or on a single silicon chip.

Packaging *does*, however, have an impact because of the partitioning that it enforces. Nobody would dare to insert a million discrete devices into a large chassis using discrete point-to-point wiring. Large systems built from discrete devices are broken down into subchassis, mother boards, and module boards carrying the actual components. This physical partitioning encourages careful consideration of the logical partitioning and of the interfaces between the modules at all levels of the hierarchy. Since such systems are typically designed by large teams, early top-down decisions concerning the partitioning and the interfaces must be made and enforced rather rigidly—for better or for worse. This keeps the total complexity in the scope of each individual designer limited in magnitude, and thus manageable.

VLSI permits the whole system to be concentrated in the basically unstructured domain of a single silicon chip which does not *a priori* force any partitioning or compartmentalization. On the positive side, this freedom may be exploited for significant performance advantages. On the negative side, it may result in a dangerous situation where the complexity within a large, unstructured domain simply overwhelms the designer. A similar crisis was faced by software engineers when unstructured programs started to grow to lengths in excess of 10 000 lines of code. The crisis was alleviated by the development of suitable design methodologies, structuring techniques, and documentation styles. Many of the lessons learned in the software domain are also applicable to the design of VLSI systems.

Much as programming was done throughout the 1950's by small groups of highly qualified people solving problems in their own style, integrated circuit design in the 1970's was still done by small clusters of layout wizards. This often led to intricate, if not mysterious, circuits. This "cottage industry" style of creating systems [3] starts to fail once programs approach tens of thousands of lines of code or once circuits end up with 100 000 devices. More formal and organized methods are required to create larger systems.

Sections II and III explore the nature of complexity in general and analyze methods to deal with complexity that were developed in the framework of software engineering. In Sections IV and V the point is made that the problems associated with VLSI designs are even harder than those encountered in large programs.

Section VI presents general developments in the emerging VLSI system design tools that address the discussed problems, and Section VII tries to predict the evolution of such tools for three different time frames. In Sections VIII and IX the role of the designer is re-evaluated in the face of this changing design environment, and some recommendations are made for the education of the next generation of designers.

Most sections are relatively self-contained and can be read as a sequence of minipapers. On the whole, they should not be viewed as the final word on the subject of VLSI complexity, but rather as food for thought and a nucleus for further discussion.

II. The Nature of Complexity

VLSI is more than just "a lot of LSI"—in the same sense that a city is not just a large village and the human brain is not just a large collection of ganglions. The 256-kbit RAM's

---

Manuscript received April 23, 1982; revised November 8, 1982. This work was sponsored by the Defense Advance Research Projects Agency (DOD) under ARPA Order 3803; monitored by the Naval Electronic System Command under Contract N00039-81-K-0251.

The author is with the Computer Science Division, Electrical Engineering and Computer Sciences, University of California, Berkeley, CA 94720.

random-access memories) of 1982 are quite different from the first 1-kbit RAM's that appeared on the market a decade ago. As the scale of any system is increased by orders of magnitude, its organization typically changes because different issues become relevant. Internal structuring starts to appear, specialized subparts emerge, communication between the different parts gains importance, and, at some level, the *complexity* of the system becomes an explicitly stated concern.

The introduction of new design methodologies to deal with this concern requires an understanding of complexity and its special demands. Webster's Dictionary gives the following definition:

> complex: referring to that which is made up of many elaborately interrelated or interconnected parts, so that much study or knowledge is needed to understand or operate it.

This section tries to provide some insight into the notion of system complexity. It will emerge that it is useful to distinguish between explicit (apparent) complexity and implicit (hidden) complexity.

A. Explicit Complexity

Unlike the information content of a transmitted message, there is no formal absolute measure for complexity. In general one tends to call those systems more complex that need more words or more bytes to be adequately described or specified. Thus a comparison of the lengths of the corresponding descriptions can be used as a measure for the *relative* complexities of two or more *similar* systems. In order for such a comparison to make sense, the systems to be compared must be described in the same language and in terms of the same primitive elements.

This *apparent* or *explicit* complexity of a system depends strongly on the language or notation used for its description. By using suitable abstractions, the details of the subcomponents of the system can be hidden, and the apparent complexity can be reduced dramatically. It is equally important to exploit any structure or regularity to reduce the length of the description and to express more succinctly the "essence" of a particular system. This can be achieved with the use of suitable notations designed to express repetition, symmetry, and other regularities in an efficient and compact manner. In this light, an early microprocessor such as the 8008, with only a few thousand transistors, is more complex than one of the large memory chips with, say, 16 384 bits. Because of its regularity, the latter can be grasped and described more easily.

B. Implicit Complexity

However, complexity is not simply proportional to the number of different parts or the length of an optimally encoded layout description. The interactions between the parts play an important role. Steward [4] defines complexity in the following way:

> Given the parts and their behaviors, complexity is the difficulty involved in using the relations among the parts to infer the behavior of the whole.
> Or phrased another way:
> Complexity is how much more the whole is than just the sum of its parts.

This *hidden* or *implicit* complexity of a system involves issues such as the behavior of the system and the way in which it achieves a particular function. A collection of gates that implement the well-defined Boolean function of an adder is far less complex than some historical radio receiver stage with a single vacuum tube which simultaneously amplifies the incoming HF signal, demodulates the LF component, and preamplifies the sound signal.

These intricate aspects of a system's behavior could, in principle, be reduced to the previously defined measure of explicit complexity. One would have to take into account the complete systems documentation, including: all its functional specifications, a description of its operation, the plans for its physical realization, and the set of repair instructions. Since it is normally impractical to sum up all this diverse information, one needs some other guidelines to evaluate this more elusive notion of implicit complexity. Looking at the way in which systems are composed from smaller components is a suitable approach.

C. Complexity in a Structured System

Few systems are described or documented in a hierarchically flat manner. Large systems are viewed as being composed of subsystems, which, in turn, consist of components on a lower hierarchical level. *Cohesion* and *coupling* are two useful concepts for a more detailed evaluation of the complexity of a module at a particular level of the system hierarchy. They address the nature by which components are combined into modules at the next higher level of the description.

Cohesion: Cohesion is a measure of how closely the internal parts that make up a module belong together when seen from different perspectives. In general, a high level of cohesion is desirable, since it leads to a simpler description at the next higher level and thus reduces its explicit complexity. High cohesion can be derived from functional similarity, from logical or physical grouping, or from the fact that all subparts work on the same data. Yourdon and Constantine [5] define several different criteria by which the cohesion of a compound software module can be judged. They can readily be extended to the domain of integrated circuits as demonstrated by the examples in Table I.

Spatial cohesion is something that is unique to the hardware domain and plays a big role in analog VLSI components such as capacitor arrays or optical sensor systems. Temporal, logical, or coincidental cohesion are very weak and are primarily used to keep things ordered from an organizational point of view.

Coupling: Coupling measures how much the submodules interact. Good structuring and clever systems partitioning aim at reducing implicit complexity by minimizing the amount of interaction between subparts. The interaction patterns should be simple and regular. This leads to independence of design, simple layout, high testability, and easy modifiability. Table II gives some measures by which the degree of coupling can be judged.

TABLE I
MEASURES OF COHESION IN COMPOUND MODULES

Name	Explanation	Example
functional	all parts contribute to function.	operational amplifier
sequential	portion of a data-flow diagram	filter cascade
communicational	data abstraction, same data used	LIFO stack
spatial	need to be physically close	sensor array
procedural	same procedural block	bus controller
temporal	used at a similar time	start-up circuitry
logical	a group of similar functions	I/O-pad library
coincidental	random collection	miscellaneous wiring

TABLE II
MEASURES OF COUPLING BETWEEN SUBMODULES

Perspective	Degree of Coupling	Example
Topology:	linear flow-through	filter chain
	hierarchical tree	carry look-ahead
	lattice modularity	memory array
	irregular mosaic	processor control
Interaction:	continuous	light sensor
	periodic	timer
	occasional	reset
Bandwidth:	high	video signal
	bursty	disk head
	low	voice signal

Minimal coupling, using explicit communications between submodules rather than parasitic interactions, supports modularity and makes possible plug-compatible replacements. The topology of the interactions should be kept as simple as possible. Clean modularization reduces implicit complexity since self-contained modules are easier to specify than a collection of highly interacting components.

III. Dealing with Complexity

Complexity is an integral part of large, "interesting" systems. In this section, general techniques for managing this complexity will be discussed. Most of them have been developed or refined in the context of large programs and form the basis of good software engineering. They are also applicable to the design of complex VLSI systems.

A. Design Equals Documentation

A complex system often can be understood more easily when viewed from the proper perspective. This is the role of good documentation. *Design* and *documentation* should become synonymous.

Software maintenance costs usually exceed initial production costs by a significant factor [6]. Software systems are constantly being modified, upgraded, and expanded [7]. This means that other people must be able to thoroughly understand what the original developer(s) had in mind. A similar issue is emerging in VLSI design as the development cycle of many products starts to exceed the average time an employee spends with the same company, and thus designs have to be passed from one designer to the next one.

Among good software designers, it is now common practice to make the documentation an integral part of the program. It has been realized that a separately kept documentation folder is never up-to-date and, before long, describes something quite different from the actual code. Documentation for a typical integrated circuit, however, is still normally scattered over several different media and/or several organizational groups, including the systems architects, the logic designers, and the layout crew. Often some pieces of the documentation are lacking entirely, making the circuit a mystery to all but the original design group. Modifying or even just debugging of such a system by anybody but the original designer(s) tends to be a real nightmare.

Notations and Representations: An important step in minimizing the *apparent* complexity of a system at any level of its description is to use suitable languages and/or notations, tailored to the particular issue that needs to be documented. Hoare emphasizes the importance of the proper notation by pointing out that the history of mathematics is, to a large degree, the story of improvements of notation [8]. The use of bad, hard-to-decipher notation is not only unproductive, it breaks the thought process aimed at finding problem solutions.

One should use all possibilities of the representation medium to enhance the clarity of the constructs that are being used: suitable mnemonic labels, symbolic shapes, differentiating colors, sound.... To simply ask for more extended English descriptions for each item is not the right way to go. Abstract, terse symbols are crucial to making the "grand picture" visible.

B. Use of Abstraction

Humans can effectively deal with only one problem at a time and can concern themselves with only a few constraints simultaneously [9]. This necessitates judicious use of hierarchy and abstraction. If the representation of a system can be rendered more succinct, one has taken a big step towards managing its complexity. Abstraction in the form of *user-defined macros* or *parameterized subroutine or cell calls* permits one to focus on the few qualities that are essential at a particular phase of the design process. It hides internal interactions and irrelevant details that contribute nothing to the solution but would clutter the picture and thus hamper the relevant thought processes.

Examples from the world of VLSI are: lumped devices hiding the complicated physical processes inside a contiguous piece of silicon; logic gates abstracting from a particular realization; register-transfer modules showing the basic data operations while hiding details of timing; and instruction-set descriptions for computer architectures hiding all but the primitives that the programmer has to deal with.

In addition, abstraction permits the creation of *generic modules*, usable with different data types or implementable with different technologies. Abstract encapsulation of modules in well-defined interfaces is the basis for efficient design of large systems. Well-defined modules, properly documented and made available through module libraries, can be used repeatedly by many designers. Design systems for integrated circuits often provide several plug-compatible versions of such modules which vary in a few parameters such as power and speed, or the availability of an explicit testing interface. Efficient construction of large systems is not possible without a library of well-designed modular components.

C. High-Level Descriptions

High-level languages lead to shorter and more readable programs. They make it easier to understand the organization and behavior of programs or systems and result in designs that are more likely to be free of errors, easier to debug and modify.

Programmer Productivity: Long-term average productivity in a commercial software production environment is about 10 lines of code per day. One arrives at this number by dividing the total number of lines in the final program by the amount of programmer time used for its development. The number is surprisingly independent of the language used—but the resulting net programmer effectiveness is not. A high-level language statement such as CASE or a complicated algebraic expression assignment is more powerful than an assembler-level LOAD instruction. Thus the net programmer productivity in terms of the functional effect produced by his/her code scales strongly with the power of the primitives employed. Similarly, in integrated circuit design, designers can place on average only five to ten *items* per day [10]. If ten *processors* rather than ten *transistors* are inserted into the layout per day, the effective productivity is increased dramatically.

Compilers versus Code Efficiency: High-level descriptions must be mapped down to the level of the actual implementation. Traditionally, coding in lower level languages has resulted in more compact and faster-executing code than what can be produced by a compiler. However, compiler technology is improving. Present-day compilers for high-level programming languages will do an adequate job for most applications and will do it far more quickly and reliably than any programmer. Of course, skilled programmers, willing to devote a large amount of time, can still outperform them by a factor 2-3 in small program segments, but this effort is only justified if there are hard constraints such as real-time requirements or the need to fit a program into a limited amount of on-chip memory.

High-level languages are now being introduced to the development of VLSI systems. There are several long-term research efforts in VLSI design aimed at the construction of a *silicon compiler* [11]-[13]. Driven by high-level specification of the desired circuit, these automated design systems are expected to translate high-level systems descriptions into correct circuit implementations in a particular technology. These layout compilers are still in their infancy and produce relatively inefficient results for all but narrowly defined, highly structured subsystems.

However, when the rising development costs for VLSI systems become comparable to the fabrication costs of ten thousands of devices, a faster development cycle at the price of a less efficient implementation becomes an attractive alternative for many systems. The increasing market share of gate arrays [14] and macro cells [15] must be seen in this light.

D. Partitioning

Problems that are too large to be handled as a whole must be partitioned into smaller, more manageable parts. Section III-E reviews some formal methods for finding suitable partitionings. In any such subdivision, the interactions between the subparts must be given particular attention.

Intermediate Decisions and Specifications: When a module is partitioned into submodules, the latter must be specified in enough details so that one can verify that the partitioning step does not violate any of the original specifications. In carrying out the successive refinement steps, one must also assure that the implementation of every module agrees with the abstract view used at the higher level. The properties of the composition of the new submodules must be assured for *all* possible concerns, e.g., functional correctness, timing constraints, signal levels, and even testability.

If *formal* verification is not possible—and it may be a long time before this becomes practical in the domain of circuit design—another solution has to be found. Intermediate system simulation, using proper models for the submodules, is a usable approach. VLSI system design tools must have the means to create, retain, and make use of such intermediate models. They should readily permit partial and mixed-mode simulations [16], [17] of a system in a top-down manner, making use of the abstractions for the subparts. Such a simulation can cut off all branches of the design hierarchy for which suitable simulation models exist. The resulting savings in simulation efforts are considerable.

Separation of Functionality and Implementation: The specifications of a system are normally separated into a description of what the system is supposed to do plus a list of performance objectives. This provides another boundary along which a problem can be partitioned.

During the design of a system it is also useful to separate the concern for functional correctness from the details of the actual implementation, which must often include an optimization step; the number of packages is minimized or a set of packages is distributed evenly on a few printed circuit boards. Constantine refers to this optimization and packaging step as clustering pieces of a problem solution into physical constraints without unduly compromising the integrity of the original design [5]. Such packaging optimizations should be made as the final step and only after the algorithmic optimizations have been exploited to the fullest. Once the design has been compromised by low-level packaging optimizations, it loses a lot of its generality, portability, and durability in the face of emerging technologies. Such optimization tricks are difficult to comprehend and to reverse at a later time.

Conceptually, the design process should thus contain the following three concerns:

1) functional design: guaranteeing proper behavior;
2) implementation: finding a suitable structure;
3) optimization: fine tuning the physical arrangement.

These basic steps should also be taken in the design of a VLSI system. However, several iterations through these three steps may be required to make sure that the nature of the implementation medium has been properly taken into account in the high-level design decisions.

E. Structuring Methods

Structuring is a key approach to large programs, systems, or problems in general. Good structuring implies partitioning. Many of the concepts suitable for judging the structure of programs [18] are also applicable to VLSI systems. As a start, one can demand that a well-designed subsystem implement a single, independent function, have only a few "low-bandwidth" entry or exit points, be separately testable from the module boundaries, and be itself constructed of a limited number of submodules that obey the same criteria.

Unfortunately it is often not clear *how* one should partition a particular problem or system; it can involve the most important and most difficult decisions a designer must face. This section reviews some program structuring strategies and discusses their applicability for VLSI systems design.

Functional Decomposition: Functional Decomposition is an application of the very general and popular *divide-and-conquer technique*. It consists of a recursive subdivision into parts, so that the joint operation of the properly interconnected parts performs the specified operation of the whole.

It can be used in a top-down manner, splitting a large system into cohesive blocks [19]. The recursive refinement step to be executed in this design strategy is as follows. Start from the definition of the function to be performed and its desired interface. Identify logical subparts, define the function and interfaces of these parts, and specify their interactions. Then verify that when the subparts are interconnected in the specified manner, their overall behavior corresponds to the original specification at the previous level. Unfortunately, this last step is nontrivial in programming and even harder in VLSI systems design.

Any significant design effort will have to iterate between top-down and bottom-up techniques because the characteristics of the medium used in the final implementation might not be visible at the top. Functional decomposition also plays a role in conjunction with the bottom-up phase. The natural building blocks of a particular technology and effective implementations thereof are derived in a bottom-up manner. The overall problem specification is then scanned for the usage of such blocks and partitioned accordingly.

The advantage as well as the problem with functional decomposition lie in its generality. If there is a large gap between the top-level problem specification and the natural building blocks dictated by the technology, the methodology will not provide guidelines for reproducible implementations and an excessive variety of possible decompositions results. The quality of the final result will then depend strongly on the intuitive insights of the designer. This is equally true for VLSI systems and for large programs.

Flow Graph Design: A more specific way to decompose a problem is to follow its *flow of data or of control*. In the first approach, the *data* flowing through the system are viewed as the primary ingredients, and all functional blocks are viewed as filters transforming the data as they pass through [20]–[22]. This approach suggests a further decomposition of each module into input circuitry, transformation circuitry, and output circuitry. Particular concern has to be given to synchronization, flow control, and possibly to buffering of intermediate data at the interfaces.

This is a particularly important and suitable approach for integrated circuits since it deals explicitly with the communication between blocks and puts proper emphasis on spatial ordering as dictated by the flow of data. It is being used in signal-processing devices and is particularly well exemplified in systolic arrays [23]. The formal data-flow graph of a system is a good starting point for a layout. Even in laying out the datapath of a microprocessor, the flow of data and of control often form the basis for the floor plan. Typically, they are assigned to the directions of the two coordinate axes so that the two flows cross at right angles [24], [25], [13].

The price to formally partition each "filtering module" into input circuitry, transformation circuitry, and output circuitry appears to be too high for present-day VLSI systems. In regular structures, such as systolic arrays [23] implemented as an iteration of identical cells, it is preferable to design outputs and inputs as matched pairs. Significant savings in power and layout area result from this optimization step. For modules that are used more randomly and are combined with a variety of other modules, the extra overhead of buffering provides abstraction and modifiability. So far, the control flow of a system has been used less directly to produce a structured circuit layout, except possibly for the case of a microcoded control section.

Data-Structure Design: Another successful structuring method in software engineering is to start with the *data structures*. The idea is to describe a problem through its associated objects, and to specify these objects through suitable abstractions and their interactions [20], [26]. This technique applies particularly well to small subsystems performing a specific function. Because there are fewer arbitrary and different ways to find data structures that correspond to a particular model of a subsystem, the resulting system structuring is more reproducible than with other methods. However, it is not so clear how this approach can be used for large systems.

The application of this method to VLSI systems design is also difficult. Direct mappings into hardware have been realized for only a rather limited number of data structures. The parallelism desired in VLSI circuits is a further difficulty. The interactions between objects and their spatial ordering are crucial. There is a high price on interactions between submodules, which is normally not considered in the setup of data structures.

There is plenty of room for innovation. VLSI will provide ever more sophisticated and efficient implementations for such data structures as trees, queues, and content-addressable memories. Such structures can either be implemented directly with special hardware primitives, e.g., an array of associative memory cells, or they can be built using standard RAM's with suitable control structures. In either case, these VLSI data structures should be properly encapsulated so that the user sees only the desired external behavior, regardless of implementation.

Programming Calculus: Perhaps the most intellectually satisfying approach to the construction of correct and reliable software is to prove correctness during construction of the program [27], [28]. The design freedom is deliberately limited so that program and proof can be constructed hand-in-hand. This design strategy requires that the system to be built has solid, unambiguous specifications. One way to do this formally is to state the required result as an assertion in predicate calculus. Each top-down refinement step is then carried out with rigorous strength to guarantee that the specifications are still met and to produce the specifications for the next lower level of modules. Till now this approach has been limited to fairly small program segments.

There is some expectation that in the long run the circuits resulting from silicon compilers (Section III-C) can be shown to be correct by construction. However, the day is still far in the future when such generators will produce correct and economically viable solutions for arbitrary systems for all sensible combinations of input parameters, and formal proofs of correctness should not be expected in this century.

F. Restriction to a Limited Set of Constructs

Böhm and Jacopini [29] have shown that the logic of any program can be represented as a hierarchical arrangement of three basic constructs: *sequence, iteration,* and *selection.* Others [30], [31], [20] have also recommended that the number and type of control flow constructs in programs be limited to these few well-defined constructs and that unstructured control flow, as resulting from indiscriminate use of the controversial GOTO statement, should be minimized. For software systems, Jackson [20], Brooks [32], and Dijkstra [33] recommend the use of a rather limited set of allowable structuring topologies and even suggest restriction to pure *tree structures.*

For the structuring of integrated circuits, such a formal restriction to tree structures leads to implementations that are too inefficient. However, in general, restriction to few robust, well-understood constructs has many advantages. In particular, the use of standard modules that have been proven to work previously can make a system more modular and easier to understand, and reduces the possibility for errors. This comes at the price of some inefficiency and associated performance loss; thus the degree to which one can adhere to that philosophy depends on the application. Here is a sample of possible restrictions in VLSI design:

1) At the geometrical layout level, one might restrict oneself to only "Manhattan" geometry, i.e., only rectilinear features. This simplifies the required design tools and results in substantial speedup of such operations as layout rule checking for the price of some loss in layout density.

2) At the logic level, one could rely exclusively on pretested standard cells interconnected in a standardized automatic or

semiautomatic way. This leads quickly to relative risk-free implementations; but the chips are large and dominated by wiring area.

3) At the register level, one can avoid problems with hazards and race conditions by using strictly synchronous sequential circuits with no unclocked feedback loops.

4) At the large system level, hung flip-flops and similar synchronization problems can be avoided with a self-timed sequencing approach.

At all levels, the improvement in testability and understandability gained from such restrictions in the types of constructs used will make the introduced inefficiencies well worthwhile, if overall systems constraints can still be met.

G. Testing the Design

Testing is an integral part of the design process. Every design decision must be checked for its appropriateness. Any large system designed in an "open-loop" manner has a very small chance to work correctly.

There are many different ways in which people can convince themselves that a part or a design "works." Two extreme approaches are:

1) by running the part in its intended application environment;
2) by formally proving the correctness of the design from its specifications.

Both approaches have obvious drawbacks. The first one corresponds to driving a car around the block to see whether it works. In a complicated system this can never give full confidence that there are no undiscovered flaws; there is nothing to tell the designer what has *not* been tested yet. The second approach is like taking a lawyer's affidavit based on an inspection of the car assembly line; unless the inspection is complete and the conclusions are derived in an impeccable manner, the result is just as questionable.

Any test requires exact knowledge of what the system is supposed to do. However, at the onset of the design of a system, its specifications are seldom completely defined. As the design is refined and the constraints become better understood, specifications get added and fine tuned. In the end, specifications must become an exact description of what the designer believes that the system will do so that this information can then become the basis for tests or design verifications. The formulation of exact systems specifications is thus another important part of the *design* effort.

Testing should be kept in mind during the design of any system, program, or VLSI part. The designer should constantly ask questions of the kind: What is this part supposed to do? How will it be used? What are the extremes of the input parameters or operating conditions it will have to withstand? How can it be asserted that the completed part will be doing what it is supposed to do? Asking these tough questions and trying to provide answers to them will lead to cleaner and safer designs, which are also more amenable to debugging and testing.

H. Tools and Design Methodologies

Of crucial importance in the construction of large and complex systems is a good set of tools and a suitable design method.

Design Tools: The user interface is an important part of any design tool, be it for program development or the design of VLSI circuits. As computer time becomes relatively less expensive than the designer's time, the tools become more interactive in nature. Batch jobs should be used only for large program runs where the interaction of the designer is not required. Program development, on the other hand, should be done in an environment where the programmers get all the help they need to be most productive, such as syntax directed editors [34] and powerful high-level debuggers [35]. An equivalent set of tools in the domain of integrated circuit design would include design stations with built-in design rule checking, automatic layout compaction, and tight coupling to a set of simulators.

Most currently available tools fall short of the designer's expectations in many ways. Here is a list of frequently heard complaints about today's commercially available design tools for integrated circuits. Similar criticism has been voiced for software development tools [36].

1) *Cumbersome commands.* Too many actions (key strokes or cursor moves) are required to invoke frequently used operations such as placing a rectangle in a layout.

2) *Not responsive enough.* Manipulations on a graphics screen should be at least as fast as the corresponding manipulations with pencil and paper.

3) *Lack of uniformity.* Design tools reside on several machines with quite different accessing methods and interfacing protocols; this discourages the effective and frequent use of these tools.

4) *Poor integration with other tools.* Tools use different data formats or different hierarchical structuring; this makes it cumbersome to use different tools on the same design.

5) *Capabilities too narrow.* Some tools, efficiently tailored to a specific approach such as Gate Arrays [14], [37] do not permit the integration of other useful devices or functional blocks, even when these devices are compatible with the fabrication technology.

Support for Good Methodologies: An even more serious, general shortcoming is that most tools do not support a clean design methodology. Layouts done on a graphics editor can be rather ad hoc, since there is nothing to enforce high-level structuring disciplines. Normally there is no place where the designer can capture semantic information, either for the benefit of other designers re-using a particular cell or for other tools such as circuit extractors or logic simulators. Graphics editors and design tools centered around layout information alone are not sufficient for the design of complex VLSI systems.

The analysis of tradeoffs, resolution of conflicting demands, and finding the best structure are at the heart of the design process. According to Parnas [38], this leads to three key criteria by which different design methodologies should be compared:

1) In what order are decisions made?
2) How are these decisions recorded?
3) When is their correctness verified?

The search continues for good formal structuring methods giving consistent results. The search for the right design methodology for VLSI systems and for tools that properly support the latter has only just begun.

Consistency of Style: A good design methodology is one that reproducibly generates a particular system implementation, independent of whoever is applying it. So far, no design method, even in the field of software engineering, automatically produces unique solutions. As a result, one can find widely differing programming or layout styles. Even if the advantages of one style over another are hard to quantify, adherence to a consistent style in the design of a large system is important from the point of view of clarity and maintainability.

TABLE III
DIFFERENT REPRESENTATIONS OF A SUBSYSTEM

Particular Aspect	Description Format
Semantic behavior:	English text
Test specification:	text, tables
Logic function:	equations, state diagram
Timing behavior:	waveforms
Geometrical area:	graphics display
Connection diagram:	display, wire list
Power consumption:	text, tables

Representations, too, should use a consistent notation to make understanding easy. The use of familiar constructs greatly enhances understandability. The pattern-matching processes in the human brain are very powerful and can be trained to quickly find relevant patterns. This should be exploited in the selection of the notations for the various representations of a problem.

For similar reasons, interactive tools also should have consistent user interfaces. A particular command such as *delete* should have a corresponding effect in all tools in a design environment, regardless whether it is used in a graphics layout tool, in a text editor, or in a file manager. A uniform design environment will greatly enhance designer productivity.

IV. ADDITIONAL DIFFICULTIES IN VLSI SYSTEMS DESIGN

The general problems associated with complex systems that were discussed in the previous sections also occur in VLSI systems design. The related techniques developed in the field of software engineering can thus readily be applied to VLSI design. However, VLSI systems cause additional problems based primarily on the two-dimensional nature of integrated circuits and the need for physical interconnections between modules, i.e., the inability to "jump" to a "submodule." This section will outline problems that go beyond those that are normally faced by the software engineer.

A. Multiple Views and Representations

In VLSI systems, the number of different concerns is quite large and diverse. This necessitates a much richer set of views and representations than is normally used to document a software system. First, there is a larger spectrum of hierarchical abstractions, ranging from functional specifications to machine-level layout descriptions. Secondly, for a particular level of detail, several different representations may be used to optimally capture a particular concern of the designer. For instance, at the module level of a digital system, a pipelined piece of logic might be described by:

1) its logic function and test patterns for switch-level functional simulation;
2) suitable delay models to be used in signal-independent worst case timing verification;
3) separate logic gates with timing parameters useful for timing simulation and checking for glitches.

The different representations cannot always be fit into a single hierarchical ordering. Table III gives a sampling of various concerns in VLSI systems and possible notations to capture these concerns. This set of representations is richer than the one normally encountered in software engineering.

Unfortunately, these many different representations do not easily follow from one another. In a VLSI chip, most information can be computed from a final layout, but this is a very time-consuming and computation-intensive task. In order to provide an interactive design environment, most derived information should be stored explicitly in a database once it has been computed, so that it is readily available to the designer. Other information, such as functional specification, is provided top-down before any layout exists. This must be stored and later compared against the information derived from the actual implementation.

Keeping the various representations up-to-date and consistent in the face of daily changes is a major task. The search continues for a centralized representation from which most of the other representations can be derived quickly and inexpensively. Conversions between the different representations are particularly difficult if the designers do not relinquish part of their freedom and do not restrict themselves to a subset of well-defined modules and constructs.

B. Partitioning a VLSI System

Section III-E dealt with the difficulties in top-down design methods related to the fact that it is not clear *how* a large system should be partitioned into smaller blocks. In highly integrated systems, such as a VLSI chip, there are additional difficulties. Because of the lack of explicit interfaces, the partitioning step is often done in an *ad hoc* manner, improperly documented, and never formally verified to meet the original specifications of the overall system. Specifications that might have been precise at the higher level become more fuzzy or may even be violated in the intermediate partitioning steps. It is often not until the whole partitioning hierarchy has been instantiated, and a global systems simulation is being performed, that such errors get noticed. At this late stage it is often no longer clear *who* was responsible for the original violation, and an innocent group of people may have to sweat to correct the problem.

Clean and precise specifications are needed *at all levels* of the design hierarchy and in all transformations of representations. This is tedious and difficult, but unless it is done properly, the original problem has not *really* been partitioned; at best, a more terse and economical notation has been found for something that is still functionally unstructured. As an example, consider a switched-capacitor filter comprising some switches and capacitors and five identical operational amplifiers, each with, say, 30 transistors. In any reasonable description, the amplifier will be factored out as some kind of macro cell. But unless a simpler and more abstract model is used to represent it in the systems simulation, i.e., as long as the filter circuit is presented to the simulator with all 200-odd components visible, the circuit has not been partitioned functionally. Proper functional partitioning is accomplished only if each amplifier is viewed as a new lumped element with its own higher level simulation model.

There is a lot more work involved in creating satisfactory models in *all* relevant representations of a physical component than what it takes to give an abstract description of a cleanly designed program subroutine.

C. Optimization and Packaging

Because of constraints on implementation, physical and logical hierarchy need not have a direct one-to-one correspondence. The logical design hierarchy results from a suitable partitioning into significant functional blocks. The physical hierarchy may be determined by packaging constraints or by geometrical placement restrictions.

In present-day integrated circuits, functional solution and technical implementation are rather narrowly intertwined.

The performance loss resulting from the separation of the two concerns and from the use of automated implementation has been considered too high by most companies producing high-volume integrated circuits. The emergence of integrated, customized systems of VLSI complexity will force us to rethink this issue. The intertwining of functional solution and implementation should be considered an optimization step that has to be justified, rather than being the default approach.

To make these tradeoffs wisely, a better understanding needs to be developed of where the boundary between functional solutions and implementation can be drawn. The special nature of VLSI technology and its physical realities must be considered already at the system level. Algorithmic changes may be called for when one switches from a software solution to a printed-circuit board implementation and later to a VLSI chip. Yet, at the same time, one cannot afford to tie a design exclusively to a single fabrication process.

Many of the high-level partitioning steps are relatively implementation independent; e.g., the decomposition of a large processor into register-transfer level modules is independent of whether the system will be implemented in NMOS or in CMOS technology. Thus the top end of such a design system can be common to many different technologies. In principle, one could decompose a computer down to the gate level and then simply use either the NMOS or CMOS gate implementations. However, that would lead to highly inefficient realizations of such functions as an operand shifter for which compact, technology-specific realizations exist. The more one wants to push performance and optimize an implementation, the higher up the technology-specific concerns have to enter the selection of the algorithm and the structuring process. This optimization step is crucial for the fabrication of present-day VLSI chips and will result in a certain loss of clarity in the design.

D. Layout

A particularly important set of constraints stems from the two-dimensional nature of integrated circuits.

Electrical Connectivity: Whereas high-level language programmers normally need not concern themselves with the overhead of jump instructions, communication between modules in a physical machine is expensive, both in terms of space and time. For components interacting with high frequency, physical proximity is crucial since the penalties for long, high-bandwidth interconnections are severe. In large integrated circuits, the partitioning into blocks and their location on the chip must be chosen carefully since the physical distances between components and the total bandwidth between them remains frozen once the chip is made; there is no equivalent of a memory hierarchy managed by an operating system which can move frequently used information closer to the point where it is used, e.g., into a cache memory.

The communications overhead is particularly severe if the signals must travel from chip to chip or from one printed-circuit board to another. This results in a substantial expense in power and in loss of bandwidth. In addition, the number of pins is often strictly limited. (In the software domain this would correspond to the situation where the number of parameters that can be passed to a subroutine is limited to, say, two integers or one floating-point variable!) The physical packaging hierarchy must thus be chosen with particular care.

Two-Dimensionality: The concern with interconnectivity and geometrical placement is particularly intensive in VLSI systems since the implementation is restricted to the two-dimensional space of the surface of a silicon wafer. The signal paths to other modules compete with the computational elements themselves for the rather limited chip surface. This places severe topological restrictions on implementation. A particular module can have only a very small set of close neighbors, and it is often hard to decide which of several contenders should get the preferred spot. By comparison, the sequence and ordering of the definitions of subroutines is rarely a concern to a programmer (except when forward references are disallowed).

Furthermore, changing the placement of a module in a layout is a much harder task than rearranging software modules. Very few design tools reroute the attached wires automatically when a circuit block is moved to a different corner of the chip floor plan; i.e., the equivalent of a link editor for two-dimensional layouts is still in the research stage. Moreover, no really good language exists for expressing topology.

E. Exploiting Concurrency

To realize their true potential, VLSI systems should be designed for as much concurrent action as possible. Ideally, one would like to see all the gates on a chip do useful work most of the time. But so far, the concurrency exploited on VLSI chips has been of a relatively simple and straightforward nature. That is not surprising since the problem of exploiting concurrency in a general manner is not even solved in the software domain.

A first goal must be to find adequate high-level descriptions for highly parallel but irregular operations. Timing sheets with multiple traces, a method typically employed by integrated circuit designers, does not scale effectively to the VLSI level. Table IV gives an overview over various levels of concurrency, ordered by their generality and, alas, by their difficulty of implementation.

TABLE IV
DEGREES OF CONCURRENCY

Type	Example
Bit concurrency:	n-bit parallel adder
Vector operations:	matrix multiplication
Pipelining:	overlapped instruction execution
Set concurrency:	evaluation of alternatives
Specialist functions:	co-processors
Task concurrency:	communicating processes
Random concurrency:	everything else ...

The simplest kind is bit concurrency. Parallel operation on the bits of fixed-sized operands, such as parallel addition, is routinely employed in microprocessors and normally implemented in a straightforward manner by iteration of the proper 1-bit cells.

The next level, the lock-step concurrency, as employed in single-instruction multiple-data vector machines, is conceptually not much harder. Problems arise when the number of data components exceeds the available hardware resources, and the job has to be done partly parallel, partly serial.

Control for the third level, pipelining, is more complicated. The execution times in the various stages have to be matched carefully in order to efficiently process different data elements through the same basic functional blocks. Particular problems arise at discontinuities in the data stream where the pipeline has to be refilled with relevant information.

Really difficult problems arise with the higher levels of concurrency, such as set, task, or random concurrency. At this level, one tries to extract concurrency from a problem in such a manner that different operations may be performed simultaneously on different data. The potential improvements in functional throughput resulting from an exploitation of parallelism in this most general form are tremendous. This problem is still unsolved even at the algorithmic level and remains one of the major challenges for this decade.

F. Technology Changes

VLSI designs are aimed at a moving target. Because of the length of the development cycle, a new design is often aimed at a presently emerging, still rather speculative implementation process, with the hope that the process will be mature by the time that the design is finished. As the envisioned process changes, the emerging layout has to be adjusted, demanding modular and modifiable designs.

Because the technology changes so fast, the adaptation to a new process must occur in a reasonably short time span. If the design takes longer than this period, it will have missed a fast moving target. This also speaks for the usage of higher level, and thus more technology-independent descriptions. If the system design has been carried out at too low a level of representation, it will not be usable with new fabrication processes. High-level language descriptions with suitable layout compilers will be more adaptable.

G. Debugging and Testing a VLSI Chip

Debugging the prototype design of a VLSI system involves all the problems that the software engineer faces and then some additional ones. Because a VLSI chip is a rather monolithic system, it cannot be easily tested in bits and pieces, i.e., one subroutine at a time—unless explicit measures have been taken to partition the system for debugging or testing purposes. Because of the small physical features on the chip, it is very tedious and costly to gain access to the inner parts of a VLSI circuit. Only signals that are brought out to the terminals of the chip are readily observable. In other words, the equivalent of a debugger with breakpoints and monitoring of variables has not yet been invented for VLSI chips.

Checking a monolithic system becomes exponentially harder with increasing size [39] and the amount of complexity that can be packed into a single VLSI chip is far beyond the maximum reasonable size of a testable block. Special measures are thus necessary to break the overall system into smaller, testable parts, such as providing internal access points or scan-in scan-out registers [40].

Imperfect Implementation: In addition to debugging, i.e., verifying that the *design* of a chip is correct, there is also the problem of testing each copy to make sure that it is acceptable. This problem can usually be ignored in software, since a simple redundancy check is normally sufficient to guarantee that the copy is a perfect replica of the original.

When writing a program, one typically assumes that there will be a system that guarantees, often by complicated means, that all instructions are properly executed in the sequence specified. The VLSI designer faces a less ideal world. Signal voltages are only defined in certain *ranges;* in addition, they are subject to *noise* that may well exceed the stated tolerances. Alpha particles can change the contents of a memory cell. Temporary voltage surges occur on the supply lines when large numbers of signal lines change state simultaneously.

The overall system has to be immune against such erratic behavior. Overdesign, redundancy, checking, and recovery mechanisms have to be built into larger functional blocks and subsystems so that they can present a more nearly ideal picture to the observer at the next higher level of abstraction. In debugging the design as well as in testing the final product, a particular operation has to be checked over a range of "environmental" conditions, such as supply voltage variations, temperature ranges, and processing parameter deviations. This adds an extra dimension to the debugging/testing problem.

V. Tradeoffs in VLSI Chip Complexity

In software systems there is a clear trend to higher level languages. With ever faster processors and cheaper memories, the inefficiencies in code density and execution speed in a compiled program become insignificant in comparison to the advantages associated with the usage of high-level languages (Section III-C). Hand optimization at the assembly language level is reserved for critical inner loops, often-used routines, or for applications with real-time constraints.

The issue of layout density is more important for VLSI chips than code density is for programs because of the severe technological restrictions on maximum chip size. Efficiency of implementation cannot be ignored. It will be necessary to sacrifice a certain amount of modularity and modifiability in order to fit a system on an economically viable chip. This section considers these tradeoffs.

A. Hardware versus Software

In designing a complete system based primarily on some VLSI chips, a new tradeoff has to be considered: what part of the function should be implemented in silicon and what part can be done with software? This tradeoff actually spans a whole spectrum of possibilities. Consider the example of a microprocessor: should a complicated instruction be wired into the decoding logic, should it be programmed into the microcode memory, or should the compiler compose this function from a collection of more primitive instructions? These decisions are of particular importance in VLSI system design, since the individual chip is an entity that must work as a whole. If the chip design is too ambitious, the chip cannot be fabricated economically. Thus at any given point in time, the number of active devices of a certain type that can be used economically on a single chip represents a rather rigid resource limitation. One must decide very carefully how to best use this resource.

B. Simplicity to Get the Job Done

A very direct approach to managing complexity is to question whether the complexity asked for in a set of specifications is indeed reasonable and necessary. Some recent studies into instruction sets and organizations of microprocessors showed that evolving products need not necessarily follow a trend to ever increasing complexity [41]. Complexity introduced to provide rarely used instructions may not be worth the extra costs in terms of increased chip area, a slower overall machine cycle due to the delays through more complicated decoders or micro stores, and delayed market entry because of a prolonged design cycle. Worse yet, such additional circuitry may run against the hard limit of resources on the chip and thus take away devices from other functions that might have made a larger contribution to the overall performance of the system. In the domain of integrated circuits, unnecessary functionality as well as irregularity [42] comes at a much higher price than in software engineering.

C. Complexity for Higher Yield

Beyond a certain point, the yield of good integrated circuits emerging from the fabrication process drops off exponentially with increasing active chip area [43]. Due to these hard constraints on chip size, a reduction in the number of devices on the chip is well worth some increase in complexity.

In large regular arrays such as memory blocks, groups of two or four adjacent cells are often designed jointly so that proper rotation and mirroring can be used to share contacts or power bus lines. Such a cluster of two or four cells is then repeated to produce the overall array. Similarly, large amounts of random logic are typically not implemented as a single, large, and sparsely populated programmed logic array (PLA), but are broken down into several smaller PLA's [44]. In both cases, while apparent complexity is slightly increased, considerable area savings may result.

As integrated circuit technology evolves, chip functionality will increase. But the limit of how many fully functional devices can be placed on a single chip cannot rise indefinitely. To make significant improvements in the total functionality of a chip, new methods have to be exploited. The emerging 64-kbit memory parts already employ redundant memory cells and circuit elements which take over for the devices that have fabrication flaws [45]. So far, the parts have to be tested individually, and the proper rewiring or reprogramming has to be done from the outside. It is reasonable to expect that at some point in the future there will be more complex, self-testing circuits that reconfigure themselves internally to portray to the outside a flawless function of a given specification.

D. Complexity for Better Performance

Among the reasons for casting systems in silicon is the aim of compactness and performance. This implies a certain degree of system optimization. Simply taking the most regular and straightforward implementation might result in an intolerably bulky or slow chip.

Even RISC [41] is not just a simple Turing machine. Instruction fetch and execution are overlapped. This pipelining increases the system complexity, but no microprocessor designer would give up the resulting performance gain at the present time. New commercial microprocessors go even further and use more sophisticated pipelining schemes. Increased complexity and increased performance need to be carefully evaluated to find the optimum approach. Similarly, increasing the richness of the instruction set leads to a point of diminishing return. A general guideline on this issue has been given by Wulf [46]. He demands that instruction sets offer *primitives, not solutions.*

A look into the future sometimes justifies cramming more complexity and functionality onto a particular chip than is justified by present-day resource limitations. First, one expects that these limits will expand, and that the decision to add extra features will be right in the light of future developments. Secondly, computer architectures tend to be around for a long time. A company that plans to launch a whole family of processors of various performance ranges needs to accommodate the basic instruction set even in the lowest performer; this might distort the proper balance of resource allocation for this particular member of the family.

VI. General Trends in VLSI Systems Design

Many of the techniques for complexity management reviewed in Section III are slowly being integrated into presently emerging VLSI circuit design systems. This section summarizes the main trends and explores how the discussed methods are adapted to the special problems of VLSI systems design.

TABLE V
Comparison of Languages

IC DESIGN	and	PROGRAMMING
Functional specifications		Specification languages
Self-generating layouts		APL, ...
Protected, abstract modules		Ada, ...
Subcircuits, explicit connections		Pascal, ...
Symbolic description		Basic, ...
λ-based conceptual features		Intermediate code
Mask geometry		Assembler language
Mann or MEBES format		Machine language

A. Towards High-Level Languages for VLSI Systems

There is a rich spectrum of possible views which can be used to discuss the design of a VLSI system. Table V shows the most important hierarchical levels in a high-to-low order and attempts to put them in perspective by pairing these levels with representations from the world of programming languages.

In the 1970's, the practice of laying out integrated circuits corresponded to assembly language programming. Even today, most designers still have an overriding concern for density in order to end up with a chip that is small enough to be manufactured economically. Traditional designers feel that the compactness and performance achievable with good hand layout are worth the tediousness of the approach.

In the domain of software engineering, high-level languages are now used routinely for system developments of substantial size. Assembly language code is reserved for small, high-volume systems with limited memory, such as games and controllers, or for real-time systems with absolute timing constraints. The same trends towards high-level languages seen in software engineering are also apparent in the domain of integrated circuit design. As integrated circuit technology progresses, it will soon be possible to place more than a million devices on a single chip. For random control circuits, this is an inordinate amount—far more than any designer can reasonably handle manually. In such complicated systems, efficiency is a lesser issue than achieving correct operation.

High-level languages also reduce development time. This is even more important in the fast-changing field of integrated circuit fabrication. Earlier market entry will translate into large economical advantages in this highly competitive field. The evolution of single-chip VLSI systems creates a demand for more personalized systems that fit the customer's needs exactly; multiyear development times are no longer tolerable in this context. Minimizing the size of integrated circuits will only pay off for high-volume products (more then 100 000 samples). For customized VLSI applications, some layout inefficiencies can normally be tolerated; the most important consideration is to get the job done correctly and on time.

B. Partitioning of Concerns and Problem-Specific Representation

The development of a VLSI system involves even more diverse considerations than writing a large software system. If possible, the various concerns should be addressed independently. Jackson recommends the separation of problem-oriented concerns from machine-oriented ones [20]. The use of two complementary languages is suggested: a programming language to solve the problem, and an execution language to specify how to compile and execute with efficiency. The notation for the two languages should be optimized for the task at hand.

Similarly, in VLSI design, different languages and notations should be used at different points of the abstraction hierarchy.

Representations and notations must be matched to the problems that one tries to solve. In the design of VLSI systems there is a large variety of different concerns; a rich spectrum of notations is thus necessary. The proper definition of these representations is important because notations not only determine the form of the final solution, but also shape the way we think about a problem [47].

This issue has recently been addressed by Stefik *et al.* [48]. They define several representations and develop appropriate notations and composition rules. Each one deals with only a few concerns and is thus most effective in avoiding a particular class of bugs:

"*Linked Module Abstraction*" deals with event sequencing at the systems or subsystem level. Modules get started by accepting a token; they do exactly one task at a time; when finished they emit a token to the next module. This representation helps avoid deadlocks and the sampling of data that are not ready.

"*Clocked Registers and Logic*" are concerned with the details of timing. This abstraction tries to eliminate errors such as race conditions in unclocked feedback loops, e.g., by using strictly alternating two-phase clocking.

"*Clocked Primitive Switches*" try to assure proper digital behavior by addressing the ratios of digital NMOS inverters and avoiding passive charge sharing on intermediate nonrestoring nodes.

"*Layout Geometry*" is concerned with geometrical layout rules and tries to prevent spacing errors or unrealizable mask features.

Stefik *et al.* [48] point out that such problem-specific representations need not fall into a strict hierarchical ordering. The relative ranking of the abstractions along different axes of concern might well be permuted.

C. Emphasis on Structure and Abstraction

Without the introduction of some structure, i.e., hierarchy and regularity, the problem of VLSI system design would be unmanageable. As the organizational problems start to overshadow the fabricational difficulties, the use of a *constrained hierarchy* [49] becomes more attractive. Additional restrictions are imposed on the structuring process:

1) The hierarchical partitioning is truly functional with proper abstractions at each node.

2) The hierarchies in the various representations (geometrical, logical) must correspond to one another.

These restrictions on the designer's freedom result in a higher degree of clarity and in more efficient operation of most design tools, at the price of some loss of layout density.

Regularity, i.e., reusing the same submodules as often as possible, also increases effectiveness. Lattin has introduced the *regularity factor*, derived by dividing the total number of features (transistors, rectangles) by the number of features actually drawn by the designer. Present-day microprocessor chips have regularity factors ranging around 5 for commercial products [10] and reaching about 20 for experimental devices [50].

Traditional Logic Cell Macros: A traditional approach to structuring large digital circuits relies on the logical abstractions of the well-known small- and medium-scale components, such as gates, flip-flops, and registers. This approach is used in the *standard cell* and *gate array* systems. The *standard cell* or *polycell* approach [15] relies on a library of layouts of such logic components and on some automatic or semiautomatic procedure to place the cells and to route the interconnections. The *macro cell* or *gate array* approach [14], [37] uses a combination of preprocessed wafers and libraries of predefined wiring options to generate the same standard logic functions. The design systems normally contain powerful routing algorithms that wire most of the chip automatically. The designer has to help only when there are special timing or area constraints. The increasing market share of such products indicates that these are indeed practical approaches. In both cases, engineers can approach the VLSI design in the traditional manner and map their logic designs onto a silicon chip with relatively little effort.

The drawback with these approaches is that they use the wrong weighting factors. They encourage the designer to minimize the number of logic gates, assuming that wires are free, whereas on VLSI chips, they dominate the layout area as well as the timing delays. Suitable representations for VLSI must properly address the interconnections at all levels of the partitioning hierarchy.

Advanced systems permit the integration of custom-designed macro cells into the final layout like any other library cell. This approach has the advantage that higher level functional blocks, such as the datapath of a microprocessor or a control PLA, can be designed by alternative design methods, leading to more compact implementations.

Hierarchical Analysis Tools: As regularity and hierarchy is empasized in the synthesis of our designs, the design tools also need to be restructured to exploit this new methodology in the analysis phase. Design rule checkers or circuit extractors that walk rectangle by rectangle through a regular array are no longer tolerable for 64-kbit memory chips. The generating cells and their generic constellations in the array should be checked *once*; repeated instances of checked constellations can then be skipped. This requires some sophisticated book-keeping, but the performance advantages of hierarchical circuit extractors and design rule checkers are well worth it [51]-[54].

A well thought-out abstraction hierarchy must also be exploited in the simulation tools. Once a low-level module, such as a 1-bit adder, has been constructed and checked to behave correctly, that information must be tightly linked to the definition of that cell. System simulations then no longer need to work all the way to the level of the physical differential equations but can stop at the module level and use a more abstract model describing the behavior of the module at that level.

The use of hierarchical analysis tools can only succeed if it can be guaranteed that the introduced abstractions remain valid. In particular, the cell has to be protected so that its behavior cannot be changed accidentally. In the layout domain, some kind of protection frame can be employed so that accidental interference into inner parts of a cell can be detected. There are some inefficiencies associated with this approach; cells cannot be packed as tightly as they could be in an unrestricted environment. However, this mechanism should only be invoked for cells of a reasonable size, where introducing a new level of abstraction is appropriate. The protection frames serve the same roles as the *module* construct in some high-level languages; across module boundaries all interactions have to be declared explicitly. At that level, the loss in packing density can be tolerated.

D. Separation of Design and Implementation

Another application of abstraction and partitioning of concerns occurs in the transition from design to implementation. Traditionally, the layout engineer had to stay in close contact with the fabrication line and the mask makers in order to understand what exact geometrical patterns (what polarity, suitably grown or shrunk) had to be submitted in order to receive the desired physical features on the fabricated silicon wafer. Certain process variations force the designer to rework one or more mask levels. In addition, the selection of the proper processing test structures and alignment marks was also a responsibility of the designer.

The experiments with multiproject chips shared among many designers at several different universities [55] demonstrated that design and implementation can indeed be separated. The designer is responsible only for his or her own design and need not worry about mask polarity, alignment marks, process compensation, or monitors for critical dimensions. The designs are submitted in a standardized low-level geometrical descriptive form such as CIF2.0, the Caltech Intermediate Form [56]. A centralized service organization will then merge the different designs onto several reasonably sized chips together with a *starting frame* containing the alignment marks and some process control monitors. These chips, in turn, are assembled into a suitable tiling pattern, which also contains a few slots for gross registration marks and more extensive test structures, to cover a whole wafer. The same service organization will arrange and coordinate mask making and the actual fabrication of the silicon wafer and will subsequently redistribute the fabricated and possibly packaged chips to the original designers. This approach distributes the costs of mask making and wafer fabrication among all the designers and brings the cost per design down to an affordable range for small volume silicon systems; such services have been called "silicon foundries" [57].

This approach shields the occasional designer from the time-consuming and confusing details that one has to consider when submitting a design for fabrication. The silicon foundry approach abstracts the implementation process to a single "module" with a well-defined interface. The designer specifies the physical features that should appear on the final chip. All the expand/shrink operations necessary to compensate for the actual processing are performed by the implementation service. Only chips originating from wafers that pass the process control tests will be returned to the customers. This approach can only be successful with well-established, stable fabrication technologies. Again, the additional abstractions and the corresponding reduction in complexity in the design process comes at the price of somewhat lower performance.

E. Procedural Generation of Modules and Systems

While a silicon compiler for complete systems is still some time off in the future, procedural generators for special, frequently used, functional blocks become commonplace. The most important ones produce PLA's and read-only memories (ROM's). These module generators take a set of parameters or, in the case of ROM's and PLA's, the Boolean equations of the desired logic function, and then compose the complete layout from predefined low-level cells. Because these blocks are so important and used so frequently, a lot of effort has been spent to optimize the basic cells from which the final layout is composed.

Other, more complicated functional blocks, are special data processing elements, such as the datapath of a typical microprocessor forming a linear arrangement of registers, shifter elements, incrementers, or complete arithmetic/logic units [24], [13]. Hewlett-Packard claims [58] that about 90 percent of its data processing chips could be cast in the format of such a generalized datapath.

The first set of these module generators is aimed at a specific process and a specific set of layout rules. In later versions, more fabrication independence will be gained by performing the cell composition at the logic level and passing the result through a fabrication-rule-dependent circuit compactor [59]–[61].

Eventually there will be module generators that handle many *diverse* fabrication technologies (NMOS, CMOS, SOS, . . .). However, because the most suitable topology for a PLA is different for these different implementation technologies, only the higher levels of these generators, e.g., the algebraic minimization and the optimizations through folding and splitting, can be shared. To be most effective, these module generators must be fully integrated with the layout system. Their output must have the complete documentation of a typical macro module. This requires that the module generator produce all necessary representations, ideally including test specifications.

F. Enhanced Use of Graphics and Interactive Tools

VLSI system design is strongly tied to the two-dimensional nature of the implementation medium. Because of the lack of suitable languages to express the relative geometrical arrangements of individual modules and the wiring between them, the use of graphics plays an important role. Presently graphics display hardware is becoming inexpensive enough, so that there is no reason not to provide every tool that can profit from it with a graphics interface. A highly responsive, interactive graphics display is going to be the core of every integrated circuit design station.

The success of the module generators discussed in the previous section has led many people to postulate that *all* design work should be done in a procedural manner. This approach would have two major drawbacks:

1) The description of the geometry of low-level cells is rather tedious.

2) Without a two-dimensional representation the designer lacks an important element in the feedback from the design system.

Graphics is an invaluable aid at many levels of the design. For the geometrical layout of complicated leaf cells, graphics is clearly the preferred medium of interaction. At the higher levels, where a procedural description of the modules is more appropriate, graphics still plays an important role for checking the results of the procedural constructions. A good interactive graphics system can also play an important role in the strategic planning of the chip layout. It permits the presentation of the problem to a human in a format where the designer's intuition can be tapped most effectively.

On the other hand, it seems more appropriate to use a procedural approach to specify the placement of the "1"-cells in a large PLA. Because of these different tradeoffs, there is an ongoing dispute of the use of "Pictures versus Parentheses" [62]. Both approaches are optimal for certain tasks, and a good design system thus cannot ignore either one of them.

A major disadvantage with present-day procedural design tools stems from their "batch-processing" nature. It is often necessary to make small changes in a nearly finished design. If placement or routing is specified in a procedural manner, the introduction of a small change in the specifications may result in a completely different layout. This will invalidate the simulation and performance verification efforts expended on the 95 percent of the chip that were correct beforehand. Restating the constraints in such a manner that only the *desired* changes take place is normally not possible.

In the future, these procedural tools need to be integrated into a design system with interactive interfaces on which desired changes in placement or routing can be specified in a natural manner. New algorithms are required that make only *incremental* changes and derive the next solution as a *minimal change* from the previous solution. Many problems need to be solved to create such an integrated system in which the designer gets the best benefits from both approaches.

VII. Evolution of Design Tools

To a large degree, the evolution of design environments for systems on silicon will follow the evolution of programming environments [63]-[66]. However, future tools need to take into account the special demands of VLSI design discussed in Section IV. This section attempts to project how this will be achieved in three different time frames. Tomorrow's design tools already exist in prototype form in many research laboratories, and so one can have a fairly clear idea what they will look like. The remarks concerning future design environments are based on work that is currently in active pursuit in several research institutions with the goal of creating prototypes within the next couple of years. The last subsection on ultimate ways of creating solid-state systems is a rather speculative extrapolation of some ideas currently being discussed in research papers and workshops.

A. Tomorrow's Design Tools

The trends discussed in Section VI will gradually make their way into the emerging design tools. Based on the work currently in progress in many research and development organizations, one can have the following expectations about tomorrow's design systems: they will have strongly improved human interfaces, addressing most of the issues discussed in Section III-H. The tools for layout planning, module placement, and routing will make use of ever more powerful algorithms to do a large fraction of the task, but retain a strong interactive nature to permit the designers to make full use of their intuition and judgment.

There will be an increased use of higher level symbolic descriptions coupled with more and more automatic generation of frequently needed modules. Explicit design methodologies, such as the use of a constrained hierarchy, will gain preference, and the analysis tools will start to exploit the hierarchy used in the design process.

The two main components of tomorrow's design system will be a modular set of tools and a database.

A Modular Set of Tools: The same considerations that apply to large programs or VLSI systems also apply to the development of the design environment itself: the latter should also be highly structured and composed of modular components with simple, well-defined interfaces.

The art of VLSI design is still not fully understood, and new methodologies are still evolving. It is thus too early to specify a rigid design system that performs the complete design task; quite likely, such a system would be obsolete by the time it becomes available to the user. It is more desirable to create a framework that permits the usage of many common tools in different approaches and that supports a variety of different design methods and styles. In short, the environment should provide *primitives* rather than *solutions* or, in other words, *mechanisms* rather than *policies*.

Intricate interaction between the various tools must be avoided. Every tool should do one task well and with reasonable efficiency [36]. Compatible data formats are needed to make possible ready exchange of information among the tools or direct piping of data through a whole chain of tools. These formats should be comprehensive enough to carry all relevant information for various tools working on a particular design.

At Berkeley such a collection of tools [67] is being developed, embedded in the UNIX [64] operating system. UNIX already provides many of the facilities needed for our planned environment; a suitable hierarchical *file structure*, a powerful monitor program in the form of the UNIX *shell* [68] and convenient mechanisms for *piping* the output of one program directly into the input of a successor program.

Database and Data Management: All information concerning a particular design should be kept in one place, ideally in some *database* readily accessible to designers and design tools. In an emerging system at Berkeley [67], a VLSI design is mapped onto the structure of the UNIX file and directory system. Each module of the system under development is represented as a directory of files on the computer's storage system. Submodules are represented as immediate subdirectories or as files in special library directories. Each such directory may contain a varying number of files corresponding to different representations of this module. There may be files describing layouts, circuit diagrams, symbolic representations, suitable models for high-level simulation, sets of test vectors for this module, and possibly some plain English documentation. The number of files at each node is potentially unlimited.

Currently, many design systems for custom circuitry use the geometrical layout information to "glue" everything together. It is from this low-level description that other representations are derived; many of the analysis tools work from that level, e.g., circuit extraction and design rule checking. This is an unsatisfactory approach. Too much of the designer's intent has been lost in that low-level representation and has to be pieced together again by the analysis tools. If there is to be a "core" description from which other representations are derived, it has to be at a higher level. The trend is to move to a symbolic description [59] that is still close enough to the actual geometry so that ambiguities in the layout specification can be avoided. Yet at the same time, this description must have provisions to specify symbolically the electrical connections and functional models of subcircuits [69], [70].

Regardless of the exact structure of the database, the various different representations of a design should be at the fingertips of the designer so that one can readily go back and forth to the one representation that best captures the problem formulation with which the designer is grappling at the moment.

B. Future Design Environments

Based on the improved tools discussed in the previous section, five to ten years from now one will see the emergence of integrated yet flexible development *environments* for solid-state systems. Major parts of such future environments are currently in the active research stage at several institutions.

Design Management: An ever larger fraction of VLSI systems will be generated in a semiautomatic manner in which the tedious low-level operations are performed automatically. Just as future programming environments will tend to deemphasize programming by end users [71], the system design environments of the 1990's will encourage most users to work at a higher level. These environments will contain parameterized components in a variety of the technologies available at that time and the necessary tools to place and interconnect them semiautomatically. The designs will be described at much higher levels and significant parts of them will be compiled into the layouts for the chosen technology. "Design *management* tools" rather than design tools will play an increasingly important role.

Tools to help with the management of the overall design effort, including task distribution, activity scheduling, and design reviews, will gain importance with respect to the tools that just support the technical aspects of the design. Some utilities that have proven beneficial in the domain of software engineering have started to make inroads into solid-state systems design. One such utility, the Source Code Control System (SCCS) [72], plays a role when a system is too big for a single designer. It maintains a record of versions of a file or a system. This is useful when several people work together on a project and thus need to make changes to a document that is jointly used by all of them. SCCS keeps a record with each set of changes of what the changes are, why they were made, by whom, and when. Old versions of the document can be recovered, and different versions can be maintained simultaneously. It also assures that two persons are not editing the same file at the same time, which could lead to the loss of one person's modifications.

In the realm of VLSI system design, a more formal *configuration management* [73] discipline should be adopted to control modification of the specification files, the emerging microcode, or the lists of the intermodule connections. These are typical examples of documents that change continuously throughout the design phase and which may have to receive input from several people.

Design Data Base and Module Libraries: The file system containing the information about a design will evolve into a full database. This database containing the emerging custom design will be complemented by a library of cells that will be shared by many designers working on different designs. For custom designs using hand layout, these libraries have only been moderately successful. There is a lot of reinventing the wheel. Weinberg [74] noted that the problem with program libraries is that "everyone wants to put something in, but no one wants to take anything out." Most of this is caused by cumbersome access methods, inadequate documentation, poor adaptability of the modules, and a "not invented here" mentality [75], [76].

The libraries of the next decade must evolve into database systems that include convenient, interactive search procedures, employing hierarchical menus or the possibility of querying in a subset of a natural language to find the desired function. For the selected cells, the system must produce layers of more and more detailed documentation so that the designer can unambiguously determine whether a particular cell will suit his needs or will have to be modified. All relevant design information must also be available on-line so that modifications can be done easily.

Such a library should contain all regularly used, generic, functional units, such as memory blocks, stacks, ALU's, analog-to-digital converters—to name just a few. These components should be suitably parameterized to span a wide range of possible applications and should be available in a range of different technologies. In addition, they should be properly encapsulated and provided with unambiguous, easy-to-use, standardized interfaces by which they can be fit together without a need for the designer to pay attention to the device-level details of this interface. The effort needed to create, document, maintain, and adapt all these modules to an evolving technology can be prohibitive for small companies. It is conceivable that such libraries will be sold or leased on a commercial basis by special companies.

Automatic Consistency Checking: As outlined in the previous section, the various pieces that make up the whole VLSI system will be designed at a higher level and stored in that form to increase the designer's productivity and to maintain some technology independence. A rich set of representations is thus associated with each module. Special software is needed to maintain the consistency of the various views of each module. While such consistency checks could be built into the database, it is preferable to put the know-how of this task into special tools. As more sophisticated tools become available, they can readily be introduced without affecting the structure of the database.

By 1990 silicon compilers will have made a lot of progress. They will then hopefully be able to handle fairly complex subsystems. However, they will still not be mature enough to produce guaranteed correct results from a set of functional specifications. It may not even be possible to prove that they produce proper results for all sensible parameter combinations. A substantial amount of checking will, therefore, still be needed.

A mechanism to alleviate this situation and to catch many possible errors at compile time is to rely on built-in assertions. These assertions verify that certain restrictions hold; they can perform bounds checking on parameters or combination of parameters, monitor current densities, or control module dimensions. Today, assertions are frequently used to verify timing constraints in large systems [77], [78]. They will play an ever increasing role in future systems.

In the design of a large system, individual modules are separately constructed. For a specific implementation, these descriptions have to be compiled into a representation that unambiguously describes the information to be passed to the fabrication line. Some simulation and performance checks may have to be done after this transformation. Rather than doing these steps jointly for a whole VLSI system, it will become more practical to do it on a per-module basis. This corresponds to the notion of separate compilation of the modules of large programs; the required facilities for a suitable environment for VLSI design [79] would be rather similar. A good model is the *make* facility [80] in UNIX. A *makefile* at each node in the file system describes the dependencies between the various data descriptions which may exist at that node, as well as the rules by which a desired description can be obtained from the information on which it depends. In addition, all files carry a time stamp. When the *make* facility is invoked, all files which have predecessors with more recent time stamps will be regenerated. In this manner, incremental updates propagate through the hierarchy.

At all levels, submodules can be tested or verified independently and then protected against further accidental changes. If a deliberate change is made inside such a module, then the flags indicating that the module has been verified are reset to false. An automatic sequence of test programs will recheck the new version and assert the corresponding flags if the specified tests are passed. Thus recompilation or rechecking of modules is required only if something changes inside. Small corrections in a large layout may thus be contained to a small fraction of the overall design.

Testing and Debugging: The areas that are most in need of

advancement are debugging and testing of VLSI systems. Because of the monolithic structure of integrated circuits, more than ever before, debugging and testing must be carefully planned during the design of the system [81]. The key issue is to subdivide the VLSI system into smaller parts that are easier to test. The controllability and observability of internal nodes can be greatly improved with such scan-in scan-out techniques as LSSD [40]. This methodology can be integrated into the design systems that rely on logic gates and register-transfer components. In such a system, a special type of latch is used for every flip-flop, unless explicitly told otherwise, which can then be strung together into scan-in scan-out registers. This approach is particularly powerful if for every encapsulated module the design environment automatically generates a set of test vectors, or, if that is not possible, prompts the designer for the delivery of such a set to encourage him or her to address the issue.

In the same context, better and more technology-specific fault models will have to be developed. It is too simplistic just to deal with stuck-at faults in a solid-state system that is dominated by interconnects. *Shorts* between different wires need to be considered. The simulation of the effect of such faults *per se* is no problem. However, it is impractical to test all possible pairs of nodes for the effect of a short between them. With circuit extractors operating from the layout or from a suitable symbolic description that defines the topology unambiguously, all adjacencies can be determined, and short simulation can thus be limited to relevant pairs. Broken wires, creating *open circuits* can be reduced to corresponding stuck-at faults.

C. Ultimate Ways for Creating Systems

This section is rather speculative; it is not expected that the concepts discussed herein will become reality in this century.

Specification-Driven System Generation: There is a possibility that ultimately routine system design will be driven directly by very-high-level language system specifications. This goal has been pursued in the software domain for the last two decades; but automatic, specification-driven software generation has yet to be demonstrated on a *general* program of substantial size.

In the VLSI domain, such specification-driven designs might emerge as an outgrowth of ever more sophisticated module generators. PLA generators currently require only a few specifications concerning their intended logic functions, constraints on the ordering of inputs and outputs, and some indication to what degree minimization, splitting, and folding should be performed. As these generators evolve, one may soon be able to select from several different structural variants to cover a wide range of delay/power tradeoffs, and to map the resulting design into most common integrated circuit fabrication processes. Once such facilities have been created, a specification-driven system is not too far away. The provision of such facilities for *arbitrary* VLSI systems, however, will lag many years behind. Here, as in software engineering, the difficulty of providing exact, unambiguous specifications is a major part of the problem.

Formal verification for programs [82]-[84] has been discussed for more than a decade. However, it is still impractical, too cumbersome, and mathematically too difficult for most programmers [85]. Application to the domain of VSLI design lies even further in the future because of the more complicated medium and its "analog" behavior. However, most inventions that have revolutionized our lives, have required 20 to 50 years from first demonstration to the point where they started to have a significant impact [71]. This is true for the telegraph, the steam engine, photography, as well as for television. More readable specification languages will be developed, and most of the tedium of grinding through the formal verification steps will be automated to the point where the computer is taking care of the details under high-level directions by the designer. So, while verification methods have not yet become a smashing success in their first 15 years, it may be well before the turn of the century that they will affect the way in which large designs will be approached.

Knowledge Engineering and Expert Systems: Techniques developed in the field of Artificial Intelligence are expected to have a major influence on our development environments for software engineering as well as for VLSI design. The goal is to capture the essence of the design process and establish a formal design methodology. Once VLSI design is well enough understood that we can start to express explicitly some of the plans followed by good designers, some of that reasoning and background knowledge can be incorporated into *Expert Systems* [86]. Such expert systems combined with rich databases of suitable software tools or electronic subsystems will then be able to piece together reasonable solutions to routine problems. The specially constrained, high-performance systems will be left to the human experts for a long time to come.

Self-Testing and Self-Reorganizing Systems: Even on that day far in the future, when the emerging designs will be provably correct because of the methodology that produced them, fabrication yields will still be less than 100 percent. Testing of the fabricated circuits will thus still be a necessity. When done in the traditional way, it can become a major, if not the dominant, part of their costs. More of the testing chore has to be off-loaded to the VLSI system itself.

Here is a possible scenario. The wafer carries, in addition to the desired VLSI chips, a grid of power and control lines that connects all chips through some suitable circuitry to a few external connectors. In the testing phase, the whole array is powered up and some control signal will initiate a self-test in all the chips on the wafer. Bad chips that draw an excessive amount of power will automatically trip a circuit that disconnects them from the power grid. The remaining circuits will start a small internal state machine that tests a local ROM. If the first test is successful, a second phase will be initiated, in which these registers and the tested ROM from a more sophisticated testing machine that now can test systematically all major parts of the associated chip. If necessary, more layers of this testing hierarchy can be introduced. The final result of all those tests, a single bit of information, will be written nondestructively into an easily readable spot; this might be done by selectively blowing a small fuse, so that good and bad chips can be identified by visible inspection.

This general idea can be extended to systems that coninue to perform self-checks while they are in operation. A collection of multiple redundant blocks, constantly monitoring themselves and each other, may be a way to move into wafer scale integration. Failing blocks will be disconnected from the power supplies and spare parts will be turned on instead. This approach is only viable if the blocks themselves are large enough (more than a 100 000 devices) so that the control and monitoring overhead can be tolerated. The size of these blocks, in turn, requires a more mature technology so that blocks of a few hundred thousand devices can be built with reasonable yield.

VIII. THE ROLE OF THE DESIGNER

A. The Design Tool Manager

Just as evolving programming environments show a trend away from programming, designers of solid-state systems will spend an ever smaller fraction of their time designing at the solid-state level. The general trend has already started. The low-level layout functions get off-loaded to tools of increasing sophistication. The "tall, thin designer" [87] who carries the design from the top-level down to the layout will be a short-lived phenomenon. Solid-state systems engineers who work too much at the layout level neglect the more important design decisions that need to be made at the higher levels of abstraction. A lot of the low-level technical tasks can soon be left to computer-based design tools. The systems designer will be able to rely on tools and on prototype modules generated by expert designers. He or she will thus change from being a technical designer to being a *manager of design tools*. At the same time, *computer-aided design* will evolve into synergistic *designer-directed semiautomatic design*.

A good manager, rather than doing the job himself, will concentrate on creating an environment in which the job can get done most efficiently. This requires an attitude change on the part of the typical engineer. The most leverage out of human ingenuity can be obtained if the latter is used to build new and better tools, which then can do the job more or less automatically. In this mode, the impact of the work of individual engineers can be compounded.

B. The Expert Designer

In all the excitement about the potential of these sophisticated design environments and the power of the proposed tools, one should not overlook the role of the technical expert designer. As technology scales down, the medium gets more difficult, devices are dominated by edge effects and parasitics, and more design parameters have to be considered. It takes an expert to know whether the constrained design space provided for the average designer should be left, where it pays to cross abstraction boundaries, and when large advantages can be gained by violating the rules. Feedback from various tools can be used to check the intuitive insights of the expert.

Experts, too, will rely heavily on design tools. Even a less sophisticated tool can be used to survey the design space in order to decide which approaches are worth studying in detail. The layout expert who plans to hand-pack a certain array structure to the limit possible in a given technology, will be well advised to use a less skillful, automatic compaction program first to quickly find promising topologies that can then be optimized further by hand. Without such an exploratory search of the design space, the expert designer might get "trapped in a local minimum," overlooking a deeper global minimum resulting from a different topology. Seeing a succession of many possible solutions also leads to a better understanding of the tradeoffs, and develops heuristics that permit homing-in on desirable solutions more quickly.

Designers most interested and skilled in low-level design will make their contributions by improving the basic cells on which the solid-state systems compilers must rely. They will formalize their knowledge and incorporate it into the systems that drive the design tools. The support by the expert designer will be vital in keeping the automatic design environment from becoming obsolete.

C. Multi-Designer Teams

Even with a very powerful set of almost fully automatic design tools, large VLSI systems will very likely still require the help of more than one designer. A big problem, well known in the area of software engineering, lies in the management of a design team. Too large a team can become completely ineffective [32] if so much time is spent in meetings discussing *what should be done*, that there is no time left for *actually doing* something. Computers can alleviate this situation in two ways. First, by providing more sophisticated tools, computers will allow the individual designer to achieve more, and this will reduce the size of design teams. Secondly, properly set up electronic communication can also increase the efficiency of the interactions between designers.

The foundation is provided by a sophisticated electronic mail system. A lot of time is wasted when busy people try to get in touch with each other in real time. Forwarding messages to an electronic mailbox that is read when the other party next uses the computer permits more frequent interaction. In addition, electronic messages tend to be relatively terse and devoid of a lot of the additional overhead encountered in many telephone conversations. Most people can deal with an order of magnitude more computer messages than phone conversations per unit time.

Such a system can be expanded readily to contain interest-oriented mailing lists for semipublic announcements and electronic bulletin boards that store relevant information about an emerging system design or a family of design tools. By browsing through these bulletin boards periodically, every member of a large design department can keep informed about issues of general concern.

Electronic mailboxes and bulletin boards can be integrated with the design system itself. Analysis tools, such as geometrical design rule checkers, or update tools such as the described *make* facility, may do some of their tedious work during the night and then send the results and suitable diagnostics to the mailbox of the designer. Changes to cells in a common library may be announced on the bulletin board, and reported to every designer employing these cells in a system that is currently in the active design stage.

Computers can also help formalize the tools used in the management of the design team. They can keep records of the formal distribution of tasks and responsibilities, and of planned and actual project schedules. They can record design decisions, preserve the specifications of the various subsystems and their interfaces, and use them to verify the design in functional simulations. Such a system could even initiate group meetings for design reviews and structured walk-throughs.

IX. RECOMMENDATIONS FOR ENGINEERING CURRICULA

Currently the more innovative design systems are being developed by people with a strong background in computer science, software engineering, or artificial intelligence. The tools required as well as the VSLI systems to be built require an explicit methodology to cope with the complexity involved; ad hoc methods are no longer sufficient.

Hands-on Experience with Complexity Management: It is thus recommended that courses be introduced into the engineering curriculum that explicitly and implicitly teach some of the techniques already well established in software engineering. Courses that concentrate on one big project, demonstrating how the techniques can be used to subdivide a large task into manageable parts, will prove highly beneficial. At the University of California at Berkeley several courses that give the students experience in complexity management exist in the area of Computer Science, even at the undergraduate level. These are courses in which the students build major parts of a compiler, modify significant pieces of an operating system, or design complete subsystems in a VLSI layout course. In an advanced course sequence, a small group of students has recently built a complete microprocessor [41]. On the other hand, such courses are much sparser in the traditional engineering schools, where the emphasis is typically more on design at the "component" level. There is, for instance, rarely a course on how to design a complete television set or a frequency spectrum analyzer.

Emphasis on Specification and Testing: Functional partitioning goes hand in hand with properly specifying the resulting submodules. To emphasize this point, all designers should be forced to specify precisely what it is that they are going to design or implement, be it hardware or software. Before they write a single line of code or draw a single circuit schematic, they should write down exactly how they are going to check whether their product fits certain specifications.

They should also have prepared a complete debugging plan and hand in their test files with their project proposal. Such a debugging plan starts with the simplest possible test, e.g., measuring the power consumption, for the case when chips come back in which not a single recognizable signal sequence appears at the output. Then, in case that this simple test works, there is a sequence of incrementally more sophisticated experiments building on the results of earlier tests. Thinking about testing early in the design process and doing the thorough preparations outlined above will strongly influence the design. The changes that such discipline enforces will be most appreciated when the prototype chips come back from fabrication.

X. Conclusion

Many of the techniques that are needed to manage the complexity of VLSI system design are already known. During the last decade, several of them have reached a stage of maturity in the domain of software engineering. However, they are only slowly entering the domain of integrated circuit design; they have not yet appeared as integral parts of commercial VLSI design tools.

The introduction of these new design techniques and their emergence in VLSI design tools will change the style in which VLSI systems will be produced in the future. The bulk of the products will no longer be designed by circuit wizards who, by ad hoc methods based on their experience and intuition, will come up with chip designs that do miraculous and mysterious things. More circuits will be specified by high-level descriptions and by functional specifications from which the lower level descriptions and ultimately the layout will be compiled in an automatic or semiautomatic manner. The time will soon come where a systems engineer can draw on a set of sophisticated semiautomatic tools that will produce a properly structured, understandable, and testable chip that will adhere to specifications and has a good chance to work the first time.

To make this happen sooner rather than later requires primarily a change in attitude. Such a change is taking place in many research laboratories. It can gain momentum if the engineering schools in their curricula put proper emphasis on explicit and practical instruction in complexity management.

Acknowledgment

The thoughts expressed in this paper have clearly been influenced by my environment: the many colleagues at UC Berkeley and other academic and industrial institutions with whom I interact during my work, at conferences, and indirectly through publications and correspondence. I am sure that many of my friends will recognize their models and metaphors appearing implicitly or explicitly in this text. I would like to express to all of them my sincere gratitude for the stimulating environment and the fruitful personal interactions that make it so exciting to work in this rapidly evolving field. Special thanks go to S. C. Johnson, H. T. Kung, J. K. Ousterhout, R. L. Russo, S. Trimberger, P. W. Verhofstadt, and A. I. Wasserman who have given me a lot of constructive criticism on earlier drafts.

References

[1] G. E. Moore, "Progress in digital integrated electronics," presented at the IEEE Int. Electron Devices Meet., Talk 1.3 (Washington, DC, Dec. 1975).

[2] G. E. Moore, Quote at the First Caltech Conference on VLSI, Pasadena, CA, Jan. 1979.

[3] J. N. Buxton, "Software engineering," in *Programming Methodology*, D. Gries, Ed. New York: Springer, 1978, pp. 23-28.

[4] D. V. Steward, "Analysis and complexity," in *Systems Analysis and Management*. New York: Petrocelli Books, 1981, p. 2.

[5] E. Yourdon and L. L. Constantine, *Structured Design*. Englewood Cliffs, NJ: Prentice-Hall, 1979.

[6] P. Freeman and A. I. Wasserman, *Tutorial: Software Design Techniques*, 3rd ed. Los Alamitos, CA: IEEE Computer Society, 1980.

[7] L. A. Belady and M. M. Lehman, "Characteristics of large systems," in *Research Directions in Software Technology*, P. Wegner, Ed. Cambridge, MA: MIT Press, 1979, pp. 106-142.

[8] C.A.R. Hoare, "Hints on programming language design," Memo AIM 224, Stanford Artificial Intelligence Laboratory (Oct. 1973). Reprinted in *Tutorial: Programming Language Design*, A. I. Wasserman, Ed. Los Alamitos, CA: IEEE Computer Society, 1980.

[9] G. A. Miller, "The magical number seven, plus or minus two: Some limitations on our capacity for processing information," *Psychol. Rev.*, vol. 63, pp. 81-97, 1956.

[10] W. W. Lattin, J. A. Bayliss, D. L. Budde, J. R. Rattner, and W. S. Richardson, "A methodology for VLSI chip design," *Lambda*, vol. 2, no. 2, pp. 34-44, 2nd quarter, 1981.

[11] D. L. Johannsen, "Bristle blocks: A silicon compiler," in *Proc. Caltech Conf. VLSI* (Pasadena, CA), pp. 303-310, Jan. 1979.

[12] J. M. Siskind, J. R. Southard, and K. W. Couch, "Generating custom high performance VLSI designs from succinct algorithmic descriptions," in *Proc. Conf. on Adv. Research in VLSI* (MIT, Cambridge, MA), pp. 28-40, Jan. 1982.

[13] H. E. Shrobe, "The data path generator," in *Proc. Conf. on Adv. Research in VLSI* (MIT, Cambridge, MA), pp. 175-181, Jan. 1982.

[14] R. J. Blumberg and S. Brenner, "A 1500 gate, random logic, large-scale integrated masterslice," *IEEE J. Solid-State Circuits*, vol. SC-14, pp. 818-823, 1979.

[15] B. W. Kernigham, D. G. Schweikert, and G. Persky, "An optimum channel routing algorithm for polycell layouts of integrated circuits," in *Proc. 10th Design Automation Workshop* (Portland, OR), pp. 50-59, June 1973.

[16] V. D. Agrawal et al., "The mixed mode simulator," in *Proc. 17th Design Automation Conf.*, pp. 618-625, June 1980.

[17] A. R. Newton, "Timing, logic, and mixed mode simulation for large MOS integrated circuits," NATO Advanced Study Institute on Computer Design Aids for VLSI Circuits, Sogesta-Urbino, Italy, 1980.

[18] G. D. Bergland, "A guided tour of program design methodologies," *Computer*, vol. 14, no. 10, pp. 13-36, Oct. 1981.

[19] N. Wirth, "Program development by stepwise refinement," *Commun. ACM*, vol. 14, no. 4, pp. 221-227, Apr. 1971.

[20] M. A. Jackson, *Principles of Program Design*. New York: Academic Press, 1975.

[21] T. DeMarco, *Structured Analysis and System Specification*. Englewood Cliffs, NJ: Prentice-Hall, 1979.

[22] C. Gane and T. Sarson, in *Structured Systems Analysis: Tools and Techniques*. Englewood Cliffs, NJ: Prentice-Hall, 1979.

[23] H. T. Kung and C. E. Leiserson, "Algorithms for VLSI processor arrays," in *Introduction to VLSI Systems*. C. A. Mead and L. A. Conway, Eds. Reading, MA: Addison-Wesley, 1980.

[24] D. L. Johannsen, "Our machine: A microcoded LSI processor," Display File 1826, Dept. Comp. Science, Caltech, Pasadena, CA, July 1978.

[25] R. W. Sherburne, M.G.H. Katevenis, D. A. Patterson, and C. H. Séquin, "Datapath design for RISC," in *Proc. Conf. on Adv. Research in VLSI* (MIT, Cambridge, MA), pp. 53-62, Jan. 1982.

[26] J. D. Warner, *Logical Construction of Programs*. New York: Van Nostrand, 1974.

[27] E. W. Dijkstra, *A Discipline of Programming*. Englewood Cliffs,

NJ: Prentice-Hall, 1976.
[28] D. Gries, "An illustration of current ideas on the derivation of correctness proofs and correct programs," *IEEE Trans. Software Eng.*, vol. SE-2, no. 4, pp. 238-244, Dec. 1976.
[29] C. Böhm and G. Jacopini, "Flow diagrams, turing machines and languages with only two formation rules," *Commun. ACM*, vol. 9, no. 5, pp. 366-371, May 1966.
[30] H. D. Mills, "Mathematical foundations for structured programming," IBM Tech. Rep. FSC 72-6012, Federal Syst. Div., Gaithersburg, MD, 1972.
[31] D. E. Knuth, "Structured programming with goto statements," *Computing Surveys*, vol. 6, no. 4, pp. 261-301, Dec. 1974.
[32] F. P. Brooks, Jr., *The Mythical Man-Month*. Reading, MA: Addison-Wesley, 1975.
[33] E. W. Dijkstra, "The humble programmer," *Commun. ACM*, vol. 15, no. 10, pp. 859-866, Oct. 1972.
[34] T. Teitelbaum and T. Reps, "The Cornell program synthesizer: A syntax-directed programming environment," *Commun. ACM*, vol. 24, no. 9, pp. 563-573, Sept. 1981.
[35] M. Linton, "A debugger for the Berkeley Pascal system," Master's Report, U.C. Berkeley, June 1981.
[36] S. Gutz, A. I. Wasserman, and M. J. Spier, "Personal development systems for the professional programmer," *Computer*, vol. 14, no. 4, pp. 45-53, Apr. 1981.
[37] D. Hightower and F. Alexander, "A mature I^2L/STL gate array layout system," in *Dig. Papers, COMPCON*, pp. 149-155, Feb. 1980.
[38] D. L. Parnas, "The use of precise specifications in the development of software," in *Information Processing 77 (Proc. IFIP Congress)*. Amsterdam, The Netherlands: North Holland, 1977, pp. 861-867.
[39] G. J. Myers, *The Art of Software Testing*. New York: Wiley, 1980.
[40] E. B. Eichelberger and T. W. Williams, "A logic design structure for LSI testability," *J. Des. Automat. Fault-Tolerant Comput.*, vol. 2, pp. 165-178, May 1978.
[41] D. A. Patterson and C. H. Séquin, "RISC I: A reduced instruction set VLSI computer," in *Proc. 8th Int. Symp. on Computer Architecture* (Minneapolis, MN), pp. 443-457, May 1981.
[42] W. A. Wulf, "Compilers and computer architecture," *Computer*, vol. 14, no. 7, pp. 41-47, July 1981.
[43] A. B. Glaser and G. E. Subak-Sharpe, "Failure, reliability and yield of integrated circuits," in *Integrated Circuit Engineering*. Reading, MA: Addison-Wesley, 1978, pp. 746-799.
[44] R. Ayres, "Silicon compilation—Hierarchical use of PLAs," in *Proc. Caltech Conf. on VLSI* (Pasadena, CA), pp. 311-326, Jan. 1979.
[45] S. S. Eaton, D. Wooton, W. Slemmer, and J. Brady, "Circuit advances propel 64-k RAM across the 100 ns barrier," *Electronics*, vol. 55, no. 6, pp. 132-136, Mar. 24, 1982.
[46] W. A. Wulf, "Keynote address," presented at the Symp. on Arch. Supp. for Prog. Lang. and Operat. Syst., Palo Alto, CA, Mar. 1982.
[47] —, "Trends in the design and implementation of programming languages," *Computer*, vol. 13, no. 1, pp. 14-25, Jan. 1980.
[48] M. Stefik, D. G. Bobrow, A. Bell, H. Brown, L. Conway, and C. Tong, "The partitioning of concerns in digital systems design," in *Proc. Conf. on Adv. Research in VLSI* (MIT, Cambridge, MA), pp. 43-52, Jan. 1982.
[49] M. Tucker and L. Scheffer, "A constrained design methodology for VLSI," *VLSI Des.*, vol. 3, no. 3, pp. 60-65, May 1982.
[50] D. T. Fitzpatrick, J. K. Foderaro, M.G.H. Katevenis, H. A. Landman, D. A. Patterson, J. B. Peek, Z. Peshkess, C. H. Séquin, R. W. Sherburne, and K. S. VanDyke, "VLSI implementation of a reduced instruction set computer," in *Proc. CMU Conf. on VLSI Systems and Computations* (Pittsburgh, PA), pp. 327-336, Oct. 1981.
[51] M. E. Newell and D. T. Fitzpatrick, "Exploiting structure in integrated circuit design analysis," in *Proc. Conf. on Adv. Research in VLSI* (MIT, Cambridge, MA), pp. 84-92, Jan. 1982.
[52] L. K. Scheffer, "A methodology for improved verification on VLSI designs without loss of area," in *Proc. 2nd Caltech Conf. on VLSI* (Jan. 19-21, 1981).
[53] T. Whitney, "A hierarchical design analysis front end," in *Proc. VLSI81, Int. Conf. on Very Large Scale Integration* (Edinburgh, Scotland), pp. 217-225, 1981.
[54] S. C. Johnson, "Hierarchical design validation based on rectangles," in *Proc. Conf. on Adv. Research in VLSI* (MIT, Cambridge, MA), pp. 97-100, Jan. 1982.
[55] L. A. Conway, A. Bell, and M. E. Newell, "MPC79: A large-scale demonstration of a new way to create systems in silicon," *Lambda*, vol. 1, no. 2, pp. 10-19, 2nd quarter, 1980.
[56] B. Hon and C. H. Séquin, *Guide to LSI Implementation* (2nd revised and extended edition), Xerox PARC, Palo Alto, CA, Jan. 1980.
[57] W. D. Jansen and D. G. Fairbairn, "The silicon foundry: Concepts and reality," *Lambda*, vol. 2, no. 1, pp. 16-26, 1st quarter, 1981.
[58] W. J. Haydamack, Public Lecture, University of California, Berkeley, Fall 1981.
[59] J. D. Williams, "STICKS—A graphical compiler for high level LSI design," in *AFIPS Conf. Proc., NCC*, vol. 47, pp. 289-295, 1978.
[60] A. Dunlop, "SLIP—Symbolic layout of integrated circuits with compaction," *Computer-Aided Des.*, vol. 10, pp. 387-391, Nov. 1978.
[61] M. Y. Hsueh and D. O. Pederson, "Computer-aided layout of LSI circuit building blocks," in *Proc. IEEE Int. Solid-State Circuits Conf.* (Tokyo, Japan), pp. 474-477, 1979.
[62] C. H. Séquin, "Pictures versus parentheses: Design methodologies of the 1980's," Panel Discussion at COMPCON, San Francisco, CA, Feb. 1982.
[63] A. I. Wasserman, "Automated development environments," *Computer*, vol. 14, pp. 7-10, Apr. 1981.
[64] B. W. Kernighan and J. R. Mashey, "The UNIX programming environment," *Computer*, vol. 14, no. 4, pp. 12-22, Apr. 1981.
[65] W. Teitelman and L. Masinter, "The INTERLISP programming environment," *Computer*, vol. 14, no. 4, pp. 25-33, Apr. 1981.
[66] A. I. Wasserman and S. Gutz, "The future of programming," *Comm. ACM*, vol. 25, no. 3, pp. 196-206, Mar. 1982.
[67] A. R. Newton, D. O. Pederson, A. L. Sangiovanni-Vincentelli, and C. H. Séquin, "Design aids for VLSI: The Berkeley perspective," *IEEE Trans. Circuits Syst.*, vol. CAS-28, no. 7, pp. 666-680, July 1981.
[68] S. R. Bourne, "UNIX time-sharing system: The UNIX shell," *Bell Syst. Tech. J.*, vol. 57, no. 6, pp. 1971-1990, Jul.-Aug. 1978.
[69] C. H. Séquin and A. R. Newton, "Description of STIF 1.0," in *Design Methodologies for VLSI*. Groningen, The Netherlands: Noordhoff, Jan. 1982, pp. 147-171.
[70] S. A. Ellis, K. H. Keller, A. R. Newton, D. O. Pederson, A. L. Sangiovanni-Vincentelli, and C. H. Séquin, "A symbolic layout design system," presented at Int. Symp. on Circuits and Systems, Rome, Italy, May 1982.
[71] A. K. Graham, "Software design: Breaking the bottleneck," *IEEE Spectrum*, vol. 19, no. 3, pp. 43-50, Mar. 1982.
[72] M. J. Rochkind, "The source code control system," *IEEE Trans. Software Eng.*, vol. SE-1, no. 4, pp. 364-370, Dec. 1975.
[73] E. H. Bersoff, V. D. Henderson, and S. G. Siegel, "Software configuration management: A tutorial," *Computer*, vol. 12, no. 1, pp. 6-14, Jan. 1979.
[74] G. M. Weinberg, *The Psychology of Computer Programming*. New York: Van Nostrand-Reinhold, 1971.
[75] L. A. Belady, "Evolved software for the 80's," *Computer*, vol. 12, no. 2, pp. 79-82, Feb. 1979.
[76] A. I. Wasserman and L. A. Belady et al., "Software engineering: The turning point," *Computer*, vol. 11, no. 9, pp. 30-41, Sept. 1978.
[77] T. M. McWilliams, "Verification of timing constraints on large digital systems," in *Proc. 17th Design Automation Conf.* (Minneapolis, MN), pp. 139-147, June 1980.
[78] W. E. Cory and W. M. VanCleemput, "Development in verification and design correctness," in *Proc. 17th Design Automation Conf.* (Minneapolis, MN), pp. 156-164, June 1980.
[79] A. R. Newton, "The VLSI design challenge of the 80's," in *Proc. 17th Design Autom. Conf.*, pp. 343-344, June 1980.
[80] S. I. Feldman, "Make—A program for maintaining computer programs," *Software—Practice and Experience*, vol. 9, pp. 255-265, Apr. 1979.
[81] J. Grason and A. W. Nagle, "Digital test generation and design for testability," in *Proc. 17th Design Automation Conf.* (Minneapolis, MN), pp. 175-189, June 1980.
[82] P. Naur, "Proof of algorithms by general snapshot," *BIT*, vol. 6, pp. 310-316, 1966.
[83] R. Floyd, "Assigning meaning to programs," in *Proc. Symp. in Applied Mathematics*, vol. 19, pp. 19-32, American Math. Soc., Providence, RI, 1967.
[84] C.A.R. Hoare, "An axiomatic basis for computer programming," *Commun. ACM*, vol. 12, no. 10, pp. 576-583, Oct. 1969.
[85] R. DeMillo, R. Lipton, and A. J. Perlis, "Social processes and proofs of computer programs," *Commun. ACM*, vol. 22, no. 5, pp. 271-280, May 1979.
[86] M. Stefik, J. Aikins, R. Balzer, J. Benoit, L. Birnbaum, F. Hayes-Roth, and E. Sacerdoti, "The organization of expert systems: A perspective tutorial," Tech. Rep., Xerox PARC, Palo Alto, CA, Jan. 1982.
[87] C. A. Mead, "VLSI and technological innovation," Keynote Address, presented at the Caltech Conf. on VLSI, Pasadena, CA, Jan. 1979.

A Knowledge-Based System Using Design For Testability Rules

Paul W. Horstmann*

Dept. E34, Bldg. 901
IBM Corp., General Technology Division
P.O. Box 390
Poughkeepsie, NY 12602

ABSTRACT

This paper describes a prototype system developed at Syracuse University for automating the design for testability by using artificial intelligence techniques. The system's objective is to use logic programming for the creation of a Design for Testability (DFT) expert system. For a limited set of test cases, this expert system functions as a "testability" expert, or as a designer's assistant, in that it will check for DFT rule violations, and should these exist, transform the design to remove them. The results of preliminary experiments are given and and some future development on this system is discussed.

INTRODUCTION

The number and use of various Design For Testability (DFT) techniques has greatly proliferated due to the increase in VSLI design complexity and the attendant testing problems (1). These techniques, which can be used at both the logical and physical levels of design, include Level Sensitive Scan Design (LSSD), Scan/Set, Scan-Path, Random Access Scan, etc. These approaches essentially structure the memory elements (flip/flops, latches and registers) of the design to avoid uncontrolled feedback loops, and assure the controlability/observability of the memory elements. In addition to these enhancements of testability, other techniques under development allow the efficient self-testing (or built-in test (BIT)) of IC chips and modules. These techniques include methods for pseudo-random pattern generation and signature analysis (BILBO, AAFIS, etc.) and procedures that specifically take advantage of certain forms of regularity in the design (Programmable Logic Arrays (PLAs), Random Access Memory (RAM), etc.).

These DFT techniques are driven by the cost of implementing the design, the testing approach to be used, and the allowable defect level of the shipped product. The trade-offs between the various DFT techniques, and the preceding factors often require the use of a testability expert. This paper addresses the development of such a testability expert as a knowledge-based system using logic programming. This expert system represents initial work (2) done at Syracuse University on the mechanization of techniques which designers employ in modifying their designs so as to satisfy given design rules.

The expert system discussed in this paper will check a design for DFT rule violations using a simplified set of rules that define LSSD, and show how these violations can be analyzed and corrected by the expert system. The system can allow the extension of these rules to include non-LSSD approaches as required. However, the initial project did not include these extensions. In addition to checking the DFT rules, the system also generates control and observation information, which can be used by an automatic test pattern generator. The expert system uses a design model based upon a logic programming approach and is essentially technology independent, although some provisions have been made for constructs such as "wired-ORs," etc. It is the intent of this system to provide as rule-based approach to the checking of testability and not to provide an explicit testability measure, e.g. TMEAS, SCOAP, etc. (1). At the time of this paper, experiments have been performed using a gate-level description of an up/down counter, a functional description of a DMA control chip, and a bus/operator description of the M6800 microprocessor.

OVERVIEW OF EXPERT SYSTEMS

An expert system (3) is essentially a computer program (system) based on a collection of task specific knowledge (or data) which mimics the reasoning of human experts. The system uses expert knowledge to solve problems, such as medical diagnosis and computer system configuration, that normally do not yield numerical solutions. The use of expert systems in Computer Aided Design (CAD) is in the initial stages of development. Two notable examples of these sytems are: 1) PALLADIO - This system, currnetly in use at XEROX-PARC (4) is being used to investigate new design methodologies and techniques. 2) DAA - This system, developed at Carnegie-Mellon Univeristy (5)   , uses knowledge based expert programs to synthesize VLSI designs (expressed in terms of registers, operators, data paths, and control siganls) from algorithmic expressions of the design. These rules, which describe this system, define the various synthesis operations and the contexts in which they are applicable. These examples were chosen to be illustrative and are by no means a complete set of the expert (or rule-based) systems in existance.

An expert system usually consists of a knowledge base and inference mechanisms, which manipulate the

---

* This work was supported by an IBM Graduate Fellowship while the author completed his degree requirements at Syracuse University.

knowledge base. The knowledge base is a set of rules and facts, which describe the expert's knowledge. These rules are usually in the form of IF-THEN rules (Fig. 1), which indicate actions that may be taken if certain situations arise. The inference mechanisms describe the means by which new knowledge is generated from the current knowledge base. These mechanisms, which can also be rules, guide the expert system through its search for the solution of a particular goal. A simple strategy for inference rules is to search the rules data base to find a rule whose antecedents (conditional elements) match assertions in the data base. This form of inference mechanism is used by the logic programming language, PROLOG, as implemented on the DEC KL-10 (6). PROLOG is described in more detail in the next section. A more complicated strategy is to organize the rules into networks (often called semantic networks), which determine when certain rules get applied.

IF:
    The current context is design transformation,
    and the current DFT violation is clock observation,
    and a multiplexer is to be added,

THEN:
    Pick a design output w/o a currently attached multiplexer,
    and create a multiplexer node,
    and connect control input to the design input for test control,
    and connect the terminal node in the given clock path to the test data input of the multiplexer,
    and disconnect the output from its source node,
    and connect output to output pin of the multiplexer,
    and connect the source node to the system data input of the multiplexer.

Figure 1. Example of Knowledge Base Rule

The expert knowledge (or knowledge base) used in the implementation at Syracuse resides in the DFT, control sequence, and transformation rules. The DFT rules state what the expert believes to be an easily testable design. The rules defining the control sequence generation describe the information the expert feels is needed for further processing in automatic test pattern generation and tester operation. The transformation rules describe how an expert may transform the design to achieve the degree of testability defined by the DFT rules. One of the short-comings of expert systems is their inability to efficiently handle more than one level of abstraction or domain of representation (system architecture, data-path descriptions, functional descriptions, circuit level descriptions, etc.) By limiting the domain of knowledge to the functional level of description in both the design model and the DFT rules, this problem with multiple levels of knowledge or abstraction is avoided.

LOGIC PROGRAMMING AS APPLIED TO THE EXPERT SYSTEM

The logic programming language PROLOG was chosen to implement this DFT expert system. As a language, PROLOG has three major attributes that influenced the decision. 1) All of the information needed for the expert system could be entered as PROLOG clauses. This included the design model, the functional behaviors, the DFT expert knowledge, the inference and system clauses, and the actual proof mechanisms used. 2) PROLOG is an easy language to learn and debug. Complicated systems can be written, and easily debugged using PROLOG. For example, one person built a nominal delay functional simulator using PROLOG in an elapsed time of one month (7). 3) Syracuse University has a significant development activity in logic programming. It was expedient to make use of this activity in the system development.

Logic programming, a technique used for some artificial intelligence (AI) systems, combines logic clauses (assertions) and a form of automatic deduction as the inference mechanism. The language PROLOG has gained acceptance throughout the AI community (especially Europe and Japan) and may be a base for future high-level programming languages and functions. For example, both Japan and the United Kingdom have chosen PROLOG as part of their fifth generation computer research (8). These efforts have as one of their goals the implementation of a PROLOG processor in VLSI.

The basis for logic programming is first-order predicate calculus, where the logical relations are expressed as sets of assertions or clauses. These clauses are both declarative in that they describe objects and relationships, and procedural in that they are executed as functions. The details of the operation and use of PROLOG will not be further described here.

For the prototype system, the meta-PROLOG system developed by Bowen and Kowalski (9) is used as the proof (or inference) engine. Meta-PROLOG sits on top of the PROLOG base system and allows experimentation with a clause set that describes a specific theory. A major component of this meta-PROLOG system is the DEMO theorem prover. DEMO will take a given goal and prove it true or false against a specified input (or base) set of facts and rules. In the DFT expert system, the input (or base) theory consists of the PROLOG design model, the functional behaviour clauses, and the inference clauses (defined in the next section). The goals given to DEMO are the actual DFT rules to be checked. In effect, the DEMO prover is iteratively invoked by the expert system for each individual rule: with each success (return of true) the next rule is tried, until all of the rules have been used. When a rule fails DEMO, the attempted proof (along with the rule and input theory) is used as input data by the violation analyzer. The attempted proof is the set (search tree) of selected clauses that were used during the failing proof process. In addition to the clauses, the variable bindings (substitutions) are also saved as part of the attempted proof. Each DFT rule is tried (proved) over all of the nodes in the design model. By limiting the scope of the resolution process to only the design model and the inference clauses, inadvertant execution of the DFT system clauses is avoided. In addition to clause execution, the DEMO prover allows more control than the regular PROLOG system over both the order of clause resolution and the direction the proof process takes. Regular PROLOG will proceed in a left-to-right depth-first clause order in its proof process. The DEMO system can allow any arbitrary order for the proof process as specified by the user of DEMO.

DESIGN FOR TESTABILITY EXPERT SYSTEM

The representation of a digital design in PROLOG consists of a set of assertions (facts), which describe the node interconnections, and the functions of the nodes. The functional descriptions can be either structural or behavioral, and are also implemented using PROLOG. A simple expample of a design, along with PROLOG design clauses is shown in Fig. 2.

The PROLOG design description (called the design model) in Fig. 2 contains two general types of statements. The first (labelled (1)) describes the node functions and their interconnections. Each of these clauses gives a node name, node function, a list of the come-from nodes (and pin names), and a list of the go-to nodes (and pin names). The second type of clause (labelled (2)) states the special behavior of the node indicated. For example, each of the clock inputs has a clause defining that input pin to be a clock.

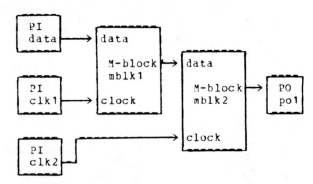

a) Schematic of Design

```
data(pi,(),(c(mblk1,data,_))).        (1)
clk1(pi,(),(c(mblk1,clock,_))).
clk2(pi,(),(c(mblk2,clock,_))).
clk1(clock).                          (2)
clk2(clock).
mblk1(m-block,(c(data,data,_),
              c(clk1,clock,_)),
       (c(mblk2,data,_))).
mblk2(m-block,(c(mblk1,data,_),
              c(clk2,clock,_)),
       (c(po1,_,_))).
po1(po,(c(mblk2,_,_)),()).
```

b) PROLOG Design Clauses

Figure 2. Simple Design PROLOG Clauses

Given this design model, the DFT rules checker and analyzer operates in the following fashion:

1. A design preprocessor program (under development) or a designer, enters the design model to the system in PROLOG logic clauses. Any memory elements (flip/flops, latches, registers, etc.) in the design are assumed to be represented by "m-blocks." These blocks (functions) serve to indicate the placement of the respective memory elements in the design structure. The m-block is essentially a single-clocked polarity-hold data latch.

2. The set of DFT rules (including the control sequence and transformation rules) to be used for this design are specified and given to the DFT rules checker/analyzer. These rules are predefined and accessed automatically during the system operation. More than one set or version of the rules may be used for the DFT check and analysis since a single design methodology may not be optimal for all designs.

3. The DEMO system is invoked by the expert system to check the DFT rules.

4. Those rules that find insufficient behavior specification in the design model will make appropriate assumptions about the design. Designer assistance may be required, depending upon the situation. For example, if only one input can be found for a required clock input, the system assumes there is only one possible choice. However, if a choice between inputs is possible, then the designer may be called upon to make the decision.

5. Any rule that notes structural violations in the design (by failure to find a proof by DEMO) will pass to the analysis portion of the DFT checker. This analyzer will attempt to identify the section of the design that needs correction in order to pass the given rule.

6. After the analysis is complete, the system will transform the design by the addition of more design clauses to conform to the DFT rules. This transformation may either be automatically performed (with nominal assistance from the designer when design choices must be made), or be manually performed by the designer.

7. Iteration of steps 3-6 continues until all of the rules have been checked and proved true. After checking is complete, the required control sequence information for an automatic test pattern generator, or tester, will be derived from the results of the rule checking. This derivation will be dependent upon a successful check of the involved DFT rules.

DESIGN RULES IN THE DFT EXPERT SYSTEM

The following paragraphs describe some of the DFT rules and clauses that have been developed for the expert system. These rules are loosely based on the Level Sensitive Scan Design (LSSD) DFT approach developed at the IBM (10). Currently there are a total of 11 actual rules defined in the DFT system and these are comprised of approximately 40 PROLOG clauses. In addition to the rules themselves, PROLOG inference clauses anaylze the design model in reference to the DFT rules. These clauses essentially tell the PROLOG analyzer how to manipulate the nodal description of the design. The expert knowledge used in this system resulted from an analysis of the LSSD methodology, a study of some of the automatic test pattern generators in use at IBM and the author's work on DFT techniques (2). For a system to be put in production use, this knowledge may be inadequate, and a larger collection of expert knowledge may be required to describe different variations and implementations of the LSSD concepts.

Two key examples of the inference clauses are the CONNECT and FCTMATCH clauses. The CONNECT clause has the form:

CONNECT(Node1,Node2,Control).

Its basic operation is either to return a value of true if Node1 is connected to Node2, or to return a nodename by instantiating the variable specified as either Node1 or Node2. The Control variable is used to limit the set of potential path candidates for the path between Node1 and Node2. This is used to prohibit path tracing through memory elements for some of the design rules, and for other types of path constraint. By using an unassigned (unbound) variable as either Node1 or Node2, the come-froms or go-tos of the given Node2 or Node1 can be found. This is an example of one of logic programming's useful features that allows the same clause definition to be used to check the occurance of a particular situation and if any information is unspecified, fill in the required values in order to satisfy the logic relationships. Unspecified (uninstantiated) variables are treated as "don't cares" by the PROLOG match (unification) process. When a valid match is found in the clause data-base, and unification occurs, any unbound variables are bound to the values found in the found clause head. For example, use the design shown in Fig. 2 and the invocation of CONNECT given below:

CONNECT(clk1,Goto,noControl).

This invocation of CONNECT will return with the Goto variable instantiated (bound) to the value "mblk1," which is the goto (or successor) node for "clk1." The term "noControl" is used to indicate that no constraints are to placed on the path set used by CONNECT. The backtracking capbility of PROLOG is also used by the inference clauses to find other occurrences of nodes, which will satisfy the given values. Since PROLOG will uninstantiate any bound variables during the backtrack operation, this can be used to try different variable bindings. Continuing with the example given above, should backtracking occur and the CONNECT clause be retried, the clause will return in the "Goto" variable the value "mblk2," then "po1," and then, upon subsequent attempts CONNECT will fail since it has exhausted all possibilities for the successor path of the "clk1" node.

The second clause, FCTMATCH, is used in a similiar fashion and has the following format:

FCTMATCH(Nodename,FunctionName).

This clause will determine if the given node (via Nodename) has the indicated function (via FunctionName) or as in the case of CONNECT it will return the appropriate instantiated (bound) variable if either Nodename or FunctionName is unassigned and a potential match can be found in the design model.

The DFT rules used in the checking operation use the inference clauses (those above and others) to express the required relationships for a structured design. One example of the DFT rules is the m-block observability (MBO) rule as shown in Fig. 3. The set of system clauses that define the DFT rule checker is constructed such that each node of the design is checked for each rule statement. The DFT rule checker attempts to show for this rule that for a given node: 1) The first clause succeeds immediately unless the given node is an m-block, i.e., the MBO rule is automatically true for all design nodes that are not m-blocks. 2) If the node is an m-block, it must connect to a design output via a path which does not traverse any m-blocks (the control value for CONNECT removes all m-block nodes from consideration in this search for successor nodes); 3) If it is an m-block, and it does not connect to an output, then it must connect directly (the control will restrict the CONNECT operation to paths of length one) to another m-block whose clock is different from the current node's clock (via the EXCLK predicate). In effect, each node in the design must succeed for one of the three MBO clauses (which describe the MBO rule) if the design is to be considered violation-free for this rule.

```
mbo(Node):-not(fctmatch(Node,m-block)).
mbo(Node):-fctmatch(Node,m-block),
          connect(Node,Outnode,((mblock,1)),
          fctmatch(Outnode,po).
mbo(Node):-fctmatch(Node,m-block),
          connect(Node,Outnode,((any,1)),
          fctmatch(Outnode,m-block),
          exclk(Node,Outnode).
```

NOTE: ":-" means "implied by"

Figure 3. Example of m-block Observability Rule

The use of the EXCLK predicate in the second clause for the MBO rule is an example of the nesting of DFT rules. EXCLK is a DFT rule that tells the expert system exactly what is meant by "exclusive" clocks for the testability rules. The following statements show how PROLOG processes the MBO rule for the design shown in Fig.2

-Pick first node, data,
-Fail on FCTMATCH against an m-block, implies that not(FCTMATCH) succeeds.
-Therefore MBO (clause 1) succeeds.
-Pick node, clk1,
-Fail on FCTMATCH against an m-block, implies not(FCTMATCH) succeeds.
-Therefore MBO (clause 1) succeeds.
-Pick node, clk2,
-Fail on FCTMATCH against an m-block, implies not(FCTMATCH) succeeds.
-Therefore MBO (clause 1) succeeds.
-Pick node, mblk1,
-Succeed for FCTMATCH, implies not(FCTMATCH) fails.
-Therefore MBO (clause 1) fails; try MBO (clause 2).
-Succeed for FCTMATCH for m-block,
-Connect m-blk1 to successor node, mblk2,
-Fail on FCTMATCH against a po block for mblk2,
-Backtrack on CONNECT for continuation of the successor path,
-CONNECT fails since there exist no other paths that do not encounter an m-block,
-Backtrack to FCTMATCH, which fails since there is

no choice,
-Therefore MBO (clause 2) fails; try MBO (clause 3).
-Succeed on FCTMATCH against an m-block,
-CONNECT output pin to successor node,
-Succeed on FCTMATCH against m-block for node mblk2,
-Invoke EXCLK to check for exclusive clocks on the original and successor m-blocks.
-The EXCLK rule succeeds, therefore the MBO rule succeeds for the mblk1 node,
-Pick node mblk2,
-Succeed on FCTMATCH for m-block, implies not(FCTMATCH) fails,
-Therefore clause 1 fails; try clause 2,
-Succeed on FCTMATCH for m-block,
-CONNECT output pin to successor node (po1),
-Succeed for FCTMATCH on po block for node po1,
-Therefore succeed for MBO clause 2,
-Pick node po1,
-Fail for FCTMATCH on m-block, implies not(FCTMATCH) succeeds,
-Therefore MBO clause 1 succeeds, and since all of the design nodes have been processed, the design passes the MBO rule.

The MBO rule used in the preceding example was kept simple to illustrate the PROLOG operation. However, it has two limitations for actual use in VLSI designs. The first is that the check only guarantees observability for one level of shift register. It is possible for the ends of the shift register to connect to each other (as in a ring) and have this configuration pass the MBO rule without having true observability. The second limitation is that the rule statement prevents any intervening logic between the stages of the register. This is obviously too severe a restriction for practical designs. Both of these limitations can be removed by making the MBO clauses more sophisticated, and by adding additional options on the definition of observability.

Similar rules also exist for the other requirements of a structured design following the LSSD rules. In addition to these DFT rules, we have defined rules that allow the extraction of certain control information from the design structure. These rules determine the sensitizing patterns for the m-block clocks for control and observation. The patterns so generated can be used at a later date for automatic test pattern generation (ATPG). An example of the m-block control sequence information rule is given in Fig. 4. This rule will succeed whenever a sensitized path can be found between the controllable node the m-block in question.

```
csgmbc(Node):-
    fctmatch(Node,mblock),
    connect(Innode,(Node,data),((m-block,1)),
    control(Innode),
    sensitize(Innode,Node,Nodelist),
    csgoutput(Innode,Node,control,Nodelist).
```

Figure 4. Control Sequence for M-block Data Input

The design transformation rules (one of which is shown in Fig. 1) are a collection of rules that tell the DFT system what to do with the design in the presence of specific DFT rule violations. These violations occur when the system fails to prove one of the DFT rules true for the given design model. These rules act diffently than the checking rules. Instead of operating in the mode where, "If this condition is true, and if that condition is true, etc., implies the check for this node succeeds, or fails," the transformation uses the violated DFT rule and the information generated during its attempted proof. Given these two items; the transformation rules will determine the node and interconnections to be changed to avoid a violation, and then make these changes to the design. In the current system, this merely consists of adding extra, or new, logic to the design. Thus the intended behavior of the design is preserved without regard to the exact cause of the DFT rule violation. For the prototype system these transformation rules are limited to local transformations in that they operate only on small well-defined portions of the design, and do not attempt to restructure the entire design in a global fashion.

To illustrate the transformation process, the PROLOG implementation of the multiplexor given in Fig. 1 is shown in Fig 5. This rule is invoked whenever a clock observation violation is encountered and the preferred transformation is to add an output multiplexor. This rule will use the last node in the attempted proof as the source of the DFT rule violation. Given that node, it will determine its immediate predecessor (via CONNECT and the 2nd use of PICKNODE) and then proceed to find a design primary output (po). The rule then checks to see if this output has already been used for test multiplexor purposes (via a check of the immediate predecessor node's function). If all of the preceeding sub-goals have been shown to be true, a new node name is generated (via GENNAME), and the multiplexer is inserted into the design. This multiplexer will serve to provide clock observability when the control input to the multiplexer indicates that the design is in "test mode." During normal system operation of the design, the multiplexer is controlled such that the original path to the output is preserved.

```
synthesis(clko,Searchtree):-
    picknode(Searchtree,Node,terminal),
    connect((Innode,Inpin),Node,((any,1)),
    picknode(Searchtree,Innode,_),
    fctmatch(Outnode,po),
    connect((Datanode,Datapin),Outnode,((any,1)),
    not(fctmatcht(Datanode,mux)),
    genname(Innode,Newname),
    addmux(Datanode,Datapin,Innode,Inpin,Outnode,_,Newna
```

Figure 5. Clock Observation Transformation Rule

An important consideration in the development of these DFT and transformation rules has been overall system efficiency. In the description of the PROLOG interpreter we mentioned that the interpreter will backtrack whenever it reaches a point where a goal is non-empty and other sub-goal choices remain. This is an area for potential inefficiency in a PROLOG expert system. The implementation was designed to limit the possible PROLOG backtracking to that portion of the operation where the system is searching for good (or bad) paths. Since a check for certain types of connectivity involves some form of path tracing, backtracking can be avoided here. For the remainder of the operations, violation analysis, and design transformation, little backtracking is performed.

DESIGNER INTERFACE

One of the advantages of using logic programming for the DFT rule checker and analyzer is that it is very easy to allow a designer to interupt the operation of the checker at any time and ask for an explanation of how it arrived at a particular violation (or non-violation). A similar type of interface is also available during the transformation processing. The DFT checker and analyzer is able to do this type of interface since it keeps all of the information used in the attempted proof until the next proof is attempted. For example, when the designer asks "Why can't you observe CLK1?," the system will look at the last failing clause in its proof tree and find that it was attempting to show that the picked node's function was an output. The system knows it started with a clock input and was supposed to terminate at an output. From this information the response will be "Failed to connect this clock pin to an output". The general use of natural language in PROLOG is a non-trivial area of reasearch and is being pursued by a number of logic programmers world-wide. The expert system uses a very simple I/O technique which rest primarily upon a good choice of PROLOG clause and variable names. The example uses a limited natural language interface currently under development.

RESULTS

The examples given in the figures of this paper represent cases of the current implementation of this DFT expert system at Syracuse University. Approximately 60 production type rules now describe the DFT expert's knowledge. This includes both those rules that describe the attributes of a structured design, and those rules that describe the allowed transformations in the presence of design rule violations. With this rule set, we performed experiments using a 4-bit binary up/down counter (at the gate level of description), a DMA control chip (at the functional level of description), and the M6800 microprocessor design (using a bus level description). By limiting the scope of each design experiment as well as shifting the level of functional abstraction for the design, approximately 45 to 60 PROLOG clauses were required to represent the design model. In all three cases, the system was able to find DFT rule violations and to make design transformations within the limits of its DFT knowledge by adding test points and multiplexers.

The entire expert system uses approximately 400 PROLOG clauses (excluding the design model). We are currently using the version of PROLOG implemented on the DEC KL-10 (with 18-bit addressing) for which a total of 500 clauses is the limit of operation (for interpretted PROLOG). Since this is a relatively small expert system (and design model) the performance is well within the limits for an interactive system, i.e., less than 2-3 seconds response for the rule checks and transformations. For larger collections of PROLOG clauses, the compiled version of PROLOG clauses is perferrable to the interpreted version and is considered in some of the follow-on activities for this research. The use of the compiled version should allow approximately a 25 percent increase in the allowed model size. The real gain in the use of the compiled version is in performance. Here the gain is approximately a factor of 10 (from 3K-5K LIPS to 40K LIPS). The term LIPS stands for Logic Inferences Per Second and is the accepted measure of logic programming systems. It is expected that the fifth generation computer systems will be capable of at least 500K LIPS (early versions) to greater than 1000K LIPS for more mature versions.

These results indicate that it is feasible to implement an expert system for the design for testability (DFT) problem. One of the immediate follow-on activities will be to move the current DFT system to either the PROLOG system that is currently running on an IBM System 4341, or one that is currently being implemented in the C programming language on a Data General MV8000. The actual choice of PROLOG implementations is still being investigated at this time. This move will allow a substantial increase in the size of both the DFT production rule set, and the design models used due to the change in CPUs.

CONCLUSIONS

The use of logic programming approach to the DFT problem is definitely feasible, and can be used as a starting point in the production use of the system. During the work on this system the following general observations were made: 1) It was found to be easy to add rules, or otherwise modify the system, since the expert knowledge is a separate entity from the reasoning mechanisms of the system. By keeping the various sections of inference and knowledge in the system separate from each (in an abstract sense) it is possible to modify one area without affecting another, such as the different design rules. 2) The system could easily interact with the designer, since a large amount of information is available on why certain decisions were made, etc. Essentially the entire clause data-base is accessible to interogation. 3) The system can indirecly indicate areas where it needs further information for efficient processing. This is especially true for the design transformations. These rules are both the largest set of rules, and the ones most susceptable to modification or addition. Some of the rule extensions or additions involve the use of heuristics, which can indicate what might be good modifications to try for the given design model. 4) One can add (or change) rules while using the system. Although a designer can do this, the best use of this capability is during system debug and test. This capability exists in PROLOG since the user has direct access to the clause data-base and can make modifications directly. 5) PROLOG was good language (system) to use in developing such an expert system: however, for the best system performance, some of the operations done with PROLOG clauses should be implemented in a procedural language, e.g., PASCAL, C, etc. These operations would include some of the path tracing (CONNECT) and table look-up (FCTMATCH) types of processing.

Our follow-on research includes expanding the learning aspects of the expert system as well as incorporating more expert knowledge, more sophisticated DFT rules and transforms, improving the designer interface with some natural language processing, and using more efficient PROLOG implementations on different processors.

A very interesting extension to the DFT expert system is to integrate it into the overall VSLI design methodolgy. In this way, the design for test would be part of the natural design flow and, if considered early in the design cycle, the actual overhead for implementing DFT structures can be minimal. One way to perform this integration would be to combine synthesis systems with the DFT system.

ACKNOWLEDGEMENTS

The author thanks Drs. Edward P. Stabler and Kenneth A. Bowen at Syracuse University for their assistance in the development of this system, and Mr. William L. Huston for his patient review of this manuscript.

REFERENCES

1) T. W. Williams, and K. P. Parker, "Design for testability: a survey," IEEE Transactions on Computers, Vol. C31, No. 1, January 1982, pp. 2-15.

2) P. W. Horstmann, Automation of the Design for Testability Using Logic Programming, Ph.D. Dissertation, Dept. of Electrical and Computer Engineering, Syracuse University, October, 1983.

3) D. Nau, "Expert Systems," IEEE Computer, February, 1983, pp. 63-85.

4) H. Brown, C. Tong, and G. Foster, "PALLADIO: an exploratory envirnment for circuit design," IEEE Computer, December 1983, pp. 41-56.

5) T. Kowalski, and D. Thomas, "The design automation assistant: a prototype system," The Proceedings of the 20th Design Automation Conference, June 1983, pp. 479-481.

6a) W. F. Clocksin, and C. S. Mellish, Programming in PROLOG, Springer-Verlag, 1981.

6b) R. Kowalski, Logic for Problem Solving, Elsevier - North Holland, 1982.

7) P. W. Horstmann, "Expert systems and logic programming for CAD," VLSI Design Magazine November, 1983, pp. 37-46.

8) E. A. Feigenbaum., and P. McCorduck, The Fifth Generation, Addison-Wesley, Reading, MA, 1983.

9) K. A. Bowen, and R. Kowalski, "Meta-level PROLOG," Logic Programming, Clark, K. L., and S.-A. Tarnlund (eds.), Academic Press, 1982, pp. 153-172.

10) E. B. Eichelberger, and T. W. Williams, "A logic design structure for LSI testability," The Proceedings of the 14th Design Automation Conference, June 1977, pp.462-467.

ARTIFICIAL INTELLIGENCE

Diagnostic Reasoning Based on Structure and Behavior*

Randall Davis
The Artificial Intelligence Laboratory, Massachusetts Institute of Technology, Cambridge, MA 02139, U.S.A.

ABSTRACT

We describe a system that reasons from first principles, i.e., using knowledge of structure and behavior. The system has been implemented and tested on several examples in the domain of troubleshooting digital electronic circuits. We give an example of the system in operation, illustrating that this approach provides several advantages, including a significant degree of device independence, the ability to constrain the hypotheses it considers at the outset, yet deal with a progressively wider range of problems, and the ability to deal with situations that are novel in the sense that their outward manifestations may not have been encountered previously.

As background we review our basic approach to describing structure and behavior, then explore some of the technologies used previously in troubleshooting. Difficulties encountered there lead us to a number of new contributions, four of which make up the central focus of this paper.

– We describe a technique we call constraint suspension *that provides a powerful tool for troubleshooting.*

– We point out the importance of making explicit the assumptions underlying reasoning and describe a technique that helps enumerate assumptions methodically.

– The result is an overall strategy for troubleshooting based on the progressive relaxation of underlying assumptions. The system can focus its efforts initially, yet will methodically expand its focus to include a broad range of faults.

– Finally, abstracting from our examples, we find that the concept of adjacency *proves to be useful in understanding why some faults are especially difficult to diagnose and why multiple representations are useful.*

1. Introduction

The overall goal of this research is to develop a theory of reasoning that exploits knowledge of structure and behavior. We proceed by building programs that use such knowledge to reason from first principles in solving

---

*This report describes research done at the Artificial Intelligence Laboratory of the Massachusetts Institute of Technology. Support for the laboratory's Artificial Intelligence research on electronic troubleshooting is provided in part by the Digital Equipment Corporation and in part by the Defense Advanced Research Projects Agency.

Artificial Intelligence **24** (1984) 347–410
0004-3702/84/$3.00 © 1984, Elsevier Science Publishers B.V. (North-Holland)

problems. The initial focus is troubleshooting digital electronic hardware, where we have implemented a system based on a number of new ideas and tools.

Troubleshooting digital electronics is a good domain for several reasons. First, troubleshooting seems to be one good test of part of what it means to 'understand' a device. We view the task as a process of reasoning from behavior to structure, or more precisely, from misbehavior to structural defect: given symptoms of misbehavior, we are to determine the structural aberration responsible for the symptoms. Second, the task is interesting and difficult because the devices are complex and because there is no established theory of diagnosis for them. Third, the domain is appropriate because the required knowledge is readily available from schematics and manuals. Finally, the application itself is relevant and tractable.

Work with a similar intent has been done in other domains, including medicine [23], computer-aided instruction [6], and electronic troubleshooting [7, 18], with the 'devices' ranging from the gastro-intestinal tract, to transistors and digital logic components.

This work is novel in a number of respects, some of which have been reported in an earlier publication [8]. As noted there:
- We have developed languages that distinguish carefully between structure and behavior, and that provide multiple descriptions of structure, organizing it both functionally and physically.
- We have argued that the concept of *paths of causal interaction* is a primary component of the knowledge needed to do reasoning from structure and behavior.
- We have developed an approach to troubleshooting based on the use of a layered set of categories of failure and demonstrated its use in diagnosing a bridge fault.

This paper substantially expands and develops this line of work[1], reporting several new contributions:
- We describe a new technique called *constraint suspension*, capable of determining which components can be responsible for an observed set of symptoms.
- We show that the categories of failure, previously derived informally, can be given a systematic foundation. We show that the categories can be generated by examining the assumptions underlying our representation.
- This in turn gives us a new way to view our approach to troubleshooting: the methodical enumeration and relaxation of assumptions about the device.
- Finally, we describe the concept of *adjacency* and argue that it helps in

---

[1]For the sake of continuity and ease of presentation, parts of [8] are reprinted here, including the bridge-fault example and some of the background material (machinery for describing structure and function, and the introduction to troubleshooting). This material is reprinted with permission from [8], Copyright Academic Press Inc. (London) Ltd.

understanding both what it means to have a good representation, and why multiple representations can be useful.

Section 2 is introductory, supplying a brief review of the overall concerns, expanding on the ideas listed above as a preview of what is central to this work.

If we are to reason from first principles, we require: (i) a language for describing structure, (ii) a language for describing behavior, and (iii) a set of principles for troubleshooting that use the two descriptions to guide their investigation. The central part of the paper describes each of these components, paying particular attention to the nature of the reasoning that underlies troubleshooting.

Sections 3 and 4 explore our approach to describing structure and behavior, while Sections 5 through 9 deal with troubleshooting. In Section 5 we consider the machinery traditionally used to reason about circuits and show that it fails to solve the problem we face. We then show how a technique called discrepancy detection offers a number of useful advances, but claim that it must be used with care. We suggest that the potential difficulties center around implicit assumptions typically made when using the technique. We find a solution to the problems encountered by making those assumptions explicit, organizing them appropriately, and providing for a way to surrender them one by one, methodically expanding the scope of failure the program considers. Section 9 provides an example of these ideas in operation, showing how they guide the diagnosis of a bridge fault (an unintended electrical connection, often caused by extra solder running between two adjacent pins on a chip).

In Sections 10 and 11 we draw back from the specific example presented and attempt to generalize our methods. We argue that the concept of paths of interaction offers a useful framework for asking questions about a domain. We claim that the notion of 'adjacency' both explains some of the difficulty in reasoning about bridges and provides guidance in selecting and using multiple representations.

Section 12 reviews the machinery we have presented, pointing out is limitations and suggesting directions for future work. Finally, Section 13 compares our work to a number of previous approaches to similar problems.

2. Central Concerns

Some of the examples in later sections of this paper require a substantial amount of detail, yet the principles they illustrate are often relatively uncomplicated. In order to be sure that the important points are not lost in the detail, we describe the fundamental principles here briefly, enlarging on them in the remainder of the paper.

Candidate generation can be achieved by using a new technique we call constraint suspension.

Given the symptoms of a malfunction, we wish to generate candidate components, i.e., determine which components could plausibly be responsible for the malfunction. We model the intended behavior of a device as a network of interconnected constraints, where each constraint models the behavior of one of the components. If the device is malfunctioning, then the outputs predicted by the entire constraint network will not match the actual outputs. Normally, contradictions in constraint networks are handled by retracting one or more of the input values. Constraint suspension takes the dual view and asks instead: "is there some *constraint* (component behavior) whose retraction will leave the network in a consistent state?" Each such constraint that we find corresponds to a component whose misbehavior can account for all the observed symptoms. In many cases the technique will also supply the details of the component misbehavior (i.e., show how it is misbehaving). We describe the technique in detail in Section 6.1.

Paths of causal interaction play a central role in troubleshooting.

An important part of the basic knowledge needed for troubleshooting is understanding the mechanisms and pathways by which one component can affect another. Electronic components typically interact because they are wired together, but they can also interact due to heat, capacitive coupling, etc. We argue that viewing the problem in terms of paths of interaction is more useful and revealing than the fault models (e.g., stuck-ats) traditionally used.

In doing troubleshooting we are faced with an unavoidable problem of complexity vs. completeness.

To be good at troubleshooting, we need to handle many different kinds of paths of interaction. But this presents a serious problem. If we include all possible paths, candidate generation becomes indiscriminate: every component could somehow be responsible for the observed symptoms. Yet omitting any one kind of path would make it impossible to diagnose an entire class of faults.

One technique for dealing with this dilemma is enumerating and layering the categories of failure.

As experienced troubleshooters know, some things are more likely to go wrong than others. But what are the 'some things'? And what does 'more likely' mean? We make both of these notions more precise.

We show how to generate the 'things that can go wrong' by making explicit the assumptions underlying our representation; the result is a list of 'categories of failure'. The categories are organized by using the most likely first and adding additional, less likely categories only in the face of contradictions.

Associated with each category of failure is a collection of paths of interaction. Hence the ordered listing of categories produces an ordering on the paths of interaction to be considered. This allows us to constrain the paths we consider initially, making it possible to constrain candidate generation. But no path is permanently excluded, hence no class of faults is overlooked.

The result is a strategy for troubleshooting based on the methodical enumeration and relaxation of underlying assumptions.

This technique allows us to deal with a progressively wider range of different faults without being overwhelmed by too many candidates at any one step.

The categories of failure can be generated systematically by examining the assumptions underlying our representation.

As we demonstrate in Section 10, the categories can be derived by listing the assumptions implicit in the representation and considering the consequences of violating each in turn. Because the representation employs little more than the traditional notion of black boxes, the categories of failure deal with information transmission, and the results of this exercise have relevance broader than digital electronics. We suggest that the results may be applicable to software and speculate about wider applications.

The concept of 'adjacency' proves to be useful in both troubleshooting and the selection of representations.

We noted above that the possible pathways for device interaction are an important part of the knowledge required for troubleshooting. We take the view that devices interact because they are in some sense *adjacent*: electrically adjacent (wired together), physically adjacent (hence 'thermally connected'), electromagnetically adjacent (not shielded), etc. Each of these definitions can be used as the basis for a different representation of the device, different in its definition of adjacency. The multiplicity of possible definitions helps to explain why some faults are especially difficult to diagnose: *they result from interactions between components that are adjacent in a sense (representation) that is unusual or subtle.*

This concept appears to generalize in several ways. We view faults as modifications to the original design, and claim that changes that appear small and local in one representation may not appear small and local in another. We also suggest that adjacency can help determine what makes a 'good' representation: one in which the change can be seen as compact. It may also explain some of the utility of using multiple representations: they offer multiple different definitions of adjacency.

Adjacency is a useful principle to the extent that the 'single-initial-cause heuristic' holds true.

A commonsense heuristic suggests that malfunction of a previously working device generally results from a single cause rather than a number of simultaneous, independent events. When there is indeed a single cause, there will be some representation in which a compact change accounts for the difference between the good and faulty device.

The concept of a 'single point of failure' is not well defined without specification of the underlying representation.

A failure in a single physical component, for example, may manifest as multiple points of failure in the behavior of the overall device. Consider a single chip with four AND gates in it. If the chip is damaged (via over-voltage, heat, mechanical stress, etc.), we will have a single point of failure in the physical representation, but may find four different points of failure (the four AND gates) in the functional view of the device. We claim as a result that 'single point of failure' is by itself an under-determined concept, and that to make precise sense of the term we need to indicate both the nature of and the level of abstraction of the underlying representation to which 'point' refers. We claim further that the use of multiple representations can add diagnostic power by offering different ways of resolving apparently independent failures into a single cause.

3. Describing Structure

If we wish to reason about structure and behavior, we clearly need a way of representing both. We consider each of these in turn.

By structure we mean information about the interconnection of modules. Roughly speaking, it is the information that would remain after removing all the textual annotation from a schematic.

Two different ways of organizing this information are particularly relevant to troubleshooting: functional and physical. The functional view gives us the machine organized according to how the modules interact; the physical view tells us how it is packaged. We thus prefer to replace the somewhat vague term 'structure' by the slightly more precise terms *functional organization* and *physical organization*. As we will see, every device is described from both perspectives, producing two distinct, interconnected descriptions.

Our aim is to provide a means of encoding in one place all of the information about a circuit that is typically distributed across several different documents. To this end our approach provides a way of representing much of the information traditionally found in a schematic, as well as the hierarchical description often encoded in block diagrams.

The most basic level of our structure description is built on three concepts: *modules*, *ports*, and *terminals* (Fig. 1). A module can be thought of as a standard black box; ports are the places where information flows into or out of a module. Every port has at least two terminals, one terminal on the outside of the port and one or more inside. Terminals are primitive elements; they are the places we can 'probe' to examine the information flowing into or out of a

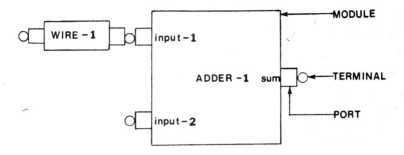

Fig. 1. The basic terms used in structure description.

device through a port, but they are otherwise devoid of interesting substructure.

Two modules are attached to one another by superimposing their terminals. In Fig. 1, for example, wire-1 is a (wire) module that has been attached to the input-1 port of adder-1 (an adder module) in this fashion.

The language is hierarchical in the usual sense: modules at any level may have substructure. In practice, our descriptions terminate at the gate level in the functional hierarchy and the chip level in the physical hierarchy, since for our purposes these are black boxes—only their behavior (or misbehavior) matters. Fig. 2 shows the next level of structure of the adder and illustrates why ports may have multiple terminals on their inside: ports provide the important function of shifting level of abstraction. It may be useful to think of the information flowing along wire-1 as an integer between 0 and 15, yet we need to be able to map those four bits into the four single-bit lines inside the adder. Ports are the places where such information is kept. They have machinery (described below) that allows them to map information arriving at their outer terminal onto their inner terminals.

Since our ultimate intent is to deal with hardware on the scale of a mainframe computer, we need terms in the vocabulary capable of describing levels of organization more substantial than the terms used at the circuit level. We can, for example, refer to *horizontal*, *vertical*, and *bit-slice* organizations, describing a memory, for instance, as "two rows of five 1K RAMs". We use these specifications in two ways: as a description of the organization of the device and a specification for the pattern of interconnections among the components.

Our eventual aim is to provide an integrated set of descriptions that span the levels of hardware organization ranging from interconnection of individual modules, through higher level of organization of modules, and eventually on up through the register transfer and PMS [4] levels. Some of this requires inventing vocabulary like that above, in other places we may be able to make use of existing terminology and concepts.

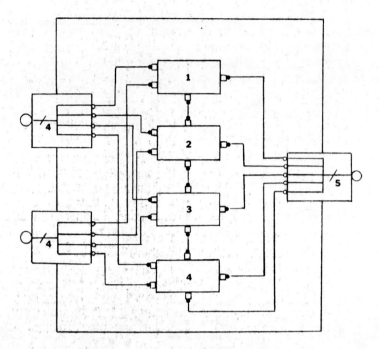

FIG. 2. Next level of structure of the adder, showing that it is implemented as a ripple-carry.

The structural description of a module is expressed as a set of commands for building it. The adder of Fig. 2, for example, is described by the instructions shown in Fig. 3. With NBitsWide bound to 4, the first expression indicates that we should repeat the following sequence of operations four times:
– create an adder slice;
– run a wire from the first input of the adder to the first input of the slice;
– run a wire from the second input of the adder to the second input of the slice;
– run a wire from the output of the slice to the output of the adder.
The next expression builds the carry chain and the last wires up the high-order bit.

```
(definemodule adder NBitsWide
  (repeat NBitsWide i
    (part slice-i adder-slice)
    (run-wire (input-1 adder) (input-1 slice-i))
    (run-wire (input-2 adder) (input-2 slice-i))
    (run-wire (output slice-i) (sum adder))
  (repeat (− NBitsWide 1) i (run-wire (carry-out slice-i)
                                      (carry-in slice-[i + 1])))
  (run-wire (carry-out slice-[NBitsWide − 1]) (sum adder)))
```

FIG. 3. Parts are described by a pathname through the part hierarchy, e.g., (input-1 adder).

These commands are executed by the system, resulting in the creation of data structures that model all the components and connections shown. These data structures are isomorphic to the diagram shown in Fig. 2. That is, the data structures are connected in the LISP sense in the same ways that the objects are connected in Fig. 2. As in real devices, information flow occurs as a result of these interconnections: slice-1 can place information on its end of the data structure modeling the carry-out wire, for example, because the two are superimposed. This information will then be propagated to the other end of the wire and pass into the data structure modeling slice-2 because those two are superimposed.[2] The utility of this approach is explored in Section 12.

The definition in Fig. 3 is thus a specification of a prototypical ripple-carry adder; invoking it with a specific value for the width parameter produces one particular instance. We have found it useful to maintain this prototype/instance distinction for several reasons. It allows the standard economy of representation, since information common to all instances can be stored once with the prototype. It also makes possible the parameterization illustrated above, allowing us to describe the overall structure of the device, capturing the generalization inherent in the standard pattern of interconnections. Finally, it allows us to build up a library of module definitions available for later use. There is considerable utility in assembling such a library, utility beyond the ease of describing more complicated devices. By requiring that we define modules outside of the context of their use in any particular circuit, we encourage an important form of 'mental hygiene' described further in Section 12.2.

Since our eventual aim is troubleshooting of devices as complex as a computer, the complete description of the device could become quite large. To deal with this we have made use of 'lazy instantiation': when an instance of a module is created, only the 'shell' is actually built at that time. For the adder in Fig. 2, for example, the outer 'box' and ports are built, but the substructure—the slices—are not built until they are actually needed. Only when we need to 'look inside' the box, i.e., drop down a level of structural detail, is additional structure actually built.

As a result, the system maintains a compact description of the device, expanding only where necessary. This can save a considerable amount of space. Even for the simple adder example (Fig. 2), there is a lot to describe: each slice is built from two half-adders and an OR-gate, each half-adder is built from an XOR- and an AND-gate. The initial instantiation is simply an adder with two inputs and an output, rather than four levels of detail and twenty gates. For more complex devices the savings can of course be greater.

The examples illustrate how we describe functional organization: an adder

---

[2] We do not, by the way, have any special mechanism for dealing with wires. They are simply another module, albeit one with a particularly simple behavior: information presented to either port will be propagated to the other.

composed of slices, which are in turn composed of half-adders, etc. Exactly the same module, port and terminal machinery is used to describe physical organization, but now the hierarchy of modules is cabinets, boards, and chips.

Since, as noted earlier, every component has both a functional and physical description, we have two interconnected description hierarchies. We represent this in our system by having the two hierarchies linked at their terminal nodes (Fig. 4).

We determine the physical location of any non-terminal entry in the functional organization by aggregating the physical locations of all of its leaves. These descriptions are available at various levels of abstraction. The physical location of slice-1, for example, is the list of all the locations of its gates (e.g., E1, E4, E7, etc.); alternatively we can also say simply that it is on board-1, if all of these chips are in fact on the same board.

By having available cross-links between the two descriptions, it becomes possible to answer a range of useful questions. For example, we can ask, "where physically do I find the (part that functions as the) address translation register", or, "what function(s) does this quad-and-gate chip perform?"

Our description language has been built on a foundation provided by a subset of DPL [3]. While DPL as originally implemented was specific to VLSI design, it proved relatively easy to 'peel off' the top-level of language (which dealt with chip layout) and rebuild on that base the new layers of language described in Fig. 4.

Since pictures are a fast, easy and natural way to describe structure, we have developed a simple circuit-drawing system that permits interactive entry of

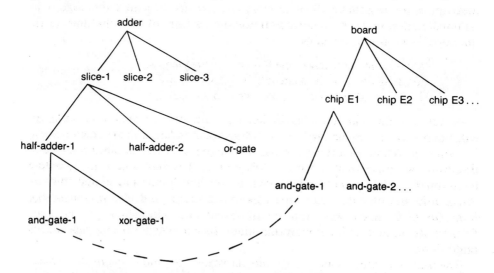

FIG. 4. Sample interconnection of functional and physical descriptions.

pictures like those in Figs. 1 and 2. Circuits are entered with a combination of mouse movements and key strokes; the resulting structures are then 'parsed' into the language shown in Fig. 3.

4. Describing Behavior

By behavior we mean the black box description of a component: how is the information leaving the component related to the information that entered it? A variety of techniques have been used in the past to describe behavior, including simple rules for mapping inputs to outputs, Petri-nets, and unrestricted chunks of code. Simple rules are useful where device behavior is uncomplicated, Petri-nets are useful where the focus is on modeling parallel events, and unrestricted code is often the last resort when more structured forms of expression prove too limited or awkward. Various combinations of these three have also been explored.

4.1. Simulation rules and inference rules

Our initial implementation is based on a constraint-like approach [30, 32]. Conceptually a constraint is simply a relationship. The behavior of the adder of Fig. 1, for example, can be expressed by saying that the logic levels of the terminals on ports input-1, input-2 and sum are related in the obvious fashion.

In practice, this is accomplished by writing a set of expressions covering all the different individual relations (the three for the adder are shown below) and setting them up as demons that watch the appropriate terminals. A complete description of a module, then, is composed of its structural description as outlined earlier and a behavior description in the form of rules that interrelate the logic levels at its terminals.

```
to get sum from (input-1 input-2) do (+ input-1 input-2)
to get input-1 from (sum input-2) do (- sum input-2)
to get input-2 from (sum input-1) do (- sum input-1)
```

A set of rules like these is in keeping with the original conception of constraints, which emphasized the non-directional, relationship character of the information. When we attempt to use it to model behavior of physical devices, however, we have to be careful. This approach is well suited to modeling behavior in analog circuits, where devices are largely non-directional. But we can hardly say that the last two rules above are a good description of the *behavior* of an adder chip—the device doesn't do subtraction; putting logic levels at its output and one input does not cause a logic level to appear on its other input.

The last two rules really model the *inferences we make about the device.* Hence our implementation distinguishes between *simulation rules* that represent flow of electricity (digital behavior, the first rule above) and *inference rules* representing flow of inference (conclusions we can make about the device, the next two rules).

We find this distinction useful in part for reasons of 'mental hygiene' (the two kinds of rules deal with different kinds of knowledge) and in part because it contributes to the robustness of the simulation in the face of unanticipated events. Consider for example a circuit that included an adder, and imagine that a fault in that circuit resulted in some other component trying to drive the output of the adder. If we treated all rules the same, our simulation would suggest that the device did the subtraction described above, yet this simply doesn't happen.

4.2. Implementation

Our implementation accomplishes this distinction by using two parallel but separate networks, one containing the simulation rules modeling causality, the other containing the inference rules. We keep them distinct by giving each terminal two slots, one holding the value computed by the simulation rules, the other holding the value computed by the inference rules (Fig. 5). Each network then works independently, using the standard demon-like style of propagating values through its collection of slots.

In addition to firing rules to propagate information around the network, this rule-running machinery also keeps track of dependencies, allowing us to determine how the value in a slot got there. This is done by having each slot keep track of (i) every rule that uses this slot as an input, (ii) every rule that can place a value in this slot (i.e. uses it as an output) and (iii) which rule did, in fact, provide the current value (since, in general, more than one rule can set the value of a slot).[3]

This dependency network offers several advantages. As we explore below, it is one of the foundations of the discrepancy detection approach to trou-

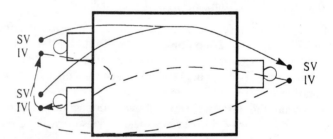

FIG. 5. Adder with one simulation rule (solid) and two inference rules (dashed). SV: simulation value; IV: inference value.

[3]The user can also set a value; this is most commonly done at the primary inputs.

bleshooting. As the work in [6] demonstrated, such a network also makes it easy to explore hypothetical situations. We can place a value in some terminal and observe the consequences, i.e., see what values propagate from it. When done with the exploration, removing the value at the terminal causes the dependency network to remove everything that depends on it as well. This mechanism makes it easy to discover the answer to some questions by simulation: we can get the device to a particular state, then explore multiple alternative futures from that point.

While all of the examples given thusfar deal with combinatorial devices, we can also represent and model simple devices with memory. This requires three simple augmentations of the approach described above. First, we use a global clock and timestamp all values. Second, we extend the behavior-rule vocabulary so that it can refer to *previous* values. For example, one simulation rule for a D flipflop would be

> To get output from (input previous-output clock)
>   do (if clock is high then input else previous-input)

Finally, our propagation machinery is extended to keep a history of values at each terminal (details are in [10]). This model of time is still very simple, but has allowed us to represent and reason about basic flipflops, memories, etc.

5. Troubleshooting

Having provided a way of describing functional organization, physical organization and behavior, we come now to the important third step of providing a troubleshooting mechanism that works from those descriptions. We develop the topic in three stages. In the first stage we consider test generation—the traditional approach to troubleshooting—and explain how it falls short of our requirements.

We consider next the style of debugging known as *discrepancy detection* and demonstrate why it is a fundamental advance. Further exploration, however, demonstrates that this approach has to be used with care in dealing with some commonly known classes of faults. We suggest that the difficulties arise from a number of implicit assumptions typically made when troubleshooting.

In discussing how to deal with the difficulties uncovered, we argue for the primacy of *models of causal interaction*, rather than traditional fault models. We point out the importance of making these models explicit and separate from the troubleshooting mechanism. The result is a strategy for troubleshooting based on the systematic enumeration and relaxation of underlying assumptions. This approach allows us to deal with a progressively wider range of different faults, without being overwhelmed by too many candidates at any one step.

We demonstrate the power of this approach on a bridge fault, a traditionally difficult problem, showing how our system locates the fault in a focused process that generates only a few plausible candidates.

5.1. The traditional approach

The traditional approach to troubleshooting digital circuitry (e.g., [5]) relies primarily on the process of path sensitization in a range of forms, of which the D-Algorithm [26] is one of the most powerful. A simple example of path sensitization will illustrate the essential character of the process. Consider the circuit shown in Fig. 6 and imagine that we want to determine whether the wire labeled A is stuck at 1. We try to put a zero on it by setting both x1 and x2 to 0. Then, to observe the actual value on the wire we set x3 to 1 thereby propagating the value unchanged through G2, and set x4 to 0, making the output of G3 = 0 allowing the value of A to propagate unchanged through G1.

Troubleshooting with this approach is then accomplished by running a complete set of such tests, checking for all stuck-ats on all wires.

For our purposes this approach has a number of significant drawbacks. Perhaps most important, it is a theory of *test generation*, not a theory of *diagnosis*. Given a specified fault, it is capable of determining a set of input values that will detect the fault (i.e., a set of values for which the output of the faulted circuit differs from the output of a good circuit). The theory tells us how to move from faults to sets of inputs; it provides little help in determining what fault to consider, or which component to suspect. Such questions are a central issue in our work because of complexity: the size of the devices we want to work with in the long run precludes the use of diagnosis trees—complete decision trees for all possible faults.

A second drawback of the existing approach is its lack of sharp distinction between *diagnosis* in the field and *verification* at the end of the manufacturing line. As a result of economic and historical forces, diagnostics written to verify the correct operation of a newly manufactured device have been pressed into service in the field. Yet the tasks are sharply different. We are not requesting verification that a machine is free of faults. The problem facing us is "Given

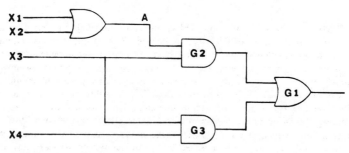

FIG. 6. Simple example of path sensitization.

the observed misbehavior, determine the cause". We know that the device has worked in the past, we know that some part of it has failed, and we know the symptoms of that failure. Given the complexity of the device, it is important to be able to use all of this information as a focus for further exploration. Only if this fails might it make sense to fall back on a set of diagnostics that, by design, start with no information and test exhaustively.

A final drawback of the existing theory is its unavoidable use of a set of explicitly enumerated fault models.[4] Since the theory is based on boolean logic, it is strongly oriented toward faults whose behavior can be modeled as some form of permanent binary value, typically the result of stuck-ats and opens. One consequence of this is the paucity of results concerning bridging faults.[5]

In solving the diagnosis problem, though, we have a significant luxury: we can treat as an error any behavior that differs from the expected correct behavior. The misbehavior need not be modelable in terms of any fixed set of faults, it need only be different from what should have resulted.

In summary, the technology often used for troubleshooting is oriented toward test generation and the task of verification. We, on the other hand, are concerned with troubleshooting, a process that makes use of test generation (for an example, see [27]), but which requires as necessary precursors the processes of candidate generation (determining which components may be failing) and 'symptom generation' (how they may be failing). The next section explores an approach that supports both of these.

6. Discrepancy Detection and Candidate Generation

One response to these problems has been the use of 'discrepancy detection' (e.g., [6, 11]). The two basic insights of the technique are (i) the substitution of violated expectations for specific fault models and (ii) the use of dependency records to trace back to the possible sources of a fault. Instead of postulating a possible fault and exploring its consequences, the technique looks for mismatches between the values it expected from correct operation and those actually obtained. This allows detection of a wide range of faults because

---

[4]Fault models are necessary in verification if we want to avoid the exponential effort of exhaustive testing. If we treat a device as a black box (e.g., saying only that it is an adder), we are forced to verify all of its behavior, a task that is potentially exponential in the number of inputs and amount of state. The most common way of avoiding this is by combining knowledge of the substructure of the device (e.g., that it is a carry-chain adder) with a specific set of faults to consider (e.g., stuck-ats on all wires), to produce tests for all such faults on all specified parts of the substructure. This task is at worst a product of the number of faults and wires.

[5]While the theory underlying verification may be limited in the range of faults it can describe, in practice it turns out to handle a large part of the problem: experience indicates that a large percentage of all faults turn out to be *detected* (but not diagnosed) by checking just for stuck-ats. Hence we can determine that something is wrong (satisfying the *verification* task); determining the identity and location of the error (*diagnosis*), however, is a different problem.

misbehavior is now simply defined as anything that isn't correct, rather than only those things produced by a stuck-at on a line.

The inspiration behind using dependency networks is that any component on the path from an input to an incorrect output could conceivably have been responsible for the faulty behavior. As we have seen, the simulator builds a dependency network by recording the propagation of values as it simulates the circuit. Using those records (or, equivalently, the original schematic) as a guide to which components to examine appears to be an effective way to focus attention appropriately. As we will see, it is in fact an interesting trap.

We work through a simple example to show the basic approach of discrepancy detection, then add to it the idea of constraint suspension. We comment on the strengths of the resulting procedure for candidate generation, then use the same example to illustrate difficulties that can arise.

Consider the circuit in Fig. 7.[6] If we set the inputs as shown, the system will use the behavior descriptions to simulate the circuit, constructing dependency records as it does so, and indicating that we should expect 12 at F.

If, upon measuring, we find the value at F to be 10, we have a conflict between observed results and our model of correct behavior. We trace back through the dependency network to enumerate the possible sources of the problem [11]. The dependency record at F indicates that the value expected there was determined using the behavior rule for the adder and the values emerging from the first and second multiplier. One of those three must be the source of the conflict, so we have three possibilities to pursue: either the adder-behavior rule is inappropriate (i.e., the first adder is broken), or one of the two inputs did not have the expected values (and the problem lies further back). Consideration of the first possibility immediately generates Hypothesis 1: adder-1 is broken.

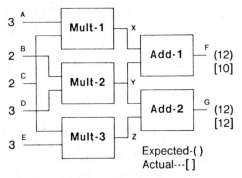

FIG. 7. Troubleshooting example using discrepancy detection.

[6]As is common in the field, we make the usual assumptions that there is only a single source of error and the error is not transient. Both of these assumptions are important in the reasoning that follows; we comment further on them below.

To pursue the second possibility, we assume that the second input to adder-1 is good. In that case the first input must have been a 4 (reasoning from the result at F, valid behavior of the adder, and one of the inputs), but we expected a 6. Hence we now have a discrepancy at the input to adder-1; we have succeeded in pushing the discrepancy one step further back along the dependency chain. The expected value there was based on the behavior rule for the multiplier and the expected value of its inputs. Since the inputs to the multiplier are primitive (supplied by the user), the only alternative along this line of reasoning is that the multiplier is broken. Hence Hypothesis 2 is that adder-1 is good and mult-1 is faulty.

Pursuing the third possibility: if the first input to adder-1 is good, then the second input must have been a 4 (suggesting that the second multiplier might be bad). But if that were a 4, then the expected value at G would be 10 (reasoning forward through the second adder). We can check this and discover in this case that the output at G is 12. Hence the value on the output of the second multiplier can't be 4, it must be 6, so the second multiplier can't be causing the current problem.

This style of reasoning can be described as the *interaction of simulation and inference*: simulation generates expectations about correct outputs based on inputs and knowing how devices work (simulation rules); inference generates conclusions about actual behavior based on observed outputs and device inference rules. The comparison of these two, in particular differences between them, provides a foundation for troubleshooting.[7]

6.1. Constraint suspension

In the discussion above, we glossed over the machinery used to determine when a candidate is consistent (as adder-1 was) and when it is inconsistent with all the available evidence (as mult-2 was). One of the novel contributions of this work is the development of the constraint-suspension technique as a way of providing an answer to this question.

To see how it works, consider once again the first step of the problem. When we examine the dependency record at F, we find that the value there resulted from the behavior rule for adder-1 and the values coming from mult-1 and mult-2. As above, the first possibility is that adder-1 is broken.

But this is a *local inference* (i.e., it is based only on the dependency record at F) and we have to be sure it's *globally consistent* with all the symptom data. More precisely, we want to ask whether there is *some* assignment of values to the ports of the adder that is consistent with all of the inputs and observed outputs. Is there any way in which adder-1 alone could be broken and produce the symptoms noted?

[7]The guided probe technique, in common use in industry, is based on a set of ideas that is closely related, though not identical. We discuss the differences in Section 13.

We can do this conveniently by using the 'constraint-like' character of our representation and the notion of *constraint suspension*. While the simulation and inference rules are usually kept distinct, for the moment we use the whole collection of them together, in effect a network of constraints that can indicate whether we have a consistent set of assignments to the inputs and outputs. If, for example, we were to try to assign to the network the inputs and *observed* outputs of Fig. 7, it would report a contradiction: there is no way for all the rules to be active (i.e., all the components working as expected) and for those inputs to have produced the observed outputs.

Normally contradictions in constraint networks are handled by retracting one of the values inserted into the network. But here we are sure of the values (the inputs we sent in to the circuit and the outputs we measured); what we are unsure of is the constraints (component behaviors). We therefore take a dual view, and rather than looking for a value we can retract, we look for a *constraint* whose retraction will leave the network in a consistent state. This is the basic idea behind constraint suspension.

To check the global consistency of adder-1, for example, we *suspend* (disable) all the rules in adder-1, assign the input values to input ports A through E, and assign the *observed* values to output ports F and G. We then allow the whole collection of rules to run to quiescence, determining for us whether there is *some* set of values on the ports of adder-1 consistent with the inputs and observed outputs.

If the network does reach a consistent state, we know that the candidate can account for the symptoms. In addition, we can examine the resulting state to see what values the candidate must have at its ports. In the case of adder-1, for example, the network indicates that the inputs must be 6 and 6, and the output 10. Thus in the process of determining the global consistency of a candidate, constraint suspension also produces symptom information about the misbehavior.

If the network reports an intractable contradiction, there is no assignment of values to the component that is consistent with all the symptoms, and hence no way for that component to account for the observed malfunction. For example, when we disable the rules of mult-2 and insert the inputs and observed outputs, an inescapable contradiction results. This demonstrates that mult-2 cannot account for all the observed values.

Fig. 8 provides a complete description of the candidate-generation process in a code-like notation (the procedure has been made easier to follow by ignoring the simulation/inference distinction for the moment and assuming we have a traditional constraint network). Candidate generation occurs in three basic steps: simulate the circuit and collect discrepancies, determine potential candidates using the dependency records; and finally, for each potential candidate determine global consistency and symptom values by using constraint suspension.

CANDIDATE GENERATION PROCEDURE

STEP 1: COLLECT DISCREPANCIES
  1.1 Insert device inputs into the constraint network inputs
      ; *e.g. insert 3, 2, 2, 3, and 3 at primary inputs A through E*
      ; *simulation predicts values at F and G*
  1.2 Compare predicted outputs with observed and collect discrepancies
      ; *e.g., prediction and observation differ at F.*

STEP 2: DETERMINE POTENTIAL CANDIDATES VIA DEPENDENCY RECORDS
  2.1 For each discrepancy found in Step 1:
      follow the dependency chain back from the predicted value
      to find all components that contributed to that prediction
      ; *these are all the components "upstream" of the discrepancy*
      ; *e.g., if we follow the dependency chain back from the 12*
      ; *at F, we find adder-1, mult-1, and mult-2*
  2.2 Take the intersection of all the sets found by Step 2.1
      ; *this yields the components common to all discrepancies (and*
      ; *hence potentially able to account for all discrepancies)*
      ; *(in the example above there is only one discrepancy)*

STEP 3: DETERMINE CANDIDATE CONSISTENCY VIA CONSTRAINT SUSPENSION
  3.1 For each component found in Step 2.2:
     3.1.1 Turn off (suspend) the constraint modeling its behavior
     3.1.2 Insert observed values at outputs of constraint network
         ; *(inputs were inserted earlier at Step 1.1)*
     3.1.3 If the network reaches a consistent state
        – the component is a globally consistent candidate
        – its symptoms can be found at its ports
        – add the candidate and its symptoms to the candidate list
         ; *e.g., adder-1 and its values of 6, 6, and 10*
      otherwise
        the candidate is not globally consistent, ignore it
         ; *e.g., mult-2*
     3.1.4 Retract the values at constraint network outputs
     3.1.5 Turn on the constraint turned off in Step 3.1.1
        ; *(these last two just get ready for the next iteration of 3.1)*

FIG. 8. Candidate generation via constraint suspension (assuming single point of failure).

As we explore further below, there are a number of important assumptions underlying this reasoning. But constraint suspension provides a mechanism that is both very useful and characteristic of a basic theme underlying this work: the careful management of assumptions. The traditional approach to diagnosis proceeds by assuming that it knows how the component might be failing: it is displaying one of the known misbehaviors found in the set of fault models. We, on the other hand, *proceed by simply suspending all assumptions about how a component might be behaving*. We then allow the symptoms to tell us what the component might be doing. By suspending the constraint in the adder, for example, we are in effect withdrawing all preconceptions about how that component is behaving. We then let the symptoms and the rest of the network tell us whether there is any behavior at all that is consistent with all our observations.

6.2. Advantages of discrepancy detection and constraint suspension

The combination of discrepancy detection and constraint suspension provides a very useful mechanism with a number of advantages:

(1) It is fundamentally a *diagnostic* technique, since it allows systematic isolation of the possibly faulty devices, and does so without having to pre-compute fault dictionaries, diagnosis trees, or the like.

(2) It reasons from the structure and behavior of the device: the candidate-generation process works from the schematic itself to determine which components might be to blame.

(3) Since it defines failure behaviorally, i.e., as anything that doesn't match the expected behavior, it can deal with a wide range of faults, including any systematic misbehavior. This is more widely applicable than a fixed set of models like stuck-ats.

(4) As we saw above, the technique yields symptom information about the malfunction: if adder-1 is indeed the culprit, then we know a little about how it is misbehaving. As will become clear, this information turns out to be useful in several ways.

(5) The approach allows natural use of hierarchical descriptions, a marked advantage for dealing with complex structures.

In the example above, for instance, we determined the relevant candidates at the current, fairly high level of description, never having to deal with lower-level descriptions (e.g., gate-level devices). We could now continue the process 'inside' either candidate, using the next level of description in exactly the same fashion, to determine what subcomponents might be responsible.

(6) Continuing the process at the next level might indicate that no subcomponent could be responsible, ruling out that candidate.

We might, for example, find that, given how the adder is implemented, there may be no subcomponent of it that can logically account for the '6 plus 6 equals 10' symptom that the adder would have to be displaying. Thus the same candidate-generation machinery will either provide a set of candidate subcomponents at the next level, or indicate that none can account for the inferred misbehavior, exonerating this candidate.

(7) This approach keeps knowledge about logical plausibility distinct from knowledge about physical plausibility. This helps simplify construction of the system.

Constraint suspension answers the question of logical plausibility of a candidate; it determines whether there is *any* set of values the component might display that could account for all the symptoms. The technique (by design) knows nothing about whether that set of values is in fact physically plausible.

Our candidate-generation machinery would, for example, consider a forked wire to be a plausible candidate if it inferred that the values at its three ports were 1 at the 'input' (the point where some device is driving the wire), and 0 and 1 at the two 'outputs' (where the wire in turn drives two other devices). Viewed at the black-box level, the wire is a three-port device that could well display the symptoms noted. To know that this is implausible requires understanding the physics of a specific technology: a wire will display different values at its ends as a result of breaks, but a broken wire in TTL will manifest as a high. Hence the pattern of values given can be ruled out by using knowledge of the particular technology.

The candidate generator thus provides a list of logically plausible components; further pruning of this list can then be done by invoking a distinct body of technology-specific knowledge. Keeping the two distinct simplifies the construction of both.

(8) *The technique extends in straightforward fashion across multiple tests.*

Each test of the overall device provides one set of symptoms which in turn yields one or more components with their suspected misbehaviors. The test in Fig. 7, for instance, gives us one misbehavior for adder-1 and one for mult-1; subsequent tests can provide additional misbehaviors. Two kinds of knowledge then relate results across tests.

The non-intermittency assumption indicates that if a component is misbehaving, that misbehavior is at least consistent and reproducible. Hence if a candidate has identical inputs in a later test, it must have an identical output. If, however, a later test were to indicate that adder-1 was a candidate with inputs 6 and 6, and an output of 13, those two test results (and the non-intermittency assumption) exonerate adder-1.

The second source of knowledge comes from additional information concerning faults physically plausible within a specific technology. Consider a situation in which Test 1 indicates that a particular wire is a candidate because it is getting a 0 but propagating a 1, while Test 2 indicates the wire is a candidate because it is getting a 1 but propagating a 0. Considered simply as a two-port device, this is logically consistent and the candidate generator will report it as such. But knowledge about TTL circuits tells us that there is no physically plausible fault which will cause a wire to start behaving as an inverter, hence the wire can be exonerated.

(9) *This approach makes the $\langle N \rangle$-point-of-failure assumption both explicit and easily modified.*

In Fig. 8 above the single-point-of-failure assumption appears in two places: at Step 2.2 where we intersected the sets, and at Step 3.1 where we disable the rules for exactly one component when checking candidate consistency. To deal with any specific $N > 1$, the simplest change is to take the *union* at Step 2.2 and take pairs, triples, etc., at Step 3.1.

While this does not in any sense solve the problem of multiple points of failure, it does demonstrate two important points. First, it shows that, unlike many approaches, our constraint-suspension technique is not limited to dealing with a single point of failure. Second, it does provide some aid in grappling with multiple failures, since we can methodically try possible single failures, then pairs, triples, etc.[8]

Unfortunately, as we demonstrate, the power of all of this mechanism is only part of the story.

6.3. Subtleties in candidate generation

Consider the slightly revised example shown in Fig. 9. Reasoning as before, we would discover in this case that there is only one hypothesis consistent with the values measured at F and G: the second multiplier is malfunctioning, outputting a 0.

Yet there is another quite reasonable hypothesis: the third multiplier might be bad (or the first).

But how could this produce errors at both F and G? The key lies in being wary of our models. The thought that digital devices have input and output ports is a convenient abstraction, not an electrical reality. If, as sometimes happens (due to a bent pin, bad socket, etc.), a chip fails to get power, its inputs are no longer guaranteed to act unidirectionally as inputs. If the third multiplier were a chip that failed to get power, it might not only send out a 0 along wire Z, but it might also pull down wire C to 0. Hence the symptoms result from a single point of failure (mult-3), but the error propagates along an 'input' line common to two devices.

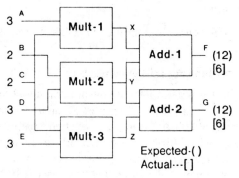

FIG. 9. Troublesome troubleshooting example.

---

[8]While *checking* pairs, etc. using constraint suspension (Step 3.1.3) is computationally simple, *generation* of appropriate pairs, triples, etc., (Step 3.1) is still an exponential process. This has not yet proved to be untenable, since both the number of faults considered and the number of components at any given level of description are relatively small. We are nevertheless exploring ways of improving the process. For example, some simple bookkeeping tricks can be used to rule out some of the pairs, triples, etc., very quickly.

The most immediate problem lies in our implicit acceptance of unidirectional ports—when checking the inputs to multi-1, we assumed that the inputs were machinery. We implicitly assumed that wires get information only from output ports—when checking the inputs to mult-1, we assumed that the inputs were 'primitive'. We looked only at terminals A and C, never at the other end of the wire at mult-3.

Bridges are a second common fault that illustrates another place where we need to be careful: the reasoning style used above can *never* hypothesize a bridging fault, again because of implicit assumptions and their subtle reflection in the reasoning. Bridges can be viewed as wires that don't show up in the design. But we traditionally make an implicit closed-world assumption: the structure description is assumed to be complete and anything not shown there 'doesn't exist'. Clearly this is not always true. Bridges are only one manifestation; wiring errors during assembly are another possibility.

Let's review for a moment. The traditional test-generation technology suffered from a number of problems: among others, it is a technology for test generation, not diagnosis, and it uses a limited fault model. The use of discrepancy detection and constraint suspension improves on this substantially by providing a diagnostic ability, by defining a fault as anything that produces behavior different from that expected, and by working directly from descriptions of structure and behavior. This seems to be perfectly general, but, as we illustrated, it has to be used with care:

Put simply, the virtue of the technique is that it reasons from the schematic; the serious flaw in the technique is that it reasons from the schematic *and the schematic might be wrong*.

We believe it is instructive to examine the basic source and nature of this problem.

7. Mechanism and Knowledge

In the example above we encountered some interesting situations because we failed to make explicit a number of important assumptions underlying the reasoning. In the power-failure example, we were assuming implicitly that there was only one possible direction of causality at an input port, and thus never examined the other end of wire C. Similarly, tracing back through the circuit from input X of adder-1, we looked only at mult-1, because that was the only apparent connection at that point. We never looked at, say mult-3, because there was no wire leading there, hence no reason to believe one might affect the other.

Note carefully the character of these assumptions: they concern the *existence of causal pathways, the applicability of a particular model of interaction*. In the power failure example we assumed implicitly that there was no way for mult-3 to affect mult-1 through wire C, yet such a path is possible. As we noted above in discussing the possibility of a bridge fault, there seemed to be no path in the schematic that would allow mult-3 to affect adder-1, yet a pathway is in fact possible.

The problem is not in the existence of such assumptions—they are in fact crucial to the reasoning process. The problem lies instead in the careful and explicit management of them. To see the necessity of having assumptions about causal pathways, consider the nature of the candidate-generation task. Given a problem noticed at some point in the device, candidate generation attempts to determine which modules could have caused the problem. To answer the question we must know by what mechanisms and pathways modules can interact. Without *some* notion of how modules can affect one another, we can make no choice, we have no basis for selecting any one module over another.

In this domain the obvious answer is 'wires': modules interact because they're explicitly wired together. But that's not the only possibility. As we saw, bridges are one exception; they are 'wires' that aren't supposed to be there. But we also might consider thermal interactions, capacitive coupling, transmission-line effects, etc.

Generating candidates, then, should not be thought of in terms of tracing wires (or dependency records). Rather, we claim, it should be thought of in terms of *tracing paths of causality*. Wires are only the most obvious pathway. In fact, given the wide variety of faults we want to deal with, we need to consider many different pathways of interaction.

And that leaves us on the horns of a classic dilemma. If we include every interaction path, candidate generation becomes *indiscriminate*—there will be some (possibly convoluted) pathway by which every module could conceivably be to blame. Yet if we omit any pathway, there will be whole classes of faults we will *never* be able to diagnose.

What can we do? We believe that two steps are important. First, we have to recognize that our inference mechanisms—in this case dependency detection and constraint suspension—are not the source of problem-solving power. The power is instead in the knowledge that we supply those techniques, i.e., the pathways of interaction.

And therein lies the second step: there is an important task in enumerating and organizing the pathways of interaction to be considered. How can we do this? We believe that human performance supplies a useful clue.

8. Organizing the Pathways of Interaction

We appear to be faced with an unavoidable dilemma, caught between the desire to be complete and the need to constrain the possibilities we consider. But people face exactly the same dilemma and seem to handle it. What do they do?

The answer seems be an instantiation of Occam's razor: an experienced engineer knows that some things are more likely to go wrong than others. He will, as a result, attempt to generate solutions that employ simpler and more likely hypotheses first, falling back on more elaborate possibilities only in the face of an intractable contradiction (i.e., given the current set of assumptions, there is no way to account for the observed misbehavior). There are three important points here.

(1) The engineer has a notion of 'the kinds of things that can go wrong'.

(2) There is an ordering criterion that indicates which category of hypotheses to entertain first.

(3) The categories are ordered but none is permanently excluded.

To capture this same sort of behavior in our program we need to (i) make precise the notion of 'what can go wrong', and (ii) determine what constitutes a 'simple' explanation. We consider both of these briefly here, as background for the example that follows, then address the issue in detail in Section 10.

To address the first of these, we need some methodical way to define and generate the possible kinds of failures. This is accomplished by enumerating the assumptions built into our 'module and information path' representation and then characterizing the variety of failure that results from violating each assumption. We refer to the resulting list as the *categories of failure*.

One such category is illustrated by the problem presented in Fig. 9. The implicit assumption there was that information flows in only one direction at an input (or output) port. If this assumption is violated, we get a category of failure we term an 'unexpected-direction' failure. The other categories generated in this way are described in Section 10.

Given such a notion of 'what can go wrong', we now need an appropriate metric for ordering the list. This is currently accomplished by relying on the experience of expert troubleshooters, who tell us which categories of failure are encountered more frequently than others. Stuck-ats are more likely than assembly errors, for example. While the ordering criterion may eventually need be more elaborate, its precise content is less an issue here than its character: it is a summary of empirical experience that helps us to order the kinds of hypotheses we consider. For our current domain, this approach produces the following list:
– localized failure of function (e.g., stuck-at on a wire, failure of a RAM cell),
– bridges,
– unexpected direction (e.g., the power-failure problem),
– multiple point of failure,
– intermittent error,
– assembly error,
– design error.

We start by attempting to generate candidates in the localized-failure category, assuming that the structure is as shown in the schematic, that there was only a single point of error, that information flowed only in the predicted directions, etc. Only if this leads to a contradiction are we willing to surrender an assumption (e.g., that the schematic was correct) and entertain the notion that a bridge might be at fault. If this too leads us down a blind alley then we would surrender additional assumptions and consider ever more elaborate hypotheses, eventually entertaining the possibility of multiple errors, an assembly error (every individual component works but they have been wired up incorrectly) and even design errors (the implementation is correct but cannot produce the desired behavior).

This mimics what we believe a good engineer will do: make all the assumptions necessary to simplify a problem and make it tractable, but be prepared to discover that some of those simplifications were incorrect. In that case, surrender some of those assumptions and be willing to consider additional kinds of failure.

In terms of the dilemma noted above, the categories of failure serve as a set of *filters*. They restrict the paths of interaction we are willing to consider, thereby preventing candidate generation from becoming indiscriminate. In using the localized-failure-of-function category, for example, we are assuming for the moment that the structure is as shown in the schematic, hence there are no additional paths of interaction (and thus no bridges). But these are filters that we have carefully ordered and consciously put in place. If we cannot account for the observed symptoms with the current set of filters in place, we remove one, leaving us with a set that is less restrictive, allowing us to consider additional interaction paths and hence more elaborate hypotheses.

There are of course no guarantees that this will lead us to the correct category of failure without any false steps. It is possible that all the evidence at a given point is consistent with, say, a localized failure of function in a single component, yet replacing that component may make it clear that the fault is elsewhere. As always with Occam's razor, our only assurance is that we are generating a hypothesis that is by some measure the simplest and most likely, and that's the best we can do. We may subsequently discover that the problem is in fact more complex. By making the ordering criteria explicit and accessible, we have at least provided a place for embedding knowledge that can make the choice of hypothesis category as informed as possible at each step.

9. Example: a Bridge Fault

As we have noted, traditional automated reasoning about circuits works from a predefined list of fault models and uses the mathematical style of analysis exemplified by boolean algebra or the D-Algorithm. As a result, it is strongly oriented toward faults that can be modeled as a permanent binary value. One problem with this is its inability to provide useful results concerning bridge faults.

In this section we show how our system works when faced with a bridge fault, illustrating a number of the ideas described above. While the example has been simplified for presentation, there is still unfortunately a fair amount of detail necessary. A summary of the basic steps will help make clear how the problem is solved.

(1) The device is a 6-bit carry-chain adder. The problem begins when we notice that the attempt to add 21 and 19 produces an incorrect result.

(2) The candidate-generation process outlined above generates a set S_1 of three candidates, any of whose malfunction can explain this result.

(3) A new set of inputs (1 and 19) is chosen in an attempt to discriminate among the three possibilities. The adder's output is incorrect for this set of inputs also. The candidate generator indicates that there are two candidates capable of explaining this new result.

(4) Neither of these two candidates are found in S_1. Thus we reach a contradiction: no component is capable of explaining the data from both sets of inputs.

(5) Put slightly differently, we have a contradiction *under the current set of assumptions and interaction models*. We therefore have to surrender one of our assumptions and use a different interaction model.

(6) The next model—bridge faults—surrenders the assumption that the structure is as shown in the schematic and considers one class of modifications to the structure: the addition of one wire between physically adjacent pins.

(7) The combination of functional information (the expected pattern of values produced by the fault) and physical adjacency provides a strong constraint on the set of connections which might be plausible bridges.

(8) The first application of this idea produces two hypotheses that are functionally plausible, but both are ruled out on physical grounds.

(9) Dropping down a level of detail in our functional description reveals additional bridge candidates, two of which prove to be both functionally and physically plausible. One of these proves to be the actual error.

A key point is the utility of ordering the paths of interaction to be considered. Starting with a very restricted category of failure, we discover that it leads us to a contradiction. We surrender an assumption, consider an additional category and hence an additional pathway of interaction: bridge faults. We show how knowledge of both structural and functional organization allows us to generate a select few bridge-fault hypotheses, eventually discovering the underlying fault.

9.1. The example

Consider the six-bit adder shown in Fig. 10 and imagine that the attempt to add 21 and 19 produces 36 rather than the expected value of 40. Invoking the candidate-generation process described above, we would find that there are three devices (SLICE-1, A2 and SLICE-2), any one of whose malfunction can explain the misbehavior.[9]

[9]The example has been simplified slightly for presentation.

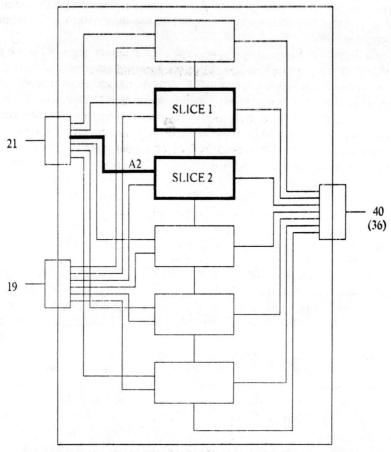

FIG. 10. Six-bit adder constructed from single-bit slices. Heavy lines indicate components implicated as possibly faulty.

A good strategy when faced with several candidates is to devise a test that can cut the space of possibilities in half. In this case changing the first input (21) to 1 will be informative: if the output of SLICE-2 does not change (to a 0) when we add 1 and 19, then the error must be in either A2 or SLICE-2.[10]

---

[10]This and subsequent test generation is currently done by hand. Work on automating test generation is in progress [27]. The reasoning behind this test relies on the single-fault assumption: if the malfunctioning component really were SLICE-1, both A2 and SLICE-2 would be fault-free. Hence the output of SLICE-2 would have to change when we changed one of its inputs. (Notice, however, if the output actually does change, we don't have any clear indication about the error location: SLICE-2, for example, might still be faulty.)

DIAGNOSTIC REASONING

As it turns out, the result of adding 1 and 19 is 4 rather than 20. Since the output of SLICE-2 has not changed, it appears that the error must be in either A2 or SLICE-2.

But if we invoke the candidate generator, we discover an oddity: the only way to account for the behavior in which adding 1 and 19 produces a 4 is if one of the two candidates highlighted in Fig. 11 (B4 and SLICE-4) is at fault.

Therein lies our contradiction. The only candidates that account for the behavior of the first test are those in Fig. 10, the only candidates that account for the second test are those in Fig. 11. There is no overlap, so there is no single candidate that accounts for all the observed behavior.

Our current category—the localized failure of function—has thus led us to a

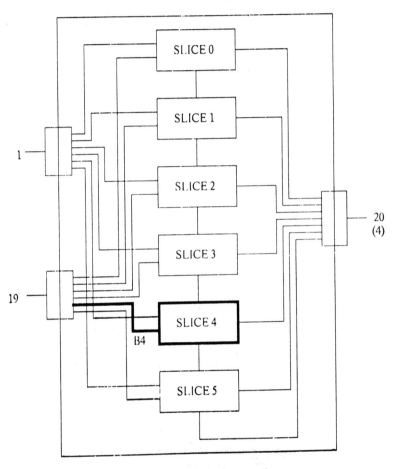

FIG. 11. Components indicated as possibly faulty by the second test.

contradiction.[11] We therefore surrender it and consider the next, less restrictive category, one that allows us to consider an *additional kind of interaction path*—bridging faults. The problem now is to see if there is some way to unify the test results, some way to generate a single bridge-fault candidate that accounts for all the observations.

Much of the difficulty in dealing with bridging faults arises because they violate the rather basic assumption that the structure of the device is in fact as shown in the schematic. But admitting that the structure may not be as pictured says only that we know what the structure *isn't*. Saying that we may have a bridge fault narrows it to a particular class of modifications to consider, but the real problem here remains one of *making a few plausible conjectures about modifications to the structure*. Between which two points can we insert a wire and produce the behavior observed?

To understand how we answer that question, consider what we have and what we need. We have test results, i.e., observations of *behavior*, and we want conjectures about modification to *structure*. The link from behavior to structure is provided by knowledge of electronics: in TTL, a bridge fault acts like an AND-gate, with ground dominating.[12]

From this fact we can derive a simple pattern of behavior indicative of bridges. Consider the simple example of Fig. 12 and assume that we ran two tests. Test 1 produced one candidate, module A, which should have produced a 1 but yielded a 0 (the zero is underlined to show that it is an incorrect output).

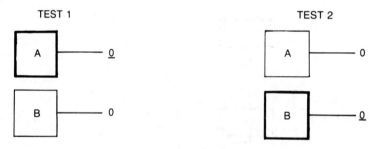

FIG. 12. Pattern of values indicative of a bridge. Heavy lines indicate candidates.

---

[11] Note that dropping down another level of detail in the functional description cannot help resolve the contradiction, because our functional description is a tree rather than a graph: in our work to date, at least, no component is used in more than one way. (If the functional description were in fact a graph, we could easily continue down it to see if the two candidate sets did indeed have a subcomponent in common).

[12] This is an oversimplification, but accurate enough to be useful. In any case, the point here is how the information is used; a more complex model could be substituted and carried through the rest of the problem. Note also that for notational convenience, we assume in the rest of the description that ground is equivalent to a 0.

DIAGNOSTIC REASONING

Module B was working correctly and produced a 0 as expected. In Test 2 this situation is exactly reversed, A was performing as expected and B failed.

The pattern displayed in these two tests makes it plausible that there is a bridge linking the outputs of A and B: in the first test the output of A was dragged low by B, in the second test the output of B was dragged low by A.

We have thus turned the insight from electronics into a pattern of values on the candidates. It is plausible to hypothesize a bridge fault between two modules A and B from two different tests if: in Test 1, A produced an erroneous 0 and B produced a valid 0, while in Test 2, A produced a valid 0 while B produced an erroneous 0. Note that this can resolve the contradiction of non-overlapping candidate sets: it hypothesizes one fault that involves a member of each set and accounts for all the test data.

Thus, if we want to account for all of the test data in the original problem with a single bridge fault, we need a bridge that links one of the candidates from the first test (SLICE-1, A2, SLICE-2) with one of the candidates from the second test (B4, SLICE-4), and that mimics the pattern shown in Fig. 12.

Fig. 13 shows the candidate-generation results from both tests in somewhat more detail. As noted earlier, the candidate-generation procedure can indicate for each candidate the values that would have to exist at its ports for that candidate to be the broken one. For example, for SLICE-1 to be at fault in Test

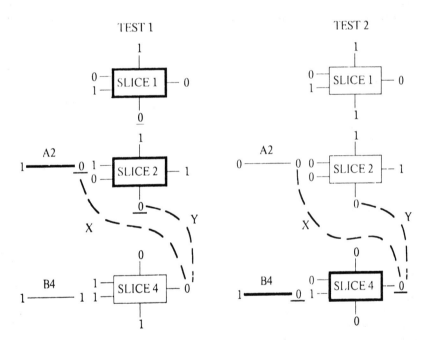

FIG. 13. Candidates and values at their ports.

1, it would have to have the three inputs shown, with its sum output a zero (as expected) and its carry output also a zero (the manifestation of the error, underlined).

In Fig. 13 there are two pairs of devices that match the desired pattern, yielding two functionally plausible bridge hypotheses:
– dotted line X, bridging wire A2 to the sum output of SLICE-4;
– dotted line Y, bridging the carry output of SLICE-2 to the sum output of SLICE-4.

But the faults have to be physically plausible as well. For the sake of

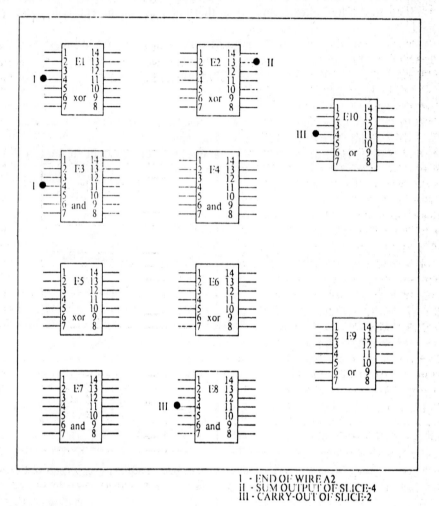

I - END OF WIRE A2
II - SUM OUTPUT OF SLICE-4
III - CARRY-OUT OF SLICE-2

FIG. 14. Physical layout of the board with first bridge hypotheses indicated. (Slices 0, 2, and 4 are in the upper 5 chips, slices 1, 3, and 5 are in the lower 5.)

DIAGNOSTIC REASONING

simplicity, we assume that bridge faults result only from solder splashes at the pins of chips.[13] To check physical plausibility, we switch to our physical representation, Fig. 14. Wire A2 is connected to chip E1 at pin 4 and chip E3 at pin 4; the sum output of SLICE-4 emerges at chip E2, pin 13. Since they are not adjacent, the first hypothesis is not physically reasonable. Similar reasoning rules out Y, the hypothesized bridge between the carry-out of SLICE-2 and the sum output of SLICE-4.

So far we have considered only the top-level of functional organization. We can run the candidate generator at the next lower level of detail in each of the non-primitive components in Fig. 13. (Dropping down a level of detail proves useful here because additional substructure becomes visible, effectively revealing new places that might be bridged.)

We obtain the components and values shown in Fig. 15. Checking here for the desired pattern, we find that either of the two wires labeled A2 and S2 could be bridged to either of the two wires labeled S4 and C4, generating four functionally plausible bridge faults.

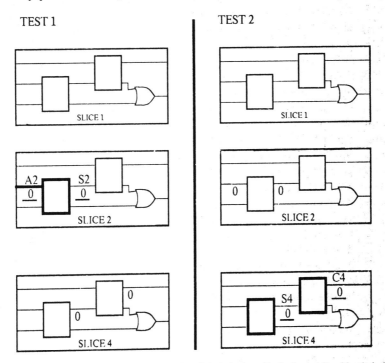

FIG. 15. Candidates at the next level of functional description. Each single-bit adder is built from two 'half-adders' and an OR-gate. (To simplify the figure, only the relevant values are shown.)

[13] Again this is correct but oversimplified (e.g., backplane pins can be bent or bridged), but as above we can introduce a more complex model if necessary.

Once again we check physical plausibility by examining the actual locations of A2, S2, S4, and C4 (Fig. 16).[14] As illustrated there, two of the possibilities are physically plausible as well: A2–S4 on chip E1 and S2–S4 on chip E2.

Switching back to our functional organization once more, Fig. 17, we see that the two possibilities correspond to (X) an output-to-input bridge between the XOR-gates in the rear half-adders of SLICE-2 and SLICE-4, and (Y) a bridge between two inputs of the XOR-gates in the forward half-adders of slices 2 and 4.

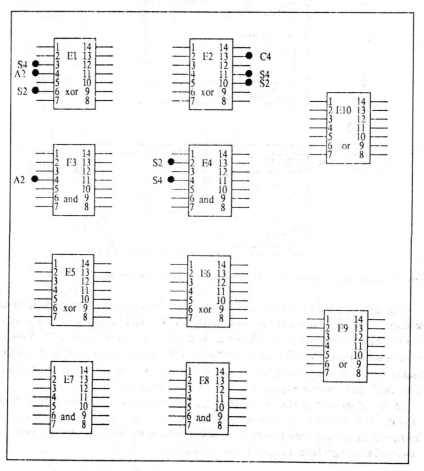

FIG. 16. Second set of bridge-fault hypotheses located on physical layout.

[14]Note that the erroneous 0 on wire S2 can be in any of three physical locations because S2 fans out (inside the module it enters on its right).

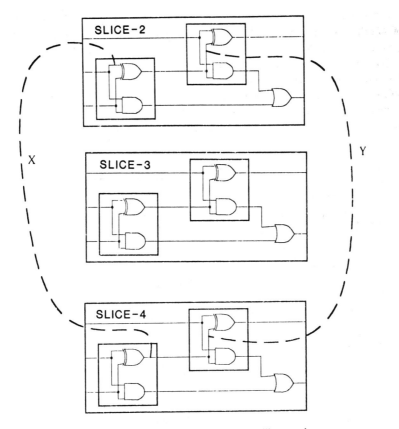

FIG. 17. Functional representation with bridge-fault hypotheses illustrated.

It is easy to find a test that distinguishes between these two possibilities:[15] adding 0 and 4 means that the inputs of SLICE-2 will be 1 and 0, with a carry-in of 0, while the inputs of SLICE-4 will both be 0, with a carry-in of 0. This set of values will show the effects of bridge Y, if it in fact exists: the sum output of SLICE-2 will be 0 if it does exist and a 1 otherwise. When we perform this test the result is 1, hence bridge Y is not in fact the problem.

Bridge X becomes the likely answer, but we should still test for it directly. Adding 4 and 0 (i.e., just switching the order of the inputs), is informative: if bridge X exists the result will be 0 and 1 otherwise. In this case the result is 0, hence the bridge labeled X is in fact the problem.[16]

[15]As above, tests are currently generated by hand.
[16]Had both been ruled out by direct test, then we would once again have had a contradiction on our hands and would have had to drop back to consider a still more elaborate model with additional paths of interaction.

9.2. The example: summary and comments

A fundamental point illustrated by the example is the utility of a layered set of models as a device for making explicit our simplifying assumptions and for dealing with complexity in candidate generation. Our original model, localized failure of function, incorporated the largest set of simplifying assumptions and was the most restrictive. It worked initially, but we eventually found ourselves unable to account for all the observations. At that point we surrendered one of our assumptions, adopted a less restrictive model, and considered an additional path of interaction. This allowed us to generate several bridge-fault hypotheses, one of which eventually proved to be correct. The notion of layers of interaction models thus provided an important overall framework guiding the problem solving.

A second source of problem-solving power came from using multiple representations. This was particularly important in constraining the generation of bridge candidates. Faced with the original, apparently insoluble problem, we admitted the possibility of a bridge. But this left us with the difficult task of deciding where the bridge might be. The problem is quite similar to adding a new line to complete a proof in geometry and the difficulty is analogous: new constructions are difficult in general because they are relatively unconstrained [22].

The system's search for likely candidates turned out to be focused because we were able to derive useful constraints from each of our multiple representations. Consider the physical representation, where the constraint is physical adjacency. We started by choosing a particular variety of adjacency, contiguity of pins on chips. This produced a significant reduction in the potential search space, but still left us with too many choices to produce an effective hypothesis generator: trying each pair of pins would be too unwieldy.

We then found a way to reduce the search using the functional representation. We used knowledge about electronics to derive a link from behavior to structure, producing a pattern characteristic of bridges. This reduced the search space considerably, since the pattern had to include one candidate from each of the two sets (each of which is itself typically small), and since the pattern of 'alternating zeros' is relatively rare.

The result was sufficiently constrained to be an effective generator of bridge-fault hypotheses. We were then able to use the physical representation as a filter on the hypotheses generated.

In general then, our system produces a focused development of its solution by relying on several keys:

(1) The layered set of interaction models produced by methodical enumeration and relaxation of underlying assumptions constrained the categories of errors we were willing to consider, yet allowed us to consider more elaborate hypotheses when simpler ones failed.

(2) The availability of multiple representations allowed us to take advantage of constraints associated with each representation.

DIAGNOSTIC REASONING

(3) We were able to use one representation as the basis for a constrained generator of hypotheses and use the other to filter the hypotheses generated.

In Section 11 we speculate on ways of generalizing the set of ideas used here. We consider the character of the representations used and ask what made them effective generators and filters. Our goal there is to produce a set of principles that will function as guidelines in selecting representations, making it possible to carry over this approach to other problems in other domains.

10. Categories of Failure

Since the categories of failure and paths of interaction play a significant role in our approach, two obvious questions concern their origin and ordering. Where do we get them, and how can we determine an appropriate ordering?

We want a methodical way to generate the categories so that we have some reason to believe that the result is systematic and hence reasonably complete. We need to define the criterion for simplicity so that we know how to order the categories to produce a sequence of successively more elaborate hypotheses.

10.1. Origins

The list in Section 8 of possibilities to consider is of course specific to our current domain. We believe that our overall framework does, however, offer a useful set of questions that we can ask about any domain to generate an analogous listing. Appropriately enough, many of the questions derive from examining carefully our simple 'module and information path' representation, asking what assumptions the representation makes about the world, and then asking what the consequences are of violating those assumptions.

The simplest assumption (indicated schematically in Fig. 18(a)) lies in believing that a module can be said to have a particular behavior. To say this is violated means, as we saw earlier, that the module isn't behaving in accordance with the assigned behavior, i.e., it is broken. This is the simple 'localized-failure-of-function' category that heads the list in Section 8. Simple examples in our current domain include the traditional stuck-at (a particular kind of failure of a wire module), as well as failures of primitive gates (e.g., a NOR-gate acting as an inverter), or even the use of an incorrect part during assembly (e.g., a NOR-gate chip instead of a NAND-chip).

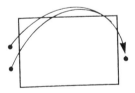

FIG. 18(a). Assigning behavior to a module.

Our representation also assumes that modules have ports, each with a specified direction (Fig. 18(b)). Yet as we saw above, this too can be violated. We term it an 'unexpected-direction' error, with the power-failure example as one instance.

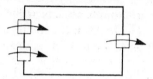

FIG. 18(b). Assuming that ports have a specified direction.

Treating a module as a black box means committing only to its behavior; when we drop down a level in the description, we are assuming that it has a specific substructure (Fig. 18(c)). But that may not be true; we refer to this as a 'structure error'. How could the substructure be different from what we expect? In our current domain there are a wide range of answers, i.e., a wide range of paths of interaction: bridges, thermal or transmission-line effects, as well as out-of-date schematics or a wiring error during assembly. Each of these has the effect of producing a substructure different from what we expected.

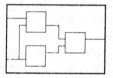

FIG. 18(c). Assuming that a substructure is as specified.

Finally, our representation assumes that the overall behavior of a module should be matched by the aggregate behavior of its substructure (Fig. 18(d)). Violating this assumption means there is a design error: all components work as specified, but they cannot produce the desired behavior.

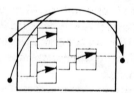

FIG. 18(d). Assuming that the overall behavior matches the aggregate of its substructure.

Thus, examining carefully the character of the assumptions built into our representation yields a set categories of failure. These in turn provide a number of questions we can ask about the domain, questions whose answers are the pathways of interaction to be considered. While we have answered these questions here for our digital circuits, we believe the questions are more broadly applicable, and may prove to be a useful way to think about other domains where the 'module and information path' view is appropriate.

One additional category derives from examining our candidate generator. As noted in Section 6.2, the single-point-of-failure assumption is embodied in the fact that we disabled the rules for exactly one component when checking candidate consistency. This too may be an incorrect assumption, yielding yet another category, multiple faults.

10.2 Ordering

The example in Section 9 used a very simple approach to ordering the categories: a fixed order based on frequency of occurrence as reported by experienced troubleshooters. While the program currently uses this approach, it is easy to imagine more sophisticated ordering schemes that still fit naturally into this framework. The remainder of this section speculates about a number of such possibilities.

More elaborate static-ordering criteria might take account of the age of the machine or the age of the design. We might want to indicate that assembly errors are more common in machines received from the production line recently, and design errors more likely if the device is a new model or has undergone substantial redesign.

Note that, were we willing to model enormously more of the world, we might be able to *infer* that design errors are less common in well-established designs, but this is beyond the scope of our efforts and the investment would be very large for a relatively modest return. As it stands, we are in this case willing to rely on some simple experiential observations for guidance, without bothering to model the causality in any detail.

The character of the reported misbehavior can be revealing as well. Small changes in external behavior are often suggestive of the local-failure-of-function category, while substantial variations in behavior can be suggestive of assembly and wiring errors. We might thus borrow from medicine the notion of a 'presenting complaint' and use information about the character of the misbehavior to reorder the categories to be considered.

An obvious further extension would make the ordering dynamic. We might start out as above, but reorder the categories in response to information gained as inference and testing proceeds.

Whatever the ordering criteria chosen, the overall point is that having an ordered set of fault categories provides a number of advantages. First, it offers a means of expressing and managing simplifying assumptions. Second, it provides a simple mechanism for dealing with complexity, by limiting the set of interaction paths to consider. Finally, it supplies a relatively natural site for embedding and using the empirical knowledge of an experienced engineer that might be difficult to represent at the level of first principles.

11. The Adjacency Principle

The bridge-fault example raises two interesting questions:
  (1) Why are bridge faults so difficult?
  (2) Why does the physical representation prove to be so useful?

To see the answer, we start with the trivial observation that all faults are the result of some difference between the device as it is and as it should be. With bridge faults the difference is the addition of a wire between two physically adjacent points.

Now recall the nature of our task: we are typically presented with a device that misbehaves, not one with obvious structural damage. Hence we reason from behavior, i.e., from the functional representation. And the important point is that for a bridge fault, the difference in question—the addition of a single wire—is not small and local in that representation. As the comparison of Figs. 16 and 17 makes clear, the new wire connects two points that are adjacent in the physical representation but widely separated in the functional representation.

The difference is also not as simple in that representation: if we include in our functional diagram the AND-gate implicitly produced by the bridge (Fig. 19), we see that a single added wire in the physical representation maps into an AND-gate and a fanout in the functional representation.

This view helps to explain why bridge faults produce behavior that is difficult to understand. Bridge faults are modifications that are simple and local in the physical description, but our reasoning is done using the functional description. Hence the dilemma:

The desire to reason from behavior requires us to use a representation that does not necessarily provide a compact description of the fault.

This non-locality and complexity should not be surprising, since devices physically adjacent are not necessarily functionally related. Hence there is no guarantee that a change that is compact in one will produce a change that is compact in the other. More generally, *changes compact in one representation are not necessarily compact in another.*

We can turn this around to put it to work for us:

Part of the art of choosing the right representation(s) for diagnostic reasoning is finding one in which the suspected change is *compact.*

This explains the utility of the physical representation: it's the 'right' one

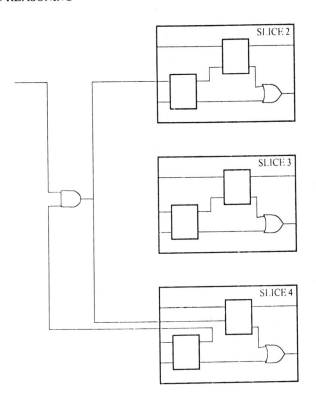

FIG. 19. Full functional representation of bridge fault X.

because it's the one in which the change is compact—the affected components are adjacent.

Going one step further, we might ask why adjacency is the relevant organizing principle. We believe the answer follows from two facts: (a) devices interact through physical processes (voltage on a wire, thermal radiation, etc.) and (b) physical processes occur locally, or more generally, causality proceeds between devices that are in some sense adjacent: there is no action at a distance. To make this useful, we turn it around:

The paths of interaction are one way to define adjacency. That is, each kind of interaction path can define a representation.

Bridge faults arise from *physical adjacency* and hence are local in the physical representation. The notion of *thermal adjacency* and a corresponding representation would be useful in dealing with faults resulting from heat conduction or radiation, *electromagnetic adjacency* would help with faults dealing with transmission line effects, etc.

Each of these produces a different representation, different in its definition of adjacency. And each will be useful for understanding and reasoning about a category of failure.

The paths of interaction are in effect a set of 'representation generators'. They can tell us what kinds of views to take of the device; how to represent it in order to make use of the fundamental belief that physical action proceeds between components that are adjacent in some way.

Finally, we can push this one step further, asking why we expect that there will in fact be some representation, some definition of adjacency, that 'makes sense' of the problem at hand. Why should there be some representation in which the fault appears to be a compact perturbation? The belief appears to be based on what we might call the 'single-initial-cause heuristic':

It is often the case that the malfunction of a previously working device results from a single cause, rather than a number of simultaneous, independent events.

When there is indeed a single cause, there will be some representation in which a small and local change accounts for the difference between the good and faulty device. To the extent that a malfunction results from multiple, *unrelated* events (e.g., multiple failures in an old machine; multiple problems inserted deliberately), it becomes far less likely that we can find a single representation that offers a unifying perspective for every fault at once.[17]

One additional illustration of the utility of multiple representations with multiple definitions of adjacency comes from considering devices with shared components. In some designs, one physical component (e.g., an AND-gate) may be used for two different purposes. We can model this by having two different modules in the functional representation that point to the same module in the physical representation (using the cross-links shown in Fig. 4). If the physical gate is bad, then the candidate generator may find that it needs two points of failure to account for the symptoms. But what looks like two failures in the functional representation can be resolved to a single point of failure by shifting to the physical representation.

Similarly, if a chip containing four AND-gates is completely bad, our candidate generator will find four apparently distinct candidates. Once again, shifting to the physical representation gives us a view in which the problem can be understood as a single failure.

In both cases shifting to a different definition of adjacency helps resolve the issue and provides alternate definitions of the notion of a single point of failure. In the first case two modules distant in the functional representation are adjacent (in fact coincident) in the physical representation; in the second case four

---

[17]This work has not yet explored the problem of multiple, related failures, where one fault causes a cascade of other failures.

different functional modules become adjacent (same package) in the physical representation.

As this illustrates, the basic concept of a 'single point of failure' is not in fact well defined until we specify the representation. We should really speak of a single point with respect to a particular representation, since, as we have seen, what appears as multiple distinct points in one representation may be a single point in a different representation.

We still have substantial additional work to do on these topics, but we seem at least to be asking the right questions. The concept of pathways of interaction and the corresponding definitions of adjacency appear to make sense for a wide range of hardware faults.

The ideas also appear to work in other domains. When debugging software, for example, the pathways of interaction differ (e.g., procedure call, mutation of data structures), but the resulting perspectives appear to make sense and there are some interesting analogies. Unintended side effects in software, for example, are in some ways like bridge faults. An unexpected direction of information flow can result during a procedure call, by assigning a value to a parameter that turns out to be called by name rather than by value.

More generally still, there appears to be substantial breadth to this whole perspective of viewing different representations as embodying different definitions of adjacency. Consider software once again. Our typical view of software is functional, in flowcharts or dataflow diagrams. Yet there are faults where the notion of 'physical adjacency' is crucial in understanding the bug. Consider for example an out-of-bounds array addressing error. Full understanding of the bug may require knowing how the software is mapped physically into memory, so that we can determine which cell was actually referenced.

A similar phenomenon is apparent in speech: the classic 'recognize speech' vs. 'wreck a nice beach' confusion is entertaining in part because the two are so 'far apart' in meaning (distant in any semantic representation) yet adjacent in their phonetic representation. The idea even supports 'troubleshooting' in this domain. In reading an abstract recently, I encountered a number of phrases that made no sense in an otherwise coherent text. Then, considering the origin of the document, it became plausible that the text might have been dictated over a noisy (transatlantic) phone line. This suggested that a phonetic representation might be the 'correct' one for debugging, i.e., the one in which the errors could be seen as small and local changes. It was then easy to debug phrases like 'decorative representations' ('declarative representations') and 'rule space systems' ('rule-based systems').

Two widely accepted bits of wisdom about AI suggest that having the 'right' representation is an important key in problem solving and that multiple representations often contribute substantial power. We believe that the notion of adjacency as a way of defining representations is useful in understanding more about both of these ideas. In the effort to define what is meant by the 'right' representation, for example, a number of general guidelines have emerged, at the level of suggestions that appropriate representations 'make the important things explicit', and 'expose the natural constraints' (e.g., [34]). The notion of adjacency offers one additional level of detail to these catchphrases. It suggests that natural constraints arise because things are adjacent 'in some sense', and that 'in some sense' can have multiple different instantiations. We can generate a variety of representations by seeking out different definitions of adjacency. Finally, this in turn may help explain the utility of multiple representations: part of the power in using them arises because they provide us with multiple different definitions of adjacency.

12. Evaluation and Future Work

Since this work is still in its formative stages, there are still a number of interesting questions about its advantages and limits and appropriate directions for future work.

12.1. Advantages of reasoning from structure and behavior

There are several advantages in basing reasoning on knowledge of structure and behavior, rather than previous approaches like the empirical associations used in a number of earlier programs (e.g., MYCIN [9, 28], INTERNIST [24], or PROSPECTOR [17]). Reasoning from first principles provides a strong degree of machine independence, makes the system easier to construct and maintain, facilitates defining the program's scope of competence, and makes it capable of dealing with bugs that display novel symptoms.

Reasoning from first principles offers a significant degree of machine independence. The tools and techniques described above work directly from the descriptions of structure and behavior and make explicit their assumptions about those descriptions. This allows us to cover a wide range of devices built from digital logic components. Even more fundamentally, concepts like modules linked by causal pathways, expected behavior, etc., are widely useful, extending perhaps to include handheld calculators, digital watches, automobiles and software.

Rules, on the other hand, are typically strongly machine-specific. There is distressingly little carry-over from one machine to the next when we capture troubleshooting knowledge at the level of symptom/disease associations. Even simple changes to a single machine (e.g., design upgrades) can mean substantial changes in the rules. Diagnosing a new machine means even more difficulty. Each new system would require a new knowledge base with all of the difficulties that entails.

A system based on reasoning from first principles is easier to construct because there is a way of systematically enumerating the required knowledge: the structure and behavior of the device. A system based on empirical associations is more difficult to construct because the character of the knowledge makes it necessary to extract the rules on a case-by-case basis. To the extent that the knowledge is a distillation of the expert's experience, the best we can do is to assemble a representative collection of cases and ask an expert for the rules dealing with each case. No more systematic method of collecting rules is available and the process often continues for an extended period of time. This time lag is a particular problem in dealing with electronic hardware: the time necessary to accumulate the relevant experience is beginning to be longer than the design cycle for the next model of the machine.

A system based on reasoning from first principles is also easier to maintain, since modifications to the machine design are relatively easy to accommodate. We can update the structure and behavior specifications for each modified component, rather than having to determine how each change should modify the overall behavior and the troubleshooting strategy.

The ability to enumerate the knowledge systematically also aids in defining the program's scope and competence. We know what parts of the machine have been modeled and to what level of detail.

When building a system from empirical associations, it is more difficult to define precisely what such a program 'knows'. A precise answer to the question is often possible only for a problem that can be formalized, so almost every AI system suffers from this to some extent. But the more systematic our mechanism for enumerating the required knowledge, the more precise we can be about defining the program's competence. Deriving the knowledge strictly from a collection of case studies is one of the less systematic mechanisms, yet it is in such cases that rule-based systems are appropriate. As a result, often the best we can say about the system is that it has a set of rules dealing with one or another class of problems, with only a very informal measure of how thoroughly that class has been covered.

Finally, reasoning from first principles offers the possibility of dealing with novel faults. As we have seen, our system does not depend for its performance on a catalog of observed error manifestations. Instead it takes the view that any discrepancy between observed and expected behavior is a bug, and it uses knowledge about the device structure and behavior to determine the possible sources of the bug. As a result, it is able to reason about bugs that are novel in the sense that they are not part of the 'training set' and are manifested by symptoms not seen previously.

Since rules are a distillation of an expert's experience, a program built from them will be reasonably sure of handling only cases quite similar to those the human expert has already seen, solved, and communicated to the program. We have little reason to believe that the program will handle a bug whose outward manifestation is unfamiliar, even if the root cause is within the claimed scope of the system.

At times there may not be a choice: in domains where knowledge truly is anecdotal and experiential, a rule-based encoding may be appropriate. Where knowledge in the domain does admit some more basic model, however, we find the reasons above present a strong argument for capturing that level of understanding in the program.

12.2. Describing structure and behavior

12.2.1. *Previous representations*

There is a long history of attempts to represent structure and behavior in general and computer hardware in particular. Our approach has a number of features in common with that work. The primary points of overlap are the black-box view of modules and the use of hierarchical descriptions (both of which are found in [16, 33], for example). Major points of difference include (i) our distinction between, and explicit representation of, functional, physical and behavioral information, (ii) emphasis on the creation of a domain-specific language, and (iii) the ability to make multiple uses of our representations.

Many previous languages fail to include information about physical organization at all. This seems to have resulted from their origin in work on machine design, where physical packaging is sometimes considered only after functional design has been accomplished.

In some other description languages there is a significant intermixing of structure and behavior information, arising apparently from the use of traditional programming languages as foundations. There are many advantages to implementing a new language as a variant on some existing language (like ALGOL), including the availability of a compiler and other language development tools. But this makes it all too easy to carry over a set of habits that may be inappropriate. The temptation exists, for example, to use datatype declarations as a mechanism for expressing the existence of functional modules. Hence we see such things as

```
STRUCTURE:

DECLARE
MEMORY[0: 1023]⟨0 : 15⟩
ACCUMULATOR⟨0 : 16⟩
PC⟨0: 11⟩
IR⟨0: 15⟩;

BEHAVIOR:

IR := MEMORY[PC]
IF IR⟨8:11⟩ = 13   ; add instruction
   . . .
etc.
```

There is a significant intermixing of structure and behavior here. The declarations are supposed to contain the structure information, but where is the indication that the instruction register (IR) is connected to the memory? It is indicated indirectly in the behavior section of the code, where we discover that the IR is capable of being loaded from memory. We prefer that such information be made explicit. Among other things, this permits more systematic fault insertion. In the current example, we would be able to simulate the effects of faulty behavior in the bus that links the memory and instruction register.

A second major difference in our approach lies in our creation of a domain-specific language, one that incorporates the concepts and vocabulary of computer architecture. As noted, several hardware description languages have used the black-box view and the module/port approach as a starting point. Simple interconnections of devices are easily accommodated in this scheme, but the attempt to describe even small real circuits soon presents problems. One common insight is that many circuits contain a substantial amount of regularity. The pattern of interconnections in memory or between slices of an ALU, for example, is often easily expressed as some form of iteration. One common response (see, e.g., [21]) is to adopt more of the traditional programming language constructs, like iterative loops.

But this presents problems. A bit-slice CPU, for example (Fig. 2), has a form of regularity different from that displayed by an array of memory chips. Using a **for** loop to express both of them ignores, or at best makes obscure, the important difference. The result is often difficult to interpret as well. It can take a considerable amount of study to determine that a tightly coded iterative loop in fact expresses a well-known organization.

We are pursuing an alternate approach, described in Section 3, of assembling a vocabulary of terms common to architects. Each such term labels a particular kind of organization and has associated with it information on how to wire up that configuration. The result is in effect a new, high-level language, with all of the standard benefits of such. It allows us to make explicit the nature of the organization in the circuit, as well as providing a compact, efficient, and easily understood language.

A third major difference in our approach is our ability to make use of descriptions in several different ways. The same description is used (i) as a basis for the troubleshooting module, (ii) as a database of facts about connectivity, part identity, etc., (iii) as a body of code that can be run to simulate the device, and (iv) as the basis for a display program for observing the device. This is possible because we have avoided the temptation to write code oriented toward a single purpose like simulation, and have instead produced a set of data structures (described in Sections 3 and 4) that can be used in all the different ways noted. This approach is not unknown in the hardware description language world—ISPS has been used quite profitably in this fashion [2]—but it is rare. We find that it offers sufficient leverage to be worth the additional effort.

12.2.2. Advantages

Our approach offers a number of features which, while not necessarily novel, do provide useful performance. For example, there is a unity of device description and simulation, since the descriptions themselves are 'runnable'. That is, the behavior descriptions associated with a given module allow us to simulate the behavior of that module; the interconnection of modules specified in the structure description then causes results computed by one module to propagate to another. Thus we don't need a separate description or body of code as the basis for the simulation, we can simply 'run' the description itself. This ensures that our description of a device and the machinery that simulates it can never disagree about what to do, as can be the case if the simulation is produced by a separately maintained body of code.

Our use of a hierarchic approach and the terminal, port, module vocabulary makes multi-level simulation very easy. In simulating any module we can either run the behavior associated with that module (simulating the module in a single step), or 'run the substructure' of the module, simulating the device according to its next level of structure. Enabling the behavior that spans the entire module gives us a one-step simulation; enabling the abstraction-shifting machinery that implements a port gives us a detailed simulation (by allowing information to propagate into the next lower level). Since the abstraction-shifting behavior of ports is also implemented with the constraint-like mechanism described in Section 4, we have a convenient uniformity and economy of machinery.

Varying the level of simulation is useful for speed (no need to simulate verified substructure), and provides as well a simple check on structure and behavior specification: we can compare the results generated by the module's behavior specification with those generated by the next lower level of simulation. Mismatches typically mean a mistake in structure specification at the lower level.

Work in [6] described the importance of the 'no function in structure' concept, essentially a point of methodology and 'mental hygiene', suggesting that device behaviors be defined independent of their use in any specific circuit. We have adopted this perspective and built a small library of component descriptions (adders, wires, AND-gates, etc.); a set of prototypical modules with structure and behavior descriptions independent of any particular circuit. Devices are then constructed by assembling and interconnecting instances of those prototypes, using the language described in Section 3.

A wire, for example, is a device whose behavior is a simple bi-directional propagation: information appearing at either one of its terminals will be propagated to the other. Any particular use of a wire typically has a single direction of propagation in mind, but our simulator runs the behavior description that reflects the 'actual' (bi-directional) behavior.

While this approach offers no formal guarantees that the behavior definitions are free of implicit assumptions, it does provide an environment that strongly encourages attention to the issue by (i) distinguishing clearly between behavior definition of an individual module and its intended use in a circuit, and (ii) by forcing every module of a particular type of share the same behavior specification. This approach is especially important in troubleshooting, since some of the more difficult faults to locate are those that cause devices to behave not as we know they 'should', but as they are in fact electrically capable of doing.

Finally, our approach offers a convenient mechanism for fault insertion. A wire stuck at zero, for example, is modeled by giving the wire a behavior specification that maintains its terminals at logic level 0 despite any attempt to change them. Bridges, opens, etc., are similarly easily modeled.

12.2.3. *Limitations*

The behavior-specification mechanism we have described is quite straightforward and some of its limits are well known. A set of constraints is, for example, a relatively simple mechanism for specifying behavior, in that it offers no obvious support for expressing behavior that falls outside the 'relation between terminals' view. A bus protocol, for example, would require additional machinery to represent the state-transition network describing the protocol. We might also want to describe and reason about behavior in higher-level terms like *enables*, or *inhibits*, suggesting the need for a vocabulary similar to the one developed in [25].

In addition, our current propagation mechanism works well when dealing with simple quantities like numbers or logic levels, but cannot deal with more elaborate symbolic expressions. What, for example, do we do if we know that the output of an OR-gate is 1 but we don't know the value at either input? We can refrain from making any conclusion about the inputs, which makes the rules easy to write but misses some information. Or we can write a rule which express the value on one input in terms of the value on the other input. This captures the information but produces problems when trying to use the resulting expression elsewhere. A simple but effective propagation of symbolic expressions is accomplished in [19], suggesting that the approach taken there may be a good starting point.

12.3. Limitations in candidate generation

This limitation in our propagation machinery is also responsible for the primary limitation in the candidate-generation facility. As we have seen, the basic technique works by looking for a contradiction: we 'turn off' the behavior rules for a single device and see if there is any set of assignments to its terminals that

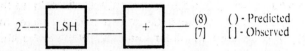

FIG. 20. Reconvergent error. There is no way for the input wire alone to account for the symptoms. (The first device shifts its input left one place and sends the result to both outputs.)

is consistent with the inputs and observed output symptoms. If no contradiction is reported, we consider the component to be a candidate.

But sometimes no contradiction is reported because the propagation machinery is too weak to discover it. One simple case can arise from reconvergent errors. In the simple circuit shown in Fig. 20, for example, our current system is unable to determine that the input wire could not be a consistent candidate. When, in using constraint suspension, we 'turn off' the behavior of the input wire, and insert the input (2) and observed output (7), several terminals end up with no values assigned. This is not considered a contradiction, so the input wire is put on the list of plausible candidates.

In principle, we should have been able to rule it out, but including it as a candidate does not present any serious problems. The candidate set will be larger than it should be, so our system may end up doing some additional work in narrowing the set later (e.g., running more tests than should strictly have been necessary). The net result is some inefficiency, but no intractable difficulties. Enhancing the propagation machinery to handle this situation would involve difficult problems, including propagating and manipulating symbolic expressions [32], and reasoning about such properties as integer solutions to equations.

12.4. Comments on the example

We chose a bridge example because they have traditionally been difficult; we considered the problem in TTL because of its common use. But this turns out to be a convenient choice. Bridges in TTL are easily modeled and the pattern easily checked for in the symptoms; we noted earlier the power this supplies in constraining the search space. In other technologies, unfortunately, the behavior of bridge faults is not so easily described. The reasoning is correspondingly more difficult and our system would be less focused in its generation of hypotheses.

A second problem highlighted by the example is the nature of the overall control of problem solving. Some parts are reasonably clear. The use of the categories of failure, for example, is to date at least fairly straightforward, starting with the most likely and moving toward the more exotic categories. We were also able to describe the overall strategy behind using and switching between multiple representations (using the functional as a hypothesis generator and the physical as a filter), but this is currently hardwired into the system. It would of course be better to have the system able to make this choice. Finally, we have yet to develop a globally defined strategy concerning the use of the description hierarchy. In general we work at the higher description levels before moving down, but this may not always be appropriate. It is not yet clear, for example, how 'deep' to pursue candidate generation before stopping to generate distinguishing tests.

Consider the example of Fig. 7. There we determined that either the first multiplier or first adder could be at fault. Should we now drop down a level and try generating candidates inside the adder and multiplier, or should we try another test vector to provide more symptom data? Either of these strategies might indicate which of the two candidates contains the broken component. As noted earlier, dropping down a level may demonstrate that no subcomponent of (say) the adder can account for the original set of symptoms, hence it cannot be at fault: running a second test provides additional symptom data that may distinguish between the candidates. The appropriate strategy presumably depends on the costs in the specific case at hand: when we are very near the 'bottom' of the description hierarchy, it may cost relatively little to go down the one final level; when tests are relatively cheap to generate and run, that may be the preferred approach.

12.5. The example: implementation

As we have noted, test generation in the example is currently done by hand, but all the rest has been implemented, in FRANZ LISP running on a VAX 780. The example shown requires approximately 3 minutes of CPU time, but since this is an early prototype, no attention has been given yet to producing efficient code.

The system is still a simple feasibility demonstration and as such lacks a number of design features necessary before it can be used as a serious tool. The current control structures, for example, are too deeply hardwired into the system. The sequencing through various categories of failure is currently embodied in a collection of procedure calls. Yet, given our emphasis on enumerating and keeping careful track of assumptions (as in Section 10), the selection of a failure category would more appropriately be accomplished with a general TMS system of the sort described in [15]. We do have a simple form of TMS in the dependency networks maintained by the simulation and inference rules, but it would be useful to construct a more general version and use it to keep track of the assumptions underlying the failure categories.

We are also working on a graphics interface that will allow dynamic display of the reasoning. As the sequence of figures shown earlier suggests, much of the process of candidate generation (for both the individual components and bridges) is easily understood in terms of diagrams. We are developing a system that will allow us to do this, displaying the results as they are generated.

12.6. Scaling: device complexity and time

With any initial demonstration of this size, the scaling issue is always of concern. We have demonstrated the feasibility of a particular technique on a small combinatorial circuit assembled from simple devices. What happens when the circuit gets considerably larger and the devices get more complex?

We believe that size alone is unlikely to be a disabling problem, arguing that the design task imposes a limit on complexity. In order to make the design task tractable, a circuit with several hundred or several thousand components must have some sort description more compact than a simple listing of the components. Without such a hierarchy, it is unlikely it could have been designed successfully.

Complex behavior is likely to present a more difficult problem. Describing the behavior of devices at the scale of gates and adders is relatively straightforward. Describing the behavior of a disk controller is likely to be considerably more challenging, but, by the design argument once again, we speculate that it will not be overwhelmingly so.

The significant problem with complex devices is likely to lie in the inference rules. It is one thing to describe what a disk controller should do, it may be quite another task to infer what some of its inputs were, given its outputs. One important subproblem we will encounter here is propagating symbolic expressions, since, as noted, many devices are not uniquely invertible.

A second problem lies in extending our work to deal with more elaborate models of time and more complex devices with state. Our current system uses only the simplest model of time, enabling us to deal with simple synchronous devices. There are several steps we need to take to elaborate this. We have to model propagation delays so we can deal with races. Since protocols play an important role in communication, we need to be able to represent and reason about them.

Reasoning about devices with memory will also require elaboration of our candidate generation machinery. Our planned approach will be in the spirit of the current effort: where discrepancy detection currently moves us backward in space through the circuit, we intend to extend it to move backward in time as well, inferring values at previous time slices.

In facing the problem of scaling this feasibility demonstration up to problems of practical size, then, we find two issues of central concern: how can we generate inference rules for complex devices, and how can we model time in a way that allows us to reason over a significant number of slices? Work on both of these is currently under way.

12.7. Limits of modeling: analog devices and incomplete models

Though we have set our sights here on reasoning from first principles, as with any representation we eventually encounter a level of detail not incorporated in our model. We have been working strictly in the digital world and as such cannot model or reason about analog phenomena. This is most obvious in the power-failure example: our representation makes it easy to *incorporate* the insight that input ports can behave as output ports, but reasoning at the level of the digital abstraction precludes *deriving* that insight. Marginal signals are a second example: the effects might be captured at the digital level but some of the reasoning would be outside our current abilities.

Since every representation incorporates some level of abstraction, the issue is not that we encounter limits, but rather how important those limits are to the current domain and how difficult it would be to press beyond them. In answer to the first of these, we simply need more experience to determine how much of the problem can be captured at the level of detail we use.

Concerning the second, previous work (e.g., [13]) suggests that dealing with analog circuits and their continuous variables requires confronting significant additional problems. One difficulty arises because most analog devices are bi-directional and this often makes candidate generation considerably less constrained. There are more paths of interaction to be considered and hence more components in the circuit that could have caused the fault. In the example of Fig. 9, for instance, allowing ports to be bi-directional widens the set of candidates (as it should) to include all the multipliers. Working in the digital domain means that most of our devices are uni-directional, keeping the set of candidate components smaller.

A second problem arising in analog circuits is the use of designs that rely on aggregate properties. The feedback or hysteresis in a device, for example, cannot be said to reside in or result from any one component. This produces difficulties when troubleshooting because some of the more interesting faults are those that are local to a single component but that disable the desired aggregate property. Tracing the fault from the disappearance of the aggregate property back to a single component can be difficult.

Finally, dealing with the continuous values present in analog devices presents its own set of problems, motivating much of the work on qualitative physics (e.g., [13]). We have to determine whether it will be possible to quantize the domain, developing and reasoning from a small vocabulary of labels like *high, low, float, rising,* and *falling,* or whether more complex machinery is necessary.

An equally pressing issue concerns the completeness of our models. Our system is focused in its efforts in part because its models of structure and behavior are complete: we knew, for example, exactly how the adder was constructed and how it behaved. Yet much real-world troubleshooting is done with incomplete models. Experienced engineers employ much the same sort of reasoning shown here even when the device is a mainframe computer and the behavior is on the scale of an operating system, yet they clearly are not using complete models of either. The ability to specify and use incomplete models would thus help address the scaling issue and would increase the likely scope of utility of our work to fields like medicine, where complete models are simply unavailable.

12.8. But that's not how it's *really* done

The performance of the program we have developed is in some ways noticeably different from the standard practice of a human expert. As expert systems work rather than cognitive science, the intent here is to be inspired by human performance without modeling it in detail. It is nevertheless useful to consider how our system differs from real practice.

While we argued earlier for the difficulty of troubleshooting based solely on empirical associations, it is clear that rules can serve useful roles. They might, for example, offer a form of memory to shortcut the process for problems previously encountered. This would clearly be an improvement on our current system, which will solve a problem from first principles every time, no matter how many times it is encountered. Such rules can also help focus the process by recognizing symptoms characteristic of particular kinds of malfunctions (e.g., power-supply failures), characteristic of particular locations of failures (e.g., memory, I/O bus, etc.), or characteristic of particular machines (e.g., the disk controllers on this model tend to fail sooner).

Real troubleshooting also typically involves extensive use of logic probes, an issue this work has not addressed at all yet. While it will clearly be important in the long run, we would argue that it is appropriately delayed for several reasons. First, we claim that inference is 'free', while measurement is often quite expensive. It's comparatively expensive in time: many inferences can be drawn in the time it takes to place a probe. More important, it's expensive in potential loss of information: it is often necessary to put cards on extenders or otherwise disturb the current state of the machine to make the measurement and the information lost can be crucial. The current trend in hardware speed makes this imbalance likely to continue.[18] As a result, we claim that it's well worth it to 'think as much as you can' before taking another measurement.

The real issue here is the longstanding problem of information gathering. Where, contrary to our current assumption, we do not have complete data available, the real problem is reasoning from the current stock of information and deciding what measurement to take next. Our current examples are small enough that complete information is reasonable, but we will clearly encounter the issue as we scale the problem up to larger devices.

Finally, we might ask what fraction of the troubleshooting problem is not currently handled by simpler, existing technology of the sort characterized by stuck-at models and state-of-the-art diagnostics. A significant percentage of the problem is solved in this fashion, but the fraction left unsolved turns out to be

---

[18]There is some countervailing trend in the design of hardware that offers visibility of internal machine state and the possibility of automated probing.

quite expensive. It is not unusual to find a strongly bi-modal distribution in which problems are commonly solved in either two hours or two days. There is thus a significant problem here, which, because of issues like decreasing design lifetimes, is likely to become worse.

13. Related Work

13.1. Hardware diagnosis

Two lines of work developed in the hardware-diagnosis community have some interesting overlap with the work described here: the guided probe [5] and the effect-cause analysis of Abramovici and Breuer [1].

The guided-probe technique has been in use in industry for some time and shares some basic ideas with our approach to troubleshooting. Both are based on the notion of discrepancy detection; both trace an error at an output back to its source by following the wiring of the circuit, and both use simulation to produce the correct values.[19]

One important difference arises because the guided-probe approach does not have anything analogous to our inference rules (it uses a logic probe to measure voltages at nodes interior to the circuit). Having inference rules is important because it allows us to separate candidate generation from probing. That is, we can determine the entire set of plausibly broken components before making any additional measurements on the circuit. This can be an advantage because, given the entire candidate set, we may be able to select a few (or even one) places to probe that best reduce the size of the candidate set (e.g., the usual half-split strategy). The standard guided-probe approach, by interleaving discrepancy detection with probing, in effect requires us to consider every candidate when it is first encountered.[20]

One further practical concern suggests additional utility of inference rules. Given the tendency toward increasingly exotic packaging technology, it is becoming more difficult to probe at random on a board. In such cases the inference rules become especially useful.

Finally, since the guided probe works directly from the schematic, it also inherits the fundamental problem noted earlier: the schematic may be wrong, and there is nothing in this approach capable of dealing with that problem.

Abramovici and Breuer have independently developed an approach to diagnosis that has a number of the features of a constraint-based system. Their deduction algorithm is similar to the use of inference and simulation rules in discrepancy detection, and they use multiple tests as we do to further prune the

---

[19]Some automatic testers use a board known to be good as the source of the correct values.

[20]The lack of inference rules and corresponding need to probe suggests that, where we earlier characterized our approach as the interaction of simulation and inference, the guided probe is analogously characterized as the interaction of simulation and measurement.

candidate set. One interesting result of their work is in its application to synchronous sequential circuits, where they show that this approach can diagnose a fault that prevents initialization (i.e., the initial states of the flipflops are unknown). This has long been a difficult roadblock for traditional fault-dictionary style diagnostics.

Drawbacks in the approach include its extensive use of inferences drawn from the fact that a wire can be labeled as 'normal' (i.e., it has been observed to take on all possible values). The label is easy to establish in the world of binary gates but clearly gets considerably more difficult for circuits modeled at higher levels. More seriously, the approach has been extensively explored for faults modelable as wires struck at 1 or 0, but does not appear to go beyond this. One of the examples in [1], for instance, produces a unique diagnosis for a circuit under the stuck-at model, but does not indicate that, among other possibilities, a malfunctioning OR-gate (*not* modelable as a stuck-at) is another, equally valid diagnosis. Finally, like other approaches that work solely from the schematic, it has no mechanism for considering that the schematic may be incorrect.

13.2. Fault models and categories of failure

In reviewing some of the traditional approaches to troubleshooting (Section 5), we noted several problems. We pointed out the historical trend toward using code originally designed for verification (proof that a device is totally free of faults) to do troubleshooting. We noted that code designed for verification requires fault models—if we are to avoid exhaustive testing, we need to limit our tests to errors produced by a pre-specified list of faults. We then claimed that we do not need traditional fault models when the task is diagnosis and when a fault is defined as anything different from the correct behavior.

Yet in doing discrepancy detection we found it necessary to come up with an apparently similar sort of list: we had to enumerate the categories of failure in order to limit the paths of interaction we considered. Have we in fact made any progress or have we simply substituted one list for another?

We claim that progress has been made along two fronts. First, by focusing on troubleshooting and defining a fault as any discrepancy, we can deal with a wider range of faults, including any systematic misbehavior. Second, we have taken a step toward providing a somewhat more formal way of generating the entries on the list. The traditional fault-model list is typically an informally generated listing of erroneous behaviors that have been observed in practice. Carefully examining the assumptions underlying our representation, as we did in Section 10, is in effect a 'generator' of categories. It produces a number of well-known categories (e.g., local failures like stuck-ats, assembly errors, design errors) and suggests some less obvious ones as well (e.g., direction of information flow errors). While it is still not a formally complete generator, we at least have some relatively systematic basis for enumerating categories of failure to consider.

13.3. Other AI work

As we have noted at several points above, a number of ideas developed in previous AI research have proved very useful in this undertaking.

13.3.1. Troubleshooting

From the work of Sussman and Steele on constraints, for example, we take the local propagation style of computation [32] and the maintenance of dependency networks [29]. Work by De Kleer first demonstrated [11] that a first-order theory of troubleshooting could be based on examining the dependency records left behind by a local propagator: any device on the path to a discrepancy should be considered a potential candidate; any device that participates in a 'corroboration' (a place where predicted and measured values agree) can be ruled out. We used the first half of this approach in our candidate generator, when Step 2 of Fig. 8 used the dependency network leading to a discrepancy to determine the potential candidates.

Our work moves beyond this in two ways: by viewing troubleshooting in the framework of methodical relaxation of assumptions, and by the use of constraint suspension. Relaxation of assumptions handles bridge faults, while the approach in [11] specifically excludes all such errors in topology. The approach there works by examining dependency records produced by the local propagator, but the propagator worked from the original schematic. Nothing in this approach provides a way of entertaining the idea that the original schematic was incorrect; there is no mechanism for hypothesizing additional paths of interaction not shown in the original description.

Constraint suspension provides two additional advantages. First, like the approach in [11], it determines which components might be at fault, but then it also determines symptom values for the candidates. This provided important information in solving the bridge fault problem in particular, and in general allows candidate generation to continue at successively lower levels of description.

Second, as noted in [11], the use of corroborations in the first-order theory above runs into trouble with any device whose behavior can inhibit propagation of an error (e.g., an AND-gate or multiplier whose other input is 0). Consider for example a slight modification to Fig. 7. If adder-2 were a fourth multiplier and input E were 0, we would get a corroboration at G (0 expected and observed), apparently exonerating mult-2. But in that circuit mult-2 would in fact be a valid candidate. Our constraint-suspension approach handles this situation without difficulty: by running the behavior descriptions of all but one component (the one being tested for candidacy), it effectively determines the appropriate consequences of all discrepancies and corroborations.

The approach in [11] was developed further in SOPHIE [6] to include knowledge of component-fault modes (e.g., resistors can be shorted, open, high or low), and to include knowledge about higher-level modules that was specific to the particular circuit and module. Our approach differs in using a uniform approach to device modeling and troubleshooting, where SOPHIE's troubleshooting of the high-level modules was handled using the circuit-specific knowledge (the first principles approach was used at the level of primitive components).

That system also noted the potential difficulties arising from implicit assumptions, and speculated about the use of a general truth-maintenance system to deal with the problem. This does not appear to have been implemented.

We have drawn from the work in [6] in some other respects, both for specific techniques and overall approach. That work demonstrated the use of simulation as a fundamental tool in troubleshooting (using it for example to predict observations from inputs and to predict consequences of failures). It showed how dependency chains could support hypothetical reasoning. The work also made clear the importance of 'mental hygiene' in building simulation models. Ideas like the 'no function in structure' concept were both directly useful (as noted in Section 12) and helped sensitize us to the importance of implicit assumptions.

That work also kept distinct the kinds of knowledge used in doing troubleshooting. The system relied on dependency records to generate the candidate set first, and only then used knowledge about component fault modes to prune the set.

13.3.2. Adjacency

The notion of adjacency has been pursued from several different directions, some of which proved useful in our own approach to the concept. The original conception of constraints [32] helped define the issue and characterize local propagation as a way of thinking about computation. Work by De Kleer [12] and De Kleer and Brown [13, 14] developed a number of principles involving local propagation when reasoning about cause and effect in physical devices.

A slightly different conception of adjacency underlies some of the work of Lenat [20]. There, syntactic changes to LISP descriptions produced useful new mathematical concepts because LISP and mathematics are languages that are 'close' together. That is, the primitives and method of composition are similar enough that the languages are structurally similar: concepts that are 'adjacent' mathematically often have LISP definitions that are quite similar. As a result, making small syntactic changes to a LISP expression that captured one mathematical concept often produced an expression embodying another meaningful mathematical concept.

This underscores the importance of the choice of representation language when making changes to a description: the changes will produce meaningful results to the extent that the languages share a definition of adjacency. A similar inspiration lies behind our analysis of the need for the 'right' representation in understanding different kinds of faults: the physical definition was appropriate for bridge faults because it provided the appropriate definition of adjacency.

In Section 11 we suggested that it can be useful to have several different definitions of adjacency, provided by the multiple representations (functional, physical, etc.) employed there. A related notion, multiple views of a circuit, shows up in the concept of slices [31] used in design. A central observation in that work was that simplifying the design task requires being able to have several different views of a circuit, each one packaging up things differently. The basis for the packaging is typically teleological: a particular section of a circuit is packaged up in a slice because it has an identifiable behavior that accomplishes a particular purpose. The design task can then be simplified by reasoning about the device using that abstraction, without being concerned about how that behavior is actually accomplished. Multiple slices can provide multiple views of the same section of the circuit; slices thus offer a mechanism for expressing multiple abstractions.

Our emphasis has been on using different definitions of what adjacent can mean in order to produce fundamentally different representations for troubleshooting. Slices may offer multiple different views of the circuit, but they are all functional views and share the same basic representational vocabulary (that of functional modules). We have both a functional representation with its definition of adjacency and its vocabulary, and a physical representation with its own distinct definition of adjacency and vocabulary (cabinets, boards, chips). Our focus is thus not on having multiple views of the device, but on having fundamentally different representations, and on carefully enumerating the different criteria (definitions of adjacency) that can be used as the basis for each representation. The definitions are in turn derived from the different pathways of interaction, since those pathways reflect the mechanisms by which faults manifest.

Finally, our categories of failure are similar in spirit to the class-wide assumptions in [13]. As pointed out there, it is impossible to write assumption-free descriptions of behavior. The problem manifests itself in the most basic terms in simply choosing a vocabulary. Any behavior description will be built from a finite vocabulary of terms, yet there is no obvious limit to the set of terms that *might* prove relevant (weight, color, flexibility, etc.). Since the vocabulary must be finite, it is important to provide at least an explicit, defensible set of selection criteria. The problem arises also in using a given vocabulary to describe a device. In describing a door bell, for example, the model in [13] needs to refer to the effect of the coil's magnetic field on the clapper, but it invokes the common assumption that the field is not strong enough to induce currents in nearby wires. Again there should be explicit, defensible grounds for making such assumptions.

The authors suggest that introductory physics provides a convenient and routinely, if tacitly, used set of criteria; they term them 'class-wide assumptions'. These are assumptions about behavior whose utility (and credibility) arises in part because they apply to whole sets of devices. Ignoring the effect of the magnetic field on nearby wires, for example, is a reasonable assumption because it applies to a wide class of devices.

Our categories of failure are similar in spirit. We are motivated by the same original consideration, namely that it is impossible to create assumption-free descriptions of structure and behavior. We have made the identification and handling of such assumptions a primary concern and have been able to provide a somewhat more systematic generation of the categories. Since De Kleer and Brown are concerned with qualitative physics, they appropriately look to that field for inspiration and have the task of identifying and accumulating the assumptions. Our more tightly focused concern—troubleshooting—allows us to examine what kinds of things can go wrong, and makes possible the relatively systematic generation of assumptions accomplished by perturbing the representation, as we did in Section 10.

This also reflects our concern with enumerating the pathways of interaction as completely as possible. While class-wide assumptions reflect important knowledge about the domain, that knowledge is often in the form of a reason not to include a particular pathway of interaction (e.g., omitting the path from the coil to nearby wires in the door-bell example). We want to press one step further on by building a list of possible pathways that is as systematic as possible, and then consider the variety of fault characterized by the existence (or omission) of such a pathway.

14. Summary

We seek to build a system that reasons from first principles in understanding how devices operate. We find troubleshooting of digital hardware to be a tractable and fertile ground for exploration. We have developed languages describing structure and behavior that distinguish carefully between them, and provide information about structure that is organized both functionally and physically.

We find that the traditional machinery for troubleshooting focuses primarily on test generation and its use in verification of device behavior. Our problem is better characterized as diagnosis in the presence of known symptoms, in devices complex enough that it is important to use the symptoms to help guide the troubleshooting.

We view the process as the interaction of simulation and inference, with discrepancies between them driving the generation of candidates. Discrepancy detection and tracing through dependency records gives us a foundation for troubleshooting that identifies potential candidates. The technique of constraint suspension extends this by supplying symptom values for the candidates and handling both discrepancies and corroborations with a single mechanism.

In exploring this approach further, we find that the concept of paths of causal interaction plays a key role, supplying the knowledge that makes the machinery work. We need an explicit model of causal interactions in order to determine which components to consider. The amount of such knowledge then leads us to a fundamental dilemma: the desire to deal with a wide range of faults seems to force us to choose between an inability to discriminate among candidates and the inability to deal with some classes of faults.

In response we have developed a troubleshooting strategy based on the methodical enumeration and relaxation of underlying assumptions about the device. We were able to generate the assumptions in a relatively systematic fashion by examining the module and information path representation of hardware. By considering the consequences of violating each assumption, we were able to generate a collection of categories of failure.

We then invoked a version of Occam's razor, noting that some categories of failure are more likely than others. This provides a criterion for ordering the categories and pathways of interaction to be considered. We start with the simplest category of failure first and consider only one class of paths of interaction initially. If this fails to generate a consistent hypothesis, we surrender one of our underlying assumptions, adding the next category of failure, and consider an additional pathway of interaction.

We illustrated this approach by diagnosing a bridge fault. When our initial categorization—local failure of function—encountered a contradiction, we surrendered the assumption that the schematic was correct and considered one additional path, bridges. This staged relaxation of assumptions permitted a constrained generation of hypotheses. Within the bridge-fault category, additional restriction was then provided by using constraints associated with both the physical and functional representations.

Drawing back from this specific example, we explored several possible generalizations of the work. We found applicability in software, for example, for the notion of enumerating the assumptions in the representation and using this to generate categories of failure. While the particular pathways of interaction were different, the technique appears to phrase a relevant set of questions. We found that some errors in software are usefully thought of in terms of a set of pathways of interaction from that domain.

Our overall approach also suggests that part of the expertise of a domain lies in knowing how to simplify a problem, i.e., knowing what simplifying assumptions can be made, recognizing when an assumption has failed to help, and knowing how to recover and get the solution back on course. Enumerating and ordering the categories of failure provides one simple mechanism for expressing such knowledge.

A further generalization of this work came from examining the difficulty involved in dealing with bridge faults. We found that an important property of a representation is its definition of adjacency. In this view bridge faults are difficult because they are simple and local changes to the physical representation, but neither simple nor local in our original, functional representation. We started with the functional representation because we were presented with behavioral manifestations of a fault, and hence needed to reason from a representation organized according to behavior. But this proved to have an inappropriate definition of adjacency for the problem at hand, so we shifted representations. This in turn lead us to a useful guideline in choosing representations for diagnostic reasoning: we should attempt to find a representation in which the suspected change can be viewed as a compact modification affecting adjacent devices.

The underlying rationale for focusing on defining adjacency is a belief that faults manifest through processes that act locally, i.e., there is no action at a distance. We then found that we could define a number of useful kinds of adjacency by considering the paths of interaction. Each of these defines a different metric (Euclidean, thermal, electromagnetic, etc.), each generating a different representation.

We pursued this one additional step, asking why we believe there will in fact be some representation in which the fault can be seen as compact. The belief appears to rest on the heuristic that a malfunction in a previously operational device often results from a single cause rather than a number of independent events. Like any heuristic, this may not always be correct, since multiple, independent failures do occur. But it is true often enough and there are substantial advantages to choosing a representation with a good definition of adjacency.

Because our basic representation machinery employs little more than the traditional black-box notion of information transmission, we suggest that some of the central concepts in this work may have relevance of considerable breadth. The examination of assumptions underlying the representation and constraint suspension, for instance, appear to be applicable in a number of different areas. We noted above the potential use in software of examining underlying assumptions. We conjecture that both techniques may apply to any system that might be modeled in terms of information transmission, ranging from hardware, to software, to organizations. Organizations may have pathways of interaction other than those on the personnel chart, for example, and we might consider 'organization troubleshooting' via constraint suspension.

Finally, we observe that there has been growing focus on the power contributed by choosing a 'good' representation and the utility of multiple representations. We suggest that one of the characteristics of a good representation is that it provides a definition of adjacency that makes the fault appear compact, and that the utility of multiple representations arises in part from the multiple different definitions of adjacency they provide.

ACKNOWLEDGMENT

Contributions to this work were made by members of the Hardware Troubleshooting project at MIT, including: Howie Shrobe, Walter Hamscher, Mark Shirley, Harold Haig, Art Mellor, John Pitrelli, and Steve Polit.

Some of the initial inspiration for this work came from conversations with Ed Feigenbaum; periodic arguments with Mike Genesereth have helped sharpen vague intuitions and were an early source of specific examples. The presentation in this paper was improved by comments from John Seeley Brown, Johan De Kleer, Ken Forbus, Doug Hofstader, Mark Shirley, and Patrick Winston.

REFERENCES

1. Abramovici, M. and Breuer, M.A., Fault diagnosis in synchronous sequential circuits based on an effect-cause analysis, *IEEE Trans. Comput.* **31** (1982) 1165–1172.
2. Barbacci, M.R., Instruction set processor specifications (ISPS): The notation and its applications, Carnegie-Mellon University Tech. Rept. CMU-CS-79-123, Pittsburgh, PA, 1979.
3. Batali, J. and Hartheimer, A., The design procedure language manual, MIT AI Memo 598, Cambridge, MA, 1980.
4. Bell, G. and Newell, A., *Computer Structures: Readings and Examples* (McGraw-Hill, New York, 1971).
5. Breuer, M.A. and Friedman, A., *Diagnosis and Reliable Design of Digital Systems* (Computer Science Press, Rockville, MD, 1976).
6. Brown, J.S., Burton, R. and De Kleer, J., Pedagogical and knowledge engineering techniques in SOPHIE I, II and III, in: D.H. Sleeman and J.S. Brown (Eds.), *Intelligent Tutoring Systems* (Academic Press, New York, 1982).
7. Davis, R., Shrobe, H., Hamscher, W., Wieckert, K., Shirley, M. and Polit, S., Diagnosis based on structure and function, in: *Proceedings National Conference on Artificial Intelligence*, Pittsburgh, PA (August, 1982) 137–142.
8. Davis, R., Reasoning from first principles in electronic troubleshooting. *Internat. J. Man–Mach. Stud.* **19** (1983) 403–423.
9. Davis, R., Buchanan, B.G. and Shortliffe, E.H., Production rules as a representation in a knowledge-based consultation system, *Artificial Intelligence* **8** (1977) 15–45.
10. Davis, R. and Shrobe, H.E., Representing structure and behavior of digital hardware, *IEEE Trans. Comput.* **32** (1983) 75–82.
11. De Kleer, J., Local methods for localizing faults in electronic circuits, MIT AI Memo 394, Cambridge, MA, 1976.
12. De Kleer, J., The origin and resolution of ambiguities in causal arguments, in: *Proceedings Sixth International Joint Conference on Artificial Intelligence*, Tokyo, Japan (August, 1979) 197–203.
13. De Kleer, J. and Brown, J.S., Assumptions and ambiguities in mechanistic mental models, Xerox PARC Rept. CIS-9, Palo Alto, CA, 1982.
14. De Kleer, J. and Brown J.S., Naive physics based on confluences, Xerox Parc Rept., 1983.
15. Doyle, J., A truth maintenance system, *Artificial Intelligence* **12** (1979) 231–272.
16. Estrin, G., A methodology for design of digital systems—supported by SARA at the age of one, in: *Proceedings NCC* (1978) 313–324.

17. Gasching, J., Preliminary evaluation of the performance of the PROSPECTOR system for mineral exploration, in: *Proceedings Seventh International Joint Conference on Artificial Intelligence*, Vancouver, BC (August, 1981) 308–310.
18. Genesereth, M., The use of hierarchical models in the automated diagnosis of computer systems, Stanford HPP Memo 81-20, Stanford, CA, 1981.
19. Kelly, V. and Steinberg, L., The CRITTER system—Analyzing digital circuits by propagating behaviors and specifications, in: *Proceedings National Conference on Artificial Intelligence*, Pittsburgh, PA (August, 1982) 284–289.
20. Lenat, D., Heuretics: theoretical and experimental study of heuristic rules, in: *Proceedings National Conference on Artificial Intelligence*, Pittsburgh, PA (August, 1982) 159–163.
21. Lim, W.Y-P., HISDL—A structure description language, *Comm. ACM* **25** (1982) 823–830.
22. Newell, A. and Simon, H., *Human Problem Solving* (Prentice-Hall, Englewood Cliffs, NJ, 1972).
23. Patil, R., Szolovits, P. and Schwartz, W., Causal understanding of patient illness in medical diagnosis in: *Proceedings Seventh International Joint Conference on Artificial Intelligence*, Vancouver, BC (August, 1981) 893–899.
24. Pople, H., Heuristic methods for imposing structure on ill-structured problems, in: P. Szolovits (Ed.), *Artificial Intelligence in Medicine*, AAAS Selected Symposium 51, 1982.
25. Reiger, C.R. and Grinberg, M., A system for cause-effect representation and simulation for computer-aided design, in: J.-C. Latombe (Ed.), *Artificial Intelligence and Pattern Recognition in Computer-Aided Design* (North-Holland, Amsterdam, 1978) 299–334.
26. Roth, J.P., Diagnosis of automata failures: A calculus and a method, *IBM J. Res. Develop.* **10** (1966) 278–291.
27. Shirley, M. and Davis, R., Digital test generation from hierarchical models and symptom information, in: *Proceedings IEEE International Conference on Computer Design*, November, 1983.
28. Shortliffe, E., *Computer-Based Medical Consultations: Mycin* (American Elsevier, New York, 1976).
29. Stallman, R.M. and Sussman, G.J., Forward reasoning and dependency-directed backtracking in a system for computer-aided circuit analysis, *Artificial Intelligence* **9** (1977) 135–196.
30. Steele, G., The definition and implementation of a computer programming language based on constraints, MIT TR-595, Cambridge, CA, 1980.
31. Sussman, G.J., Slices: At the boundary between analysis and synthesis, in: J.-C. Latombe (Ed.), *Artificial Intelligence and Pattern Recognition in Computer-Aided Design* (North-Holland, Amsterdam, 1978) 261–299.
32. Sussman, G.J. and Steele, G., Constraints—a language for expressing almost-hierarchical descriptions, *Artificial Intelligence* **14** (1980) 1–40.
33. Van Cleemput, W.M., An hierarchical language for the structural description of digital systems, in: *Proceedings Fourteenth Design Automation Conference*, New Orleans, LA (1977) 377–385.
34. Winston, P., *Artificial Intelligence* (Addison-Wesley, Reading, MA, 2nd ed., 1984).

Received August 1983; revised version received November 1983

Bibliography

1. General References 580
2. Surveys and Tutorials 581
3. Symbolic Layout and Artwork Analysis 581
4. Partitioning, Placement, and Routing 582
5. Process Modeling 583
6. Device Modeling 585
7. Faults and Fault Modeling 586
8. MOSFET Device/Circuit Reliability and Testing 587
9. Fault Masking and Multiple Fault Detection 588
10. Multivalued Logic and Fault Detection 589
11. Bridging Faults 589
12. Circuit, Timing, Logic, and Fault Simulation 589
13. Functional Testing 591
14. Testing and Fault Tolerance of Interconnection Networks 592
15. Memory Testing 592
16. PLA Testing 593
17. Design for Testability and Built-in Self-Test 594
18. Spectral Analysis and Syndrome Testability 594
19. Probabilistic Test and Signature Analysis 595
20. Application of AI Techniques to Design and Test 595

1. General References

1. J. Millman, *Microelectronics: Digital and Analog Circuits and Systems*, McGraw-Hill, New York, NY, 1979.
2. C.A. Mead and L. Conway, *Introduction to VLSI Systems*, Addison-Wesley, Reading, MA, 1980.
3. S. Muroga, *VLSI System Design*, Wiley, New York, NY, 1982.
4. D.A. Hodges and H.G. Jackson, *Analysis and Design of Digital Integrated Circuits*, McGraw-Hill, New York, NY, 1983.
5. S.M. Sze (Ed.), *VLSI Technology*, McGraw-Hill, New York, NY, 1983.
6. M.I. Elmasry (Ed.), *Digital MOS Integrated Circuits*, IEEE Press, New York, NY, 1981.
7. P.R. Gray et al. (Eds.), *Analog MOS Integrated Circuits*, IEEE Press, New York, NY, 1980.
8. J. Newkirk and R. Mathews, *The VLSI Designer's Library*, Addison-Wesley, Reading, MA, 1983.
9. C. Seitz, *Structured VLSI Design*, Addison-Wesley, Reading, MA, 1986 (to appear).
10. N. Weste and K. Eshraghian, *Principles of CMOS VLSI Design: A Systems Perspective*, Addison-Wesley, Reading, MA, 1985.
11. M.A. Breuer and A.D. Friedman, *Diagnosis and Reliable Design of Digital Systems*, Computer Science Press, Rockville, MD, 1976.
12. E.J. McCluskey, *Logic Design Principles*, Prentice-Hall, Englewood Cliffs, NJ, 1986 (to appear).
13. R.G. Bennets, *Introduction to Digital Board Testing*, Crare-Russak, New York, NY, 1982.
14. R.G. Bennets, *Design of Testable Logic Circuits*, Addison-Wesley, Reading, MA, 1984.
15. A.C. Stover, *ATE: Automatic Test Equipment*, McGraw-Hill, New York, NY, 1984.
16. J.D. Lenk, *Handbook of Electronic Test Procedures*, Prentice-Hall, Englewood Cliffs, NJ, 1983.
17. J.D. Lenk, *Handbook of Advanced Troubleshooting*, Prentice-Hall, Englewood Cliffs, NJ, 1983.
18. J.P. Roth, *Computer Logic, Testing, and Verification*, Computer Science Press, Rockville, MD, 1980.
19. D.J. Pradhan (Ed.), *Fault-Tolerant Computing*, Prentice-Hall, Englewood Cliffs, NJ, 1986 (to appear).
20. D.P. Siewiorek and R.S. Swarz, *The Theory and Practice of Reliable System Design*, Digital Press, Maynard, MA, 1982.
21. J.E. Arsenault and J.A. Roberts (Ed.), *Reliability and Maintainability of Electronic Systems*, Computer Science Press, Rockville, MD, 1980.
22. P. Antognetti et al. (Eds.), *Computer Design Aids for VLSI Circuits*, Sifthoff and Nordhoff, Rockville, MD, 1981.
23. G. Rabbat (Ed.), *Hardware and Software Concepts in VLSI*, Van Nostrand Reinhold, New York, NY, 1983.
24. M.A. Breuer (Ed.), *Digital Systems Design Automation Languages, Simulation and Data Base*, Computer Science Press, Rockville, MD, 1975.
25. M.A. Breuer (Ed.), *Design Automation of Digital Systems*, Prentice-Hall, Englewood Cliffs, NJ, 1972.
26. J.D. Ullman, *Computational Aspects of VLSI*, Computer Science Press, Rockville, MD, 1984.
27. R.F. Ayres, *Silicon Compilation and the Art of Automatic Microchip Design*, Prentice-Hall, Englewood Cliffs, NJ, 1983.
28. J. Wakerley, *Error Detecting Codes, Self-Checking Circuits and Applications*, Elsevier North Holland, New York, NY, 1978.
29. W.W. Peterson and E.J. Weldon, *Error Correcting Codes*, Cambridge, MA, MIT Press, 1972.
30. IEEE, *IEEE Standard Dictionary of Electrical and Electronics Terms*, Wiley-Interscience, New York, NY, 1984.
31. H.Y. Chang, E.G. Manning, and G. Metze, *Fault Diagnosis of Digital Systems*, Wiley-Interscience, New York, NY, 1970.
32. A.D. Friedman and P.R. Menon, *Fault Detection in Digital Circuits*, Englewood Cliffs, NJ, Prentice-Hall, 1971.
33. S.L. Hurst, *The Logical Processing of Digital Signals*, Crane-Russak, New York, 1978.
34. M.G. Karpovsky, *Finite Orthogonal Series in the Design of Digital Devices*, Wiley, New York, NY, 1976.
35. J. Mavor et al., *Introduction to MOS LSI Design*, Addison-Wesley, Reading, MA, 1983.
36. F. van de Wiele et al. (Eds.), *Process and Device Modeling for Integrated Circuit Design*, Leyden, The Netherlands, Noordhoff, 1977.
37. P. Lala, *Fault-Tolerant and Fault-Testable Hardware Design*, Prentice-Hall, Englewood Cliff, NJ, 1985.
38. S.W. Golomb, *Shift Register Sequences*, Aegean Park Press, Laguna Hills, CA, 1982.
39. B.T. Brown and J.H. Miller (Eds.), *Numerical Analysis of Semiconductor Devices and Integrated Circuits*, Dublin, Ireland, Boole Press, 1981.
40. R. Rice (Ed.), *VLSI Support Technologies: Computer-Aided Design, Testing and Packaging*, IEEE Computer Society Press, Washington, D.C., 1981.
41. C.C. Timoc (Ed.), *Selected Reprints on Logic Design*

for Testability, IEEE Computer Society Press, Washington, D.C., 1984.

42. H. Fuchs (Ed.), *Selected Reprints on VLSI Technologies and Computer Graphics,* IEEE Computer Society Press, Washington, D.C., 1983.

43. R. Rice (Ed.), *VLSI—The Coming Revolution in Applications and Design,* IEEE Computer Society Press, Washington, D.C. 1980.

44. D.J. McGreiry and K.A. Pickar, *VLSI Technologies—Through the 80's and Beyond,* IEEE Computer Society Press, Washington, D.C., 1982.

2. Surveys and Tutorials

1. P.G. Kovijanic, "A New Look at Test Generation and Verification," *Proc. 14th Design Automation Conf.,* IEEE, June 1977, pp. 58–63.

2. E.I. Muehldorf, "Designing LSI for Testability," *Dig. Papers, 1976 Ann. Semiconductor Test Symp.,* IEEE, Oct. 1976, pp. 45–49.

3. E.I. Muehldorf and A.D. Savkar, "LSI Logic Testing—An Overview," *IEEE Trans. Comput.,* Vol. C-30, No. 1, Jan. 1981, pp. 1–17.

4. A.K. Susskind, "Diagnostics for Logic Networks," *IEEE Spectrum,* Vol. 10, Oct. 1973, pp. 40–47.

5. T.W. Williams and K.P. Parker, "Testing Logic Networks and Design for Testability," *Computer,* Vol. 12, Oct. 1979, pp. 9–21.

6. W.G. Fee (Ed.), *Tutorial: LSI Testing,* IEEE Computer Society Press, Washington, D.C., 1978.

7. H.K. Reghbati, "VLSI Testing and Validation Techniques," *Advances in Computers,* Vol. 26, Academic Press, NY, 1987 (to appear).

8. M.G. Buehler and M.W. Sievers, "Off-Line, Built-in Test Techniques for VLSI Circuits," *IEEE Computer,* Vol. 15, No. 6, June 1982, pp. 69–82.

9. J.P. Hayes and E.J. McCluskey, "Testability Considerations in Microprocessor-Based Design," *IEEE Computer,* Vol. 13, No. 3, March 1980, pp. 17–26.

10. S.B. Akers, "Test Generation Techniques," *IEEE Computer,* Vol. 13, No. 3, March 1980, pp. 9–16.

11. J. Grason and A.W. Nagle, "Digital Test Generation and Design for Testability," *ACM-IEEE Design Automation Conference,* 1980, pp. 175–189.

12. W.A. Clark, "From Electron Mobility to Logical Structure: A View of Integrated Circuits," *ACM Computing Surveys,* Vol. 12, No. 3, Sept. 1980, pp. 325–356.

13. M.S. Abadir and H.K. Reghbati, "LSI Testing Techniques," *IEEE Micro,* Vol. 3, No. 1, Feb. 1983, pp. 34–51.

14. A.R. Newton, "Computer-Aided Design of VLSI Circuits," *Proc. IEEE,* Vol. 69, No. 10, Oct. 1981, pp. 1189–1199.

15. E.H. Frank and R.F. Sproull, "Testing and Debugging Custom Integrated Circuits," *ACM Computing Surveys,* Vol. 13, No. 4, Dec. 1981, pp. 425–452.

16. G.D. Hachtel and A.L. Sangiovanni-Vincentelli, "A Survey of Third-Generation Simulation Techniques," *Proc. IEEE,* Vol. 69, No. 10, Oct. 1981, pp. 1264–1280.

17. A.E. Ruehli and G.S. Dittow, "Circuit Analysis, Logic Simulation, and Design Verification for VLSI," *Proc. IEEE,* Vol. 71, No. 1, Jan. 1983, pp. 34–48.

18. M.S. Abadir and H.K. Reghbati, "Functional Testing of Semiconductor Random Access Memories," *ACM Computing Surveys,* Vol. 15, No. 3, September 1983, pp. 175–198.

19. J.P. Hayes, "Fault Modeling, *IEEE Design & Test,* Vol. 2, No. 2, April 1985, pp. 88–95.

3. Symbolic Layout and Artwork Analysis

1. M.C. Revell and P.A. Ivey, "ASTRA—A CAD System to Support a Structured Approach to IC Design," *Proc. of VLSI-83,* 1983, pp. 413–422.

2. M.Y. Hsueh, "Symbolic Layout and Compaction of Integrated Circuits," *Ph.D. thesis,* EECS-Dept., Univ. of California, Berkeley, 1979.

3. S.M. Rubin, "An Integrated Aid for Top-Down Electrical Design," *Proc. of VLSI-83,* 1983, pp. 63–72.

4. R.A. Auerbach, B.W. Lin, and E.A. Elsayed, "Layouts for the Design of VLSI Circuits," *Computer Aided Design,* Vol. 13, No. 5, 1981, pp. 271–276.

5. S.C. Johnson, "Hierarchical Design Validation Based on Rectangles," *Conference on Advanced Research in VLSI, M.I.T.,* 1982, pp. 97–100.

6. T. Lenguaer, "The Complexity of Compacting Hierarchically Specified Layouts of Integrated Circuits," *Proc. of 23rd FOCS,* 1982, pp. 358–368.

7. J.P. Schoelkopf, "LUBRICK: A Silicon Assembler and Its Application to Data Path Design for FISC," *Proc. of VLSI-83,* 1983, pp. 435–446.

8. J.J. Levy, "On the Lucifer System," *Advanced Course on VLSI Architecture,* University of Bristol, Great Britain, 1981.

9. N.H.E. Weste, "MULGA—An Interactive Symbolic Layout System for the Design of Integrated Circuits," *Bell Syst. Tech. J.,* Vol. 60, No. 6, 1981, pp. 823–857.

10. B. Ackland and N. Weste, "An Automatic Assembly Tool for Virtual Grid Symbolic Layout," *Proc. of VLSI-83,* 1983, pp. 457–466.

11. T.W. Whitney and C.A. Mead, "POOH: A Uniform Representation for Circuit Level Designs," *Proc. of VLSI-83,* 1983, pp. 401–412.

12. A.E. Dunlop, "SLIM—The Translation of Symbolic Layouts into Mask Data," *Proc. 17th Design Automation Conf.,* IEEE, 1980, pp. 595–602.

13. J.D. Williams, "STICKS—A Graphical Compiler for High Level LSI Design," *Proc. of Nat. Comp. Conf.*, 1978, pp. 289–295.
14. L. Cardelli and G. Plotkin, "An Algebraic Approach to VLSI Design," *Proc. of VLSI-81*, 1981, pp. 173–182.
15. J. Rosenberg et al., "A Vertically Integrated VLSI Design Environment," *Proc. 20th Design Automation Conf.*, IEEE, 1983.
16. R. Goldin, B. Juran, and M. Lotvin, "Amoeba: A Symbolic VLSI Layout System," *Proc. 21st Design Automation Conf.*, 1984, pp. 294–300.
17. J.A. Gordon, "Generation of Sticks Micro-Cell Layouts from Schematics," *M.S. Thesis*, Department of Electrical Engineering, MIT, 1984.

4. Partitioning, Placement, and Routing

1. J. Soukup, "Global Router," *Proc. 16th Design Automation Conf.*, 1979, pp. 481–484.
2. B.T. Preas and C.W. Gwyn, "Methods for Hierarchical Automatic Layout of Custom LSI Circuit Masks," *J. Des. Automat. Fault-Tolerant Comput.*, Vol. 3, No. 1, 1978, pp. 41–58.
3. U. Lauther, "A Min-cut Placement Algorithm for General Cell Assemblies Based on Graph Representation," *Proc. 16th Design Automation Conf.*, 1979, pp. 41–58.
4. S. Goto, "A Two-Dimensional Placement Algorithm for the Master Slice LSI Layout Problem," *Proc. 16th Design Automation Conf.*, 1979, pp. 11–17.
5. N. Nan and M. Feuer, "A Method for Automatic Wiring of LSI Chips," *Proc. IEEE Int. Symp. Circuits and Systems*, 1978, pp. 11–15.
6. D.N. Deutsch, "A Dog-leg Channel Router," *Proc. 13th Design Automation Conf.*, 1976, pp. 425–433.
7. F. Rubin, "The Lee Connection Algorithm," *IEEE Trans. Comput.*, Vol. C-23, 1974, pp. 907-914.
8. J. Soukup, "Fast Maze Router," *Proc. 15th Design Automation Conf.*, 1978, pp. 100–102.
9. T. Yoshimura and E.S. Kuh, "Efficient Algorithms for Channel Routing," *Memo. UCB/ERL M80/43*, Electronic Research Laboratory, College of Engineering, Univ. of California, Berkeley, CA, Aug. 1980.
10. W.R. Heller, W.F. Mikhail, and W.E. Donath, "Prediction of Wiring Space Requirements for LSI," *J. of Des. Automat. Fault-Tolerant Comput.*, Vol. 2, No. 2, 1978, pp. 117–144.
11. M. Hanan, P.K. Wolff, and B.J. Anguili, "Some Experimental Results on Placement Techniques," *J. Des. Automat. Fault-Tolerant Comput.*, Vol. 2, No. 2, May 1978, pp. 145–164.
12. C.Y. Lee, "An Algorithm for Path Connections and its Application," *IRE Trans. Electron. Comput.*, Sept. 1961, pp. 346–365.
13. D.G. Schweikert, "A 2-Dimensional Placement Algorithm for the Layout of Electrical Circuits," *Proc., 13 Design Automation Conf.*, 1976, pp. 408–416.
14. G. Persky, D.N. Deutsch, and D.G. Schweikert, "LTX—A Minicomputer-Based System for Automatic LSI Layout," *J. Des. Automat. Fault-Tolerant Comput.*, Vol. 1, No. 3, May 1977, pp. 217–256.
15. M. Breuer, "Min-Cut Placement," *J. Des. Automa Fault-Tolerant Comput.*, Vol. 1, No. 4, Oct. 1977, pp. 343-362.
16. W. Heyns, W. Sansen, and H. Beke, "A Line Expansion Algorithm for the General Routing Problem with Guaranteed Solution," *Proc. 17th Design Automation Conf.*, June 1980, pp. 243–249.
17. M. Hung and W.O. Rom, "Solving the Assignment Problem by Relaxation," *Operations Res.*, Vol. 28, No. 4, Aug. 1980, pp. 969–982.
18. J. Soukup and J.C. Royle, "On Hierarchical Routing," *J. Digital Systems*, Vol. 5, No. 3, Sept. 1981.
19. M. Weisel and D.A. Mlynski, "An Efficient Channel Model for Building Block LSI," *Proc. 1981 IEEE Int. Symp. Circuits and Systems*, Apr. 1981, pp. 118–121.
20. P.P. Chaudhuri, "An Ecological Approach to Wire Routing," *Proc. 1979 IEEE Int. Symp. Circuits and Systems*, July 1979, pp. 854-857.
21. K.J. Loosemore, "Automated Layout of Integrated Circuits," *Proc. 1979 IEEE Int. Symp. Circuits and Systems*, 1979, pp. 665–668.
22. R.H. Krambeck, D.E. Blahut, H.F.S. Law, B.W. Colbry, H.C. So, M. Harrison, and J. Soukup, "Top-Down Design of a One-Chip 32 Bit CPU," *VLSI-81 Conf.*, Edinburgh, Scotland, Aug. 1981.
23. B.W. Kernighan and S. Lin, "An Efficient Procedure for Partitioning Graphs," *Bell Syst. Tech. J.*, 1970, pp. 291–307.
24. M. Breuer and K. Shamsa, "A Hardware Router," *J. of Digital Syst.*, Vol. 4, No. 4, 1980, pp. 393–408.
25. K. Sato and T. Nagai, "A Method of Specifying the Relative Locations Between Blocks in a Routing Program for Building Block LSI," *Proc. 1979 IEEE Int. Symp. Circuits and Systems*, 1979, pp. 673–676.
26. H. Shiraishi and F. Hirose, "Efficient Placement and Routing Techniques for Master-slice LSI," *Proc. 17th Design Automation Conf.*, 1979, pp. 673–676.
27. B.T. Preas and W.M. Van Cleemput, "Placement Algorithms for Arbitrarily Shaped Blocks," *Proc. 16th Design Automation Conf.*, 1979, pp. 474–480.
28. M. Hanan, "Net Wiring for Large Scale Integrated Circuits," *IBM Research Rep. RC-1375*, 1965.

29. C.M. Fidducia and R.M. Matthyses, "A Linear-Time Heuristic for Improving Network Partitions," *Proc. 19th Design Automation Conf.*, June 1982, pp. 175-181.
30. K. Kimura et al., "An Automatic Routing Scheme for General Cell LSI," *IEEE Trans. CAD*, Vol. CAD-2, No. 4, Oct. 1983, pp. 285-292.
31. T. Yoshimura, "An Efficient Channel Router," *Proc. 21st Design Automation Conf.*, 1984, pp. 38-44.
32. G. Hamachi and J. Ousterhout, "A Switch Box Router with Obstacle Avoidance," *Proc. 21st Design Automation Conf.*, 1984, pp. 173-179.
33. T. Yoshimura and E.S. Kuh, "Efficient Algorithms for Channel Routing," *IEEE Trans. CAD*, Vol. CAD-1, No. 1., Jan. 1982, pp. 25-35.
34. R. Rivest and C. Fidducia, "A 'Greedy' Channel Router," *Proc. 19th Design Automation Conf.*, 1982, pp. 418-424.
35. M. Burstein and R. Pelavin, "Hierarhical Channel Router," *Proc. 20th Design Automation Conf.*, 1983, pp. 591-597.
36. A. Sangiovanni-Vincentelli and M. Santomauro, "YACR: Yet Another Channel Router," *Proc. Cust. Int. Circ. Conf.*, 1983, pp. 327-331.
37. H. Rothermel and D. Mlynski, "Automatic Variable-Width Routing for VLSI," *IEEE Trans. CAD*, Vol. CAD-2, No. 4, Oct. 1983, pp. 271-284.
38. I.H. Kirk et al., "Placement of Irregular Circuit Elements on Non-Uniform Gate Arrays," *Proc. 20th Design Automation Conf.*, 1983, pp. 637-643.
39. T. Kozawa, C. Miura, and H. Terai, "Combine and Top Down Placement Algorithm for Hierarchical Logic VLSI Layout," *Proc. 21st Design Automation Conf.*, 1984, pp. 667-669.
40. H.E. Krohn, "An Over-Cell Gate Array Channel Router," *Proc. 20th Design Automation Conf.*, 1983, pp. 665-670.
41. C.O. Newton and P.A. Young, "Optimisation of Global Routing for the UK5000 Gate Array by Iteration," *Proc. 20th Design Automation Conf.*, 1983, pp. 651-657.
42. S. Sastry and A. Parker, "On the Relation of Wire Length Distributions and Placement of Logic on Master Slice ICs," *Proc. 21st Design Automation Conf.*, 1984, pp. 710-711.
43. C.L. Wardle et al, "A Declarative Design Approach for Combining Macrocells by Directed Placement and Constructive Routing," *Proc. 21st Design Automation Conf.*, 1984, pp. 594-601.

5. Process Modeling

1. B.E. Deal and A.S. Grove, "General Relationship for the Thermal Oxidation of Silicon," *J. Appl. Phys.*, Vol. 36, 1965, p. 3770.
2. M.M. Atalla and E. Tannenbaum, "Impurity Redistribution and Junction Formation in Silicon by Thermal Oxidation," *Bell Syst. Tech. J.*, Vol. 39, July 1960, p. 933.
3. S.M. Hu, "Oxygen, Oxidation Stacking Faults, and Related Phenomena in Silicon," *Proc. Materials Research Soc. Meet.*, Apr. 1981.
4. D.A. Antoniadis, M. Rodoni, and R.W. Dutton, "Impurity Redistribution in SiO_2-Si During Oxidation: A Numerical Solution Including Interfacial Fluxes," *J. Electrochem. Soc.*, Vol. 126, No. 11, Nov. 1979, p. 1939.
5. M.Y. Tsai, F.F. Morehead, J.E. Baglin, and A.E. Michel, "Shallow Junctions by High-Dose as Implants in Si: Experiments and Modeling," *J. Appl. Phys.*, Vol. 15, July 1980, p. 3230.
6. H. Ryssel, K. Haberger, K. Hoffman, G. Prinke, R. Dumcke, and A. Sack, "Simulation of Doping Processes," *IEEE Trans. Electron Devices*, Vol. ED-27, Aug. 1980, p. 1484.
7. D.A. Antoniadis, M. Rodoni, and R.W. Dutton, "Models for Computer Simulation of Complete IC Fabrication Processes," *IEEE J. Solid State Circuits*, Vol. SC-14, Apr. 1979, p. 412.
8. R. Tielert, "Two-Dimensional Numerical Simulation of Impurity Redistribution in VLSI Processes," *IEEE Trans. Electron Devices*, Vol. ED-27, Aug. 1980, p. 1479.
9. R. Reif, T.I. Kamins, and K.C. Saraswat, "A Model for Dopant Incorporation into Growing Silicon Epitaxial Films: I. Theory," *J. Electrochem. Soc.*, Vol. 126, No. 4, Apr. 1979, p. 644.
10. R. Reif and R.W. Dutton, "Computer Simulation in Silicon Epitaxy," *J. Electrochem. Soc.*, 1981.
11. L. Mei, S. Chen, and R.W. Dutton, "A Surface Kinetics Model for Plasma Etching," *IEEE Int. Electron Devices Meet. Tech. Dig.*, Dec. 1980, p. 831.
12. H.J. De Man and R. Mertens, "SITCAP—A Simulator of Bipolar Transistors for Computer-Aided Circuit Analysis Programs," *ISSCC Dig. Tech. Papers*, Feb. 1973, p. 104.
13. H.G. Lee, "Two-Dimensional Impurity Diffusion Studies: Process Models and Test Structures for Low-Concentration Boron Diffusion," *Ph.D. Dissertation*, Stanford Electronics Lab. Tech. Rep. G201-8, Aug. 1980.
14. Y. Nakajima and Y. Fukuawa, "Simplified Expression for the Distribution of Diffused Impurity," *Japan J. Appl. Phys.*, Vol. 10, 1971, p. 162.
15. R.B. Fair and J.C. Tsai, "Profile Parameters of Im-

planted Diffused Arsenic Layer in Silicon," *J. Electrochem. Soc.*, Vol. 123, 1976, p. 583.

16. R.B. Fair, "Boron Diffusion in Silicon-Concentration and Orientation Dependence Background Effects, and Profile Estimation," *J. Electrochem. Soc.*, Vol. 122, No. 6, June 1975, p. 800.

17. R.W. Barton, S.A. Schwarz, W.A. Tiller, and C.R. Helms, "Segregation Effects on the Phosphorous Distribution at a Moving Si-SiO^2 Interface," *Thin-film Interfaces and Interaction Ions*, Electrochem. Soc., 1980.

18. D.A. Antoniadis, A.G. Gonzalez, and R.W. Dutton, "Boron in Near-Intrisic 100 and 111 Silicon under Inert and Oxidising Ambients—Diffusion and Segregation," *J. Electrochem. Soc.*, Vol. 125, No. 5, May 1978, p. 813.

19. D.C. D'Avanzo, R.W. Rung, A. Gat, and R.W. Dutton, "High Speed Implementation and Experimental Evaluation of Multilayer Spreading Resistance Analysis," *J. Electrochem. Soc.*, Vol. 125, No. 7, July 1978, p. 1170.

20. R.W. Dutton, D.A. Divekar, A.G. Gonzalez, S.E. Hansen, and D.A. Antoniadis, "Correlation of Fabrication Process and Electrical Device Parameter Variations," *IEEE J. Solid State Circuits*, Vol. SC-12, Aug. 1977, p. 349.

21. L.C. Parillo, G.W. Reutlinger, R.S. Payne, A.R. Tretola, and R. Kreatsch, "The Sensitivity of Transistor Gain to Process Variations in an all Implanted Bipolar Technology," *IEEE Int. Electron Devices Meet. Tech. Dig.*, Dec. 1977, p. 265A.

22. D.C. D'Avanzo, "Modeling and Characterisation of Short Channel Double Diffused MOS Transistors," *Ph.D. Dissertation*, Stanford University, Tech. Rep. G-201-6, Mar. 1980.

23. K. Yokoyama, A. Yoshii, and S. Horiguchi, "Threshold-Sensitivity Minimization of Short-Channel MOSFET's by Computer Simulation," *IEEE Trans. Electron Devices*, Vol. ED-27, Aug. 1980, p. 1509.

24. R.F. Motta, P. Chang, J.G. Chern, and N. Godinho, "Computer-Aided Device Optimization for MOS/VLSI," *IEEE Trans. Electron Devices*, Vol. ED-27, Aug. 1980, p. 1559.

25. R.R. Troutman, "Ion-Implanted Threshold Tailoring for Insulated Gate Field-Effect Transistors," *IEEE Trans. Electron Devices*, Vol. ED-24, Mar. 1977, p. 182.

26. J.W. Slotboom, "Analysis of Bipolar Transistors," *Ph.D. Dissertation*, Eindhoven University, Eindhoven, The Netherlands, 1977.

27. A. De Mari, "An Accurate Numerical One-Dimensional Steady State Solution of the *p-n* Junction," *Solid State Electron.*, Vol. 11, 1968, p. 33.

28. G.D. Hachtel, R.C. Joy, and J.W. Cooley, "A New Efficient One-Dimensional Analysis Program for Junction Device Modeling," *Proc. IEEE*, Vol. 60, Jan. 1972, p. 86.

29. D.C. D'Avanzo, M. Vanzi, and R.W. Dutton, "One-Dimensional Semiconductor Device Analysis (SEDAN)," Stanford Electronics Lab., Stanford, CA, *Tech. Rep. SEL 79-033*, Oct. 1979.

30. J.A. Greenfield and R.W. Dutton, "Nonplanar VLSI Device Analysis Using the Solution of Poisson's Equation," *IEEE Trans. Electron Devices*, Vol. ED-27, Aug. 1980, p. 1520.

31. S. Selberherr, A. Schutz, and W. Potzl, "MINIMOS—A Two Dimensional MOS Transistor Analyzer," *IEEE Trans. Electron Devices*, Vol. ED-27, Aug. 1980, p. 1540.

32. W. Oldham, S. Nandgoankar, A. Neureuther, and M. O'Toole, "A General Simulator for VLSI Lithography and Etching Processes: Part I—Applications to Projection Lithography," *IEEE Trans. Electron Devices*, Vol. ED-26, Apr. 1979, p. 717.

33. W.G. Oldham, A.R. Neureuther, C. Sung, J.L. Reynolds, and S.N. Nandgoankar, "A General Simulator for VLSI Lithography and Etching Processes: Part II—Applications to Deposition and Etching," *IEEE Trans. Electron Devices*, Vol. ED-27, Aug. 1980, p. 1455.

34. A.R. Neureuther, C.H. Ting, and C.-Y. Liu, "Application of Line Edge Profile Simulation to Thin-Film Deposition Processes," *IEEE Trans. Electron Devices*, Vol. ED-27, Aug. 1980, p. 1459.

35. A.R. Neureuther, "Simulating VLSI Wafer Topography," *IEEE Trans. Electron Devices Meet. Tech. Dig.*, Dec. 1980, p. 214.

36. H.G. Lee, R.W. Dutton, and D.A. Antoniadis, "On Redistribution of Boron During Thermal Oxidation of Silicon," *J. Electrochem. Soc.*, Vol. 126, No. 11, Nov. 1979, p. 2001.

37. H.G. Lee, J.D. Sansbury, R.W. Dutton, and J.L. Moll, "Modeling and Measurement of Surface Impurity Profiles of Laterally Diffused Regions," *IEEE J. Solid State Circuits*, Vol. SC-13, Aug. 1978, p. 455.

38. K. Tanguchi, M. Kashiwagi, and H. Iwai, "Two-Dimensional Computer Simulation Models for MOS LSI Fabrication Process," *IEEE Trans. Electron Devices*, Vol. ED-28, May 1981, p. 574.

39. M.S. Mock, "A Two-Dimensional Mathematical Model of the Insulated Gate Field-Effect Transistors," *Solid-State Electron.*, Vol. 16, 1973, p. 601.

40. B.R. Penumalli, "Process Simulation in Two Dimensions," *ISSCC Dig. Tech. Papers*, Feb. 1981, p. 212.

41. D.J. Chin, M.R. Kump, H.G. Lee, and R.W. Dutton,

"Process Design Using Coupled 2D Process and Device Simulators," *IEEE Int. Electron Devices Meet. Tech. Dig.*, Dec. 1980, p. 223.

42. T.F. Hasan, S.U. Katzman, and D.S. Perloff, "Automated Electrical Measurements of Registration Errors in Step-and-Repeat Optical Lithography," *IEEE Trans. Electron Devices*, Vol. ED-27, Dec. 1980, p. 2304.

43. B.J. Gordon, "On-line Capacitance-Voltage Doping Profile Measurement of Low-Dose Ion Implants," *IEEE Trans. Electron Devices*, Vol. ED-27, Dec. 1980, p. 2268.

44. M.G. Buehler, "Effect of the Drain-Source Voltage on Dopant Profiles Obtained from the DC MOSFET Profile Method," *IEEE Trans. Electron Devices*, Vol. ED-27, No. 12, Dec. 1980, p. 2273.

45. H.G. Lee and R.W. Dutton, "Measurement of Two-Dimensional Profiles Near Locally Oxidized Regions," *IEEE Int. Electron Devices Meet. Tech. Dig.*, 1979.

46. A.M. Lin, R.W. Dutton, and D.A. Antoniadis, "The Lateral Effect of Oxidation on Boron Diffusion in 100 Silicon," *Appl. Phys. Lett.*, Vol. 35, Nov. 1979, p. 799.

6. Device Modeling

1. K.M. van Vliet and A.H. Marshak, "The Shockley-Like Equations for the Carrier Densities and the Current Flow in Materials with a Nonuniform Composition," *Solid-State Electron.*, Vol. 23, 1980, pp. 49–53.

2. C.R. Crowell and S.M. Sze, "Current Transport in Metal Semiconductor Barriers," *Solid-State Electron.*, Vol. 9, 1966, p. 1035.

3. D.M. Caughey and R.E. Thomas, "Carrier Mobilities in Silicon Empirically Related to Doping and Field," *Proc. IEEE*, Vol. 55, 1967, pp. 2192–2193.

4. G. Baccarani and P. Ostoja, "Electron Mobility Empirically Related to the Phosphorous Concentration in Silicon," *Solid-State Electron.*, Vol. 18, 1975, pp. 579–580.

5. K.K. Thornber, "Relation of Drift Velocity to Low-Field Mobility and High-Field Saturation Velocity," *J. Appl. Phys.*, Vol. 15, No. 4, 1980, pp. 2127–2136.

6. B.V. Gokhale, "Numerical Solution for a One-Dimensional Silicon n-p-n Transistor," *IEEE Trans. Electron Devices*, Vol. ED-17, 1970, pp. 594–602.

7. K. Yamaguchi, "Field-Dependent Mobility Model For Two Dimensional Numerical Analysis of MOSFET's," *IEEE Trans. Electron Devices*, Vol. ED-26, 1979, pp. 1068–1074.

8. J. Cooper and D.F. Nelson, "Measurement of the High-Field Velocity of Electrons in Inversion Layers on Silicon," *IEEE Electron Devices Lett.*, Vol. EDL-2, 1981, pp. 171–173.

9. S. Selberherr, A. Schutz, and H.W. Potzl, "Minimos-A Two Dimensional MOS Transistor Analyzer," *IEEE Trans. Electron Devices*, Vol. ED-27, 1980, pp. 1540–1549.

10. A.H. Marshak and K.M. van Vliet, "Carrier Densities and Emitter Efficiency in Degenerate Materials With Position Dependent Band Structure," *Solid-State Electron.*, Vol. 21, 1977, pp. 429–434.

11. M.S. Mock, "Transport Equations in Heavily Doped Silicon and the Current Gain of a Bipolar Transistor," *Solid-State Electron.*, Vol. 16, 1973, pp. 1251–1259.

12. A. Nakagawa, "One-Dimensional Device Model of the n-p-n Bipolar Transistor Including Heavy Doping Effects Under Fermi Statistics," *Solid-State Electron.*, Vol. 22, 1979, pp. 943–949.

13. J.W. Slotboom and H.C. DeGraaff, "Measurement of Bandgap Narrowing in Si Bipolar Transistors," *Solid-State Electron.*, Vol. 19, 1976, pp. 857–862.

14. A.W. Wieder, "Emitter Effects in Shallow Bipolar Devices Measurements and Consequences," *IEEE Trans. Electron Devices*, Vol. ED-27, 1980, pp. 1402–1408.

15. R.P. Mertens, J.L. van Meerbergen, J.F. Nijs, and R.T. van Overstaeten, "Measurement of the Minority-Carrier Transport Parameters in Heavily Doped Silicon," *IEEE Trans. Electron Devices*, Vol. ED-27, No. 5, 1980.

16. J.G. Fossum and D.S. Lee, "A Physical Model for the Dependence of Carrier Lifetime on Doping Density in Nondegenerate Silicon," *Solid-State Electron.*, Vol. 25, No. 8, 1982, pp. 741–747.

17. J. Dziewor and W. Schmid, "Auger Coefficients for Highly Doped and Highly Excited Silicon," *Appl. Phys. Lett.*, Vol. 31, 1977, pp. 346–348.

18. J.A. Greenfield and R.W. Dutton, "Nonplanar VLSI Device Analysis Using the Solution of Poisson's Equation," *IEEE Trans. Electron Devices*, Vol. ED-27, 1980, pp. 1520–1532.

19. E.M. Buturl, P.E. Cottrell, B.M. Grosman, and K.A. Salsburg, "Finite-Element Analysis of Semiconductor Devices: The Fielday Program," *IBM J. Res. Dev.*, Vol. 25, 1981, pp. 218–231.

20. J.W. Slotboom, "Computer Aided Two Dimensional Analysis of Bipolar Transistors," *IEEE Trans. Electron Devices*, Vol. ED-20, 1973, pp. 669–679.

21. H.H. Heimeier, "A Two Dimensional Numerical Analysis of a Silicon n-p-n Transistor," *IEEE Trans. Electron Devices*, Vol. ED-20, 1973, pp. 708–714.

22. O. Manck, H.H. Heimeier, and W.L. Engl, "High Injection in a Two-Dimensional Transistor," *IEEE Trans. Electron Devices*, Vol. ED-21, 1974, pp. 403–409.

23. G.D. Hachtel, M. Mack, and R.R. O'Brien, "Semiconductor Device Analysis Via Finite Elements," *Proc. 8th Asilomar Conf. on Circuits, Systems and Computers,* 1974, pp. 332–338.

24. M.S. Mock, "Time-Dependent Simulation of Coupled Devices," in *Numerical Analysis of Semiconductor Devices and Integrated Circuits,* (B.T. Browne and J.J.H. Miller, Eds.), Dublin, Ireland: Boole Press, 1981.

25. A. De Mari, "An Accurate Numerical One-Dimensional Solution of the p-n Junction Under Arbitrary Transient Conditions," *Solid-State Electron.,* Vol. 11, 1968, pp. 1021–1053.

26. G.D. Hachtel, R.C. Joy, and J.W. Coole, "A New Efficient One-Dimensional Analysis Program for Junction Device Modeling," *Proc. IEEE,* Vol. 60, 1972, pp. 86–98.

27. O. Manck and W.L. Engl, "Two Dimensional Computer Simulation for Switching a Bipolar Transistor Out of Saturation," *IEEE Trans. Electron Devices,* Vol. ED-22, 1975, pp. 339–347.

28. G.D. Hachtel, M.H. Mack, R.R. O'Brien, and B. Speelpenning, "Semiconductor Analysis Using Finite Elements; Part I: Computational Aspects," *IBM J. Res. Dev.,* Vol. 25, 1981, pp. 232–245.

29. R.K. Brayton, F.G. Gustavson, and G.D. Hachtel, "A New Efficient Algorithm for Solving Differential-Algebraic Systems Using Implicit Backward Differentiation Formulas," *Proc. IEEE,* Vol. 60, 1972, pp. 98–108.

30. R. Laur and H.P. Strohband, "Numerical Modeling Technique for Computer Aided Circuit Design," *Proc. IEEE Int. Symp. Circuits and Systems,* 1976, pp. 247–250.

31. W.L. Engl and H. Dirks, "Functional Device Simulation by Merging Numerical Building Blocks," *Numerical Analysis of Semiconductor Devices and Integrated Circuits,* (B.T. Brown and J.J.H. Miller, Eds.), Dublin, Ireland: Boole Press, 1981.

32. D. Vandope, J. Borel, G. Merckel, and P. Saintot, "An Accurate Two-Dimensional Numerical Analysis of the MOS Transistor," *Solid-State Electron.,* Vol. 15, 1972, p. 547.

    H.K. Gummel, "A Self-Consistent Iterative Scheme for One Dimensional Steady State Transistor Calculations," *IEEE Trans. Electron. Devices,* Vol. ED-11, 1964, pp. 455–465.

    E.M. Buturla and P.E. Cotrell, "Simulation of Semiconductor Transport Using Coupled and Decoupled Solution Techniques," *Solid-State Electron.,* Vol. 23, 1980, pp. 331–334.

    J.D. Arcy, E.J. Prendergast, and P. Lloyd, "Modeling of Bipolar Device Structures: Physical Simulations," *IEDM Dig. Tech. Papers,* 1981, pp. 516–519.

36. T. Adachi, A. Yishii, and T. Sudo, "Two-Dimensional Semiconductor Analysis Using Finite-Element Method," *IEEE Trans. Electron Devices,* Vol. ED-26, 1979, pp. 1026–1031.

37. T. Wada and R.L.M. Dang, "Modification of ICCG Method for Application to Semiconductor Device Simulators," *Electron. Lett.,* Vol. 18, 1982, pp. 265–266.

38. W.L. Engl, R. Laur, and H.K. Dirks, "MEDUSA: A Simulator for Modular Circuits," *IEEE Trans. Computer-Aided Des.,* Vol. CAD-1, 1982, pp. 85–93.

7. Faults and Fault Modeling

1. R. Boute and E.J. McCluskey, "Fault Equivalence in Sequential Machines," *Proc. Symp. on Computers and Automata,* Polytech. Inst. Brooklyn, Apr. 13–15, 1971, pp. 483–507.

2. R.T. Boute, "Optimal and Near-Optimal Checking Experiments for Output Faults in Sequential Machines," *IEEE Trans. Comput.,* Vol. C-23, No. 11, Nov. 1974, pp. 1207–1213.

3. R.T. Boute, "Equivalence and Dominance Relations Between Output Faults in Sequential Machines," *Tech. Rep. 38, SU SEL-72-052,* Stanford Univ., Stanford, CA, Nov. 1972.

4. F.J. O. Dias, "Fault Masking in Combinatorial Logic Circuits," *IEEE Trans. Comput.,* Vol. C-24, May 1975, pp. 476–482.

5. J.P. Hayes, "A NAND Model for Fault Diagnosis in Combinational Logic Networks," *IEEE Trans. Comput.,* Vol. C-20, Dec. 1971, pp. 1496–1506.

6. E.J. McCluskey and F.W. Clegg, "Fault Equivalence in Combinational Logic Networks," *IEEE Trans. Comput.,* Vol. C-20, Nov. 1971, pp. 1286–1293.

7. K.C.Y. Mei, "Fault Dominance in Combinational Circuits," *Tech. Note 2, Digital Systems Lab.,* Stanford Univ., Stanford, CA., Aug. 1970.

8. K.C.Y. Mei, "Bridging and Stuck-at Faults," *IEEE Trans. Comput.,* Vol. C-23, No. 7, July 1974, pp. 720–727.

9. R.C. Ogus, "The Probability of a Correct Output from a Combinational Circuit," *IEEE Trans. Comput.,* Vol. C-24, No. 5, May 1975, pp. 534–544.

10. K.P. Parker and E.J. McCluskey, "Analysis of Logic Circuits with Faults Using Input Signal Probabilities," *IEEE Trans. Comput.,* Vol. C-24, No. 5, May 1975, pp. 573–578.

11. K.K. Saluja and S.M. Reddy, "Fault Detecting Test Sets for Reed-Muller Canonic Networks," *IEEE Trans. Comput.,* Vol. C-24, No. 5, Oct. 1975, pp. 995–998.

12. D.R. Schertz and G. Metze, "A New Representation for Faults in Combinational Digital Circuits," *IEEE Trans. Comput.*, Vol. C-21, No. 8, Aug. 1972, pp. 858–866.

13. J.J. Shedletsky and E.J. McCluskey, "The Error Latency of a Fault in a Sequential Digital Circuit," *IEEE Trans. Comput.*, Vol. C-25, No. 6, June 1976, pp. 655–659.

14. J.J. Shedletsky and E.J. McCluskey, "The Error Latency of a Fault in a Combinational Digital Circuit," *Proc. FTCS-5*, 1975, pp. 210–214.

15. K. To, "Fault Folding for Irredundant and Redundant Combinational Circuits," *IEEE Trans. Comput.*, Vol. C-22, No. 11, Nov. 1973, pp. 1008–1015.

16. D.T. Wang, "Properties of Faults and Criticalities of Values Under Tests for Combinational Networks," *IEEE Trans. Comput.*, Vol. C-24, No. 7, July 1975, pp. 746–750.

17. G.L. Schnable and R.S. Keen, "On Failure Mechanisms in Large-Scale Integrated Circuits," *Advances in Electronics and Electron Physics*, Vol. 30, Academic Press, New York, NY, 1971, pp. 79–138.

18. E.A. Doyle, Jr., "How Parts Fail," *IEEE Spectrum*, Vol. 18, No. 10, Oct. 1981, pp. 36–43.

19. G.R. Case, "Analysis of Actual Fault Mechanisms in CMOS Logic Gates," *Proc. 13th Design Automation Conf.*, June 1976, pp. 265–270.

20. J. Galiay, Y. Crouzet, and M. Vergniault, "Physical Versus Logical Fault Models in MOS LSI Circuits: Impact on Their Testability," *IEEE Trans. Comput.*, Vol. C-29, No. 6, June 1980, pp. 527–531.

21. R.L. Wadsack, "Fault Modeling and Logic Faults," *Proc. 15th Design Automation Conf.*, June 1978, pp. 124–126.

22. R.L. Wadsack, "Fault Modeling and Logic Simulation of CMOS and MOS Integrated Circuits," *Bell System Tech. J.*, Vol. 57, No. 4, May–June 1978, pp. 1449–1474.

23. P. Banerjee and J.A. Abraham, "Fault Characterization of MOS VLSI Circuits," *Proc. 1982 Int. Conf. on Circuits and Computers*, New York, NY, 1982, pp. 564–568.

24. P. Banerjee and J.A. Abraham, "Characterization and Testing of Physical Failures in MOS Logic Circuits," *IEEE Design & Test*, Vol. 1, No. 3, August 1984, pp. 76–86.

8. MOSFET Device/Circuit Reliability and Testing

1. S.W. Levin, "Alpha Emission Measurement of Lids and Solder Preforms on Semiconductor Packages," *IEEE Trans. Components, Hybrids, and Manuf. Technol.*, Dec. 1979, pp. 391–395.

2. J.A. Wooley et al., "Low Alpha-Particle-Emitting Ceramics: What is the Lower Limit? (For Dynamic RAM and CCD Memories Encapsulation)," *IEEE Trans. Components, Hybrids, and Manuf. Technol.*, Dec. 1979, pp. 388–391.

3. T.C. May, "Soft Errors in VLSI: Present and Future," *IEEE Trans. Components, Hybrids, and Manuf. Technol.*, Dec. 1979, pp. 377–387.

4. P.W. Peterson, "The Performance of Plastic-Encapsulated CMOS Microcircuits in a Humid Environment," *IEEE Trans. Components, Hybrids, and Manuf. Technol.*, Dec. 1979, pp. 422–425.

5. E.E. King, "Radiation-Hardening Static NMOS RAMs," *IEEE Trans. Nucl. Sci.*, Dec. 1975, pp. 5060–5064.

6. C.S. Guenzer et al., "Single Event Upset of Dynamic RAMs by Neutrons and Protons," *IEEE Trans. Nucl. Sci.*, Dec. 1975, pp. 5042–5047.

7. G.J. Brucker, "Application of Simple Hardening Techniques in System Design and Verification of Survival by EMP and Gamma Tests (CMOS/SOS Memory Circuits)," *IEEE Trans. Nucl. Sci.*, Dec. 1979, pp. 4949–4952.

8. W. Nasswetter, "Failure Analysis on a 65K MOS RAM with a New Type of Memory Display," *IEEE Fifth European Solid State Circuits Conf., ESSCIRC 79*, 1979, pp. 113–115.

9. D.J.W. Noorlag et al., "The Effect of Alpha-Particle-Induced Soft Errors on Memory Systems with Error Correction," *IEEE Fifth European Solid State Circuits Conf., ESSCIRC 79*, 1979, pp. 86–88.

10. R.W. Stevens and J.R. Brailsford, "Cost Effective Semiconductor Memory Testing," *IEEE Fifth European Solid State Circuits Conf., ESSCIRC 79*, 1979, pp. 82–85.

11. D.L. Crook, "Techniques of Evaluating Long Term Oxide Reliability at Wafer Level (MOS Memory Devices)," *IEEE 1978 Int. Electron Devices Meet.*, 1978, pp. 444–448.

12. R. Kondo, "Test Patterns for EPROMs," *IEEE J. Solid-State Circuits*, Aug. 1979, pp. 730–734.

13. C.W. Green, "PMOS Dynamic RAM Reliability: A Case Study," *IEEE 17th Annual Proc. Reliability Physics*, 1979, pp. 213–219.

14. G. Schindlbeck, "Analysis of Dynamic RAMs by Use of Alpha Irradiation," *IEEE 17th Annual Proc. Reliability Physics*, 1979, pp. 30–34.

15. C.C. Huang et al., "Component/System Correlation of Alpha Induced Dynamic RAM Soft Failure Rates," *IEEE 17th Annual Proc. Reliability Physics*, 1979, pp. 23–29.

16. H.P. Gibbons and J.D. Pittman, "Alpha Particle Emissions of Some Materials in Electronic Packages," *IEEE 29th Electronics Components Conf.*, 1979, pp. 257-260.

17. J.W. Peeples, "Electrical Bias Level Influence on the Results of Plastic Encapsulated NMOS 4K Dynamic RAM," *IEEE Trans. Electron Devices*, Jan. 1979, pp. 72-77.

18. H.A. Batdorf et al., "Reliability Program and Results for a 4K Dynamic RAM," *IEEE Trans. Electron Devices*, Jan. 1979, pp. 52-56.

19. D.S. Yaney et al., "Alpha-Particle Tracks in Silicon and their Effect on Dynamic MOS RAM Reliability," *IEEE Trans. Electron Devices*, Jan. 1979, pp. 10-16.

20. J.M. Waddell, "Achieving Quality and Reliability in Memories," *New Electron.*, March 1979, pp. 30-40.

21. H. Roder, "Quality Test of Semiconductor Memories by Jitter Measurement," *Electron. Lett.*, July 1979, pp. 428-429.

22. J.A. Roberts, "A Design Review Approach Toward Dynamic RAM Reliability," *Microelectron. and Reliability*, Jan.-Feb. 1979, pp. 97-105.

23. G.J. Brucker and L. Thurlow, "Memory Behaviour in a Radiation Environment," *Electron. Eng.*, Jan. 1979, pp. 67-71.

24. J.E. Arsenault and D.C. Roberts, "MOS Semiconductor Random Access Memory Failure Rate," *Microelectron. and Reliability*, Jan.-Feb., 1979, pp. 81-88.

25. L.J. Gallace and H.L. Pujol, "Improving COS/MOS Reliability," *Electron.*, Nov. 1978, p. 27.

26. R. Freeman and A. Holmes-Siedle, "A Simple Model for Predicting Radiation Effects in MOS Devices," *IEEE Trans. Nucl. Sci.*, Dec. 1978, pp. 1216-1225.

27. M. Schlenther et al., "In-Situ Radiation Tolerance Tests of MOS RAMs," *IEEE Trans. Nucl. Sci.*, Dec. 1978, pp. 1196-1204.

28. P.J. Vail, "A Survey of Radiation Hardened Microelectronic Memory Technology," *IEEE Trans. Nucl. Sci.*, Dec. 1978.

29. T.P. Haraszti, "Radiation Hardened CMOS/SOS Memory Circuits," *IEEE Trans. Nucl. Sci.*, Dec. 1978, pp. 1187-1195.

30. R.W. Tallon, "Ionizing Radiation Effects on the Sperry Rand Nonvolatile 256-Bit NMOS RAM Array (SR2256)," *IEEE Trans. Nucl. Sci.*, Dec. 1978, pp. 1176-1180.

31. A. London and R.C. Wang, "Dose Rate and Extended Dose Characterization of Radiation Hardened Metal Gate CMOS Integrated Circuits," *IEEE Trans. Nucl. Sci.*, Dec. 1978, pp. 1172-1175.

32. J.C. Pickle and J.T. Blandford, "Cosmic Ray Induced Errors in MOS Memory Cells," *IEEE Trans. Nucl. Sci.*, Dec. 1978, pp. 1166-1171.

33. J.H. Fusselman, "Memory Tester Correlation," *IEEE 1978 Semiconductor Test Conf.*, 1978, pp. 240-242.

34. R. O'Keefe et al., "Test Patterns for Static RAMs," *IEEE 1978 Semiconductor Test Conf.*, 1978, pp. 86-88.

35. T.P. Haraszti, "CMOS/SOS Memory Circuits for Radiation Environments," *IEEE J. Solid-State Circuits*, Oct. 1978, pp. 669-676.

36. J.W. Peeples, "Influence of Electrical Bias Level on 85/85 Test Results of Plastic Encapsulated 4K RAMS," *IEEE Reliability Physics 16th Annual Conf.*, 1978, pp. 154-160.

37. H.P. Feuerbaum et al., "Quantitative Measurement with High Time Resolution of Internal Waveforms on MOS RAMs Using a Modified Scanning Electron Microscope," *IEEE J. Solid-State Circuits*, June 1978, pp. 319-325.

38. R.M. Alexander, "Accelerated Testing in FAMOS Devices: 8K EPROM," *IEEE Reliability Physics 16th Annual Conf.*, 1978, pp. 229-232.

39. H.A. Batdorf et al., "Reliability Evaluation Program and Results for a 4K Dynamic RAM," *IEEE Reliability Physics 16th Annual Conf.*, 1978, pp. 14-18.

40. R.C. Foss, "Electrical Testing for Design Verification and Operating Margins (MOS Dynamic RAM)," *IEEE Reliability Physics 16th Annual Conf.*, 1978, pp. 14-18.

41. A.J. Gonzales and M.W. Powell, "Resolution of MOS One-Transistor Dynamic RAM Bit Failures Using SEM Stroboscopic Techniques," *J. Vac. Sci. and Technol.*, May-June 1978, pp. 1043-1046.

9. Fault Masking and Multiple Fault Detection

1. V.K. Agarwal and A.S.F. Fung, "Multiple Fault Testing of Large Circuits by Single Fault Test Sets," *IEEE Trans. Comput.*, Vol. C-30, No. 11, Nov. 1981, pp. 855-865.

2. V.K. Agarwal and G.M. Masson, "Generic Fault Characterizations for Table Look-up Coverage Bounding," *IEEE Trans. Comput.*, Vol. C-29, No. 4, Apr. 1980, pp. 288-299.

3. C.W-Y. Cha, "Multiple Fault Diagnosis in Combinational Networks," *Report No. UIUC-74-2215*, Dept. of Elect. Eng., Univ. of Illinois, June 1974.

4. F.J.O. Dias, "Fault Masking in Combinational Logic Circuits," *IEEE Trans. Comput.*, Vol. C-24, No. 5, May 1975, pp. 476-482.

5. M. Fridrich and W.A. Davis, "Minimal Fault Tests for Combinational Networks," *IEEE Trans. Comput.*, Vol. C-23, No. 8, Aug. 1974, pp. 850-859.

6. J.E. Smith, "On Necessary and Sufficient Conditions for Multiple Fault Undetectability," *IEEE Trans. Comput.*, Vol. C-28, No. 10, Oct. 1979, pp. 801-802.

10. Multivalued Logic and Fault Detection

1. Y.M. Ajabnoor and M.H. Abd-Elbarr, "Stuck-Type Fault Detection in Multi-Valued Combinational Circuits," *Proc. of the 11th Int. Symp. on Multi-Valued Logic*, May 1981, pp. 275–282.
2. W. Coy and K. Moraga, "Description and Detection of Faults in Multi-Valued Logic Circuits," *Proc. of the 9th Int. Symp. on Multi-Valued Logic*, May 1979, pp. 74–81.
3. T.T. Dao, "Recent Multi-Valued Circuits," *Proc. of 1981 COMPCON (Spring)*, Feb. 1981, pp. 194–203.
4. T.T. Dao, E.J. McCluskey, and L.K. Russell, "Multivalued Integrated Injection Logic," *IEEE Trans. Comput.*, Vol. C-26, No. 12, Dec. 1977, pp. 1233–1241.
5. J.L. Huertas and J.M. Carmona, "Low Power Ternary CMOS Circuits," *Proc. of the 9th Int. Symp. on Multi-Valued Logic*, May 1979, pp. 170–174.
6. H-Y. Lo, "Generalized Fault Detection for Multi-Valued Logic Systems," *Proc. of the 10th Int. Symp. on Multi-Valued Logic*, May 1980, pp. 178-186.
7. H-Y. Lo and S.C. Lee, "A Map-Partition Method for the Fault Detection of Multi-Valued and Multi-Level Combinational Circuits," *Proc. of the 11th Int. Symp. on Multi-Valued Logic*, May 1981, pp. 283-289.
8. E.J. McCluskey, "Logic Design of MOS Ternary Logic," *Proc. of the 10th Int. Symp. on Multi-Valued Logic*, May 1980, pp. 1–5.
9. K.C. Smith, "The Prospects of Multivalued Logic: A Technology and Application View," *IEEE Trans. Comput.*, Vol. C-30, No. 9, Sept. 1981, pp. 619–634.
10. R.J. Spillman and S.Y.H. Su, "Detection of Single Stuck-Type Failures in Multivalued Combinational Networks," *IEEE Trans. on Comp.*, Vol. C-26, No. 12, Dec. 1977, pp. 1242-1251.
11. M. Stark, "Two Bits per Cell ROM," *Proc. of 1981 COMPCON (Spring)*, Feb. 1981, pp. 209–212.
12. Z.G. Vranesic, "Applications and Scope of Multi-Valued LSI Technology," *Proc. of 1981 COMPCON (Spring)*, Feb. 1981, pp. 213–215.

11. Bridging Faults

1. M. Karpovsky and S.Y.H. Su, "Detection and Location of Input and Feedback Bridging Faults among Input and Output Lines," *IEEE Trans. Comput.*, Vol. C-29, No. 6, June 1980, pp. 523–527.
2. C.H. Lin and S.Y.H. Su, "Feedback Bridging Faults in General Combinational Networks," *Proc. 8th Int. Symp. Fault-Tolerant Computing*, June 1978.
3. K.C.Y. Mei, "Bridging and Stuck-at Faults", *IEEE Trans. Comput.*, Vol. C-23, No. 7, July 1974, pp. 720–727.

12. Circuit, Timing, Logic, and Fault Simulation

1. W. Johnson et al., "Mixed-Level Simulation from a Hierarchical Language," *J. Digital Systems*, Vol. 4, Fall 1980, pp. 305–335.
2. R.E. Bryant, "MOSSIM: A Switch-Level Simulator for MOS LSI," *Proc. 18th Design Automation Conf.*, June 1981, pp. 786–790.
3. D. Holt and D. Hutchings, "A MOS/LSI Oriented Logic Simulator," *Proc. 18th Design Automation Conf.*, June 1981, pp. 280–287.
4. A.K. Bose et al., "A Fault Simulator for MOS LSI Circuits," *Proc. 19th Design Automation Conf.*, June 1982, pp. 400–409.
5. J.P. Hayes, "A Unified Switching Theory with Applications to VLSI Design," *Proc. IEEE*, Vol. 70, Oct. 1982, pp. 1140–1151.
6. E.G. Ulrich and T. Baker, "Concurrent Simulation of Nearly Identical Networks," *Computer*, Vol. 7, No. 4, April 1974, pp. 39–44.
7. B.R. Chawla, H.K. Gummel, and P. Kozak, "MOTIS—An MOS Timing Simulator," *IEEE Trans. Circuits Syst.*, Vol. CAS-22, Dec. 1975, pp. 901–910.
8. L.W. Nagel, "SPICE2: A Computer Program to Simulate Semiconductor Circuits," *ERL Memo ERL-M520*, Univ. of California, Berkeley, May 1975.
9. W.T. Weeks, A.J. Jiminez, G.W. Mahoney, D. Mehta, H. Qassemzadeh, and T.R. Scott, "Algorithms for ASTAP: A Network Analysis Program," *IEEE Trans. Circuit Theory*, Vol. CT-20, Nov. 1973, pp. 628–634.
10. G.D. Hachtel, R.K. Brayton, and F.G. Gustavson, "The Sparse Tableau Approach to Network Analysis and Design," *IEEE Trans. Circuit Theory*, Vol. CT-18, Jan. 1971, pp. 101–113.
11. C. Ho, A.E. Ruehli, and P. Brennan, "The Modified Nodal Approach to Network Analysis," *IEEE Trans. Circuits Syst.*, Vol. CAS-25, June 1975, pp. 504–509.
12. B.R. Chawla, H.K. Gummel, and P. Kozak, "MOSIS—An MOS Timing Simulator," *IEEE Trans. Circuits Syst.*, Vol. CAS-22, Dec. 1975, pp. 901–909.
13. H.N. Nham and A.K. Bose, "A Multiple Delay Simulator for MOS LSI Circuits," *Proc. 17th Design Automation Conf.*, June 1980.
14. V.D. Agrawal et al., "A Mixed Mode Simulator," *Proc. 17th Design Automation Conf.*, June 1980.
15. A.R. Newton, "Techniques for the Simulation of Large-Scale Integrated Circuits," *IEEE Trans. Circuits Syst.*, Vol. CAS-26, Sept. 1979, pp. 741–749.
16. G. Arnout and H. De Man, "The Use of Threshold Functions and Boolean-Controlled Network Elements for Macromodeling of LSI Circuits," *IEEE J. Solid-State Circuits*, Vol. SC-13, June 1978, pp. 326–332.

17. K. Skallah and S.W. Director, "An Activity-Directed Circuit Simulation Algorithm," *IEEE ICCC'80 Conf. Proc.*, Oct. 1980, pp. 1032-1035.

18. P.M. Trouborst and J.A.G. Jess, "Macromodeling by Systematic Code Reduction," *IEEE ICCC'80 Conf. Proc.*, Oct. 1980, pp. 337-340.

19. W.M.G. van Bokhoven, "Macromodeling and Simulation of Mixed Analog-Digital Networks by a Piecewise-linear Systems Approach," *IEEE ICCC'80 Conf. Proc.*, Oct. 1980, pp. 361-365.

20. I.N. Hajj, "Sparsity Considerations in Network Solution by Tearing," *IEEE Trans. Circuits Syst.*, Vol. CAS-27, May 1980, pp. 357-366.

21. S.A. Szygenda and E.W. Thompson, "Digital Logic Simulation in a Time-Based, Table Driven Environment; Part 1: Design Verification," *IEEE Computer*, Mar. 1975, pp. 24-36.

22. E.G. Ulrich, "Time Sequenced Logical Simulation Based on Circuit Delay and Selective Tracing of Active Network Path," *Proc. ACM 20th Nat. Conf.*, 1965, pp. 437-448.

23. A.R. Newton, "Timing, Logic, and Mixed Mode Simulation for Large MOS Integrated Circuits," *Proc. NATO Advanced Study Institute on Computer Design Aids for VLSI Circuits*, Aug. 1980.

24. E.G. Ulrich, "Exclusive Simulation of Activity in Digital Net Networks," *Commun. ACM*, Vol. 12, No. 2, Feb. 1969, pp. 102-110.

25. N.B. Rabbat, A.L. Sangiovanni-Vincentelli, and H.Y. Hsieh, "A Multilevel Newton Algorithm with Macromodeling and Latency for Analysis of Large-Scale Nonlinear Networks in the Time Domain," *IEEE Trans. Circuits Syst.*, Vol. CAS-26, No. 9, Sept. 1979.

26. A.L. Sangiovanni-Vincentelli and N.B. Rabbat, "Techniques for the Time Domain Analysis of VLSI Circuits," *Proc. Inst. Elect. Eng.*, Vol. 127, Part G, 1980, pp. 292-301.

27. H. De Man, "Computer Aided Design for Integrated Circuits: Trying to Bridge the Gap," *IEEE J. Solid-State Circuits*, Vol. SC-14, June 1979, pp. 613-621.

28. R.K. Brayton, F. Gustavson, and G.D. Hachtel, "A New Efficient Algorithm for Solving Differential-Algebraic Systems Using Implicit Backward Differentiation Formulas," *Proc. IEEE*, Vol. 60, Jan. 1972, pp. 98-108.

29. G. Guardabassi, "A Note on Minimal Essential Sets," *IEEE Trans. Circuit Theory*, Vol. CT-18, 1971, pp. 557-560.

30. G.W. Smith and R.B. Walford, "The Identification of a Minimal Feedback Vertex Set of a Directed Graph," *IEEE Trans. Circuits Syst.*, Vol. CAS-22, 1975, pp. 9-15.

31. H.Y. Chang, G.W. Smith, Jr., and R.B. Walford, "LAMP: System Description," *Bell Syst. Tech. J.*, Vol. 53, Oct. 1974, pp. 1431-1449.

32. J. Katzenelson, "An Algorithm for Solving Nonlinear Resistive Networks," *Bell Syst. Tech. J.*, Vol. 44, 1965, pp. 1605-1620.

33. T. Fujisawa and E.S. Kuh, "Piecewise-Linear Theory of Non-Linear Networks," *SIAM J. Appl. Math.*, Vol. 22, No. 2, March 1972.

34. A. Siagiovanni-Vincentelli and T.A. Bickart, "Bipartite Graphs and an Optimal Bordered Triangular Form of a Matrix," *IEEE Trans. Circuits Syst.*, Vol. CAS-26, No. 10, Oct. 1979, pp. 880-890.

35. M.F. Moad, "A Sequential Method of Network Analysis," *IEEE Trans. Circuit Theory*, Vol. CT-17, Feb. 1970, pp. 99-104.

36. G. Guardabassi and A. Sangiovanni-Vincentelli, "A Two-Level Algorithm by Tearing," *IEEE Trans. Circuits Syst.*, Vol. CAS-23, 1976, pp. 783-791.

37. J.D. Crawford, M.Y. Hsueh, A.R. Newton, and D.O. Pederson, "MOTIS-C User's Guide," *Electronics Research Laboratory*, Univ. of California, Berkeley, CA, June 1978.

38. G. De Micheli and A. Sangiovanni-Vincentelli "Numerical Properties of Algorithms for Analysis of MOS VLSI Circuits," *Proc. European Conf. Circuit Theory and Systems*, Aug. 1981.

39. H. De Man, G. Arnout, and P. Reyneart, "Mixed Mode Circuit Simulation Techniques and Their Implementation in DIANA," *Proc. NATO Advanced Study Inst.*, Aug. 1980.

40. A.E. Ruehli, A. Sangiovanni-Vincentelli, and N.B. Rabbat, "Time Analysis of Large Scale Circuits Using One-Way Macro-Models," *Proc. IEEE Int. Symp. Circuits and Systems*, 1980, pp. 766-770.

41. V.D. Agrawal, A.K. Bose, P. Kozak, H.N. Nham, and E. Pascas-Skewes, "A Mixed-Mode Simulator," *Proc. 17th Design Automation Conf.*, June 1980, pp. 618-625.

42. A.R. Newton, "The Simulation of Large Scale Integrated Circuits," *Mem. No. UCB/ERL M78/52*, Electronics Research Lab., Univ. of California, Berkeley, CA, July 1978.

43. H. De Man, L. Darcis, I. Bolsens, P. Reynaert, and D. Dumlugol, "A Debugging Guided Simulation System for MOS VLSI Design," *Digest of Technical Papers, ICCAD-83*, Sept. 1983, pp. 137-138.

44. R.A. Saleh, J.E. Kleckner, and A.R. Newton, "Iterated Timing Analysis in SPLICE1," *Digest of Technical Papers, ICCAD-83*, Sept. 1983, pp. 139-140.

45. E. Lelarasmee, A.E. Ruehli, and A.L. Sangiovanni-Vincentelli, "The Waveform Relaxation Method for Time-Domain Analysis of Large Scale Integrated Circuits," *IEEE Trans. on CAD*, Vol. CAD-1, No. 3, July 1982, pp. 131–145.

46. C.F. Chen and P. Subramaniam, "The Second Generation MOTIS Timing Simulator—An Efficient and Accurate Approach for General MOS Circuits," *Proc. 1984 Int. Symp. Circuits and Systems*, 1984.

47. R.E. Bryant, "An Algorithm for MOS Logic Simulation," *Lambda*, Vol. 1, No. 3, 1980, pp. 46–53.

48. R.E. Bryant, "A Switch-level Model and Simulator for MOS Digital Systems," *Caltech Report 5065:TR:83*, 1983.

49. D. Dumlogol, H.J. De Man, P. Stevens, and G.G. Schrooten, "Local Relaxation Algorithms for Event-Driven Simulation of MOS Networks Including Assignable Delay Modeling," *IEEE Trans. on Computer-Aided Design*, Vol. CAD-2, No. 3, July 1983, pp. 193–202.

50. P. Subramaniam, "Table Models for Timing Simulator," *Proc. 1984 Custom Integrated Circuits Conf.*, 1984.

51. J. White and A.L. Sangiovanni-Vincentelli, "RELAX2: A New Waveform Relaxation Approach for the Analysis of LSI MOS Circuits," *Proc. 1983 Int. Symp. Circuits and Systems*, May 1983.

52. C.-Y. Lo, H.N. Nham, and A.K. Bose, "A Data Structure for MOS Circuits," *Proc. 20th Design Automation Conf.*, June 1983, pp. 619–624.

53. B.R. Chawla, H.K. Gummel, and P. Kozak, "MOTIS—An MOS Timing Simulator," *IEEE Trans. Circuits Syst.*, Vol. CAS-22, No. 12, Dec. 1975, pp. 901–910.

54. H.N. Nahm and A.K. Bose, "A Multiple Delay Simulator for MOS LSI Circuits," *Proc. 17th Design Automation Conf.*, June 1980, pp. 610–617.

55. L.W. Nagel, "ADVICE for Circuit Simulation," *Proc. 1980 Int. Symp. Circuits and Systems*, 1980.

13. Functional Testing

1. S.B. Akers, "Binary Decision Diagrams," *IEEE Trans. Comput.*, Vol. C-27, No. 6, June 1978, pp. 509–516.

2. S.B. Akers, "Functional Testing with Binary Decision Diagrams," *Proc. 8th Int. Symp. on Fault-Tolerant Computing*, June 1978, pp. 75–82.

3. D. Siewiorek and K. Lai, "Testing of Digital Systems," *Proc. IEEE*, Oct. 1981, pp. 1321–1333.

4. M. Lin and K. Rose, "Applying Test Theory to VLSI Testing," *Proc. Test Conf.*, 1982.

5. T. Middleton, "Functional Test Vector Generation for Digital LSI/VLSI Devices," *Proc. Test Conf.*, 1983, pp. 682–691.

6. C. Robach and G. Saucier, "Microprocessor Functional Testing," *Proc. Test Conf.*, 1980.

7. J. Crafts, "Techniques for Memory Testing," *IEEE Computer*, Oct. 1979, pp. 23–31.

8. M. Annarartone and M. Sami, "An Approach to Functional Testing of Microprocessors," *Proc. 12th Symp. on Fault-Tolerant Computing*, June 1982, pp. 158–164.

9. S. Su and Y. Hsieh, "Testing Functional Faults in Digital Systems Described by Register Transfer Language," *J. Digital Systems*, 1981, pp. 447–457.

10. Y. Min and S. Su, "Testing Functional Faults in VLSI," *Proc. 19th Design Automation Conf.*, 1982.

11. S. Thatte and J. Abraham, "Test Generation for Microprocessors," *IEEE Trans. Comput.*, June 1980.

12. K. Lai, "Functional Testing of Digital Systems," *Ph.D thesis*, Comp. Sc. Dept., Carnegie-Mellon Univ., Dec. 1981.

13. J. Oakley, "Symbolic Execution of Formal Machine Descriptions," *Ph.D. Thesis*, Comp. Sci. Dept., Carnegie-Mellon Univ., 1979.

14. Y. Levendel and P. Menon, "Test Generation Algorithms for Computer Hardware Description Language," *IEEE Trans. Comput.*, July 1982, pp. 577–588.

15. M. Breuer and A. Friedman, "Functional Level Primitives in Test Generation," *IEEE Trans. Comput.*, March 1980, pp. 223–235.

16. J. Abraham and K. Parker, "Practical Microprocessor Testing: Open and Close Loop Approaches," *Proc. Test Conf.*, 1981, pp. 308–311.

17. M. Breuer, A. Friedman, and A. Iosupovicz, "A Survey of the State-of-the-Art of Design Automation," *IEEE Computer*, Oct. 1981, pp. 58–75.

18. C. Bellon et al., "Automatic Generation of Microprocessor Test Programs," *Proc. 19th Design Automation Conf.*, 1982, pp. 566–573.

19. R. Wadsack, "The BELLMAC CPU's Design Verification, Debugging, and Fault Coverage Tests," *1982 Bell Systems Conf. on Electronic Testing*, 1982.

20. T. Sridhar and J. Hayes, "A Functional Approach to Testing Bit-Sliced Microprocessors," *IEEE Trans. Comput.*, Aug. 1981, pp. 563–571.

21. K. Saluja, L. Shen, and S. Su, "A Simplified Algorithm for Testing Microprocessors," *Proc. Test Conf.*, 1983, pp. 608–615.

22. L. Shen and S. Su, "A Functional Testing Method for Microprocessors," *Proc. 14th Int. Symp. on Fault-Tolerant Computing*, 1984.

23. M.S. Abadir and H.K. Reghbati, "Test Generation for LSI: A Case Study," *Proc. of ACM-IEEE Design Automation Conf.*, 1984, pp. 180–195.

24. M.S. Abadir and H.K. Reghbati, "Functional Test Generation for LSI Circuits Described by Binary Decision Diagrams," *Proc. of IEEE's International Test Conference*, 1985.

25. M.S. Abadir and H.K. Reghbati, "Functional Specification and Testing of Logic Circuits," *International Journal of Computers and Mathematics*, 1985.

14. Testing and Fault Tolerance of Interconnection Networks

1. D.P. Agrawal, "Graph Theoretic Analysis and Design of Multistage Interconnection Networks," *IEEE Trans. Comput.*, Vol. C-32, No. 9, Sept. 1983, pp. 637–648.

2. D.C. Operferman and N.T. Tsao-Wu, "On a Class of Rearrangeable Switching Networks; Part II: Enumeration Studies and Fault Diagnosis," *Bell Syst. Tech. J.*, May/June 1971, pp. 1601–1618.

3. J.P. Shen, "Fault Tolerance Analysis of Several Interconnecting Networks," *Proc. 1982 Int. Conf. on Parallel Processing*, 1982, pp. 102–112.

4. K.M. So and J.J. Narraway, "On-line Diagnosis of Switching Networks," *IEEE Trans. Circuits Syst.*, Vol. CAS-21, No. 7, July 1979, pp. 575–583.

5. J.J. Narraway and K.M. So, "Fault Diagnosis in Inter-Processor Switching Networks," *Proc. Int. Conf. on Circuits and Computers*, Oct. 1980, pp. 741–753.

6. C.L. Wu and T.Y. Feng, "Fault-Diagnosis for a Class of Multistage Interconnection Networks," *1979 Int. Conf. on Parallel Processing*, Aug. 1979, pp. 269–278.

7. D.P. Siewiorek et al., "A Case Study of Cmmp, CM*, and CVMP: Part I—Experiences with Fault Tolerance in Multiprocessor Systems," *Proc. IEEE*, Vol. 64, No. 10, Oct. 1978, pp. 1178–1200.

8. D.P. Agrawal, "Automated Testing of Computer Networks," *1980 Int. Conf. on Circuits and Computers*, Oct. 1980, pp. 717–720.

9. K.M. Falavarjani and D.K. Pradhan, "Fault Diagnosis of Parallel Processor Interconnection Networks," *Proc. 1981 Fault-Tolerant Comp. Symp.*, June 1981.

10. S. Sowrirajan and S.M. Reddy, "A Design for Fault-Tolerant Full Connection Networks," *1980 Conf. on Information Sciences and Systems*, pp. 536–540.

11. D.P. Agrawal, "Testing and Fault-Tolerance of Multistage Interconnection Networks," *IEEE Computer*, Vol. 15, No. 4, April 1982, pp. 41–53.

12. D.P. Agrawal and D. Kaur, "Fault Tolerant Capabilities of Redundant Multistage Interconnection Networks," *Proc. Real-Time Systems Symp.*, Dec. 1983, pp. 119–127.

13. D.C.H. Lee, "Fault Diagnosis of (N,K) Shuffle/Exchange Networks," *M.S. thesis*, EE Dept., CMU, Feb. 1983.

14. K.N. Levitt, M.W. Green, and J. Goldberg, "A Study of the Data Communications Problems in a Self-Repairable Multiprocessor," *Proc. SJCC*, 1968, pp. 515–527.

15. Memory Testing

1. W.E. Sohl, "Selecting Test Patterns for 4K RAMs," *IEEE Trans. Manuf. Techn.*, Vol. MFT-6, No. 1, 1977, pp. 51–60.

2. D.S. Suk and S.M. Reddy, "Test Procedures for a Class of Pattern-Sensitive Faults in Semiconductor Random Access Memories," *IEEE Trans. Comput.*, Vol. C-29, No. 6, June 1980, pp. 419–429.

3. D.S. Suk and S.M. Reddy, "A March Test for Functional Faults in Semiconductor Random Access Memories," *IEEE Trans. Comput.*, Vol. C-30, No. 12, Dec. 1981, pp. 982–985.

4. S.M. Thatte and J.A. Abraham, "Testing of Semiconductor Random Access Memories," *Proc. 7th Annual Int. Conf. on Fault-Tolerant Computing*, 1977, pp. 81–87.

5. J.P. Hayes, "Detection of Pattern-Sensitive Faults in Random Access Memories," *IEEE Trans. Comput.*, Vol. C-24, No. 2, Feb. 1975, pp. 150–157.

6. J.P. Hayes, "Testing Memories for Single-Cell Pattern-Sensitive Faults," *IEEE Trans. Comput.*, Vol. 29, No. 3, Mar. 1980, pp. 249–254.

7. A.R. Klayton, "Fault Analysis for Computer Memory Systems and Combinational Logic Networks," *Ph.D thesis*, Lehigh Univ., 1971.

8. J. Knaizuk and C.R.P. Hartmann, "An Algorithm for Testing Random Access Memories," *IEEE Trans. Comput.*, Vol. C-26, No. 4, April 1977, pp. 414–416.

9. J. Knaizuk and C.R.P. Hartmann, "An Optimal Algorithm for Testing Stuck-at Faults in Random Access Memories," *IEEE Trans. Comput.*, Vol. C-26, No. 11, Nov. 1977, pp. 1141–1144.

10. R. Nair, "Comments on an Optimal Algorithm for Testing Stuck-at Faults in Random Access Memories," *IEEE Trans. Comput.*, Vol. C-28, No. 3, Mar. 1979, pp. 258–261.

11. R. Nair, S.M. Thatte, and J.A. Abraham, "Efficient Algorithms for Testing Semiconductor Random Access Memories," *IEEE Trans. Comput.*, Vol. C-27, No. 6, June 1978, pp. 572–576.

12. C.V. Ravi, "Fault Location in Memory Systems by Program," *Proc. AFIPS Spring Joint Computer Conf.*, Vol. 34, May 1969, pp. 393–401.

13. S.C. Seth and K. Narayanswamy, "A Graph Model for Pattern-Sensitive Faults in Random Access Memories," *IEEE Trans. Comput.*, Vol. C-30, No. 12, Dec. 1981, pp. 937–977.

14. J. Crafts, "Techniques for Memory Testing," *IEEE Computer*, Vol. 12, Oct. 1979, pp. 23–31.
15. M.S. Abadir and H.K. Reghbati, "Functional Testing of Semiconductor Random Access Memories," *ACM Comput. Surv.*, Vol. 15, No. 3, Sept. 1983, pp. 175–198.
16. Z. Sun and L.T. Wang, "Self-Testing of Embedded RAMs," *IEEE Int. Test Conf.*, 1984, pp. 148–156.
17. W.H. McAnney et al., "Random Testing for Stuck-at Storage Cells in an Embedded Memory," *IEEE Int. Test Conf.*, 1984, pp. 157–166.

16. PLA Testing

1. V.K. Agrawal, "Multiple Fault Detection in Programmable Logic Arrays," *IEEE Trans. Comput.*, Vol. C-29, 1979, pp. 518–522.
2. P. Bose and J.A. Abraham, "Test Generation for Programmable Logic Arrays," *Proc. 19th Design Automation Conf.*, 1982, pp. 574–580.
3. S. Bozorgui-Nesbat and E.J. McCluskey, "Lower Overhead Design for Testability of PLAs," *Proc. of Int. Test Conf.*, 1984, pp. 856–865.
4. M.A. Breuer and X. Zhu, "A Knowledge-Based System for Selecting a Test Methodology for a PLA," *Proc. of DAC*, 1985, pp. 259–265.
5. C.W. Cha, "A Testing Strategy for PLAs," *Proc. 15th Design Automation Conf.*, 1978, pp. 326–334.
6. W. Daehn and J. Mucha, "A Hardware Approach to Self-Testing of Large Programmable Logic Arrays," *IEEE Trans. Comput.*, Vol. C-30, 1981, pp. 829–833.
7. E.B. Eichelberger and E. Lindbloom, "A Heuristic Test-Pattern Generator for Programmable Logic Arrays," *IBM J. Res. Dev.*, Vol. 24, 1980, pp. 15–22.
8. H. Fleisher and L.I. Maissel, "An Introduction to Array Logic," *IBM J. Res. Dev.*, Vol. 19, March 1975, pp. 98–109.
9. H. Fujiwara and K. Kinoshita, "A Design of Programmable Logic Arrays with Universal Test," *IEEE Trans. Comput.*, Vol. C-30, 1981, pp. 823–828.
10. H. Fujiwara and K. Kinoshita, "A New PLA Design for Universal Testability," *IEEE Trans. Comput.*, Vol. C-33, August 1983, pp. 37–48.
11. S.Z. Hassan and E.J. McCluskey, "Testing PLAs Using Multiple Parallel Signature Analyzers," *Proc. of 13th FTCS*, 1983, pp. 422–425.
12. K.A. Hua, J.Y. Jou, and J.A. Abraham, "Built-in Tests for VLSI Finite State Machines," *FTCS-14*, 1984, pp. 292–297.
13. J. Khabaz, "A Testable PLA Design with Low Overhead and High Fault Coverage," *FTCS-13*, 1983, pp. 426–429.
14. J. Khabaz and E.J. McCluskey, "Concurrent Error Detection and Testing for Large PLA's," *IEEE J. Solid-State Circuits*, Vol. SC-17, 1982, pp. 386–394.
15. K.L. Kodanadapani and D.K. Pradhan, "Undetectability of Bridging Faults and Validity of Stuck-at Faults Test Sets," *IEEE Trans. Comput.*, Vol. C-29, pp. 55–59.
16. G.P. Mak, J.A. Abraham, and E.S. Davidson, "The Design of PLAs with Concurrent Error Detection," *Proc. of 12th Annu. Symp. Fault-Tolerant Computing (FTCS-12)*, 1982, pp. 303–310.
17. D.L. Ostapko and S.J. Hong, "Fault Analysis and Test Generation for Programmable Logic Arrays," *IEEE Trans. Comput.*, Vol. C-28, pp. 617–627.
18. D.K. Pradhan and K. Son, "The Effect of Undetectable Faults in PLAs and a Design for Testability," *Dig. 1980 Test Conf.*, 1980, pp. 359–367.
19. J. Rajski and J. Tyszer, "Easily Testable PLA Design," *Proc. of Euromicro*, 1984, pp. 139–146.
20. J. Rajski and J. Tyszer, "Combinational Approach to Multiple Contact Faults Coverage in PLAs," *IEEE Trans. Comput.*, Vol. C-34, No. 6, June 1985, pp. 549–552.
21. K.S. Ramanatha and N.N. Biswas, "A Design for Complete Testability of Programmable Logic Arrays," *IEEE Test Conference*, 1982, pp. 67–74.
22. H.K. Reghbati, "Fault Detection in Programmable Logic Arrays," *SFU-CMPT Tech. Report 85-11*, 1985.
23. K. Saluja, K. Kinoshita, and H. Fujiwara, "An Easily Testable Design of Programmable Logic Arrays for Multiple Faults," *IEEE Trans. Comput.*, Vol. C-32, Nov. 1983, pp. 1038–1045.
24. J.E. Smith, "Detection of Faults in Programmable Logic Arrays," *IEEE Trans. Comput.*, Vol. C-28, 1979, pp. 845–853.
25. F. Somenzi et al., "PART: Programmable Array Testing Based on a Partitioning Algorithm," *IEEE Trans. CAD*, Vol. CAD-3, No. 2, 1984, pp. 142–149.
26. K. Son and D.K. Pradhan, "Completely Self-Checking Checkers in PLAs," *IEEE Test Conf.*, 1980, pp. 231–237.
27. Y. Tamir and C.H. Séquin, "Design and Application of Self-Testing Comparators Implemented with MOS PLA's," *IEEE Trans. Comput.*, Vol. C-33, No. 6, June 1980, pp. 493–506.
28. R. Treuer, H. Fujiwara, and V.K. Agrawal, "Implementing a Built-in Self-Test PLA Design," *IEEE Design & Test*, Vol. 2, No. 2, Apr. 1985, pp. 37–48.
29. R. Treuer, H. Fujiwara, and V.K. Agrawal, "A Low-Overhead, High Coverage, Built-In Self-Test PLA Design," *Proc. of FTCS-15*, 1985, pp. 112–117.

30. S.L. Wang and A. Avizienis, "The Design of Totally Self Checking Circuits using Programmable Logic Arrays," *Proc. of 9th Annu. Symp. Fault-Tolerant Computing (FTCS-9)*, 1979, pp. 173–180.

31. R.-S. Wei and A. Sangiovanni-Vincentelli, "A PLA Test Pattern Generation Tool," *Proc. of Design Automation Conf.*, 1985, pp. 197–203.

32. S. Yajima and T. Aramaki, "Autonomously Testable PLAs," *Proc. FTCS-11*, 1981, pp. 41–43.

17. Design for Testability and Built-in Self-Test

1. S. Dasgupta, R.G. Walther, and T.W. Williams, "An Enhancement to LSSD and Some Applications of LSSD in Reliability, Availability, and Serviceability," *11th Int. Symp. on Fault-Tolerant Computing*, 1981, pp. 32–34.

2. H. Eiki, K. Inagaki, and S. Yajima, "Autonomous Testing and its Application to Testable Design of Logic Circuits," *10th Int. Symp. on Fault-Tolerant Computing*, 1980, pp. 173–178.

3. E.B. Eichelberger and T.W. Williams, "A Logic Design Structure for LSI Testability," *Proc. 14th Design Automation Conf.*, 1977, pp. 462–468.

4. E.B. Eichelberger and E. Lindbloom, "Random-Pattern Coverage Enhancement and Diagnosis for LSSD Logic Self-Test," *IBM J. Res. Dev.*, Vol. 27, No. 3., May 1983, pp. 265–272.

5. P.P. Fasang, "BIDCO: Built-in Digital Circuit Observer," *IEEE Test Conf.*, 1980, pp. 261–266.

6. S. Funsatsu, N. Wakatsuki, and A. Yamada, "Designing Digital Circuits with Easily Testable Consideration," *IEEE Test Conf.*, 1978, pp. 98–102.

7. D. Komonytsky, "LSI Self-Test Using Level-Sensitive Scan Design and Signature Analysis," *IEEE Test Conf.*, 1982, pp. 414–424.

8. D. Komonytsky, "Synthesis of Techniques Creates Complete System Self-Test," *Electronics*, Mar. 1983, pp. 110–115.

9. B. Konemann, J. Mucha, and G. Zwiehoff, "Built-in Logic Block Observation Technique," *IEEE Test Conf.*, 1979, pp. 37–41.

10. B. Konemann, J. Mucha, and G. Zwiehoff, "Built-in Test for Complex Digital Integrated Circuits," *IEEE J. Solid-State Circuits*, Vol. SC-15, No. 3, June 1980, pp. 315–318.

11. E.J. McCluskey and S. Bozorgui-Nesbat, "Design for Autonomous Test," *IEEE Trans. Comput.*, Vol. C-30, No. 11, Nov. 1981, pp. 866–875.

12. E.J. McCluskey, "Verification Testing," *Proc. 19th Design Automation Conf.*, 1982, pp. 495–500.

13. E.J. McCluskey, "Built-in Verification Test," *IEEE Test Conf.*, 1982, pp. 183–190.

14. E.J. McCluskey, "Verification Testing—A Pseudo-exhaustive Test Technique," *IEEE Trans. Comput.*, Vol. C-33, No. 6, June 1984, pp. 541–546.

15. D.R. Resnick, "Testability and Maintainability with a New 6K Gate Array," *VLSI Design*, March/April 1983, pp. 34–38.

16. J. Savir, "Syndrome-Testable Design of Combinational Circuits," *IEEE Trans. Comput.*, June 1980, pp. 442–451.

17. J. Savir, "Syndrome-Testing of 'Syndrome Untestable' Combinational Circuits," *IEEE Trans. Comput.*, Vol. C-30, No. 8, Aug. 1981, pp. 606–608.

18. J. Savir, G. Ditlow, and P.H. Bardell, "Random Pattern Testability," *13th Annual Int. Symp. on Fault-Tolerant Computing*, June 1983, pp. 80–89.

19. R.M. Sedmak, "Design for Self-Verification: An Approach for Dealing with Testability Problems in VLSI-based Designs," *IEEE Test Conf.*, Oct. 1979, pp. 112–124.

20. R.M. Sedmak, "Implementation Techniques for Self-Verification," *IEEE Test Conf.*, Nov. 1980, pp. 267–278.

21. Z. Sun and L.T. Wang, "Self-Testing of Embedded RAMs," *IEEE Test Conf.*, 1984.

22. K.D. Wagner, "Design for Testability in the Amdahl 580," *COMPCON Spring 83*, 1983, pp. 384–388.

23. S. Strom, "ROM-Resident Software Self-Tests Microcomputers," *Electronic Design*, Vol. 28, No. 13, June 1980, pp. 131–134.

24. V.K. Agrawal, "Increasing Effectiveness of Built-In-Testing by Output Data Modification," *13th Annual Fault-Tolerant Computing Symp.*, 1983, pp. 227–233.

25. S.B. Akers and B. Krishnamurthy, "On the Application of Test Counting to VLSI Testing," *Proc. Chapel Hill Conf. on VLSI*, 1985, pp. 343–362.

26. S.B. Akers, "On the Use of Linear Sums in Exhaustive Testing," *Proc. of FTCS-15*, 1985, pp. 148–153.

18. Spectral Analysis and Syndrome Testability

1. J.C. Muzio and D.M. Miller, "Spectral Fault Signatures for Internally Unate Combinational Networks," *IEEE Trans. Comput.*, Vol. C-32, No. 11, Nov. 1983, pp. 1058–1062.

2. A.K. Susskind, "Testing by Verifying Walsh Coefficients," *IEEE Trans. Comput.*, Vol. C-32, No. 2, Feb. 1983, pp. 198–201.

3. J. Savir, "Syndrome Testable Design of Combinational Circuits," *IEEE Trans. Comput.*, Vol. C-29, No. 6, June 1980, pp. 442–451.

4. V.H. Tokmen, "Disjoint Decomposability of Multi-Valued Functions by Spectral Means," *Proc. 10th Int. Symp. on Multiple Valued Logic*, 1980, pp. 88-93.

5. J.C. Muzio, "Composite Spectra and the Analysis of Switching Circuits," *IEEE Trans. Comput.*, Vol. C-29, 1980, pp. 750-753.

6. G. Markowsky, "Syndrome-Testability can be Achieved by Circuit Modification," *IEEE Trans. Comput.*, Vol. C-30, 1981, pp. 604-606.

7. E. Eris and D.M. Miller, "Syndrome-Testable Internally Unate Combinational Networks," *Electron. Lett.*, Vol. 19, No. 16, 1983.

8. W.C. Carter, "The Ubiquitous Parity Bit," *Proc. 12th Int. Symp. on Fault-Tolerant Computing*, 1982, pp. 289-296.

9. J. Savir, "Syndrome-Testing of 'Syndrome Untestable' Combinational Circuits," *IEEE Trans. Comput.*, Vol. C-30, No. 8, 1981, pp. 606-608.

10. J. Savir and J.P. Roth, "Testing for and Distinguishing Between Failures," *Proc. 12th Int. Symp. on Fault-Tolerant Computing*, 1982, pp. 165-172.

11. Z. Barzilai, D. Coppersmith, and A.L. Rosenberg, "Exhaustive Generation of Bit Patterns with Applications to VLSI Self-testing," *IEEE Trans. Comput.*, Vol. C-32, No. 2, Feb. 1983, pp. 190-193.

19. Probabilistic Test and Signature Analysis

1. S.W. Golomb, *Shift Register Sequences*, Aegean Park Press, Laguna Hills, CA, 1982.

2. G. Gordon and H. Nadig, "Hexadecimal Signatures Identify Troublespots in Microprocessor Systems," *Electronics*, March 3, 1977, pp. 89-96.

3. S.Z. Hassan, "Algebraic Analysis of Parallel Signature Analyzers," *Centre for Reliable Computing, Tech. Rep. No. 82-5*, Computer Systems Lab., Stanford Univ., Stanford, CA, June 1982.

4. S.Z. Hassan, D.L. Lu, and E.J. McCluskey, "Parallel Signature Analyzers: Detection Capability and Extensions," *COMPCON Spring 83*, San Francisco, CA, 1983.

5. J.P. Hayes, "Transition Count Testing of Combinational Logic Circuits," *IEEE Trans. Comput.*, Vol. C-27, No. 6, June 1976, pp. 613-620.

6. J. Losq, "Efficiency of Random Compact Testing," *IEEE Trans. Comput.*, Vol. C-27, No. 6, June 1978, pp. 516-525.

7. J.J. Shedletskey, "Random Testing: Practicality vs. Verified Effectiveness," *7th Int. Conf. on Fault-Tolerant Computing*, June 1977, pp. 175-179.

8. J.E. Smith, "Measure of the Effectiveness of Fault Signature Analysis," *IEEE Trans. Comput.*, Vol. C-29, No. 6, June 1980, pp. 510-514.

9. T. Sridhar et al., "Analysis and Simulation of Parallel Signature Analyzers," *IEEE Test Conf.*, 1982, pp. 656-661.

10. L.T. Wang and E.J. McCluskey, "A New Condensed Linear Feedback Shift Register Design for VLSI/System Testing," *14th Int. Conf. on Fault-Tolerant Computing*, 1984, pp. 360-365.

11. J.L. Carter, "The Theory of Signature Testing for VLSI," *Proc. 14th Annu. ACM Symp. on Theory of Computing*, May 1982, pp. 66-76.

12. R. David and G. Blanchet, "About Random Fault Detection in Combinational Networks," *IEEE Trans. Comput.*, Vol. C-25, June 1976, pp. 659-664.

13. J.J. Shedletsky and E.J. McCluskey, "The Error Latency of a Fault in a Sequential Digital Circuit," *IEEE Trans. Comput.*, Vol. C-25, June 1976, pp. 655-659.

14. J. Savir and P.H. Bardell, "On Random Pattern Test Length", *IEEE Trans. Comput.*, Vol. C-33, June 1984, pp. 467-474.

15. P. Thevenod-Fosse and R. David, "A Method to Analyze Random Testing of Sequential Circuits," *Digital Processes*, Vol. 4, 1978, pp. 313-332.

16. W.C. Carter, "Signature Testing with Guaranteed Bounds for Fault Coverage," *IEEE Test Conf.*, 1982, pp. 75-82.

17. R. David, "Signature Analysis of Multi-Output Circuits," *14th Annual Fault-Tolerant Computing Symp.*, June 1984, pp. 366-371.

20. Application of AI Techniques to Design and Test

1. H.K. Reghbati and N. Cercone, "Knowledge Representation for Computer-Aided Design and Test of VLSI Systems," *Knowledge Representation*, Springer Verlag, 1986 (to appear).

2. M.J. Bending, "Hitest: A Knowledge-Based Test Generation System," *IEEE Design & Test*, Vol. 1, No. 2, 1984, pp. 83-92.

3. P.W. Horstmann, "Automation of the Design for Testability Using Logic Programming," *Ph.D. thesis*, Dept. of Elect. Eng., Syracuse Univ., Oct. 1983.

4. J.H. Kim et al., "Exploiting Domain Knowledge in IC Cell Layout," *IEEE Design and Test*, Vol. 1, No. 3, 1984, pp. 52-64.

5. P.W. Horstmann, "Expert Systems and Logic Programming for CAD," *VLSI Design*, Nov. 1983, pp. 37-46.

6. R. Zippel, "An Expert System for VLSI Design," *Proc. Int. Symp. Circuits and Systems*, May 1983.

7. B.C. Williams, "Qualitative Analysis of MOS Circuits," *Artificial Intelligence*, Vol. 24, 1984, pp. 281-346.

8. R. Stallman and G. Sussman, "Forward Reasoning and Dependency-Directed Backtracking in a System for Computer-Aided Circuit Analysis," *Artificial Intelligence*, 1977, pp. 135–196.

9. J. De Kleer, "How Circuits Work," *Artificial Intelligence*, Vol. 24, 1984, pp. 205–280.

10. R. Davis, "Diagnostic Reasoning Based on Structure and Behavior," *Artificial Intelligence*, Vol. 24, 1984, pp. 347–410.

11. M.R. Genesereth, "The Use of Design Descriptions in Automated Diagnosis," *Artificial Intelligence*, Vol. 24, 1984, pp. 411–436.

12. H.G. Barrow, "VERIFY: A Program for Proving Correctness of Digital Hardware Designs," *Artificial Intelligence*, Vol. 24, 1984, pp. 437–471.

13. W.S. Adolph and H.K. Reghbati, "Application of AI Techniques to VLSI Design," *S.F.U. CMPT Tech. Rep. 85-12*, 1985.

14. W.S. Adolph and H.K. Reghbati, "Application of AI Techniques to VLSI Testing," *S.F.U. CMPT Tech. Rep. 85-14*, 1985.

15. M.S. Abadir and M.A. Breuer, "A Knowledge-Based System for Designing Testable VLSI Chips," *IEEE Design & Test*, Vol. 2, No. 4, August 1985.

16. T.J. Kowalski and D.E. Thomas, "The VLSI Design Automation Assistant," *Proc. 22nd Design Automation Conf.*, 1985, pp. 252-258.

17. M.A. Breuer and Xian Zhu, "A Knowledge-Based System for Selecting A Test Methodology for a PLA," *Proc. 22nd Design Automation Conf.*, 1985, pp. 259–265.

18. L.I. Steinberg and T.M. Mitchell, "The Redesign System: A Knowledge-Based Approach to VLSI CAD," *IEEE Design & Test*, Vol. 2, No. 1, Feb. 1985, pp. 45–54.

Glossary

Accelerated test: A test in which certain factors, such as voltage, temperature, and so forth, are chosen to exceed those stated in the reference conditions, in order to shorten the time required to observe the stress response of the item, or magnify the response in a given time. To be valid, an accelerated test should not alter the basic modes and/or mechanisms of failure, or their relative prevalence.

Alias: A faulty circuit test output response signature that is identical to the fault-free signature.

Biased random test: A test in which the random input vectors are not equally probable.

BILBO: Also known as the built-in logic block observer. It is a chip-level self-test scheme, in which the test pattern source is an LFSR, and the response evaluator is an MISR.

Bit-sliced: A digital system is said to be bit-sliced if it is realized as a one-dimensional array or cascade of identical modules.

Bit-sliced (micro-) processor: A bit-sliced system that can perform the function of the execution unit of a computer's central processing unit.

Boundary scan path: Also known as external scan path, it consists of a scan path flip-flop connected to each of the chip bonding pads.

Built-in logic block observer: *See* BILBO.

Built-in self-test: Also known as BIST, refers to the capability of a device to generate its own test pattern set and to compress the test results into a compact pass/fail indication.

Built-in tests: Built-in tests provide access to internal circuitry and can be designed to test the circuits at speed.

Burn-in: The operation of components or systems prior to their ultimate application, intended to stabilize their characteristics and to identify early failures.

Characterization test: The series of tests designed to test a device as fully as possible. These tests will include not only a functional check-out but also parametric tests and voltage-supply tolerance tests. Implicit in the last two tests is an element of device burn-in.

Concurrent self-test: A concurrent self-test is aimed at detecting all faults of interest during the normal circuit operation.

Concurrent error detection: A technique that allows the symptoms of failure to be observed on-line.

Controllability: The ability to force the internal nodes into specified logic states.

Debugging: The art of determining the exact nature and location of suspected errors and removing them.

Design for testability: Producing designs that are easily testable.

Design rules: Geometric constraints on layout artwork.

Dynamic testing: AC (dynamic or parametric) testing verifies the time-related behavior of a circuit and involves the measurement of actual voltage and current levels.

Equivalent faults: Two or more faults are equivalent if all are detected by the same set of tests.

Error: *See* Physical fault.

Escape probability: The probability that the circuit will be declared fault-free when it is, in fact, faulty.

Exhaustive testing: The application of all 2^n input combinations to the (combinational) circuit under test.

External scan path: *See* Boundary scan path.

Fault: *See* Logical fault, Physical fault, *and* Parametric fault.

Fault cover: The total set of faults detected by a test.

Fault dominance: If a fault a dominates another fault b, then any test that detects b will also detect a. Note that a and b may be distinguishable.

Fault model: The manifestation of physical defects are captured via fault models.

Fault simulation: Simulating the behavior of logic circuits in the presence of faults.

Golden unit: The system known to be fault free.

Intermittent fault: A fault is intermittent if it is only occasionally detectable due to unstable physical failures.

Level-sensitive scan design: *See* LSSD.

LFSR: *See* Linear feedback shift register.

Linear feedback shift register: Informally, a shift register with feedback, interspersed with exclusive ORs.

Logical fault: A logical fault causes the logic function of a module or an input signal to be changed to some other function. Stuck-at-one and stuck-at-zero are classical logic faults.

LSSD: A design for testability technique, which translates the problem of testing sequential circuits into one of testing combinational circuits, which is more tractable.

Margining: Putting the device under extra stress, to ensure integrity over the long term.

Multiple-input signature register: The MISR is basically a signature accumulator, where an EXOR gate has been added between any two consecutive latches. As far as its behavior is concerned it can be transformed to an equivalent single-input signature accumulator.

Non-concurrent testing: It requires that normal operation be suspended before the test can be applied.

Observability: The ability to measure externally the internal logic states of a design.

Off-line testing: *See* Non-concurrent testing.

On-line testing: *See* Concurrent error detection.

Parametric fault: Parametric faults often change the value of a circuit parameter causing an alteration in some factor such as circuit speed, current, or voltage.

Permanent fault: A fault is said to be permanent if it is continuous and stable, reflecting an irreversible physical change in the circuit.

Physical fault: Any discrepancy between the observed and specified behavior of the circuit causes an error, the cause of which is said to be a physical fault.

Prime fault: A fault on a primary input or a fan-out branch.

Pseudoexhaustive testing: Applies all possible inputs to the blocks of the circuit under test, rather than to the entire circuit.

Pseudorandom test: An approximation to a truly random test. The major difference between a random test and a pseudorandom test is that the latter is fully repeatable whereas the former is not.

Random test: A random selection of input vectors.

Scan test: The test performed to verify the correct operation of the memory elements of the LSSD approach. In this test, patterns are fed through the scan-in port and collected via the scan-out port.

Signature analyzer: A portable tester, originally introduced by Hewlett-Packard, in which the output response is compressed through an LFSR.

Static testing: In DC (static or functional) testing, the digital circuit is exercised by applying binary input patterns and analyzing the corresponding steady-state outputs to determine correct functional behavior.

Stick diagram: At the transistor level, the symbolic layout is often called a stick diagram, since interconnections are represented by their center lines, and hence resemble sticks.

Symbolic layout: A description which forms a bridge between a schematic view of the circuit and its mask-level layout.

Syndrome analysis: An exhaustive testing strategy that requires counting the number of 1s in the output response stream.

Test sequence: A specific order of related tests.

Test software: Software used by automatic test equipment to control the unique stimuli and measurement parameters used in testing the unit under test.

Test validity: The degree to which a test accomplishes its specified goal.

Test vector: A test pattern that specifies a primary input stimulus plus a known fault-free primary output response.

Testing: The process of detecting errors.

Transient fault: A fault is transient if it is caused by temporary environmental conditions.

Two-port flip-flop: A flip-flop having two control inputs, with the data source determined by which of the controls is pulsed.

Unidirectional error: An error due to a fault in which some lines that are supposed to be at logic 0 are at logic 1, or some of the lines that are supposed to be at logic 1 are at logic 0, but not both.

Uniform random test: The random input vectors are equally probable.

Unit under test: The object being tested or debugged.

Universal test set: A testable programmable logic array (PLA) can be designed in such a way that the test patterns do not depend on the personality of the PLA. The required test set is said to be universal.

Verification testing: *See* Exhaustive testing.

Yield: The fraction of total die locations on a wafer that result in working circuits.

Author Index

Abadir, M.S. 6
Aboulhamid, E.M. 453
Abraham, J.A. 356
Ackley, D. 150
Agrawal, D.P. 337
Agrawal, V.D. 35, 298, 305, 311, 401
Baker, C.M. 59
Barber, M.R. 482
Barzilai, Z. 425
Bierman, H. 467
Blank, T. 90
Bosch, R. 429
Buehler, M.G. 202
Cerny, E. 453
Cohen, H. 150
Crouzet, Y. 283
Davis, R. 515
Ditlow, G.S. 75
Friedman, A.D. 320
Fujiwara, H. 350
Galiay, J. 283
Hamachi, G.T. 47
Hayes, J.P. 289, 327
Henley, F.J. 195
Hnatek, E. 471
Horstmann, P.W. 508
Hua, K.A. 356
Jain, S.K. 35, 298, 311
Johnson, D. 477
Jou, J.-Y. 356
Kinoshita, K. 350
Koenemann, B. 430
Kubalek, E. 175

LeBlanc, J.J. 457
Lipsett, R. 150
Mahoney, M.V. 165, 362
Marcus, R.B. 249
Markowsky, G. 421, 425
Marschner, E. 150
Mayo, R.N. 47
McCluskey, E.J. 437, 445
McCormick, S.P. 67
Menzel, E. 175
Mercer, M.R. 401
Newton, A.R. 24
Ousterhout, J.K. 47, 109
Parker, K.P. 383
Reddy, M.K. 305
Reddy, S.M. 305
Reghbati, H.K. 6
Ruehli, A.E. 75
Savir, J. 398, 407, 419, 421, 425
Scott W.S. 47
Séquin, C.H. 490
Shahdad, M. 150
Sheehan, K. 150
Shirachi, D.K. 371
Singer, D.M. 35
Smith, M.G. 425
Sridhar, T. 324
Taylor, G.S. 47
Terman, C. 59
Vergniault, M. 283
Waxam, R. 150
Williams, T.W. 383
Zwiehoff, G. 430

Subject Index

Abadir-Reghbati Technique 15, 162
Advanced Micro Devices 381
Analog Testing 362
Artificial Intelligence Techniques 465, 515
Automatic Test Equipment 465
BILBO 379
Binary Decision Diagrams 16
BIST 379, 437, 445
Bridging Faults 589
Built-in Logic Block Observer 379
Built-in Self-Test 379, 437, 445
C-Testability 317
Checker 7
Circuit Extraction 42
CMOS Testing 305
CODEC Testing 371
Compact Testing 19
Comparison Testing 18
Concurrent Error Detection 3
Concurrent Testing 7
Controllability 398
Cross-Point Faults 350
CRYSTAL 45
CWC 380
D-Algorithm 12, 163, 298
Design for Testability 379, 383
Design Rule Checking 42
Design Rules 41
Device Simulation 42
DFT 379, 383
Diagnostic Reasoning 515
Diagnostic Techniques 249
Digital Signal Processing 318, 362
EDIF 45
Electronic-Beam Testing 175
Expert Systems 465
Explicit Testing 7
Failures 161
Fairchild 466
Fault Coverage 8
Fault Location 8
Fault Masking 588
Fault Modeling 284, 289
Fault Simulation 162
Fault Tolerance 3
Flaws 161
Functional Failure 9
Functional Testing 317, 591
Golden Unit 16
Hewlett Packard 381
IBM 381
IC Fabrication 1
Interconnection Networks Testing 337
Iterative Arrays 317
Knowledge-Based Systems 465
Laser-Beam Testing 195
Level-Sensitive Scan Design 379
LOCST 381, 457
Logic Programming 508
LSSD 379
Magic 44, 47
Memory Testing 318
Microprocessor Testing 318
MISR 379
Monolithic Memories 381
Multivalued Logic 589
Observability 398
Path Sensitization 12
PLA Testing 317
Point Defects 162
Probabilistic Test 595
Process Monitoring 1
Process Simulation 42
Production Testing 2
PrototypeTesting 2
Register-Transfer Level Simulation 42
Scanning Electron Microscopy 250
SDC 380
Self-Checking Circuits 3
Serial Shadow Register 381
Signature Analysis 19
Simulation 42
Simulation Engine 45, 90
SPICE 43
STAFAN 311
Stored-Response Testing 18
Stuck-at Fault 9
STUMPS 379
Switch-Level Simulation 42
Syndrome Testing 380, 407

Syndrome Untestable	421	Thatte-Abraham Technique	13
Test Chips	202	Timing Simulation	42
Test Equipment	465	TITUS	35
Test Generation	162	Transition Counting	19
Test Heads	482	Transmission Electron Microscopy	250
Test Structures	202	VHDL	45, 150
Test Systems	465	VHSIC	45, 150
		VLSI Circuits	1
Testability Measures	401	Weighted Syndrome Sums	425
Testing	162	Yield	1

Author's Biography

Hassan K. Reghbati was born in Kerman, Iran, on January 18, 1949. He studied Engineering Physics at Arya-Mehr University of Technology, Tehran, Iran, Physics at M.I.T., Cambridge, Massachusetts, U.S.A.; and Electrical Engineering at the University of Toronto, Toronto, Canada.

Dr. Reghbati has worked in the area of integrated circuits for the last fifteen years. He was with the Department of Engineering Physics, Arya-Mehr University of Technology, from 1970 to 1973, with the Department of Physics, M.I.T., from 1973 to 1974, and with the Department of Electrical Engineering, University of Toronto, from 1974 to 1978, all as a lecturer or research assistant. Since 1978 he has been an Assistant Professor of Computer Science at the University of Saskatchewan (1978-1982) and Simon Fraser University (1982-1985).

Dr. Reghbati has published widely in journals, books, and conference proceedings. He is a contributing author of the *Encyclopedia of Computer Science and Engineering, Advances in Computers,* and *Knowledge Representation*. One of his papers has been reprinted in *Auerbach Annual 1980: Best Computer Papers*. A member of the IEEE and the Canadian Society for Computational Studies of Intelligence, he has been active professionally, serving on the program committee of conferences, as a consultant to the industry, and as a member of standards committees.

Dr. Reghbati is the Guest Editor of the forthcoming Special Issue of the *International Journal of Computers & Mathematics,* on "Diagnosis and Reliable Design of VLSI Systems."

BCC TK 7874 .T8855 1985

Tutorial--VLSI testing &
 validation techniques